PHYSICAL HYDRODYNAMICS

Physical Hydrodynamics

Second Edition

Etienne Guyon

Honorary director of Ecole normale supérieure. Ecole supérieure de Physique et Chimie (Paris)

Jean-Pierre Hulin

Senior CNRS Research Scientist, Emeritus, FAST Laboratory, Paris-Saclay University.

Luc Petit

Professor of Physics, ILM Laboratory, Université Claude Bernard-Lyon1

Catalin D. Mitescu

Professor of Physics, Emeritus at Pomona College

OXFORD
UNIVERSITY PRESS

OXFORD
UNIVERSITY PRESS

Great Clarendon Street, Oxford, OX2 6DP,
United Kingdom

Oxford University Press is a department of the University of Oxford.
It furthers the University's objective of excellence in research, scholarship,
and education by publishing worldwide. Oxford is a registered trade mark of
Oxford University Press in the UK and in certain other countries

Translated with revisions and extensions from the French language edition of:
Hydrodynamique physique – 3e édition
De Etienne Guyon, Jean-Pierre Hulin et Luc Petit

First Edition published in 2001

Published in the United States of America by Oxford University Press
198 Madison Avenue, New York, NY 10016, United States of America

British Library Cataloguing in Publication Data
Data available

Library of Congress Control Number: 2014936053

ISBN 978–0–19–870244–3 (hbk.)
ISBN 978–0–19–870245–0 (pbk.)

Printed and bound by
CPI Group (UK) Ltd, Croydon, CR0 4YY

Foreword

Fluid mechanics is a subject with a long history, but yet is young with regard to recent discoveries, and has many applications which affect everyday life. Its history is a parade of the great names of science: from the eighteenth century the Bernoullis, Euler and Lagrange; from the nineteenth century Cauchy, Navier, Stokes, Helmholtz, Rayleigh, Reynolds and Lamb; and from the twentieth century Couette, Prandtl, G.I.Taylor and Kolmogorov.

In the natural environment, we can now rely on the 5-day weather forecasts and tornado warnings; early success in predicting tides is today applied to automated tsunami warnings; understanding the circulation in the oceans and atmosphere is applied to pollution, the ozone hole and climate change. In the interior of the earth, fluid mechanics is important in mantle convection, volcanoes and their dust clouds, oil reservoirs and possible CO_2 sequestration.

Fluid mechanics is central to many industries. The design of aircraft developed from simple ideas in the early twentieth century to low-drag forms of wings with winglets at their tips and improved body profiles by the end of the century. At the same time, jet-noise was dramatically reduced with the introduction of wide entrance bypass fans which shield the fast jet. Simple and complex fluids are processed in various manufacturing industries: for glass and other materials, chemical engineering and food processing.

Recent research in fluid mechanics includes: microfluidics at the micron scale with the possibility of multiple simultaneous tests of small biological samples; similar scales and effects of wettability in ink-jet-printing; the ideas of convection allowing the design of energy-efficient buildings by using natural convection; and the control of instabilities and turbulence.

With such a wealth of ideas and applications, there is a major challenge of how to teach the subject. Some material is best left to specialized Masters courses. But the basic core has to be taught in a way to help students' progress to the advanced topics, current and future. The authors of this book have adopted in my opinion an approach and style which should interest and educate students, preparing them for the future. I fear that some alternative approaches fail on this: some engineering courses have an over-reliance on Computational Fluid Dynamics, which can be unsafe in novel applications; some mathematical courses are lost in the enormous difficulty of proving the governing equations have, or do not have, solutions in the simplest of situations, an open Clay prize problem. The approach in this book is grounded in experiment and reality. The chosen structure of the presentation helps students come to deep insights into the subject.

In my opinion, the subject of fluid mechanics has benefited in the last 30 years from the contributions of French physicists such as the authors of this book, bringing a fresh approach to the subject along with novel experimental techniques and an appreciation of practicalities.

John Hinch
University of Cambridge

Contents

Introduction

The place of fluid mechanics has been poorly defined in the scientific world. On the one hand, it has a strong connection with applied mathematics, in particular in France where the impact of such research has been extremely important from the nineteenth century onward. In recent times, this tendency has been reinforced by significant developments of computational science in such domains as turbulence and instabilities. On the other hand, engineering communities deal with sophisticated technical problems regarding flows and transfer of heat and matter, which require approximate solutions, but where some basic understanding is often missing. A consequence of this state of affairs is that both physicists and chemists, who are also involved in these issues, have remained on the sidelines. This was the case in the basic training of the authors of the present book. Yet another possible reason has been the strong polarization of physicists towards problems on the quantum scale, and relativity, although the pioneering giants of the field (Einstein, Bohr, Heisenberg . . .) had a good background in continuum mechanics. For a long time, physicists did not follow closely the important developments in the field, or followed them only indirectly: for example, the information on perfect flows and vorticity, of three of the authors of the present volume was triggered by their experimental work on superfluid Helium. The study of hydrodynamic instabilities was approached by analogies with phase transitions. Physical chemists dealt with complex materials such as liquid crystals, colloids, or polymers: they had to apply mechanics but also other approaches like scaling laws.

This schematic description of a mutual ignorance applies less and less frequently nowadays. Over the last decades, the interaction between scientists trained in physics and in mechanics has continued to increase through a great deal of joint research. The first edition of *Physical Hydrodynamics*, which was published in English 13 years ago, is presented in a completely renewed and expanded form and content while keeping the same physical, pedagogical approach. Moreover, 36 exercises have been added at the end of selected chapters and their correction is provided at the end of the book.

Indeed, in the course of the last 30 years, physicists and chemists have approached the subject in a manner complementary to classical mechanics. This implied the use of experimental and theoretical tools which had been developed for domains of physics such as condensed matter, statistical physics and material science. International conferences such as the APS-DFD or Euromech meetings led to interactions between participants with very diverse backgrounds. A broader field of applications has also been considered, such as in the life sciences, in addition to more classical engineering ones. But this also implies less formal ways of reasoning, such as for example, scaling approaches making extensive use of the classical tools of experimental physics.

The present book is based on our experience as experimental physicists. Actually, the word "hydrodynamics" is somewhat misleading and we use it in the sense that was given in *Hydrodynamica* by Daniel Bernoulli. We deal here not only with liquids but also with gas flows such that compressibility effects cannot be ignored. We have therefore excluded such problems as high-velocity gas flows, which involve a coupling between the equation of motion and thermodynamics. One specificity of our approach is to attempt to tie in, as often as possible, the macroscopic behavior of fluids to their microscopic properties. We also rely, as often as is reasonable, on order of magnitude arguments rather than just on formal derivations. Thus, in introducing dimensionless numbers, as is customary in the field, we stress the fact they are ratios (in Greek, αναλογια, *analogy*, which means ratio) of similar quantities such as characteristic lengths, times, energies, and so on, rather than relying on algebraic manipulations based on dimensional analysis. This was the approach of Osborne Reynolds in introducing the dimensionless quantity which bears his name. Rather than "cookbook recipes", we look for a deeper physical understanding of the mechanisms at play in a field extremely rich in experimental observations.

The book can be roughly divided into three parts:

The first five chapters (1 to 5) are devoted to the basic and classical elements of fluid mechanics. In the first chapter, we give a schematic description typical of elementary out-of-equilibrium statistical physics on non-equilibrium transport processes and of the spectroscopic tools used for these studies. Diffusive versus convective processes are discussed in the following chapter. Kinematics and dynamics of flows as well as the use of conservation laws provide the classical foundation of the book.

Different regimes of flows are analyzed in the following four chapters (6 to 9): potential flows, flows governed by vorticity (with an extension to rotating flows as encountered in geostrophic conditions); quasi-parallel flows and low Reynolds numbers flows.

The last chapters (10 to 12) are devoted to more detailed and complex phenomena which simultaneously involve different mechanisms: boundary-layer flows lead to many applications as, in particular, those found in chemical engineering. Unstable flows are presented in a simple way emphasizing the coupling between the different mechanisms involved and, finally, the chapter on turbulent flows in which mathematical developments have been kept to a minimum.

As stated above, our physicists' approach was initiated by our research work on various themes on fluid mechanics not often dealt with by specialists of classical mechanics (such as, sound in superfluid He^4, instabilities in liquid crystals, percolation in porous media). Our basic training made use of references such as Landau's *Fluid Mechanics*. However, over the course of time, we benefited from other communities and, in particular, the British school of G.I. Taylor (the exquisite four volumes of his complete works) and direct contacts with G.K. Batchelor who provided us an access to a pragmatic approach to fluids where ingenious experimental discoveries are accompanied with rigorous reasoning and more mathematical treatment well connected with experimental reality. The teaching of John Hinch, who has accepted to write a "Foreword" to this book, and Keith Moffatt, who introduced fluid mechanics to a number of us in a famous summer institute in Les Houches, in 1973, have provided us with a number of fine tools suitable both for our research as well as for undergraduate and graduate teaching at the origin of this book. In the USA, connections with A. Acrivos, J. Brady, G. Homsy, J. Koplik, L. Mahadevan, H. Stone and many other colleagues have broadened our vision of the field. Our initial curiosity was stirred up by several films of the NCFMF (National Committee for Fluid Mechanics Films), and by the Album of Fluid Motion of M. Van Dyke (which inspired our "Ce que disent les fluides" in French). More recently, the Multimedia Fluid Mechanics project directed by G. Homsy has broadened the range of documents available for teaching purposes. Our English speaking colleagues which have used the recent French edition of the book in their classes have encouraged us to produce this new, second edition of *Physical Hydrodynamics*.

Above all, it is in the everyday life of class and laboratory activities that the book has been constructed. The research and teaching of P. Bergé, B. Castaing, C. Clanet, Y. Couder, M. Farge, M. Fermigier, P. Gondret, E. Guazzelli, J.F. Joanny, F. Moisy, B. Perrin, Y. Pomeau, M. Rabaud, D. Salin, B. Semin, J.E. Wesfreid and many others are at the origin of elements of this book. We wish to thank particularly several colleagues who have directly contributed to specific parts of this book: C. Allain (non-Newtonian flows),

A. Ambari (polarogaphy), A.M. Cazabat (dynamics of wetting), M. Champion (flames), C. Clanet and D. Quéré (capillarity), P. Gondret (exercises), F. Moisy (text and exercises on turbulence and rotating flows), C. Nore (MHD), N. Ribe (free liquid jets) and J. Teixeira (spectroscopy of fluids).

The translation, as well as the necessary adjustment to the English scientific form, has been made essentially by C.D. Mitescu. We express our gratitude to Dr. Natalie Reinert, DVM, who has edited the book during the last year. She has also contributed greatly to the style of the translation. Our gratitude goes to Mrs Nicole Mitescu, wife of C.D. Mitescu and mother of Nathalie for proof reading the book. We thank her as well as Marie Yvonne, Danièle and Christine for their patience during the preparation of this new edition.

E. Guyon, J.P. Hulin, L. Petit, C.D. Mitescu

The Physics of Fluids

1

From a microscopic viewpoint, the study of the physics of fluids can be considered as a branch of thermodynamics. In a classical thermodynamic approach, we study the equilibrium states of pure substances – solids, liquids and gases – and the changes of state between these phases. A generalization of this approach is the study of fluctuations in the immediate neighborhood of an equilibrium state; these fluctuations are not only characteristic of the state, but also or the way in which equilibrium is reached. Thus, for a physical system with a large number of particles, which has undergone a "small disturbance" relative to its state of thermodynamic equilibrium, there exist straightforward proportionality relations between the fluxes that tend to restore equilibrium and the amplitude of the displacement.

The study of these relations and the definition of the transport coefficients which characterize them constitute the core of this first chapter. The discussion first emphasizes a macroscopic viewpoint (Section 1.2), and then a microscopic one (Section 1.3). We also analyze (Section 1.4) some of the surface phenomena which appear when two fluids have a common boundary (interface). Finally, we provide a brief overview of the application of optical spectroscopy, of X-ray techniques and of neutron scattering to the study of liquids (Section 1.5); such measurements permit the study of fluctuations about the equilibrium point and the subsequent evaluation of the transport coefficients. At the outset, though, we present in Section 1.1 a simple description of the microscopic nature of a fluid, and attempt to describe the influence of its microscopic characteristics on its macroscopic properties.

1.1 The liquid state

The periodic arrangement of atoms in a crystal is quite familiar to us from X-ray studies of its microscopic structure, or from the observation of its external shape. In this – the *solid* – state of matter, atoms remain fixed relative to one another except for small amplitude vibrations resulting from thermal motion. In the other extreme limit, *gases* at low pressure are nothing but a dilute system of particles with mutual interactions, weak except at the moment of a collision. Kinetic theory models of gases allow one to understand, from a microscopic viewpoint, the evolution of their equilibrium variables, such as temperature or pressure, except in the neighborhood of a critical point. On the other hand, the precise description of a *liquid*—with characteristics midway between those of gases and solids—is much more delicate; should we consider it a very dense gas or a disordered solid? Microscopic models of liquids often combine features from these two extremes. In particular, model two-dimensional systems, both microscopic and macroscopic, provide a powerful tool for the analysis of both the structure and the static properties of the various states of matter.

Physical Hydrodynamics. Second Edition. Etienne Guyon *et al.*

1.1.1 The different states of matter: model systems and real media

Visual representation of different states of matter by means of an air table

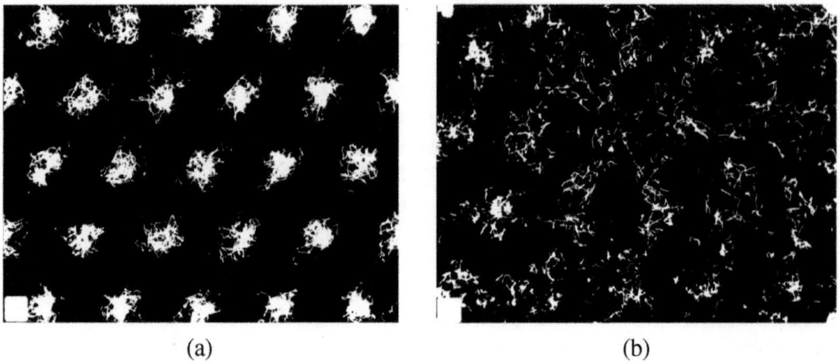

Figure 1.1 *Configurations observed as a result of the movement of disks on a vibrating air table, corresponding to different values of the packing fraction, C: (a) model for the solid state,* $C = 0.815$; *(b) liquid-state simulation,* $C = 0.741$ *(plates, courtesy Piotr Pieranski)*

An air table consists of a large horizontal plate drilled with a pattern of identical, small diameter and uniformy spaced holes, through which air at high pressure is forced upwards. A set of identical disks of radius R, placed on the table, and levitated by the air flow, can thus move around with negligible friction. "Thermal motion" of these disks can be simulated by vibrations of the supporting horizontal plate, or of its lateral boundaries. Depending on the mean concentration of the disks, we observe the characteristics of the different states of matter. Figures 1.1a and 1.1b were obtained by fastening a small light source on each disk and recording photographically the corresponding trajectories (with an exposure time much longer than the mean time between collisions). The trajectories of the disks appear as white traces on the figures.

The concentration is characterized by the ratio of the surface area covered by the disks to the total area of the table—a ratio defined as the *packing fraction, C*.

Maximal packing fraction: $C = C_M$ The maximum value C_M for two-dimensional systems of disks, is obtained with a compact triangular packing. The disks then form a perfect, two-dimensional crystal lattice (Figure 1.2); this state represents that of a perfect crystal with no thermal vibrations.

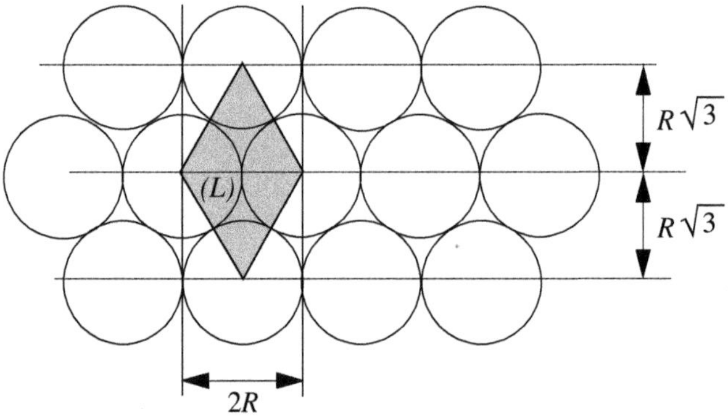

Figure 1.2 *Maximally compact configuration for a packing of disks of uniform diameter: their centers form a plane, triangular, crystalline lattice*

Proof

In this configuration, the elementary pattern (unit cell) which, repeated periodically, leads to the triangular lattice, is a diamond-shaped figure (L) of surface area $S_0 = (2R)(2R)\sqrt{3}/2$. This unit cell contains exactly the equivalent area $S_p = \pi R^2$ of one disk. We hence find a packing fraction:

$$C_M = \frac{S_p}{S_0} = \frac{\pi R^2}{2R^2\sqrt{3}} = 0.901. \tag{1.1}$$

Some people refer to this, somewhat incorrectly, as a hexagonal lattice, which would correspond to having an element at the vertices of a hexagon, with no element in the center.

High packing fraction So long as the packing fraction remains close to C_M, the disks, as observed from the trajectories of the luminous points (Figure 1.1a), undergo limited displacements around their equilibrium positions. Their mean positions remain, however, constant and the resultant average structure is periodic: we have there an image of atomic vibrations in a solid. Such vibrations are associated with sound propagation; the displacements of individual particles are transmitted from one neighbor to the next in response to a disturbance applied at the other end of the solid, leading to propagation modes known as *phonons*. But, in this packing, it is virtually impossible for two neighboring rows to undergo a relative displacement greater than R and, consequently, each particle retains the same neighbors. The amplitude of any resulting slippage is thus quite limited, and elastic restoring forces result.

Medium packing fraction For a packing fraction smaller than $C_0 \approx 0.8$, we have a transition to a different regime: an individual particle is now able to escape from the "cage" created by its neighbors. In this instance, particles no longer have a fixed position relative to immediately adjacent ones; the system now simulates a two-dimensional "liquid" (Figure 1.1b). Simultaneously, the periodicity of the crystal has vanished. The resulting fluidity of the system leads to the occurrence of global displacements of the disks with respect to each other in response to a relative motion of the side walls of the container.

Low packing fraction $C \ll C_0$ In this final instance, we have a "gas" of particles. The relative distance between "nearest neighbors" can now be quite large (of order $R/\sqrt{C}$), whereas it was of order $2R$ for the "liquid."

Numerical simulations in terms of a hard-disk model

Results similar to those of Section 1.1.1 can be obtained by means of numerical simulations in which the interaction between particles is of the hard-disk type; in this case, the interaction potential between pairs of particles is zero when the distance r between their centers is greater than $2R$, and infinitely repulsive when $r \leq 2R$. These calculations confirm and extend the results of the analog model we have presented above. These simulations can be improved by introducing more realistic interaction potentials—such as the Lennard–Jones potential,—which, for three-dimensional systems, is of the form:

$$V(r) = V_0\left[\left(\frac{2R}{r}\right)^{12} - \left(\frac{2R}{r}\right)^{6}\right]. \tag{1.2}$$

This potential allows one to take into account the very slight interpenetration between pairs of particles, strongly limited by the Pauli exclusion principle, when $r \leq 2R$; it also introduces a weak, attractive, *van der Waals interaction* between particles, which becomes dominant at large distances ($r \gg R$). Equation 1.2 predicts the existence of a minimum in the potential $V(r)$ for a value r_0 of the order of 2.2 R, thus indicating a potentially stable equilibrium state absent in the hard-disk or hard-sphere model. By introducing this potential, we alter slightly the equation of state of the two-dimensional ideal gas, and obtain a result similar to the van der Waals equation for pure substances. More specifically, there appears a domain where liquid–gas coexistence is possible.

Three-dimensional models

In Section 1.1.1, we simulated the structure of solids and the solid–liquid transition by a system of flat circular disks. Can we also create similar models in three-dimensions by stacking beads of uniform diameter, which we rearrange by shaking, or by *fluidization* techniques (keeping them temporarily apart by forcing through them an upward stream of fluid from the bottom of the stack)? We find that it is indeed possible to represent certain structural forms of matter with the help of periodic packings of beads; such is the case, for example, for the *face-centered cubic* (FCC) lattice (Figure 1.3a), of packing fraction 0.74.

Packing fraction of a face-centered cubic lattice

In this periodic packing, the basic building block is a cube of side $a = \left(4R/\sqrt{2}\right)$ (Figure 1.3b), since the diagonal of the faces of this cube has a length $4R$; this unit cell contains the equivalent of four complete spheres of radius R. The packing fraction C (fraction of the volume taken up by the spheres) is thus:

$$C = \frac{4\,(4/3)\,\pi R^3}{a^3} = \frac{4\,(4/3)\,\pi R^3}{\left(4R/\sqrt{2}\right)^3} = \frac{\pi}{3\sqrt{2}} \simeq 0,74 \qquad (1.3)$$

The value of 0.74 for the packing fraction of the FCC structure is the highest known for any packing of spheres of uniform diameter R. Once created, this packing ensures long-range periodic order just as does the triangular lattice in two dimensions. Kepler's conjecture, which stipulated that this limit could not be exceeded for a packing of uniform spheres, was proved mathematically only in . . . 1998!

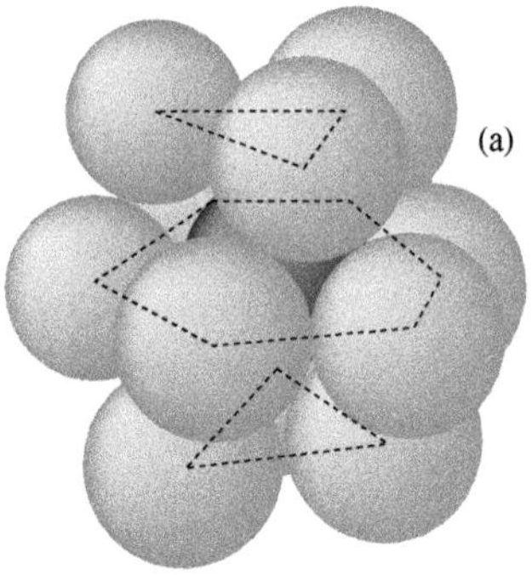

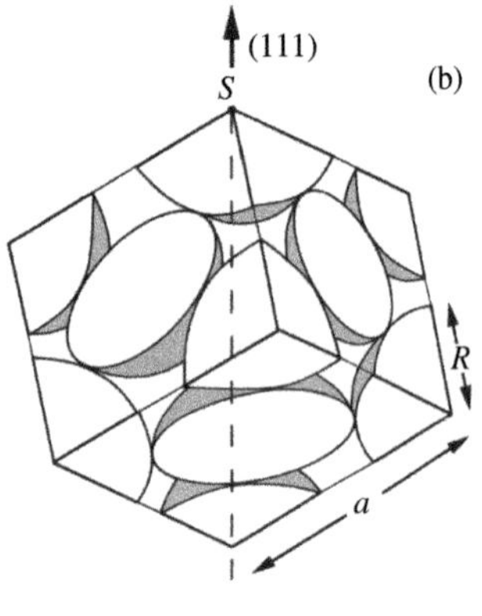

Figure 1.3 *(a) Compact face-centered cubic (FCC) packing for a system of beads of uniform radius R. (b) The FCC lattice of inter-touching spheres has been sliced into the cube which forms the unit cell of the lattice. This cube contains the equivalent of four complete spheres. We can more easily visualize the correspondence with Figure 1.3a if we observe that the sphere S at the ends of the* (111) *axis, the principal diagonal of the cube, plays the role of the (partly hidden) sphere in the center of the packing of Figure 1.3a*

The difference from the two-dimensional case is that this periodic packing is not spontaneously generated when a container is filled with identical spheres, even if shaken around. There is, in fact, an infinite variety of ways of arranging 12 identical spheres around a single one; the icosahedron of Figure 4 in the next page is just one particular example. It is thus not possible to construct a crystal starting from an arbitrary local filling. This is quite different from the two-dimensional case where the pattern for putting six disks in contact with a central one is unique: they must be located at the vertices of a hexagon which then constitutes the seed of the two-dimensional triangular crystal of Figure 1.2.

In real life, when we fill a container at random with uniform spheres we create a disordered packing with a packing fraction ranging from 0.59 to 0.64. The resulting structure represents fairly well that of an amorphous metal which can be created by rapid deposition of metallic liquid or vapor onto a very cold substrate. It is also a good representation of the instantaneous position of atoms in a simple liquid (Figure 1.1b). Finally, this packing is also a good representation of granular or porous media (sand, sandstone) discussed in Section 9.7.

We can equally obtain three-dimensional crystal models by using uniform-diameter, micron-sized spherical latex particles in an ionic solution. As long as Coulomb repulsion between spheres is sufficiently strong, these interactions lead to the formation of a periodic lattice of particles (*a colloïdal crystal*). With increased concentration of ions in the solution, the interaction is screened more and more strongly; the periodic structure ultimately disappears and the particles aggregate.

1.1.2 The solid–liquid transition: a sometimes nebulous boundary

The limit between a solid and a liquid is not always as simple as it appears in rest state; it depends closely on the amplitude and duration of applied stresses. The branch of science studying the evolution of the deformation of materials under stress is known as *rheology*. In Section 4.4, we discuss the different kinds of response observed in various fluids; here, we confine our discussion to two examples.

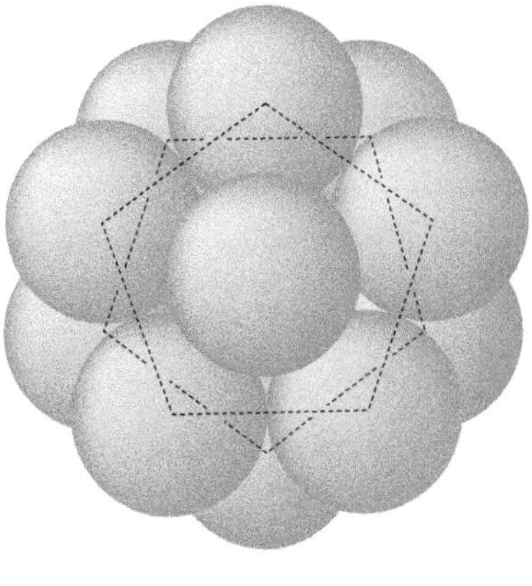

Figure 1.4 *This packing, in the form of an icosahedron is made up, just as in Figure 1.3a, starting with 12 identical spheres uniformly placed around one in the center: it has a higher probability of occurrence than the FCC packing. Because it has fivefold symmetry, it cannot be indefinitely repeated to create a crystal lattice*

Modelling plastic flow in two dimensions

The change in shape of solids, such as metal bars or springs, resulting from the action of forces, is no longer reversible beyond a certain threshold stress, known as the elastic limit. Thus many substances, though apparently solid, flow or creep when subjected to high stresses (e.g., glaciers, the earth's crust, metal sheets being cold pressed). This is referred to as *plasticity*. The crucial role that defects in crystalline packings play in this phenomenon can be illustrated by means of the two-dimensional models. These defects might be, for example, the appearance of a new row of particles starting from a point in the lattice (*dislocation*), or perhaps a contact line between two lattices of different orientation (*grain boundary*). We can easily observe this by looking at a flat plane on which a monolayer of beads has been laid out. If the plane is slightly inclined, we see that the motion of beads in the neighborhood of defects allows for global deformations of the system. Thus, defects account for the flow of solid matter. This observation is the basis of modern metallurgy.

Effect of the rate of change of the stresses on the deformation of a medium

The rate at which stresses vary plays a role as important as their magnitude in the behavior of a substance. We see in Section 4.4.4 that the response of some substances to variable-frequency perturbations displays a transition from a solid-like state (at high frequencies) to a liquid-like one (at lower frequencies). The crossover between these two regimes occurs in the neighborhood of a time constant τ_{De} characteristic of the substance: the kind of behavior observed depends on the ratio of τ_{De} and of the characteristic time (or period) of the excitation. This ratio, defined by Equation 4.44, is called *the Deborah number*.

1.2 Macroscopic transport coefficients

Let us now discuss transport phenomena which appear in fluids due to small deviations from equilibrium conditions, deviations small enough that the system response continues to be approximately linear. Three types of transport can then be studied:

- the transport of *thermal energy* resulting from spatial variations of temperature,
- the transport of *matter* due to variations in concentration,
- the transport of *momentum* in a moving fluid.

Though it is this last transport property which is treated in particular depth in this textbook, in the present chapter we consider only the first two types listed above.

These several kinds of transport phenomena frequently coexist in the physics of fluids, as indicated in the following examples:

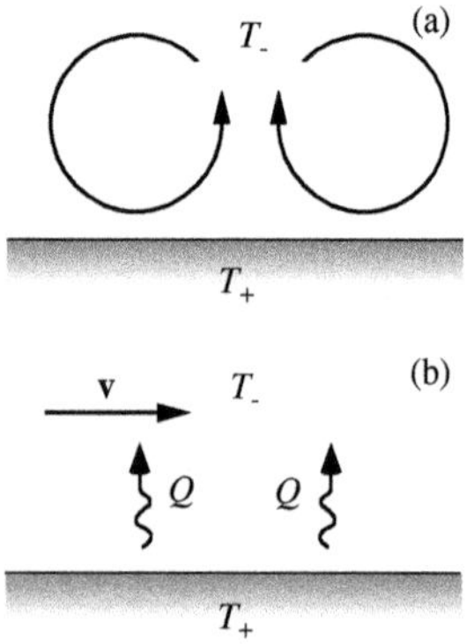

Figure 1.5 *Heat exchange between a fluid and a heated plate: (a) in the presence of spontaneous convective fluid motion induced by the existing temperature difference (discussed in detail in Chapter 11); (b) between an air stream and a warm body, resulting in the formation of a boundary layer (See Chapter 10)*

- a warm object (at temperature T_+) placed in a motionless fluid at lower temperature T_-, frequently generates a *convective* circulation in the upper region of the fluid (Figure 1.5a). In turn, the fluid motion increases the heat exchange between object and fluid;
- when we blow air on a glowing wooden ember, we affect simultaneously the transport of mass, of matter and of heat (Q); the kinetics of the exothermic burning process is correspondingly accelerated (Figure 1.5b).

We see, in these examples, a superposition of a number of exchange mechanisms: not only *convective* (drag by the moving fluid), *radiative* and *chemical* (associated with the reactions occurring), but also *diffusive* or *conductive* exchange. These depend only on the microscopic properties of the fluid and can be analyzed in terms of small deviations from the equilibrium state. In this chapter, we discuss mainly the diffusive effects.

We first consider, from a macroscopic viewpoint, the familiar example of heat conduction, and proceed from there to look at mass diffusion (Section 1.2.2); we provide in Section 1.3 a microscopic picture of these effects in gases and liquids. In Chapter 2, we explain how the viscosity of a fluid results from a similar diffusion of momentum. The reason for our simultaneous presentation of diffusive transport of these three physical quantities (heat, mass and momentum) is the fact that the equations describing each process and its observed characteristics are mathematically identical. The results obtained for heat transport are thus easily translated to that of mass, or momentum (requiring only a change of definition of the associated coefficients and, for the momentum, a change from scalar to vector quantities).

1.2.1 Thermal conductivity

Definition of thermal conductivity: the governing equation for heat transport under stationary conditions

A semi-infinite homogeneous body (solid, liquid or gas) occupying the space corresponding to positive values of x is subjected to a temperature gradient $\mathrm{d}T/\mathrm{d}x$ in the direction of the x-axis. This gradient is obtained by applying a temperature difference $T_1 - T_2$ between two planes P_1 and P_2 a distance L apart (Figure 1.6). We shall consider heat conduction across a cross-sectional area S perpendicular to the x–axis.

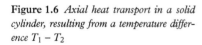

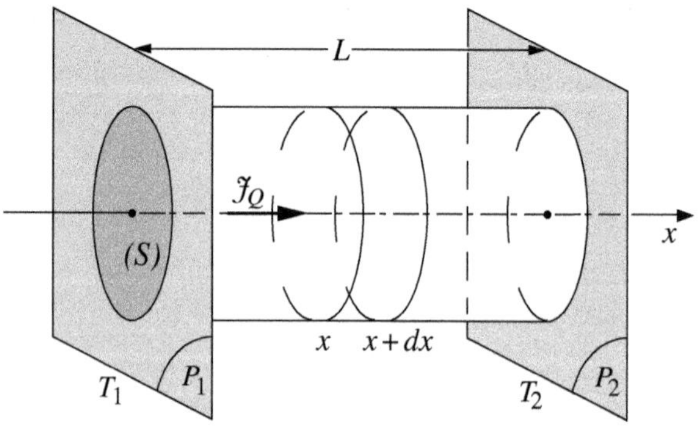

Figure 1.6 *Axial heat transport in a solid cylinder, resulting from a temperature difference $T_1 - T_2$*

The heat flux $\mathcal{J}_Q$ per unit area and per unit time, is directly proportional to the temperature difference $T_1 - T_2$, as given by:

$$\mathcal{J}_Q = \frac{\delta Q}{S\,\delta t} = k\frac{T_1 - T_2}{L} = -k\frac{\mathrm{d}T}{\mathrm{d}x}. \tag{1.4}$$

The negative sign indicates that heat flow occurs in the direction opposite to the temperature gradient. The last equality in Equation 1.4 results from the fact that, in the equilibrium stationary state, the temperature varies linearly between the two boundary values T_1 and T_2, a result which we justify below. The coefficient k (known as the *thermal conductivity*) is a function only of the properties of the material. It satisfies the dimensional equation:

$$[k] = \frac{[M][L]^2[T]^{-3}}{[L]^2.[\Theta]/[L]} = [M][L][T]^{-3}[\Theta]^{-1} \tag{1.5}$$

(here $[L]$ stands for length, $[M]$ for mass, $[T]$ for time and $[\Theta]$ for temperature). Typical values of this coefficient may be found in Table 1.2 in the appendix of this chapter.

The result of Equation 1.4 can be generalized to the case where the temperature T varies with all three coordinates in three-dimensional space, to assume the vector form:

$$\mathbf{J}_Q(\mathbf{r}) = -k\,\nabla T(\mathbf{r}). \tag{1.6}$$

This result is mathematically identical to the local form of Ohm's Law $\mathbf{j}(\mathbf{r}) = -\sigma\,\nabla V(\mathbf{r})$ which relates the electric current density $\mathbf{j}(\mathbf{r})$ to the potential $V(\mathbf{r})$ in electrodynamics (σ being here the electrical conductivity of the medium). We can now easily understand why, under stationary conditions, the temperature varies linearly with distance in the geometry of Figure 1.6, just as the potential varies linearly between two parallel electrodes (at different potentials) immersed in a conducting fluid. The *linear* dependence in Equation 1.6 expresses a proportionality between a *flux* (of heat) and a *thermodynamic force* (the temperature gradient); we find similar relations for the other transport phenomena in the neighborhood of equilibrium conditions.

Application of the thermal conductivity equation to a cylindrical geometry

The experiment which we describe below allows one to measure the thermal conductivity of a solid material in the shape of a hollow cylinder of external radius a, as shown in Figure 1.7. Initially, the cylinder is placed in an isothermal bath at uniform temperature T_1 in such a way that $T(r = a) = T_1$ at all times. The hollow region inside the cylinder ($r < b$) is completely insulated from the external bath. In this inside space, we place a thermometer, to measure the interior temperature $T_i(t)$, as well as a heater resistance R delivering a constant power P uniformly along the entire height H of the cylinder. If the aspect ratio H/a for the cylinder is sufficiently large, we can assume that the heat flux $\mathbf{J}_Q(r)$ is effectively radial and does not vary along the axis of the cylinder. At time $t = 0$, the heater is turned on, and we observe the variation of the interior temperature $T_i(t)$ of the cylinder as measured by the thermometer (Figure 1.7). At the end of a sufficient time lapse, this temperature stabilizes at a value T_2. Measurement of the temperature difference T_2-T_1 provides us with the value of the thermal conductivity k of the cylinder as follows: If we calculate the radial heat flux $\mathbf{J}_Q(r)$ per unit area (for $b<r<a$) between the two infinitesimally close cylinders at radii r and $r+dr$, the magnitude of the total heat flux is:

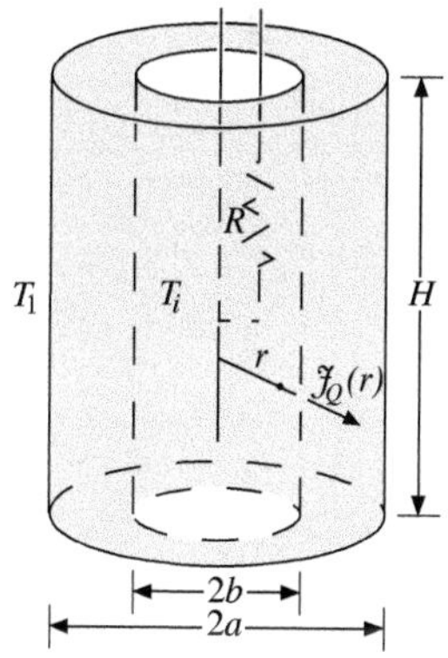

Figure 1.7 *Schematic representation of the experimental set-up for measuring the thermal conductivity of a hollow, solid cylinder*

$$P = 2\pi\, r H \mathcal{J}_Q(r) \tag{1.7}$$

where:

$$\mathcal{J}_Q(r) = -k\frac{\mathrm{d}T}{\mathrm{d}r}. \tag{1.8}$$

This calculation, valid in a stationary regime, is identical to that for the electrical conductivity of a conductor formed by two co-axial cylinders under a potential difference $(V_2 - V_1)$ (corresponding to the temperature difference $T_2 - T_1$) and carrying a current I (which corresponds to the heater power P).

Under stationary conditions, the heater power P is constant. Combining these two results, we obtain the differential equation:

$$\frac{dr}{r} = -\frac{2\pi H k}{P} dT. \quad (1.9)$$

Integrating this equation, subject to the boundary conditions on the temperature, we find:

$$k = \frac{P}{2\pi H(T_2 - T_1)} \mathrm{Log}\frac{a}{b}. \quad (1.10)$$

Thermal exchanges under non-stationary conditions - the Fourier equation

The previous example was discussed under the assumption that a stationary condition had been achieved, with the temperature at every point independent of time. We then measured heat diffusion under a fixed gradient. Let us now look at a more general case, where the temperature T is a function of both position and time. We will first examine the simpler case of temperature variation in the one-dimensional geometry of Figure 1.6. The evolution of the temperature $T(x,t)$ is described by what is known as Fourier's equation in one dimension, an equation which provides the relationship between the partial derivatives of the temperature T with respect to time and to the spatial coordinate x:

$$\rho C \frac{\partial T(x,t)}{\partial t} = k \frac{\partial^2 T(x,t)}{\partial x^2} \quad (1.11a) \qquad \text{or:} \qquad \frac{\partial T(x,t)}{\partial t} = \kappa \frac{\partial^2 T(x,t)}{\partial x^2} \quad (1.11b)$$

where we write:

$$\kappa = \frac{k}{\rho C} \quad (1.12)$$

here ρ is the density of the material, and C its specific heat; the coefficient κ is the *thermal diffusivity* (a large value of the thermal diffusivity corresponds to a large thermal conductivity k, and to a small inertia with respect to heat transfer, as measured by the product ρC). Dimensionally, Equation 1.11b implies that the ratio of κ to the square of a length is inversely proportional to time; we thus obtain $[\kappa] = [L]^2[T]^{-1}$ in units of square meters per second. Values of k, ρ and C for a number of typical fluids are given in the table at the end of this chapter.

Derivation

If we express the energy balance for an infinitesimal elementary volume of cross-section S, bounded by the adjacent planes at x and $x + dx$ (Figure 1.6), the thermal energy entering through S during the time dt is, according to Equation 1.6:

$$q_x(x) S\, dt = -k \frac{\partial T(x,t)}{\partial x} S\, dt. \quad (1.13)$$

The heat flux exiting is:

$$q_x(x + dx)\, S\, dt = -k \frac{\partial T(x + dx, t)}{\partial x} S\, dt. \quad (1.14)$$

The difference between these two fluxes is the net increase in energy content in the material between the two planes, i.e:

$$q_x(x)\, S\, dt - q_x(x + dx)\, S\, dt = k\, S\, dt \left[-\frac{\partial T(x,t)}{\partial x} + \frac{\partial T(x + dx, t)}{\partial x} \right] = k\, S\, dt \frac{\partial^2 T(x,t)}{\partial x^2} dx, \tag{1.15}$$

using a Taylor series expansion for computing the right-hand side of the above equation. Equating this result to the increase in thermal energy as a function of time leads to:

$$\rho\, C\, S\, dx \frac{\partial T(x,t)}{\partial t} dt = k\, S\, dt \frac{\partial^2 T(x,t)}{\partial x^2} dx. \tag{1.16}$$

Simplifying Equation.1.16, we obtain Equations 1.11.

The form of Equation 1.11b is very general and characteristic of all diffusion phenomena: we find it in identical form, but with different variables, in problems of mass or momentum transport. It is frequently advantageous to make use of this correspondence where the diffusivity coefficients have identical dimensions but different numerical values. We note that, under stationary conditions in the one-dimensional geometry, Equation 1.11b reduces to $\partial^2 T(x,t)/\partial x^2 = 0$. We have then, in that particular case, a temperature gradient constant with respect to the space coordinate x and consequently, as originally assumed in Equation 1.4, a linear variation of the temperature with position.

More generally, in a problem where the temperature is a function of all three coordinates (x, y, z), the term $\partial^2 T(x,t)/\partial x^2$ in Equation 1.11 is replaced by the Laplacian $\nabla^2 T$ so that:

$$\frac{\partial T(\mathbf{r},t)}{\partial t} = \frac{k}{\rho C} \nabla^2 T = \kappa \nabla^2 T, \tag{1.17}$$

which is the three-dimensional heat-diffusion equation (*Fourier's equation*).

Application to the one-dimensional propagation of temperature variations

We consider now an homogeneous material occupying a semi-infinite half space $x > 0$ and, initially, at a uniform temperature $T = T_0$. At time $t = 0$, one imposes a temperature $T = T_1$ on the boundary $x = 0$. The temperature profiles in the material at different times $t > 0$ show that the perturbation imposed in the plane $x = 0$ spreads progressively by thermal diffusion (Figure 1.8a below).

The solution of this problem sheds light on numerous other examples of diffusion. We might note first of all that, by using the thermal diffusivity κ and the time variable t, we can construct a new variable $\sqrt{\kappa t}$ which has the dimensions of a length. It is therefore, quite reasonable to normalize the space coordinate x in terms of this characteristic length, defining a new dimensionless variable $u = x/\sqrt{\kappa t}$; similarly, instead of calculating $T(x,t)$, we introduce the normalized variable $(T - T_0)/(T_1 - T_0)$, which therefore takes on values 1 (for $x = 0$), and 0 (for $x \to \infty$). We find then, as derived below, the solution:

$$\frac{T - T_0}{T_1 - T_0} = \left[1 - \mathrm{erf}\left(\frac{u}{2}\right)\right], \tag{1.18}$$

where erf(u) is the error function, defined by $\mathrm{erf}(u) = \frac{2}{\sqrt{\pi}} \int_0^u e^{-\zeta^2} d\zeta$; replacing z by $u/2$, erf($u/2$) equals 0 for $u = 0$, and approaches unity as u tends to infinity. Thus, for the particular boundary conditions which we have considered here, we have obtained a solution which does not depend separately on x and t but only on the reduced variable $u = x/\sqrt{\kappa t}$. Figure 1.8b indicates that the spatial profiles of the normalized temperature, corresponding to different moments in time, coincide when they are displayed as a function of this dimensionless

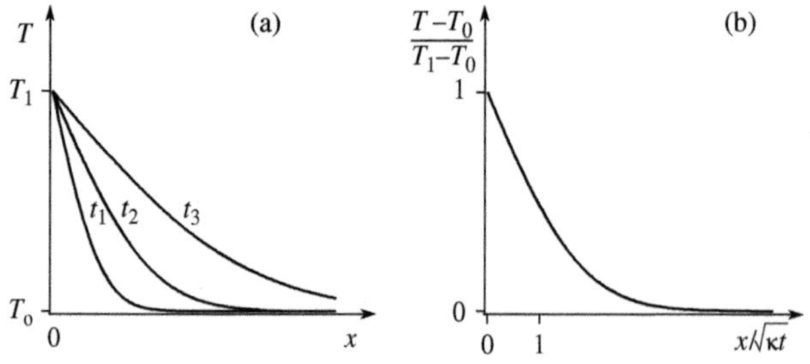

Figure 1.8 *Profiles of the variation of temperature with distance along an "infinitely long" bar; (a) profiles at different times $t_1 < t_2 < t_3$ upon applying a step function in temperature at the origin $x = 0$; (b) profiles for different times as a function of the normalized distance $x/\sqrt{\kappa t}$. The corresponding universal curve is predicted by Equation* 1.18

variable. For other boundary conditions, it might have been necessary to look for more complicated solutions which depend separately on x and t.

Qualitatively, we see that, at a given time t, the effect of a temperature perturbation has spread to a distance of order $\sqrt{\kappa t}$, (within a scale factor of the order of unity) in good agreement with the dimensional form of κ. We note here the essential characteristic of diffusive phenomena, a *proportionality between the mean diffusion distance* and *the square root of the time*. This result specifically explains the ineffectiveness of diffusive mechanisms at great distances: if it takes a time t_L for a temperature perturbation to diffuse through a distance L, it will need a time $t_{10L} = 100\ t_L$ for it to diffuse through a distance ten times as large! For air, for example, $\sqrt{\kappa t} = 1$ cm for $t = 10$ s; $\sqrt{\kappa t} = 10$ cm for $t = 10^3$ s $\approx$ 16 mins.

Derivation of Equation 1.18

We seek trial solutions of the form:

$$T(x,t) = f\left(\frac{x}{\sqrt{\kappa t}}\right) = f(u). \tag{1.19}$$

Substituting the expression (1.19) into Equation 1.11b and writing the corresponding partial derivatives of T with respect to t and x:

$$\left(\frac{\partial T(x,t)}{\partial t}\right)_x = -\frac{1}{2}t^{-3/2}\frac{x}{\sqrt{\kappa}}f'(u) \quad \text{and:} \quad \left(\frac{\partial^2 T(x,t)}{\partial x^2}\right)_t = \frac{1}{\kappa t}f''(u),$$

we obtain the relation:

$$f''(u) + \frac{1}{2}uf'(u) = 0. \tag{1.20}$$

Defining $F(u) = f'(u)$, the differential equation for $F(u)$ in terms of the reduced variable u is:

$$F'(u) + \frac{u\,F(u)}{2} = 0, \quad (1.21) \qquad \text{with solution:} \quad F = F_0\,e^{-u^2/4}. \quad (1.22)$$

We conclude:

$$f(u) = A\,\mathrm{erf}\left(\frac{u}{2}\right) + B, \qquad \text{where:} \quad u = \frac{x}{\sqrt{\kappa t}}. \tag{1.23}$$

Constants A and B are evaluated from the temperature boundary conditions; for $x = 0$: $T = T_1 = B$, while, for $x \to \infty$: $T = T_0 = A + B$.

Transient heat diffusion in a cylindrical geometry

We have discussed earlier the radial heat transport in a hollow cylinder under stationary conditions. Let us now consider the transient temperature variations in the limiting case where the cylinder is no longer hollow but solid. We suppose that the cylinder of outer radius a, initially at a temperature T_0, is suddenly plunged at time $t = 0$ into a fluid of temperature $T_0 + \delta T_0$. We will not carry out in detail the required calculation in terms of series of functions, but we confine ourselves to displaying the results obtained.

Figure 1.9 shows the temperature profiles for the reduced temperature $\delta T(r/a)/\delta T_0$ at different values of the time t (r is the distance from the axis of the cylinder, $\delta T(r/a)$ is the variation of the local temperature relative to the initial temperature T_0). The letters which label each profile correspond to different values of the ratio t/τ_D, where τ_D $(= a^2/\kappa)$ represents, as shown earlier, the order of magnitude of the diffusion time over a distance a. We observe that, for short times, the temperature variation is confined to a layer close to the surface ($r = a$) of thickness of the order of $\sqrt{\kappa t}$. The temperature profiles are similar to those which we have shown previously in Figure 1.8 for a plane boundary, i.e., for an infinitely large radius of curvature. For longer times, of the order of $0.1\tau_D$, the perturbation in temperature begins to be observed on the axis of the cylinder. It is everywhere equal to $T_0 + \delta T_0$ when t becomes of order τ_D. We see in Section 2.1.2 that these profiles equally describe the temporal and spatial variation of the fluid velocity in a cylinder filled with fluid and which is suddenly set into rotation.

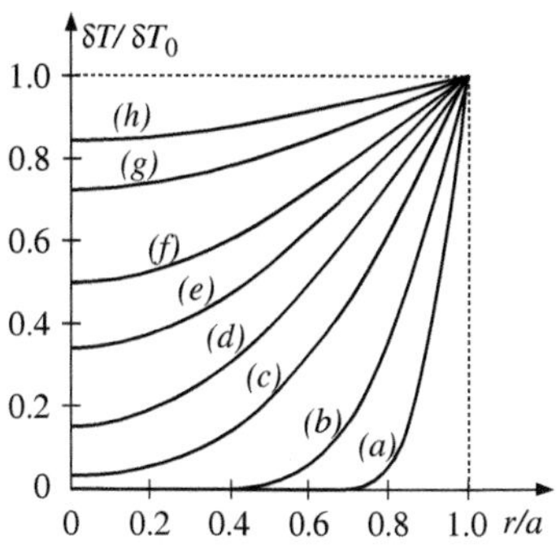

Figure 1.9 *The diffusion of heat in a cylinder initially at temperature T_0 which is suddenly immersed, at time $t=0$, in a heat bath such that its surface is then maintained at temperature $T_0 + \delta T_0$. The distribution of the reduced temperature $\delta T/\delta T_0$ is displayed as a function of the normalized distance r/a from the axis of the cylinder. The several curves correspond to different times, measured in terms of the reduced-time variable* ($t/\tau_D = \kappa t/a^2$)*: (a)* $t/\tau_D = 0.005$; *(b)* 0.01; *(c)* 0.04; *(d)* 0.1; *(e)* 0.15; *(f)* 0.2; *(g)* 0.3; *(h)* 0.4

Propagation: diffusion versus wave motion

In order to estimate the effectiveness of various processes, let us compare the diffusion Equation 1.17 to the classical equation of propagation of a wave of amplitude A travelling in the x-direction:

$$\frac{\partial^2 A}{\partial t^2} = c^2 \frac{\partial^2 A}{\partial x^2}, \tag{1.24}$$

where c is the velocity of the wave. We know that, for Equation 1.24, the most general solution is of the form: $A(x,t) = f_1(x - ct) + f_2(x + ct)$. The functions $f_1(x - ct)$ and $f_2(x + ct)$ describe wave propagation, respectively in the positive and negative x-directions, at a constant velocity c; the distance d covered by the wave is proportional to the time. We have seen that Equation 1.17 has solutions for which the propagation distance x varies like the square root of the time t; thus the "effective velocity of propagation" x/t decreases with distance. Physically, this result is due to the fact that the flux of the variable which is diffusing (temperature, concentration ...) is proportional to the gradient of this variable: the more the variation front spreads out, the slower the propagation.

Thus, a simple change in the order of the derivative with respect to time yields completely different behaviors for the case of wave propagation and for diffusion. Diffusive phenomena are efficient at short times, or for relatively small distances. On the other hand, wave propagation and fluid convection resulting from the fluid motion, which also leads to a displacement linear in time, dominate in every other situation.

1.2.2 Mass diffusion

Conservation of mass for a diffusing substance

Let us replace, in the one-dimensional example of Figure 1.6, the temperature T by the concentration of a substance diluted by the main fluid: this tracer substance might be another gas, or smoke within the gas, ions, dye molecules, or radioactive isotopes in the liquid. In this experiment of *gradient diffusion*, we are interested in determining the flux of tracer which results from a gradient in its concentration. We could also investigate the phenomenon of *self-diffusion* which describes the redistribution of "tagged" molecules (even in the absence of a gradient) among other unmarked molecules of the same type. In a dilute solution, the gradient diffusion and self-diffusion coefficients have identical values. For the case where diffusing particles interact, these two coefficients can take on different values.

The concentration can be measured by the number density $n(x,t)$ of tracer particles per unit volume, or by the mass density ρA of the tracer A, per unit volume of the mixture. In this case, the expression equivalent to Equation 1.6 (sometimes known as *Fick's equation*) is:

$$\mathbf{J}_m = -D_m \nabla \rho_A, \tag{1.25}$$

where $\mathbf{J}_m$ is the current density (mass per unit area per unit time), and D_m the *molecular diffusion coefficient* of the tracer. Its value is a function of the properties of both the tracer and the substance through which it diffuses. D_m satisfies the dimensional equation

$$[D_m] = \frac{[M][L]^{-2}[T]^{-1}[L]}{[M][L]^{-3}} = [L]^2[T]^{-1},$$

thus having the same dimensions as the thermal diffusivity coefficient (Equation 1.16). Table 1.2, in the appendix to this chapter, also provides a few, typical, numerical values for this coefficient.

We can derive the partial differential equation which relates the variations of the density with position and time by following, step by step, the same procedure used for the heat conductivity problem; i.e., by expressing in two different ways the mass conservation of the tracer within a given volume element. We thus obtain for the one-dimensional case, an equation analogous to (1.11b):

$$\frac{\partial \rho_A}{\partial t} = D_m \frac{\partial^2 \rho_A}{\partial x^2}. \tag{1.26}$$

In Section 1.2.1, we discussed examples of heat conduction with fixed temperature boundary conditions on the walls of the sample. These examples would correspond here to a problem where the boundary values of the concentrations of the chemical substances are fixed. In that case, the mathematical solutions carry over exactly, merely substituting ρ_A for the variable T and replacing κ with D_m.

In practical terms, it is simpler, in the case of the mass diffusion problem, to prescribe a particular initial concentration of tracer, and to observe how it evolves with time. The equivalent thermal problem would consist in generating, over a rather short initial time interval, a certain amount of heat at specific, localized points, and then observing the resultant temperature distribution.

Spreading of a tracer initially localized in a plane

Let us assume that we introduce uniformly into the interior of a fluid a mass M_A of tracer, initially localized throughout a very narrow layer in the plane $x = 0$. Mathematically, this distribution is of the type of a Dirac δ-function with $\rho_A = M_A\,\delta(x)$. Provided the tracer does not interact with the fluid, the conservation of mass can be written:

$$\int_{-\infty}^{+\infty} \rho_A(x)\,\mathrm{d}x = M_A = \text{constant}. \tag{1.27}$$

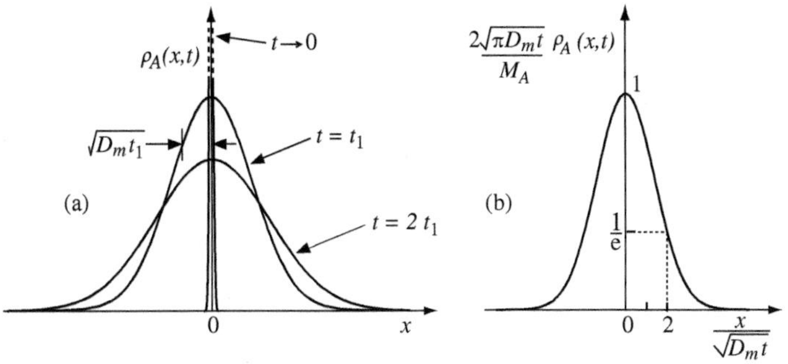

Figure 1.10 *Spreading, by means of molecular diffusion, of a drop of tracer initially localized in the plane $x = 0$ at time $t = 0$; (a) concentration profiles at times $t = 0$, $t = t_1$ and $t = 2t_1$; (b) universal curve representing the superposition of all curves of Figure 1.10a when we use the normalized variable $x/\sqrt{D_m t}$ as abscissa and the amplitude $(2\sqrt{\pi D_m t}/M_A)\,\rho_A(x,t)$ as ordinate. The graph corresponds to the Gaussian of Equation 1.29*

To check these constraints, we can take the derivative *g(x,t)*, with respect to the variable x, of Equation 1.23 for thermal diffusion obtained above in Section 1.2.1. One has :

$$g(x,t) = \frac{A}{2\sqrt{\pi \kappa t}} e^{-\frac{u^2}{4}}, \tag{1.28}$$

with: $u = x/\sqrt{\kappa t}$. The function *g(x,t)* is the solution of Equation 1.11b: effectively, in taking the derivative of this equation with respect to x, we find that, if a function is a solution of the equation, its derivative with respect to x is also a solution. In replacing κ by D_m we therefore obtain:

$$\rho_A(x,t) = \frac{M_A}{2\sqrt{\pi D_m t}} e^{-\frac{x^2}{4D_m t}}. \tag{1.29}$$

The solution of this problem is a Gaussian (Figure 1.10). Its width increases proportionally to the square root of the time: this is illustrative of the variation of the propagation distance characteristic of diffusive phenomena. At the same time the amplitude decreases as $1/\sqrt{t}$ so that the area under the curve is conserved, this area representing the total mass of tracer injected. As was the case for the graph of Figure 1.8b, we find that, by using the coordinate $u = x/\sqrt{\kappa t}$ as abscissa and the product $2\sqrt{\pi D_m t}\,\rho_A(x,t)/M_A$ as the ordinate, we can superimpose the different results on the universal curve displayed in Figure 1.10b.

The variation given by Equation 1.23 is the response to a step-like variation of the initial temperature (or of the concentration of tracer) and not, as here, to a very localized injection of a finited amount of tracer. The corresponding solution given by Equation 1.28 differs therefore significantly from Equation 1.23: it is actually its spatial derivative.

1.3 Microscopic models for transport phenomena

The microscopic laws which we have derived for heat and mass diffusion have a very general applicability to fluids and solids. Up to this point, we have considered neither the microscopic nature of these transport processes nor the relation between the diffusion coefficients and the structure of the medium in which the transport occurs. We are going to give a simplified description of the microscopic mechanisms of transport. In the sections that follow we proceed to discuss mass diffusion in terms of a random walk, we apply kinetic theory to transport processes in gases and, finally, we treat the case of liquids.

1.3.1 The random walk

We study here again the spread of a drop of tracer initially localized at a point. Let us analyze the movement of the tracer particles which might be molecules, or very small particles, much less than a micron in size, sometimes called Brownian particles. Because of thermal motion, these particles are not stationary, but follow rather complex trajectories which entail a sequence of random changes in direction (*Brownian motion*).

To analyze this process, we use the model of a random walk (sometimes also known as the "staggering drunkard" problem!). Starting from an origin O at initial time $t = 0$, the walker takes steps of constant length ℓ and time duration τ setting out, at each subsequent step, in a new direction completely independent of the preceding one. Physically, ℓ corresponds to the mean free path of the particles; τ to the mean time between two collisions; and $\bar{u}$ to the thermal motion velocity (in this case, just as for random walk, the trajectory between two collisions is a straight line). We wish to evaluate the mean distance from the origin O at which the walker (representing one of the tracer molecules) is to be found at the end of a number of steps $N = t/\tau$. In statistical physics, we are concerned with averages: we therefore need to calculate the mean square displacement $< R(t)^2 >$ where R is the magnitude of the displacement of the walker from his starting point: in order to obtain a statistically significant value, the result is averaged over an ensemble of independent random walks.

The average of the vector displacement $\mathbf{R}(t)$ is, however, zero because all directions are independent and equivalent.

We then find that:

$$< R(t)^2 > = \frac{t}{\tau}\ell^2 = \frac{\ell^2}{\tau}t = D_m t \quad (1.30) \qquad \text{where:} \qquad D_m = \frac{\ell^2}{\tau} = \bar{u}\ell = \bar{u}^2\tau. \quad (1.31)$$

The root-mean-square (rms) displacement $\sqrt{< R(t)^2 >}$ relative to the initial position increases therefore like the square root of the time, a result which is thus equivalent to the law of diffusive spreading.

Proof

At the end of the first step, $<R(\tau)^2> = R(\tau)^2 = \ell^2$. We demonstrate by mathematical induction that, after N steps:

$$< R^2\,(N\tau) > = N\ell^2. \quad (1.32)$$

Let us assume that, after $(N\text{-}1)$ steps: $<R^2((N\text{-}1)\tau)> = (N\text{-}1)\,\ell^2$. For a given walk such as the one illustrated in Figure 1.11, let us denote by $\mathbf{M}_{N\text{-}1}$ the position at the end of step N-1, and by $\mathbf{M}_N$ that after the next (Nth) step. Thus, we have the vector equality:

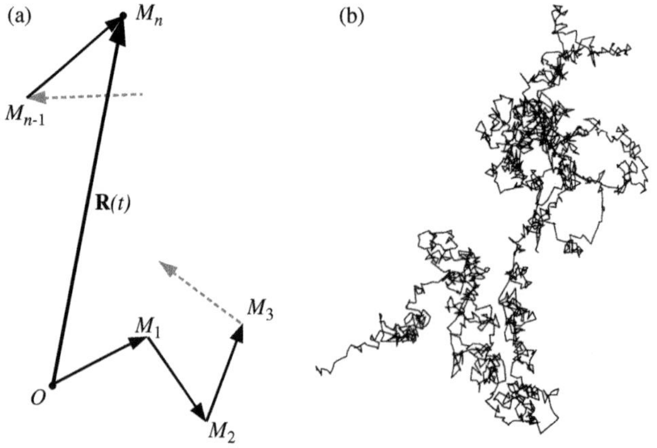

Figure 1.11 *(a) Displacement of a tracer particle during a Brownian random walk; we assume that each step has the same size. (b) A real life Brownian trajectory obtained by image analysis for a polystyrene sphere which was followed for* 2000 *s. An elementary displacement lasts on average* 1 *s.* [1 *cm on the graph corresponds to* 3.5 *μm*] *(doc. G. Bossis)*

$$(\mathbf{OM}_N)^2 = (\mathbf{OM}_{N-1} + \mathbf{M}_{N-1}\mathbf{M}_N)^2 = (\mathbf{OM}_{N-1})^2 + (\mathbf{M}_{N-1}\mathbf{M}_N)^2 + 2\,(\mathbf{OM}_{N-1})\,.\,(\mathbf{M}_{N-1}\mathbf{M}_N)\,. \tag{1.33}$$

Consider now the average value of Equation 1.33 over a large set of independent walks each starting from the origin O at time $t = 0$. Because of the fact that each step is completely independent of the previous one, the mean value of the scalar product $(\mathbf{OM}_{N\text{-}1}).(\mathbf{M}_{N-1}\,\mathrm{M}_N)$ must be zero (this product naturally has, with equal probability, positive and negative values). We find therefore the expected result:

$$(\mathbf{OM}_N)^2 = (N-1)\,\ell^2 + \ell^2 + 0 = \mathrm{N}\,\ell^2, \tag{1.34}$$

which is equivalent to Equation 1.30 because $t = N\tau$.

We should note that the proof of Equation 1.30 does not depend on the dimensionality of the space in which the random walk or the diffusion occurs. In the particular case where this walk occurs along a straight line, it is relatively easy to demonstrate that the law giving the probability of spreading from a point along this line is a Gaussian one. One recovers in this way independently the result of Equation 1.29 obtained by a purely macroscopic approach. This proof is based on the binomial theorem applied to a sequence of steps:

$$(p+q)^N = p^N + N\,p^{N-1}q + \ldots + C_r^N p^r q^{N-r} + \ldots + q^N. \tag{1.35}$$

If p (or q) (with $p + q = 1$) represents the probability of stepping to the right (or to the left), the general term $C_r^N p^r q^{N-r}$ gives the probability that we have taken r steps to the right and $N - r$ steps to the left. In the limit of a very large number of steps (N tending to infinity), we can show that the distribution of the binomial terms approaches a Gaussian, centered at the origin if $p = q = 1/2$.

In the previous analysis, we treated the case where every step of the random walk was of uniform length ℓ. It can be shown that the Gaussian model obtained under these conditions remains valid even when the distribution of the sizes of the random walk steps is not too broad (more precisely, it is sufficient that the probability of taking a step of size ℓ decreases more rapidly than $1/\ell^2$). If this is not the case, then the *rms* displacement of the random walk will increase with time more rapidly than $\sqrt{t}$, due to the fact that rare but very long steps will dominate the diffusive behavior. In this case, we talk about *abnormal diffusion* (here, *hyperdiffusion*).

If the random walk takes place along a plane which is slightly inclined, each step is slightly biased in the downhill direction. At the end of a number of steps N, the distribution of the walkers is no longer centered at the origin O, as in the case of the previous Gaussian, but is centered around a point O' such that $\mathbf{OO}'$ points downward. We thus observe a convective phenomenon which is superimposed on the diffusive phenomenon of the random walk: this convective phenomenon will dominate over long periods of time because the displacement OO' varies linearly with time while the diffusive transport distance increases only as $\sqrt{t}$. We can study this problem in the one-dimensional model wich we have just discussed, by taking probabilities p and q different for the right and left directions. We then find that the displacement of the center of gravity of the spot varies like $|OO'| = N\,(p\text{-}q)\,\ell$, but the spread remains Gaussian. This result represents an example of the diffusion of tracer in the presence of a permanent field of volume forces, such as in the case of gravity on heavier particles (sedimentation), or for an electric field on ions.

1.3.2 Transport coefficients for ideal gases

Representative elementary volume

Before trying to describe the macroscopic interpretation of transport coefficients, we will give here the definition of a *representative elementary volume* ($\mathcal{REV}$). In the mechanics of continuous media, we are called upon to define macroscopic quantities: pressure, temperature, velocity, density. These quantities are averages over corresponding microscopic quantities, on a scale (the $\mathcal{REV}$) large compared to that of the microscopic variations (m) but small by comparison with the macroscopic scale ($\mathcal{M}$) (Figure 1.12). This elementary

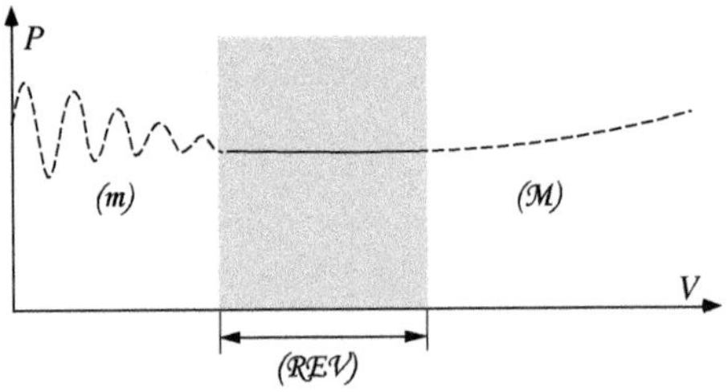

Figure 1.12 *Definition of a representative elementary volume REV) used for defining macroscopic quantities characterizing continuous media; V represents the volume over which the physical variable P is being averaged*

The range of length scales involved is indeed extremely variable from one system to another. In a simple liquid, the microscopic scale is of the order of a nanometer while macroscopic variations take place over millimetric distances (or more) in usual channels, but micrometers in microfluidic devices: the *REV* is then easily defined. In porous media studied in Section 9.7, the size of the pores is generally a few 10 μm in porous rocks and a *REV* can still be defined in homogeneous rocks. It will be much more difficult if the rock displays heterogeneities of millimetric size or more such as fractures.

representative volume is not found in all cases, particularly in very inhomogeneous media for which macroscopic changes occur over very short distances.

Molecular diffusion for an ideal gas

We consider again the case of the one-dimensional geometry of Section 1.2.1 (Figure 1.6) and we will study a dilute gas of tracer molecules, for which the number density $n(x)$ varies uniformly in the direction of the x-axis (Figure 1.13) The tracer molecules are accompanied by other molecules which redistribute themselves in order to compensate for the pressure variations associated to those of the density $n(x)$. The molecules move about with an effective velocity $\bar{u}$ due to thermal motion. We then obtain (see below) the following value for the diffusion coefficient:

$$D_m = \frac{1}{3}\bar{u}\,\ell. \tag{1.36}$$

We observe that Equation 1.36 coincides (within a factor of 1/3) with the model obtained for the random walk in Section 1.3.1 if we use, as the velocity between collisions, the thermal motion velocity of the molecules and if we take the length of each step to be the mean free path. In this particular case, this demonstrates the equivalence of the two phenomena, that of the spreading of a drop of tracer and the flux created by a gradient in the concentration: the two treatments, that of the kinetic theory of gases and that of the random walk, are thus equivalent.

Proof

In order to determine D_m, we evaluate the total flux of tracer particles across a plane cross-section located at x_0 and of unit area (Figure 1.13). We designate by $\mathcal{J}_+$ (or $\mathcal{J}_-$) the flux of tracer particles coming from the left (or from the right) of this plane (the flux represents the number of particles which cross a unit area per unit time). Because of the effect of the gradient in the concentration of particles $n(x)$, these two fluxes are different. The ratio between the difference $(\mathcal{J}_+ - \mathcal{J}_-)$ and $(-\partial n/\partial x)$ then gives us the molecular diffusion coefficient D_m. An exact calculation would use the distribution in magnitude and direction of thermal molecular velocities. We will make here a few simplifying assumptions:

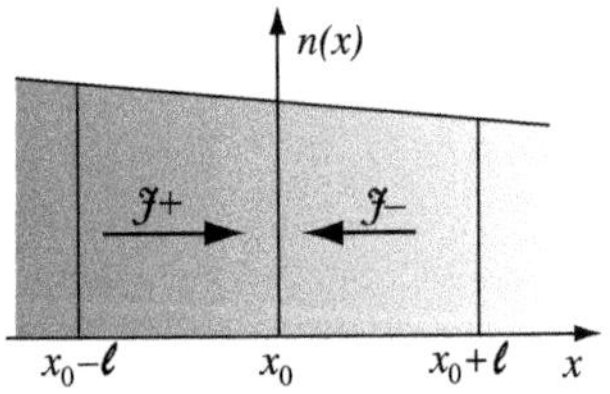

Figure 1.13 *Schematic representation of a one-dimensional, kinetic theory, diffusive transport model for tracer particles in a gas*

- We consider that only a third of the molecules move along the $\pm x$ direction with the thermal velocity $\bar{u}$; the remaining two-thirds will move in the y- and z-directions and, because of this, will not contribute to transport in the direction that we are considering.
- In order to evaluate the fluxes coming from the right or from the left, we divide the space into cells of side equal to the mean free path ℓ of the tracer particles. It is at this microscopic scale, which, however, is large compared to the particle size that exchanges occur. The mean free path will hereafter be considered as the minimum scale for the variations of the average properties: concentration, temperature and velocity.

Based on these assumptions, the flux $\mathcal{J}_+$ can be written as:

$$\mathcal{J}_+ = \frac{1}{2}\frac{1}{3} n\,(x_0 - \ell)\,\bar{u}. \tag{1.37a}$$

The extra factor 1/2 accounts for the fraction of particles which move in the positive x direction; $n(x)$ is the number density of molecules in the plane located at $x_0 - \ell$. Equation 1.37a represents actually the classical definition of flux and the factor 1/6 allows us to take into account the different directions of the molecular velocities. The main assumption is the introduction of the location $x_0 - \ell$ which amounts to assuming that particles which arrive at x_0 have taken a step exactly equal to ℓ. We write in the same way the flux $\mathcal{J}_-$ of particles coming from the right.

$$\mathcal{J}_- = \frac{1}{6} n\,(x_0 + \ell)\,\bar{u}. \tag{1.37b}$$

The net flux $\mathcal{J} = \mathcal{J}_+ - \mathcal{J}_-$ which results from the concentration gradient dn/dx is thus:

$$\mathcal{J} = \frac{1}{6}\bar{u}[n(x_0 - \ell) - n(x_0 + \ell)] = -\frac{1}{3}\bar{u}\,\ell\frac{dn(x)}{dx} = -D_m\frac{dn(x)}{dx}. \tag{1.38}$$

The resultant Equation 1.36 matches exactly (including the coefficient 1/3) the result of a more exact calculation based on the ideal-gas model.

Quantitative application to helium gas We now put into Equation 1.36 the properties of helium gas at standard temperature and pressure (STP). The mass of a helium atom is $m = \mathcal{M}/\mathcal{N}$ in which $\mathcal{M} = 4 \times 10^{-3}$ kg/mole and $\mathcal{N} = 6.02 \times 10^{23}$ atoms/mole is Avogadro's number. The rms thermal velocity of the helium atom is given by:

$$\frac{1}{2}m\,u_{rms}^2 = \frac{1}{2}m\,\overline{u^2} = \frac{2}{3}k_B T \quad (1.39a) \qquad \text{i.e.:} \quad u_{rms} = \sqrt{\frac{3k_B T}{m}}, \quad (1.39b)$$

where $k_B = 1.38 \times 10^{-23}$ J/K is Boltzmann's constant. We obtain the value $u_{rms} = 1{,}370$ m/s.

We get the mean free path ℓ from the classical derivation of this result in the kinetic theory of gases:

$$\ell = \frac{1}{\sqrt{2}n\sigma_c}, \tag{1.40}$$

where $\sigma_c = 1.5 \times 10^{-19}$ m^2 is the effective cross section of a helium atom and the number density of atoms is $n = \mathcal{N}/\mathcal{V}$ with a molar volume $\mathcal{V} = 22.4 \times 10^{-3}$ m^3.

We obtain: $\ell = 1.8 \times 10^{-7}$ m and $D_m = \frac{1}{3}u_{rms}\ell = 8 \times 10^{-5}\text{m}^2.\text{s}^{-1}$.

A more rigorous kinetic theory approach makes a distinction between $u_{rms} = \sqrt{\overline{u^2}} = \sqrt{3k_BT/m}$ and the mean velocity $\bar{u} = |\bar{u}| = \sqrt{8k_BT/(\pi m)}$. The former ($u_{rms}$) occurs in the microscopic derivation of the pressure of an ideal gas of non-interacting particles. The latter, $\bar{u}$, is the one which must be used for calculating transport properties. However, we note that these two speeds differ by only 8%, a totally negligible difference, as we are mainly interested in orders of magnitude.

Explanation of Equation 1.40

It indicates that, in a cylinder of cross-section σ_c and of length ℓ, we find, on the average, a single molecule; such a cylinder represents the volume "swept out" by the molecule as it moves through a distance ℓ. At least another particle must be present inside this volume for a collision to take place.

We should observe that, in this model valid for ideal gases (without particle interactions), the diffusivity D_m increases with temperature as $\sqrt{T}$ if the number density n is constant (since $\bar{u}$ has the same dependence).

Calculation of the thermal diffusivity of an ideal gas

By an argument similar to that used for molecular diffusion, we find the following values for the thermal conductivity and thermal diffusivity:

$$k = \frac{1}{2}\rho C_V \bar{u}\,\ell \qquad (1.41a) \qquad \text{and:} \qquad \kappa = \frac{k}{\rho C_V} = \frac{\bar{u}\,\ell}{2}. \qquad (1.41b)$$

where the thermal velocity $\bar{u}$ is given by Equation 1.39b, ρ is the density and C_V is the specific heat at constant volume. Equation 1.41b for the thermal diffusivity κ is very similar to Equation 1.36 for D_m. The mechanism which controls transport, diffusion through thermal motion, is essentially the same in the two cases.

We note that the thermal conductivity k is independent of the number density n of molecules so long as the approximation of an ideal gas is valid. In fact, ρC_V is proportional to the number density n and Equation 1.41a contains therefore the product $\ell n = n/(\sqrt{2}n\sigma)$ which is independent of n. This result, at first somewhat surprising, is understandable if we realize that, in increasing the number density n of particles, we increase the frequency of collisions between them but we reduce by a similar amount the mean free path and, consequently, the effectiveness of the transport.

Proof

Here, we assume that we have, on the one hand, particles of only one kind and that their number density n is constant. On the other hand, we also assume that there exists a temperature gradient constant in the x-direction. Keeping in mind Equation 1.39b, we will then also have a gradient of the velocity $\bar{u}$. Furthermore equations like 1.37 and 1.38, which evaluate the particle flux, will now describe the thermal fluxes desinated by $\mathcal{J}_Q$, such that:

$$\mathcal{J}_{Q+} = \frac{1}{6}\rho C_V T(x_0 - \ell)\,\bar{u}(x_0 - \ell) \qquad (1.42a)$$

and:

$$\mathcal{J}_{Q-} = \frac{1}{6}\rho C_V T(x_0 + \ell)\,\bar{u}\,(x_0 + \ell). \qquad (1.42b)$$

The quantity $\rho\, C_V T$ is the energy per unit volume associated to the thermal motion of the molecules.

In contrast to the previous case of mass diffusion, spatial variations of the thermal fluxes $\mathcal{J}_{Q+}$ and $\mathcal{J}_{Q-}$ result from those of both factors T and $\bar{u}$ (the average molecular velocity $\bar{u}$ due to the thermal motion also varies with T); since $\bar{u} \propto \sqrt{T}$, one has: $\partial\bar{u}/\partial x = (\bar{u}/2T)\partial T/\partial x$. Using this relation, the net thermal flux $\mathcal{J}_Q$ becomes, by summing the contributions of the variations of $\bar{u}$ and T:

$$\mathcal{J}_Q = \mathcal{J}_{Q+} - \mathcal{J}_{Q-} = -\frac{1}{2}\rho C_V \bar{u}\,(x_0)\,\ell\frac{\mathrm{d}T\,(x)}{\mathrm{d}x} = -k\frac{\mathrm{d}T\,(x)}{\mathrm{d}x}, \qquad (1.43)$$

from which we instantly derive Equations 1.41a–b.

Validity of ideal-gas models

The results obtained above do not apply either to gases at very low pressures or to dense gases which have properties approaching those of liquids.

- In the first case, if the mean free path ℓ becomes much larger than the size L of the container, there are essentially more collisions with the walls, and many fewer collisions between particles; but such collisions are the necessary condition to establish the statistical equilibrium state described by the kinetic theory of gases. This particular regime, called the *Knudsen* regime, occurs in containers at very low pressures: in fact, for pressures of the order of 0.1 Pascal, the mean free path of the molecules of a gas becomes of the order of tens of centimeters. We can therefore reach *Knudsen* regime in macroscopic-sized tubes.
- The second limit is that for which the average distance between particles (of the order of $n^{-1/3}$) is comparable to the mean free path. This is the situation in liquids, which we are about to discuss.

1.3.3 Diffusive transport phenomena in liquids

In contrast to gases, we cannot understand the different transport coefficients for liquids starting from a simple and unique model such as that of kinetic theory. We discuss here very briefly the cases of mass diffusion and of thermal diffusion.

Molecular diffusion coefficient in liquids

When we discussed transport properties in gases, we assumed that the interactions between molecules are negligible in the time interval between two collisions. In liquids, however, interactions always remain very important. Thus, we are going to begin by studying the diffusion of spherical particles of radius R used as tracers: in this case, if the particle moves relative to the liquid with a velocity $\mathbf{v}$, the force of interaction with the liquid is the *Stokes force:*

$$\mathbf{F}_1 = -6\pi\,\eta\,R\mathbf{v}, \tag{1.44}$$

where η is a coefficient known as the viscosity, characteristic of the fluid, which will be defined in Chapter 2. We justify this particular form of the interaction force in Section 9.4.2. If the particle is small enough (typically for R smaller than a micron), thermal motion effects are significant enough for us to calculate the corresponding molecular diffusion coefficient. We extrapolate then this result to smaller diameters in order to *estimate* the diffusion coefficient for a molecular tracer. We can put Equation 1.44 in the form:

$$\mathbf{F}_1 = -\frac{\mathbf{v}}{\mu} \quad (1.45) \qquad \text{with:} \qquad \mu = \frac{1}{6\,\pi\,\eta\,R} \quad (1.46)$$

(μ is called the *mobility*). In a famous paper on Brownian motion published in 1905, Einstein derived the following general relationship between the molecular diffusion coefficient D_m and the mobility μ:

$$D_m = \mu\,k_B T. \quad (1.47) \qquad \text{i.e., after Equation 1.46:} \quad D_m = \frac{k_B T}{6\,\pi\,\eta\,R}. \quad (1.48)$$

We observe that the coefficient D_m indicates a spreading, in the absence of external forces but in the presence of thermal motion, while the mobility μ is defined in the presence of an external force $\boldsymbol{F}_1$.

Equation 1.48 applies reasonably well for particle radii even down to molecular dimensions; we can therefore use it to estimate an order of magnitude for the diffusion coefficient in liquids. Thus, for a molecule 1 nm in diameter and a liquid of viscosity $\eta = 10^{-3}$ Pa.s, we find: $D_m = 2.2 \times 10^{-10} \mathrm{m}^2/s$– a value much smaller than that obtained for gases.

Proof of the Einstein relation

In order to prove the Einstein relation, we will write, for a particular case, the result of equilibrium between the effects of thermal motion and of an external force. Let us assume that we have a group of particles of some tracer (molecular or not), of mobility μ and diffusion coefficient D_m submitted to a constant force field f oriented along the x-direction. Practically, this model might correspond to the behavior of a group of Brownian particles sedimenting as a result of gravity. We assume that they are in thermal equilibrium with a thermostat at temperature T. The force field leads to a potential $U = -fx$. This potential induces a gradient dn/dx of the number density n of particles; locally, $n(x)$ obeys the Boltzmann distribution law.

$$n(x) = n_0 \, \mathrm{e}^{-\frac{U}{k_B T}} = n_0 \, \mathrm{e}^{\frac{fx}{k_B T}} \qquad (1.49) \qquad \text{whence:} \quad \frac{1}{n}\frac{\mathrm{d}n}{\mathrm{d}x} = \frac{f}{k_B T}. \qquad (1.50)$$

The gradient dn/dx leads to a diffusive particle flux $\mathcal{J}_m$ such that:

$$\mathcal{J}_m = -D_m \frac{\mathrm{d}n}{\mathrm{d}x} = -D_m \, n \frac{f}{k_B T}. \qquad (1.51)$$

The force f leads to an average drift velocity of the particles $v_d = \mu f$. We observe that v_d is an average velocity for the particles taken as a group; it is generally much smaller than the individual particle velocities due to the thermal motions which are, instead, randomly oriented in all directions. v_d results in a particle flux $\mathcal{J}_d$ such that:

$$\mathcal{J}_d = n \, v_d = n \, \mu f. \qquad (1.52)$$

For statistical equilibrium, these two fluxes must cancel each other out so that $\mathcal{J}_m + \mathcal{J}_d = 0$. In combining equations (1.51) and (1.52) into this result, we find the Einstein relation (1.47).

Thermal conductivity of liquids

In a liquid there are two mechanisms of heat transfer. The first results from the nearest neighbor propagation of individual particle vibrations in the liquid (represented approximately [Figure 1.1] in the model experiment using a dense ensemble of vibrating disks). The second occurs for liquids metals (Hg, Na, ...) and involves the electrons which are also responsible for electrical conductivity. These two mechanisms are very similar to those of the thermal conductivity in a crystalline solid; we will not discuss them here. One notes, however, the great effectiveness of the mechanism of electronic transfer which results in the fact that liquid metals are very good conductors of heat, while also having a high specific heat. Liquid sodium is, for example, used as a thermal conduction liquid in the heat exchangers of breeder nuclear reactors.

Values of diffusive transport coefficients

In Table 1.2, we indicate the values of diffusive transport coefficients for a few pure substances. In addition to the coefficients D_m and κ which we have discussed in this chapter, we have added the ***kinematic viscosity*** coefficient ν. It represents the diffusivity of momentum and its physical meaning will be discussed in detail in Chapter 2 (ν is proportional to the dynamic viscosity η introduced in Equation 1.44 and inversely proportional to the density ρ, so that $\nu = \eta/\rho$). The coefficients D_m, κ and ν each have the same dimension $[L]^2[T]^{-1}$. In many processes, the two diffusion phenomena might occur at the same time. Their relative effect will then be a very important parameter. We characterize this by means of a dimensionless number which represents a ratio of the corresponding time constants evaluated along a distance of the order of the characteristic length L of the flow. We will see several examples of such dimensionless numbers in Section 2.3. They play an important role in a number of phenomena involving heat and mass transfer.

1.4 Surface effects and surface tension

The interfaces between a liquid and another body, be it another gas, another liquid or a solid, play a very important role in the equilibrium and flow of films or of liquid layers. We will first introduce the coefficient called the surface tension, and discuss its relationship with the energy of these interfaces.

1.4.1 Surface tension

Evidence of the effects of surface tension

Let us consider the experiment shown in Figure 1.14: a liquid film (soapy water, for example) is held on a rectangular frame in which one side is movable. If we allow this side to move freely, we observe that it moves in such a way as to minimize the surface area of the film. In order to hold it stationary, we must exert a force F which is proportional to the length L of the movable side. If we want to increase the surface area of the film by a value $dS = L\,d\ell$, we must supply an energy dW which corresponds to the work done by force F and is expressed by:

$$dW = F\,d\ell = 2\gamma L\,d\ell = 2\gamma\,dS. \tag{1.53}$$

γ is called the *surface tension* coefficient between the liquid in question and air; the factor 2 results from the fact that the liquid film is made up of two liquid–air interfaces. Equation 1.53 indicates that γ corresponds to an energy per unit area for each interface; it also represents the value of the force per unit length exerted by each interface on the side of the

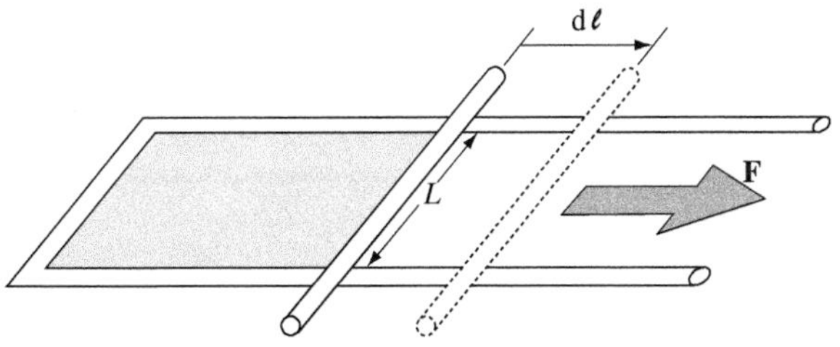

Figure 1.14 *A simple experimental illustration of surface tension*

frame. Its units are therefore Newtons per meter (N/m). Its value for pure water at room temperature is approximately 70×10^{-3} N/m. In Table 1.1, we give numerical values of the surface tension and of its dependence on temperature for a few common liquids; the values given correspond to the surface tension between liquids and air; they are quite different in the case of interfaces between two liquids, or between a liquid and solid, as we will see in the next paragraph.

Table 1.1 *Order of magnitude of a few physical parameters characterizing the interfacial properties of some common liquids.*

	Surface tension γ (N/m)	Temperature dependence of γ $-\frac{d\gamma}{dT}$ (mN/(m.K))	Capillary length $l_c = \sqrt{\frac{\gamma}{\rho g}}$ (m)
Liquid metals	$7\ 10^{-2}$ to 2.5	10^{-2} to 10^{-1}	2 to 5×10^{-3}
Organic liquids	50×10^{-3}	10^{-2} to 10^{-1}	1 to 3×10^{-3}
Molten salts	10^{-1}	10^{-2}	2 to 3×10^{-3}
Silicone oils	20×10^{-3}	10^{-2}	10^{-3}
Water	70×10^{-3}	10^{-1}	3×10^{-3}
Molten glass	10^{-1}	10^{-2}	5×10^{-3}

Surface tension phenomena result in minimizing the area of the interface, while taking into account the other constraints placed on the system (gravity, pressure . . .). Particularly, in the absence of gravity, a drop takes on a spherical shape because this results in a minimized surface for a given volume.

Physical origin of the surface tension forces

Surface tension is associated with the internal attractive forces between the molecules of a fluid: van der Waals forces, hydrogen bonds (for example in water), ionic bonds, metallic bonds (in metals such as mercury). Within the volume of a fluid, the forces exerted by each molecule are compensated by those exerted by its neighboring molecules. If we introduce an interface, for example in a vacuum, the forces exerted in that direction are not compensated: this is the origin of surface energy. The value of the surface tension will vary greatly according to the nature of the forces which occur between atoms or molecules. The high surface tension of liquid metals (0.48 N/m for mercury and up to 2.5 N/m for osmium at 3000 K) is explained by the large value of the energy associated with metallic bonds; in contrast, the van der Waals forces which play a dominant role in numerous molecular substances, only give values of the order 20×10^{-3} to 25×10^{-3} N/m.

Surface tension, which we have just defined, involves an interface between a liquid and vacuum. In the case of contact between two bodies, the surface energy of each is modified by the presence of the other, and we talk about *interfacial tension*. The latter depends on the surface tension of each of the two components, as well as on the energy of interaction between them. In order to indicate that we are talking about an interfacial tension, we give a subscript corresponding to the bodies involved; thus γ_{AB} represents the interfacial tension (i.e., the energy per unit area between a substance A and a substance B).

Effect of temperature on surface tension

The reduction of the cohesion of a liquid with increasing temperature leads to a reduction of the surface tension. For moderate changes in temperature, we can use the following linear dependence:

$$\gamma(T) = \gamma(T_0)\,[1 - b(T - T_0)]\,. \tag{1.54}$$

The coefficient b, of the order of 10^{-2} to 10^{-1} K^{-1}, is positive, which indicates that the surface energy decreases as the temperature increases. Thus, the surface tension becomes zero at the critical point. We will see in Section 8.2.4 that these variations can result in flows (Marangoni effect).

Changes in surface tension in the presence of a surfactant

The presence of a third substance (or surfactant) at the interface between two fluids can lead to a reduction of the surface tension. Without going into great detail about the reciprocal interaction between three bodies at the same place, we can explain qualitatively this reduction in surface tension: consider a fatty-acid surfactant (such as stearic acid, which is the main component of candles). It is a compound formed by a polar, acid, partially ionized head, and of a long tail made up of CH_2 radicals. In the presence of water, the molecule aligns in such a way that its polar head is on the water side (we talk about a *hydrophilic* head) and the aliphatic *hydrophobic* tail toward the external medium. For this reason we talk about amphiphilic compounds. This interface of surfactant molecules between two fluids reduces the direct interaction between their molecules, and accordingly reduces the interfacial tension as a result of the affinity of the surfactant for each of the two fluids. It then becomes energetically more favorable for the system to increase the surface area between the two fluids.

Surfactants play a key role in numerous problems of physical chemistry, particularly in the area of detergents. These molecules attach to drops of fatty material with the hydrophobic tail at the interior and the hydrophilic head toward the exterior: this allows the fats to dissolve in water. Surfactants also affect the mobility of liquids near interfaces.

1.4.2 Pressure differences associated with surface tension

Young–Laplace's Law

Consider now, as in Figure 1.15a, a spherical drop (1) of radius R immersed in another fluid (2). For this droplet to be in equilibrium, the pressure of the interior fluid must exceed that on the outside by an amount:

$$p_1 - p_2 = 2\frac{\gamma}{R}. \tag{1.55}$$

For a soap bubble, the pressure difference measured between the inside and the outside of the bubble would be twice this value, again due to the presence of two liquid–air interfaces, each of which gives a contribution $2\gamma/R$ (Figure 1.15b).

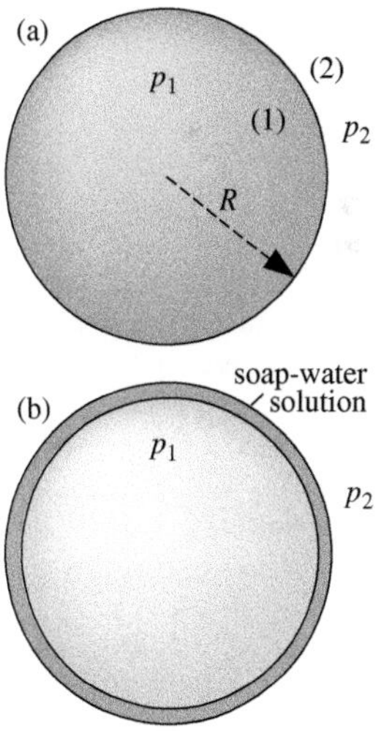

Figure 1.15 *(a) Capillary pressure difference p_1 - p_2 between the interior (1) and outer (2) regions of a spherical fluid droplet immersed in an external fluid; (b) the case of a soap bubble where there are two interfaces between soapy water and air*

Proof

The radius of the drop corresponds to the equilibrium between the surface tension effects (minimizing the area of the interface) and the excess pressure inside the drop relative to the external one (this excess pressure can, for example, be maintained by connecting the inside of the droplet to an external reservoir at fixed pressure p_1).

To derive Equation 1.55, let us apply the principle of virtual work, corresponding to an increase dR of the radius of the drop, under constant pressure difference pression $\Delta p = p_1 - p_2$. The value of Δp for mechanical equilibrium is such that the variation of the total energy dW_1 vanishes. Two contributions make up dW_1:

- one, dW_s, is due to the change in surface energy of the sphere:

$$dW_s = d\left(4\pi\,\gamma R^2\right) = 8\pi\,\gamma R\,dR, \tag{1.56}$$

- the second, dW_p, corresponds to the work done by pressure forces:

$$dW_p = -\Delta p\,dV = -(p_1 - p_2)\,d\left[(4/3)\,\pi R^3\right] = -(p_1 - p_2)\,4\pi R^2 dR. \tag{1.57}$$

In writing $dW_s + dW_p = 0$, we obtain Equation 1.55, also known as the Young–Laplace's law.

For the case where the boundary surface between the two fluids has arbitrary shape, the Young–Laplace's law takes on the more general form:

$$p_1 - p_2 = \gamma\left(\frac{1}{R} + \frac{1}{R'}\right) = \gamma C, \tag{1.58}$$

where R and R' are the principal radii of curvature of the surface at the point in question; the principal radii are defined as the extreme values of the radii of curvature of the curves, sectioned from the surface by a pair of mutually orthogonal planes, each containing the normal **n** to the surface (Figure 1.16a). The sum $C = (1/R) + (1/R')$ is called the *local mean curvature* of the interface. The radii of curvature R and R' have algebraic signs and are positive when the corresponding center of curvature is located on the side of fluid (1). In Figure 1.16b we see a liquid film bounded by a frame in a configuration where the average curvature at each point is zero: R and R' are different from zero but of opposite sign. This allows the film to satisfy Young–Laplace's law, Equation 1.58, when, like in this case, the pressure on the two sides of the interface is the same.

The Young–Laplace's law affects quite a number of physical phenomena, among which we might mention in particular the nucleation of bubbles in a boiling liquid. For a liquid in equilibrium being gradually heated, it is impossible for minute vapor bubbles to appear at the normal boiling point, because this would require a correspondingly high excess pressure inside the bubble at the moment that the bubble is formed ($\Delta p \propto 1/R$); as a result, boiling is delayed. Boiling generally occurs only when microscopic bubbles, already existing within the fluid, begin to grow. To minimize this delay in boiling, bubble generators, such as glass beads, may be placed in the liquid; these then provide a minimum size scale for vapor bubbles to nucleate and grow. The larger this latter scale, the smaller the superheating.

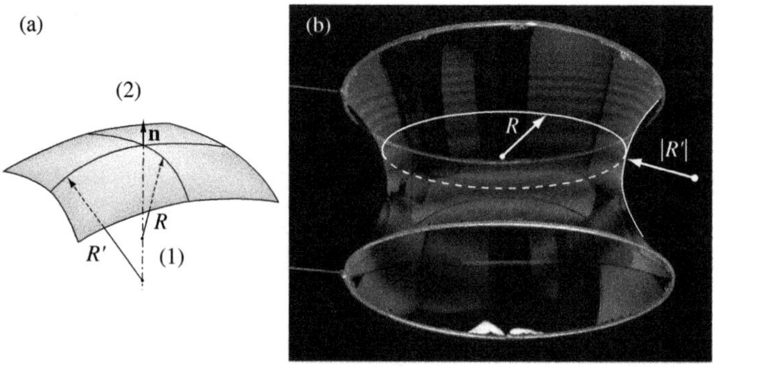

Figure 1.16C *(a) Geometry of an arbitrary boundary surface between two fluids (1) and (2), illustrating the corresponding definition of the principal radii of curvature R and R′; (b) the surface of a soap film stretched between two circular metal rings (at the top and bottom of the figure) is open and, thus, the pressure on the two sides of the film is the same. The local mean curvature C = [(1/R) + (1/R′)] is thus zero at every point. The surface generated in this way, a catenoid, achieves a minimum in the surface area of the film, while taking into account the boundary conditions imposed by the two rings. (after a photograph by S. Schwartzenberg, © Exploratorium, www.exploratorium.edu)*

1.4.3 Spreading of drops on a surface – the idea of wetting

In this section, we study the conditions under which a drop of liquid *wets* a solid surface—that is spreads out on the surface.

Spreading parameter

In order to see whether the spreading of a liquid (l) on a solid plane (s) is energetically favorable, we compare the interface energy values for a solid–vacuum interface when covered and not covered by a liquid layer (Figure 1.17). We call γ_{sl} the surface energy per unit area in the presence of a liquid, γ_{so} the energy relative to vacuum, and $\gamma = \gamma_{lo}$ the *interfacial energy* between the liquid and vacuum. The total surface energies in these two situations are respectively $F_{\sigma f} = \gamma_{lo} + \gamma_{sl}$ and $F_{\sigma} = \gamma_{so}$. We will assume that the thickness of the film is sufficiently large compared to the range of the intermolecular forces so that we can neglect the interaction between the two interfaces (we refer to this to as a macroscopic film). There exists, however, microscopic films of submicron thickness for which this condition is not satisfied.

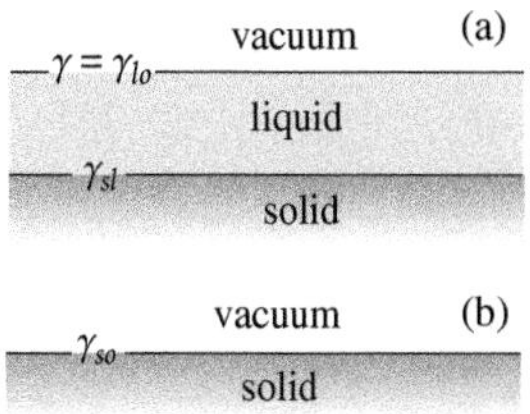

Figure 1.17 *Calculation of the interfacial energies for a surface which is covered (a), or not covered (b) by a liquid film*

The presence of a macroscopic film will therefore be energetically favorable if the difference

$$S_0 = F_{\sigma} - F_{\sigma f} = \gamma_{so} - \gamma_{sl} - \gamma \tag{1.59}$$

is positive. We call S_0 the *spreading parameter*. In most cases, the external environment of the surfaces consists of a gas, and we must substitute for the surface energies γ_{lo} and γ_{so} relative to the vacuum the energies γ_{lg} and γ_{sg} relative to the gaseous environment. These differences can be significant, particularly between γ_{sg} and γ_{so} because of the absorption phenomena, and, more specifically, if the gaseous phase contains liquid vapor (equilibrium condition in the case of a saturated vapor). In that case we make practical use of the spreading parameter S.

$$S = F_{\sigma} - F_{\sigma f} = \gamma_{sg} - \gamma_{sl} - \gamma, \tag{1.60}$$

where γ refers here to the interfacial tension γ_{lg} between the liquid and the gas.

Partial and total wetting

When the spreading parameter S is negative, we are in a situation of partial wetting; when S is positive, we have total wetting and there might exist a liquid film on the solid surface. From a molecular point of view, "hard" solids, having strong bonds (ionic, covalent or metallic), have large surface energies (γ_{so} = 0.5-1 N/m). Examples of these are glass, silica and metallic oxides: these are generally wetted by most molecular liquids. They are also easily contaminated by impurities present in the environment, which could alter the wetting behavior.

On the other hand, the solids with lower energy bonds (van der Waals, hydrogen bonds) will have weaker surface energies ($\gamma_{sg} = 5 \times 10^{-2}$ N/m). This will be the case for polymers, for Teflon or for paraffin: such surfaces are far more difficult to contaminate. Wetting will be total or partial, depending on the liquid involved.

Equilibrium of a drop on a plate

Let us first consider the case of small droplets for which gravity is negligible (Figure 1.18a). We can write the requirements for the equilibrium of forces on the triple contact line of the

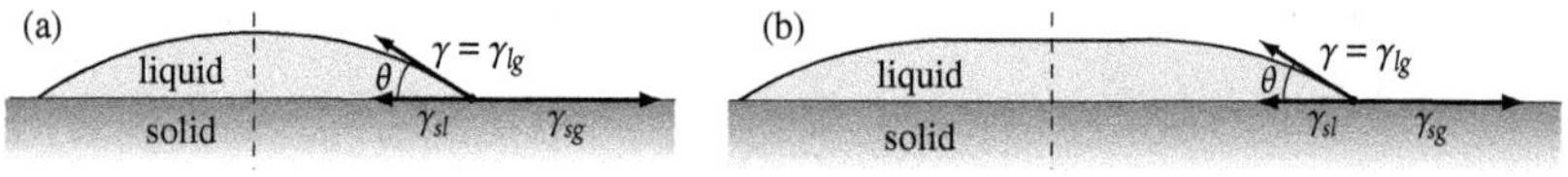

Figure 1.18 *(a) Equilibrium of a droplet wetting a flat solid plane, when the droplet is small enough for gravity effects to be negligible; (b) case of a large droplet flattened by the effect of gravity*

liquid interface with the plate. The vertical component is compensated by the elastic reaction of the solid plate (for spreading on a liquid substrate, we would observe a deformation of the liquid). We can write the equilibrium of the horizontal components of the force in the form:

$$\gamma_{sl} + \gamma \cos\theta = \gamma_{sg}. \tag{1.61}$$

Combining this result with Equation 1.60 for the spreading parameter, we obtain the law of Young–Dupré:

$$\gamma(\cos\theta - 1) = S, \tag{1.62}$$

where θ is called the *static contact angle*. We note that, for $S = 0$, we have $\cos\theta = 1$. Larger drops are flattened (Figure 1.18b) but the contact angle at the interface with the surface keeps the same value.

Movement of an interface

Let us now consider the case of moving contact lines. We are talking about motion at infinitesimally small velocities under quasi-static conditions. The case of interfaces moving at finite velocities will be dealt with in Section 8.2.2. Even when the motion is extremely slow, we observe that there is a difference between the contact angles of a liquid with the wall, depending on the direction of the motion of the contact line. We observe this effect in the spreading or retraction of drops on a plane (Figure 1.19a and b), as well as in the case of spontaneous motion of a drop on an inclined plane (Figure 1.19c) or when a hanging drop moves within a tube (Figure 1.19d).

At very low speeds of movement considered in this chapter, this effect is due to the presence of surface roughness or of chemical inhomogeneity on the solid surface. We talk about an advancing θ_a or receding θ_r contact angle the limiting values in each of these situations, where the speed of the contact line tends toward zero.

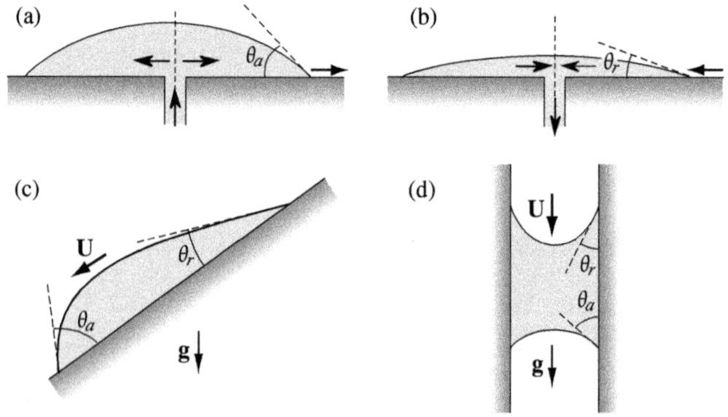

Figure 1.19 *Difference between advancing (a) and receding (b) contact angles for droplets placed in a situation of partial wetting when (a) their volume is increased or (b) decreased. (c) Advancing and receding contact angles at the front and rear of a droplet moving at very low speed down an inclined plane. (d) Advancing and receding contact angles for the case of a small liquid bubble which moves very slowly downward in a capillary tube*

1.4.4 Influence of gravity

Shape of droplets; the Bond number

The effects of capillarity, which are directly related to the curvature of interfaces, are important whenever we look at small-scale phenomena. For large objects, they are overridden by the effects of volume forces such as gravity. As an illustration of this, let us examine the sequence of mercury droplets of different sizes placed on a horizontal surface (Figure 1.20). The smallest are spherical, while the larger ones display a flattened shape due to the effects of gravity.

Figure 1.20 *Flattening, under the action of gravity, of mercury droplets of various sizes lying on a flat glass plate. The smallest droplet has a diameter of the order of 2 mm. We note that the mercury does not wet the glass, i.e. near the glass the mercury surface is convex. The idea of wetting has been discussed in a previous paragraph (Photograph by the authors.)*

Let us estimate the order of magnitude of the differences between the capillary and hydrostatic forces affecting a droplet in the shape of a spherical cap placed on a flat surface (Figure 1.21).

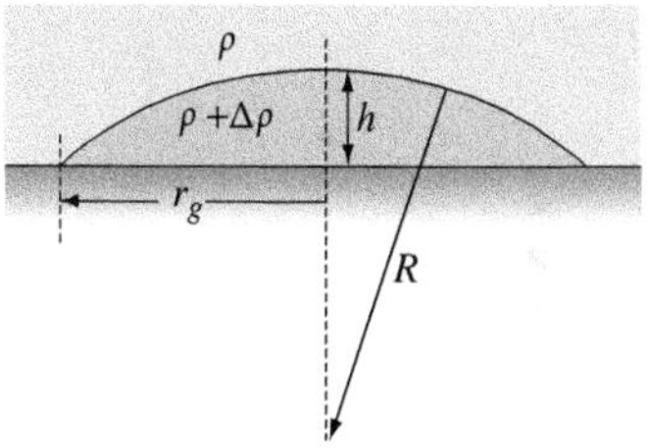

Figure 1.21 *Geometry of a partially wetting droplet placed on a flat plate*

In more general terms, let us consider a droplet of fluid of density $\rho + \Delta\rho$ immersed in a fluid of density ρ. The difference between the capillary and hydrostatic pressures over the height h of the droplet is respectively:

$$\Delta p_{\mathrm{cap}} = 2\frac{\gamma}{R} \quad (1.63a) \qquad \text{and:} \qquad \Delta p_{\mathrm{grav}} = \Delta\rho\, g\, h. \quad (1.63b)$$

Also, if the droplet is not too strongly deformed relative to a spherical shape, the radius r_g of the contact line can be written as $r_g^2 \simeq 2Rh$. We can therefore characterize the relative importance of the effects of gravity and capillarity by the ratio of the corresponding pressure differences:

$$Bo = \frac{\Delta\rho\, g\, h}{2\gamma/R} \approx \frac{\Delta\rho\, g\, r_g^2}{\gamma}. \qquad (1.64)$$

This ratio is called the *Bond number*. A large value of the Bond number corresponds to the effects of gravity dominating those of surface tension. The value $Bo = 1$ allows us to define a critical scale l_c, called the *capillary length*. It is expressed as:

$$l_c = \sqrt{\frac{\gamma}{\Delta\rho\, g}}. \qquad (1.65)$$

Typical values of l_c are given in Table 1.1. In the case of mercury, it is of the order of $l_c \simeq 2$ mm; for water, $l_c \simeq 3$ mm. Small values of the Bond number (for which capillary effects dominate) are also observed in hydrodynamic experiments under microgravity conditions. We can also reduce the effective density in experiments which use two non-miscible fluids with a small difference $\Delta\rho$ in their density; we then talk about compensated gravity (due to the effect of Archimedes' buoyancy). To gauge the relative importance of surface tension for given flow conditions, one compares l_c to the characteristic dimensions of the flow.

Capillary rise along a wall

The capillary length l_c appears in numerous phenomena. It represents, for example, the order of magnitude of the capillary rise of a fluid along a vertical wall in the case where the fluid is wetting, i.e. when the center of curvature of the interface is outside the liquid (Figure 1.22).

We will determine the height to which a liquid rises in the neighborhood of the wall. We should point out that there might exist a very thin, microscopic, film (a few hundred Angstroms in thickness) above the contact line of the meniscus with the solid wall. We will neglect the existence of this film in the present discussion.

The pressure inside the liquid at an ordinate $y(x)$ below the interface can be written: $p_s(x) = p_{at} + \rho\, g\, y(x) - \gamma/R(x)$ where $R(x)$ represents the local radius of curvature of the interface (it is positive in the present case) and p_{at} is the air pressure above the interface. The term γ/R represents the effects of surface tension, while the term $\rho g y(x)$ represents those of the hydrostatic pressure. On the other hand, just below the interface, in its flat region, the pressure in the liquid is p_{at}. We therefore arrive at the equation:

$$\rho\, g\, y(x) = \frac{\gamma}{R(x)}. \tag{1.66}$$

Using the geometrical relationship $R(x) = \left[\left(1 + y'(x)^2\right)^{3/2}\right]/y''(x)$, we obtain the differential equation:

$$\rho g y = \gamma \frac{y''}{(1 + y'^2)^{3/2}} \quad (1.67) \qquad \text{i.e.:} \quad d(y^2) = -2\frac{\gamma}{\rho g} d\left(\frac{1}{\sqrt{1 + y'^2}}\right). \quad (1.68)$$

We observe that there appears the scaling length characteristic of the problem, i.e. the capillary length $l_c = \sqrt{\gamma/\rho g}$. By integrating, using the asymptotic conditions for very large

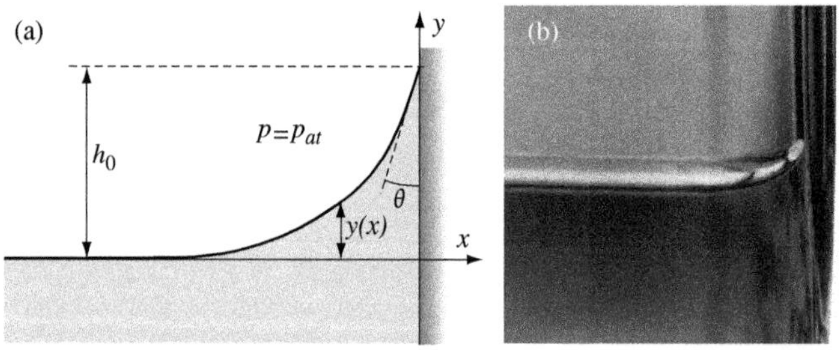

Figure 1.22 *Rise of a wetting fluid on a vertical wall: (a) schematic diagram; (b) experiment conducted with water in a glass container. The rise is of the order of a millimeter. (Photograph C. Rousselin, Palais de la Découverte)*

x: $y \to 0$ and $y' \to 0$, and observing that $y'(x = 0) = -\cot\theta_0$, we obtain the capillary rise along the wall:

$$h_0^2 = 2l_c^2(1 - \sin\theta_0). \tag{1.69}$$

This capillary rise is, as expected, of the order of magnitude of the capillary length, corrected by a factor involving the wetting properties of the wall by the liquid (contact angle θ_0). The case of Figure 1.22 is that of a wetting fluid ($\theta_0 < 90°$). For a non-wetting fluid such as mercury ($\theta_0 > 90°$), the interface drops in the neighborhood of the wall instead of rising, and h_0 represents the height to which the liquid goes down near the wall, below its free surface far away from this wall.

Capillary rise in a tube: Jurin's law

When a capillary tube is placed in a liquid (Figure 1.23a), we observe, as in the previous case, a surface deformation near the walls, but also a rise of the meniscus in the tube relative to the external fluid if the tube is of a small diameter and if the fluid wets the walls of the tube (contact angle θ with the wall less than 90°); if the fluid is non-wetting, as in the case of mercury, the level of the interface in the tube drops below the external fluid.

We calculate now the difference in levels between the inside and outside of the capillary tube. We write the equilibrium condition for the column of liquid located in the tube (column of diameter d and height h) under the effect of its weight ($\pi d^2 h\rho g/4$) and of the pressure difference at the level of the interface (Figure 1.23b). Let us assume that the interface between the liquid and air at the interior of the capillary is spherical and denote by R its radius of curvature. The pressure P_1 just below the interface is equal to ($P_{at} - 2\gamma/R$). On the other hand, R is related to the diameter d of the capillary tube and to the contact angle θ by the equation $d = 2R\cos\theta$. For the pressure difference between the two sides of the interface, we therefore have:

$$\Delta p = P_{at} - P_1 = \frac{4\gamma\cos\theta}{d}. \tag{1.70a}$$

The pressure P_1 also equals $P_{at} - \rho\, g\, h$. Substituting this expression for P_1 in Equation 1.70a, we obtain:

$$h = \frac{4\gamma\cos\theta}{\rho g d} = 4\frac{l_c^2}{d}\cos\theta. \tag{1.70b}$$

This is Jurin's law. It is all the better verified since we can neglect relative to h the capillary rise on the walls given by Equation 1.69. This implies that r should be very small compared to h. For a value of the surface tension $\gamma = 7\times10^{-2}$ N/m (water), $\cos\theta = 0.6$ and $d = 1$ mm, we find a value $h \approx 20$ mm, much greater than l_c ($l_c = 3$ mm for water). For a non-wetting fluid (contact angle $\theta > 90°$), Equation 1.70b is still valid but gives a negative value of h; this implies a lowering of the level of the interface in the tube.

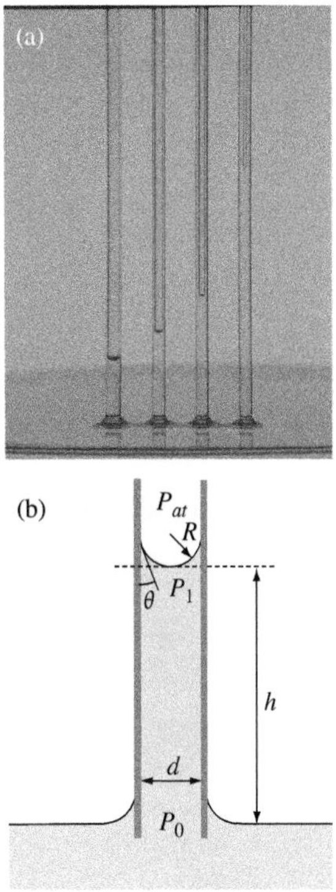

Figure 1.23 *(a) Capillary rise of fluid in a set of tubes of diameters d decreasing from left to right (document K. Piroird); (b) schematic diagram for the calculation of the rise height*

1.4.5 Some methods for measuring the surface tension

Several of these methods are based on the phenomena which we have described above. Among the first, surface tension is measured directly without requiring some knowledge of the contact angle of the liquids with the corresponding solid surfaces.

Measures of the geometry or of other characteristics of droplets

- ***Method of the hanging droplet***
  Influenced by gravity, a droplet is formed at the end of a capillary tube and its shape is analyzed by image processing. It is then compared to theoretical profiles by adjusting the value of the surface tension until the agreement is satisfactory. The accuracy of this estimate can be of the order of 1 percent, or even better in ideal cases.
- ***Weight of droplets formed at the end of a capillary tube*** (dropper)
  The weight of droplets produced by a capillary tube is approximately proportional to the surface tension, and depends on the radius of the tube. In averaging a sufficient number of measurements and applying appropriate corrections, one can obtain an accuracy of the order of a few percent on this type of measurement.

We will now describe methods which involve the contact angle of liquids with solid walls. Combined with the methods just described, they allow us to determine at the same time the surface tension and the contact angle.

Measurement of forces on solid surfaces

These methods often display problems in reproducibility in the preparation of the surfaces, and depend on the manner in which the surfaces are put in contact with fluids.

- ***Measurement of the height of capillary rise***
  This method consists of measuring the height to which a liquid rises in a capillary tube. We determine the surface tension by using Jurin's law, established in Equation 1.70b.
- ***Measurement of the pulling force of the interface on a solid object.***
  We draw vertically upward a hollow cylinder of radius R immersed in a liquid, and we measure the pulling force exerted by the liquid on the cylinder at the moment at which the latter detaches from the interface. The force F obeys:

$$F = P_{\text{Arch.}} + 2\,(2\pi R)\,\gamma\cos\theta$$

where $P_{\text{Arch.}}$ represents the Archimedes' buoyancy exerted on the cylinder (factor 2 results from the presence of two liquid–solid–air contact lines at the level of the cylinder). At the moment when the cylinder detaches from the surface, as shown in Figure 1.24, Archimedes' buoyancy is no longer effective and the liquid–air interface is vertical at the point of contact with the cylinder. We therefore have $\cos\theta = 1$ and:

$$F = 4\pi\,R\,\gamma \tag{1.71}$$

We should note that, in order to have reliable measurements, one must work with very clean surfaces.

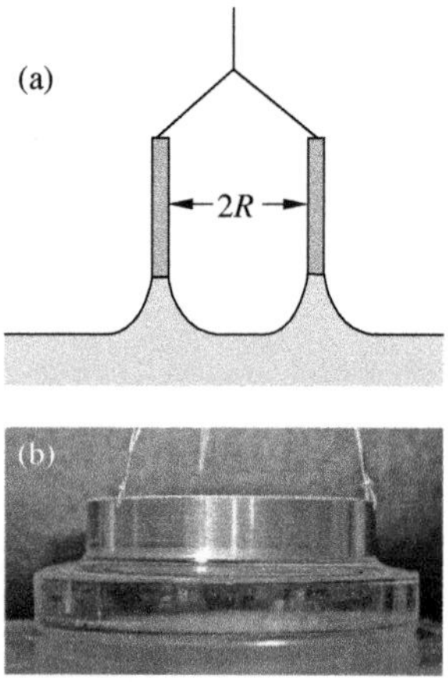

Figure 1.24 *Measurement of the surface tension by the method of pulling a hollow cylinder (a) Schematic diagram of the measurement (b) an actual experimental setup (courtesy of P. Jenffer)*

This method of determining surface tension by measuring the force is known as Wilhelmy's method When by when the solid object is a plate and du Noüy's method when it is a ring. In practical devices, correction factors must be introduced in Equation 1.71.

Measurement of the contact angle

The contact angles of a liquid surface with a wall can first of all be determined by combining measurements of the force or of the capillary rise on interfaces with direct measurements of the surface tension. They can also be determined directly by observations with a microscope in favorable cases. Other techniques allow for more rapid and precise measurements in particular cases.

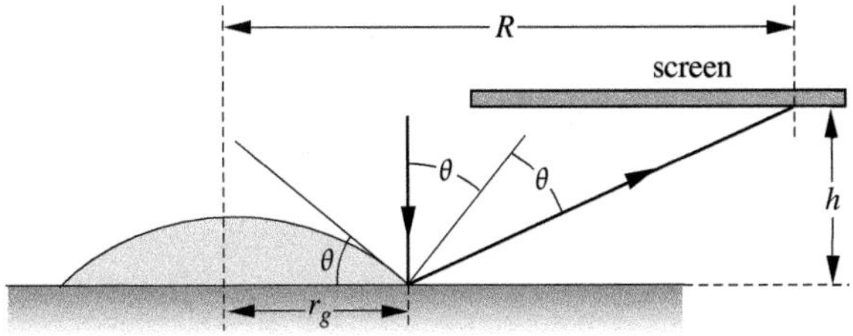

Figure 1.25 *Schematic diagram of the measurement of the contact angle between a droplet and a flat plane by reflection from the upper surface of the droplet*

- ***Interferometric measurements***. Using monochromatic light, the measurement of fringes of constant width near a contact line, in the case of a very small wetting angle (<2°), allows for the measurement of that angle;
- ***Reflection measurements*** (Figure 1.25) from the upper surface of a drop (contact angles less than 20°).

The limit of the illumined area of the screen corresponds to rays reflected at the contact line, and the corresponding radius R is given by:

$$\tan 2\theta = \frac{R - r_g}{h}. \tag{1.72}$$

1.4.6 The Rayleigh–Taylor instability

An example of competition between surface tension and gravity effects is the Rayleigh–Taylor instability. We have seen that the effect of surface tension is to minimize the surface area of the interface between two fluids—one way of visualizing this is to imagine the interface as an elastic membrane. Let us now consider the case of an horizontal interface separating two fluids of different densities, with the lighter fluid located underneath the heavier one. Such a system is gravitationally unstable; indeed any fluctuation of the surface from its initial flatness leads to a pressure imbalance tending to amplify the disturbance. However, in this instance, surface tension tends to flatten the surface to its original shape, thus attempting to restore the equilibrium (Figure 1.26a). This latter effect will be the more marked the smaller the interface. We can evaluate the parameter governing the instability by considering more carefully these competing mechanisms.

The driving mechanism for this instability is gravity, with which we associate a hydrostatic pressure difference δp_1 between one side and the other of the interface:

$$\delta p_1 \approx \left(\rho' - \rho\right) g\varepsilon, \tag{1.73}$$

where ε is the infinitesimal vertical displacement of the interface and ρ and ρ' are the densities of the two fluids (Figure 1.26a). The order of magnitude of the "stabilizing" pressure differences created by surface tension is:

$$\delta p_2 \approx \frac{\gamma}{R}, \tag{1.74}$$

where R is the radius of curvature of the interface. By assuming that this interface is similar to a spherical bubble (first order approximation in ε/R), this radius R is related to the vertical

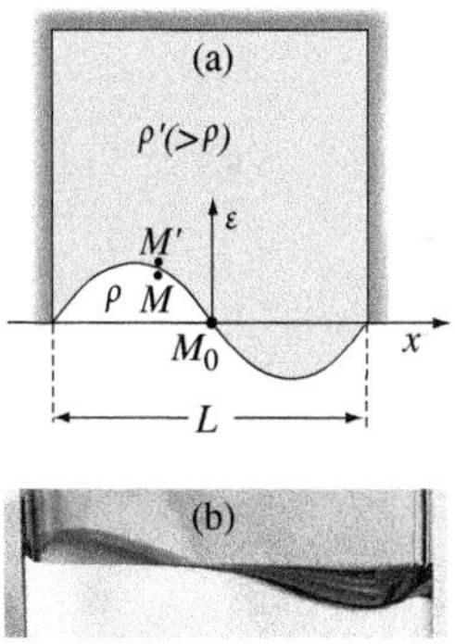

Figure 1.26 *Rayleigh–Taylor instability; (a) geometry of the experiment; (b) picture of a beaker containing very viscous oil (5000 times more viscous than water) after being suddenly turned upside down (plate by C. Rousselin - Palais de la Découverte)*

displacement ε by the equation $\varepsilon R \approx (L/4)^2$. The relative magnitude of these two effects will then be measured by the ratio:

$$\frac{\delta p_1}{\delta p_2} \approx \frac{\Delta\rho\, g\, \varepsilon}{\gamma/R} \approx \frac{\Delta\rho\, g\, L^2}{\gamma}. \tag{1.75}$$

with $\Delta\rho = \rho' - \rho$. It is the value of this ratio which determines the stability of the interface. In this parameter, we are not surprised to find the form of the Bond number (Equation 1.64), since we are considering in both these cases the relative importance of gravity and surface tension.

We can easily observe this instability by using a sufficiently viscous fluid, in order to slow down the growth of the interface deformations, and to facilitate their observation (Figure 1.26b). For example, by turning rapidly upside down a pot of viscous honey, we can observe the appearance of a swelling of the interface on one side and a depression on the other.

Discussion of the Rayleigh–Taylor instability

Let us consider a container filled with a fluid of density ρ' and displaying at its lower end an opening of length L in the x-direction (Figure 1.26a) and very long in the perpendicular direction z, so that we can neglect curvature effects in this latter direction. This system is placed above and in contact with a second fluid of lower density ρ'. Let us assume that any deformation of the interface occurs in the plane of the figure, and let us denote by $\varepsilon(x,t)$ the vertical displacement of this interface relative to its initial plane configuration. Let us denote by M and M' two infinitely close points located on both sides of the interface in each of the two fluids. If $R(x)$ denotes the radius of curvature of the interface at the level of these two points (counted positively if the center of curvature is inside the upper fluid), we can write according to Laplace's law:

$$p_{M'} - p_M = \frac{\gamma}{R(x)}. \tag{1.76}$$

The fundamental principle of hydrostatics applied inside each of these two fluids (principle which we can apply since the fluids are at their limit of stability with a vanishingly low velocity) allows us to write:

$$p_{M'} = p_{M_0} - \rho' g\, \varepsilon \quad (1.77a) \qquad \text{and:} \qquad p_M = p_{M_0} - \rho g\, \varepsilon. \quad (1.77b)$$

At the point M_0, the radius of curvature of the interface is indeed zero and the pressure has the same value on the two sides of this interface. Eliminating $p_{M'}$, p_M and p_{M0} from the three equations above, and taking into account the fact that the interface has small deformations because we are close to the equilibrium point (so that $R(x) \simeq 1/(\mathrm{d}^2\varepsilon/\mathrm{d}x^2)$), we obtain:

$$\frac{\mathrm{d}^2\varepsilon}{\mathrm{d}x^2} = -\frac{\Delta\rho\, g}{\gamma}\varepsilon(x,t). \tag{1.78}$$

This equation has a general solution:

$$\varepsilon(x,t) = A\cos kx + B\sin kx \quad (1.79) \qquad \text{with:} \qquad k = \sqrt{\frac{\Delta\rho g}{\gamma}}. \quad (1.80)$$

If we further assume that the interface is fixed at a point of contact with the lateral walls, $\varepsilon(x,t)$ satisfies the boundary conditions: $\varepsilon\,(x{=}0) = \varepsilon(x{=}L) = 0$. Also, the average displacement of the interface $\int_0^L \varepsilon(x,t)\,\mathrm{d}x$ must be zero in order to ensure the conservation of fluid in the container. All these conditions together restrict the solution to:

$$\varepsilon(x,t) = B\sin kx, \tag{1.81}$$

where $k = (2\, n\, \pi/L)$ and n is an integer. The threshold is obtained for the smallest value of k satisfying this condition ($n = 1$) with: $2\pi/L = \sqrt{\Delta\rho\, g/\gamma}$ or, also:

$$\frac{\Delta\rho\, g L^2}{\gamma} = 4\pi^2. \tag{1.82}$$

In this equation we find the form of the parameter obtained in Equation 1.75. An order of magnitude calculation with an air–water interface ($\Delta\rho = 10^3$ kg/m^3 and $\gamma \approx 70 \times 10^{-3}$ N/m) gives a threshold:

$$L_c = \sqrt{\frac{4\pi^2\gamma}{\Delta\rho g}} \approx 1.7 \cdot 10^{-2}\text{m}. \tag{1.83}$$

The condition: $L > L_c$ for observing the instability will therefore be very frequently satisfied.

The approach used here is very general in the study of hydrodynamic instabilities (studied in detail in Chapter 11). We have assumed a type of deformation or *instability mode* (Equation 1.79), and we have looked for the conditions under which we have the limit of stability (i.e., it represents a solution of equations of motion independent of time). It is that of the unstable modes which corresponds to the smallest threshold which determines the appearance of a global instability.

1.5 Scattering of electromagnetic waves and particles in fluids

1.5.1 Some probes of the structure of liquids

Scattering probes

We have demonstrated above (Section 1.3), the close relationship between the transport properties of fluids and their microscopic structure. This structure, as well as motion on a very small scale, can be studied by analyzing the diffraction of a wave incident on the substance. The wavelength involved must be comparable to, or larger by a few orders of magnitude than inter-atomic distances. We can use not only *electromagnetic waves* (X-rays or visible light), but also beams of particles (*electrons* or *neutrons*). For atomic particles, the equivalent probe wavelengths are given by the *de Broglie relation*:

$$\lambda = \frac{h}{p} = \frac{h}{mv} \tag{1.84}$$

(where h is Planck's constant, while p, m and v, are respectively the momentum, mass and velocity of the particle, assumed to be non-relativistic).

These three types of measurements are currently used in the study of liquids. Neutron scattering techniques use either neutrons generated by the fission of uranium in nuclear reactors (*High Flux Istotope Reactor* at the Oak Ridge National Laboratory (ORNL, USA), *Institut Laue-Langevin* in Grenoble, *Laboratoire Léon Brillouin* in Saclay), or those expelled by bombarding a target with high energy particles (the *Spallation Neutron Source* (ORNL, USA)), *Isis* (Rutherford Appleton Laboratory, UK), the future *European Spallation Source*(Lund, Sweden) which are built exclusively for this purpose. These have taken a very important role in the last decades. By comparison with X-ray diffraction, neutron scattering is relatively more sensitive to the light elements, H, O, C, N which we find in many organic liquids. This results from the fact that neutrons do not interact with the electron cloud surrounding atoms, but only with the atomic nuclei. Also, for the same element, this interaction varies from one isotope to another. The two other kinds of spectroscopic probes "see" the electrons around atoms, and have accordingly a sensitivity which increases rapidly with the atomic number: these techniques are obviously more sensitive to heavy elements.

A few orders of magnitude

The techniques which we have just mentioned allow one to analyze the structure of a liquid or its *elementary excitations* depending on the wavelength and the instruments used. These excitations could be pressure waves, or elementary diffusive waves excited thermally. We are now going to list a few orders of magnitude of the parameters corresponding to the techniques in current use.

In order to gather information on the structure of a liquid on the scale of *interatomic distances*, we use probes corresponding to wavelengths of the same order:

X-rays: these are classically generated by bombarding a metallic anode with electrons accelerated by an electron gun. The corresponding wavelengths are of the order of a few Ångstroms (e.g., 1.54 Å for the Kα -line of copper). Current studies are also carried out by means of very strong *synchrotron radiation* obtained from the acceleration of electrons in *storage rings: Advanced Light Source* in Berkeley *(USA), Diamond light source* in the Rutherrford Appleton Laboratory (UK), *SOLEIL synchrotron* (Source Optimisée de Lumière d'Energie Intermédiaire) in Saclay (France), *ESRF* (European Synchrotron Radiation Facility) in Grenoble (France), ...

Electrons: for electrons accelerated through a potential difference of 200 V, the de Broglie wavelength is 0.87 Å. These are *slow, low-energy, non-relativistic electrons.*

Neutrons: They are obtained from nuclear reactors or from spallation sources (see previous page) by slowing high energy neutrons in a hydrogen environment, either at room temperature (*thermal neutrons*) or at very low temperatures (*cold neutrons*). After a sufficient number of collisions with the nuclei of the medium, the velocity distribution of the neutrons approaches thermal equilibrium with the liquid. At 300 K, the corresponding peak wavelength is 1.78 Å.

Scattering of visible light: Because of its longer wavelength, we use such light to analyze density or composition fluctuations of the fluid at *large scales.* It represents a very useful tool for studying transport phenomena in liquids, but at scales larger than a micron.

In the two sections that follow, we discuss *elastic* and *inelastic scattering of X-rays and neutrons*, and then continue to *Rayleigh* and *Brillouin scattering of light.*

1.5.2 Elastic and inelastic scattering

In this section, we first discuss the characterization of the structure of liquids on the scale of a few atoms (a few Ångstroms). However, the techniques of analysis that we use can frequently be carried over to the scattering of light.

Pair correlation function in an atomic liquid

The essential information about the distribution of interatomic distances in a liquid is contained in its pair correlation function, $g(r)$ (Figure 1.27), defined as follows: if we take an atom centered at O, $g(r)$ is related to the number $n(r)\,dr$ of atoms whose centers can be found at a distance from O between r and $r + dr$ by the relation:

$$4\pi r^2 g(r)\, \mathrm{d}r = n(r)\, dr. \tag{1.85}$$

The value of $g(r)$ is practically zero inside a sphere of radius $r = 2r_0$, where r_0 is the atomic radius, because of the impenetrability of atoms. It displays a first peak for a value slightly

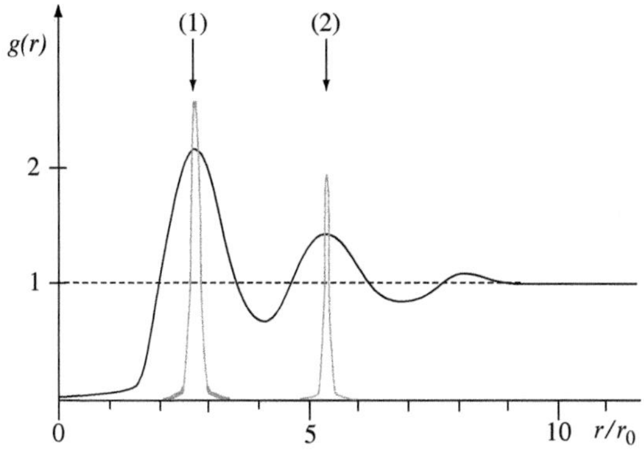

Figure 1.27 *Comparison of the variation of the radial distribution function with distance from the atom taken as the origin O, for a metal just above (solid line) and just below (dotted line) its melting point. Markers 1 and 2 indicate respectively the positions of the first and second nearest neighbors*

greater than $2r_0$, which corresponds to the presence of nearest neighbor atoms (Figure 1.27). As r increases, the function $g(r)$ then displays several damped oscillations which correspond to the effects of second and third neighbors. Beyond a distance of a few atomic radii, the function $\rho(r)$ is constant and equal to the average density of atoms: this indicates the absence of long range order in a liquid.

The distribution function $\rho(r)$ corresponds to an average on all possible configurations of atoms in a liquid. We will show that it can be inferred from scattering experiments, for example using X-rays of wavelengths of the order of Ångstroms.

Elastic scattering

Figure 1.28 shows schematically a scattering experiment using wave vectors $\mathbf{k}_i$ and $\mathbf{k}_d$ corresponding to the incident wave, the scattered wave and of the transfer wave vector $\mathbf{q}$:

$$\mathbf{q} = \mathbf{k}_d - \mathbf{k}_i \tag{1.86}$$

$\mathbf{q}$ represents the momentum transfer between the wave and the medium. We assume in this case an *elastic scattering*, i.e., without change in energy. The magnitudes of $\mathbf{k}_d$ and $\mathbf{k}_i$ are therefore equal.

The scattered amplitude $A(\mathbf{q})$ in the direction of $\mathbf{k}_d$ is

$$A(q) = C\,D(\mathbf{q})\left(1 + \iiint_V g(r)\,e^{i\mathbf{q}.\mathbf{r}}d^3\mathbf{r}\right). \tag{1.87}$$

We assume, in this case, that the hypothesis of elastic scattering is approximately satisfied for X-ray photons of wavelength of the order of an Ångstrom. The corresponding energy is of the order of 12 KeV and is much greater than that of excitations created in the fluid: this energy will therefore be virtually unchanged during the scattering. This will not be the case for probes such as thermal neutrons of the same wavelength for which the energy is much lower ($kT \approx 1/40$ eV at 300 K).

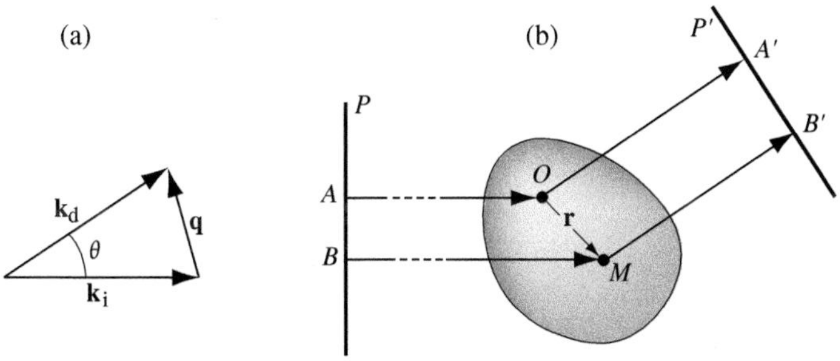

Figure 1.28 *Schematic representation of elastic X-ray diffraction by a liquid; (a) representation in terms of wave vectors in Fourier transform space; (b) representation in physical space (the planes P and P′ are at very large distance relative to the scale of the sample)*

Proof

Let us assume that the complex amplitude of the wave scattered by a molecule is $D(q)$ (we assume molecules with spherical symmetry, so that the angle of the wave vectors with the characteristic directions of the molecule has no particular effect). Let us take a molecule O as the origin, and assume there is a second molecule at a point M such that $\mathbf{OM} = \mathbf{r}$ (Figure 1.28b). Let us calculate the phase change $\Delta\varphi$ between the waves scattered by O and M from the optical path difference AOA' and BMB' ($\mathbf{AB}$ and $\mathbf{A'B'}$ are respectively perpendicular to the incident and scattered wave vectors). In taking the projection of $\mathbf{OM}$ on $\mathbf{OA}$ and $\mathbf{OA'}$, we find $\Delta\varphi = (\mathbf{k}_\mathrm{i} - \mathbf{k}_\mathrm{d}).\mathbf{r} = -\mathbf{q}.\mathbf{r}$. The amplitude of the two resultant waves is therefore:

$$A_{O+M} = D(\mathbf{q})(1 + \mathrm{e}^{-\mathrm{i}\,\Delta\varphi}) = D(\mathbf{q})\,(1 + \mathrm{e}^{\mathrm{i}\,\mathbf{q}.\mathbf{r}}). \tag{1.88}$$

In order to obtain a significant value, we must integrate this expression on the ensemble of vectors $\mathbf{r}$ corresponding to the scattering volume by weighting that by the probability of the presence $g(r)$ on the second molecule: this leads to Equation 1.87.

The expression for the factor $S(\mathbf{q})$ uses the Fourier transform of $\rho(r)$ to go from the real space to that of the wave vector $\mathbf{q}$. The inverse transform allows us to come back from measuring $S(\mathbf{q})$ to the result for $\rho(r)$; this is what makes the method particularly interesting. As it is the product $\mathbf{q}.\mathbf{r}$ which occurs in the integral, the information on the large scale structure will be obtained from wave vectors $\mathbf{q}$ which are small (we talk about small angle scattering); for a given value of $\mathbf{k}_\mathrm{i}$, this implies that the direction of the wave vector $\mathbf{k}_\mathrm{d}$, where we make the measurement, will make a small angle with the incident wave. When the function $\rho(r)$ is periodic, as is the case for a crystalline solid, $S(\mathbf{q})$ will be also periodic in $\mathbf{q}$. We will observe diffraction peaks at angles where the Bragg condition is satisfied for reflection from crystal planes.
The passage from real space to Fourier space for wave vectors and the resulting properties are a common characteristic of all the scattering techniques, which are used in the study of the scattering of light.

In Equation 1.87, the total scattered amplitude in the direction $\mathbf{k}_\mathrm{d} = \mathbf{k}_\mathrm{i}+\mathbf{q}$ depends on two factors:

- one $D(\mathbf{q})$, which is related to the structure of individual molecules;
- the other $S(\mathbf{q}) = \left(1 + \iiint_{\mathcal{V}} \rho(r)\, \mathrm{e}^{\mathrm{i}\mathbf{q}.\mathbf{r}} \mathrm{d}^3\mathbf{r}\right)$, known as the *structure factor*, is determined by the distribution of the relative positions of molecules.

The elastic scattering of X-rays is thus seen as a powerful method to analyze the correlations between the position of the molecules in the liquid (when these are all identical and of spherical symmetry). The wavelength of the X-rays used is well adapted to the analysis of phenomena at the scale of atomic distances.

Inelastic scattering

If, in the previous example, we have considered scattering as elastic, there is, more generally, always a transfer of momentum and of energy during the scattering: the magnitude $|\mathbf{k}_\mathrm{d}|$ of the scattered wave vector is then different from that $|\mathbf{k}_\mathrm{i}|$ of the incident wave vector. This consideration will be particularly important in the case of thermal neutrons where we have seen above that the energy is much smaller than that of X-rays of the same wavelength ($kT \approx 1/40$ eV at 300 K for a wavelength of 1 Å): the relative value of the change in this energy is therefore much more significant and we then have easily measurable *inelastic scattering*. The latter provides very worthwhile information on the internal excitation modes of a fluid.

Let us now generalize the previous discussion to whatever type of particles (neutrons . . .) or waves (X-rays, light . . .) which are incident on or scattered from a particle A; the latter could be an atom or a fictitious particle representing the complex effect of the interaction of the wave with the fluid. In using Equation 1.84 to determine wave vectors, we can write the energy and the momentum of these different objects in the form:

	Particle or wave i	Particle A	Particle or wave d
Energy	$\hbar\,\omega_\mathrm{i}$	$\hbar\,\Omega$	$\hbar\,\omega_\mathrm{d}$
Impulse	$\hbar\,\mathbf{k}_\mathrm{i}$	$\hbar\,\mathbf{q}$	$\hbar\,\mathbf{k}_\mathrm{d}$

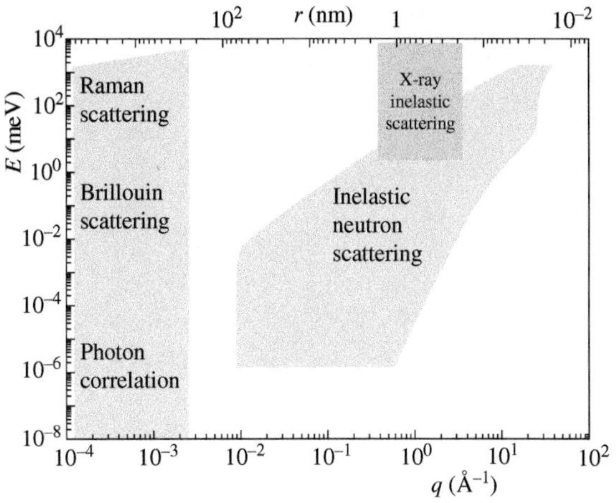

Figure 1.29 *Regions of energy and wave-vector transfer for the various experimental probes used in the spectroscopic analysis of liquids*

Among these different variables we have the following relationships, which imply the conservation of energy and of momentum:

$$\Omega = \omega_d - \omega_i \qquad \mathbf{q} = \mathbf{k}_d - \mathbf{k}_i. \tag{1.89}$$

These relationships will be valid for local probes as well as for light for which the wavelength is much greater. Light allows for the study of phenomena at a characteristic scale much greater than the others, as we will see in the following sections.

Figure 1.29 displays the domains of energy and wave vector transfer for different classical probes used in the analysis of liquids.

1.5.3 Elastic and quasi elastic scattering of light: a tool for studying the structure and diffusive transport in liquids

An example of elastic scattering of light: Rayleigh scattering by a dilute emulsion

Let us observe (Figure 1.30) a long tube containing a dilute solution of small scale particles (for example, water to which we have added a very small amount of milk). If we illumine the tube through one of its ends, we will see reddish light coming out the opposite end; on the other hand, at right angles to the beam, we observe scattered light which is bluish. This experiment is analogous to the more common observation of the color of the sky, which appears red at sunset, when we look in the direction of the rays of the sun (light transmitted after absorption), and blue in the perpendicular direction (light scattered from the molecules in the air). In the above experiment, the scattering occurs from density (and, as a result, refraction index) fluctuations, due to the presence of the milk droplets. The method of calculating the scattered light is analogous to that used in Section 1.5.2 for the elastic scattering of X-rays. If the solution is dilute, the intensities of the light scattered by the droplets simply add up, because $g(r)$ is practically a constant: the variation of the scattered intensity with the wave vector corresponds then to that of $D(\mathbf{q})$.

The change of color observed results from absorption and scattering which are different for the different components of the spectrum of white light. Monochromatic light of a given

Figure 1.30C *Rayleigh scattering from a beam of white light (coming from the left) incident on a container filled with water to which a tiny amount of milk has been added. The light scattered at right angles to the beam has a bluish tint while the light transmitted to the screen is reddish. (document courtesy of B. Valeur)*

wavelength keeps the same wavelength: this phenomenon, called *Rayleigh scattering*, appears then as elastic.

In the previous example, Rayleigh scattering results from concentration fluctuations due to the presence of the dilute emulsion of milk; in a pure liquid, its amplitude would be much lower, and is associated to variations of the index of refraction due to the fluctuations of local temperature. These fluctuations propagate diffusively, as we have seen in Section 1.2 of this chapter, and not as waves.

If we illuminate a suspension of Brownian particles by coherent and monochromatic laser light, the time dependent fluctuations in the light allow us to characterize the dynamics of these particles. This technique, called *dynamic scattering of light* (or photon correlation), has at present numerous applications.

Forced Rayleigh scattering

We will now analyze a model scattering experiment: *forced Rayleigh scattering*, where large amplitude temperature variations are artificially created.

In a static liquid, we create a figure of interference fringes by means of a high-power, but very short, laser pulse: the monochromatic beam is separated into two components, which converge at the same spot inside the liquid under study (Figure 1.31). The spatial period Λ of the resulting pattern of interference fringes (the grid spacing) is determined from the angle φ between the interfering beams, by the relation

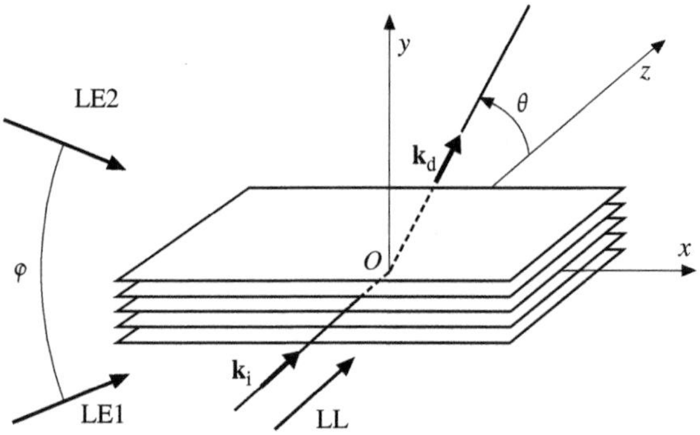

Figure 1.31 *Principle of forced Rayleigh scattering experiment: a refractive index grating is "written" inside the liquid by the interference of two beams* LE *generated by a same pulsed laser. This grating is then "read" by the beam* LL *of a low power continuous laser which gets diffracted*

$$\Lambda = \frac{\lambda_0}{2n \sin \varphi/2}, \tag{1.90}$$

where λ_0 denotes the free space wavelength of the light, and n the index of refraction of the liquid. The rise in temperature resulting from the laser pulse evolves with time and distance y, according to a law of the type:

$$\Delta T(y, t) = \Delta T_0 \quad e^{-\kappa(2\pi/\Lambda)^2 t} \cos \frac{2\pi\, y}{\Lambda}. \tag{1.91}$$

Proof
Taking into account the high power of the laser LE, which forms a temperature grid (*writing* beam), a spatial modulation $\Delta T(y)$ of the temperature is associated with this interference pattern as:

$$\Delta T(y, t) = \Delta T_0(t) \cos \frac{2\pi y}{\Lambda}.$$

The amplitude $\Delta T_0(t)$ of this variation in temperature attenuates with time because of thermal diffusion; the wavelength Λ of the modulation remains, however, almost constant in time. The modulation $\Delta T(y,t)$ satisfies the thermal diffusion Equation 1.17. In applying this equation to the preceding expression $\Delta T(y, t)$, we obtain:

$$\frac{\partial \Delta T_0(t)}{\partial t} = -\frac{4\pi^2 \kappa}{\Lambda^2} \Delta T_0(t), \qquad \text{i.e.:} \qquad \Delta T_0(t) = \Delta T_0\, e^{-\kappa(2\pi/\Lambda)^2 t}.$$

In substituting this equation in the expression for $\Delta T(y, t)$, we get back Equation 1.91.

In order to measure the amplitude of these temperature variations, we illuminate the system of fringes in the z-direction, perpendicular to the plane of the interfering beams, with a continuous beam from a second, low power, laser LL with wavelength λ. This *reading* beam is diffracted by the modulations in the index of refraction due to the temperature modulation. Because the grid is periodic, we observe diffraction maxima for certain wave vectors. This phenomenon is equivalent, in the case of visible light to the elastic scattering of X-rays with the momentum transfer described in Figure 1.28. The transfer wave vector $\mathbf{q} = \mathbf{k}_d - \mathbf{k}_i$ corresponds to the most intense maximum with magnitude $q = 2\pi/\Lambda$. In fact, the incident wave vector $\mathbf{k}_i$ and the scattered wave vector $\mathbf{k}_d$ have the same magnitude k (a characteristic of elastic scattering) and make an angle θ to each other such that:

$$\sin \frac{\theta}{2} = \frac{q}{2k} \approx \frac{\lambda}{2\Lambda} \tag{1.92}$$

In Section 1.5.2, we have seen that the scattered amplitude is proportional to the amplitude of the variations in density. The diffracted intensity $I_d(t)$ is thus proportional to the square of the amplitude $\Delta T_0(t)$, which describes the temperature modulation. According to Equation 1.91, it decreases in time as:

$$I_d(t) \propto e^{-2\kappa q^2 t} = e^{-2(t/\tau_Q)}. \tag{1.93}$$

This decrease allows us to measure κ since q is known ($q = 2\pi/\Lambda = (4\pi/\lambda)\sin(\theta/2)$); note that, according to Section 1.2.1, $\tau_Q = 1/(\kappa\, q^2)$ is of the order of the thermal diffusion time along a distance equal to the wavelength Λ=$2\pi/q$ of the grid.

Spontaneous Rayleigh scattering

Spontaneous Rayleigh scattering corresponds to scattering of light by fluctuations in the temperature or concentration due to thermal motion; these are naturally present within a liquid. The technique of forced Rayleigh scattering, which we have studied above, represents an excellent model for scattering by such spontaneous fluctuations. The latter can be broken down into elementary perturbations, such as those that we have discussed; their wave vector can assume all magnitudes in all possible directions. The scattered amplitude is a combination of the corresponding scattered amplitudes. The light intensity detected is generally very weak: for a laser of 0.1 W, it varies from a few photons to 10^7 photons per second.

In choosing a direction of observation for the scattered light, one selects at the same time the direction and the wavelength of the grid which scatters the light. We thus achieve a filtering of the spontaneous fluctuations from the ensemble of wavelengths, keeping only the components which provide maximum light in the chosen direction.

The width $\Delta\omega$ of the spectrum around the frequency ω_0 of the incident wave allows us to determine the coefficients of the diffusive processes present in the fluid. This width varies greatly: it can be as small as a fraction of a hertz if we observe the (very slow) scattering by large particles, but it can also reach several tens of megahertz for fast processes.

1.5.4 Inelastic scattering of light in liquids

When the spectrum of light scattered from a fluid is analyzed with high enough resolution, a pair of satellite lines (B) are found on either side of the central unshifted Rayleigh-scattered line (R). These side peaks, illustrated in Figure 1.32, and known as *Brillouin scattering*, result from the diffraction of the beam by spontaneous density (and, correspondingly, pressure) fluctuations, propagating as acoustic waves in the fluid. The shift between the Rayleigh and Brillouin lines is of a few gigahertz. The width of the lines provides information about the time constants of the relaxation and the fluctuations in the mass and concentration. In order to understand the physics of this phenomenon, we proceed, just as in the case of Rayleigh scattering, to describe a model experiment where density fluctuations are externally created.

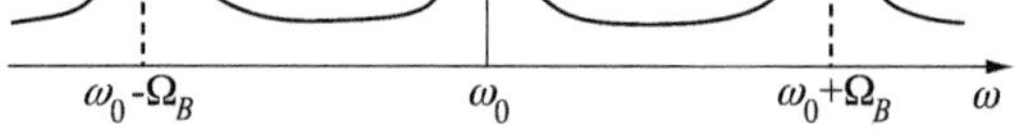

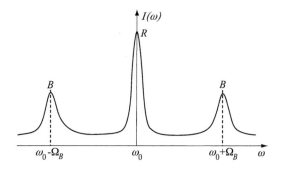

Figure 1.32 *Expanded-scale schematic of the scattered-light spectrum with two Brillouin-scattering lines (B) $\omega_0 \pm \Omega_B$ symmetrically located on either side of the central, unshifted Rayleigh-scattering line (R). The width of each spectral line provides a measurement of the relaxation time constants for mass and concentration fluctuations. The separation between the Rayleigh and Brillouin lines is of the order of a few gigahertz*

Forced Brillouin scattering

This experiment provides a model for inelastic scattering through the Doppler effect. A high-frequency, acoustic travelling wave of frequency f_s is generated within a liquid by means of a quartz transducer (Figure 1.33). This wave consists of a modulation of the pressure and density of wavelength Λ, propagating through the liquid. Visible light of free-space wavelength λ and frequency ω_0, is beamed at an angle $\theta/2$ to the planes of the acoustic wave. The optical beam is reflected at an equal angle from the acoustic wave-planes, and

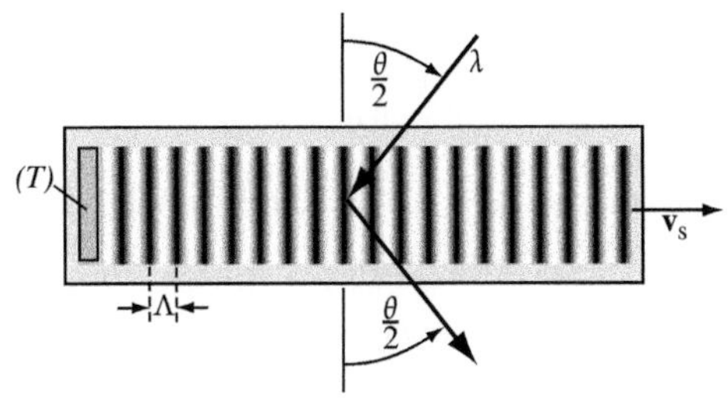

Figure 1.33 *Forced Brillouin scattering of a monochromatic light beam of wavelength λ from an acoustic wave of wavelength Λ induced in a fluid by the transducer T*

the beams reflected at each plane interfere. Since the optical path difference from one plane to the next is $2\Lambda\sin(\theta/2)$, we have constructive interference, provided that the equation:

$$2\Lambda\sin\frac{\theta}{2} = p\frac{\lambda}{n} \tag{1.94}$$

is satisfied; p is any integer, (assumed hereafter to be unity, corresponding to the first-order diffraction peak), and n represents the index of refraction of the liquid. It should be observed that this expression is precisely identical to the *Bragg condition* for the reflection of X-rays from the planes of a crystal. It can be derived by following each of the steps in Section 1.5.3 for Equation 1.92. At the same time, because the reflecting planes are not stationary, the frequency ω_d of the diffracted wave is Doppler-shifted by an amount Ω_B such that:

$$\frac{\Omega_B}{\omega_0} = \frac{\omega_d - \omega_0}{\omega_0} = 2\frac{v_s}{c/n}\sin\frac{\theta}{2} = 2n\frac{v_s}{c}\sin\frac{\theta}{2} = \frac{v_s}{c}\frac{\lambda}{\Lambda}; \tag{1.95}$$

here, v_s is the velocity of the acoustic wave in the fluid ($f_s = v_s/\Lambda$ is then its frequency). Since $c = \omega_0\,\lambda/2\pi$, we see that the Brillouin frequency shift Ω_B from ω_0 equals the angular frequency $\omega_s = 2\pi\, f_s$ of the acoustic wave. The measurement of this frequency shift ω_B from the spectrum of the scattered light (Figure 1.32), and the experimental measurement of the scattering angle θ, for which we have a maximum in the scattered intensity, allows us to determine the velocity v_s from the relation $v_s = (\Omega_B/2\pi)(\lambda/[2n\sin(\theta/2)])$. The sign of the Doppler shift depends on the direction of propagation of the travelling wave relative to the incident light beam. If the travelling wave is replaced by a more easily generated standing wave – which, we recall, is nothing but a linear superposition of two waves travelling in opposite directions – two symmetrically shifted lines are observed.

Orders of magnitude

Let us consider an acoustic wave with speed of sound v_s = 1500 m/s and frequency f_s = 150 MHz. The free-space wavelength of the light from a He–Ne laser is 6328 Å and the index of refraction of the liquid is n = 1.5. The acoustic wavelength is thus: $\Lambda = v_s/f_s = 10^{-5}$ m. Applying Equation 1.94, we find a Bragg angle $\theta = 6.3 \cdot 10^{-2}$ radians. The angular frequency ω_0 of the light is given by $\omega_0 = 2\pi c/\lambda = 3 \times 10^{15}$ s^{-1}. From Equation 1.95, we obtain $\Omega_B = 9.4 \times 10^{8}$ s^{-1}. The resultant frequency split is sufficiently large to be easily measurable, even using incoherent light.

Spontaneous Brillouin scattering

As in the case of forced Brillouin scattering, we observe in the spectrum of light scattered from a liquid, Brillouin scattering lines, symmetrically located on either side of the central Rayleigh line. In the present case, these lines are due to thermodynamic pressure fluctuations which occur and propagate *spontaneously* in the liquid (as opposed to the case just discussed, in which they are externally generated).

Experimentally, one uses a collimated emitter beam together with a detector sufficiently small in size, so that the angle θ between the incident and the reflected beams is well defined. This is equivalent to selecting out those components of the pressure fluctuations for which the Bragg reflection condition (Equation 1.94) is satisfied, and such that the angles of incidence and reflection are equal. Equation 1.95 then gives immediately the propagation velocity v_s for these fluctuations. As implied by Equation 1.94, a fixed θ value is equivalent to a given wave vector **q** of the fluctuations under study; by varying θ, one can measure the dependence of v_s on **q**, i.e., the dispersion law for the acoustic wave. It should be noted, however, that the frequencies involved in such measurements are outside the normal range of classical ultrasonic techniques.

As a generalization of these ideas, Brillouin scattering can be used directly to characterize the velocity distribution of a system of particles in suspension in a fluid. In such a case, the frequency variations that occur during scattering from the moving objects are measured.

In the diffraction spectrum, other lines more distant from the Rayleigh scattering peak than the Brillouin lines can be seen; these are due to the excitation of (rotational and vibrational) internal modes of the molecules, and correspond to the phenomenon of Raman scattering, which we shall not discuss here.

A very similar technique is *laser Doppler anemometry*, described in detail in Section 3.5.3. This method uses light, suspended particles to obtain information about the movement of a fluid. By illuminating the particles by means of a laser beam, and observing the frequency shift in the scattered light, the velocity of the particles can be determined.

In conclusion, we find that the Brillouin and Rayleigh scattering techniques are highly complementary for studying transport phenomena in liquids:

- Rayleigh scattering leads to the analysis of diffusive transport phenomena, and the measurement of the related coefficients.
- Brillouin scattering provides information about convective and wave-transport modes; it allows us to measure the speed of sound in materials at high frequencies.

1A Appendix - Transport coefficients in fluids

Table 1.2 *Heat, mass and momentum transport coefficients in a number of common fluids. For the molecular diffusion coefficients D_m, we estimate the order of magnitude of the self-diffusion coefficients (i.e., of the fluid in itself), in order to display the great numerical difference between gases and liquids.*

	Thermal conductivity k (J/m s K)	Specific heat at constant volume C_v (J/K)	Density ρ (kg/m^3)	Thermal diffusivity $\kappa = k / \rho C_v$ (m^2/s)	Molecular diffusivity D_m (m^2/s)	Kinematic viscosity ν (m^2/s)	Prandtl number $Pr = \nu / \kappa$	Dynamic viscosity $\eta = \rho\ \nu$ (Pa s)
Liquid metals	$1 - 10^2$	$\approx 10^3$	$2 \times 10^3 - 2 \times 10^4$	$10^{-6} - 10^{-4}$	$10^{-9} - 10^{-8}$	$10^{-8} - 10^{-6}$	$10^{-3} - 10^{-1}$	$10^{-4} - 10^{-3}$
Organic liquids	≈ 0.15	$10^3 - 3 \times 10^3$	10^3	$10^{-8} - 10^{-7}$	$10^{-10} - 10^{-7}$	$10^{-7} - 10^{-6}$	$1 - 10$	$10^{-4} - 10^{-3}$
Molten salts	$10^{-7} - 10^{-6}$	$10^3 - 4 \times 10^3$	$\approx 2 \times 10^3$	10^{-7}	$\approx 10^{-10}$	10^{-6}	10	10^{-3}
Silicone oils	0.1	2×10^3	$\approx 10^3$	10^{-7}	$10^{-13} - 10^{-9}$	$10^{-5} - 10^{-1}$	$10 - 10^7$	$10^{-2} - 10^3$
Water	0.6	4×10^3	10^3	10^{-7}	$10^{-10} - 10^{-8}$	10^{-6}	10	10^{-3}
Molten glass (800 K)	10^{-2}	$\approx 10^3$	$\approx 3 \times 10^3$	10^{-6}	$\approx 10^{-12}$	10^{-2}	$10^3 - 10^4$	10
Air (gas at p = 1 atm and T = 300 K)	2.6×10^{-2}	10^3	1.29	2.24×10^{-5}	$10^{-5} - 10^{-4}$	1.43×10^{-5}	0.71	1.85×10^{-5}

Momentum Transport Under Various Flow Conditions

2

In the first chapter, we have seen how heat, or miscible tracers, are transported by diffusion. The flux of heat, or of tracer, is then proportional to the gradient of the quantity being transported (temperature or tracer concentration) and directed along this gradient, the flux being oriented so as to tend to attenuate these gradients. There exists another mechanism (often much more effective) of heat or tracer transport: convection through flow. Thus, in a region of rapidly flowing fluid, a drop of dye is displaced on average at the velocity of the fluid, all the while displaying some spreading due to the effect of molecular diffusion or of the transverse velocity gradients.

At the beginning of this chapter (Section 2.1), we show that the momentum of a moving fluid can be transported, like heat or tracer concentration, simultaneously by diffusion and by convection. However, an important difference from the cases just mentioned is that momentum is a vector quantity, whereas temperature and concentration are scalars. Section 2.2 includes a simplified discussion of the microscopic models of the related coefficient, the viscosity, paralleling that of Chapter 1 for the other transport coefficients. We then compare in Section 2.3 the relative effectiveness of the convection and diffusion mechanisms, which will lead us to the definition of the Reynolds number. Finally, in Section 2.4, we illustrate by the cases of flow in a tube or around a cylinder or a sphere the changes occurring in the flow regime as the Reynolds number increases.

2.1 Diffusive and convective transport of momentum in flowing fluids

2.1.1 Diffusion and convection of momentum: two illustrative experiments

It is easy to understand the transport of momentum by convection, if we first consider a liquid in uniform parallel flow with a constant vector velocity $\mathbf{U}$. Every element of fluid carries along (convects) its momentum while being displaced at its own velocity, the local velocity $\mathbf{U}$ of the flow. The momentum flux, per unit area, per unit time, is equal in this case to the product of $\mathbf{U}$ with the quantity being transported, the momentum $\rho\,\mathbf{U}$ (where ρ is the density of the fluid). The resulting term $\rho\, U^2$ has dimensions of a pressure; $\rho\, U^2/2$ is often called the *dynamic pressure* of the fluid and we shall see it appearing naturally in the discussion of the conservation of energy (Section 5.3.2).

Momentum transfer by diffusion is, as we shall see, also an effective mechanism, but it is frequently overshadowed by convective transport. Since the latter occurs in the direction of

Physical Hydrodynamics. Second Edition. Etienne Guyon *et al.*

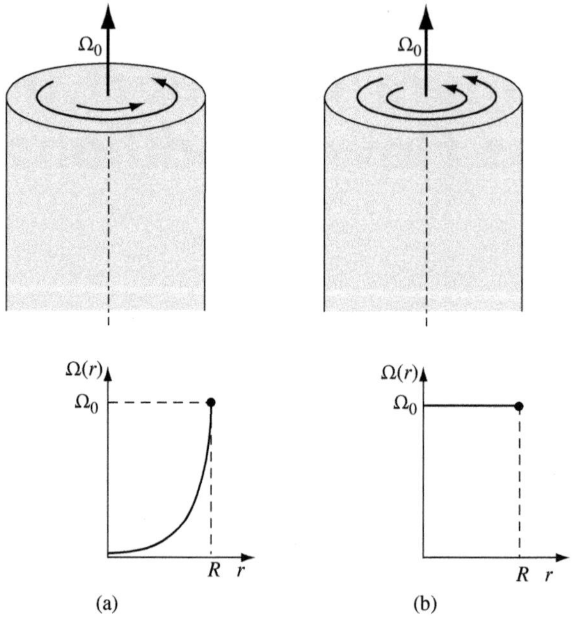

Figure 2.1 *(a) Initiation of the motion of a viscous fluid, located in a cylindrical container of which the wall is suddenly set into rotation at a constant angular velocity Ω_0; (b) final steady-state motion. The graphs shown below each figure display the corresponding radial dependence of the angular velocity profile at the upper surface of the fluid*

the flow, it is in fact easier to identify diffusion in the direction normal to the flow as outlined in the experiment described in Figure 2.1. A long cylinder with vertical axis and radius R is filled with a liquid for which we can visualize the flow by means of particles dropped on the surface. The system is initially at rest. At the very beginning, we start the cylinder in motion with a constant angular velocity Ω_0. First of all, only the liquid layers in the immediate vicinity of the cylinder begin to move with its angular velocity (Figure 2.1a). This fluid flow is characterized by the angular velocity $\Omega(r,t) = v(r,t)/r$, where the local velocity $\mathbf{v}$ of the flow is perpendicular to the radius $\mathbf{r}$. The flow propagates from one layer to the next toward the innermost layers; after a very long time, the whole fluid rotates with a uniform angular velocity equal to that of the cylinder (Figure 2.1b). This phenomenon is remarkably similar to the problem of thermal diffusion which we have discussed in Chapter 1 (Section 1.2.1): there, we considered a solid cylinder made up of material with thermal diffusivity κ at an initial uniform temperature T_0; at the first moment, we change the temperature of the outer wall of the cylinder to a value $T_0 + \delta T_0$. The perturbation in temperature propagated by diffusion toward the inner layers and the thickness of the affected region increased with time as $(\kappa t)^{1/2}$; the very same propagation law as $t^{1/2}$ is observed in the hydrodynamic experiment. Moreover, we will see that, in this flow experiment in a rotating cylinder, we can also define a diffusion coefficient for the momentum: this allows us to observe a rigorous correspondence between the angular velocity profiles $\Omega(r)$ at different times and the corresponding thermal diffusion profiles $\delta T(r)$ in Figure 1.9.

The description given above (Figures 2.1a–b) would be rigorously true for an infinitely long cylinder. In a real situation, the bottom of the container creates an effect which eventually dominates the manner in which the velocity profile evolves after a long time, as a result of the creation of a *secondary flow* due to the rotation of the bottom of the container (see Section 7.7.2).

We observe a transfer of the "momentum" information by radial diffusion; the convection due to hydrodynamic flow cannot contribute to this propagation because the fluid moves in a "tangential" direction, perpendicular to the radius. A second important result of this experiment is the fact that the velocity of the solid wall is precisely equal to that of the adjacent fluid. This is characteristic of all normal viscous fluids. There occurs a kind of frictional force between layers of fluid in contact with the solid which causes the fluid immediately next to the wall to move at the same velocity. This diffusive transport of momentum is characterized by a property which depends on the fluid, the *viscosity*, which we will now discuss from a macroscopic viewpoint.

2.1.2 Momentum transport in a shear flow – introduction of the viscosity

Macroscopic definition of the viscosity

The example which we have just discussed corresponded to a non-stationary problem in which the velocity at a given point was time dependent. Let us now analyze the case of the stationary flow of a fluid located between two infinite parallel plates and separated by a distance a in the normal direction y (Figure 2.2). One plate is fixed, while the other moves parallel to itself at constant velocity V_0 in the x-direction. The fluid flow results from the motion of the upper plate. Under stationary conditions (i.e., after enough time has passed since the upper plate was set in motion) we observe that the velocity of the fluid varies linearly from 0 to V_0 from one plate to the other according to the relationship:

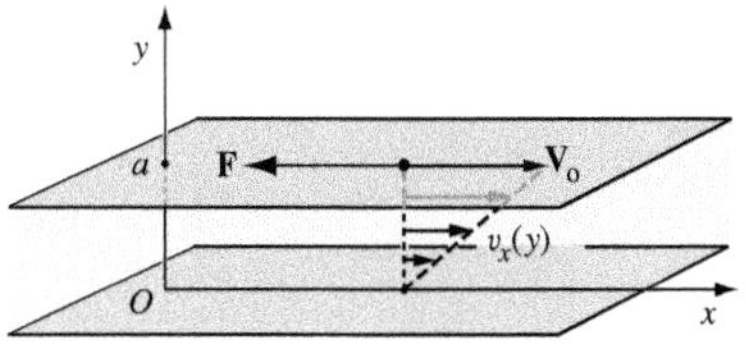

Figure 2.2 *Geometry of a simple shear flow*

$$v_x(y) = V_0 \frac{y}{a} \tag{2.1}$$

The resulting flow is known as *simple shear flow* or *plane Couette flow*. This type of flow can be compared to thermal conductivity between two parallel plates at different temperatures (Section 1.2.1): under stationary conditions, the temperature varied linearly between the boundary values T_1 and T_2 at the plates. Here, the vector field $\mathbf{v}(y)$ replaces the scalar temperature field $T(x)$. Equation 1.6 relating the thermal energy flux and the temperature gradient corresponds to proportionality between the frictional force $\mathbf{F}$ (for an area S of the plate), which opposes the relative motion of the plates, and the velocity gradient between them ($\mathbf{F}$ is in the direction of the negative x-axis).

$$\frac{F_x}{S} = \eta \frac{V_0}{L} = -\eta \frac{\partial v_x}{\partial y} \tag{2.2}$$

The ratio F_x/S is known as the *shear stress*, with dimensions of a pressure.

The relationship between the characteristics of the shear stress F_x/S and those of the thermal energy flux will be more clear as we study the mechanism at a molecular level.

The constant η is known as the *dynamic viscosity* of the fluid (because it is related to a force), or simply as the *viscosity*. Its dimensional form is, according to Equation 2.2:

$$[\eta] = \frac{[M]\,[L]\,[T]^{-2}[L]^{-2}}{[L]\,[T]^{-1}[L]^{-1}} = [M]\,[L]^{-1}[T]^{-1}$$

Its unit in the SI system is the Pascal-second (Pa.s) (1 Pa.s = 1 kg/(m.s)).

Another usual notation for the viscosity is μ. We find in the literature other historical units such as the Poiseuille (Pl), which is the old name of the SI unit, and the CGS unit which is the Poise (Po) and is equal to 0.1 Pa.s. We find in Table 1.2 at the end of Chapter 1, values for the viscosity of several common fluids.

Equation for the diffusion of momentum

Going back to the non-stationary initial problem discussed in Section 2.1.1, let us consider the simple situation of a plane geometry (Figure 2.3): we will assume a flow in the x-direction, where the component $v_x(y,t)$ of the velocity is only a function of the y-coordinate in the perpendicular direction. The partial differential equation governing the variation of the velocity $v_x(y,t)$ with position y and time t is as follows:

$$\frac{\partial v_x}{\partial t} = \frac{\eta}{\rho}\frac{\partial^2 v_x}{\partial y^2} = \nu \frac{\partial^2 v_x}{\partial y^2}. \tag{2.3}$$

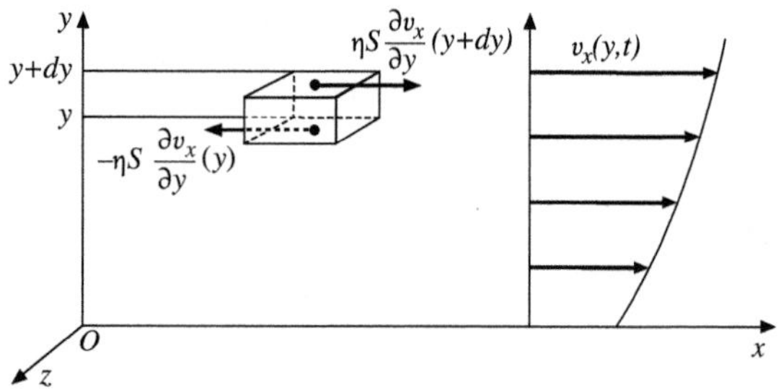

Figure 2.3 *Balance of the shear forces acting on an element of volume of the fluid located between two planes in relative shear motion*

It represents the equivalent for the velocity (or the momentum if we multiply both sides by the fluid density ρ) of the thermal diffusion Equation 1.11b and of the mass diffusion Equation 1.26. Temperature, or tracer concentration, are replaced by the components of the velocity (here, v_x), or of the momentum per unit volume ($\rho\, v_x$). This equation involves the coefficient ν, which depends on the properties of the fluid, and is known as the ***kinematic viscosity***; it obeys the relationship:

$$\nu = \frac{\eta}{\rho} \tag{2.4}$$

and has dimensions $[L]^2/[T]$. The kinematic viscosity ν represents a diffusion coefficient for momentum, quite analogous to the thermal diffusion coefficient κ and the mass diffusion coefficient D_m which we introduced in the previous chapter. The correspondence allows for a better understanding of the analogy, introduced at the beginning of this chapter, between thermal diffusion and the propagation of a velocity perturbation toward the center of a cylinder (Section 2.1.1).

Proof

Consider the balance of forces acting on a volume element bounded by two plane parallel surfaces of cross section S and sides y and $y + \mathrm{d}y$. The wall at position y undergoes a shear force $-\eta S[\partial v_x/\partial y](y)$ due to the fluid below it, and in the negative x-direction, as shown in Figure 2.3. In the same way, the wall at position $y + \mathrm{d}y$ is acted on by a force $+\eta S[\partial v_x/\partial y](y + \mathrm{d}y)$ exerted by the fluid above it and in the positive x-direction. Consequently, we have a resultant force on the volume $S\,\mathrm{d}y$:

$$-\eta\, S\frac{\partial v_x}{\partial y}(y) + \eta\, S\frac{\partial v_x}{\partial y}(y + \mathrm{d}y) = \eta\, S\frac{\partial^2 v_x}{\partial y^2}\,\mathrm{d}y.$$

This force causes the volume to accelerate according to Newton's law:

$$\rho\, S\,\mathrm{d}y\frac{\partial v_x}{\partial t} = \eta\, S\,\frac{\partial^2 v_x}{\partial y^2}\mathrm{d}y.$$

We obtain Equation 2.3 by dividing both sides by $\rho\ S\ \mathrm{d}y$.

Equation 2.3 can be generalized to geometries in two or three dimensions so long as we do not have convective terms (a concept which we will discuss in Section 3.1.3); for this, we may substitute in Equation 2.3 the derivative with respect to the spatial coordinate by a Laplacian. Each component of the velocity vector will then obey the resulting equation. We can then write the result in vector form:

$$\frac{\partial \mathbf{v}}{\partial t} = \nu \nabla^2 \mathbf{v} \tag{2.5}$$

(the vector Laplacian $\nabla^2\mathbf{v}$ is a vector such that its x component, for example, is equal to $\nabla^2 v_x$). We will prove and study this three-dimensional form in Chapter 4, where we will include pressure terms in the complete formulation of this equation.

Flow near a flat plane caused to move suddenly parallel to itself

This problem is an application of Equation 2.5 to a plane version of the motion of the fluid in the neighborhood of the wall of a cylinder, which we discussed in Section 2.1.1. We assume that, at time $t = 0$, an infinite solid plane at position $y = 0$, is suddenly made to move parallel to itself with constant velocity V_0 in the x-direction (Figure 2.4). We are interested in the motion of the fluid with velocity $v_x(y,t)$ in the semi-infinite space located above the $y = 0$ plane.

Figure 2.4a shows the evolution of the velocity profile with time. The solution of Equation 2.3 for this problem is rigorously identical to that of the thermal diffusion problem next to an infinite plane whose temperature is kept constant (Section 1.2.1). It is sufficient to replace κ by ν and the reduced temperature by $v_x(y,t)/V_0$. The kinematic viscosity $\nu = \eta/\rho$ thus represents the diffusion coefficient for momentum. By a change of variable $u = y/(\nu t)^{1/2}$, analogous to that used for thermal diffusion, we find for Equation 2.3 the same solution as for Equation 1.18:

$$v_x\left(\frac{y}{2\sqrt{\nu t}}\right) = V_0\left(1 - \frac{1}{\sqrt{\pi \nu t}}\int_0^y e^{-\left(\frac{\xi^2}{4\nu t}\right)} d\xi\right) = V_0\left(1 - \mathrm{erf}\left(\frac{u}{2}\right)\right) \tag{2.6}$$

where the function erf(u) is, like in Chapter 1, the integral $(2/\sqrt{\pi})\int_0^u \exp(-z^2)\mathrm{d}z$. The ratio v_x/V_0 then depends uniquely on the variable u. All the profiles can then be deduced from one another by expanding the length scale in the y-direction by a factor proportional to $\sqrt{t}$: we say then that those are *self-similar*. The dependence of v_x/V_0 on y and t, given by Equation 2.6, is identical to the temperature dependence found in Chapter 1 for the thermal

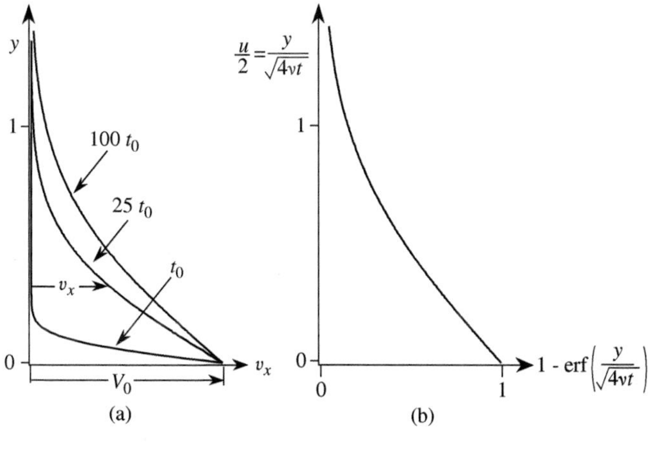

Figure 2.4 *(a) Time dependence of the velocity profile $v_x(y,t)$ generated by the sudden displacement of a plane boundary wall parallel to itself (the Rayleigh problem). (b) Normalised coordinate representation of the universal function* 1− erf $(u/2)$ $(u = y/(\nu t)^{1/2})$ *corresponding to the previous figure. The two figures displayed here are exact analogues of Figures* 1.8 a,b *in Chapter* 1 *(but with ordinates and abscissae interchanged)*

diffusion problem. Figure 2.4b is analogous to Figure 1.8b and shows the variation of the ratio v_x / V_0 as a function of u: the region where the influence of the perturbation is observed has a thickness of the order of $\delta \approx \sqrt{\nu t}$ (diffusion length), but the velocity v_x is not exactly zero at greater distances.

The initiation of the flow of a fluid by viscous coupling is the more effective if its viscosity η is large and its density ρ (and thus, its inertia) is small. In the case of water, for example, the diffusion length for a velocity perturbation is of the order of 3 mm for $t = 10$ s, and about 10 cm for $t = 10^4$ s, i.e., about three hours; this illustrates the poor effectiveness of diffusion over long time periods, as we have already pointed out in Chapter 1 for the case of thermal diffusion.

One observes that the kinematic viscosity of air is of the same order of magnitude as that of water: this is due to a compensation between the density and the dynamic viscosity which are both smaller by a factor of one thousand for air.

The $\sqrt{t}$ dependence of the diffusion length (common to all diffusion mechanisms) makes viscous diffusion an ineffective mechanism at large distances. As a result, convective mechanisms, often quite complex, dominate whenever we consider flow in containers of sufficiently large size; such mechanisms have, as do waves, propagation distances which increase linearly with time. The terminology "sufficiently large" is unfortunately quite imprecise: we will discover, further down, that an analysis based on certain dimensionless numbers will allow us to make our reasoning much more quantitative.

2.2 Microscopic models of the viscosity

Just as for the cases of mass and heat transport, understanding the microscopic mechanisms responsible for viscosity leads to a better grasp of the phenomenon. Here again, these mechanisms turn out to be quite different depending on whether we are discussing gases (where our treatment parallels rather closely that used for the other transport coefficients) or liquids.

2.2.1 Viscosity of gases

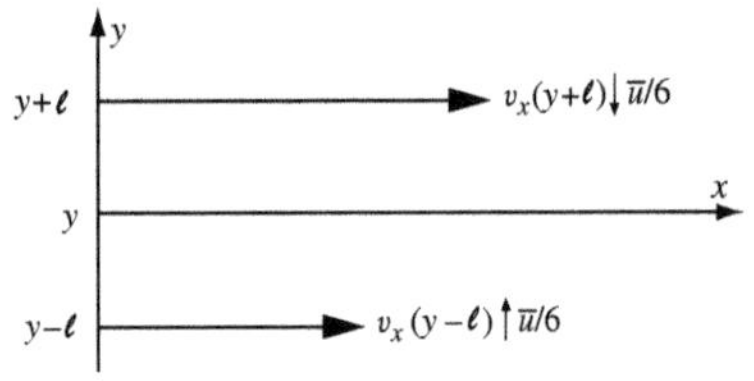

Figure 2.5 *Schematic representation of the simplified calculation from which the viscosity of a gas may be obtained*

In this section, we adapt the kinetic-theory model (Section 1.3) to analyze momentum transport in a shear flow, and to evaluate the viscosity coefficient defined in the previous section. We consider a stationary shear flow in a gas, with the plane geometry illustrated in Figure 2.5. The streamlines are parallel to the x-axis and a velocity gradient exists along the y-direction normal to the flow.

We evaluate here the transfer, by way of the existing components of the thermal motion in the y-direction, of the average x-component of the momentum, $m\, v_x(y)$ (where m is the mass of a molecule). We encounter two completely different velocity scales:

- $\bar{u}$, the *average* magnitude of the thermal velocity of individual molecules,
- the bulk velocity v_x, which represents a slight bias of the thermal motion in the positive x-direction; v_x exists because of the relative motion of the boundary planes.

In defining, as we did in Chapter 1, ℓ as the mean free path of the molecules and n as their number density, we obtain the expression for the *dynamic viscosity*:

$$\eta = \frac{1}{3} m\, n\, \bar{u}\, \ell \qquad (2.7)$$

The *kinematic viscosity* ν is defined by Equation 2.4:

$$\nu = \frac{\eta}{\rho} = \frac{1}{3}\, \bar{u}\, \ell \qquad (2.8)$$

For ideal gases, ν takes on, therefore, an expression almost identical to that of the corresponding thermal and mass, diffusion coefficients, κ and D_m, introduced in Chapter 1. This underscores the close similarity with these two other diffusive transport mechanisms that our microscopic analysis had already suggested. For ideal gases, kinetic theory gives us, by combining Equation 2.7 with the results of Section 1.3.2:

$$\eta \propto \frac{\sqrt{m\,T}}{\sigma_c}$$

where σ_c is the effective collision cross-section of the molecules, and T the temperature of the gas.

Just as in the case of the thermal conductivity k, the viscosity η is independent of the density of the gas (and of the pressure) since the product $n\,\ell \approx 1/\sigma_c$ does not depend on it. As we pointed out in Section 1.3.2, this reasoning does not apply to gases at very low or very high pressures.

Proof of the result for the viscosity

Let us evaluate the flux of momentum per unit time, crossing unit area of a plane at elevation y and coming from "above" (in Figure 2.5). This flux equals $-(1/6)m\,v_x(y+\ell)n\,\bar{u}$, the geometric factor 1/6 comes in because we account for the different directions of the molecular velocities by assuming that they are uniformly distributed. The flux of momentum corresponding to molecules coming from "below" is similarly written as $(1/6)m\,v_x(y-\ell)n\,\bar{u}$ (if we have a uniform temperature, $\bar{u}$ is independent of y). Thus, in the presence of a non-zero shear velocity gradient $\partial v_x/\partial y$, we have a net transfer of momentum across the plane with coordinate y for the component in the x-direction.

The resulting flux of momentum per unit area and per unit time (taking as positive the contribution coming from the region below the plan y) is thus:

$$-\frac{1}{6}m\,n\,\bar{u}\,[v_x(y+\ell)-v_x(y-\ell)].$$

This finite flux of momentum can be interpreted as a frictional force $\mathbf{F}$, which acts between two liquid layers located on either side of the plane y. The direction of this force, which acts in the x-direction, corresponds to the drag of the layers moving at lower velocity by those of higher velocities.

By applying the law of conservation of momentum $\mathbf{F} = \mathrm{d}\mathbf{p}/\mathrm{d}t$ to the force per unit area F_x/S and combining the result with the expression 2.2, we find:

$$\frac{F_x}{S} = -\frac{1}{6}m\,n\,\bar{u}\,[v_x(y+\ell)-v_x(y-\ell)] = -\eta\frac{\partial v_x}{\partial y},$$

which leads immediately to Equation 2.7, after having used Taylor series expansion.

2.2.2 Viscosity of liquids

In Chapter 1, we studied the diffusion of particles in a liquid: we then assumed that it was governed by forces of the type "viscous friction," exerted by the fluid on the particles.

The viscous forces within a liquid can be analyzed by extending this first study, but we will use here a different model. We assume that the molecules in the liquid are all of the same size and move in a manner similar to the grains of a powder. The relative motion of "grains" associated with the shear flow occurs by the passing of individual grains from the cell

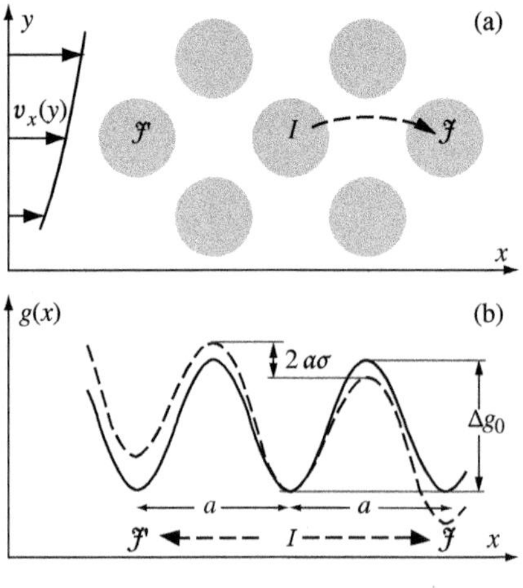

Figure 2.6 *Illustration of the principle underlying the calculation of the viscosity of liquids. The energy barrier Δg_0 that a particle must overcome to pass from cell I into the potential wells J and J' (a) is symmetric in the absence of a flow (solid line in (b)). It becomes asymmetric (dashed line) in the presence of a shear stress; this asymmetry then allows shear flow to occur*

bound by their nearest neigbors (which we assume are in contact with one of the grains) to the next cell; we show this in Figure 2.6, where the curve $g(x)$ represents the variation of potential energy of the particle as a function of the distance x in the direction of the motion.

In the absence of flow, motion from a cell I to the neighboring cell $\mathcal{J}$ requires an activation energy Δg_0 in order to pass over the potential barrier separating the two sites. The application of a shear stress σ results in an asymmetry of the profile $g(x)$ (dashed curve): this favors the jump of a particle from cell I toward the neighboring cell $\mathcal{J}$ on its right (for $\sigma > 0$) rather than toward the cell on its left $\mathcal{J}'$. In the case illustrated in the figure, a shear flow with $G = \partial v_x/\partial y > 0$ will thus appear (each layer slips over the previous one, resulting in the increase of the absolute velocity of each plane with its co ordinate y). In the limit of weak stresses, this description leads to the following expression for the viscosity:

$$\eta \approx \frac{h}{\alpha} \mathrm{e}^{\frac{\Delta g_0}{k_B T}}. \tag{2.9}$$

This equation indicates that viscosity decreases as the temperature increases, according to a law of the Arrhenius type describing the activation process of the jump across the potential barrier. This dependence is opposite to that for gases, where the viscosity increases with the temperature as $\sqrt{T}$.

Proof of Equation 2.9 for the viscosity

The passage from one potential well to the other occurs because of the thermal activation energy $k_B T$; as a result, frequency of jumps from a cell I to the neighboring cell $\mathcal{J}$ obeys a Maxwell–Boltzmann relation (h is Planck's constant):

$$f \approx \frac{k_B T}{h} \mathrm{e}^{-\frac{\Delta g_0}{k_B T}}. \tag{2.10}$$

The presence of a shear stress σ results in an asymmetry of the energy barrier heights in going towards sites $\mathcal{J}$ and $\mathcal{J}'$. The variations in the height of the maxima are proportional to the stress σ, i.e. to the frictional force per unit area between the layers:

$$-\Delta g = -\Delta g_0 \pm \alpha \sigma \tag{2.11}$$

One observes that this variation is first order relative to σ, the quantity $\alpha\sigma$ is a measure of the energy resulting from the shear, which lowers or increases the energy barrier in the direction of the flow (or in the opposite direction), and α is a coefficient having the dimension of a volume.

We have therefore a difference in the frequencies f_+ and f_- for the jumps from I to $\mathcal{J}$ and from I to $\mathcal{J}'$ and, as a result, a net global mobility of the particles. One has indeed:

$$I \to \mathcal{J}\ f_+ \approx \frac{k_B T}{h} e^{-\frac{\Delta g_0 - \alpha\sigma}{k_B T}} \qquad \text{and:} \qquad I \to \mathcal{J}'\ f_- \approx \frac{k_B T}{h} e^{-\frac{\Delta g_0 + \alpha\sigma}{k_B T}}.$$

Let us take as reference of zero velocity the lower layer; the average velocity v_I of the atoms in the intermediate layer is of the order of magnitude of the product of the distance a covered per jump and the net frequency difference $(f_+ - f_-)$ of these jumps:

$$v_I = a(f_+ - f_-) \approx a\frac{k_B T}{h} e^{-\frac{\Delta g_0}{k_B T}} \left(e^{\frac{\alpha\sigma}{k_B T}} - e^{-\frac{\alpha\sigma}{k_B T}} \right)$$

The velocity gradient relative to the lower layer, assumed stationary, is thus:

$$G = \frac{\partial v_x}{\partial y} \approx \frac{v_I}{a} \approx 2\frac{k_B T}{h} e^{-\frac{\Delta g_0}{k_B T}} \sinh\left(\frac{\alpha\sigma}{k_B T}\right) \tag{2.12}$$

In the limit of a weak stress, $\sinh(\alpha\sigma/(k_B T)) \approx \alpha\sigma/(k_B T)$. Equation 2.12 therefore leads to Equation 2.9 if we use formula 2.2 for computing the viscosity η $(= \sigma/G)$.

An empirical form for expressing Equation 2.12 allows us to estimate the viscosity of a liquid by using its molar volume and its boiling point T_b:

$$\eta = \frac{h}{\mathcal{V}/\mathcal{N}} e^{3.8\frac{T_b}{T}}. \tag{2.13}$$

Replacing Δg_0 by an energy term proportional to $k_B T_b$ is understandable. Boiling occurs effectively when the temperature is sufficiently high so that two neighboring particles have a significant probability to move apart. We have expressed a similar condition for the passage by thermal activation of a particle between cells I and $\mathcal{J}$ or $\mathcal{J}'$. Also, in order to derive Equation 2.13, we have assumed the volume α to be equal to the average volume $\mathcal{V}/\mathcal{N}$ for each molecule ($\mathcal{N}$ is Avogadro's number).

Example: Let us estimate the viscosity of benzene at room temperature. $\mathcal{V} = 89 \times 10^{-6}$ m^3/mole, $T_b = 353$ K, $\mathcal{N} = 6.02 \times 10^{23}$, $h = 6.62 \times 10^{-34}$ J.s. Applying Equation 2.13 leads to a value $\eta = 4.5 \times 10^{-4}$ Pa.s comparable to the experimental one, 6.5×10^{-4} Pa.s.

2.2.3 Numerical simulation of molecular trajectories in a flow

We can represent the microscopic mechanisms which lead to viscosity by means of the numerical simulation technique, known as *molecular dynamics*, which analyzes the trajectories of individual molecules (for reasons of the length of calculation times, these simulations are generally carried out with a reduced number of molecules). In such calculations, we take into account local interactions between moving molecules, the effect of walls, which are themselves represented by an ensemble of particles and the forces exerted on the fluid. Figure 2.7a shows the trajectory of one (among a large number of) molecule confined between two solid parallel planes, and subjected to a force field similar to gravity, in the direction of the x-axis. One observes a disordered movement. The changes in direction are the result of interactions with other molecules (which are not displayed in the figure). There results a superposition of a drift in the direction of applied force superimposed on the Brownian diffusion motion. The density of particles is sufficiently high so that this two-dimensional model can simulate a liquid; this explains the numerous collisions. We also observe that the acceleration in the x-direction is reduced in the neighborhood of the solid walls. The particles adjacent to the walls are at rest, as is expected for flow of a viscous fluid. Figure 2.7b shows, at different distances in the z-direction normal to the walls, the velocity averaged over a large number of particles and of different times. This parabolic drift velocity profile is an essential characteristic of the flow of viscous fluids, which we study in detail in Chapter 4.

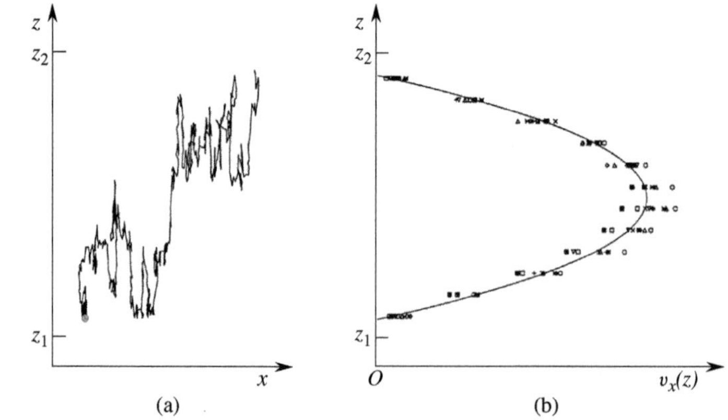

Figure 2.7 *(a) Trajectory of one individual molecule of a two-dimensional liquid, bounded by two fixed planes ($z = z_1$ and $z = z_2$) and subjected to a constant force (such as gravity) directed along the x-axis; the location of the molecule at the beginning of the simulation is near the wall. It should be noted that the scales of distances along the x and z-axes differ by a factor of 7; this explains the steep apparent slopes of the trajectories. (b) Mean, parabolic, Poiseuille-type velocity profile within the liquid deduced from the same molecular dynamics simulation (courtesy J. Koplik)*

2.3 Comparison between diffusion and convection mechanisms

2.3.1 The Reynolds number

In an arbitrary flow of fluid, the two mechanisms (convective and diffusive) for momentum transport are simultaneously active but, depending on the velocity and geometry of the flow, they will not have the same order of magnitude. Let us analyze, for example, the case of the fluid in a channel of arbitrary geometry, and let us evaluate the orders of magnitude for the transport processes.

- ***Convection***: the flux of momentum associated with convection is of the order of ρU^2 where ρ is the density of the fluid and U is the velocity characteristic of the flow (for example, the average velocity in the cross section). The order of magnitude of this flux is obtained by multiplying the value of the momentum per unit volume, of order $\rho\, U$, by the velocity U.
- ***Diffusion***: in the case of parallel flows, which we have just analyzed, the transverse flux of momentum resulting from the viscosity was $\eta\ \partial v_x/\partial y$. Most generally, it equals the product of η by combinations of the first derivatives of the velocity components; it is therefore of the order of $\eta\ U/L$. We can then evaluate a dimensionless ratio:

$$\frac{\text{convective flux of momentum}}{\text{diffusive flux of momentum}} \approx \frac{\rho U^2}{\eta\, U/L} = \frac{UL}{\nu} = Re. \tag{2.14}$$

This ratio is the definition of the *Reynolds number*, which characterizes the relative importance of the momentum transport by convection and by viscous diffusion.

The Reynolds number can also be considered as the ratio between the characteristic transport times by diffusion and convection along distances of order of magnitude L. Since ν represents the diffusivity of momentum, the characteristic diffusion time over a distance of order L is, just as for other diffusive processes, of order L^2/ν. The characteristic time for convection is of order L/U (the time for covering the distance L at the average velocity U of the flow). From this we find a ratio of the characteristic times:

$$\frac{\text{characteristic diffusion time}}{\text{characteristic convection time}} \approx \frac{L^2/\nu}{L/U} = \frac{UL}{\nu} = Re. \tag{2.15}$$

It is the fastest mechanism which propagates the perturbations which will dominate and determine the nature of the velocity field.

- In a flow at *low Reynolds number*, viscous forces and the associated diffusive transport dominate. The flow profile results from an equilibrium between the viscous friction forces and the pressure gradients (or the volume forces, externally exerted). As indicated by the equation for *Re*, such flows will be observed at low velocities and/or in very small systems (for example, bacteria or microorganisms), or also for very viscous fluids within which the frictional forces between layers are significant. These are in general very stable flows with clearly defined profiles known as *creeping flows*. We will study these in detail in Chapter 9.
- On the other hand, in the case of flow at *large Reynolds number*, the transport of momentum by convection dominates, and appears in the form of non-linear terms containing products of the velocity components and their gradients. The corresponding flows are often non-stationary: in particular, turbulent flows correspond to an infinite number of possible solutions of the equations of motion (they will be studied in Chapter 12). One observes these effects at high velocities, in fluids of low viscosity, or in very large scale systems; such flows appear as a random superposition of vortices of very variable scales. We should however observe that, at short distances of the order of the dimension of the smallest structures, momentum transfer by diffusion mechanisms plays again a major part.
- In certain instances, even in the case of large Reynolds numbers, the nature of the flow can then be the same as at low velocity. The simplest example of this is that of *parallel flows* (where only one component of the velocity is non-vanishing), such as the flow in a rotating cylinder discussed in Section 2.1.1. In this case, we can have momentum transport in the direction perpendicular to the streamlines only by viscous diffusion; the flow remains governed by viscosity, independently of the Reynolds number, so long as the velocity profile remains parallel (see chapter 8). Such flows, which we call *laminar*, will be discussed in Section 4.5.

  Nevertheless, if transverse components of the velocity are accidentally created locally, the momentum transport by convection no longer vanishes and new solutions can appear. The flow then often becomes unstable and turbulent. We will study flows of this type in Chapters 11 and 12; but in Section 2.4 of the current chapter, we will describe, for several types of flow, different sequences of the transition between flow regimes as the Reynolds number increases.

2.3.2 Convective and diffusive mass, or thermal energy, transport

Convective and diffusive transport mechanisms can occur simultaneously in problems involving the propagation of thermal energy or the dispersion of pollutants, as is the case for momentum, which we have just discussed. The analysis of convective effects is even simpler in these two cases because velocity only affects convective transport and because the quantity convected (temperature, concentration) is a scalar.

Transport of mass, Péclet number

Let us, for example, analyze the transport of a tracer substance A with local mass concentration $\rho_A(x,t)$ by a flow of characteristic velocity U. Assuming that we can neglect the effects of the difference in density introduced by the presence of the tracer, this latter is, on the one hand, carried along by the local velocity of the flow and, on the other hand, spreads by molecular diffusion. Let us call L a length scale characteristic of the variations in the concentration of the tracer; the diffusive flux of tracer obeys:

$$\mathbf{J}_m = -D_m \nabla \rho_A \approx D_m \frac{\rho_A}{L},$$

where D_m is its molecular diffusion coefficient (Equation 1.25). For the convective flux we have, by an argument similar to that used for momentum in the previous paragraph:

$$\mathcal{J}_{conv} \approx \rho_A U.$$

The ratio of the fluxes associated with these two mechanisms can be expressed by:

$$\frac{\mathcal{J}_{conv}}{\mathcal{J}_m} \approx \frac{\rho_A U}{D_m \, \rho_A / L} = \frac{UL}{D_m} = Pe. \tag{2.16}$$

The ratio Pe, known as the *Péclet number*, represents for the phenomena of mass (tracer) dispersion the equivalent of the Reynolds number for momentum. We can also, just as we did for the Reynolds number, define the Péclet number, as the ratio of the characteristic times for transport of mass along a distance of order L by diffusion and by convection.

Thermal energy transport, Prandtl number

For thermal diffusion, we can carry out similar comparisons between convective and diffusive transport. This leads us to define, just as in the previous paragraph, a thermal Péclet number by:

$$Pe_\theta = \frac{UL}{\kappa},$$

where κ is the thermal diffusivity introduced in Chapter 1. In order to evaluate the effectiveness of diffusive transport of thermal energy relative to that of momentum, we calculate a *Prandtl number*, which is the ratio between the coefficients of kinematic viscosity and thermal diffusion:

$$Pr = \frac{\nu}{\kappa} = \frac{Pe_\theta}{Re}. \tag{2.17}$$

In other words, the Prandtl number Pr represents the ratio of the characteristic times L^2/κ and L^2/ν for the diffusion of temperature and velocity fluctuations over the same distance L. Typical values of this number are listed in Table 1.2 at the end of Chapter 1. In the same way, the *Lewis number*, $Le = D_m/\kappa$, represents a ratio between the thermal diffusion and mass diffusion times; it occurs in the phenomena involving combustion, which we will study in Section 10.9. Table 2.1 at the end of this section summarizes the definition of the various dimensionless numbers which involve diffusion coefficients.

For gases, the coefficients of thermal diffusion κ, of mass diffusion D_m, and of kinematic viscosity ν are of the same order of magnitude as we have seen in Section 2.2.1: the Prandtl number and the Lewis number are therefore of the order of unity. In a gas dynamic problem, where we simultaneously observe mass and momentum transport, the Reynolds and Péclet numbers are also of the same order of magnitude. On the other hand, in liquids,

Reynolds number	$Re = \frac{UL}{\nu}$	$\frac{\text{momentum diffusion time}}{\text{momentum convection time}}$
Péclet number	$Pe = \frac{UL}{D_m}$	$\frac{\text{mass (tracer) diffusion time}}{\text{mass (tracer) convection time}}$
Thermal Péclet number	$Pe_\theta = \frac{UL}{\kappa}$	$\frac{\text{thermal diffusion time}}{\text{thermal convection time}}$
Prandtl number	$Pr = \frac{\nu}{\kappa}$	$\frac{\text{thermal diffusion time}}{\text{momentum diffusion time}}$
Schmidt number	$Sc = \frac{\nu}{D_m}$	$\frac{\text{mass (tracer) diffusion time}}{\text{momentum diffusion time}}$
Lewis number	$Le = \frac{D_m}{\kappa}$	$\frac{\text{thermal diffusion time}}{\text{mass (tracer) diffusion time}}$

Table 2.1 *Dimensionless numbers characterizing the relative magnitude of the various mechanisms governing diffusion and convection.*

the Prandtl number can take on very different values depending on the thermal conduction mechanisms which are observed. Thus, for liquid metals, it is the transport of thermal energy by the conduction electrons which dominates; this leads to a high thermal diffusion and a correspondingly low value of the Prandtl number. On the other hand, in electrically insulating liquids with high viscosities (such as organic oils), the thermal diffusivity varies very little from one oil to another, while the viscosity and, consequently, the Prandtl number, can take on very large values.

We can also define a mass Prandtl number $Sc = \nu / D_m$, which is called the *Schmidt number*. If we consider the value of the coefficient D_m for liquids discussed in Section 1.3.3, we see that, as the viscosity ν of the medium increases, D_m decreases at the same time. The values of the Schmidt number Sc in liquids of high viscosity may therefore be very high (for water, which has a low viscosity, Sc is already of the order of 10^3). In such instances, even for low Reynolds numbers, the spread of a local concentration of tracer (such as a dye) due to the effect of velocity gradients is much more important than that due to molecular diffusion.

The spreading of tracers in porous media is an example of the application of phenomena of convective dispersion. In such media (studied in Section 9.7), the size of the flow channels is very small, and the Reynolds number Re is generally small. Because of the large variations of the velocity from one point to another, resulting from the random geometry of the medium, the spread of a tracer acted on by velocity gradients is characterized by a dispersion coefficient which takes on the role of the diffusivity for this problem: this coefficient is normally much larger than that, D_m, of molecular diffusion. More importantly, for turbulent flows where the velocity at a given point varies with time, convective transport is dominant except over very small distances, for which mixing by molecular diffusion takes place much faster than mixing by convection.

We will analyze a few examples of this problem in Chapter 9, which discusses flows at low Reynolds numbers and in Chapter 10, which deals with boundary layers.

In visualizing non-stationary flows by means of tracers, the distribution of dyes does not display the instantaneous configuration of the flow, but it depends on the entire history of the evolution of the flow (Section 3.1.4).

We have defined in this section several dimensionless numbers (Reynolds, Prandtl, Péclet numbers . . .) characterizing the flows. A classical method for defining such numbers involves finding a dimensionless combination of the parameters which occur in the flow (in the form of a product of powers of these parameters). In the case of Reynolds numbers it was the viscosity η, density ρ, velocity U and length scale L. Keeping in mind the fact that the mass appears only in η and ρ, we assume a form of the type $(\eta/\rho)^\alpha U^\beta L^\gamma$. The only possible dimensionless form corresponds to $\beta = \gamma = -\alpha$ (in fact, the dimension of η/ρ is UL). This method, known as the method of Vaschy–Buckingham, provides a systematic approach in order to suggest dimensionless combinations and determine the possible number of independent ones; but it does not help our understanding of their physical significance, nor of what best characterizes a given phenomenon.

2.4 Description of various flow regimes

We have shown throughout this chapter that, depending upon the flow velocities and geometries, either diffusive or convective mechanisms dominate momentum transport in a fluid. The measure of the relative importance of these transport mechanisms is the Reynolds number, Re. We now discuss, on the basis of experimental observations, how variations in the transport processes, related to the values of the Reynolds number, affect the flow, and how transitions between the various flow regimes take place.

Everyday life provides us with numerous examples of the diversity of flows—from the perpetually fluctuating appearance of river rapids ("turbulent" flow) to the extremely smooth, stable aspect of a high-viscosity oil being poured from one container to another ("laminar" flow). There are also intermediate cases, where the flow is only intermittently turbulent, or where it varies with time in a periodic manner. This is the case for air flow past a telephone wire ("the singing wire"). The strong coupling between the deformations of a structure (chimney, bridge) and vortex emission can lead to an amplification of these deformations and result ultimately in the destruction of the structure. A well known example is the collapse, in a moderate wind (68 km/h), of the Tacoma bridge following high amplitude torsional oscillations of the deck coupled to vortex emission at the edges.

Aeroelasticity is the domain of fluid mechanics which deals with such coupling between aerodynamic forces (viscous and/or inertial ones) and the deformations of the structure submitted to the flow.

2.4.1 Flows in a cylindrical tube: Reynolds' experiment

In a seminal article published in 1883, Osborne Reynolds, taking up previous experiments done by Poiseuille, describes the transition of water flowing in a cylindrical tube from a regime which is laminar ("direct") to one which is turbulent ("sinuous"). He introduces, at the very beginning of this paper, the number which now bears his name and he carries out a very complete set of measurements in which he varies one by one the parameters of the flow: the viscosity (by changing the temperature), the diameter and the velocity. He shows experimentally, as a complement to his theoretical analysis, that the laminar-turbulent transition occurs in these various cases for the same value Re_c of this parameter. He also shows that this transition is accompanied by a change in the relation governing the variation of the pressure difference between the ends of a tube as a function of the flow rate. Reynolds also observes that the tube needs to be longer than a certain "entry length" in order to establish a stationary velocity profile (in this case, the "Poiseuille profile"). Finally, he notes the importance of controlling the perturbations which generate the transition and affect the value of Re_c: in order to minimize them, he attaches at the entrance of the tube a kind of funnel which leads to a continuous, gradual variation of the cross-section. These different characteristics will be discussed in greater detail in Section 11.4.3.

Reynolds' paper also discusses the visualizations that can be achieved by injecting dye at given points near the entry of the tube. Observations obtained in a similar manner are displayed in Figure 2.8. At low velocity ($Re < Re_c$) (case a), the injected dye remains localized over a great distance along a straight line stream which, just as the velocity of

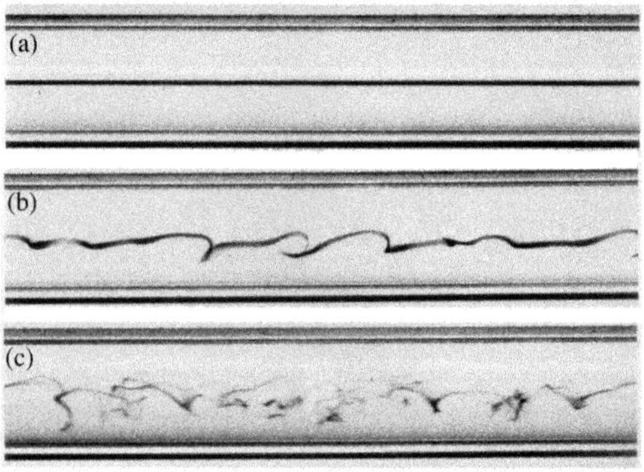

Figure 2.8 *Reproduction of the classical Reynolds experiment: visualization of a dye injected at the entrance of a circular tube in the case of (a) a laminar regime; (b) and (c) turbulent regimes (plates N.H. Johannesen and C. Lowe, An Album of Fluid Motion)*

the fluid, is parallel to the cylinder axis. It is this absence of mixing and the stability of the flow which lead to the name laminar flow. Above the critical Reynolds number Re_c (cases b, c), when the flow is turbulent, there appear transverse velocity components which vary randomly with time and distance. These result in significant mixing of the dye between the center and the regions close to the walls. Finally, for Re in the neighborhood of Re_c, Reynolds reports an intermediate regime with turbulent surges, which propagate with the flow and are separated by laminar flow phases.

2.4.2 Various flow regimes in the wake of a cylinder

One observes in Figures 2.9a–d different steps in the transition towards turbulence for the case of a flow of increasing upstream velocity U around a circular cylinder of diameter d. The flow is directed in the x-direction and the cylinder axis is oriented in the transverse z-direction. The images are obtained by illuminating the cylinder with a luminous "sheet" of light located in the x-y plane of the figure, and by injecting, upstream of the cylinder, a fluorescent dye or small, light particles. The Reynolds number characteristic of the flow is taken equal to Ud/ν. In Section 11.1, devoted to instabilities, we will discuss a model, called the Landau model, describing the behavior in the neighborhood of the critical Reynolds number ($Re_c = 47$), at which recirculating eddies appear behind the cylinder.

The same experiment can also be set up by using a container, with water a few centimeters deep, in which there is a suspension of very tiny, elongated particles; the particles are found to line up with the flow, and to reflect the light anisotropically. By displacing through the liquid a vertical cylindrical object with diameter d, of a few millimeters, we observe the different regimes displayed in Figure 2.9.

- *At low velocities* ($Re \approx 1$) (Figure 2.9a), the flow is laminar and quite symmetric between the upstream and downstream regions of the cylinder. This is characteristic of the reversibility of flows at low Reynolds numbers, which we will study in Chapter 9 (when the direction of the flow is inverted, the streamlines are unchanged).
- *Beyond a Reynolds number* $Re \approx 26$ (Figure 2.9b), one observes two fixed recirculating eddies downstream of the cylinder (recirculating flow). The length L of the recirculating region increases as Re increases.
- *Beyond a critical value* Re_c *of the order of* 47, the flow is no longer stationary and the velocity of the fluid depends on time: vortices are periodically emitted downstream of the cylinder (Figure 2.9c, for which $Re \approx 200$). They form a double row of vortices,

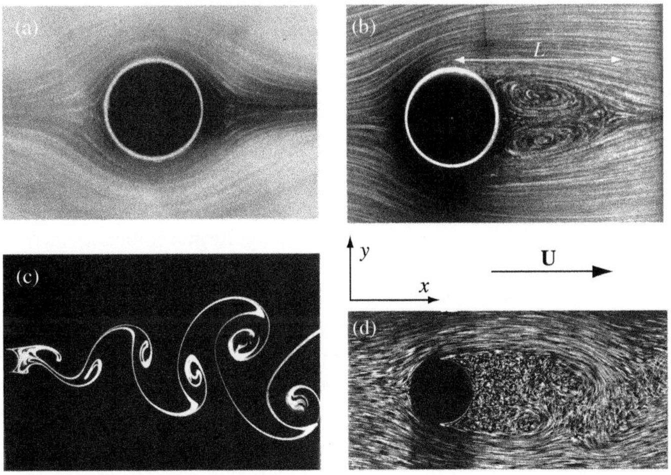

Figure 2.9 *Visualization of the flow patterns near a cylinder at various Reynolds numbers: (a) virtually symmetric upstream and downstream flow-patterns at low Reynolds number (Re* = 0.16*); (b) two fixed recirculation eddies appear downsteam, behind the cylinder (Re* = 26*); (c) periodic emission of vortices making up a Bénard–von Karman vortex street (Re* = 200*); (d) turbulent wake (Re* ~ 8000*) (plates a, b et c: S. Taneda; plate d: H. Werlé)*

known as the *Bénard–von Karman vortex street*. The frequency at which these vortices are emitted is characterized by a dimensionless number called the *Strouhal number*:

$$Sr = f\frac{d}{U}, \tag{2.18}$$

which does not depend very much on U and is of the order of unity. The frequency f is, therefore, proportional to the velocity U.

Current understanding of turbulence takes into account simultaneously the large structures which vary with the nature of the flow, the disordered motion at small scales, which is independent of its nature, as well as the transfer of the kinetic energy of the flow from larger scales to smaller ones. These important concepts of *statistical turbulence* will be evoked in Chapter 7, using examples from the dynamics of vortices, and then studied in detail in Chapter 12.

- *At very large Reynolds numbers* (Figure 2.9d), one observes incoherent turbulent motion, at smaller spatial scales, which become progressively smaller the larger the Reynolds number (in fact, their minimum size decreases as $Re^{-1/2}$). However, the periodic emission of large vortices superimposed on these fluctuations still occurs. These remain observed at extremely large Reynolds numbers in oceanographic or atmospheric flows behind very large obstacles (Figure 2.10).

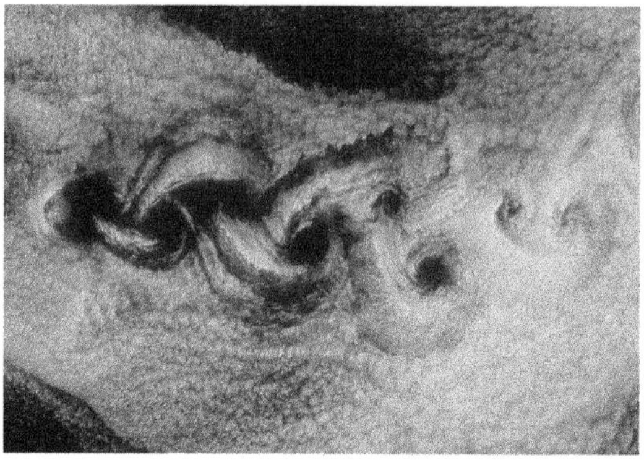

Figure 2.10C *Satellite view of cloud formations in the vortex street emitted behind the volcanic island of Rishiri-to off the northern tip of Hokkaidō in the Sea of Japan (document NASA, mission STS100)*

2.4.3 Flow behind a sphere

Figures 2.11 a–c. show that flow regimes downstream of a sphere of diameter d differ in several ways from the previous ones; these are observed in a flow at Reynolds number $Re=Ud/\nu$ (U is the velocity of the uniform flow observed sufficiently far upstream of the sphere).

- **For *Re* < 212**, one observes an axially symmetric and stationary flow with, downstream of the sphere, a toroidal vortex with its axis parallel to the velocity $\boldsymbol{U}$ (Figure 2.11a). This vortex plays the role of the two symetric recirculation zones observed for the cylinder.
- **For 212 < *Re* < 280,** there appear two vortices parallel to the flow axis, symmetric relative to the latter and counter-rotating. As the Reynolds number increases, the vortices are, at first, stationary for $Re < 267$ (Figures 2.11b and e) and then oscillate for $Re > 267$ (Figures 2.11c and f).
- Finally **for Re > 280** (Figures 2.11d and g), the flow first becomes periodic, generating regular vortices in the form of "hairpins": two longitudinal vortices appear just as in the previous case but, beyond a finite distance, they change shape and are connected by a segment transverse to the vortex. For yet larger Reynolds numbers, we have, as previously, the appearance of more complicated motion and ultimately of turbulence.

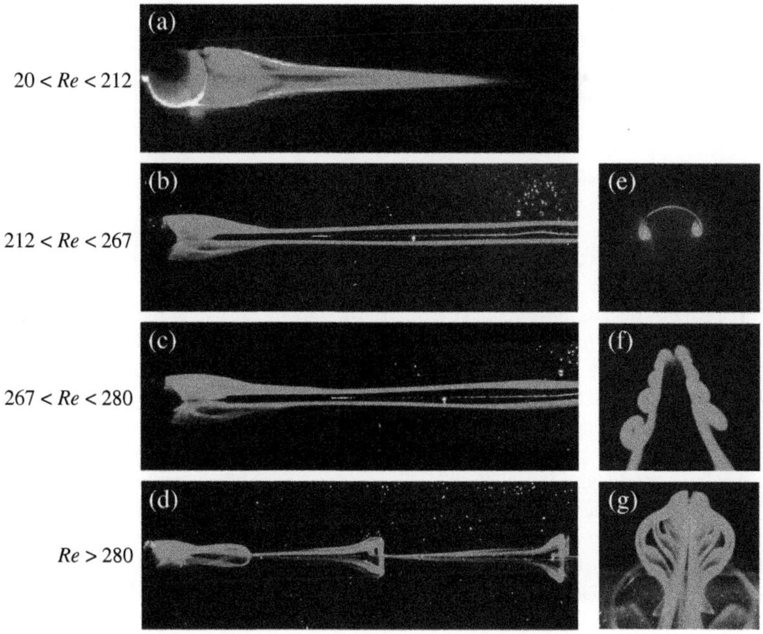

Figure 2.11C *Views from the top (a,b,c,d) and back (e,f,g) of the flow behind a sphere (at the left) visualized by injection of a fluorescent dye. For increasing Reynolds numbers Re, evaluated on the basis of the upstream velocity and the diameter of the sphere, one observes in succession (a) a toroïdal vortex attached on the downstream side of the sphere* ($20 < Re < 212$)*; (b,e) two fixed vortices with their axis parallel to the main flow* ($212 < Re < 267$)*; (c,f) two oscillating vortices* ($267 < Re < 280$)*; (d,g) "hairpin"-like structures* ($Re > 280$) *(documents A. Przadka and S. Goujon Durand, PMMH-ESPCI)*

In these two examples and, quite generally, for flows around obstacles, the increase of the inertial terms with the Reynolds number first creates stationary structures combined with the fundamental laminar flow, then induces flow components periodic with time and, finally, leads to turbulence. In other geometries, like in Poiseuille flows, turbulence appears directly above some critical Re number.

3 Kinematics of Fluids

This chapter covers the study of the motion of a fluid including, in particular, the analysis of its deformations (or strains), without being concerned here with the origin of these strains, discussed in the next chapter. We start by indicating the methods used to characterize the motion of the fluid (Section 3.1): the definition of the velocity of a particle of fluid, the Eulerian and Lagrangian descriptions, the acceleration and paths characteristic of a flow. In Section 3.2, we analyze the strains in the fluid. Section 3.3 is devoted to setting up the mathematical formulation for the law of conservation of mass, and its consequences for an incompressible fluid—this is the case for the fluids considered throughout this book. We also mention, in the same section, a number of parallels with the electromagnetic theory, covered in greater details in following chapters. We then introduce the concept of a stream function for plane (two-dimensional) or axially symmetric flows (Section 3.4), and discuss a few examples of plane flows and of the system of streamlines associated with these. Finally, in Section 3.5, we conclude by describing a number of experimental methods used to characterize the velocity field (velocity and velocity gradients) of a fluid.

3.1 Description of the motion of a fluid

3.1.1 Characteristic linear scales and the hypothesis of continuity

We define a *fluid particle* as an element of fluid of volume V such that its size, $a \approx V^{1/3}$, is:

- very small relative to the length scale L characteristic of the flow (the width of a channel, the radius of a tube, the size of an obstacle);
- very large relative to the mean free path ℓ of the molecules. If this were not the case, the molecules could cross the entire volume of the particle with no change in their energy or momentum; on a scale that small, a significant average distribution of the velocity could not then be defined.

The macroscopic length scale L of the flow might be as small as a fraction of a millimeter in the case of a capillary blood vessel, a micron for the holes in a porous medium or even less in microfluidics. The mean free path ℓ (the lower scale limit) is of the order of microns for gases at normal pressure and will, generally, be very much smaller than L in most applications. On the other hand, for gases at very low pressures (less than $\sim 10^{-4}$ Pa $\approx 10^{-6}$ Torr) or for scales L sufficiently small, the mean free path could be of the same order of magnitude as the macroscopic dimensions of the container. In this molecular region, called the *Knudsen regime*, the study of flows becomes a problem of the mechanics of discrete objects

We must be careful not to confuse the *particle* of fluid with the *molecules* (or atoms) making up the fluid; the fluid particle will always contain, as discussed above, a very large number of molecules. Figures 2.7 in Section 2.2.3 show clearly the difference between individual molecules (a) and the average velocity of the fluid (b).

Physical Hydrodynamics. Second Edition. Etienne Guyon *et al.*

which we can describe in terms of their collisions with the walls. We referred to this case in Section 1.3.2 in the discussion of the diffusion coefficient of gases.

When the model of particles of fluid is applicable, the fluid can be treated as a *continuous medium*. We then define the local velocity $\mathbf{v}$ of the fluid – i.e., the velocity of a fluid particle – as the average value of the velocities of the molecules located inside this small volume of fluid. This average is independent of the size a of the particle so long as the assumption $a \gg \ell$ holds and that a is small relative to the characteristic macroscopic length L of the flow.

3.1.2 Eulerian and Lagrangian descriptions of fluid motion

In the *Eulerian description* of fluid motion, we are concerned with the velocity $\mathbf{v}(\mathbf{r}, t)$ of a fluid particle which is located at time t at the *fixed point* M having vector position $\mathbf{r}$. At a later time t', the velocity at the same point $\mathbf{r}$ will have become $\mathbf{v}(\mathbf{r}, t')$; moreover, it will correspond to the velocity of *different* particles. This viewpoint is that of an observer at rest in the reference frame in which the velocity $\mathbf{v}$ is measured, and corresponds exactly to the experimental measurements carried out with *probes* which are *fixed* relative to the motion of the fluid; these techniques are described in Section 3.5. It is the Eulerian velocity which we observe when we look down at water flowing under a bridge. The particles of fluid we observe at a given point (indicated, for example, by dust on the surface) are different at each instant of time. Their velocity is a function both of the time at which we make the observation and of the point $\mathbf{r}$ (fixed relative to the bridge) from which we look. This Eulerian description confronts us, however, with the inconvenience of introducing non-linear terms in the expression for the acceleration as we will see in Section 3.1.3.

In the *Lagrangian description*, we follow, as it moves, a particle of fluid which had a position $\mathbf{r}_0$ ($\mathbf{r}_0 = \mathbf{OM}_0$) at a given reference time t_0. The velocity of the fluid is then characterized by the vector $\mathbf{V}(\mathbf{r}_0, t)$ which is a function of the two variables $\mathbf{r}_0$ and t. In the example of the flowing river, this point of view is that of an observer floating on a raft being moved by the current: the speed of the raft is the Lagrangian velocity. The Lagrangian point of view corresponds to measurements made with instruments which go along with the fluid as it moves, such as probe-balloons in the atmosphere, or tagged particles in a flow (Section 3.5).

We use capital letters to denote the Lagrangian velocity $\mathbf{V}(\mathbf{r}_0, t)$, as distinct from the Eulerian velocity field $\mathbf{v}(\mathbf{r},t)$.

3.1.3 Acceleration of a particle of fluid

Consider a fluid particle located at time t at the point $M_1(\mathbf{r}_1)$ (Figure 3.1); its velocity at that instant is $\mathbf{v}(\mathbf{r}_1, t)$. At a slightly later time $t' = t + \delta t$, this fluid particle is now at the point $M_2(\mathbf{r}_2)$ such that $\mathbf{r}_2 = \mathbf{r}_1 + \mathbf{v}(\mathbf{r}_1, t)\delta t + O(\delta t^2)$ and its velocity is now $\mathbf{v}(\mathbf{r}_2, t')$. The velocity change $\delta\mathbf{v} = \mathbf{v}(\mathbf{r}_2, t') - \mathbf{v}(\mathbf{r}_1, t)$ of this fluid particle in the time interval δt results:

- on the one hand, from the explicit variation of the velocity field $\mathbf{v}(\mathbf{r},t)$ with time, if the flow is non-stationary (the corresponding term is: $[\mathbf{v}(\mathbf{r}_1, t') - \mathbf{v}(\mathbf{r}_1, t)]$);
- on the other hand, from the "probing" of the velocity field by the particle; this effect contributes to the acceleration only if the field is non-uniform (the corresponding term is: $[\mathbf{v}(\mathbf{r}_2, t') - \mathbf{v}(\mathbf{r}_1, t')]$).

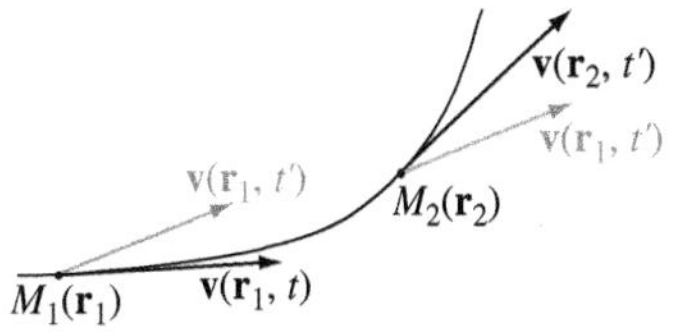

Figure 3.1 *Components of the acceleration of a particle of fluid in non-stationary flow*

The resultant velocity change $\delta\mathbf{v}$ can then be written as a first-order expansion for each of the two terms:

$$\delta\mathbf{v} = \mathbf{v}(\mathbf{r}_2, t') - \mathbf{v}(\mathbf{r}_1, t) = \frac{\partial \mathbf{v}}{\partial t}\delta t + \frac{\partial \mathbf{v}}{\partial x}\delta x + \frac{\partial \mathbf{v}}{\partial y}\delta y + \frac{\partial \mathbf{v}}{\partial y}\delta y + \frac{\partial \mathbf{v}}{\partial z}\delta z,$$

where δx, δy and δz are the components of the vector $\mathbf{r}_2 - \mathbf{r}_1$. The acceleration of the fluid particle is then:

$$\frac{\mathrm{d}\mathbf{v}}{\mathrm{d}t} = \lim_{\delta t \to 0} \frac{\delta \mathbf{v}}{\delta t} = \lim_{\delta t \to 0} \left(\frac{\partial \mathbf{v}}{\partial t} + \frac{\partial \mathbf{v}}{\partial x}\frac{\delta x}{\delta t} + \frac{\partial \mathbf{v}}{\partial y}\frac{\delta y}{\delta t} + \frac{\partial \mathbf{v}}{\partial z}\frac{\delta z}{\delta t} \right) = \frac{\partial \mathbf{v}}{\partial t} + v_x\frac{\partial \mathbf{v}}{\partial x} + v_y\frac{\partial \mathbf{v}}{\partial y} + v_z\frac{\partial \mathbf{v}}{\partial z},$$

or, in a more concise form:

From here on we will use the notation $\mathrm{d}/\mathrm{d}t$ to refer to the *Lagrangian derivative* (or *material* or yet *convective derivative*) obtained as we follow the fluid particles in their motion (this is sometimes also denoted by $\mathrm{D}/\mathrm{D}t$).

$$\frac{\mathrm{d}\mathbf{v}}{\mathrm{d}t} = \frac{\partial \mathbf{v}}{\partial t} + (\mathbf{v} \cdot \nabla)\mathbf{v}. \tag{3.1}$$

The symbolic representation in the second term in the right-hand side of the equation involves a scalar product between the vector $\mathbf{v}$ and the ∇ operator which has components $\partial/\partial x$, $\partial/\partial y$ and $\partial/\partial z$. Figures 3.2a and 3.2b illustrate the two types of contributions of the scalar product to the total acceleration in a stationary flow (such that $\partial\mathbf{v}/\partial t = 0$).

In the case of Figure 3.2a, there exists a non-vanishing component $\partial v_x/\partial x$ of the gradient of the velocity in the direction of the flow. As a result, a particle carried along by the flow undergoes an acceleration equal to $v_x\, \partial v_x/\partial x$. In Figure 3.2b, there exists a non-zero transverse gradient $\partial v_x/\partial y$ of the velocity component in the x-direction. Let us assume that there is additionally a non-zero component v_y of the velocity in the y-direction. This will result in a change of the velocity v_x *along the particle trajectory* for which the time derivative will equal $v_y \partial v_x/\partial y$. Both of these terms are included in the expression $(\mathbf{v} \cdot \nabla)v_x$ which is the component in the x-direction of the vector $(\mathbf{v} \cdot \nabla)\mathbf{v}$

We can write the equivalent of Equation 3.1 to express the variation of variables other than the velocity along the trajectory of a particle of fluid: for example, we can use this notation to express the variation of the temperature $T(\mathbf{r}, t)$ of the particle or its chemical concentration $C(\mathbf{r}, t)$. We find then that the change in temperature of a particle *along its trajectory* satisfies:

$$\frac{\mathrm{d}T}{\mathrm{d}t} = \frac{\partial T}{\partial t} + (\mathbf{v} \cdot \nabla)T, \tag{3.2}$$

where $\partial T/\partial t$ is the explicit time-derivative of the temperature of the fluid at a given fixed point. The second, right-hand-side term displays the change of T due to the fluid flow in the direction of the temperature gradient (this equation can be derived in the same manner as Equation 3.1).

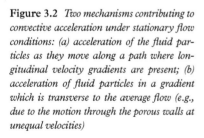

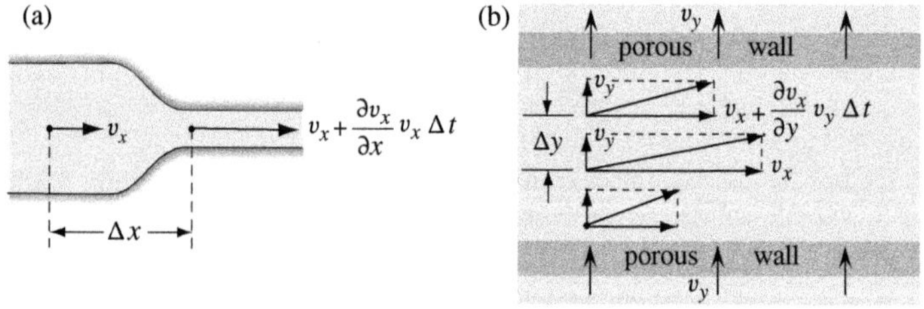

Figure 3.2 *Two mechanisms contributing to convective acceleration under stationary flow conditions: (a) acceleration of the fluid particles as they move along a path where longitudinal velocity gradients are present; (b) acceleration of fluid particles in a gradient which is transverse to the average flow (e.g., due to the motion through the porous walls at unequal velocities)*

3.1.4 Streamlines and stream-tubes, trajectories and streaklines

- ***Streamlines*** are the field lines of the vector field $\mathbf{v}(\mathbf{r}, t)$. They are defined as the tangents, at every point, to the velocity vector $\mathbf{v}(x,y,z,t_0)$ at a given time t_0. A *stream-tube* is the set of streamlines which pass through a closed space curve. We can visualize streamlines *experimentally* by taking a *short-time-exposure* photograph of a system of particles in suspension. Each particle will mark on the photograph a short segment indicating the direction of the local velocity vector; the streamlines are tangent to the segments obtained in this way, and the length of the segments is proportional to the magnitude of the velocity. We will provide in Section 3.5.1 a few additional details on the methods used to visualize and make measurements on flows.

Mathematically, a streamline is defined as a curve tangent to the local velocity vector at all points, so that a small displacement $\mathrm{d}\mathbf{M}(\mathrm{d}x, \mathrm{d}y, \mathrm{d}z)$ along the line will be co-linear with the velocity vector $\mathbf{v}$. This condition can be expressed as:

$$\mathrm{d}\mathbf{M} \times \mathbf{v} = 0, \quad \text{i.e.,} \quad \frac{\mathrm{d}x}{\mathrm{v}_x} = \frac{\mathrm{d}y}{\mathrm{v}_y} = \frac{\mathrm{d}z}{\mathrm{v}_z}.$$

We obtain the equation for the streamlines by integrating these two differential equations.

- The ***trajectory***, or ***pathline***, of a fluid particle is defined as the path which this particle follows in time – i.e., the set of successive positions through which this particle passes as it moves. We can visualize these experimentally by taking a *long-time-exposure* photograph of the displacement of a tracer emitted for a very short time at a point in the fluid (dye, light-scattering particles, hydrogen bubbles . . .). We obtain the mathematical expression for the pathline by integrating with respect to time the Lagrangian velocity field $\mathbf{V}(\mathbf{r}_0, t)$, i.e., the system of equations:

$$\frac{\mathrm{d}x}{\mathrm{d}t} = V_x(\mathbf{r}_0, t), \qquad \frac{\mathrm{d}y}{\mathrm{d}t} = V_y(\mathbf{r}_0, t), \qquad \frac{\mathrm{d}z}{\mathrm{d}t} = V_z(\mathbf{r}_0, t).$$

Indeed, if $\mathbf{r}$ is the position of a particle of fluid at time t, and $\mathbf{r} + \mathrm{d}\mathbf{r}$ its position at time $t + \mathrm{d}t$, we have, by definition of the Lagrangian velocity $\mathbf{V}(\mathbf{r}_0, t)$ (as the particle is followed along in its motion):

$$\mathbf{V}(\mathbf{r}_0, t) = \frac{\mathrm{d}\mathbf{r}}{\mathrm{d}t}, \quad \text{whence, by direct integration:} \quad \mathbf{r}(t) = \mathbf{r}_0 + \int_{t_0}^{t} \mathbf{V}(\mathbf{r}_0, t')\mathrm{d}t'.$$

- A ***streakline*** represents the set of positions, at a given time t, of a particle of fluid which coincided at an instant in the past with the point $M_0(x_0, y_0, z_0)$. It is experimentally obtained by continuous emission of a tracer (e.g., dye) at the point M_0, and provides an *instantaneous* photograph of the whole set of positions of this tracer.

For the case of stationary flow, the velocity field does not depend explicitly on the time, so that $\partial\mathbf{v}/\partial t = 0$. In this case, the streamlines, pathlines and streaklines coincide. Indeed, different "marker" particles emitted from the same point at different times follow the same trajectories: they therefore represent at the same time the streaklines. Also, the local velocity vector (independent of time) is tangent at each point to the trajectories, which therefore also display the streamlines.

In contrast, in the case of non-stationary flow, e.g., for an obstacle which moves through a container in which the fluid is otherwise at rest, these different lines are, in general, quite distinct from one another and the relationship between them is difficult to establish. We will then usually be concerned with the *streamlines* within the fluid.

3.2 Deformations in flows

In this section, we analyze the deformation (strain) of a fluid particle; this is a needed step toward the evaluation – carried out in Chapter 4 – of the force exerted by the nearby fluid. A similar procedure can be used in the analysis of strains in an elastic solid but the latter only undergo finite-amplitude deformations: as a result, in the mechanics of solids, the concepts of strain and rotation replace those, for fluids, of the rate of strain and the rate of rotation (the term "rate" used here indicates the change, per unit time, of the quantity under consideration).

3.2.1 Local components of the velocity gradient field

Consider, at a given time t, a particle of fluid located at a point $\mathbf{r}$ with velocity $\mathbf{v}(\mathbf{r}, t)$; the velocity of a nearby particle located at the point $\mathbf{r} + \delta\mathbf{r}$ is correspondingly $\mathbf{v} + \delta\mathbf{v}$. For each component δv_i $(i = x, y, z)$ of $\delta\mathbf{v}$, the velocity increment can be written to first-order relative to the components of the displacement δx_j $(j = x, y, z)$ as:

$$\delta v_i = \sum_j \left(\frac{\partial v_i}{\partial x_j}\right) \delta x_j. \tag{3.3}$$

The fact of discussing only the change in velocity between two adjacent points is equivalent to neglecting the global translational motion of the entire system of particles; such translation indeed results in no deformation. The quantities $G_{ii} = (\partial v_i / \partial x_i)$ are the elements of a second-rank tensor – the *tensor of the deformation rates* (or the *velocity gradient tensor*) – for a fluid.

It can be written as a 3×3 (or 2×2) matrix in three (or two) dimensions, – a matrix which can always be decomposed into a symmetric and an antisymmetric component, as follows:

$$G_{ij} = \frac{\partial v_i}{\partial x_j} = \frac{1}{2}\left(\frac{\partial v_i}{\partial x_j} + \frac{\partial v_j}{\partial x_i}\right) + \frac{1}{2}\left(\frac{\partial v_i}{\partial x_j} - \frac{\partial v_j}{\partial x_i}\right). \tag{3.4}$$

We can rewrite:

$$e_{ij} = \frac{1}{2}\left(\frac{\partial v_i}{\partial x_j} + \frac{\partial v_j}{\partial x_i}\right) \quad (3.5a) \qquad \text{and} \qquad \omega_{ij} = \frac{1}{2}\left(\frac{\partial v_i}{\partial x_j} - \frac{\partial v_j}{\partial x_i}\right), \quad (3.5b)$$

i.e.,

$$G_{ij} = e_{ij} + \omega_{ij}. \tag{3.6}$$

By their very definition, the tensors e_{ij} and ω_{ij} are respectively symmetric and antisymmetric. We will study their physical significance in Sections 3.2.2 and 3.2.3. To simplify our explanation, we will consider the case of deformations in two dimensions. These could be observed by seeding the free surface of the liquid with a number of fine particles. Practically, we will consider the deformation and rotation of a test object making up a small square *ABCD* shown in Figure 3.3a below.

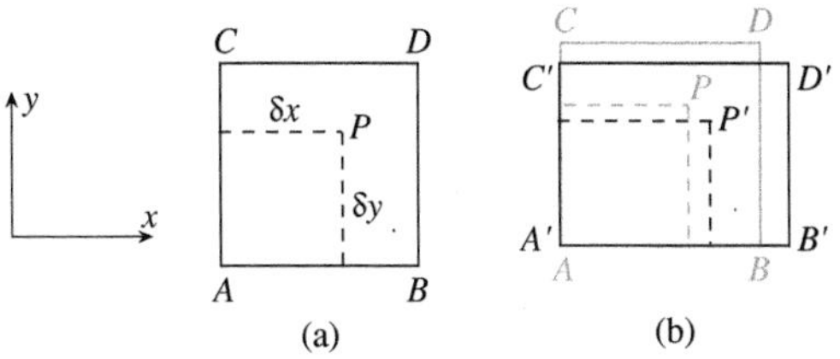

Figure 3.3 *Deformation of a square in a flow for which the velocity gradient field consists of only diagonal terms of the type $(\partial v_i/\partial x_i)$. (a) original, undeformed square at initial time t; (b) configuration of the square at the later time $t + \delta t$*

For an arbitrary point P in the test square $ABCD$, we will calculate the change in the vector **AP** $(\delta x, \delta y)$ in a time interval δt. We denote by A' and P' the position of the points A and P at the end of the time interval δt. This change can be written as:

$$\mathbf{A'P'} - \mathbf{AP} = \mathbf{PP'} - \mathbf{AA'} = \mathbf{v}(P)\,\delta t - \mathbf{v}(A)\,\delta t \approx \left(\frac{\partial \mathbf{v}}{\partial x}\delta x + \frac{\partial \mathbf{v}}{\partial y}\delta y\right)\delta t. \tag{3.7a}$$

The right side of the above equation represents the first order expansion of $\mathbf{v}(P) - \mathbf{v}(A)$ according to Equation 3.3. This approximation is valid for time intervals δt and velocity gradients $\partial v_i/\partial x_j$ sufficiently small so that the product $(\partial v_i/\partial x_j)\,\delta t$ is small compared to unity i.e., for *small deformations* (the case of *large deformations* will be examined in Section 3.2.5). By stating explicitly the components v_x and v_y of the velocity ($\mathbf{e}_x$ and $\mathbf{e}_y$ are the unit vectors along the x- and y-axes, respectively), the preceding equation can also be re-written as:

$$\mathbf{A'P'} - \mathbf{AP} \approx \left(\frac{\partial v_x}{\partial x}\delta x\,\mathbf{e}_x + \frac{\partial v_y}{\partial y}\delta y\,\mathbf{e}_y\right)\delta t + \left(\frac{\partial v_x}{\partial y}\delta y\,\mathbf{e}_x + \frac{\partial v_y}{\partial x}\delta x\,\mathbf{e}_y\right)\delta t. \tag{3.7b}$$

We will use this equation to study successively the evolution of each of the sides **AB** and **AC** of the square due to the effect of the different components G_{ij} of the *velocity gradient tensor*.

3.2.2 Analysis of the symmetric component of the rate of strain tensor: pure strain

This symmetric component is the tensor e_{ij} defined in Equation 3.5a. It contains, in the most general case, both diagonal terms ($i = j$) and off-diagonal ones ($i \neq j$). We will consider in turn the effect of each of these two terms on the test square defined in Figure 3.3a.

Deformations due to the diagonal terms of the tensor $\mathbf{e_{ij}}$

Under the action of a velocity–gradient field containing only *diagonal terms* of the kind $\partial v_i/\partial x_i$, the square $ABCD$ will undergo, after time δt, the transformation shown in Figure 3.3b. Using Equation 3.7b and taking the point B as P ($\delta y = 0$ and $\delta x = AB$), we can write for the evolution of the side **AB** to first-order in $(\partial v_x/\partial x)\,\delta t$:

$$\mathbf{A'B'} - \mathbf{AB} \approx \left(\frac{\partial v_x}{\partial x}AB\,\delta t\right)\mathbf{e}_x.$$

The side **AB** therefore undergoes no rotation (it remains parallel to the x-axis). Its relative elongation is:

$$\frac{\delta(AB)}{AB} \approx \frac{(\partial v_x/\partial x)\,AB\,\delta t}{AB} = \frac{\partial v_x}{\partial x}\delta t.$$

It is positive if $\partial v_x/\partial x$ is positive (the case for Figure 3.3). Similarly, for the evolution of the vector **AC** ($\delta x = 0$, $\delta y = AC$) we can write:

$$\mathbf{A'C'} - \mathbf{AC} \approx \left(\frac{\partial v_y}{\partial y} AC\,\delta t\right)\mathbf{e}_y.$$

The side **AC** thus doesn't rotate either under the action of the velocity gradient. Its relative elongation is:

$$\frac{\delta(AC)}{AC} \approx \frac{\partial v_y}{\partial y}\delta t.$$

Thus the sides of the square remain parallel to their original direction and, in the case of the figure it undergoes a dilation (or, respectively, a contraction) in the x- and y-directions. This illustrates the fact that the diagonal terms of the tensor of the components $\partial v_i/\partial x_j$ represent relive the rate of stretching of the fluid element in the corresponding direction (the x-direction for **AB**). Let us now estimate the relative change in the surface of the square $ABCD$.

$$\frac{\delta S}{S} = \frac{\delta(AB)}{AB} + \frac{\delta(AC)}{AC} = \left(\frac{\partial v_x}{\partial x} + \frac{\partial v_y}{\partial y}\right)\delta t = (\nabla \cdot \mathbf{v})\delta t. \tag{3.8a}$$

Effectively, the trace of the tensor [**e**] (equivalent to that of the tensor [**G**]), which is the sum of the diagonal elements ($\partial v_i/\partial x_i$), is equal to the divergence of the velocity field and represents the rate of expansion of the fluid element under consideration. In our two-dimensional example, this expansion corresponds to an increase in the surface area. More generally, for a flow where the velocity varies in three dimensions, the relative change in volume V of a parallelepiped can be written:

In order to write such relations without mentioning explicitly all terms, we shall use the (so-called) *Einstein convention*, which implies summation on repeated indices. Thus, writing ($\partial v_i/\partial x_i$) implies: $\partial v_i/\partial x_i = \sum_i (\partial v_i/\partial x_i)$.

$$\frac{\delta V}{V} = \left(\frac{\partial v_x}{\partial x} + \frac{\partial v_y}{\partial y} + \frac{\partial v_z}{\partial z}\right)\delta t = (\nabla \cdot \mathbf{v})\,\delta t. \tag{3.8b}$$

Thus, the rate of expansion of the volume of an element of fluid is also given by $\nabla\cdot\mathbf{v}$. For an incompressible fluid ($\delta V/V = 0$), this volume must remain constant; the velocity field must therefore have zero divergence.

Deformations resulting from the off-diagonal components of tensor [e]

Let us now consider the deformations of the same square $ABCD$ with its sides parallel to the axes when only the off-diagonal terms of the tensor G_{ij} (terms in $\partial v_i/\partial x_j$ with $i \neq j$) are non-zero. Let us write at this point the change in vectors **AB** and **AC**. Following the same procedure that we have just used, we obtain for the vector **AB** ($\delta x = AB$, $\delta y = 0$):

$$\mathbf{A'B'} - \mathbf{AB} \approx \left(\frac{\partial v_y}{\partial x}\right) AB\,\delta t\,\mathbf{e}_y.$$

The side **AB** is no longer parallel to the x-axis, but rotates through an angle:

$$\delta\alpha = \frac{(\partial v_y/\partial x)\,AB\,\delta t}{AB} = \frac{\partial v_y}{\partial x}\delta t, \qquad \text{so that:} \qquad \frac{\delta\alpha}{\delta t} = \frac{\partial v_y}{\partial x}. \tag{3.9a}$$

The angle $\delta\alpha$ is positive for the case of Figure 3.4 while the length of **AB** remains fixed. Calculating in the same way the change in the vector **AC**, we obtain:

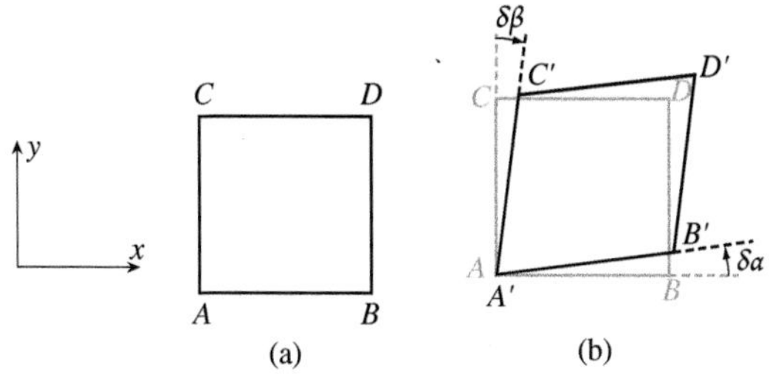

Figure 3.4 *Deformation of a square, resulting from a flow such that $e_{xx} = e_{yy} = 0$ and $\omega_{xy} = 0$. (a) at time t before any deformation has occurred; (b) geometry observed at time $t + \delta t$*

$$\mathbf{A'C'} - \mathbf{AC} = \left(\frac{\partial v_x}{\partial y} AC\,\delta t\right)\mathbf{e}_x, \qquad \text{so that:} \qquad \frac{\delta\beta}{\delta t} = -\frac{\partial v_x}{\partial y}. \tag{3.9b}$$

The angle of rotation $\delta\beta$, for the case of Figure 3.4, is negative as we assume that $\partial v_x/\partial y > 0$. If the two components $\partial v_x/\partial y$ and $\partial v_y/\partial x$ of the velocity–gradient tensor are equal, the antisymmetric component ω_{xy} will evidently vanish, while the angles $\delta\beta$ and $\delta\alpha$ will be opposite. The time-rate of change of the angle γ between the sides $\mathbf{A'B'}$ and $\mathbf{A'C'}$ then satisfies:

$$\frac{\delta\gamma}{\delta t} = -\frac{(\delta\alpha - \delta\beta)}{\delta t} = -\left(\frac{\partial v_y}{\partial x} + \frac{\partial v_x}{\partial y}\right) = -2e_{xy}. \tag{3.10}$$

Because the Equation 3.10 represents the time derivative of the angle γ, the notation $\dot{\gamma}$ is frequently used for this quantity, known as the *shear rate*. We will discuss this idea further in Chapter 4, particularly as to the manner in which a fluid is affected by the flow.

We can therefore interpret the *off-diagonal term e_{xy}* as the *rate of local angular deformation.*

Relationship between the deformations resulting from the diagonal and off-diagonal components of tensor [e]

In effect, if the trace of the tensor e_{ij} vanishes, and if we assume for the sake of simplicity that $\omega_{ij} = 0$, we have an equivalence between the deformations associated with a diagonal tensor and an off-diagonal one as we can see in the example of Figure 3.5. We assume now again that the tensor e_{ij} contains only diagonal terms, and consider an element of fluid *ABCD*, initially square but with its diagonals (not its sides) oriented parallel to the *x*- and *y*-axes. The deformations observed now are displayed in Figure 3.5 showing the square *EFGH* with its sides parallel to the *x*- and *y*-axes, and within which the first square in inscribed. The quadrangle *A'B'C'D'* has its vertices in the middle of the sides of *E'F'G'H'* and, as we see in the figure, the length of its sides does not vary (at least to first-order) while the angles between them change (resulting ultimately in a diamond shape). We obtain a deformation of the same nature as if we had selected axes parallel to the sides *AB* and *AC* and a tensor **[e]** with only off-diagonal non-zero components e_{ij}.

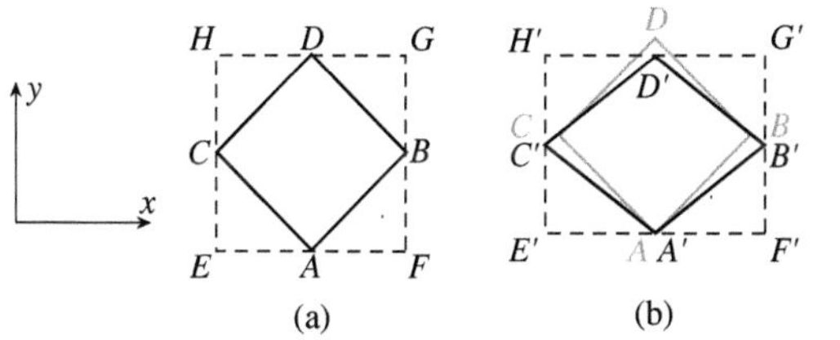

Figure 3.5 *Deformation of a square, with its diagonals aligned parallel to the x-y co-ordinate axes, resulting from a flow for which the velocity-gradient field contains only diagonal elements of the form $(\partial v_i/\partial x_i)$; (a) square with no deformation at the time t; (b) deformed square at the time $t + \delta t$*

In fact, the case of an off-diagonal tensor **[e]** is equivalent to that of a diagonal tensor if we calculate the eigenvectors of **[e]** and carry out a rotation of the axes such that **[e]** is diagonal in this new reference frame (only the components e_{ii} are non-zero).

Proof

Let us first calculate the change in the vectors **AD** and **CB** corresponding to the diagonals of the diamond. We can write, respectively:

$$\mathbf{A'D'} - \mathbf{AD} = \frac{\partial v_y}{\partial y} AD\,\delta t\,\mathbf{e}_y \quad \text{and} \quad \mathbf{C'B'} - \mathbf{CB} = \frac{\partial v_x}{\partial x} CB\,\delta t\,\mathbf{e}_x.$$

Thus, the diagonals **AD** and **CB** remain respectively parallel to the y- and x-axes even though the angles between the sides vary. The area S of the diamond, which is equal to half the product $CB.AD$, undergoes the following change:

$$\frac{\delta S}{S} = \frac{\delta(AD)}{AD} + \frac{\delta(CB)}{CB} = \frac{(\partial v_y/\partial y)\,AD\,\delta t}{AD} + \frac{(\partial v_x/\partial x)\,CB\,\delta t}{CB} = (\nabla \cdot \mathbf{v})\,\delta t.$$

We recover the result of Equation 3.8a, that the rate of change of area is proportional to the divergence of the velocity vector.

This analysis of the effects of each of the terms of the velocity–gradient field (diagonal and off-diagonal terms) suggests that we should break up the symmetric tensor **[e]**, generally known as the *tensor of the rate of change of deformation.* Let us write it as the sum of the diagonal tensor **[t]** (with three identical components t_{ii}) and of the tensor **[d]** with vanishing trace (the sum d_{ll} of its diagonal components):

$$e_{ij} = \frac{1}{3}\delta_{ij}\,e_{ll} + \left[e_{ij} - \frac{1}{3}\delta_{ij}\,e_{ll}\right] = t_{ij} + d_{ij}. \tag{3.11}$$

The diagonal tensor t_{ij} represents the *volume rate of expansion* of the elements of the fluid. The tensor d_{ij}, known as the *deviator*, is associated with the deformations that occur at constant volume.

3.2.3 Antisymmetric component of the tensor of the rate of deformation: pure rotation

We discuss now the antisymmetric tensor of components $\omega_{ij} = (1/2)(\partial v_i/\partial x_j - \partial v_j/\partial x_i)$, defined in Section 3.2.1 (Equation 3.5b) and assume that only its (off-diagonal) components can be non-zero while all those of the tensor **[e]** vanish. Let us again analyze the deformation

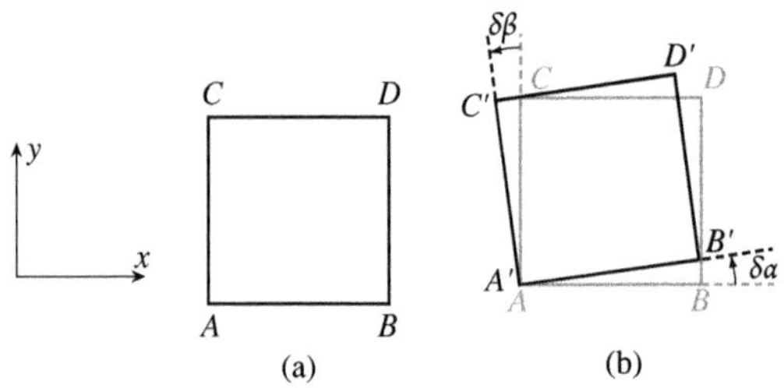

Figure 3.6 *Rotation of an elementary square resulting from the antisymmetric part ω_{ij} of the rate-of-deformation tensor G_{ij}; (a) initial position of the square at time t; (b) configuration at time $t + \delta t$*

of an elementary square (Section 3.2.2); the term ω_{ij} does not introduce any lengthening of the sides, because it would be associated with the diagonal terms of the tensor e_{ij}, all of which are assumed to be zero. Using Equations 3.9a and 3.9b, we find also that the constraint $e_{ij}=0$ implies that the changes $\delta\alpha$ and $\delta\beta$ in the angles of the sides relative to the axes are identical: the angle between each side remains $\pi/2$, and the square undergoes no change either in its shape or in its size, but only a rotation of angle $\delta\alpha = \delta\beta$ or, equivalently, $(\delta\alpha + \delta\beta)/2$. As a result of Equations 3.9a and 3.9b, the angular velocity of rotation of the square is therefore:

$$\frac{\delta\alpha}{\delta t} = \frac{\delta\beta}{\delta t} = \frac{\delta\alpha + \delta\beta}{2\,\delta t} = \frac{1}{2}\left(\frac{\partial v_y}{\partial x} - \frac{\partial v_x}{\partial y}\right) = \omega_{yx}. \tag{3.12}$$

The term ω_{ij} thus represents the angular velocity $d\alpha/dt$ of the local rotation without deformation of an element of fluid.

We could also substitute for the antisymmetric tensor of components ω_{ij} the pseudo-vector $\boldsymbol{\omega}$ such that:

$$\omega_k = -\varepsilon_{ijk}\,\omega_{ij}, \tag{3.13}$$

where $\varepsilon_{ijk} = +1$ under an even permutation of the indices i, j and k, $\varepsilon_{ijk} = -1$ for an odd permutation of these indices (and $\varepsilon_{ijk} = 0$ whenever any two indices are equal). This vector $\boldsymbol{\omega}$ (known as the *vorticity* of the flow) can then be written in the form:

$$\boldsymbol{\omega} = \nabla \times \mathbf{v}. \tag{3.14}$$

In the particular case of two-dimensional flows, the vorticity vector is normal to the plane of the flow. This idea of vorticity plays an important role in the mechanics of fluids; we will discuss it thoroughly in Chapter 7, which is entirely devoted to this topic. The pseudo-vector $\boldsymbol{\Omega} = (1/2)\,\nabla\times\mathbf{v}$, known as the *vortex vector*, represents the local angular velocity of rotation of an element of fluid; thus, in the example just discussed, we have seen that $d\alpha/dt = \Omega = \omega_{yx} = (1/2)\omega_z$.

In the same way, let us discuss the case of a three-dimensional, solid-body rotation field about the z-axis, perpendicular to the x-y plane. Denoting by $\boldsymbol{\Omega}$ the angular velocity vector, the velocity field $\mathbf{v} = \boldsymbol{\Omega} \times \mathbf{r}$ has as its components: $v_r = 0$, $v_\varphi = \Omega\, r$, $v_z = 0$. The value of the curl is:

$$\nabla \times \mathbf{v} = \left(\frac{\partial v_\varphi}{\partial r} + \frac{v_\varphi}{r}\right)\mathbf{e}_z = 2\,\Omega\,\mathbf{e}_z = 2\,\boldsymbol{\Omega}. \tag{3.15}$$

We can observe experimentally the local rotation of a fluid (and, therefore, its vorticity) by looking at the rotation of a rigid float of two crossed sticks on the surface of the fluid. More practical and precise methods of measurement will be discussed in Section 3.5.4.

To summarize the results of Sections 3.2.2 and 3.2.3, combining Equations 3.6 and 3.11, we find that the velocity–gradient tensor $G_{ij} = \partial v_i/\partial x_j$ can always be represented as the sum of three components:

$$G_{ij} = t_{ij} + d_{ij} + \omega_{ij}, \tag{3.16}$$

where:

- t_{ij} is a diagonal tensor, representing the *change in volume* (or *area* in two dimensions) of the elements of a fluid (it vanishes for an incompressible fluid);

- d_{ij} is a zero-trace, symmetric tensor. It is related to the *deformations* of the elements of the fluid, without changes in volume;
- ω_{ij} is an antisymmetric tensor, representing the solid-body *rotation* of the elements of the fluid.

The decomposition of G_{ij}, as in Equation 3.16, is essential to our understanding of the stresses induced by the flow of a fluid, a topic to be discussed in Chapter 4. We will see that the only relevant part of G_{ij} is the one corresponding to deformations (the terms t_{ij} and d_{ij}).

3.2.4 Application

Let us consider the example of a two-dimensional flow characterized by a stationary velocity field $\mathbf{v}(x,y)$ such that $v_x = \alpha\, x + 2\beta y$ and $v_y = -\alpha y$ where α and β are two constants. We will show that this field is a combination of different types of deformation and stretching of the elements of the fluid which we have just discussed. This represents a field with no average flow, for which the velocity is zero at the origin $O(0, 0)$, and with a velocity field symmetric with respect to O.

The velocity gradient tensor [**G**], defined by Equation 3.4, which characterizes these deformations, can be written:

$$[\mathbf{G}] = \begin{bmatrix} \alpha & 2\beta \\ 0 & -\alpha \end{bmatrix}. \qquad (3.17a)$$

Let us break it up into the components suggested by Equation 3.16:

$$[\mathbf{G}] = \begin{bmatrix} \alpha & 2\beta \\ 0 & -\alpha \end{bmatrix} = \underset{\text{expansion}}{\begin{bmatrix} 0 & 0 \\ 0 & 0 \end{bmatrix}} + \underset{\text{deformation}}{\begin{bmatrix} \alpha & \beta \\ \beta & -\alpha \end{bmatrix}} + \underset{\text{rotation}}{\begin{bmatrix} 0 & \beta \\ -\beta & 0 \end{bmatrix}}. \qquad (3.17b)$$

The component corresponding to a volume expansion is zero, as would be expected for incompressible flow ($\nabla\cdot\mathbf{v} = 0$). We observe that the rotation component represents the solid-body rotation of the elements of the fluid. The component corresponding to deformation results also in the angular displacement of the fluid components $\delta\mathbf{r}$ and additionally causes a change in their length; these displacements depend on the orientation of the vectors $\delta\mathbf{r}$. There will only occur expansion without rotation in two characteristic directions, which are mathematically obtained by diagonalizing the deformation tensor. If we write the condition for the determinant $(\lambda^2 - \alpha^2) - \beta^2$ of the matrix $\begin{bmatrix} \alpha-\lambda & \beta \\ \beta & -\alpha-\lambda \end{bmatrix}$ to equal zero, we obtain the eigenvalues:

$$\lambda = \pm\sqrt{\alpha^2 + \beta^2}. \qquad (3.18a)$$

The corresponding eigenvectors represent the directions in which there is no resulting rotation. These two directions are represented by mutually normal straight lines obeying the equations:

$$(\alpha - \lambda)\, x + \beta y = 0, \qquad (3.18b) \qquad \text{i.e.} \qquad y = \frac{-\alpha \pm \sqrt{\alpha^2 + \beta^2}}{\beta} x. \qquad (3.18c)$$

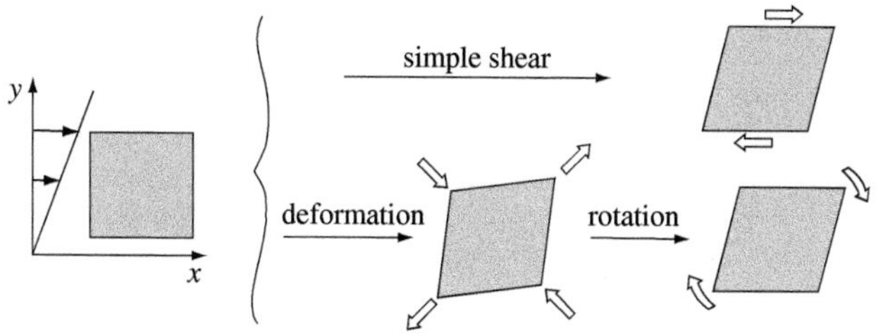

Figure 3.7 *The deformation of a volume element in simple Couette shear flow can be expressed as a superposition of a rotation and a deformation without rotation*

- The particular case $\alpha \neq 0$, $\beta = 0$ corresponds to a pure deformation without rotation of the elements of the fluid, such as shown in Figure 3.3. The eigenvectors are oriented along the initial x- and y-axes. A square aligned along the x-axis changes into a rectangle parallel to the same axes; the incompressibility condition implies that, to first order, the changes in length of the two sides are equal and opposite.
- The case $\alpha = 0$, $\beta \neq 0$ corresponds to a *simple shear flow* (or *plane Couette flow*) (Figure 3.7); it results from the superposition of a deformation and a rotation, such as those displayed in Figures 3.4 and 3.6. We then have $v_x = 2\beta y$: the velocity has a non-zero component only in the x-direction, and varies only in the direction normal to this axis. This kind of velocity field is obtained by setting into relative motion, parallel to their own plane, two parallel plates between which we put a viscous fluid. We note that the eigenvectors of the deformation component are oriented at $\pm$ 45° to the direction of the shear (the corresponding straight lines obey the equation $y = \pm x$ as shown in the general equation established above).

In order to understand better the difference between these two last flows, let us analyze the behavior of a liquid drop placed in a non-miscible fluid in which such a flow has been created (we assume that the drop is centered at the origin O):

- For the case of pure deformation, the lengthening of the drop is largest along the x-axis (if α is positive), which is one of the characteristic axes of the flow; the capillary forces, which tend to minimize the surface area of the drop, then oppose this deformation. Because the fluid velocity is zero at point O, the drop remains fixed and it will stretch (and ultimately break up by emitting small droplets from its tip) if the capillary forces which ensure its cohesion are insufficiently large. Such flow can occur in the geometry discussed in Section 3.4.2 (Figure 3.13).
- For the case of simple shear flow, the maximal stretch is at 45° to the direction of the flow (as we have just seen), and tends to deform the droplet. However, the droplet also tends to rotate because of the rotational component in the flow; the effect of the stretching is thus largely cancelled and the drop will only break up for very large velocity gradients. In the case of high flow velocities, one can even show that it can reach a stationary condition with ellipticity 0.25.

3.2.5 Case of large deformations

Let us now consider, as an example, the situation of large deformations for which the assumption, that the velocity at each point remains constant as the point is displaced, is no longer valid. This corresponds to the flow of the liquid over long periods of time, or to the

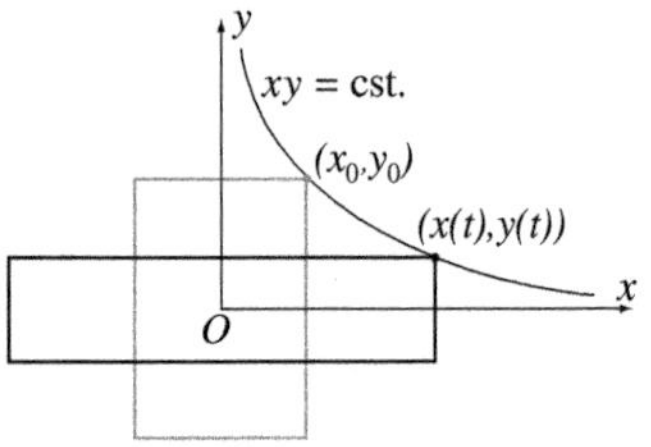

Figure 3.8 *Large deformation of an element of fluid in a pure deformation, incompressible flow*

case of extremely deformable solids. This is often the case in the deformation of geologic formations, which occur over extremely long elapsed times.

Let us consider the velocity field $v_x = \alpha x$, $v_y = -\alpha y$ (Figure 3.8). The velocity has zero divergence, and the gradient contains only diagonal terms: such a velocity field corresponds to the elongation of elements of the fluid without any change in volume.

We then find, as in the proof below, that the variation of the components $\delta x(t)$ and $\delta y(t)$ of the displacement relative to the initial position is, in matrix form:

$$\begin{bmatrix} \delta x(t) \\ \delta y(t) \end{bmatrix} = \begin{bmatrix} e^{\alpha t} - 1 & 0 \\ 0 & e^{-\alpha t} - 1 \end{bmatrix} \begin{bmatrix} x_0 \\ y_0 \end{bmatrix} = [\mathbf{D}] \begin{bmatrix} x_0 \\ y_0 \end{bmatrix}. \tag{3.19}$$

We then see that, for the general case of large deformations, the displacement no longer varies linearly either with the time, or with the velocity gradient tensor $[\mathbf{G}]$; the condition of incompressibility is still satisfied. In the approximation of small deformations ($\alpha t \ll 1$), we find by a first-order expansion of the exponential terms:

$$\begin{bmatrix} \delta x(t) \\ \delta y(t) \end{bmatrix} = t \begin{bmatrix} \alpha & 0 \\ 0 & -\alpha \end{bmatrix} \begin{bmatrix} x_0 \\ y_0 \end{bmatrix} = [\mathbf{G}]\, t \begin{bmatrix} x_0 \\ y_0 \end{bmatrix}, \tag{3.20}$$

a result which corresponds to a variation of the deformation, linear both with time and with the velocity–gradient tensor $[\mathbf{G}]$.

Proof

In order to deal with this problem, we must take into account the change of velocity of the elements of fluid during the displacement. The equation of motion of a specific point in the fluid can then be written as:

$$\frac{d\mathbf{r}}{dt} = \mathbf{v}(\mathbf{r}\,(t)), \qquad \text{instead of:} \qquad \frac{d\mathbf{r}}{dt} = \mathbf{v}(\mathbf{r}_0),$$

as was the case for small deformations seen above. We thus have:

$$\frac{dx}{dt} = \alpha x \quad (3.21a) \qquad \text{and} \qquad \frac{dy}{dt} = -\alpha y, \quad (3.21b)$$

so that, after integrating:

$$x\,(t) = x_0\, e^{\alpha t} \quad (3.22a) \qquad \text{and} \qquad y\,(t) = y_0\, e^{-\alpha t} \quad (3.22b)$$

The product of these two equations is $x(t)\, y(t) = x_0\, y_0$; the trajectories of the particles are therefore branches of hyperbolae (Figure 3.8). We note that, for a rectangle whose diagonal has an initial position $\mathbf{OM}_0(x_0, y_0)$, the area at time t is equal to $x(t)\, y(t)$ (because the point O is fixed) and therefore remains a constant. The components $\delta x(t)$ and $\delta y(t)$ of the displacement are therefore: $\delta x\,(t) = x_0\left(e^{\alpha t} - 1\right)$ and $\delta y\,(t) = y_0\left(e^{-\alpha t} - 1\right)$, from which we obtain Equation 3.19.

3.3 Conservation of mass in a moving fluid

By writing down the global balance for the mass of fluid inside a *fixed* volume, we obtain a *local* equation for the conservation of mass, also known as the equation of continuity. The same procedure will be followed in Chapter 5 where we derive *conservation laws* for the various other important quantities (e.g., energy, momentum).

3.3.1 Equation for the conservation of mass

We consider an arbitrary volume $\mathcal{V}$ fixed in the reference frame used for describing the flow of fluid, and bounded by a closed surface $\mathcal{S}$ (Figure 3.9). At each instant of time, fluid enters and leaves this volume; the rate of change of the mass m contained within the volume is the opposite of the flux leaving through the boundary surface.

We thus have:

$$\frac{\mathrm{d}m}{\mathrm{d}t} = \frac{\mathrm{d}}{\mathrm{d}t}\left(\iiint_{\mathcal{V}} \rho \,\mathrm{d}\mathcal{V}\right) = \iiint_{\mathcal{V}} \frac{\partial \rho}{\partial t}\,\mathrm{d}V = -\iint_{S} \rho\, \mathbf{v}\cdot\mathbf{n}\,\mathrm{d}S. \tag{3.23}$$

The unit vector $\mathbf{n}$ normal to the closed surface $\mathcal{S}$ is directed outward from the enclosed volume $\mathcal{V}$. Since $\mathcal{V}$ is fixed, we can interchange the order of integration and differentiation with respect to time. Further, by applying Gauss' divergence theorem to the rightmost term of Equation 3.23, we obtain:

$$\iiint_{\mathcal{V}} \left(\frac{\partial \rho}{\partial t} + \nabla\cdot(\rho\,\mathbf{v})\right)\mathrm{d}V = 0. \tag{3.24}$$

Since this equation holds for any arbitrary volume $\mathcal{V}$, the integrand must be identically zero, yielding the equation of continuity:

$$\frac{\partial \rho}{\partial t} + \nabla\cdot(\rho\,\mathbf{v}) = 0. \tag{3.25}$$

One can write the $\nabla\cdot(\rho\,\mathbf{v})$ term of Equation 3.25 as:

$$\nabla\cdot(\rho\,\mathbf{v}) = \rho\,\nabla\cdot\mathbf{v} + \mathbf{v}\cdot\nabla\rho,$$

which then allows us to rewrite this equation as:

$$\left(\frac{\partial \rho}{dt} + \mathbf{v}\cdot\nabla\rho\right) + \rho\,\nabla\cdot\mathbf{v} = 0. \tag{3.26}$$

We should note that the quantity in parentheses represents exactly the variation ($\mathrm{d}\rho/\mathrm{d}t$) of the fluid density with time for an element of fluid which is followed along with the flow (the convective derivative corresponding to the Lagrangian description). We can therefore rewrite Equation 3.26 in the form:

$$\frac{\mathrm{d}\rho}{\mathrm{d}t} + \rho\,\nabla\cdot\mathbf{v} = 0, \tag{3.27}$$

which is another representation of the equation of the conservation of mass.

In the present case, where we are concerned with the net balance of mass, there is no "source" term corresponding to the creation or annihilation of this quantity. On the other hand, in the case of a chemical reaction in a non-stoichiometric gas phase or in the presence of a solid phase, there may well exist a term of that nature.

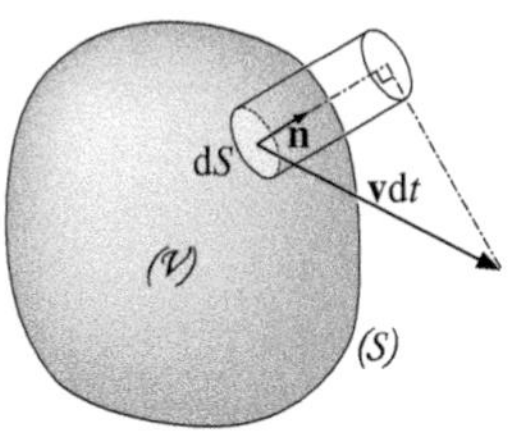

Figure 3.9 *The evaluation of the net mass change within a fixed volume ($\mathcal{V}$). The unit vector* $\mathbf{n}$ *normal to a surface element* ($\mathrm{d}S$) *is directed outward from ($\mathcal{V}$), as we evaluate the outward mass flux exiting, per unit time, from that volume element, equal to the surface integral of* $\rho\,\mathbf{v}\cdot\mathbf{n}\,\mathrm{d}S$

We observe in Equation 3.25 the exact parallel with the equation of conservation of charge in the theory of electromagnetism. The latter can be written $\partial\rho/\partial t + \nabla\cdot\mathbf{j} = 0$, where ρ and $\mathbf{j}$ represent respectively the charge and current densities.

3.3.2 Condition for an incompressible fluid

For an incompressible fluid, e.g., such that the density of each element is a constant during the flow ($\mathrm{d}\rho/\mathrm{d}t = 0$), the equation of conservation of mass has the very simple form:

$$\nabla\cdot\mathbf{v} = 0 \tag{3.28}$$

(i) Equation 3.28 at the bottom of the page is hardly surprising, since we have already seen in Section 3.2.2 that the term $\nabla\cdot\mathbf{v}$, which is the trace of the velocity–gradient tensor, represents the local rate of change of volume of an element of the fluid.

(ii) If we recall Equation 3.27, and replace $\nabla\cdot\mathbf{v}$ by the rate of change of volume $(1/V)\,(\delta V/\delta t)$, to which it is equal according to Equation 3.8b, we observe that this rate of volume expansion is the opposite of the rate of change of density, because $(1/\rho)\,(\delta\rho/\delta t) + (1/V)\,(\delta V/\delta t) = 0$.

The conditions under which a fluid can be considered as incompressible can, in most cases, be described by the inequality:

$$U \ll c, \qquad (3.29)$$

where U represents a characteristic velocity for the flow and c is the speed of pressure waves in the given fluid (i.e., *the speed of sound*). In effect, in the case of flow dominated by the effects of the inertia of the fluid, the order of magnitude of pressure changes due to the flow is $\delta p \approx \rho U^2$. This term represents the convected momentum flux through the unit area, a result that we discuss in greater detail in Chapter 5, when we consider Bernoulli's equation. If the compressibility of the fluid is χ the corresponding relative density fluctuations are $\delta\rho/\rho \approx \chi\ \delta p \approx \chi\ \rho U^2$. The compressibility of the fluid can therefore be neglected if $\delta\rho/\rho \ll 1$, i.e., $U \ll 1/\sqrt{\chi\rho}$ where $1/\sqrt{\chi\rho}$ is the speed of sound c.

We can write Equation 3.29 in a dimensionless form using the *Mach number* M defined as equal to the ratio U/c:

When the flow is dominated by the effects of viscosity (the case of flows at small Reynolds numbers which we will address in Chapter 9), Equation 3.30 is replaced by the more restrictive condition: $M \ll \sqrt{Re}$, where Re is the Reynolds number of the flow. In this instance, the pressure fluctuations in the flow are indeed of the order of $\delta p \approx \eta U/L = \rho U^2/Re$.

$$M \ll 1. \qquad (3.30)$$

This condition is clearly not satisfied in the studies of the dynamics of high velocity gases (aeronautical applications, shock waves); neglecting the compressibility of a fluid in such instances is indeed equivalent to the assumption that the speed of sound is infinite.

Consider now the case of the non-stationary flow, for which there exists a characteristic time scale T (e.g., the period of an acoustic wave). In this situation, the condition of Equation 3.29 must be supplemented by a relation between the time scale T and the characteristic time L/c over which a typical pressure perturbation is convected by the flow over a distance L. This condition can be written:

$$T \gg L/c. \qquad (3.31)$$

For a periodic flow with period T, Equation 3.31 can be written in the form of a comparison between the wavelength $\lambda = cT$ and the spatial scale L:

$$\lambda \gg L. \qquad (3.32)$$

Finally, we consider one of the consequences of incompressibility in three-dimensional flows near a stagnation point. We can write the deformation tensor e_{ij} in terms of coordinate axes coinciding with the principal axes of the tensor. Since the trace of the tensor is invariant under a rotational transformation of axes, the condition of incompressibility (the trace of the tensor equal to zero) requires that two of the eigenvalues be negative and one positive or, conversely, that only one eigenvalue be negative. Two types of local deformations correspond to very different kinds of flows satisfy this condition. In the first case (Figure 3.10a),

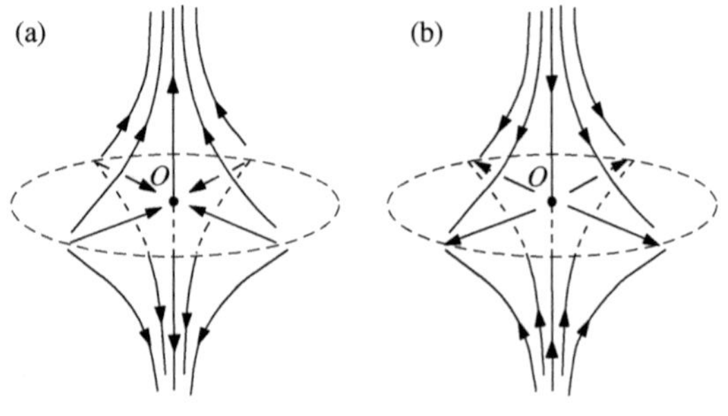

Figure 3.10 *Three-dimensional flow in the neighborhood of a stagnation point, such that two eigenvalues of the velocity gradient tensor are, respectively: (a) negative; (b) positive*

the fluid tends to "pile up" toward the origin from all directions in the plane of the negative eigenvalues, and hence to "stretch" in the third, orthogonal, direction. An elongated particle placed in this flow will therefore line up along this third axis. In the opposite case (Figure 3.10b), we might expect that a particle in the shape of a thin, flat disk would align in the plane of the positive eigenvalues. Besides this orientation effect, we can also observe a deformation by locating a drop of dye, or some non-miscible fluid, at the stagnation point O. The elongation of the drop in the first case, and its flattening out in the second, indicate the very different velocity fields in the two cases.

In the case of a two-dimensional flow discussed in Section 3.2.4, we no longer have the two different kinds of flows just discussed, because, in that situation, we have only two equal and opposite eigenvalues for the deformation tensor. There is only one possible type of deformation for a droplet placed at the stagnation point: it assumes the shape of an ellipse with its major axis oriented in the direction of the positive eigenvalue.

3.3.3 Rotational flows; potential flows

Very generally, we can represent an arbitrary vector field $\mathbf{v}(\mathbf{r})$ as the sum of three terms:

$$\mathbf{v}(\mathbf{r}) = \mathbf{v}_1(\mathbf{r}) + \mathbf{v}_2(\mathbf{r}) + \mathbf{v}_3(\mathbf{r}). \tag{3.33}$$

- The field $\mathbf{v}_1(\mathbf{r})$ has zero divergence and its curl corresponds to a local rotation with angular velocity $\boldsymbol{\Omega} = \boldsymbol{\omega}/2$ (where $\boldsymbol{\omega}$ is the vorticity of the flow). This field obeys:

$$\nabla\cdot\mathbf{v}_1(\mathrm{r}) = 0 \quad (3.34\mathrm{a}) \qquad \text{and} \qquad \nabla\times\mathbf{v}_1(\mathrm{r}) = \boldsymbol{\omega}. \quad (3.34\mathrm{b})$$

  We have here an analogy with magnetism in the approximation of quasi-permanent regimes, discussed in Chapter 7 where we consider vorticity and *rotational* flows.
- The field $\mathbf{v}_2(\mathbf{r})$ has simultaneously zero divergence and zero curl. It therefore satisfies at the same time the equations:

$$\nabla\cdot\mathbf{v}_2(\mathbf{r}) = 0 \quad (3.35\mathrm{a}) \qquad \text{and} \qquad \nabla\times\mathbf{v}_2(\mathbf{r}) = \mathbf{0}. \quad (3.35\mathrm{b})$$

  The second equation indicates that we can introduce a potential function Φ, called the *velocity potential*, such that $\mathbf{v} = \nabla\Phi$. The corresponding flows are known as *potential flows*. We will study these in Chapter 6, where we also discuss the analogy with electrostatics.
- Finally, the field $\mathbf{v}_3(\mathbf{r})$ takes into account the effects of changes in volume $(1/V)\ (\delta V/\delta t)$. It is only non-zero in the case of a compressible fluid with:

$$\nabla\cdot\mathbf{v}_3(\mathbf{r}) = \nabla\cdot\mathbf{v}(\mathbf{r}) = \frac{1}{V}\frac{\delta V}{\delta t} \quad (3.36\mathrm{a}) \qquad \text{and} \qquad \nabla\times\mathbf{v}_3(\mathbf{r}) = \mathbf{0}. \quad (3.36\mathrm{b})$$

3.4 The stream function

3.4.1 Introduction and significance of the stream function

The stream function allows us to simplify the treatment of the vector velocity of an incompressible fluid by dealing instead with a scalar field for the case where the velocity field depends only on two coordinates (two-dimensional flow, or flow with an axis of rotational symmetry).

Let us consider the flow of an *incompressible* fluid. Conservation of mass implies Equation 3.28: $\nabla\cdot\mathbf{v} = 0$. If, on the other hand, we take a more general case where none of the three vector components of the velocity vanish, we can then introduce a vector function $\mathbf{A}$ such that:

$$\mathbf{v} = \nabla \times \mathbf{A}. \tag{3.37}$$

For velocities, the vector potential $\mathbf{A}$ associated with the incompressibility condition is the analog of the vector potential introduced in the study of magnetism and resulting from the relation $\nabla \cdot \mathbf{B} = 0$ (we develop this analogy in Section 7.1.3). The practical usefulness of such a function is still somewhat limited because, here, we replace the velocity field $\mathbf{v}$ with another vector field $\mathbf{A}$. On the other hand, when the velocity depends only on two coordinate directions, the velocity field can be directly expressed by a single component of the vector potential $\mathbf{A}$; this will be the case for two-dimensional flows, and also for axially symmetric flows (with rotational symmetry about one axis). If we consider the case of a two-dimensional flow (the case of axially symmetric flows will be discussed in Section 3.4.3), the velocity vector is assumed to be translationally invariant along the z-axis and will have no component in that direction [$\mathbf{v} = (v_x(x, y), v_y(x, y), 0)$]. The condition $\nabla \cdot \mathbf{v} = 0$ is then equivalent to:

$$\frac{\partial v_x}{\partial x} + \frac{\partial v_y}{\partial y} = 0.$$

As a result of this equation, a scalar function Ψ exists such that:

$$v_x = \frac{\partial \Psi}{\partial y}, \quad (3.38a) \qquad v_y = -\frac{\partial \Psi}{\partial x}. \quad (3.38b)$$

The reader's attention is drawn to the notational convention which we follow in this text: the pair (r, φ) indicates polar coordinates in two dimensions; in three dimensions, cylindrical coordinates are denoted by (r, φ, z) while the set (r, θ, φ) is used for spherical polar coordinates. Thus r will always represent the radius vector in two and three dimensions, in both cylindrical and spherical-polar coordinates, while φ is always the azimuth angle in both two and three dimensions (i.e., the angle in the x-y- plane measured relative to the direction of the x-axis). For spherical polar coordinates, the polar axis is chosen to coincide with the Cartesian z-axis and θ indicates the (conical) polar angle, measured from the polar axis.

By inserting these results into Equation 3.37, we obtain:

$$\Psi \equiv A_z. \tag{3.38c}$$

The scalar function $\Psi(x, y)$ (known as the *stream function*) represents thus the component of the vector potential $\mathbf{A}$ in the direction perpendicular to the plane of the flow (here the z-axis). Note that we can introduce a stream function Ψ for any two-dimensional incompressible flow, whether viscous or not.

If polar coordinates (r, φ) are used in the plane of the flow, the equations for the velocity components and the stream function Ψ can be written as:

$$v_r = \frac{1}{r}\frac{\partial \Psi}{\partial \varphi}, \quad (3.39a) \qquad v_\varphi = -\frac{\partial \Psi}{\partial r}. \quad (3.39b)$$

Let us now consider the properties of the stream function. This will allow us to understand its significance.

- The lines Ψ = *constant* coincide with the streamlines.

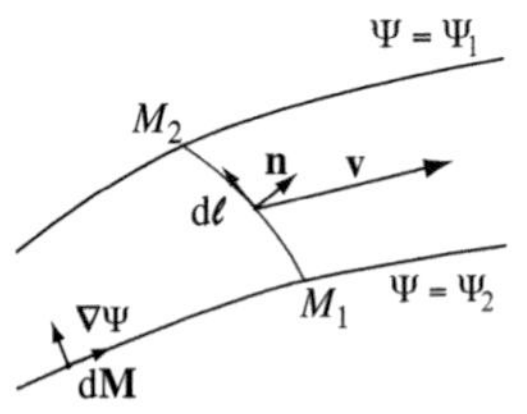

Figure 3.11 *The streamlines for two-dimensional incompressible flow (everywhere tangent to the local velocity vector) coincide with the lines along which the stream function Ψ is constant*

Proof

For the case of a two-dimensional flow, let us evaluate the scalar product $(\mathbf{v}\cdot\nabla)\Psi$ by using Equations 3.38a and 3.38b:

$$(\mathbf{v}\cdot\nabla)\,\Psi = v_x\frac{\partial \Psi}{\partial x} + v_y\frac{\partial \Psi}{\partial y} = 0. \tag{3.40}$$

Now, the lines Ψ = constant are everywhere orthogonal to the vector field $\nabla\Psi$. In fact, for a small displacement $\mathbf{dM}$, the change $d\Psi$ satisfies $d\Psi = \nabla\Psi \cdot \mathbf{dM}$; we therefore have $d\Psi = 0$ when $\mathbf{dM}$ is normal to $\nabla\Psi$ (Figure 3.11); the lines Ψ = constant are therefore parallel to $\mathbf{v}$. The same reasoning applies to axially symmetric flows.

- Again for the case of a two-dimensional flow, the quantity $\Delta\Psi = \Psi_2 - \Psi_1$ represents the rate of flow of fluid in a stream tube of rectangular cross-section located between the streamlines $\Psi = \Psi_1$ and $\Psi = \Psi_2$ and of unit depth in the z-direction.

Proof

In order to derive this result, let us evaluate the flow rate Q for the fluid in such a flow. Assuming unit depth in the z- direction, we have:

$$Q = \int_{M_1}^{M_2} \mathbf{v}\cdot(\mathbf{n}\, d\ell),$$

so that, if $\mathbf{d\ell} = (dx, dy, 0)$ and $(\mathbf{n}\, d\ell) = (dy, -dx, 0)$, we then have:

$$Q = \int_{M_1}^{M_2} \left[\frac{\partial \Psi}{\partial y} dy + \left(-\frac{\partial \Psi}{\partial x}\right)(-dx)\right] = \int_{M_1}^{M_2} d\Psi = \Psi_2 - \Psi_1. \tag{3.41}$$

It is thus established that, for this incompressible flow, the flow rate in such a stream tube is constant everywhere along the tube.

3.4.2 Stream functions for two-dimensional flows

Elementary flows (uniform flow, vortices, sources and their various combinations) will be studied in Chapter 6. We examine here, as an example, the set of two-dimensional flows for which the stream function has the form:

$$\Psi(x, y) = a x^2 + b y^2, \tag{3.42a}$$

where a and b are real constants. The components of the velocity field thus satisfy:

$$v_x = \frac{\partial \Psi}{\partial y} = 2\, b\, y, \tag{3.42b}$$

$$v_y = -\frac{\partial \Psi}{\partial x} = -2\, a\, x \tag{3.42c}$$

and the streamlines are given by the equation:

$$\Psi(x, y) = \text{constant}. \tag{3.42d}$$

Figure 3.12 on the next page illustrates the velocity profiles and the streamlines corresponding to three specific values of the ratio a/b:

- **Case $a/b = 0$** (Figure 3.12a). We then have:

$$v_x = 2\, b\, y, \tag{3.43a}$$

$$v_y = 0. \tag{3.43b}$$

We have already met such *simple shear* flows in Sections 2.1.2 and 3.2.4. This kind of flow occurs in the case of a viscous fluid located between two parallel plates, at a distance d from one another and moving parallel to their own plane with relative velocity U. The velocity field is characterized by a value of the velocity gradient, or shear rate, $\dot{\gamma} = (U/d)$ (its value here is $2b$) which has the dimensions of an inverse time. The streamlines correspond to the straight lines y = constant and are parallel to the boundary plates.

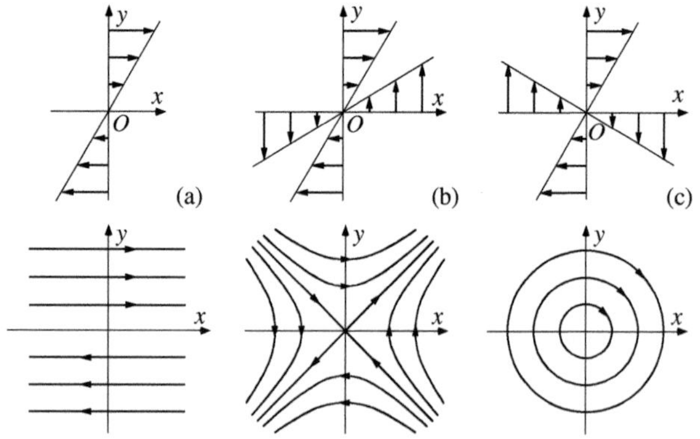

Figure 3.12 *Streamlines (below) and velocity profiles along the x- and y-axes (above) for two-dimensional, plane flows characterized by a stream function* $\Psi(x, y) = ax^2 + by^2$ *for three specific values of the ratio a/b; (a) simple shear flow a/b = 0; (b) pure shear flow a/b = −1; (c) pure rotation a/b = 1*

- **Case** $a/b = -1$ (Figure 3.12b). The streamlines obey the equation:

$$\Psi = a\left(x^2 - y^2\right) = \text{constant}. \tag{3.44a}$$

They make up a set of rectangular hyperbolae for which the asymptotes are the bisectors of each of the quadrants ($y = \pm x$). The corresponding velocity field has the components:

$$v_x = -2ay, \quad (3.44b) \qquad v_y = -2ax. \quad (3.44c)$$

This kind of velocity field is characterized by the presence of *elongation* components, which stretch out the fluid particles in the direction of the flow; an example of this is the flow of a fluid through an orifice in the form of a slit. (Figure 3.13a) and also between a set of four counter-rotating cylinders (Figure 3.13b). In this last case, we note that the central point O is a stagnation point: such a configuration is interesting if we wish to study the deformation of a non-rigid object which is not carried along by the flow.

Let us evaluate the curl of this velocity field:

$$\nabla \times \mathbf{v} = \left(\frac{\partial v_y}{\partial x} - \frac{\partial v_x}{\partial y}\right)\mathbf{e}_z = 0, \tag{3.44d}$$

in which $\mathbf{e}_z$ is the unit vector along z. This velocity field is thus irrotational; it corresponds to a pure deformation field with no rotation which we have already discussed in Section 3.2.

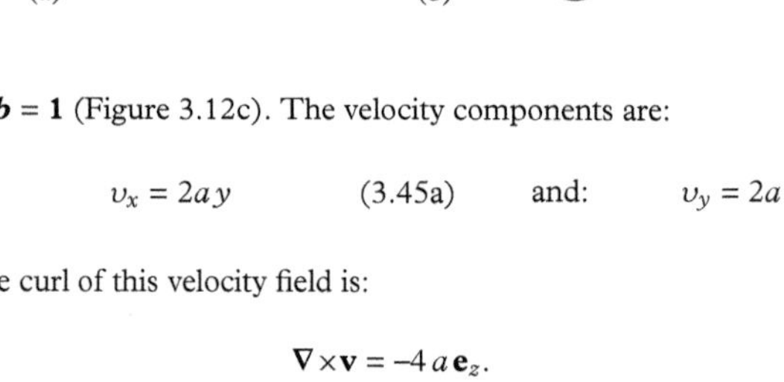

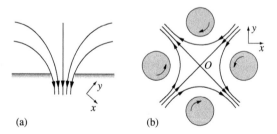

Figure 3.13 *Shape of the streamlines for the case of elongational fluid flow; (a) through a slit-shaped opening; (b) between four counter-rotating cylinders*

- **Case** $a/b = 1$ (Figure 3.12c). The velocity components are:

$$v_x = 2ay \quad (3.45a) \qquad \text{and:} \qquad v_y = 2ax. \quad (3.45b)$$

The value of the curl of this velocity field is:

$$\nabla \times \mathbf{v} = -4a\,\mathbf{e}_z. \tag{3.45c}$$

This flow corresponds to a pure rotation of the fluid around the z-axis normal to the plane of the flow; the magnitude of the angular velocity is $|\boldsymbol{\Omega}| = |(\nabla \times \mathbf{v})/2| = 2a$ (we have previously discussed this case in Section 3.2.3). The streamlines are concentric circles which obey an equation of the type:

$$\Psi = a\left(x^2 + y^2\right) = \text{constant}. \tag{3.45d}$$

In the general case for which the coefficients a and b in the equation $\Psi(x, y) = a\,x^2 + b\,y^2$ are not equal in absolute value, we can write Ψ in the form:

$$\Psi(x,y) = \frac{b+a}{2}\left(x^2 + y^2\right) + \frac{b-a}{2}\left(y^2 - x^2\right) = \Psi_{rot} + \Psi_{shear}, \tag{3.46}$$

where:

- the stream function Ψ_{rot} corresponds to *pure rotation* (the coefficients of x^2 and y^2 are equal). The magnitude Ω of the angular velocity is then equal to $b + a$.
- the stream function Ψ_{shear} corresponds to *pure shear* (the coefficients of x^2 and y^2 have opposite values).

Since the stream function Ψ and the velocity field are linearly related (as shown earlier), the latter is the sum of the velocity fields corresponding to a pure rotation and to a pure shear, i.e.:

$$v_x = (b+a)\,y + (b-a)\,y \quad (3.47a) \qquad \text{and:} \qquad v_y = -(b+a)x + (b-a)\,x. \quad (3.47b)$$

These general equations allow us, obviously, to retrieve for $b = a$ and $b = -a$, the limiting cases of pure shear flow and pure rotation which we discussed above.

3.4.3 Stream functions for axially symmetric flows

These are flows that have an axis of symmetry relative to which the velocity field is rotationally invariant. In order to satisfy automatically the equation of conservation of mass, we introduce, just as in the previous case, a scalar function – called the *Stokes stream function* – which we also denote by Ψ.

- For the case of a problem with *cylindrical symmetry*, we write the equation of conservation of mass for an incompressible fluid ($\nabla \cdot \mathbf{v} = 0$) in cylindrical coordinates (r, φ, z) (see Section 4A.2 of the appendix to Chapter 4):

$$\frac{1}{r}\frac{\partial(r\,v_r)}{\partial r} + \frac{1}{r}\frac{\partial v_\varphi}{\partial \varphi} + \frac{\partial v_z}{\partial z} = 0. \tag{3.48}$$

The velocity field of an axially symmetric flow is independent of φ, so that the previous equation becomes:

$$\frac{\partial(r\,v_r)}{\partial r} + r\frac{\partial v_z}{\partial z} = 0. \tag{3.49}$$

This equation is automatically satisfied provided we introduce a function Ψ such that:

$$v_r = \frac{1}{r}\frac{\partial \Psi}{\partial z}, \quad (3.50a) \qquad\qquad v_z = -\frac{1}{r}\frac{\partial \Psi}{\partial r}. \quad (3.50b)$$

- For a problem with *spherical symmetry*, it is natural to use spherical polar coordinates (r, θ, φ) (see Section 4A.3 of the appendix to Chapter 4). The equation of conservation of mass is then:

$$\frac{1}{r^2}\frac{\partial\,(r^2 v_r)}{\partial r} + \frac{1}{r\sin\theta}\frac{\partial\,(\sin\theta\, v_\theta)}{\partial\theta} + \frac{1}{r\sin\theta}\frac{\partial v_\varphi}{\partial\varphi} = 0. \tag{3.51a}$$

Here again, for flows with axial symmetry about the polar z-axis, the velocity components are independent of φ and the above equation becomes:

$$\frac{\partial\,(r^2\, v_r)}{\partial r} + \frac{1}{r\,\sin\theta}\frac{\partial\,(\sin\theta\; v_\theta)}{\partial\theta} = 0. \tag{3.51b}$$

This equation is also automatically satisfied by a stream function (also denoted by Ψ) such that:

$$v_r = \frac{1}{r^2\sin\theta}\frac{\partial\Psi}{\partial\theta} \quad (3.52a) \qquad \text{and} \qquad v_\theta = -\frac{1}{r\sin\theta}\frac{\partial\Psi}{\partial r}. \quad (3.52b)$$

By comparing these equations with the components of the curl of the vector potential **A** from which the velocity field (Equation 3.37) is derived, we find that the stream function Ψ is the quantity (r sin θ A_φ).

Several examples of the use of this stream function will be listed in Sections 6.2.3, 6.2.4 and 9.4.1. In appendix 6A of Chapter 6, we summarize the complete set of equations governing the components of the velocity and the stream function.

The definition of the stream function Ψ for the case of axially symmetric flows leads to a dimensionality different from that for two-dimensional flows: in this case, it has the dimensions of the product of a velocity and a surface area (m^3/s), as opposed to the product of a velocity and a length (m^2/s) for the case of a two-dimensional flow.

3.5 Visualization and measurement of the velocity and of the velocity gradients in flows

We will first describe a few techniques used to visualize flows (Section 3.5.1), and extend these to quantitative measurements of solute or tracer concentrations (Section 3.5.2). Then, we will discuss several methods for measuring the local velocity (Section 3.5.3) and, finally, measurements of the velocity field of flows and of characteristic quantities such as velocity–gradients and vorticity (Section 3.5.4).

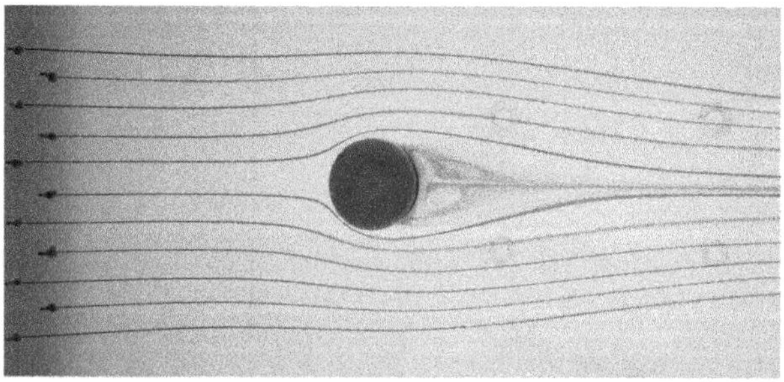

Figure 3.14C *Visualization, by means of streaklines of dye, of the flow around a cylindrical obstacle with its axis perpendicular to the plane of the figure. The flow is from left to right and the dye is injected at* 11 *discrete points (at the left of the figure). We note the recirculation zone just downstream of the cylinder. (document courtesy of L. Auffray and P. Jenffer, Université Paris-Sud)*

3.5.1 Visualization of flows

A first method for visualizing flows makes use of tracers which follow the motion of the fluid. Other techniques use the changes in the index of refraction resulting from changes in the temperature and/or the density of the flowing fluid.

Tracking by means of dyes, bubbles (liquids), smoke (gases) and particles

In the case of liquids, we frequently visualize the flow by injecting a solution of dye at different points in the liquid upstream of an obstacle. In some cases, however, the dye is emitted at the edge of the obstacle so as to observe the flow in its immediate neighborhood. We thus observe streaklines of the dye, extending over large distances in the case of laminar flows (Figure 3.14). In Section 3.1.4, we have shown that these lines coincide with the streamlines if the flow is stationary, as in the case of Figure 3.14.

Another technique involves the generation, by means of electrolysis in the flow of an ionic aqueous solution, of hydrogen micro-bubbles from a small diameter conducting wire (10 to 100 μm) which is alternately insulated, and stripped, so as to define the zones in which bubbles are emitted (Figure 3.15); the diameter of the bubbles is of the same order of magnitude as that of the wire. One can thus display the streaklines (and therefore the streamlines), in the case of stationary flow. One can, moreover, generate dyed segments of controlled length by modulating the rate of emission of bubbles (Figure 3.15): one can visualize directly in this manner the deformation of elements of the fluid as discussed in Section 3.2 (Figures 3.4 to 3.7).

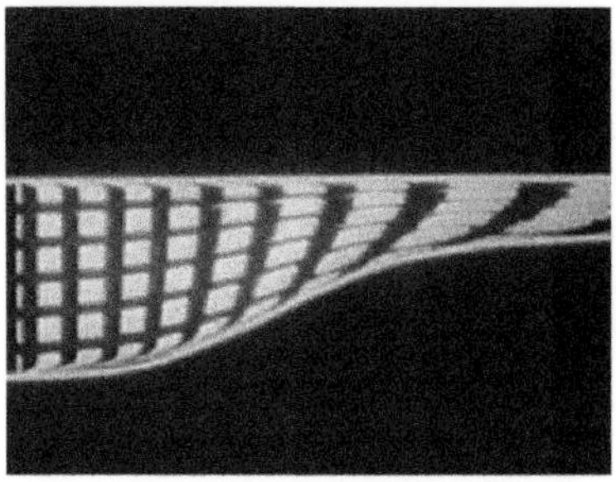

Figure 3.15 *Visualization by means of the pulsed emission of hydrogen micro-bubbles in a convergent channel. The bubbles are generated by electrolysis at a fine wire, perpendicular to the flow located just on the left of the figure (plate from NCFMF). The distortion of the pattern of squares can be considered as illustrative of the deformations discussed in Section 3.2*

One can also use, for the visualization, fine particles suspended in the fluid, which follow locally its motion; it is important to select particles of a density as close as possible to that of the fluid, in order to minimize sedimentation. Also, in order to avoid the images of particles located at different depths from the observer and being superimposed on top of one another, one illuminates the flow by means of a very thin luminous plane normal to the direction of observation, as shown in Figure 3.16: this technique is named *optical tomography*. To generate this plane of light, one uses, most often, as a light source, a laser beam which we widen in the direction of the plane by means of a cylindrical lens of very short focal length; a second lens of much longer focal length allows one then to minimize the thickness of the plane in the region of interest. One refers to this technique as visualization by a *laser plane* or *sheet* (Figure 3.20).

Figure 3.16 *Flow resulting from a small-amplitude (1.4 mm) vertical oscillation of a sphere (12 mm in diameter) located in a fluid at the center of the figure. The flow is visualized by means of glass micro-spheres in suspension in the fluid, and illuminated by a sheet of light in the plane of the photograph. The exposure time (30 s) is very long relative to the period of oscillation of the sphere (15 ms) so that the lighted lines display the streamlines of the resultant flow (document from the authors)*

A gaseous flow can be similarly tagged by certain chemical reactions between the gases present in the flow (e.g., by creating very fine ammonium chloride "smoke" by means of the reaction of ammonia with gaseous hydrogen chloride: $NH_3 + HCl \rightarrow NH_4Cl$). One can also use smoke from incense, or very fine oil droplets, and carry out, just as for liquids, optical tomography by illuminating the flow with a luminous sheet. This is the case illustrated in Figure 10.28 which displays a flame front separating a mixture of fresh gases (on the upstream side) and burnt gases (on the downstream side). Light is scattered upstream by refractory powders injected into the mixture, resulting in a bluish luminous region; downstream they are at a lower temperature and one has an orange-hued region. The conical white zone indicates the hottest region of the flame (see Section 10.9.3).

Visualization by means of anisotropic reflecting particles

This method is effective when significant variations in the direction of the flow from one point to the next exist. One can use small aluminium flakes (such as those which can be found in aluminium paints) or suspensions of particles that have the shape of long thin platelets (these fluids, known as rheoscopic fluids, are commercially available, often sold under the name "Kalliroscope^{TM}"). When such particles are in suspension in a flow, they align due to the action of velocity gradients—this results in the observation of contrasts in the reflected light, allowing us to visualize the flows. Figure 3.17 shows the application of this technique to a cellular thermal convection flow in a layer of liquid.

Figure 3.17 *Top view of cellular flow in a layer of liquid heated from underneath and bounded by a free surface above. The flow is visualized by means of flakes of aluminium in suspension in the liquid: the liquid is rising in the center of the hexagonal cells and flowing downward along their edges. The flow results from a Bénard–Marangoni instability due to changes in the surface tension with temperature, which we discuss in Section 11.3.1 (plate courtesy of J. Salan)*

Visualizing changes in refractive index by the Schlieren technique

There is a wide spectrum of optical methods which use the refraction of light rays due to local fluctuations in the index of refraction. Such variations can result from gradients in the fluid density, (and hence in the index) induced by temperature differences, as is the case in thermal convection phenomena. They may also be associated with density fluctuations due to compressibility effects, as found in high-velocity gas flows, and they may also due to variations in the chemical composition.

The *Schlieren method* is an example of these techniques which rely on variations of the index of refraction. Its principle, illustrated in Figure 3.18, consists of creating an image of the flow region by means of an optical device which blocks out the undeviated light rays which would normally propagate in the region of observation; one sees in the image only those rays which were refracted by the changes in the index, so that we can visualize the latter. Often, mirrors are used instead of lenses in these devices, because their diameter can more easily be large enough to be, as needed, comparable to the extent of the region under study. Furthermore, mirrors allow for a reduction in the effective chromatic aberrations. Figure 3.19 shows that this method allows one to visualize changes in the index of refraction of air resulting from temperature variations.

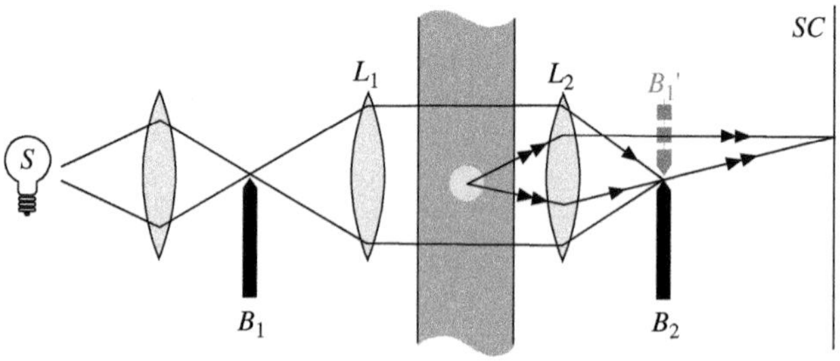

Figure 3.18 *Principle of the Schlieren technique for visualizing refraction index variations within a flow. The light beam from a source S is focused into a point source and then transformed into a parallel beam by lens L_1. The beam is then refocused by a second lens L_2 onto a sharp edge blocking out the direct beam. The only light visible on the screen SC (or recording camera) where the image of the flow is formed is due to variations in the index of refraction of the fluid*

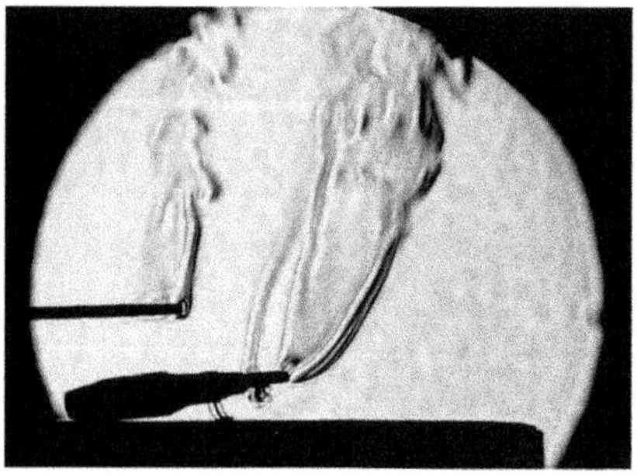

Figure 3.19C *Visualization by the Schlieren method of the variations in the index of refraction of air, resulting from temperature changes near a butane torch (lower plume) and a heated glass rod (upper plume) (document courtesy of I. Smith)*

3.5.2 Concentration measurements

Industrial or environmental applications often make use of a mixture of fluids, initially injected separately, then stirred and mixed by the flow, e.g., as in the case of chemicals which will then undergo a reaction. It is frequently essential to determine the relative concentration of each component: this can be carried out by local probes (e.g., in liquids, by electrical resistance micro probes or by electrodes measuring the concentration of specific ions).

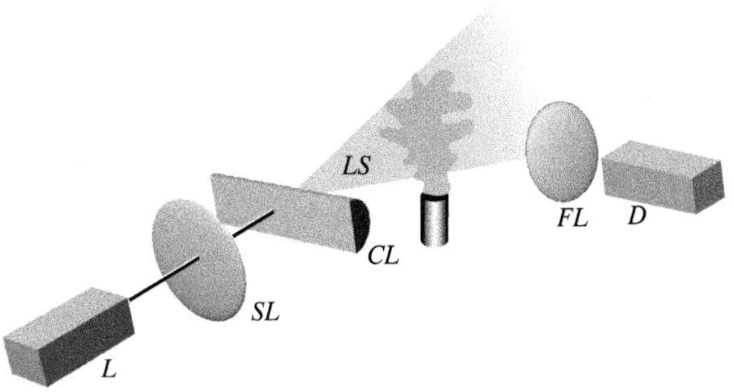

Figure 3.20 *Principle of the laser-induced fluorescence technique. The beam from a laser L is broadened into a light sheet LS by means of a cylindrical lens CL, after being focused by a spherical lens SL, so as to be as thin as possible as it passes through the region of interest in the flow (here, a vertical jet). The light scattered by the fluorescent dye is observed by a camera D, adjusted by means of a focusing lens FL to make observations in the plane light-sheet LS*

A particularly effective technique is *laser-induced fluorescence* (Figure 3.20). In the case of flowing liquids, it consists of dissolving a given concentration of a fluorescent dye in one of the components of the mixture: the flow is then illumined by a light beam of wavelength λ_0 selected to be in the region of maximum absorption of the dye. The light reemitted in all directions by fluorescence then corresponds to wavelengths $\lambda_f > \lambda_0$; its intensity is then proportional to the concentration C of the dye, so long as it is not too high, and this allows us to measure C. For this purpose, one frequently uses illumination by a *thin sheet of laser light* (Section 3.5.1). This allows one to analyze the local structure of the mixing region furthermore, because the wavelength of the laser is sharply defined, one can eliminate, by means of an interference filter, the parasitic light reflected or transmitted at the original wavelength λ_0, and visualize only the fluorescent dye.

Figure 3.21 illustrates the visualization of the mixing in a turbulent jet by means of this technique. Furthermore, the use of very short laser pulses allows for the observation of very clear and detailed views even in the case of high-velocity flows. The technique can even be extended to the observation of three-dimensional concentrations by rapidly displacing the laser beam in the direction normal to its plane, and using a high-speed camera. It can also be applied to gases when one of the gases can be rendered fluorescent.

Figure 3.21C *Display of a turbulent jet of liquid injected (from the left) into an identical fluid at rest. A fluorescent dye has been added to the jet, illuminated by a light-sheet in the plane of the figure and coinciding with the axis of the jet. The colours correspond to the local concentration of the dye in the plane of the image (document courtesy of C. Fukushima & J. Westerweel)*

3.5.3 A few methods for measuring the local velocity in a fluid

Laser Doppler anemometry (LDA)

Laser Doppler anemometry makes use of the variations in intensity of light scattered by particles carried along by the flow in a periodically spatially illuminated region. One uses two beams split from the same laser, which intersect at an angle φ within a small measurement volume (Figure 3.22). The two beams interfere there and, as a result, a network of fringes appear (the planes of light and dark fringes are normal to the plane of the two beams). When light-scattering particles are placed in a fluid and cross through the system of interference

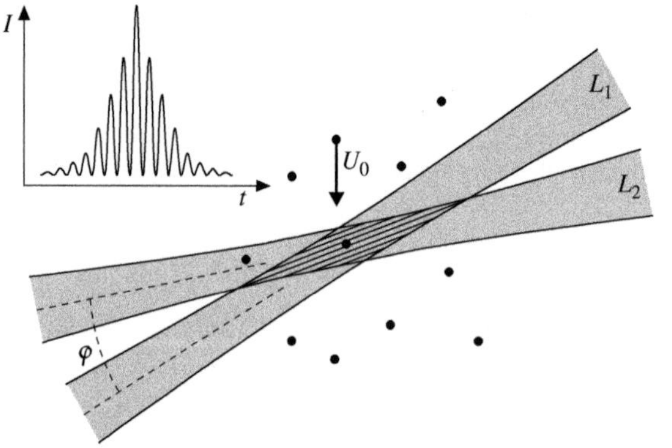

Figure 3.22 *Schematic representation of the Laser–Doppler anemometry technique, illustrating a split laser beam forming interference fringes in the region of overlap, with small particles carried along by the flowing fluid as it crosses the region; the inset figure displays the variation of the beam intensity as it is modulated by the passage of a particle through the interference region*

fringes, they are alternately bright and dark. By focusing the image of the fringes on a photomultiplier, one detects oscillations corresponding to the modulation of the intensity of the light scattered by each particle (inset of Figure 3.22): their frequency is measured by a counter or a spectrum analyzer. This frequency is related to the component of the velocity of the particle normal to the fringes by the equation:

$$f = \frac{2\,U_0}{\lambda_0}\, n \sin\frac{\varphi}{2}, \tag{3.53}$$

where φ is the angle between the beams in the fluid, n the index of refraction of the fluid, and λ_0 the wavelength of light in free space. Assuming that the velocity of the particles coincides with that of the fluid, one obtains a value for this velocity, averaged over the region of the fringes, which can be quite small (a few microns). The maximum velocities then measured can be as high as 100 m/s.

Proof
In Section 1.5.3 (Equation 1.90), we have shown that the fringe wavelength Λ, which corresponds to the geometry of the beams in Figure 3.22, satisfies:

$$\Lambda = \lambda_0 / \{2n \sin(\varphi/2)\}\,. \tag{3.54}$$

If a particle moves with a velocity component U_0 in the y-direction (thus normal to the plane of the fringes), the frequency of the changes in the illumination of the particle, thus of the intensity scattered by the particle, is $f = U_0/\Lambda$, from which we derive Equation 3.53. Typically, for U_0 = 20 mm/s, $\varphi = 30°$, n =1.33 and λ_0 = 0.5 μm, we find $f \approx 22.5$ kHz. The value of f is identical to the frequency shift for the light scattered by a moving particle: this is the reason why this technique is known as *Laser Doppler anemometry* (LDA).

Advantages of Laser Doppler anemometry This technique has the major advantage of not requiring the presence of an invasive physical probe within the flow: dust particles naturally present in a liquid are frequently sufficient to provide an easily measured signal. In the case of gases, however, it is usually necessary to seed the flow with very tiny particles.

The measurement of the velocity obtained is the absolute value, independent of the temperature fluctuations, or of the variations in the composition of the fluid (making calibration unnecessary).

One can also simultaneously determine all three velocity components with a three-coloured laser.

One can also detect the direction of the flow by allowing the fringes to move continuously by means of a variable optical retardation device (the frequency shift is related to the velocity of the particles relative to the fringes and, because of this, is not the same for two particles with opposite velocities).

Measurements can even be carried out in flames and in reactive media.

Disadvantages and limitations One can only measure the velocity of the scattering particles rather than the velocity of the fluid itself (particles up to 0.25 μm follow the changes in velocity up to frequencies around 10 kHz; particles as large as 4 μm are limited to 1 or 2 kHz); moreover, the necessity of having to seed the fluid with small particles requires that they be compatible with the fluid.

Measurements are difficult to make near walls because of the aggregation of particles and parasitic reflection of the light.

Measurements are impossible in opaque fluids.

Ultrasound anemometers

In the case of opaque liquids, one can use acoustic anemometers: these measure either changes in the velocity of sound depending on whether it moves along or in the opposite direction to the flow, or changes in frequency due to the Doppler effect when sound is back-scattered by small particles or bubbles carried by the flow. One frequently uses short bursts of soundwaves rather than a continuous sound: by choosing a time interval between the emission and the reception window of the sound, one selects the distance over which the measurement is carried out, and thus obtains a local measurement. By means of electronic measurements with several time windows, one obtains an instaneous profile of the velocity. This technique is frequently used in medicine in measurements of the flow-velocity of blood in the heart (a combination of the velocity measurement with echo-cardiography) or in blood vessels. This technique has recently been extended to industrial or laboratory measurements.

Hot-wire anemometry

Such anemometers measure the velocity of a flowing gas as a result of the cooling, by the gas, of an electrically heated wire (Figure 3.23). It is an intrusive technique (the presence of the probe can disturb the flow). Moreover, the variations of the signal with velocity are nonlinear and must be calibrated (see Section 10.8.1). On the other hand, the diameter of the wires used can be very small (a few microns) with a corresponding response time as fast as a few microseconds, a significant advantage over other methods. Therefore, such devices are frequently used in the fine-scale analysis of turbulent velocity fluctuations.

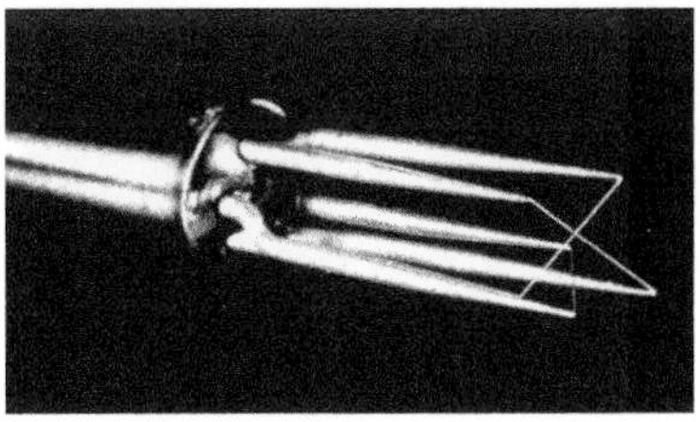

Figure 3.23 *The sensor end of a three-wire anemometer (plate courtesy of TSI)*

Principle of the measurement

One takes advantage of the variation in the resistance of the film, or of the wire, with temperature. A feedback loop applies a variable heating current so as to keep the temperature of the wire constant; the measurement of the heating current gives the velocity. This technique has the advantage, relative to a resistance measurement using heating by a constant current, of suppressing the time constant associated with the heating or cooling of the wire, and thus obtaining a frequency response of several hundred kHz. Hot-wires only detect the velocity components normal to their length: by using three wires perpendicular to each other and combining their readouts, it is then possible to determine the three components of the velocity. In the case of liquids, it is preferable to use, for reasons of robustness, heated films deposited on the surface of the tip of the probe, and protected by a thin insulating layer.

Miniature Pitot tubes

They provide a simple (though much less effective) method of measurement of the local velocity resulting from the pressure variations generated by the flow; we will describe this technique in Section 5.3.3 as an application of the Bernoulli equation.

3.5.4 Measurements of the velocity field and of velocity-gradients in a flowing fluid

Particle Image Velocimetry (PIV)

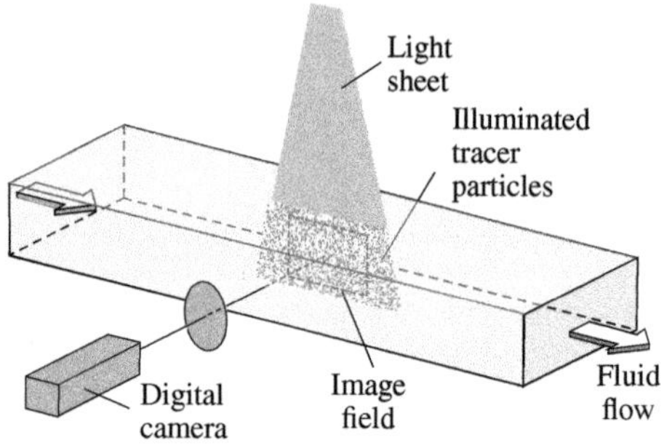

Figure 3.24 *Principle of the measurement of velocity by the PIV technique*

Principle of the measurement Thanks to recent developments in tools for analysis and image processing, it is now possible to determine the velocity flow fields from the images of particles carried along by the fluid: the main usefulness of this Particle Image Velocimetry (PIV) method is to yield in a single step the instantaneous velocity field in the region of observation, and not merely at a single location, as is the case for the techniques described in Section 3.5.3.

Just as in the optical tomography technique, the flow is illumined by a light-sheet (usually from a laser), while the fluid is seeded by fluorescent, or sometimes merely reflective, particles (Figure 3.24). A high-resolution digital camera takes two successive images of

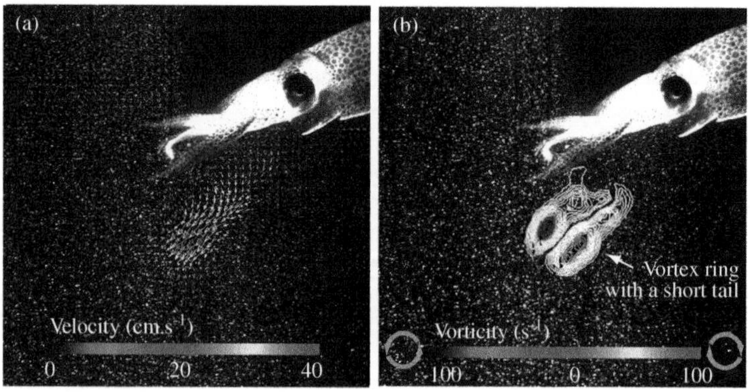

Figure 3.25C *Measurement by means of the PIV technique of (a) the velocity field and (b) the vorticity field of a vortex ring resulting from the propulsion of a squid (Lolliguncula Brevis), as it emits a series of pulsed jets (documents courtesy of I.K. Bartol, P.S. Krueger, W.J. Stewart et J.T. Thompson)*

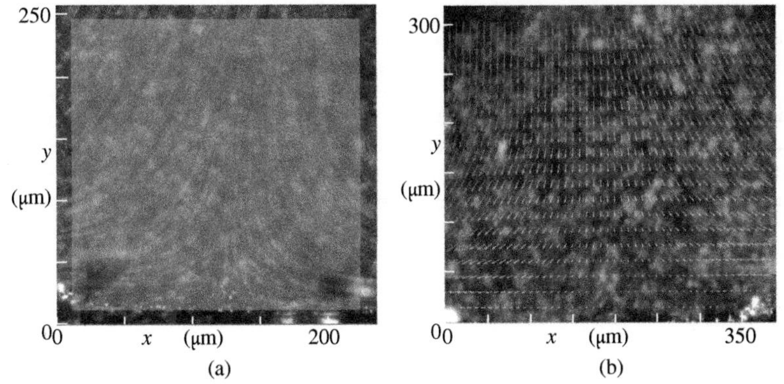

Figure 3.26C *Flow toward a plane (at the bottom of the figure) where there is a zero-velocity stagnation point in the center of the plane: (a) basic image: long exposure visualization of the streamlines by the use of particles carried along by the flow; the colours correspond to the vorticity field calculated from the velocity field of the image b. (b) velocity field obtained by the micro-velocimetry technique from the observation of the same particles (documents courtesy of C. Pirat, G. Bolognesi and C. Cottin-Bizonne, LPMCN-Lyon)*

these particles separated by a short time interval δt. Correlation analysis (described below) determines the displacement $\delta \mathbf{r}$ of particles in different regions of the flow, from which one infers the local velocities $\mathbf{v} = \delta \mathbf{r}/\delta t$. One then obtains velocity fields such as those illustrated in Figure 3.25, which correspond to the flow near a small moving squid, or that of Figure 3.26 (flow toward a stagnation point at a wall in a microchannel).

Practical measurements of the velocity field

In the case of liquids, one uses particles a few microns in size whose densities match as closely as possible that of the liquid, in order avoid sedimentation (e.g., hollow glass beads about 10 μm in diameter and with density 1.1 g/cm^3). In the case of gases, however, the density mismatch is always high, and one must, in order to slow down sedimentation, use even smaller particles such as oil droplets or spherical silica particles with sizes of the order of one micron.

In practice, in the generation of the illuminated plane, one frequently uses a pulsed laser providing short pulses of light (a few *ns*) separated by a time interval δt: one thus obtains images which are both brighter than with a continuous-wave laser and sharper thanks to the short duration of the pulse. The thickness of the illuminating plane must be sufficiently large so that particles remain inside the illuminated region during the interval between two pulses.

Digital images can be considered as matrices $I(x_i, y_j)$ where x_i and y_j are the coordinates of the pixels and I is the corresponding light intensity; they are divided into interrogation areas. For each such area in the first image I_1, one calculates a correlation function $C(\delta \mathbf{r}) = \sum_{i,j} I_1(x_i, y_j)\, I_2\, (x_i + \delta x, y_j + \delta y)$ between it and the second image I_2 (δx and δy are the components of the separation $\delta \mathbf{r}$). The value of $\delta \mathbf{r}$ for which the correlation function is a maximum then corresponds to the displacement of the fluid in the window in question during the time interval δt. $\delta \mathbf{r}$ can be determined with a resolution better than the distance between adjacent pixels provided that the size of the particles is just a tiny bit larger than this interval. One then repeats the same calculation for all the measurement windows in order to retrieve the complete velocity field. The concentration of the particles and the size of the measurement windows must be selected in such a way as to have a sufficient number of particles in each window while avoiding overlap. Sizes of windows ranging from 16×16 to 64×64 pixels are generally used, depending on whether one wishes to emphasize the number or the accuracy of the measuring points. We must also observe that, in every case, the spatial resolution of the measurement is of the order of the size of the window: since such images include at this point (in 2014) between 10^6 and 4×10^6 pixels in all, the velocity fields currently measurable contain thus typically between 10^3 and 10^4 points.

Micro-anemometry The same technique is nowadays applied to flows at very small scales, e.g., in channels used in microfluidics. In this case, we choose fluorescent particles as tracers. In fact, a problem resulting from these small devices is the importance of parasitic reflections relative to the wavelength of the illumination laser. As we have discussed in Section 3.5.2, we can eliminate these by means of an interference filter, while still preserving the light emitted by the particles, by fluorescence. Figure 3.26b compares a velocity field obtained in this manner to the streamlines obtained by a long exposure of the moving particles (Figure 3.26a).

It is difficult to obtain very thin luminous planes. Thus, in this method of micro-PIV, the measurement plane is selected by the focusing distance, making use of the very small depth-of-field of microscopes used for the observations.

Limitations and recent extensions of PIV applications A first important limitation is the fact that the PIV method measures only the components of the velocity in the plane of illumination. This difficulty can be overcome by the use of two cameras oriented obliquely relative to the plane light-beam, instead of a single normal one, as described previously (this is the *stereo-PIV* technique).

Principle of stereo-PIV

The two cameras look at the same region of measurement, their lines of sight being symmetric relative to the normal of the illuminated plane and making an angle $\pm\,\alpha$ with the latter (the plane of the lines of sight is chosen as the x-y plane in which x is perpendicular to the illuminated y-z plane). Each camera measures the velocity component of a diffusing particle normal to the line of sight. The velocity components v_z which are measured are the same for the two cameras. On the other hand, for the component in the x-y plane, one of the cameras measures a velocity $v_y \cos\alpha + v_x \sin\alpha$ and the other a velocity $v_y \cos\alpha - v_x \sin\alpha$. One therefore obtains every velocity component by combining the different measurements. However, the spatial variations of the velocity components in the direction normal to the plane of measurement can still not be determined in this way.

In order to measure a three-dimensional velocity field, one can carry out a set of measurements by displacing the luminous plane parallel to itself at each new measurement. In *PIV tomography*, a volume of fluid is illumined globally, and the three-particle coordinates are determined by combining the images from several cameras (typically four).

Initially, a further limitation of this technique, other than its limited spatial resolution, was the small repetition rate of the velocity fields (a few times per second): this limitation was due both to the rate of image acquisition by the camera and the pulse rate of the lasers. Thanks to new lasers now available, and to cameras with their own high-speed memory cards, repetition rates of a few kilohertz can be obtained at this time (2014).

Determination of the local velocity gradients

An important problem, particularly in the analysis of turbulent flows, is the determination of the velocity gradients in a flow and, more specifically, of their rotational component, which was discussed in Section 3.2.3.

When the PIV technique is used, the component of the vorticity perpendicular to the measurement plane can be easily calculated from the spatial variations of the velocity. One sees in Figure 3.25b that the presence of two high-vorticity regions is quite evident (we will see in Section 7.4.3 that this corresponds to the creation of a vortex ring). Determining other components of the vorticity requires the use of a three-dimensional system such as those described above.

We will discuss in Section 10.8.2 another technique, *polarography*, which allows for the measurement of velocity gradients in the neighborhood of a solid wall.

A direct method for measuring the vorticity consists in seeding the fluid with spherical particles which are reflective only on part of their surface area. Because of the rotational component of the flow, these particles are put into rotation at an angular velocity $\mathbf{\Omega}$ equal to half the local vorticity of the fluid ($\mathbf{\Omega} = (1/2)\ \nabla \times \mathbf{v}$) (Section 3.2.3); these particles must, however, be sufficiently light so as to match the local rotation of the fluid. They are illumined by a light beam which they reflect on a repetitive basis: one obtains the rotational velocity of the particles, and thus the vorticity of the fluid, from the frequency of these light flashes.

Finally, let us mention the technique of *forced Rayleigh scattering*, introduced in Section 1.5.3, which allows one to determine the vorticity from the rotation of the diffraction pattern of an interference grating created within a flowing fluid.

4 Dynamics of viscous fluids: rheology and parallel flows

Having introduced in Chapter 3 the concept of "rate of strain" of a flowing fluid, we discuss here the manner in which such deformations can be produced by application of a stress (external force, pressure, ...). In the mechanics of solids, there exists a proportionality between the strain (or relative deformation) and the stresses, so long as the stresses are not too strong. This relationship (Hooke's law) was first described by Robert Hooke who wrote, three centuries ago, the law "ut tensio, sic vis" ("as the strain, so is the force"). The corresponding relationship for viscous fluids, first expressed by Newton, indicates the proportionality between the rates of strain and the stresses: it applies to flows at low Reynolds number and to the so-called Newtonian fluids.

In this chapter, we first write (Section 4.1) the expression for the surface forces (pressure and viscosity) which act upon an element of the fluid. We then consider (Section 4.2) the equation of motion for a fluid (Navier–Stokes equation), under circumstances where the effects of the viscosity of the fluid cannot be neglected (i.e., for a real fluid). We will then discuss (Section 4.3) the boundary conditions at the walls which delimit the fluid flow. We will then tackle (Section 4.4) the case of non-Newtonian fluids, for which the relationship between the stress and the deformation is no longer linear and instantaneous. Finally we will analyze a few instances of parallel flow in Newtonian fluids (Section 4.5) and some non Newtonian ones (Section 4.6) for which the convective acceleration terms of the form $(\mathbf{v}.\nabla)\mathbf{v}$ introduced in Chapter 3 are identically zero.

4.1 Surface forces

4.1.1 General expression for the surface forces: stresses in a fluid

Let us consider a surface element of area dS in a fluid. We analyze the force exerted by the "piece" of fluid, located on one side of this element, on the fluid located on the other side: *stress is defined as a force per unit area*. In a fluid at rest, it is *normal* to the surface elements and its magnitude is independent of the orientation of these elements. Since this stress is isotropic, a single number is sufficient to characterize its value at each point; this is the *hydrostatic pressure*.

If the fluid is in motion, there appear additional stresses *tangential* to the element of surface dS. The latter, indicative of the frictional forces between layers of fluid sliding one relative to the other, are due to the viscosity of the fluid. We have, in fact, seen in Chapter 2 that the viscosity is a transport coefficient which characterizes the transfer of momentum from the regions of higher velocity to those of lower velocity. To determine these forces, we must know:

- the orientation of the surface dS in space, defined by means of the unit vector $\mathbf{n}$ normal to the surface ($\mathbf{dS}$ denotes the vector of magnitude dS directed along $\mathbf{n}$);
- the values of the three components of the force per unit area in the x-, y-, and z-directions for the three orientations of unit surfaces normal to these respective axes.

This leads to nine coefficients σ_{ij} which can be written as a 3×3 matrix representing the *stress tensor* $[\boldsymbol{\sigma}]$ in the fluid under consideration. The element σ_{ij} of the tensor ($i = 1, 2, 3$; $j = 1, 2, 3$) represents the component in the i-direction of the stress exerted on a surface with normal oriented in the j–direction. Accordingly:

- σ_{yx} is the y-component of the force exerted on a unit area with normal pointing in the x–direction (Figure 4.1a). This is a *tangential or shear stress.*
- σ_{xx} is the x-component of the force exerted on a surface perpendicular to the same x–direction. It is a *normal stress.*

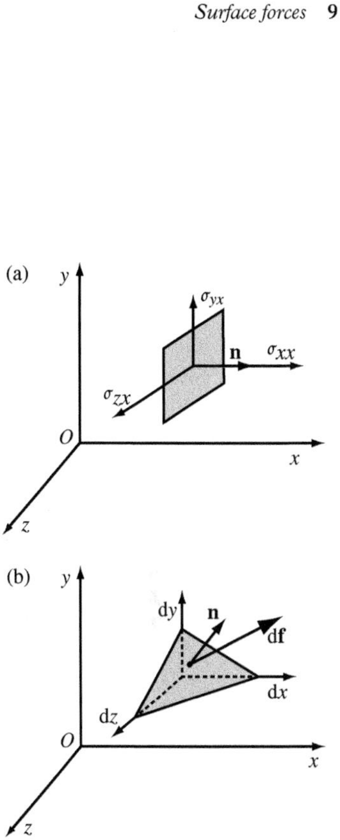

Figure 4.1 *Components σ_{xx}, σ_{yx} and σ_{zx} of the stress exerted on a surface of area dS: (a) with normal directed parallel to the x-axis; (b) with normal $\mathbf{n}$ in an arbitrary direction. Because of the existence of tangential stresses along the surface, the resultant force $\mathbf{df}$ is generally not colinear with the normal vector $\mathbf{n}$*

Let us now determine the stress $\boldsymbol{\sigma}_\mathbf{n}$ exerted on a surface dS with arbitrary normal $\mathbf{n}$ (Figure 4.1b). To this end, we must analyze the forces exerted on a tetrahedron of which three edges are in the x-, y-, and z-directions, and of respective lengths dx, dy and dz. The side of the tetrahedron bounded by the three other edges has the unit vector $\mathbf{n}$ as the outward normal with components n_x, n_y and n_z; $\mathbf{n}$ is the outward normal to the tetrahedron displayed.

We denote by $\sigma_{x\mathbf{n}}\, dS$, $\sigma_{y\mathbf{n}}\, dS$ and $\sigma_{z\mathbf{n}}\, dS$ the x-, y-, and z-components of the stress force $\mathbf{df}$ exerted on the surface dS with outward normal $\mathbf{n}$. Let us determine, for example, $\sigma_{x\mathbf{n}}$ by considering the balance of the forces exerted on the faces of the tetrahedron.

The x-components of the forces exerted on the faces perpendicular to the x-, y-, and z-directions are respectively:

$$(-\sigma_{xx})\, n_x\, dS, \qquad (-\sigma_{xy})\, n_y\, dS, \qquad \text{and} \qquad (-\sigma_{xz})\, n_z\, dS.$$

We have used here the definitions of the components of the normal and shear stresses, as well as the fact that the projections dS_x, dS_y and dS_z of the surface dS on which these stresses act are equal to the product of dS with the components n_x, n_y and n_z of the unit vector $\mathbf{n}$ along the three axes. The negative signs come about because the outward normals to these surfaces are oriented opposite to the direction of the corresponding coordinate axis, while the stresses are defined as positive if they point along the respective axis. The net x-component of all the stresses on the tetrahedron is therefore:

$$\sigma_{xn}\, dS - \sigma_{xx} n_x dS - \sigma_{xy} n_y\, dS - \sigma_{xz} n_z\, dS = \left(\sigma_{xn} - \sigma_{xx} n_x - \sigma_{xy} n_y - \sigma_{xz} n_z\right) dS. \tag{4.1}$$

Writing down Newton's second law, where dV is the volume element, ρ the density of the fluid, and d^2x/dt^2 the acceleration in the x-direction; f_x is the x-component of any volume forces present, e.g., gravity. We find:

$$(\sigma_{xn} - \sigma_{xx} n_x - \sigma_{xy} n_y - \sigma_{xz} n_z)\, dS + f_x dV = \rho\, dV \frac{d^2 x}{dt^2}. \tag{4.2}$$

Let us now take the limit as dV approaches zero by reducing proportionately each of its sides; this maintains the direction of the unit vector $\mathbf{n}$ constant. But dV approaches zero as $dS^{3/2}$. Thus the two terms with factor dV in Equation 4.2 tend to zero more rapidly than the term with factor dS, and cannot compensate for it. We therefore conclude that this latter term must be identically zero, whence:

$$\sigma_{x\mathbf{n}} = \sigma_{xx}n_x + \sigma_{xy}n_y + \sigma_{xz}n_z. \tag{4.3}$$

By cyclic permutation of the indices, we find equivalent relations for the two other components $\sigma_{y\mathbf{n}}$ and $\sigma_{z\mathbf{n}}$. This leads to the matrix equation:

$$\begin{pmatrix} \sigma_{x\mathbf{n}} \\ \sigma_{y\mathbf{n}} \\ \sigma_{z\mathbf{n}} \end{pmatrix} = \begin{pmatrix} \sigma_{xx} & \sigma_{xy} & \sigma_{xz} \\ \sigma_{yx} & \sigma_{yy} & \sigma_{yz} \\ \sigma_{zx} & \sigma_{zy} & \sigma_{zz} \end{pmatrix} \begin{pmatrix} n_x \\ n_y \\ n_z \end{pmatrix}, \tag{4.4}$$

which can also be written in the form:

$$\frac{d\mathbf{f}}{dS} = \boldsymbol{\sigma}_\mathbf{n} = [\boldsymbol{\sigma}] \cdot \mathbf{n}. \tag{4.5}$$

The term $[\boldsymbol{\sigma}]\cdot\mathbf{n}$ expresses the inner product of the second rank tensor $[\boldsymbol{\sigma}]$ with the vector $\mathbf{n}$. We also make frequent use of the notation:

$$\sigma_{i\mathbf{n}} = \sigma_{ij}\, n_j, \tag{4.6}$$

where summation is implicit on the index j (Einstein summation convention).

The diagonal terms of the stress tensor $[\boldsymbol{\sigma}]$ result both from the pressure and from the tensor $[\boldsymbol{\sigma}']$: these components, usually called *normal stresses*, are the result of the flow of the fluid and cancel out when the fluid is at rest. They occur in the case of Newtonian fluids (Section 4.1.3), but are often particularly significant in the case of viscoelastic fluids, as we will see in Section 4.4.5.

The pressure term denoted by the parameter p in Equation 4.7 must be understood as a mechanical pressure, defined in terms of the mechanical stresses which act on an element of the fluid. We cannot define the pressure in a moving fluid from thermodynamic considerations since the system is not in thermodynamic equilibrium.

Presssure forces and the shear stress tensor

We can separate out of the stress tensor $[\boldsymbol{\sigma}]$ the part which corresponds to the pressure stresses, which are the only ones acting in the absence of velocity gradients, for a fluid at rest or in uniform translational motion. This component is totally diagonal and isotropic—the stresses are all normal and all three diagonal coefficients are identical. We can therefore separate the tensor into an expression of the form:

$$\sigma_{ij} = \sigma'_{ij} - p\,\delta_{ij}, \tag{4.7}$$

where p is the pressure and δ_{ij} the Kronecker delta symbol ($\delta_{ij} = 1$ if $i = j$, and $\delta_{ij} = 0$ if $i \neq j$). The negative sign associated with p merely indicates that a fluid at rest is usually under compression, and thus the stress is acting opposite to the outward normal $\mathbf{n}$. The other term σ'_{ij} corresponds to the viscosity stress tensor: it is the part of σ_{ij} resulting from the deformation of the elements of the fluid.

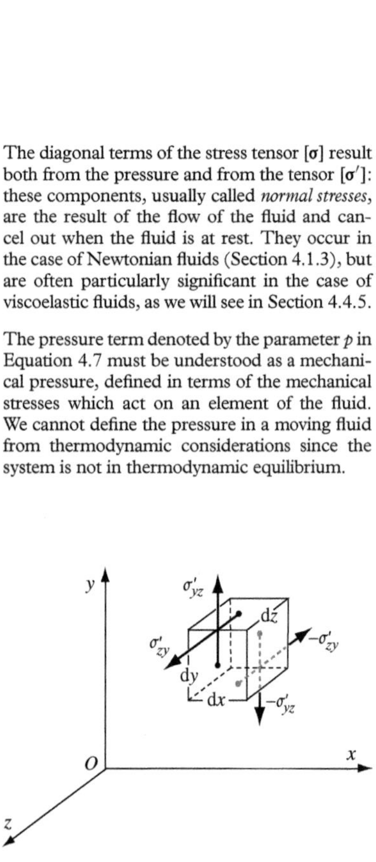

Figure 4.2 *Torques associated with the viscous forces which act on the faces of a cube located within the fluid*

4.1.2 Characteristics of the viscous shear stress tensor

We first demonstrate that the tensor $[\boldsymbol{\sigma}']$ is symmetric. This can be done by considering the balance of the torques on an infinitesimal, cubic, volume element with sides dx, dy and dz parallel to the coordinate axes (Figure 4.2).

In this demonstration, we will confine our argument to rotations around an axis in the x-direction passing through the center of the cube, such that the components σ'_{yz} and σ'_{zy} of the surface forces are the only ones contributing to the resultant torque Γ_x with respect to that axis (Figure 4.2); the same reasoning applies to the other components. The other forces exerted on the faces of a cube are either parallel to the axis of rotation or pass through it, and therefore do not contribute to Γ_x. We then obtain the expression:

$$\Gamma_x = \sigma'_{zy}(\mathrm{d}x\,\mathrm{d}z)\,\frac{\mathrm{d}y}{2} - \sigma'_{yz}(\mathrm{d}x\,\mathrm{d}y)\,\frac{\mathrm{d}z}{2} = (\sigma'_{zy} - \sigma'_{yz})\,\frac{\mathrm{d}V}{2}, \tag{4.8}$$

In calculating the torque, we have not taken into account the terms corresponding to the changes in the surface forces from one face to the other. Effectively, in Equation 4.8, these terms give a higher order contribution, of the type $(\partial\sigma'_{xy}/\partial y)\,\mathrm{d}y\,\mathrm{d}V$, which are negligible when we take the limit as the dimensions of the volume element tend to 0.

where $\mathrm{d}V = \mathrm{d}x\,\mathrm{d}y\,\mathrm{d}z$ is the volume element. If the angular acceleration of the element is written as $\mathrm{d}^2\Omega_x/\mathrm{d}t^2$ and $\mathrm{d}I$ is the moment of inertia about the axis of rotation, we have $\Gamma_x = \mathrm{d}I\ (\mathrm{d}^2\Omega_x/\mathrm{d}t^2)$. If we now take the limit (as in the previous derivation of Equation 4.3) as $\mathrm{d}V$ tends to zero, we find that $\mathrm{d}I$, which is of order $\mathrm{d}V\ (\mathrm{d}y^2 + \mathrm{d}z^2)$, tends to zero as $\mathrm{d}V^{5/3}$ (faster than $\mathrm{d}V$). We must therefore have the condition $\sigma'_{zy} = \sigma'_{yz}$ because the angular acceleration is to remain finite (even if there existed a mechanism creating a volume couple, the overall torque on an element of the volume will also be proportional to $\mathrm{d}V$ and not affect our result). This equality can be generalized to the other components of the tensor such that:

$$\sigma'_{ij} = \sigma'_{ji}. \tag{4.9}$$

This equation merely signals the equilibrium of the torques exerted on the volume elements of the fluid.

Let us now specify the relation between the viscous stresses, i.e. the tensor $[\boldsymbol{\sigma}']$, which act on a fluid and the strains of the fluid. These stresses cancel out when an element of fluid moves without strain, and, for this reason, depend neither on the velocity (global translation) nor on the local rotation. The latter is governed by the general antisymmetric tensor ω_{ij} defined in Equation 3.5b. On the other hand, the tensor $[\boldsymbol{\sigma}']$ of the viscous stresses is symmetric and must depend only on the symmetric components e_{ij} of the tensor $[\mathbf{e}]$ of the velocity gradients (rate of strain tensor), which has components:

$$e_{ij} = \frac{1}{2}\left(\frac{\partial v_i}{\partial x_j} + \frac{\partial v_j}{\partial x_i}\right). \tag{4.10}$$

4.1.3 The viscous shear-stress tensor for a Newtonian fluid

Throughout most of this text, we discuss exclusively those fluids, known as *Newtonian* fluids, for which the components σ'_{ij} of the viscosity stress tensor are assumed to depend linearly on the instantaneous values of the deformations. Assuming further that the medium is isotropic, this leads to the result:

$$\sigma'_{ij} = 2A\,e_{ij} + B\,\delta_{ij}\,e_{ll}, \tag{4.11}$$

where A and B are real constants characteristic of the fluid (and repeated indices imply summation, according to the Einstein convention).

Justifying the relation between σ'_{ij} and e_{ij}:
To first order, the components of the tensor $[\boldsymbol{\sigma}']$ are linearly dependent on the components of the deformation e_{ij}; these components must, in fact, change sign when the components of the strain change sign. The element σ'_{ij} is then written:

$$\sigma'_{ij} = A_{ijkl}\,e_{kl}, \tag{4.12}$$

where A_{ijkl} is a fourth-rank tensor, for which it can be shown that the most general form for an isotropic medium is:

$$A_{ijkl} = A\,\delta_{ik}\,\delta_{jl} + A'\,\delta_{il}\,\delta_{jk} + B\,\delta_{ij}\,\delta_{kl}.$$

Because the tensor $[\boldsymbol{\sigma}']$ is symmetric ($\sigma'_{ij} = \sigma'_{ji}$), A_{ijkl} must be symmetric with respect to interchanging of i and j, so that $A = A'$. It then follows that:

$$\sigma'_{ij} = A\,(\delta_{ik}\delta_{jl}e_{kl} + \delta_{il}\delta_{jk}e_{kl}) + B\,\delta_{ij}\delta_{kl}e_{kl} = A\,(e_{ij} + e_{ji}) + B\,\delta_{ij}e_{ll}, \tag{4.13}$$

i.e.:

$$\sigma'_{ij} = 2A\,e_{ij} + B\,\delta_{ij}e_{ll}.$$

We write Equation 4.11 in the equivalent form:

$$\sigma'_{ij} = \eta\left(2e_{ij} - \frac{2}{3}\delta_{ij}e_{ll}\right) + \zeta\,(\delta_{ij}e_{ll}), \tag{4.14}$$

The coefficient ζ appears in "incompressible" fluids only in measurements of the attenuation of sound as the propagation of sound in any fluid is necessarily accompanied by compressional effects, otherwise the velocity of sound would be infinite. For ordinary liquids, the experimental values of ζ are very small.

which uses the decomposition 3.11 of the strain tensor obtained in Section 3.2.2. The first term corresponds to deformation without change in volume, while the second one represents isotropic dilation. The latter is therefore zero for an incompressible fluid.

The first coefficient appearing in Equation 4.14 is called the *shear viscosity*. Let us first check that η is indeed the viscosity coefficient introduced in Section 2.1.2 for the particular case of simple shear flow which we used in that introduction (Figure 4.3). Assume, for example, that the velocity of the fluid $\mathbf{v}$ has only an x-component v_x which varies only along the y-direction. According to Equation 4.14, the only non-zero term of $[\boldsymbol{\sigma}']$ is then:

$$\sigma'_{xy} = \sigma'_{yx} = \eta\frac{\partial v_x}{\partial y}. \tag{4.15}$$

This expression is the same as Equation 2.2 for the term σ'_{xy} representing the tangential stresses due to the relative motion of the various layers of fluid.

The second coefficient ζ which appears in Equation 4.14 is called the *second viscosity* or the *bulk viscosity*. The corresponding stresses (diagonal terms of the tensor $[\boldsymbol{\sigma}']$ having the form $\zeta\nabla.\mathbf{v}$) are related to changes in the volume of the fluid due to compression effects. This term vanishes in the study of incompressible fluids because $\nabla.\mathbf{v} = 0$ is zero in that

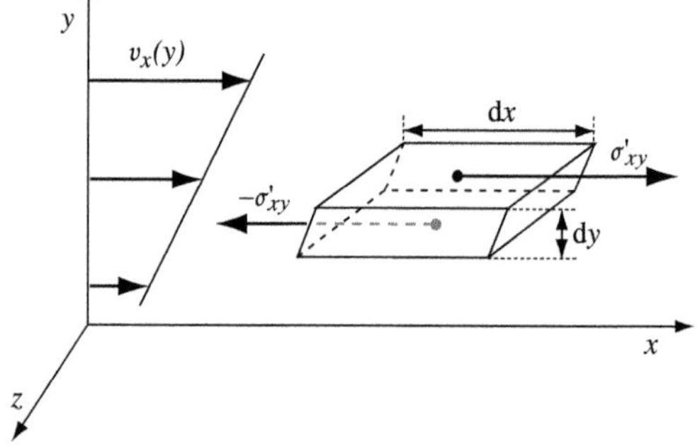

Figure 4.3 *Shear stress in a simple shear flow*

case (the circumstances under which a fluid can be treated as incompressible have been discussed in Section 3.3.2.). In such cases, Equation 4.14 simplifies to:

$$\sigma'_{ij} = 2\,\eta\, e_{ij}. \tag{4.16}$$

The viscosity coefficients η and ζ defined above are both positive. We shall prove this for the case of η in Section 5.3.1. In Appendix 4A in this chapter, the reader will find representations of the stress tensor $[\boldsymbol{\sigma}']$ in the most frequently used coordinate systems: Cartesian, cylindrical and spherical polar.

4.2 Equation of motion for a fluid

4.2.1 General equation for the dynamics of a fluid

We apply Newton's second law of motion to the dynamics of the volume of fluid $(\mathcal{V})$ by equating the time rate of change of momentum and the net forces (volume and surface) exerted on $(\mathcal{V})$. The volume $(\mathcal{V})$ is made up of a given amount of matter which moves along with the fluid and remains within $(\mathcal{V})$.

$$\frac{\mathrm{d}}{\mathrm{d}t}\left[\iiint_{\mathcal{V}} \rho\,\mathbf{v}\,\mathrm{d}\tau\right] = \iiint_{\mathcal{V}} \rho\,\mathbf{f}\,\mathrm{d}\tau + \iint_{\mathcal{S}} [\boldsymbol{\sigma}]\cdot\mathbf{n}\,\mathrm{d}\Sigma. \tag{4.17}$$

Here, $\mathrm{d}\tau$ is the differential volume of a small amount of material; $\mathrm{d}\Sigma$ is a surface element of the closed surface $(\mathcal{S})$ which bounds $(\mathcal{V})$; $[\boldsymbol{\sigma}]$ is the tensor of all the forces (pressure and viscosity) acting on $\mathrm{d}\Sigma$. The volume force $\mathbf{f}$ per unit mass of fluid might be, for instance, a gravitational force, an electrostatic force on a charged fluid, a Coriolis force in a rotating reference system, or the magnetic force on a fluid containing a suspension of magnetic particles (a ferrofluid).

We might recall that the derivative $(\mathrm{d}/\mathrm{d}t)$ is a Lagrangian derivative, evaluated in a reference system which moves with the fluid (as discussed in Section 3.1.2.). In this reference frame, the product $\rho\,\mathrm{d}\tau$, representing the mass of a small amount of fluid material is a constant: in fact, any element of the fluid always encloses, by definition, the same molecules as they move along in the local velocity field of the flow. This result allows us to differentiate with respect to time only the factor $\mathbf{v}$ in the first term of Equation 4.17, so as to write:

$$\frac{\mathrm{d}}{\mathrm{d}t}\left[\iiint_{\mathcal{V}} \rho\,\mathbf{v}\,\mathrm{d}\tau\right] = \iiint_{\mathcal{V}} \rho\,\frac{\mathrm{d}\mathbf{v}}{\mathrm{d}t}\mathrm{d}\tau. \tag{4.18}$$

Rigorously, we should undertake a detailed evaluation of the momentum balance within the volume $(\mathcal{V})$ but the simpler derivation we have used above discloses the essential physical content of the problem.

Furthermore, the total component of the surface forces in the i-direction can be written:

$$\left[\iint_{\mathcal{S}} [\boldsymbol{\sigma}]\cdot\mathbf{n}\,\mathrm{d}\Sigma\right]_i = \iint_{\mathcal{S}} \sigma_{ij}\,n_j\,\mathrm{d}\Sigma. \tag{4.19}$$

This represents the flux of the components $(\sigma_{ix}, \sigma_{iy}, \sigma_{iz})$ across surface $(\mathcal{S})$. Equation 4.19 can be transformed into a volume integral by means of Gauss' divergence theorem:

$$\iint_{\mathcal{S}} \sigma_{ij}\,n_j\,d\Sigma = \iiint_{\mathcal{V}} \frac{\partial\sigma_{ij}}{\partial x_j}\mathrm{d}\tau \tag{4.20}$$

(in the integrals of Equations 4.19 and 4.20, we use the Einstein implicit summation convention over repeated indices). Equation 4.17 can therefore be written as:

$$\iiint_{\mathcal{V}} \rho \frac{\mathrm{d}\mathbf{v}}{\mathrm{d}t} \mathrm{d}\tau = \iiint_{\mathcal{V}} \rho \mathbf{f} \, \mathrm{d}\tau + \iiint_{\mathcal{V}} \nabla \cdot [\boldsymbol{\sigma}] \, \mathrm{d}\tau. \tag{4.21}$$

The term $\nabla.[\boldsymbol{\sigma}]$ represents here the vector with components $\partial \sigma_{ij}/\partial x_j$. We further recall that, in Equation 4.21, the integrals are evaluated over a volume ($\mathcal{V}$) which moves along with the fluid. Taking the limit as this volume tends to zero and dividing by the value of the volume element, we obtain the local equation of motion for a particle of fluid:

$$\rho \frac{\mathrm{d}\mathbf{v}}{\mathrm{d}t} = \rho \mathbf{f} + \nabla \cdot [\boldsymbol{\sigma}]. \tag{4.22}$$

To be fully rigorous, as we have already mentioned in Section 3.1.1, the smallest volume over which the integral is to be evaluated, i.e. the volume of a particle of fluid, should be the smallest volume to which the assumption of a continuous medium applies. This limitation is significant only in some very specific instances, such as the case of gases at extremely low pressures.

Let us now separate out of $[\boldsymbol{\sigma}]$ the part due to the pressure from that related to the viscous forces, as we did in Equation 4.7 ($\sigma_{ij} = \sigma'_{ij} - p\,\delta_{ij}$); this expression is a reminder that pressure contributes only to normal stresses. We then have:

$$(\nabla \cdot [\boldsymbol{\sigma}])_i = (\nabla \cdot [\boldsymbol{\sigma}'])_i - \frac{\partial (p\,\delta_{ij})}{\partial x_j} = (\nabla \cdot [\boldsymbol{\sigma}'])_i - \frac{\partial p}{\partial x_i}. \tag{4.23}$$

Equation 4.22 then becomes:

$$\rho \frac{\mathrm{d}\mathbf{v}}{\mathrm{d}t} = \rho \mathbf{f} - \nabla p + \nabla \cdot [\boldsymbol{\sigma}']. \tag{4.24}$$

This equation is applicable to any fluid, as we have made no assumption about the form of the stress tensor $[\boldsymbol{\sigma}']$. Most often this equation is written by replacing the total time-derivative $\mathrm{d}\mathbf{v}/\mathrm{d}t$ by $\partial \mathbf{v}/\partial t + (\mathbf{v}.\nabla)\mathbf{v}$ just as we had done in Section 3.1.3. Thus:

$$\rho \frac{\partial \mathbf{v}}{\partial t} + \rho\,(\mathbf{v} \cdot \nabla)\mathbf{v} = \rho \mathbf{f} - \nabla p + \nabla \cdot [\boldsymbol{\sigma}']. \tag{4.25}$$

- The first term of the left-hand side expression in Equation 4.25 represents the acceleration of a particle of fluid due to the explicit time–dependence of its velocity in a fixed, Eulerian reference frame (acceleration in a homogeneous nonstationary field $\mathbf{v}(\mathbf{r},t)$);
- the second term corresponds to the changes in velocity as a particle of fluid is convected through the velocity field. This leads to an acceleration term, even if the velocity field $\mathbf{v}(\mathbf{r})$ is time–independent;
- on the right-hand side, the term $\rho\mathbf{f}$ represents the resultant of the volume forces applied to the fluid;
- the next term, $-\nabla p$, indicating the effect of the pressure, corresponds to normal stresses which are present even in the absence of fluid motion (hydrostatic pressure).

For a fluid at rest ($\mathbf{v}=0$), Equation 4.25 simplifies to the fundamental law of hydrostatics:

$$\rho\,\mathbf{f} - \nabla p = 0. \tag{4.26}$$

- finally, the last term $\nabla.[\boldsymbol{\sigma}']$ represents the viscous forces due to the strain of the fluid elements. It contains both viscous shear stresses and normal stresses which might appear during the motion of a compressible or viscoelastic fluid.

4.2.2 Navier–Stokes equation of motion for a Newtonian fluid

If we substitute for the tensor $[\boldsymbol{\sigma}']$ the explicit form obtained in Equation 4.14, we obtain for the ith component of the viscous forces:

$$(\nabla\cdot[\boldsymbol{\sigma}'])_i = \frac{\partial \sigma'_{ij}}{\partial x_j} = \eta \frac{\partial^2 v_i}{\partial x_j \partial x_j} + \left(\zeta + \frac{\eta}{3}\right) \frac{\partial}{\partial x_i}\left(\frac{\partial v_k}{\partial x_k}\right). \tag{4.27}$$

Equations 4.27 can be written in the vector form:

$$\nabla\cdot[\boldsymbol{\sigma}'] = \eta\, \nabla^2 \mathbf{v} + \left(\zeta + \frac{\eta}{3}\right) \nabla(\nabla\cdot\mathbf{v}). \tag{4.28}$$

If we substitute this in Equation 4.25, we obtain the equation of motion for a compressible or incompressible Newtonian fluid:

$$\rho \frac{\partial \mathbf{v}}{\partial t} + \rho\,(\mathbf{v}\cdot\nabla)\mathbf{v} = \rho\,\mathbf{f} - \nabla p + \eta\, \nabla^2 \mathbf{v} + \left(\zeta + \frac{\eta}{3}\right) \nabla(\nabla\cdot\mathbf{v}). \tag{4.29}$$

If the fluid flow is such that compressibility effects are negligible, $\nabla.\mathbf{v}=0$, and then the second viscosity coefficient ζ disappears. The resulting equation, known as the *Navier–Stokes equation*, finds extensive use throughout the remainder of this text:

$$\rho \frac{\partial \mathbf{v}}{\partial t} + \rho\,(\mathbf{v}\cdot\nabla)\mathbf{v} = \rho\,\mathbf{f} - \nabla p + \eta\,\nabla^2 \mathbf{v}. \tag{4.30}$$

The appendix at the end of this chapter lists explicit representations of this equation in the several coordinate systems in common use.

4.2.3 Euler's equation of motion for an ideal fluid

Euler's equation is the equation of motion for an ideal, incompressible fluid in which viscosity effects are absent. This is just a special case of the Navier–Stokes equation 4.30 in which we put $\eta = 0$, to obtain:

$$\rho \frac{\partial \mathbf{v}}{\partial t} + \rho\,(\mathbf{v}\cdot\nabla)\,\mathbf{v} = \rho\,\mathbf{f} - \nabla p. \tag{4.31}$$

This equation is rigorously applicable to an ideal fluid, i.e. one with zero viscosity. The flow of ideal fluids is frequently (but not always) *potential* (the velocity field is obtained from a potential function): this type of flow will be defined and described in Chapter 6. Some flows of viscous fluids can be described by Euler's equation, at least throughout a significant portion of their volume: we will discuss the conditions for observing such flows in Section 6.1.

We have implicitly assumed that the spatial variations $\partial\eta/\partial x_j$ and $\partial\zeta/\partial x_j$ of the viscosity coefficients η and ζ are negligible, an assumption which holds in homogeneous fluids. This is not always true; e.g., for the case of suspensions (Section 9.6), spatial variations in the concentration lead to significant local variations in the viscosity.

In Equation 4.30, it is the combination of terms $\rho\,\mathbf{f}-\nabla p$ of the volume force and the pressure gradient which causes the flow: as we have already seen, so long as $\mathbf{f}=\mathbf{g}$, setting this sum equal to zero is equivalent to writing the equation for hydrostatics. We can therefore create a flow with zero pressure gradient so long as $\rho\,\mathbf{f} \neq 0$. We will see below that this is the case, for example, for a liquid film flowing under gravity along a vertical or inclined plane, or, also, for a straight tube filled with fluid, open at both ends and at an angle to the horizontal, if we neglect capillarity.

The only ideal fluid is superfluid liquid helium. The viscosity of its isotope He^4 appears to vanish when it is cooled below a temperature of 2.172 K. The flow properties of this very peculiar fluid will be discussed in appendix 7A to Chapter 7. The isotope He^3 also becomes superfluid but at much lower temperatures of the order of 2 mK or below.

4.2.4 Dimensionless form of the Navier–Stokes equation

We can also write the Navier–Stokes equation 4.30 in terms of dimensionless parameters (that we label with "primes") for the different variables. Let L and U be the respective scale factors for the spatial characteristic length, and for the velocity of the flow. We then have:

The choice of L/U for the normalization of the time, and of ρU^2 for the pressure, is appropriate for flows where the non-linear terms dominate. For flows governed mainly by the viscosity terms, we will indicate in Section 9.2.4 another manner of writing the Navier–Stokes equation in a dimensionless form, appropriate for that regime.

$$\mathbf{r}' = \frac{\mathbf{r}}{L}, \quad \mathbf{v}' = \frac{\mathbf{v}}{U}, \quad t' = \frac{t}{L/U}, \quad p' = \frac{p - p_0}{\rho U^2}.$$

In defining p', we have subtracted out the value p_0 of the pressure in the absence of flow (hydrostatic pressure). The Navier–Stokes equation becomes after dividing both terms by $\rho\, U^2/L$:

$$\frac{\partial \mathbf{v}'}{\partial t'} + \left(\mathbf{v}' \cdot \nabla'\right) \mathbf{v}' = -\nabla' p' + \frac{\eta}{\rho\, U\, L} \nabla'^2 \mathbf{v}'.$$

We see that there appears, as a factor of the term $\nabla'^2\, \mathbf{v}'$, the inverse of the Reynolds number $Re = (UL/\nu)$ associated with the flow; this number represents the ratio of the non-linear, convective term $(\mathbf{v} \cdot \nabla)\mathbf{v}$ to the viscosity term $\nu\, \nabla^2 \mathbf{v}$ as discussed in Section 2.3.1.

From the nature of the above equation, we can state that the velocity and pressure ($\mathbf{v}'$ and p') fields, solutions of this equation, satisfying the appropriate boundary conditions for a given problem, are of the form:

$$\mathbf{v} = \mathbf{F}\left(x', y', z', t', Re\right),$$
$$p' = G\left(x', y', z', t', Re\right),$$

where $\mathbf{F}$ and G are functions dependent on the flow in the given problem.

4.3 Boundary conditions for fluid flow

The complete solution for the motion of a fluid (velocity field $\mathbf{v}(\mathbf{r}, t)$) requires both the integration of the equation of motion of the fluid particles and the specification of the boundary conditions, i.e. of the value of the variables (velocity, stress) at all boundaries of the fluid. Two different conditions apply, depending on whether the boundary medium is either a solid or another fluid.

4.3.1 Boundary condition at a solid wall

The fact that the fluid cannot penetrate into the solid requires that the components of the velocity normal to the boundary surface should be equal for the fluid and the solid:

$$\mathbf{v}_s \cdot \mathbf{n} = \mathbf{v}_f \cdot \mathbf{n}.$$

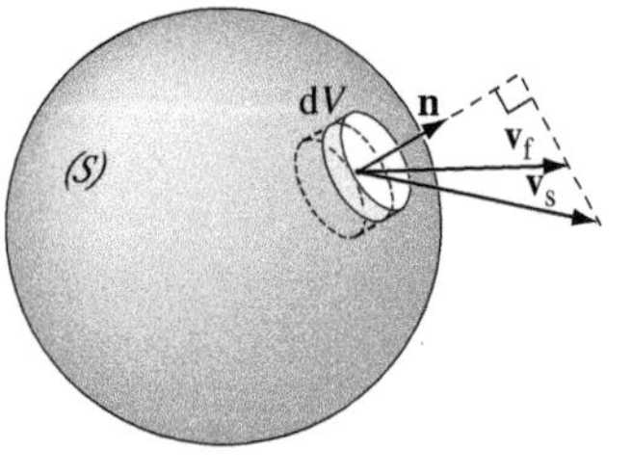

Figure 4.4 *Boundary conditions on the components of velocity at the interface between a solid and an* ideal *fluid*

This condition follows directly from the continuity equation, $\nabla \cdot \mathbf{v} = 0$, derived in Section 3.3.2. (Equation 3.28). To obtain this, we integrate this equation in a volume $\mathrm{d}V$ bounded by two neighboring surface elements located on each side of the surface of the solid and parallel to it (Figure 4.4). The boundary condition above expresses the fact that the flux of the velocity vector coming out of the solid is zero if the solid is at rest.

For an *ideal* (zero viscosity) fluid, there is no restriction on the tangential component of the velocity. This implies that the fluid can *slip* parallel to the solid surface (the case of Figure 4.4).

On the other hand, in the case of a *real* fluid, the viscosity stresses prevent any slipping of the fluid relative to the solid surface. It can be shown that such a discontinuity would lead to infinite energy dissipation at the surface as a result of the viscosity. The tangential components of the velocities of the fluid and of the solid must therefore also be equal. Coupled with the condition that the normal components must be equal, this leads to the relation:

$$\mathbf{v}_{\text{fluid}} = \mathbf{v}_{\text{solid}}, \tag{4.32}$$

i.e.: $\mathbf{v}_{\text{fluid}} = 0$ if the wall is at rest (Figure 4.5a). While this condition cannot be rigorously proved theoretically, it can, nonetheless, be considered valid for simple fluids, and on the scale of current experiments. However, significant instances of boundary slip are observable for complex fluids, and even for simple fluids, at scales smaller than a micron. They can thus have a significant influence in experiments of micro- and nano-fluidics, which are nowadays displaying important developments.

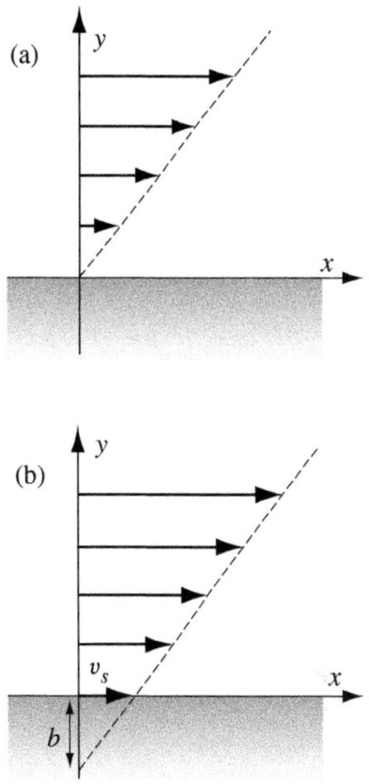

Figure 4.5 *Velocity profile for a viscous fluid near a solid wall at rest. (a) zero slip velocity; (b) finite slippage velocity* v_s

Slip length

One frequently characterizes these effects in terms of a slip length b, also known as the *Navier length*: it represents the distance from the wall where, by extrapolation, the velocity inside the solid is zero (Figure 4.5b). If the slippage velocity at the wall is v_s and if x and y are the respective coordinates parallel and perpendicular to the velocity at the wall, b is given in terms of the velocity gradient by:

$$b = \frac{v_s}{\partial v_x / \partial y}.$$

We can interpret b by assuming that the slip is accompanied by a frictional force $F_f = k\, v_s$ which balances the viscous force $\eta\ \partial v_x/\partial y$: b would then be equal to η/k and independent of v_s. This assumption is frequently in accordance with experimental data: this is the justification for using b rather than the slip velocity v_s to characterize this phenomenon.

The existence of a non-zero value for b has been known for a long time in the case of polymer solutions: a possible explanation for this is the presence of a very thin layer of solvent of thickness δ and viscosity η_s adjacent to the wall. We then have $F_f \approx \eta_s\ v_s/\delta$, i.e. $b \approx \delta\ \eta/\eta_s$; since η_s is much smaller than the viscosity η of the solution (see Section 4.4.3), we therefore have $b \gg \delta$. Macromolecules forming a type of "rug" along the surface can also play an important role. For such complex liquids, the slip lengths get to be a fraction of a micron. For ordinary liquids, slip lengths even smaller, but non-vanishing (of the order of nanometers) have also been observed, particularly along non-wetting walls. For the remainder of this text, we will neglect these effects and assume that the slip length is zero.

4.3.2 Boundary conditions at the interface between two fluids: surface tension effects

In addition to Equation 4.32 describing the continuity of the velocities, we must find a condition for the continuity of the stresses (forces per unit area) at the interface between two fluids. We need in fact to have an equilibrium between the stresses within each of the two liquids and the stresses localized at the surface.

The condition that the stresses must be equal is also valid at the surface separating a fluid and a solid. In particular, it will provide interesting information in the case of a solid wall which is easily deformed, such as a gel or rubber; it will indeed allow us to evaluate the deformation caused by flow by writing the equality between the elastic stress corresponding to the deformation of the solid, and the viscous stress exerted by the fluid.

- At the interface between two perfect fluids, the normal stresses are associated only to the pressure and their equilibrium is expressed by the Young–Laplace law (Equation 1.58). We have, between the pressures p_1 and p_2 in the two fluids, the relationship:

$$p_1 - p_2 = \gamma\left(\frac{1}{R} + \frac{1}{R'}\right), \quad (4.33a)$$

where γ is the surface–tension coefficient between fluid 1 and fluid 2, while R and R' are the principal radii of curvature of the interface. We recall that the pressure is higher on the concave side of the interface. In the more general case of a viscous fluid, the equilibrium equation must include viscous stresses and is expressed by:

The viscous stresses normal to interfaces or solid walls are very often neglected but may become important in configurations like stagnation points (Section 3.3.2. and 10.5.3), viscous jets with free surfaces (Section 8.3) or very viscous fluids.

$$\left([\sigma]^{(2)}\cdot\mathbf{n}\right)\cdot\mathbf{n} - \left([\sigma]^{(1)}\cdot\mathbf{n}\right)\cdot\mathbf{n} = \gamma\left(\frac{1}{R} + \frac{1}{R'}\right). \quad (4.33b)$$

In the above equation, the tensor products $[\sigma]^{(1)}\cdot\mathbf{n}$ and $[\sigma]^{(2)}\cdot\mathbf{n}$ represent the stresses observed on an interface with normal $\mathbf{n}$: their scalar product with the unit vector $\mathbf{n}$ normal to the interface gives the normal components of the stresses. Indices (1) and (2) refer to the two fluids in contact at the interface.

- Moreover, the equilibrium between the tangential stresses at the interface is expressed by:

Equations 4.34 and 4.35 are valid provided the surface tension coefficient γ is constant over the interface between the two fluids. We shall discuss in Section 8.2.4 the so-called Marangoni effects observed when this condition is not satisfied.

$$\left([\sigma]^{(1)}\cdot\mathbf{n}\right)\cdot\mathbf{t} = \left([\sigma]^{(2)}\cdot\mathbf{n}\right)\cdot\mathbf{t}. \quad (4.34)$$

Here, the scalar product of the stresses with the unit vector $\mathbf{t}$ tangent to the interface gives the tangential components of the stresses. Equations 4.33 and 4.34 express the equality between the action and reaction of the forces acting on the liquids at the interface.

For an incompressible Newtonian fluid, Equation 4.16 relating the stresses to the velocity gradients assumes the form $\sigma'_{ij} = \eta(\partial v_i/\partial x_j + \partial v_j/\partial x_i)$. The continuity Equation 4.34 then becomes (Figure 4.6):

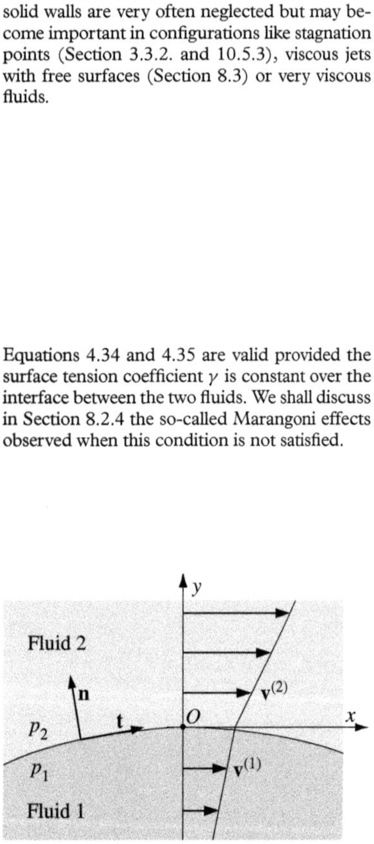

Figure 4.6 *Conditions at the interface between the viscous fluids* (1) *and* (2)

$$\eta_1\left(\left(\frac{\partial v_i^{(1)}}{\partial x_j} + \frac{\partial v_j^{(1)}}{\partial x_i}\right)n_i\right)t_j = \eta_2\left(\left(\frac{\partial v_i^{(2)}}{\partial x_j} + \frac{\partial v_j^{(2)}}{\partial x_i}\right)n_i\right)t_j, \quad (4.35)$$

where t_i (or n_i) represent respectively the components of a unit vector $\mathbf{t}$ (or $\mathbf{n}$) tangent (or normal) to the interface (Equation 4.33b may be rewritten by using the same procedure). Let us assume, specifically, that we replace the interface between the two fluids near the point O by a z-x plane, and that the velocity $\mathbf{v}$ is locally only in the x-direction and a function of y only (Figure 4.6). For this simple geometry, the expression for the tangential stress on the interface reduces to the single term $\sigma_{xy} = \eta\ \partial v_x/\partial y$ where $v_x(y)$ is the non-vanishing component of the velocity. The condition that the tangential stresses be equal leads then immediately to the condition:

$$\eta_1\frac{\partial v_x^{(1)}}{\partial y} = \eta_2\frac{\partial v_x^{(2)}}{\partial y}. \quad (4.36)$$

This simply states that the velocity gradients at the interface are inversely proportional to the dynamic viscosities of the two fluids.

Equation 4.36 turns out to be even simpler if one of the two fluids is a gas, the interface is then referred to as a *free surface*. In that case, the very low viscosity of gases in comparison of that of liquids allows us to write that the tangential stress in the liquid at the interface is essentially zero, so that (just as for an ideal fluid):

$$([\boldsymbol{\sigma}']^{(\text{liquid})} \cdot \mathbf{n}) \cdot \mathbf{t} = 0. \qquad (4.37)$$

For the specific example just treated this would imply $\partial v_x / \partial y = 0$ in the liquid at the free surface.

4.4 Non-Newtonian fluids

Up to now, we have limited our discussion to the case of *Newtonian fluids* for which there is a direct proportionality between the stresses and the rates of strain. We will now discuss the case of *non-Newtonian fluids* for which this relationship is no longer linear and can, moreover, depend on the history of the flow. These properties frequently result from the presence in the fluid of microscopic objects which are large relative to an atomic scale (while remaining small relative to the global scale of the flow): macromolecules in polymer solutions, particles in suspensions, droplets or vesicles (liquid encased in a membrane) in emulsions and biological fluids. These objects can even form yet larger structures which will have an important influence on the flow properties (aggregates of platelets in clays, clumps of particles or entangled macromolecules). Such fluids are frequently found in nature (snow, mud, blood, cream, etc.) as well as in everyday life (paints, shaving cream, mayonnaise, yoghurt, cosmetics, etc.) or in industrial settings (cement, etc.).

Understanding such flow characteristics requires the understanding of the response of such fluids to an applied stress. This is the purpose of *rheology*: this name dates only from the 1920s, and was suggested by E.C. Bingham who is considered, together with M. Reiner, as the founder of this branch of science. We might recall the statement attributed to Heraclitus: "panta rhei" (everything flows).

4.4.1 Measurement of rheological characteristics

For Newtonian or non-Newtonian fluids with time-independent characteristics, all we need to do is measure the relationship between the rate of deformation (strain) of the fluid and the stress which causes this strain. For the Newtonian case, a single experimental point is sufficient; but, in the case of non-Newtonian fluids, we need the entire curve. Furthermore, for fluids for which the response is time-dependent, and for viscoelastic fluids, we must analyse the time-dependent response of the fluid to an excitation which varies with time.

The experimental methods used can be very simple like the measurement of the flow rate through a tube at a given pressure, or of the velocity at which a small sphere falls within the fluid; measuring the time it takes for a certain volume of fluid to flow through a funnel is also an acceptable field test in many cases. However, the shear stress to which fluids are subjected in such instances is not constant. Such measurements are only quantitative in the case of Newtonian fluids (and, even for those, after careful calibration with well-known fluids).

Laboratory viscometers use geometries where the shear rate is best known and as constant as possible within the volume being measured. Figure 4.7 below illustrates a couple of the most commonly used geometries.

In the *Couette viscometer* with coaxial cylinders (Figure 4.7a), the fluid is placed between two concentric cylinders of which only one is in motion. For the simplest devices (known as constant shear), one measures the torque on one of the cylinders while fixing the speed of rotation on the second. In contrast, one can fix the torque and measure the angular velocity obtained in that manner; this last configuration, known as *applied stress*, is well adapted to fluids with a threshold (Section 4.4.2). The shear rate is almost constant throughout the volume between the cylinders, if the separation between them is small compared to their radii. Instabilities can appear at high angular velocities (Section 11.3.2), a fact that limits the domain over which we can measure shear stresses. Moreover, there can also occur parasitic effects resulting from secondary flows along the horizontal base of the cylinders; this accounts for the choice of the conical shape of the inside cylinder. Such an apparatus can be very sensitive in the case of measurements on low viscosity fluids. The most sensitive devices also allow one to make measurements in shear stress domains sufficiently small so that the internal structure of the fluids is undisturbed. Radii of the cylinders used vary from one to several centimetres and the gaps from one-tenth to a few millimetres (but larger devices are needed in order to analyze suspensions of large particles). In the case for which the inner radius R_1 and the outer radius $R_2 = R_1 + \Delta R$ are close ($\Delta R \ll R_1$) and, for cylinders of height h, the shear rate $\dot{\gamma}$ and the stress σ obey approximately:

$$\sigma = \frac{M}{2\,\pi\,R^2 h} \quad (4.38a) \qquad \text{and:} \qquad \dot{\gamma} = \frac{R}{\Delta R}\omega_0 \quad (4.38b)$$

(M is the moment of the torque applied to the cylinders, ω_0 the angular velocity of rotation, and R the average of the radii).

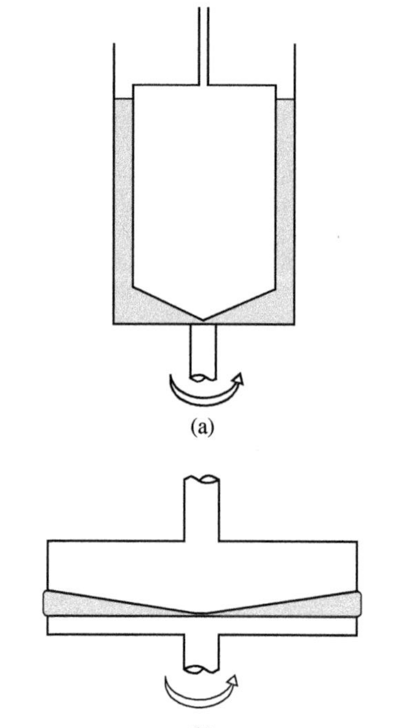

Figure 4.7 *(a) Cylindrical Couette viscometer; (b) plane-cone viscometer. A constant angular velocity is applied to one of the solids, and the resulting torque is then measured either on that surface, or on the facing one*

The *cone-plate viscometer* (Figure 4.7b) requires very little liquid and is much easier to use. The cone angles α are very small ($\alpha \leq 4°$). The upper cone is set in rotation while leaving the lower cone fixed, or vice versa: the shear rate is almost constant throughout the volume (except close to the point which is slightly truncated) if the tip of the cone coincides with the lower plane. Effectively, the thickness and the tangential velocity both increase linearly with the distance from the axis of rotation. Using the same notation as in Equations 4.38a and 4.38b, we find:

$$\sigma = \frac{3M}{2\,\pi\,R^3} \quad (4.39a) \qquad \text{and:} \qquad \dot{\gamma} = \frac{\omega_0}{\alpha}. \quad (4.39b)$$

Here again, hydrodynamic instabilities can be an impediment at high velocities, and the presence of a free surface along the sides favors evaporation. The plane-cone apparatus is in general less sensitive and only allows one to study velocities and/or shear rates larger than those for the Couette viscometers. Some viscometers allow for the measurement of stresses normal to the moving surface: such measurements allow one to detect possible elastic properties of fluids (Section 4.4.5).

4.4.2 Time-independent non-Newtonian fluids

Generally we characterize rheological properties by the manner in which the shear stress σ varies with the shear rate $\dot{\gamma}$ ($\dot{\gamma} = \partial v_x/\partial y$ for a simple shear flow). Figure 4.8 displays, on a linear scale, typical variations for different kinds of non-Newtonian fluids. In this figure, we have made the implicit assumption that the dependence of σ on $\dot{\gamma}$ is time-independent. For many of the fluids to which we refer in Section 4.4.3, this assumption is not valid.

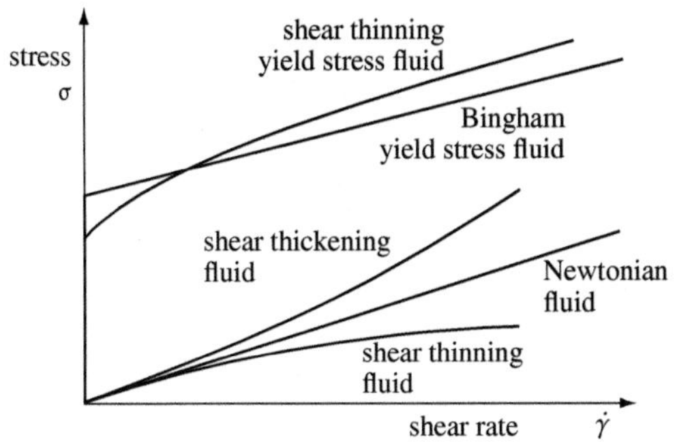

Figure 4.8 *Dependence of the stress on the shear rate for various kinds of fluids*

Shear thinning fluids

These fluids flow even under weak stress and are known as *shear thinning fluids* with an effective viscosity $\eta_{\text{eff}} = \sigma / \dot{\gamma}$ decreasing when the shear rate increases (the subscript eff is omitted in the following); these fluids are also sometimes called *pseudoplastic*. Many dilute solutions of high molecular mass polymers behave in this way, a behavior attributed to the entanglement of macromolecules which separate progressively and align as a result of the shear.

A number of dilute suspensions of solid particles are also shear thinning; the result in this case of destruction by the flow of structures originally created by attraction between particles. Other examples are shampoos, or fruit juice concentrates. Likewise, printer's inks, consisting of solid pigments in suspension in complex liquids, display similar characteristics.

Figure 4.9 shows typical rheological characteristics obtained for solutions with different concentrations of a polymer produced industrially by bacterial fermentation (scleroglucan of the polysaccharide family); this polymer is used in many chemical and agroalimentary applications. Instead of showing the variation of the stress σ as a function of the shear rate $\dot{\gamma}$, as in Figure 4.8, we display in log–log coordinates the variation of the effective viscosity $\eta = \sigma / \dot{\gamma}$ as a function of the shear rate. These curves allow us to display the deviations from Newtonian behavior. For water, this latter characteristic would appear as a horizontal line, corresponding to a constant viscosity of the order of 1 mPa.s.

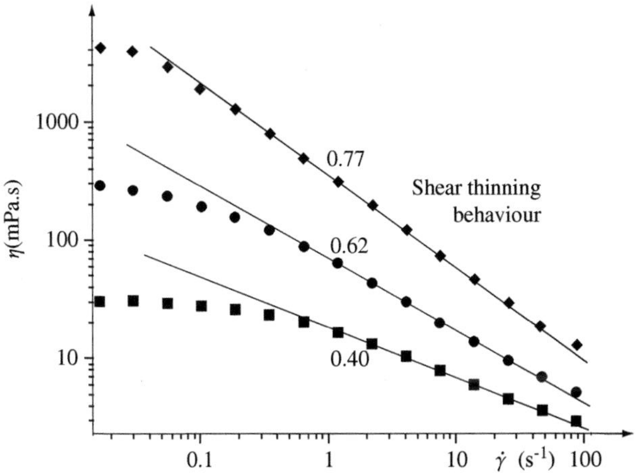

Figure 4.9 *Measurements of the effective viscosity $\eta = \sigma / \dot{\gamma}$ as a function of the shear rate $\dot{\gamma}$ carried out in steady state flow in a Couette viscometer for aqueous solutions of a polymer with relative mass concentration increasing upward (from 5×10^{-4} to 3×10^{-3}). The numbers labeling each straight line are the values of the exponent α giving the variation of the viscosity with $\dot{\gamma}$ (document courtesy C. Allain, A. Paterson, A. d'Onofrio)*

For small values of $\dot{\gamma}$, the viscosity tends toward a constant value η_0 ("*Newtonian plateau*"): η_0 increases with the polymer concentration and can be several thousand times larger than the viscosity of the solvent (water). The effective viscosity η decreases with $\dot{\gamma}$ following a power law:

$$\eta = D\,\dot{\gamma}^{-\alpha} \tag{4.40}$$

(represented by a straight line in log–log coordinates) in which $\alpha(>0)$ increases with the concentration. Moreover, the range of shear rates $\dot{\gamma}$ over which Equation 4.40 is satisfied becomes broader as the concentration of the solution increases (the lower limit of the range becomes indeed lower). At the largest shear rates, the viscosity of the solution remains, in all cases, greater than that of the solvent.

Just as for almost all non-Newtonian fluids, the variation of the viscosity reflects an evolution of the internal structure of the fluid (here, a rearranging of the macromolecules). The rheological characteristics of the solutions corresponding to Figure 4.9 vary only little when they are subjected to cyclical increases and decreases of the shear rate. Such behavior is, however, not the general case, and the rheological characteristics of complex fluids frequently evolve in a significant manner when subjected to cyclical variations of the stress.

Shear thickening fluids

These are fluids which do not have a flow threshold and for which the viscosity increases with the applied stress (they are also known as *dilatant fluids*). Some polymer solutions display this behavior: if the macromolecules are initially wound up in balls, the stresses due to the flow can unwind them into long chains, leading to an increased effective viscosity. Other fluids display a sequence of shear thinning followed by shear thickening regimes as the shear rate is increased; we can thus think of polymer chains that unwind and align, reducing the effective viscosity η, until they begin to interact which, in contrast, increases η.

Bingham fluids

Yield stress fluids do not flow until the applied stress is higher than a critical value, σ_c (these are sometimes also known as *plastic fluids*). A number of concentrated suspensions of solids in a liquid, and some polymer solutions, display a flow threshold (often called *yield stress*) beyond which the shear rate is not zero and increases with the stress. The theoretical model of a *Bingham fluid* assumes that this latter variation is linear. For real fluids, we frequently observe a variation which more closely approaches, beyond the threshold, a power law.

Let us assume an increasing pressure head is applied to such a fluid filling a cylindrical tube: we observe, immediately above a threshold value, a solid-like flow of almost all the fluid with a resultant velocity independent of the distance from the walls. The velocity gradients are localized in the immediate neighborhood of the walls, because it is the only region where the stress required to achieve shear flow is achieved. In such a case, we talk about *plug flow*. As the stress is further increased, the velocity gradient becomes progressively non-zero throughout the volume. We present, in Section 4.6.3, a quantitative calculation of such profiles.

We can interpret this behavior as the destruction of three-dimensional structures in the fluid which are generally formed when the fluid is at rest; clays have, for instance, a microscopic structure of platelets. In the absence of flow, the platelets form rigid structures, held together by weak interactions which hold until a certain threshold of stress is attained. Beyond this threshold, the structure is partly broken up, and flow becomes possible; the higher the velocity, the more the structure is broken down, while the platelets align in the direction of the flow. As a result, the increase of the stress with the shear rate is slower than that which would result from a linear relationship. We then refer to these as *shear thinning threshold fluids* (often called *Herschel–Bulkley fluids*, or *Casson fluids*). Some suspensions of colloidal particles also display these properties; in a similar manner, paints should spread easily when a paintbrush subjects them to shear, but they should not drip spontaneously after having been applied.

A precise measurement of the yield stress in such fluids can be obtained by analyzing their *creep*: we increase the applied stress in stepwise fashion, and look for the value corresponding to the step at which the shear rate in the stationary regime is no longer zero.

Figure 4.10 displays the relation between the stress and the shear rate in semi-log coordinates for a mixture of kaolin and water which corresponds to a suspension of clay particles: as the stress is gradually increased from zero, we observe no flow until a threshold value of the order of 10 Pa is reached.

Drilling muds represent a particularly illustrative example of a practical use of these fluids: these muds are injected into the bottom of the *borehole* at the level of the *wellhead* through an assembly of hollow tubes which force the rotation of the *drill bit* all the way from the surface, and then they flow back to the surface in the course of the drilling. They must be able to flow easily when a pumping pressure is applied, but they should also have sufficient resistance to shear so that they will carry along the *drill cuttings* toward the surface, and prevent their falling back down whenever the flow comes to a halt.

Among other yield stress fluids with important practical applications are creams and emulsions used in the cosmetic industry, as well as toothpastes. Another example is fresh cement (for which the yield stress is of the order of a few tens of Pascals), and quite a number of products in the food industry.

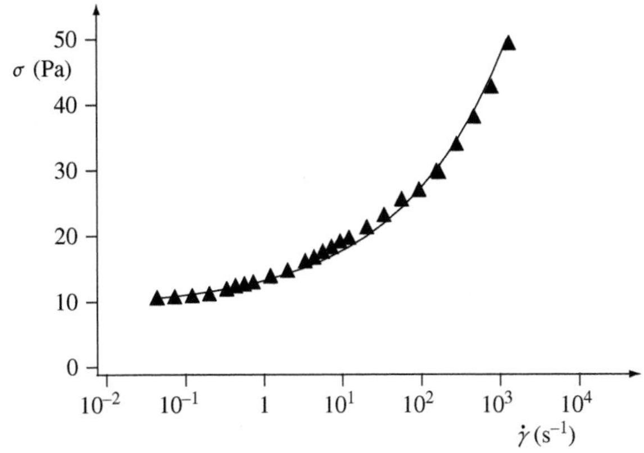

Figure 4.10 *Dependence of the shear rate on the stress (in semi-log coordinates) for a water–kaolin mixture. The solid triangles represent experimental data. The continuous curve corresponds to Equation 4.43 with the values* $\sigma_c = 9$ Pa, $n = 0.32$ *and* $K = 4.2$ Pa.s$^{-0.32}$ *(document courtesy P. Coussot)*

Approximate rheological behavior

For quite a number of practical applications, we need approximate analytical relationships between the stress and the shear rate. As we have just seen, a variation of the type of Equation 4.40, initially suggested by Ostwald, provides an approximation valid for a number of cases. An exponent $\alpha > 0$ corresponding to the shear thinning fluids (most often observed), while $\alpha < 0$ most often corresponds to shear thickening fluids. For $\alpha > 0$, Equation 4.40 predicts that η becomes infinite when the shear rate $\dot{\gamma}$ approaches zero, while we observe that η approaches instead a constant limit (often called *zero shear rate viscosity* or *Newtonian plateau*) as seen in Figure 4.9. In order to display the complete curve and, particularly, the limiting values for high and low values of $\dot{\gamma}$, we use more complex equations of the type:

$$\frac{\eta(\dot{\gamma}) - \eta_\infty}{\eta_0 - \eta_\infty} = f(\dot{\gamma}), \tag{4.41}$$

e.g., where the function $f(\dot{\gamma})$ is, for instance, of the type $f(\dot{\gamma}) = (1 + \beta^2\dot{\gamma}^2)^{-p}$ (Carreau relation).

The previous relationships predict a finite viscosity for $\dot{\gamma} = 0$, so that they are in fact not applicable to yield stress fluids. We must then use other equations: the simplest is that of Bingham, which assumes that the shear rate is proportional to the difference $\sigma - \sigma_c$, as we exceed the threshold value σ_c. Few fluids display such a simple characteristic behavior, and the equation relating stress to shear rate often displays a curvature directed downwards (stress thinning threshold fluids). One frequently represents such a behavior by equations combining the threshold effect with a power law, such as the Herschel–Bulkley relation:

$$\sigma \leq \sigma_c : \dot{\gamma} = 0 \qquad (4.42) \qquad \text{and} \qquad \sigma > \sigma_c : \sigma = \sigma_c + K\dot{\gamma}^n. \qquad (4.43)$$

This equation reproduces well the rheological characteristics of the water–kaolin mixtures displayed in Figure 4.10.

4.4.3 Non-Newtonian time-dependent fluids

Thixotropic fluids and characteristic time-scales

Thixotropic fluids display an effective viscosity which decreases with time under the action of a constant stress. Normal viscosity is restored only after some time has elapsed. Concentrated polymer solutions, and suspensions of particles, are typical examples of such behavior.

In fact, both thixotropic behavior and shear thinning behavior indicate a change of the internal structure of the fluid as it flows (quite a number of fluids are thus simultaneously *shear thinning* and *thixotropic*); the difference between these two concepts results from the relative values of the characteristic times τ_{De} for this structure to rearrange, and T, the time over which the stress is applied. The ratio:

Marcus Reiner who, together with E.C. Bingham, is considered one of the fathers of rheology, defined the *Deborah number* by referring to the bibical prophetes Deborah (Judges 5:5) who, after a victory over the Philistines sang: "the mountains flowed before the Lord." If, on a human scale, mountains appear indestructible, they nevertheless end up being deformed, and even disappearing, over the, infinitely long, time-scale of observation by the divine. It is only a question of the scale of characteristic times!

$$De = \frac{\tau_{De}}{T} \qquad (4.44)$$

of these two characteristic times is known as the *Deborah number*.

When De is small compared to unity, the fluid has time to rearrange its structure as the stress is varied (or as the applied shear rate $\dot{\gamma}$ is changed). For a shear thinning fluid, the apparent viscosity would decrease with $\dot{\gamma}$, and the value thus obtained would be independent of the measurement; moreover, if we consider a cyclical increase and decrease of the shear rate, the relationship between stress and shear rate is always the same. We can then describe this behavior by curves of the type shown in Figure 4.8. In practical terms, however, this description only applies to certain fluids, and to steady-state or slowly changing flows.

On the other hand, for a Deborah number De much greater than unity, the rheological properties change during a time of the order of τ_{De} as the structure of the fluid changes. In the same way, when one describes a sequence of increasing and decreasing shear rates, one observes hysteresis effects and ends up describing, after a transient regime, a limit cycle for which the characteristic curves corresponding to changes in one direction or the other do not coincide. Such behavior is a manifestation of thixotropic properties.

One also often encounters, in the literature describing such fluids, the *Weissenberg number We*, defined as the product $\dot{\gamma}\tau_{De}$. This number characterizes processes in which the fluid evolves as a function of time, when subjected to a constant shear rate $\dot{\gamma}$.

In the case of polymers, thixotropic behavior indicates most often, on a microscopic scale, the disentangling of clusters of macromolecules. In the case of suspensions, this can indicate the destruction of structures of particles held together by electrostatic attraction forces, or van der Waals forces. Several solutions of clays (bentonites) described in the previous section display similar behavior. A measurement of the torque required to maintain a given angular velocity shows that this torque decreases with time (on a time scale of several minutes). If the angular velocity is then reduced to zero, the fluid displays a stress-strain relation quite different from that originally observed: a change appears to have occurred in the structure of the fluid, and the original characteristics are reproduced only if the fluid is left at rest for several hours. This evolution of the internal structure of the flowing fluid is the more significant, the greater the applied velocity gradients.

Thixotropic fluids have numerous practical applications: paints and drilling muds, already described, are strongly thixotropic, an observation which reinforces the effect of the shear thinning characteristics that we have encountered above. Food products, such as ketchup, need to be shaken vigorously before they flow easily.

A few fluids display opposite characteristics to the ones we have just described, and become more viscous as a function of time when subjected to shear: one talks then about *anti-thixotropic fluids*, of which gypsum paste is an example. One also encounters the term *rheopexy*, which characterizes the progressive solidification of fluids as they are stirred.

Viscoelasticity

Viscoelasticity corresponds to a behavior intermediate between that of an elastic solid (the strain proportional to the stress, and related to it by the elastic modulus) and that of a liquid (the rate of deformation increasing with the stress). A particularly spectacular example of this is the case of the silicone *silly putty* balls (Figure 4.11) which bounce elastically off the ground (a′, b′, c′, d′), but spread out like a liquid if allowed to rest on a flat surface for a long enough time (a, b, c): when the rate at which stresses change is high (as in the case of an impact) the internal structure of the substance has no time to rearrange itself during the short time of the impact, and the material responds as an elastic solid. The corresponding energy will be stored, for example, by changes in the orientation of macromolecules in polymers.

However, if the *silly putty* is placed on a flat surface, the stress is then constant, and the putty spreads out like a liquid as a result of the rearrangements in the internal structure, on a time-scale corresponding to τ_{De}. These characteristics differ markedly from those of thixotropic fluids, which behave on short time-scales as very viscous liquids.

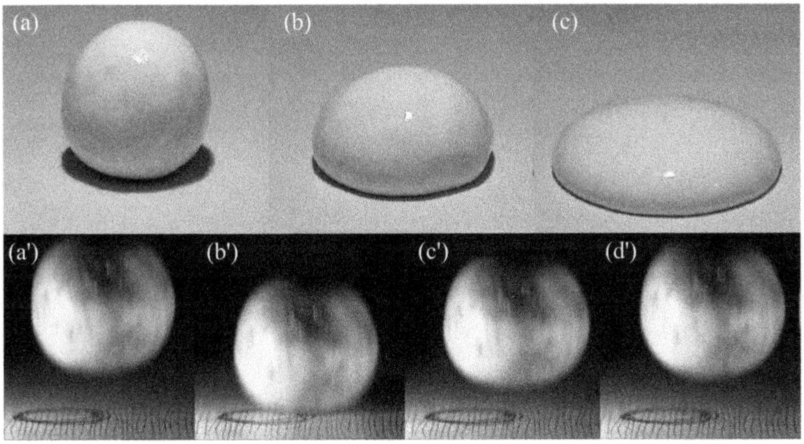

Figure 4.11 *A ball of silicone paste (silly putty) behaves very differently depending on the length of time it is subjected to a stress. (a,b,c): It spreads like a liquid if it is allowed to rest for a long while on a flat surface. (a′, b′, c′, d′): It bounces like an elastic ball if we let it drop from a height, thus subjecting it to a short, intense stress (figures from R. Lehoucq (a–c) and authors (a′–d′))*

Concentrated aqueous solutions of polymers (e.g., a solution of polyoxyethylene at mass concentrations of a few parts per thousand), bread dough and man-made textile fibers (Nylon, Kevlar, etc...) are other examples of viscoelastic fluids. Occasionally, the characteristics of viscoelastic materials approach those of solids. This is the case, for example, for jellies and for some types of foams.

Finally, viscoelastic fluids display differences between the usual components parallel and perpendicular to the plane of the shear. This effect, also seen in other kinds of fluids, will be analyzed in Section 4.4.5.

Other factors which influence the time-dependence of stresses

One also observes time-dependent behaviors in substances which have quite different rheological characteristics, not necessarily viscoelastic ones. Thus, a concentrated solution of *corn starch* in water behaves like an elastic solid when struck sharply with a spoon, while the same spoon can be inserted easily if pushed in slowly (or it will even sink, if it is just left to lie on the surface). In this example, the relaxation time of the structures is very short. It is much longer in the case of the earth's mantle, which behaves as a viscous fluid in the course of convective tectonic motion, but which is evidently solid over ordinary time-scales.

4.4.4 Complex viscosity and elasticity of viscoelastic fluids

Measurement of time-dependent characteristics of viscoelastic fluids

A first type of measurement consists of creating a stepwise change in the stress, or in the shear rate, and measuring the corresponding changes in the shear rate (or in the stress). For example, consider a sudden change (Figure 4.12a) in the torque Γ applied to one of the cylinders of a Couette viscometer. For a classical Newtonian fluid, the velocity of rotation would stabilize almost instantaneously at a new constant value after changing the torque. In contrast, in the case of a viscoelastic fluid, the angular rotation speed ω of the cylinder increases greatly at first, then undergoes a relaxation toward a smaller constant value (Figure 4.12b), or to a value which varies very slowly with time. The opposite effect is observed when the torque is suddenly dropped to zero: we might then see the cylinder rotating slightly in the opposite direction, before stopping suddenly as though it were held by a rubber torsion wire, representing the elasticity of the liquid (*recovery phenomenon*). The determination of the response time to excitations, in such measurements, allows us to estimate the time τ_{De} for the structure to adapt.

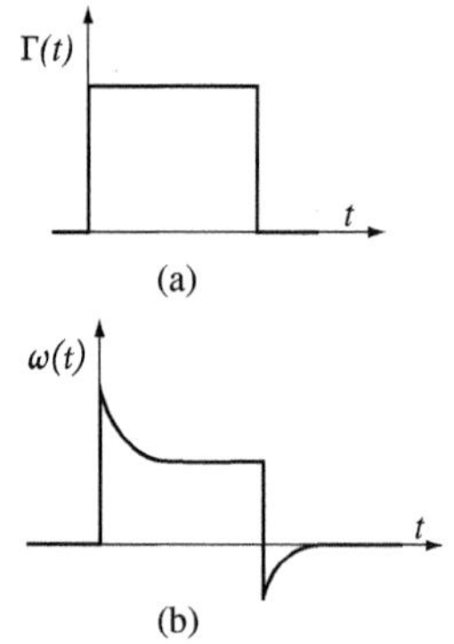

Figure 4.12 *Simplified schematic response of a viscoelastic fluid to a sudden change in the applied stress in a Couette viscometer. (a) Sudden change in the torque* Γ *applied by the viscometer; (b) corresponding changes in the angular velocity* $\omega(t)$

The kind of response observed depends on several characteristic times: the characteristic time(s) of the response of the structure of the fluid, the characteristic time for stress to change (or, for the shear rate to change), the inverse of the shear rate. According to their relative values (and the inherent non-linearities of the fluid), we might observe several kinds of behavior.

Finally, another technique for studying time-dependent fluids consists of applying to them sinusoidal variations of the stress (or of the shear rate) and measuring the corresponding changes in the complementary variable as a function of time. For small deformations, these are also sinusoidal but display a phase shift relative to the excitation: we must therefore measure both the amplitude and the phase of the signal for a given excitation. The response to these sinusoidal excitations, which we will describe in more detail below, is highly variable depending on the relative orders of magnitude of the period of the excitation, and on the time for the structure to rearrange internally (thus according to the Deborah number); it also depends on the amplitude of the deformations that result.

Definition of the characteristic coefficients

Measurements made with a sinusoidal excitation allow us to determine the complex modulus $\bar{G}(\omega)$ which relates the complex stress $\bar{\sigma}(t) = \sigma_0(\omega)\, e^{i(\omega t+\varphi)}$ and the complex strain $\bar{\gamma}(t) = \gamma_0(\omega)\, e^{i\omega t}$ with:

$$\bar{\sigma}(t) = \bar{G}(\omega)\,\bar{\gamma}(t) = (G'(\omega) + i\,G''(\omega))\,\bar{\gamma}(t) \tag{4.45}$$

(observe that $\bar{\gamma}(t)$ is the strain and not the shear rate which is, in fact, its derivative with respect to time). $\bar{G}(\omega)$ is known as the *complex modulus of rigidity*: its real part, directly related to the elasticity of the medium, is the *storage modulus* and its imaginary part, related to the viscosity, is the *loss modulus*. For an elastic solid:

$$G'(\omega) = G \qquad (4.46a) \qquad \text{and:} \qquad G''(\omega) = 0. \qquad (4.46b)$$

where G is the shear modulus of the material. For a Newtonian viscous fluid, we have the complex notation: $\bar{\dot{\gamma}}(t) = \mathrm{i}\,\omega\,\bar{\gamma}(t)$. Considering the usual equation $\bar{\sigma} = \eta\,\bar{\dot{\gamma}}\,(t)$, we have:

$$G'(\omega) = 0 \qquad (4.47a) \qquad \text{and:} \qquad G''(\omega) = \eta\,\omega. \qquad (4.47b)$$

For fluids which are essentially viscous but with a small viscoelastic component, it is more significant (but equally valid) to introduce a complex viscosity:

$$\bar{\eta}(\omega) = \eta'(\omega) - \mathrm{i}\,\eta''(\omega) = \frac{\bar{\sigma}(t)}{\bar{\dot{\gamma}}(t)} = \frac{\bar{\sigma}(t)}{\mathrm{i}\,\omega\,\bar{\gamma}(t)} = \frac{\bar{G}(\omega)}{\omega} = \frac{G''(\omega)}{\omega} - \mathrm{i}\frac{G'(\omega)}{\omega} \qquad (4.48)$$

Using these definitions we obtain the equations:

$$\eta'(\omega) = \frac{G''(\omega)}{\omega} \qquad (4.49a) \qquad \text{and:} \qquad \eta''(\omega) = \frac{G'(\omega)}{\omega}. \qquad (4.49b)$$

For a purely viscous fluid (without any elasticity), we find, as expected, $\eta'(\omega) = \eta$ and $\eta''(\omega) = 0$.

The coefficients $G'(\omega)$, $G''(\omega)$, $\eta'(\omega)$ and $\eta''(\omega)$ are frequency dependent, but they also depend on the amplitude of the excitation. A study of these fluids implies an analysis of the effects of these two parameters. Thus, the frequency of the transition between the viscous low frequency regime ($De \ll 1$) and the elastic regime at high frequencies provides an estimate of the Deborah number and, then, of the characteristic response times of the internal structure of the fluid.

Mechanical models of viscoelastic fluids

Just as in the case of the stress-strain dependence of fluids which are time-independent, it is helpful to have approximate analytic expressions for viscoelastic fluids. We can characterize the variation of G' and G'' with the excitation frequency ω by means of linear mechanical models, such as the model of the *Maxwell liquid*, which do not attempt to explain the internal structure of the material.

The Maxwell model is approximately valid for the case of a linear viscoelastic fluid which behaves as a fluid of viscosity η at small angular frequencies ω, and as a solid of elastic modulus G when the frequencies are higher (qualitatively, this corresponds to the behavior of the silly putty described above). From the viewpoint of the stress–strain relation, the material behaves like a spring (stress $\bar{\sigma}$ proportional to the strain with $\bar{\sigma} = G\bar{\gamma}$) in series with a damping mechanism (stress $\bar{\sigma} = \eta\bar{\dot{\gamma}} = \mathrm{i}\,\omega\,\eta\,\bar{\gamma}$ proportional to the shear rate $\bar{\dot{\gamma}}$) (Figure 4.13a).

For excitations at high frequencies, the damping mechanism has no time to react, and the system then behaves like a spring; at very low frequencies, the spring is hardly deformed, and only the damping mechanism is in play. Note that this is only a mechanical analog model, which in no way attempts to describe the internal structure of a material, but only to mimic its overall behavior. In this model we observe the effect of the

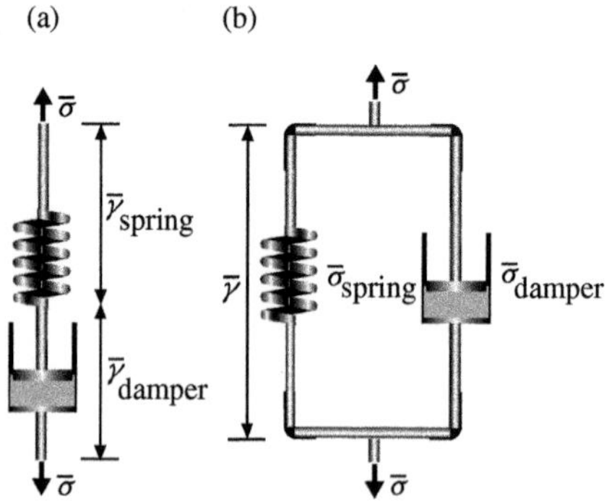

Figure 4.13 *(a) Mechanical model of a Maxwell liquid with a spring and damper in series; (b) mechanical model for a Kelvin–Voigt solid*

applied stress $\bar{\sigma}$ equally on the spring and on the damping mechanism; the two deformations $\bar{\gamma}_{spring}$ and $\bar{\gamma}_{damper}$ add up, with the result:

$$\bar{\gamma}_{Maxwell} = \bar{\gamma}_{spring} + \bar{\gamma}_{damper} = \frac{\bar{\sigma}}{G} + \frac{1}{i\omega}\frac{\bar{\sigma}}{\eta} = \frac{\bar{\sigma}}{G' + iG''}. \tag{4.50}$$

We thus have:

$$\bar{G} = G' + iG'' = \frac{i\omega\eta}{1 + i\omega(\eta/G)} = \frac{i\omega\eta}{1 + i\omega\tau}, \tag{4.51a}$$

where $\tau = \eta/G$ corresponds to a characteristic relaxation time of the fluid, which determines the transition frequency between the frequency regimes where the fluid appears as a liquid ($\omega \ll 1/\tau$; i.e. $De \ll 1$), or as an elastic solid $\omega \gg 1/\tau$; i.e. $De \gg 1$). Using expressions 4.49a-b, Equation 4.51a can be rewritten in terms of the complex viscosity:

$$\bar{\eta}(\omega) = \eta'(\omega) - i\eta''(\omega) = \frac{\eta}{1 + i\omega\tau}. \tag{4.51b}$$

The model of a *Kelvin–Voigt* solid behaves on the other hand like an elastic solid (with elastic modulus G) for slow changes in the excitation, and like a viscous liquid (viscosity η) for short-time excitations. This material can also be modeled by a spring, this time in parallel with a damper (Figure 4.13b). By an analysis similar to the preceding one, the equation for the complex elastic modulus $\bar{G}$ is now:

$$G' + iG'' = G + i\omega\eta = G(1 + i\omega\tau). \tag{4.52}$$

Real fluids rarely behave like these two limiting cases and frequently involve several characteristic times governing the rearrangement of their structures. More complex models have been developed to describe their behavior and, particularly, to take into account the effect of the amplitude of the deformations.

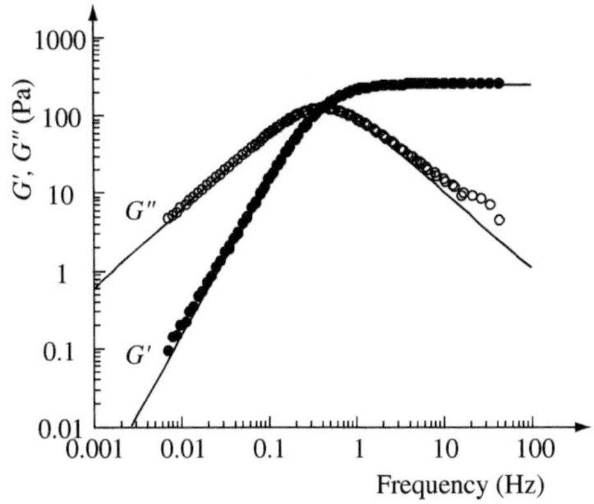

Figure 4.14 *Frequency dependence of the magnitudes G′ and G″ for a solution of giant micelles in log-log coordinates. The continuous curves correspond to the Maxwell model (Equation 4.51a) adjusted for parameter values* $\tau = 0.38$ s *and* $G = 249$ Pa *(document courtesy J.-F. Berret, G. Porte and J.-P. Decruppe)*

Example of the characteristics of a viscoelastic fluid

Figure 4.14 displays the rheological characteristics of a fluid containing giant micelles: this is the case for solutions of molecules with active surface tension compounds which gather in worm-like structures of very small diameter (two molecules wide) but very long. These structures are sometimes called "living structures" because they permanently separate and reattach dynamically. On the figure, we clearly differentiate between two types of behavior, depending on the frequency of the flow. For frequencies below 0.2 Hz, the solutions behave essentially like viscous fluids with a loss modulus G'' increasing linearly with the frequency (this corresponds to the imaginary part of Equation 4.51a); the effect of the loss modulus dominates, in this case, that of the storage modulus G' which varies as the square of the frequency. At frequencies above 2 Hz, the storage modulus G' is effectively constant, while the loss modulus G'' decreases with frequency as f^{-1}. In this region, the behavior of the fluid is similar to that of an elastic solid. At still higher frequencies, above 10 Hz, the G'' component deviates upward relative to Equation 4.51a because of the presence of new energy loss mechanisms.

4.4.5 Anisotropic normal stresses

Up to this point we have discussed different types of stress–strain relations, whether or not time-dependent, observed in the case of non-Newtonian fluids by analyzing only the behavior of the shear viscosity. We will now describe other phenomena associated to anisotropic normal stresses or to the elongational viscosity and which involve other components of the stress tensor, or other kinds of flows.

In a Newtonian fluid, there is no normal stress in parallel flows. On the other hand, if we consider a viscoelastic fluid flowing in a tube, we often observe at the exit of the tube (extrusion) an *extensional normal stress* in the direction of the flow, and an outward *pressure* in the perpendicular direction. The difference between these normal stresses accounts for the swelling of some polymer jet solutions or of molten polymers (solvent-free liquids at high temperatures), as they exit from an orifice. The final diameter of the jet increases with the rate of flow (Figures 4.15a–b) and can even reach the initial diameter several times over. Beyond a certain rate of flow, one can even observe retardation effects, where the jet only begins to swell after it has traveled some distance (Figure 4.15c). In the fabrication, by extrusion, of wires, or when plastic parts are manufactured by injection into a mold, such effects must be taken into account.

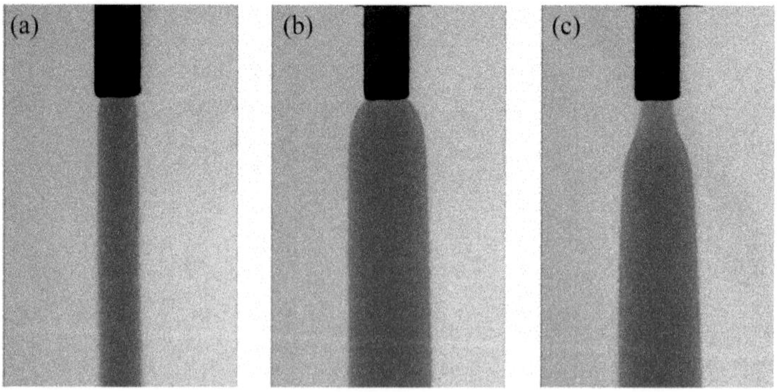

Figure 4.15 *Swelling at the exit of a capillary tube of a jet of an aqueous polymer solution injected into a salt-water solution of the same density. The flow rates are (a):* 5 ml/h, *(b)* 80 ml/h, *(c)* 120 ml/h. *(document C. Allain, P. Perrot, D. Senis, FAST)*

The *Weissenberg effect* is another classical demonstration of differences in the normal stresses. A viscoelastic fluid such as egg whites (or bread dough) rises along the axis of a rotating beater (Figure 4.16): there again, this rise is due to the nature of the normal stresses. In the case of a Newtonian fluid, one observes, in contrast, a depression of the free surface, due to the centrifugal force.

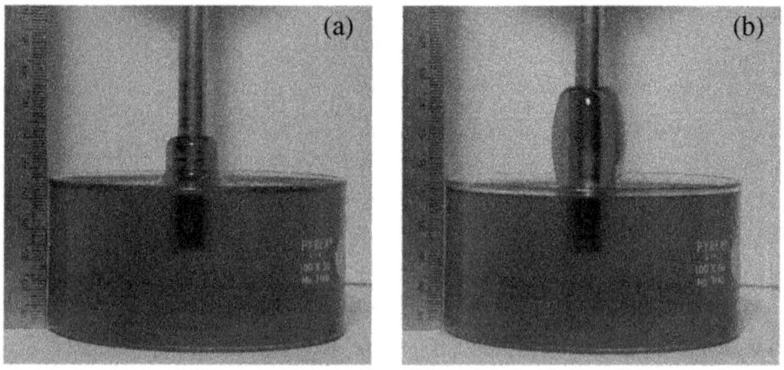

Figure 4.16C *Rise at two successive times, around a rotating axis, of a viscoelastic fluid (polystyrene dissolved in an organic solvent): this phenomenon demonstrates the Weissenberg effect (documents courtesy of J. Bico, G. Mc Kinley, MIT)*

Qualitative analysis of the Weissenberg effect: the main constituent of egg whites is albumin, which is a protein made up of a rather long chain of molecules which leads to a an elastic behavior of the egg whites; under the action of a stress, the chains are stretched and/or unrolled and require a "certain" time lapse before reaching back their equilibrium configuration. When the egg whites are caused to rotate, the macromolecules which are their constituents are stretched out around the axis of rotation as though they were rubber bands surrounding the axis. The resulting tension has the effect of "strangling" the fluid by pushing it toward the axis, a clear indication of the anisotropy of the normal stresses. If the elastic behavior of the fluid is sufficiently great, this phenomenon can compensate for, or even overcome, the effect of the centrifugal force resulting in an overpressure near the axis of rotation instead of a depression. This excess pressure results then in the fluid rising along this axis. A similar explanation can be made in the case of the swelling of jets of polymer solutions (Figure 4.15); the "rubber bands" are stretched as they flow through the extrusion orifice, and recover their length and shape by increasing their exit diameter and retaining their volume.

Finally, if normal stress anisotropies are often important in the case of viscoelastic polymer solutions, they can also be observed in other fluids such as suspensions of non-Brownian anistropic particles (e.g., rods greater than one micron in size) or solutions of giant micelles, such as those corresponding to the curves of Figure 4.14.

In quantitative terms, let us consider a two-dimensional shear flow $v_x(y)$. In the case of isotropic Newtonian fluid flow, the normal stresses σ_{xx}, σ_{yy} and σ_{zz} are all identical. On the other hand, in the flow of some non-Newtonian fluids, the difference:

$$N_1 = \sigma_{xx} - \sigma_{yy}, \tag{4.53a}$$

known as *the first normal stress difference*, is frequently large and positive (in some cases, it is as large as ten times the shear stress). The *second normal stress difference*, is defined as:

$$N_2 = \sigma_{yy} - \sigma_{zz}. \tag{4.53b}$$

Generally it has a much smaller (less than 10% of N_1) and negative value; it is taken equal to zero in some classical models (Weissenberg). N_1 and N_2 become zero in zero shear, and frequently obey a power law as a function of the shear rate $\dot{\gamma}$ (generally as $\dot{\gamma}^2$). Finally, time-dependent effects can also be observed as the normal stress varies: thus, in the experiment of Figure 4.15, the jet increases its diameter only beyond a threshold flow velocity, and at a distance downstream of the orifice which increases the greater the velocity.

Some viscometers allow for measurements of N_1 in the classical plane–cone geometry. The effects observed when the normal stresses differ can also be used to estimate indirectly these differences (the swelling of a jet, or the rise of a fluid around a rotating axis). These normal stress differences do not depend on the direction of the strain, as expected from their variation as $\dot{\gamma}^2$; the fluid is observed to rise independent of the direction of rotation.

4.4.6 Elongational viscosity

The flow of a fluid through a small opening displays elongational viscosity effects in which velocity gradients along the flow must be taken into account. Let us discuss here the case of a flow known as *elongational* between two identical disks which are parallel and coaxial, and which we pull apart. Figure 3.10a provides a schematic idea of a flow of this nature. The velocity is zero on the axis half-way between the disks (point O). Outside the median plane ($z = 0$), the axial component is directed toward the nearer disk, while the radial component is in the z-direction. A velocity field of the form:

$$v_x = -\dot{\varepsilon}\frac{x}{2}, \quad (4.54a) \qquad v_y = -\dot{\varepsilon}\frac{y}{2}, \quad (4.54b) \qquad v_z = \dot{\varepsilon}\, z \quad (4.54c)$$

is a good approximation for such a flow (it also satisfies the condition $\nabla \cdot \mathbf{v} = 0$). For such a velocity field with rotational symmetry around the z-axis, we can define an elongational viscosity (also called *Trouton viscosity*) by:

$$\eta_{el} = \frac{\sigma_{zz} - (\sigma_{xx}/2) - (\sigma_{yy}/2)}{\dot{\varepsilon}}. \tag{4.55}$$

For a Newtonian fluid, we have the simple relations $\sigma_{zz} = 2\eta\,\dot{\varepsilon}$, $\sigma_{xx} = -\eta\,\dot{\varepsilon}$, $\sigma_{yy} = -\eta\,\dot{\varepsilon}$, and we find that $\eta_{el} = 3\eta$. On the other hand, for a number of non-Newtonian fluids, specifically viscoelastic ones, we have just seen that the normal stress differences included in this definition can be significant. We call this phenomenon *strengthening* resulting from the strain: on a microscopic scale, it results, for polymer fluid, from an alignment of the macromolecules, and then of the classical behavior of these, when the molecular chains have been strongly stretched out. The viscosity η_{el} is difficult to measure because, as opposed to the previous cases, we cannot easily generate stationary elongational flows.

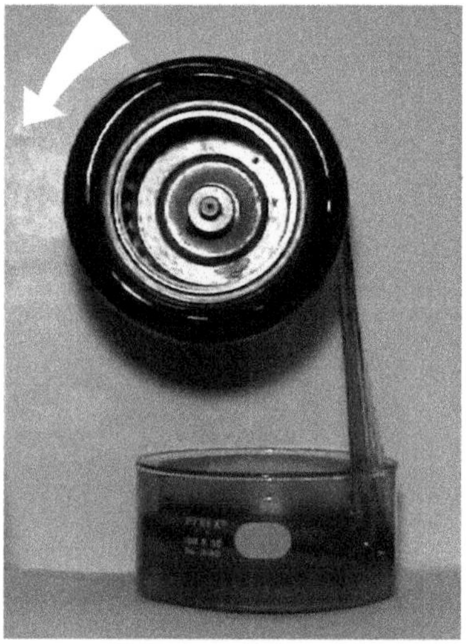

Figure 4.17C *The winding on a rotating spool of a free layer of the viscoelastic fluid shown in Figure 4.16 demonstrates the effect of its high elongational viscosity (documents courtesy of J. Bico, G. Mc Kinley, MIT)*

This result has important practical consequences for the synthetic textile industry. When we stretch a jet of ordinary liquid, capillary forces amplify the fluctuations in the diameter which appear spontaneously and eventually break up the jet into droplets (Rayleigh–Plateau instability described in Section 8.3.2). In contrast, the elongational viscosity of liquid viscoelastic polymers used in the creation of synthetic fibers is quite significant: this slows down the development of such instabilities and allows for the obtention of uniform diameter fibers, if the stretching is slowed down in the thinnest regions. A "thread" of such a fluid pulled upward from a free surface without any solid support can thus be rolled onto a spool above the liquid (Figure 4.17).

We finally point out that some flows display simultaneously the several effects described in this section: extrusion flows lead not only to the appearance of effects of the normal-stress anisotropy but also indicate an elongational component.

4.4.7 Summary of the principal kinds of non-Newtonian fluids

Table 4.1 presents a summary of the rheological properties which we have just defined in order to describe the properties of complex fluids such as (among many others): colloidal fluids, suspensions and polymers. Some of these fluids, for example yield stress and viscoelastics ones, have, as we have just seen, a behavior very close to that of a solid, under stresses which are weak and/or short-lived. On the other hand, some easily deformable solids

Table 4.1 *Various kinds of non-Newtonian fluids.*

Type	Name	Characteristic
Newtonian	Newtonian	$\eta_{\text{eff}} = \sigma/\dot{\gamma}$ constant as a function of $\dot{\gamma}$
Non-linear stress–shear rate relation	Shear thinning	η_{eff} decreases as $\dot{\gamma}$ increases
	Yield stress fluid	$\dot{\gamma} = 0$ below a threshold stress σ_c
	Shear thickening	η_{eff} increases with $\dot{\gamma}$
Time-dependent evolution under the action of a stress	Thixotropic	$\eta_{eff} \downarrow$ with time, under constant stress
	Rheopectic	$\eta_{eff} \uparrow$ with time, under constant stress
Mixed fluid-solid behavior	Viscoelastic	Elastic or viscous response dependent on the time characteristic of the excitation. High normal stresses

(such as gels) can "flow" under the action of strong and/or long-lasting stresses; they can also display viscosity when they are deformed (e.g., in the Kelvin–Voigt model discussed in Section 4.4.4) or, even, effects of surface tension.

One often talks of *soft matter*, an expression which has been introduced as a result of the seminal work of the French physicist Pierre Gilles de Gennes to describe the general family of all these materials. Their microscopic structure and their physico-chemical characteristics are generally the key to understanding their mechanical and flow properties.

4.5 One-dimensional flow of viscous Newtonian fluids

4.5.1 Navier–Stokes equation for one-dimensional flow

The Navier–Stokes Equation 4.30, which describes the motion of an incompressible Newtonian viscous fluid, results from a few simplying assumptions which we have described above. It does not have, in general, an analytic solution. This is often the case because of the presence of the non-linear term $\rho(\mathbf{v}.\boldsymbol{\nabla})\mathbf{v}$ which represents the exploring of the spatial variations of the velocity field by the particles of fluid.

This problem disappears for one-dimensional flow (often referred to as parallel flow). We will assume hereafter that the velocity is everywhere in the x-direction with:

$$v_y\,(x,\ y,\ z,\ t) = v_z\,(x,\ y,\ z,\ t)\ =\ 0. \tag{4.56}$$

Taking into account the above equations, the incompressibility condition $\boldsymbol{\nabla}\!\cdot\!\mathbf{v} = 0$ then reduces to:

$$\frac{\partial v_x}{\partial x} = 0, \quad (4.57) \quad \text{whence:} \quad (\mathbf{v}\cdot\boldsymbol{\nabla})\,\mathbf{v} = \left(v_x\frac{\partial}{\partial x}\ +\ v_y\frac{\partial}{\partial y}\ +\ v_z\frac{\partial}{\partial z}\right)\mathbf{v} \equiv 0, \quad (4.58)$$

because $\mathbf{v}$ then has only the component $(v_x(y,\,z,\,t),\,0,\,0)$. The Navier–Stokes equation then becomes:

$$\rho\frac{\partial v_x}{\partial t} = \rho\,f_x - \frac{\partial p}{\partial x} + \eta\left(\frac{\partial^2 v_x}{\partial y^2} + \frac{\partial^2 v_x}{\partial z^2}\right) \qquad \text{and:} \qquad \rho\,f_y - \frac{\partial p}{\partial y} = \rho\,f_z - \frac{\partial p}{\partial z} = 0.$$
$$\text{(4.59a)} \hspace{8cm} \text{(4.59b)}$$

The two Equations 4.59b are merely an indication of hydrostatic equilibrium (in most cases f_x, f_y and f_z are the components of the gravitational acceleration $\mathbf{g}$). In this case, $\mathbf{f} = \mathbf{g}$ is constant throughout the volume of the fluid; by taking the derivative of Equation 4.59b with respect to x, we obtain:

$$\frac{\partial}{\partial x}\left(\frac{\partial p}{\partial y}\right) = \frac{\partial}{\partial x}\left(\frac{\partial p}{\partial z}\right) = 0, \qquad \text{i.e.:} \qquad \frac{\partial}{\partial y}\left(\frac{\partial p}{\partial x}\right) = \frac{\partial}{\partial z}\left(\frac{\partial p}{\partial x}\right) = 0.$$

The pressure gradient $\partial p/\partial x$ in the flow is thus independent of y and of z; moreover, as a result of Equation 4.59a, and of the fact that v_x is independent of x, its derivative with respect to x is zero. We have therefore:

$$\frac{\partial p}{\partial x} = \text{constant} \tag{4.60}$$

throughout the volume of one-dimensional flows (but $\partial p/\partial x$ can be time-dependent for non-stationary flows).

In the case of a steady-state flow, the first term of Equation 4.22 is zero. As discussed in Section 4.2.1, this is an indication of the fact that the resultant of the forces on a volume-element of the fluid (gravity + pressure + viscous stresses) vanishes at all times for one-dimensional stationary flows: in fact, the momentum of such an element of fluid must remain constant as it moves both within space (one-dimensional characteristic) and with time (stationary condition).

The first four examples of the flows which we will analyze below are of this last kind. We will then discuss the case of cylindrical Couette flow which gives an example of a situation in which we can determine the velocity field even in the presence of non-linear terms. However, the fact that we can find a solution for such flows does not indicate that this solution is the only one which may be observed. Generally, for sufficiently large Reynolds numbers, such parallel flows become unstable: turbulent velocity fields, which are much more complicated and non-stationary, can then develop.

It is not only for parallel flows that the term $(\mathbf{v}\cdot\nabla)\,\mathbf{v}$ is negligible; this will also be the case for flows at low Reynolds numbers ($Re \ll 1$) to be discussed in Chapter 9. In this case, this non-linear term can be neglected, even for cases with arbitrary geometry, by comparison with the component $\eta\,\nabla^2\mathbf{v}$ which represents the forces of viscous friction. Such flows obey a linear equation of motion known as the *Stokes equation*. Finally, for flows which are almost one-dimensional, discussed in Chapter 8, the equation of motion again becomes linear, provided that the Reynolds number (while still larger than 1) is not too large (the limiting value then depends on the geometry of the flow). In that case we discuss *lubrification*.

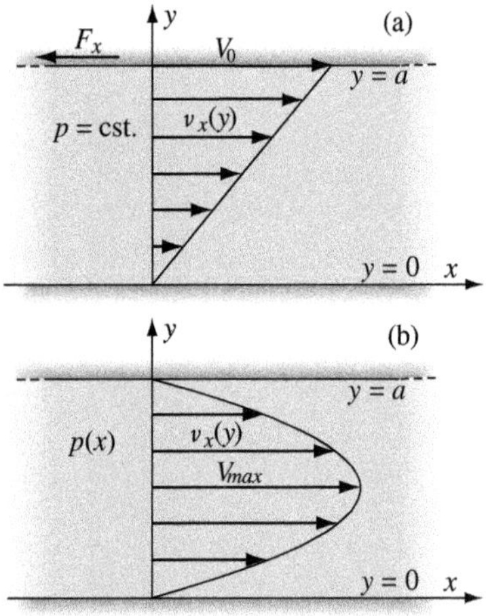

Figure 4.18 *Plane flows: (a) Couette flow; the lower plane is at rest while the upper one moves with constant velocity V_0 in the x-direction. (b) Poiseuille flow; both planes are at rest while a pressure gradient is applied (compare this flow to that of Figure 2.7b)*

4.5.2 Couette flow between parallel planes

We have previously discussed such flow between two parallel planes separated by a constant distance and in relative motion in Section 2.1.2, where we discussed the diffusion of momentum.

We calculate here the velocity profile $v_x(y)$ between a lower plane ($y = 0$) at rest, and an upper plane ($y = a$) which is displaced parallel to itself at a constant velocity $\mathbf{V}_0$ (Figure 4.18a). We assume that no pressure gradient is applied in the direction along the plates (i.e., $\partial p/\partial x = 0$ throughout the volume according to Equation 4.60), that the flow is steady and stationary ($\partial v_x/\partial t = 0$), independent of z, and that the planes are horizontal (only the component of gravity $g_y = -g$ is non-zero). Equations 4.59 become:

$$\eta\frac{\partial^2 v_x}{\partial y^2} = \frac{\partial p}{\partial x} \quad (4.61) \qquad \text{and} \qquad \frac{\partial p}{\partial y} = -\rho\, g, \quad (4.62)$$

where g is the absolute value of the acceleration of gravity (and the positive y-axis is upward). Because $\partial p/\partial x = 0$, Equation 4.61 becomes:

$$\frac{\partial^2 v_x}{\partial y^2} = 0, \quad (4.63) \qquad \text{which can be integrated:} \qquad v_x = V_0\frac{y}{a}, \quad (4.64)$$

taking into account the boundary conditions which require that the relative velocity be zero at each of the boundary planes. Gravity merely results in a vertical hydrostatic pressure gradient, which does not affect the flow. The viscous friction force per unit area on each of the planes has the value:

We note that F_x is independent of the distance y from the lower plane and that the resultant viscous force $\eta\ (\partial^2 v_x/\partial y^2)$ per unit volume of an element of the moving fluid is zero. This is required in order that the resultant overall force in the x-direction be zero on each element; in fact, the pressure and gravity forces have a zero component in the x-direction.

$$|F_x| = |\sigma_{xy}| = \eta\frac{\partial v_x}{\partial y} = \frac{\eta\, V_0}{a}. \quad (4.65)$$

4.5.3 Poiseuille-type flows

We discuss now the steady-state flow of an incompressible fluid between two fixed parallel planes, and then in a tube of circular cross-section; these flows are driven by a pressure difference applied between the boundaries of the planes, or of the tube. We assume that we are far enough from the entry points of the channel so that the velocity profile is independent of distance along the flow; the problems involving the approach to steady-state of the velocity profiles are discussed in Section 10.2, together with the downstream variation in the thickness of the boundary layers.

Flow between two fixed parallel planes

We will first consider the steady-state flow of a viscous fluid between two horizontal planes, fixed and parallel, located at $y = 0$ and $y = a$ (Figure 4.18b). The flow results from a pressure gradient $\partial p/\partial x = -K = -(\Delta p/L)$ in the x-direction. K is positive for flow in the direction of positive x and, in accordance with Equation 4.60, it is constant throughout the fluid. Equation 4.59a then becomes:

$$\eta \frac{\mathrm{d}^2 v_x}{\mathrm{d}y^2} + K = 0.$$

Integrating along the y-direction, and taking into account the boundary conditions at the walls, ($v_x = 0$ for $y = 0$ and $y = a$), we find:

$$v_x = \frac{K}{\eta}\frac{y\,(a-y)}{2} = -\frac{\partial p}{\partial x}\frac{1}{2\eta}y\,(a-y) = V_{max}\frac{4y\,(a-y)}{a^2}. \qquad (4.66)$$

The corresponding flow is known as *Poiseuille parallel flow*. It has a parabolic velocity profile with a maximum value V_{max} in the symmetry plane of the channel ($y = a/2$), which has the algebraic value:

$$V_{max} = K\frac{a^2}{8\eta} = -\left(\frac{\partial p}{\partial x}\right)\frac{a^2}{8\eta}. \qquad (4.67)$$

The flow rate Q of fluid per unit depth of the channel is the z-direction has the absolute value:

$$Q = \int_{-a/2}^{a/2} v_x(y)\,dy = K\frac{a^3}{12\eta} = \frac{\Delta p}{L}\frac{a^3}{12\eta}, \qquad (4.68)$$

where $\Delta p = p(x) - p(x+L)$ is the pressure drop over a distance L. The flow rate Q is thus proportional to a^3 for a given pressure drop Δp, thus increasing much more rapidly than the cross-section a (per unit depth) of the channel.

We can define, in terms of the flow rate Q, an average flow velocity U given by $U = Q/a$, whence:

$$U = K\frac{a^2}{12\eta} = \frac{2V_{max}}{3}. \qquad (4.69)$$

Superposition of Couette and Poiseuille flows between two parallel planes

Let us now consider a steady-state flow between two horizontal planes, separated by a distance a and parallel to the x-axis: the lower boundary $y = 0$ is taken as the reference plane and the upper plane ($y = a$) moves with velocity $v_x = V_0$ in the parallel x-direction. In contrast to the case discussed in Section 4.5.2, we apply, in addition, a pressure gradient $\partial p/\partial x$ in the direction parallel to the boundary planes and independent of the time and of the coordinates x, y and z. Integrating Equation 4.59a, we find:

$$v_x(y) = \frac{\partial p}{\partial x}\frac{y^2}{2\eta} + Cy + D. \tag{4.70}$$

The coefficient D is always zero because the velocity is everywhere zero along the plane $y = 0$; the coefficient C can be determined by recognizing that the velocity is always V_0 at the upper boundary plane. We then obtain:

$$v_x(y) = -\frac{\partial p}{\partial x}\frac{y\,(a-y)}{2\eta} + V_0\frac{y}{a}. \tag{4.71}$$

The resultant velocity field is thus the superposition of a parabolic, Poiseuille velocity-field (first right-hand term in Equation 4.71, and of a Couette-type velocity-field (second right-hand one). Figure 4.19 displays qualitatively the various resulting velocity profiles. Depending on the particular case, we might, or not, find a maximum between the two planes. The resultant flow rate per unit depth in the z-direction is obtained by integrating directly Equation 4.71, with respect to y:

$$Q = -\frac{\partial p}{\partial x}\frac{a^3}{12\,\eta} + \frac{V_0\,a}{2}. \tag{4.72}$$

We observe that it is also the sum of the respective flow rates corresponding to the Poiseuille and Couette terms. The fact that we can superimpose both the velocity fields and the flow rates is a consequence of the linearity of Equation 4.59a.

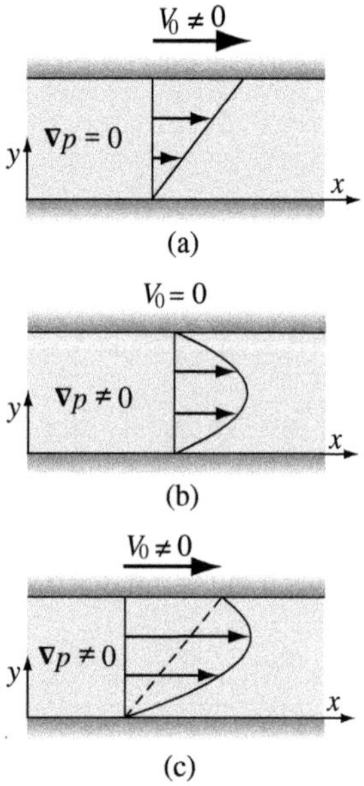

Figure 4.19 *Superposition of a shear flow (a) and of a Poiseuille flow (b); we obtain a third kind of flow (c) in which we have coexistence of shear and a pressure gradient in the direction of the flow*

We will find again examples of such flows in Chapter 8, in the discussion of the flows between two planes, in relative motion, making a small angle with respect to each other, or induced by a local temperature-gradient at the free surface of a liquid (*Marangoni effect*).

Flow in a cylindrical tube

We discuss the flow resulting from a pressure difference Δp along a length L of a horizontal cylindrical tube of radius R (Figure 4.20). We assume that the only non-zero component of the velocity is the axial velocity v_z, and that it is only a function of the radial distance r from the axis. We solve the Navier–Stokes equation by observing, as in the previous case, that the pressure head loss coefficient $K = (\Delta p/L) = -(\partial p/\partial z) = \text{cst}$.

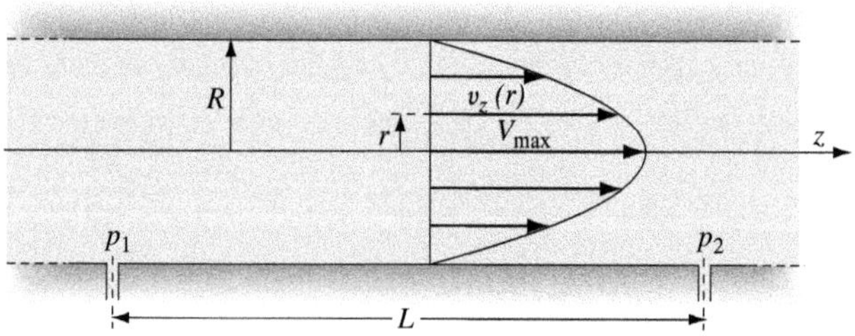

Figure 4.20 *Poiseuille flow in a cylindrical tube of radius R, resulting from a pressure difference $\Delta p = p_1 - p_2$ over a distance L*

- **(i) Velocity-field for the flow**

In cylindrical coordinates (r, φ, z) where r is the radial distance to the axis and $\varphi = 0$ for a radius vector directed upward, the Navier-Stokes equation 4.59a–b becomes:

$$0 = -\frac{\partial p}{\partial r} - \rho\, g \cos\varphi \qquad (4.73a) \qquad\qquad 0 = -\frac{1}{r}\frac{\partial p}{\partial \varphi} + \rho\, g \sin\varphi \qquad (4.73b)$$

$$0 = -\frac{\partial p}{\partial z} + \nabla^2 v_z = -\frac{\partial p}{\partial z} + \frac{\eta}{r}\left[\frac{\partial}{\partial r}\left(r\frac{\partial v_z}{\partial r}\right)\right] = K + \frac{\eta}{r}\left[\frac{\partial}{\partial r}\left(r\frac{\partial v_z}{\partial r}\right)\right]. \qquad (4.73c)$$

Equations 4.73a and 4.73b indicate that the only effect of gravity is to create throughout the cross-section of the tube, a hydrostatic pressure-gradient which leaves the flow unaffected. Just as in the case of plane Poiseuille flow, the gradient $\partial p/\partial z = -K$ is then constant throughout the flow volume, so that:

$$\frac{1}{r}\frac{\mathrm{d}}{\mathrm{d}r}\left(r\frac{\mathrm{d}v_z}{\mathrm{d}r}\right) = -\frac{K}{\eta}. \qquad (4.75)$$

By integrating this equation (eliminating the Log r term which diverges for $r = 0$) and using the boundary conditions at the tube walls ($v_z = 0$ for $r = R$) we find:

$$v_z = \frac{K}{4\eta}(R^2 - r^2) = V_{\max}\left(1 - \frac{r^2}{R^2}\right), \qquad (4.76) \qquad \text{where:} \qquad V_{\max} = \frac{K R^2}{4\eta}.$$

Here $V_{\max}$ is the maximum value of the velocity which occurs along the axis of the tube ($r = 0$).

The flow rate Q of the fluid in the tube is then:

$$Q = \int_0^R v_z(r)\, 2\pi\, r\, \mathrm{d}r = \frac{\pi K R^4}{8\eta} = -\frac{\pi R^4}{8\eta}\frac{\partial p}{\partial z}. \qquad (4.77)$$

By expressing the flow rate in terms of the tube diameter d of its length L and of the pressure drop Δp between its two ends, we obtain, finally:

$$Q = \frac{\pi}{128\,\eta}\frac{\Delta p}{L}d^4. \qquad (4.78)$$

This result, known as *Poiseuille's law*, indicates that the flow rate varies as the fourth power of the diameter of a circular tube, i.e., as the square of its cross-section. Let us compare, for example, the flow rates Q resulting from the same pressure gradient in a tube of diameter R and in one hundred tubes of diameter $R/10$ connected in parallel, representing an equivalent cross-section: the flow rate would be one hundred times smaller in this latter case. This result is quite different from that obtained in a problem of electrical current transport between two plane parallel electrodes perpendicular to the tubes: if the two sets of tubes have the same electrical conductivity, the value of the electrical resistance is the same in both cases. The major difference between these two examples results from the boundary condition of zero velocity along the tube walls, in the case of a viscous fluid, a condition which does not apply for the case of electrical current. Due to this condition, the transverse velocity gradients and, accordingly, the viscous friction forces increase greatly as the size of the channels decreases.

Assuming that v_z is only a function of r implies that the radial velocity component v_r is zero for a stationary flow. First, any v_φ component of the velocity would correspond to a transient rotation of the fluid around the axis z vanishing in the steady-state (Section 2.1.1). Taking then $v_\varphi = 0$ in the condition of incompressibility $\nabla\cdot\mathbf{v} = 0$ leads to $(1/r)\,\mathrm{d}(r\, v_r)/\mathrm{d}r = 0$. Using the boundary condition $v_r = 0$ at the tube wall of the tube leads to $v_r = 0$ everywhere.

Equation 4.73 can be derived directly by writing that the resultant force is zero on a volume bounded by two concentric cylinders of radii r and $r + \mathrm{d}r$ to which the viscous stresses are applied, and by two sections at a distance $\mathrm{d}z$ between which the pressure gradient acts. We obtain;

$$2\pi\,(r + \mathrm{d}r)\,\mathrm{d}z\,(\sigma'_{zr})_{r+\mathrm{d}r} - 2\pi\, r\,\mathrm{d}z\,(\sigma'_{zr})_r = 2\pi\, r\,\mathrm{d}r\,\mathrm{d}z\frac{\partial p}{\partial z}. \qquad (4.74)$$

This equation reduces to Equation 4.73c by a first-order Taylor expansion and a division by $2\pi\, \mathrm{d}r\, \mathrm{d}z$.

Up to now, we assumed that the acceleration of gravity had no component in the direction z of the flow (like for an horizontal flow). If this is not the case, Equation 4.73c becomes:

$$0 = -\frac{\partial p}{\partial z} + \rho\, g_z + \frac{1}{r}\frac{\mathrm{d}}{\mathrm{d}r}(r\,\sigma'_{zr}) = -\frac{\partial p}{\partial z} + \rho\, g_z + \eta\frac{1}{r}\left[\frac{\mathrm{d}}{\mathrm{d}r}\left(r\frac{\mathrm{d}v_z}{\mathrm{d}r}\right)\right].$$

Then, the sum: $-\partial p/\partial z + \rho g_z$, and not just $-\partial p/\partial z$ is the driving term of the flow.
For instance, a tube of axis z at angle θ to the horizontal and connecting two two free-surface reservoirs would have the atmospheric pressure at both ends and, therefore, a pressure gradient (constant with z) equal to zero. The flow is then only due to the component $\rho g_z = \rho g \sin\theta$, and Equation 4.78 becomes:

$$Q = \frac{\pi}{128}\frac{\rho}{\eta}\, g_z\, d^4.$$

- **(ii) Viscous-friction forces on the walls of a circular tube**

The viscous friction force exerted by the fluid on the tube can be calculated by integrating the stress over the area of the walls:

$$F = \iint_{\text{(wall)}} [\sigma] \cdot \mathbf{n} \, dS, \tag{4.79}$$

where $\mathbf{n}$ is the unit vector normal to the wall. The component F_z in the z-direction of the flow is thus:

$$F_z = \iint_{\text{(wall)}} [\sigma'_{zr} r]_{r=R} \, d\varphi \, dz = \int_0^L dz \int_0^{2\pi} [\sigma'_{zr} r]_{r=R} \, d\varphi. \tag{4.80}$$

We can also calculate f_z by writing that it must be the opposite of the difference $-\pi R^2 \, dp/dz$ between the pressure forces on two sections of the tube separated by a unit distance (we must indeed have a zero resultant of the forces acting on the fluid).

By using the expression $\sigma'_{zr} = \eta \, (\partial v_z / \partial r)$ given in appendix 4A.2 at the end of this chapter, we obtain for the force f_z per unit length of the tube:

$$f_z = 4 \, \pi \, \eta \, V_{max}. \tag{4.81}$$

One also often calculates a dimensionless viscous *drag coefficient* C_d by normalizing the force f_z by the quantity $(1/2)\rho V_{\max}^2 R$. As we will see in Section 5.3.1., the term $(1/2)\rho V_{\max}^2$ corresponds to a dynamic pressure or, with the extra factor of 1/2, to a convective momentum flux. We then obtain:

$$C_d = \frac{f_z}{(1/2)\rho \, V_{\max}^2 R} = 8\pi \left(\frac{\eta}{\rho \, V_{\max} R} \right) = \frac{8\pi}{Re}, \tag{4.82}$$

where we have defined the Reynolds number related to the flow by $Re = \rho \, V_{\max} \, R/\eta$. This kind of variation of the drag coefficient as $1/Re$ is characteristic of flows in which the convective effects associated with the terms of the type $(\mathbf{v} \cdot \nabla)\mathbf{v}$ are zero or negligible. On the other hand, in the opposite limit of large Reynolds numbers (particularly for turbulent flows), we generally obtain a drag coefficient C_d which varies little with the Reynolds number (Figures 12.9 and 12.14). The drag coefficient C_d is then a good characterization of this latter kind of flow but is less adequate for the present laminar flows.

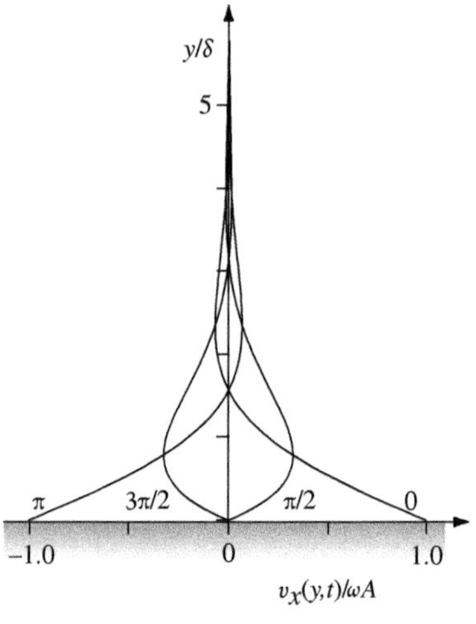

Figure 4.21 *Instantaneous profiles of the velocity $v_x(y, t)$ resulting from the oscillation of the plane $y = 0$ parallel to itself in the x-direction, with amplitude A for several phases (indicated by the labels)*

4.5.4 Oscillating flows in a viscous fluid

Shear-flow near a plane oscillating parallel to its surface

We now consider the flow of an incompressible viscous fluid above an infinite horizontal solid plane which is subjected to a sinusoidal oscillating motion parallel to itself along the the x-axis (Figure 4.21). The instantaneous displacement $\Delta x(t)$ of the plane at a time t is $A \sin \omega t$ where A is the amplitude of the motion and ω its angular frequency. This problem is the *sinusoidal* version of the sudden motion of an infinite plane at constant velocity, previously considered in Section 2.1.2.

We seek the simplest solution of the equations of motion which satisfies the symmetries of the problem and the boundary conditions both along the plate and at infinity (we assume the fluid to be at rest at very large distances). Accordingly, we assume that the flow is one-dimensional with a velocity component $v_x(y,t)$ independent of the x- and z-directions in the plane (satisfying the boundary conditions). According to Equation 4.59b, the y-component of the equation of motion reduces to the hydrostatic equation:

$$0 = -\frac{1}{\rho}\frac{\partial p}{\partial y} - g \tag{4.83a}$$

and, moreover, $\partial p/\partial z = 0$. On the other hand, $\partial p/\partial z$ must be contstant throughout the volume of the fluid (Equation 4.60), and will vanish unless there exists a horizontal pressure gradient applied from a great distance. The pressure is then independent of x and z, so that we can rewrite Equation 4.59a in the form:

$$\frac{\partial v_x}{\partial t} = \frac{\eta}{\rho}\frac{\partial^2 v_x}{\partial y^2} = \nu\frac{\partial^2 v_x}{\partial y^2}, \tag{4.83b}$$

where $\nu = \eta/\rho$. We get back the momentum diffusion equation derived in Section 2.1.2. Let us now look for a periodic solution $v_x\,(y,\,t)$ of Equation 4.83b in the form:

$$v_x\,(y,\;t) = |f(y)|\cos(\omega t + \varphi) = \Re e\left(f\,(y)\,\mathrm{e}^{\mathrm{i}\omega t}\right), \tag{4.84}$$

where $f(y)$ is a complex function (the symbol $\Re e$ indicates that we take the real part of the quantity in question). Inserting this expression into Equation 4.83b, we obtain:

$$\mathrm{i}\,\omega\,f(y) = \nu\frac{\partial^2 f}{\partial y^2}, \tag{4.85}$$

for which we have the general solution:

$$f\,(y) = C_1\,\mathrm{e}^{-(1+\mathrm{i})ky} + C_2\,\mathrm{e}^{(1+\mathrm{i})ky} \qquad \text{where:} \qquad k = \sqrt{\frac{\omega}{2\nu}}. \tag{4.86}$$

We then derive: $$v_x\,(y,\;t) = \Re e\left\{C_1\mathrm{e}^{-ky}\mathrm{e}^{\mathrm{i}(\omega t-ky)} + C_2\mathrm{e}^{ky}\mathrm{e}^{\mathrm{i}(\omega t+ky)}\right\}. \tag{4.87}$$

The velocity must approach zero as y tends towards infinity, which requires $C_2 = 0$. On the other hand, the boundary condition at the surface of the oscillating plane is written: $v_x(y = 0,\; t) = \omega A\cos\omega t$, which then allows us to evaluate the constant of integration $C_1 = \omega\,A$; we therefore obtain:

$$v_x\,(y, t) = \omega\,A\,\mathrm{e}^{-ky}\cos(\omega t - k\,y) \tag{4.88}$$

This indicates that the oscillation of the applied velocity at the surface of the plane propagates toward the interior of the fluid [the term $\cos(\omega\,t - k\,y)$] with an exponential damping, as e^{-ky}. We thus have a propagation of a transverse wave which is attenuated in the viscous fluid.

We define the *penetration depth* δ of the oscillation as the distance from the oscillating plane for which the amplitude of the velocity is decreased by a factor 1/e, such that:

$$\delta = \sqrt{\frac{2\nu}{\omega}}. \tag{4.89}$$

Example:
For an oscillatory motion with frequency 2 Hz in a fluid of kinematic viscosity $\nu = 10^{-3}\,\mathrm{m^2/s}$ (a thousand times the viscosity of water), we obtain $\delta \approx 10^{-2}\,\mathrm{m}$.

The essential result of the above analysis is the conclusion that a shear wave will not propagate over significant distances in a viscous liquid. The corresponding acoustic wave

This problem is the analog of the skin effect in electrical conduction, or of the penetration of seasonal changes in temperature into the ground. The equivalent of the viscosity coefficient is the resistivity of the conducting medium in the case of the skin effect (within a factor of μ_0), and the thermal diffusivity in the case of fluctuations in temperature in the ground. In each of these cases, we have a penetration depth δ which varies as $1/\sqrt{\omega}$ and a complex wave vector **k**, with modulus equal to $1/\delta$. Both of these variations are characteristic of all diffusive propagation phenomena.

is said to be *critically damped*. We have here an essential difference between liquids and solids: in solids, there exists the propagation, in addition to the compressional wave (ordinary sound), of two oscillation modes transverse to the direction of propagation—*shear waves* with two orthogonal polarizations. There are intermediate results with the possibility of shear waves, propagating over significant distances, which can be observed in the case of viscoelastic fluids. We will describe these in Section 4.6.4.

The frictional viscous force F_x on the oscillating plane can be written as:

$$F_x = \iint_{(\text{plane})} \sigma_{xy}\, \mathrm{d}x\, \mathrm{d}z = \iint_{(\text{plane})} \eta \left(\frac{\partial v_x}{\partial y}\right)_{y=0} \mathrm{d}x\, \mathrm{d}z. \tag{4.90}$$

Combining Equations 4.86, 4.88 and 4.90, we obtain the value f_x of the frictional force per unit area:

$$f_x = (A\sqrt{2}\,\omega\, k\eta) \cos\left(\omega t - \frac{3\pi}{4}\right) = A\,\omega^{3/2} \sqrt{\rho\,\eta} \cos\left(\omega t - \frac{3\pi}{4}\right). \tag{4.91}$$

Thus, f_x has a phase retardation of three-eighths of a period relative to the velocity $\mathbf{v}(0, t)$ of the plane. Just as we did in the case of the circular tube, we can normalize the amplitude of the force f_x by the dynamic pressure $(1/2)\,\rho A^2\omega^2$ expressed in terms of the peak value $U = \omega\, A$ of the velocity. We then obtain a coefficient of frictional drag:

$$C_d = \frac{A\,\omega^{3/2}\sqrt{\rho\,\eta}}{(\rho A^2\omega^2/2)} = 2\sqrt{\frac{\nu}{\omega A^2}} = \frac{2}{\sqrt{Re}}. \tag{4.92}$$

where $Re = (\omega\, A^2)/\nu$ is the Reynolds number for the flow, evaluated in terms of the velocity U and of the characteristic length A. This time, however, C_d is proportional to the square root of the inverse of the Reynolds number. This variation is slower than that, as $1/Re$, obtained for one-dimensional steady-state flows (Section 4.5.3.). The same type of variation will be found in Chapter 10 for boundary layers, suggesting that one has here an oscillating boundary layer.

Geophysical application:
A seismograph located at a moderate distance from a place where an earthquake has occurred normally detects three signals, corresponding to the three types of waves: two shear (S) and one pressure (P) wave, which can propagate through the solid earth. Nevertheless, if the source and the detector are located at more or less diametrically opposite ends of the earth, the seismograph detects only the signal corresponding to the pressure wave: the shear waves will not propagate indeed across the central core of the earth, the outer layer of which is liquid (for distances of 2800 to 5100 km from the center).

Flow generated between two planes by an oscillating pressure gradient

When the flow rate is modulated, the inertial effects from the fluid increase with the frequency of the modulation; then, oscillating boundary layers, very similar to the ones we have encountered in the previous example, can appear near the walls of a capillary tube. In contrast to the latter, however, the walls do not move and the pressure gradient no longer vanishes in the direction of the flow. Let us analyze such a phenomenon for the simple case of a one-dimensional flow between two fixed horizontal planes $y = \pm\, a/2$, where the local velocity $v_x(y,t)$ is along the x-axis (Figure 4.22) and independent of the z-variable.

Assume that the longitudinal pressure gradient $\partial p/\partial x\ (t)$ is sinusoidally modulated at an angular frequency ω, with a variation, in complex notation, $\partial p/\partial x\,(t) = \partial p/\partial x\,(\omega)\, e^{i\omega t}$. Let us also write Equation 4.59a to include the acceleration term $\partial v_x/\partial t$ and the gradient of the pressure. Looking for a sinusoidal solution for v_x at angular frequency ω, we obtain in complex notation:

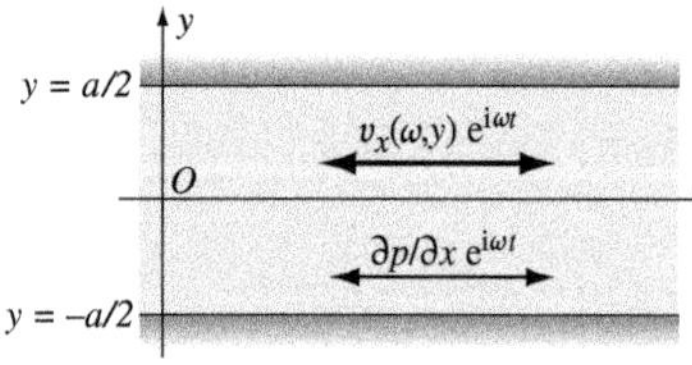

Figure 4.22 *Oscillating flow between two fixed, parallel planes, resulting from a sinusoidal pressure gradient parallel to the two planes*

$$\frac{\partial v_x}{\partial t} = \mathrm{i}\,\omega\, v_x = -\frac{1}{\rho}\frac{\partial p}{\partial x}(\omega)\ \mathrm{e}^{\mathrm{i}\omega t} + \nu\frac{\partial^2 v_x}{\partial y^2}.$$

The complex function $v_x(\omega,\, y)\ e^{i\omega\, t}$, solution of this linear differential equation which satisfies the boundary conditions $v_x = 0$ for $y = \pm\, a/2$ is then:

$$v_x(\omega, y) = \frac{\mathrm{i}}{\rho\,\omega}\frac{\partial p}{\partial x}(\omega)\left[1 - \frac{\cosh\,(k(\omega)y)}{\cosh\,(k(\omega)a/2)}\right], \tag{4.93a}$$

where:
$$k(\omega) = \sqrt{\frac{\mathrm{i}\,\omega}{\nu}} = (1+\mathrm{i})\sqrt{\frac{\omega}{2\nu}}. \qquad (4.93b)$$

We find, by integrating with respect to the variable y, the complex amplitude of the variation of the average velocity:

$$U(\omega) = \frac{1}{a}\int_{-a/2}^{a/2} v_x(\omega, y)\mathrm{d}y = \frac{\mathrm{i}}{\rho\,\omega}\frac{\partial p}{\partial x}(\omega)\left[1 - \frac{\tanh\left(k(\omega)a/2\right)}{k(\omega)a/2}\right]. \qquad (4.94)$$

Here $k(\omega)$ is the complex wave vector of the damped oscillating shear wave propagating from an oscillating wall into a viscous fluid (Equation 4.86). The expression $1/|k(\omega)| = \delta(\omega) = \sqrt{\nu/\omega}$ then gives the order of magnitude of the penetration depth $\delta(\omega)$ of a wave oscillating at angular frequency ω. The response of the system is very different depending on whether $\delta(\omega)$ is large or small compared to the channel width a between the planes:

- If $|k\ (\omega)|\,a \ll 1$ (*i.e.* $\delta(\omega) \gg a$): low-frequency regime → viscous forces dominate.

If we expand tanh ($ka/2$) in a power series for small $k\,a$, we obtain:

$$U(\omega) = -\frac{a^2}{12\,\eta}\frac{\partial p}{\partial x}(\omega)\left(1 - \mathrm{i}\frac{a^2\omega}{10\nu}\right). \qquad (4.95)$$

The first term in the expression is dominant. It merely represents the velocity we calculated for the stationary flow. The flow is only weakly affected by inertial effects, which are represented by the second, smaller, term (90° out of phase, as indicated by the factor i).

The correction term involves the ratio $Nt_\nu = a^2\omega/\nu$ of the viscous diffusion time a^2/ν over the distance a to the period $2\pi/\omega$ of the oscillation; Nt_ν (first found in 8.1.3) is also the ratio of the non-stationary characteristics of the flow ($\rho\,(\partial\mathbf{v}/\partial t)$ term in the equation of motion) to the viscous effects ($\eta\nabla^2\mathbf{v}$ term). In the case of a more complicated geometry, we would have to also introduce the term in $(\mathbf{v}.\nabla)\mathbf{v}$, so that the flow would also depend on a *Strouhal number* of the form $Sr = \omega L/U$ (U is the velocity, and L a characteristic length). This number, defined in Section 2.4.1, represents the ratio of the non-stationary and convective terms.

If $L = a$, the quotient Nt_ν/Sr corresponds to the Reynolds number $Re = UL/\nu$.

- If $|k\ (\omega)|\,a \gg 1$ (i.e. $\delta(\omega) \ll a$): high-frequency regime → inertial effects dominate.

Using the notation $\alpha = (1/2)\sqrt{\omega\,a^2/2\nu}$, and the identity $\tanh(\mathrm{i}x) = \mathrm{i}\tan(x)$ (valid for all real x), we expand the hyperbolic tangent of the complex number $(ka)/2$:

$$\tanh\frac{k(\omega)a}{2} = \tanh\left[\alpha(1+\mathrm{i})\right] = \frac{\tanh\alpha + \mathrm{i}\tan\alpha}{1 + \mathrm{i}\tanh\alpha\tan\alpha}.$$

But, for $\alpha \gg 1$, $\tanh(\alpha)$ is very close to unity, so that $\tanh(k(\omega)a/2) \approx 1$ and one obtains:

$$U(\omega) \approx \frac{\mathrm{i}}{\rho\,\omega}\frac{\partial p}{\partial x}(\omega)\left(1 - (1-\mathrm{i})\sqrt{\frac{2\nu}{\omega a^2}}\right). \qquad (4.96)$$

The governing term $(\mathrm{i}/\rho\,\omega)(\partial p/\partial x)$ corresponds to a solid body oscillation of the mass of the fluid with an amplitude determined completely by its inertial response to the oscillating pressure and 90° out of phase with it. It is only for a thickness of fluid of order $\delta(\omega)$ that viscosity effects are significant, and lead to energy dissipation which is represented by the real part of the correction term. The ratio $\sqrt{2\nu/(\omega\,a^2)}$ between the two terms of Equation 4.96 also gives the order of magnitude of the ratio of the thickness of the boundary layers to the total width of the channel.

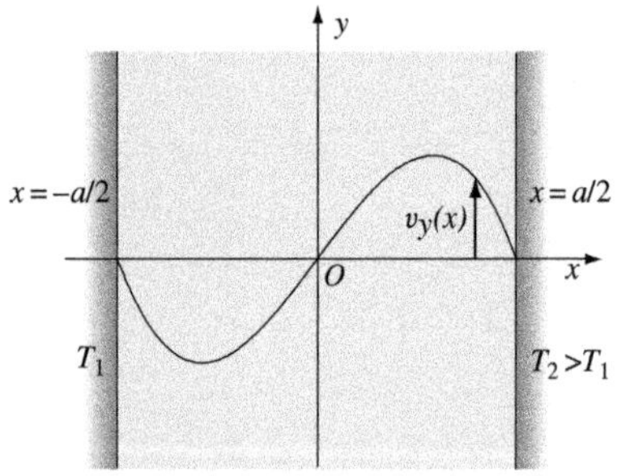

Figure 4.23 *Velocity profile for fluid flow resulting from a temperature difference between two infinite vertical plates*

4.5.5 Parallel flow resulting from a horizontal density variation

We consider here a steady-state flow between two vertical parallel plates, located at $x = \pm a/2$ and infinite in extent, resulting from a density gradient in the horizontal x-direction.

As an example, we can consider the case where the density change results from keeping the two plates at different constant temperatures T_1 and T_2 (Figure 4.23). We assume that the temperature is not a function of the y and z coordinates, and that it varies linearly along the x-direction. We then have:

$$T(x) = T_0 + \Delta T \frac{x}{a}, \tag{4.97}$$

with: $\quad T_0 = \dfrac{T_1 + T_2}{2} \quad$ and $\quad \Delta T = T_2 - T_1.$

Explanation

We confine ourselves here to the case where only the vertical component v_y of the velocity is non-zero: we do not then have any thermal transport due to the flow, from one plate to the other in the x-direction, and such transport is only a function of the thermal conductivity k of the liquid. If we assume that the themal conductivity k does not depend on the temperature and the pressure, the thermal energy flux per unit area $\mathcal{J}_Q = -\, k\, \partial T/\partial x$ and, therefore, the gradient $\partial T/\partial x$ must be constant, and not a function of x. Moreover, Equation 4.97 implies that the temperature $T(x)$ is not a function of y and z. The flow which is parallel to the plates and, therefore, to constant temperature surfaces does not then contribute to thermal energy transfer.

Let us now assume that the variation of the density ρ of the fluid with temperature obeys:

$$\rho(T) = \rho_0\,[1 - \alpha\,(T - T_0)] \tag{4.98}$$

(T_0 is the temperature at $x = 0$ and α is the thermal expansion coefficient, generally positive) and that the density ρ is independent of the pressure. We then obtain for the variation of the density with distance:

$$\rho(x) = \rho_0 + \delta\rho\,(x) = \rho_0\,[1 - \alpha\,(T(x) - T_0)] = \rho_0 \left[1 - \alpha\,\Delta T \frac{x}{a}\right]. \tag{4.99}$$

The surfaces of constant density $\rho(x)$ (or *isochores*) are thus the vertical planes $x = constant$. It then becomes impossible for hydrostatic equilibrium to be established with the fluid completely at rest ($\mathbf{v} = 0$); if that were the case, the pressure p would have to satisfy the fundamental equation of hydrostatics:

$$\nabla p = \rho\,\mathbf{g}. \tag{4.100}$$

The constant-pressure lines (*isobars*) would then be perpendicular to $\mathbf{g}$ and therefore (as usual) horizontal with $p = p(y)$. It would then be impossible for Equation 4.100 to be satisfied, because its members depend on two different variables: y for the pressure and x for the density. Accordingly, the fluid must begin to move.

Assuming a steady-state flow with vertical velocity $v_y(x)$, the equation of motion for the velocity becomes (as seen in the explanation further down):

$$[\rho(x) - \rho_0]\, g = \eta \frac{\partial^2 v_y}{\partial x^2}, \qquad (4.101a) \qquad \text{i.e.:} \qquad -\alpha\, \rho_0\, g\, \Delta T \frac{x}{a} = \eta \frac{\partial^2 v_y}{\partial x^2}. \qquad (4.101b)$$

Equation 4.101a indicates that the flow results from a volume force which varies with x and is proportional to the difference between the local density and the average density. Integrating Equation 4.101b, we then obtain:

$$v_y(x) = -\frac{\alpha\, \rho_0\, g\, \Delta T}{\eta} \frac{x}{a} \frac{x}{6} \left(x^2 - \frac{a^2}{4} \right). \qquad (4.102)$$

This profile is antisymmetric relative to $x = 0$ with zero net flow rate (Figure 4.23). The choice made for the value of $\partial p/\partial y$ corresponds then to a system closed at both ends (where we would, in fact, have near the boundaries two small perturbation zones, with a transverse recirculation flow).

Explanation:
In view of the symmetry of the problem, we can assume that v_y does not vary in the z-direction. Moreover, if ($v_x = v_z = 0$), the equation $\nabla \cdot \mathbf{v} = 0$ becomes $\partial v_y/\partial y = 0$, so that v_y will depend only on x. Under these conditions, and remembering that the velocity has a component only in the y-direction, we can rewrite Equations 4.59a and 4.59b in the form:

$$\rho(x) g + \frac{\partial p}{\partial y} = \eta \frac{\partial^2 v_y}{\partial x^2} \qquad (4.103a) \qquad \text{and:} \qquad \frac{\partial p}{\partial x} = \frac{\partial p}{\partial z} = 0. \qquad (4.103b)$$

The pressure gradient $\partial p/\partial y$ is thus independent of x and z because, taking the derivative of Equation 4.103b with respect to y, one finds:

$$\frac{\partial}{\partial y}\left(\frac{\partial p}{\partial x}\right) = \frac{\partial}{\partial y}\left(\frac{\partial p}{\partial z}\right) = 0 \qquad \text{so that:} \qquad \frac{\partial}{\partial x}\left(\frac{\partial p}{\partial y}\right) = \frac{\partial}{\partial z}\left(\frac{\partial p}{\partial y}\right) = 0.$$

Moreover, $\partial p/\partial y$ also cannot be a function of y because the other terms in Equations 4.103a do not depend on it: this gradient is therefore constant throughout the flow. Assuming that this constant is equal to the gradient of the hydrostatic pressure $- \rho_0\, g$ associated with the average density ρ_0 of the fluid which, as we have seen, corresponds to a tube closed at both ends, we obtain Equation 4.101a. Using some other value for $\partial p/\partial y$ would be equivalent to superimposing a Poiseuille type flow, corresponding to the resultant gradient $\partial p/\partial y + \rho_0\, g$.

We have assumed here that the flow resulted from a horizontal temperature gradient. The variation of the density might have other sources, such as a concentration gradient of the solute (e.g., salt in water). On the other hand, even though the velocity profile (Equation 4.102) satisfies the equation of motion for any difference of temperature ΔT, other more complicated solutions will be observed experimentally as ΔT increases; for example, one may observe recirculation cells, smaller in height than the extent of the plates and with a non-vanishing velocity component in the x-direction.

4.5.6 Cylindrical Couette flow

We consider the steady-state behavior of an incompressible fluid of viscosity η, located between two coaxial cylinders of radii R_1 and R_2 rotating about their axis with respective

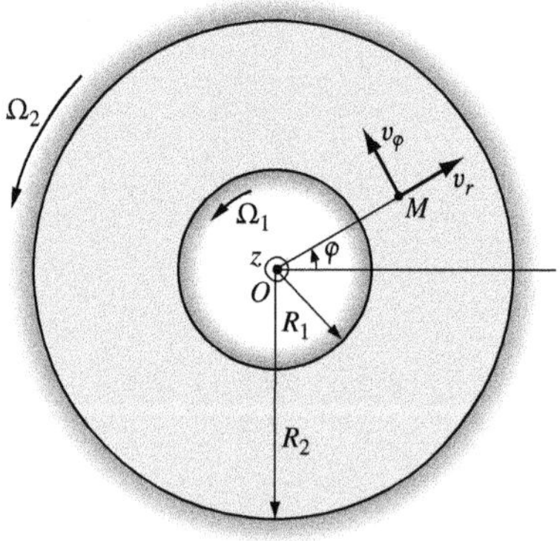

Figure 4.24 *Couette flow between two rotating concentric cylinders seen from the direction of the cylinders' axis*

angular velocities Ω_1 and Ω_2 (Figure 4.24). We assume that no external pressure gradient is applied. We choose a system of cylindrical coordinates (r, φ, z) where the z-axis coincides with the axis of the cylinders, and we evaluate the velocity components v_r, v_φ, and v_z.

Even though this is a system with a very simple geometry, quite complex flow structures can appear (depending on the angular velocity of the cylinders), as we discuss in Chapter 11. We are interested here only in the simplest possible flow, corresponding to pressure and velocity fields independent of z and φ. It is, in fact, what we observe at low velocities. In this case, keeping in mind that the system is invariant with respect to translation along the z-axis, and since there is no axial pressure gradient, we have $v_z = 0$. Moreover, in the simplest flow configuration, v_φ is independent of φ, because of the rotational invariance around the z-axis. The equation of conservation of mass for an incompressible fluid ($\nabla \cdot \mathbf{v} = 0$), given in cylindrical coordinates in Appendix 4A.2 at the end of this chapter simplifies to:

$$\frac{\partial v_r}{\partial r} + \frac{v_r}{r} = \frac{1}{r}\frac{\partial}{\partial r}(r\, v_r) = 0. \qquad (4.104)$$

Accordingly, v_r can only appear in the form $v_r = (C/r)$, where C is a constant. But the boundary conditions at the solid walls require that $v_r\ (r = R_1) = v_r\ (r = R_2) = 0$; therefore, v_r must vanish throughout the fluid.

The Navier–Stokes equation in cylindrical coordinates (see Appendix 4A.2), yields:

$$\frac{v_\varphi^2}{r} = \frac{1}{r}\frac{\partial p}{\partial r} \qquad (4.105a) \qquad \text{and:} \qquad \nu\left(\frac{\partial^2 v_\varphi}{\partial r^2} + \frac{1}{r}\frac{\partial v_\varphi}{\partial r} - \frac{v_\varphi}{r^2}\right) = 0. \qquad (4.105b)$$

In Equation 4.105a, the v_φ^2/r term corresponds directly to the inertial, centrifugal force resulting from the curvature of the trajectories of the fluid particles. Even though this term involves the square of the velocity, it does not affect the flow, as it is counterbalanced by the pressure gradient in the radial direction; it corresponds thus to a generalized form of the hydrostatic pressure, quite analogous to the effects of gravity, in the one-dimensional horizontal flows we have previously discussed. The terms of the second Equation 4.105b correspond to the φ–component of the Laplacian operator applied to the velocity in cylindrical coordinates.

Integrating Equation 4.105b (or the equivalent Equation 4.108 in the margin), we find:

$$v_\varphi = a\,r + \frac{b}{r}. \qquad (4.109)$$

The constants a and b are evaluated from the boundary conditions:

$$v_\varphi(r = R_1) = \Omega_1 R_1 \qquad \text{and} \qquad v_\varphi(r = R_2) = \Omega_2 R_2$$

which leads to the result:

$$v_\varphi = \frac{\Omega_2 R_2^2 - \Omega_1 R_1^2}{R_2^2 - R_1^2}\,r + \frac{(\Omega_1 - \Omega_2) R_2^2 R_1^2}{R_2^2 - R_1^2}\,\frac{1}{r}. \qquad (4.110)$$

The pressure distribution can be obtained by integrating Equation 4.105a. Let us now examine a few special cases among the values of the angular velocities of the cylinders and of their radii:

Equation 4.105b is also obtained by writing that the global moment of the forces on a volume bounded by two concentric cylinders of unit depth and with radii r and $r + dr$ vanishes, so that:

$$2\pi(r^2\sigma_{r\varphi})(r + dr) - 2\pi(r^2\sigma_{r\varphi})(r) = 0 \quad (4.106a)$$

i.e., as $dr \to 0$:

$$\frac{\partial}{\partial r}(r^2\sigma_{r\varphi}) = 0. \qquad (4.106b)$$

Due to the symmetry of the velocity field $\mathbf{v}(0, v_\varphi(r), 0)$, $\sigma_{r\varphi}$ is the only non-vanishing component of the stress tensor, as can be shown from the expressions of this tensor in cylindrical coordinates in Appendix 4A.2. Then:

$$\sigma_{r\varphi} = \eta\left(\frac{1}{r}\frac{\partial v_r}{\partial \varphi} + \frac{\partial v_\varphi}{\partial r} - \frac{v_\varphi}{r}\right) = \eta\left(\frac{\partial v_\varphi}{\partial r} - \frac{v_\varphi}{r}\right) = \eta\, r\frac{\partial}{\partial r}\left(\frac{v_\varphi}{r}\right) \qquad (4.107)$$

(It should be noted that $\sigma_{r\varphi}$ does vanish for solid-body rotation of the fluid, where $v_\varphi = \omega r$). By substituting the above result into Equation 4.106b, which indicates the absence of overall torque on an annular volume-element, we obtain an equivalent formulation of Equation 4.105b:

$$\frac{\partial}{\partial r}\left(r^3\frac{\partial}{\partial r}\left(\frac{v_\varphi}{r}\right)\right) = 0. \qquad (4.108)$$

- when the radii of both cylinders approach infinity, while their separation $d = R_2 - R_1$ remains constant, we recover the velocity field corresponding to the plane Couette flow;
- if $\Omega_1 = \Omega_2$, then $v_\varphi = \Omega\, r$, corresponding to solid-body rotation of the fluid;
- if $\Omega_2 = 0$ and $R_2 \to \infty$ (which amounts to eliminating the outer cylinder), then $v_\varphi = (\Omega_1 R_1^2)/r$. The result is the velocity field of a two-dimensional, irrotational vortex, which we discuss in Section 6.2.3;
- if the inner cylinder is the only one in rotation ($\Omega_1 \neq 0$, $\Omega_2 = 0$), the initial solution, discussed above, becomes unstable when the angular velocity, Ω_1, exceeds a critical value Ω_c. We show in Section 11.3.1 that, at that point, there appears a secondary flow in the form of toroidal rolls (*the Taylor-Couette instability*).

Let us now evaluate the tangential viscous torques exerted on the cylinders. Using Equation 4.107 to calculate $\sigma_{r\varphi}$, we find

$$\sigma_{r\varphi} = -\frac{2b\,\eta}{r^2}, \qquad (4.111)$$

where b is the coefficient of the $1/r$ term of the velocity field (Equation 4.109). Let us now evaluate the total viscous torque Γ_1 acting on the cylinder of radius R_1, per unit length along the z-axis. Γ_1 is equal to the product of the viscous stress $\sigma_{r\varphi}(R_1)$ by the surface $2\pi R_1$ on which the stress acts and by the radius R_1 of this surface. Thus the torque on the inner cylinder is:

$$\mathbf{\Gamma}_{R_1} = 2\,\pi R_1^2 \frac{2b\,\eta}{R_1^2}\mathbf{e}_z = -4\,\pi\,\eta\, b\,\mathbf{e}_z = 4\,\pi\,\eta\frac{(\Omega_2-\Omega_1)R_2^2R_1^2}{R_2^2-R_1^2}\mathbf{e}_z, \qquad (4.112)$$

where $\mathbf{e}_z$ is the unit vector along the z-axis. This relation shows that a solid-body rotation of the two cylinders at the same angular velocity does not result in a torque; a relative rotation ($\Omega_2 \neq \Omega_1$) is needed for a torque to be observed.

Equation 4.112 confirms the concept (already discussed in Section 4.4.1) that it is possible to determine the viscosity η from a measurement of the internal torque exerted by the fluid on one of the two cylinders when their angular velocities are different (Couette viscometer). In other devices, one applies a given torque to one of the cylinders and measures the resulting angular velocity.

In a steady-state regime, as indicated in Equation 4.111, the shear stress varies as $1/r^2$ with the distance from the axis of rotation. This result, a consequence of Equations 4.106a and 4.106b, reflects the balance of the torques between different layers of the fluid, and is also applicable to non-Newtonian fluids. The stress-strain relation can only be directly determined for the latter fluids if the stress can be assumed to be constant throughout the volume in which the measurement occurs, i.e. if the distance between the cylinders is sufficiently small relative to the average of their radii.

4.6 Simple one-dimensional, steady state flows of non-Newtonian fluids

For non-Newtonian fluids, the Navier–Stokes equation is no longer applicable. We must therefore use the general form 4.25 of the equation of motion, where the viscous component is no longer equal to $\eta\,\nabla^2\mathbf{v}$, and we must start from the general form $\nabla\cdot[\boldsymbol{\sigma}']$ of the viscous stresses.

We will first restrict our discussion to the case of fluids for which the characteristics are time-independent and isotropic, by assuming that a relationship exists between the shear rate and the viscous stresses which is well defined, and a unique function of the fluid. This is equivalent to assuming that we consider the shear thinning, or shear thickening, characteristics, but not thixotropy or visco-elasticity. The shear rate $\dot{\gamma}_{xy} = \partial v_x/\partial y$ is then not just proportional to the viscous stress but obeys a more complicated relationship $f(\sigma'_{xy})$

(Figure 4.8). [For flow with radial symmetry, we must rather use the relationship between the shear rate and the stress in polar coordinates (r, φ, z): $\partial v_z/\partial r = f(\sigma'_{zr})$].

Moreover, if we consider a flow in the x-direction between parallel planes, we can still assume that the pressure gradient $\partial p/\partial x$ is constant between the planes if a steady state flow is achieved with respect to x. In fact, even if transverse pressure gradients appear because of normal stresses, they are independent of the distance x and will not affect $\partial p/\partial x$ (the discussion of Section 4.5.1 remains valid).

4.6.1 Steady-state Couette plane flow

We will discuss a steady-state parallel flow, with the same geometry as that of Figure 4.18a: a fluid layer located between the two planes y and $y + \delta y$ must be in mechanical equilibrium; this requires that the condition $\sigma'_{xy}(y + \delta y) = \sigma'_{xy}(y)$ be satisfied for all values of y and δy, since they are the only components of the non-vanishing stresses in the x-direction. The viscous stress σ'_{xy} must therefore be constant throughout the fluid; this also means that the shear rate $\partial v_x/\partial y = f(\sigma'_{xy})$ must also be constant. We have thus, just as for a Newtonian fluid, a simple shear flow with velocity profile $v_x = V_0\, y/a$. Such a geometrical configuration would be, in theory, ideal for a direct measurement of the rheological characteristic $f(\sigma'_{xy})$, since both the shear rate and the viscous stress are well defined. Unfortunately, this condition is only approximately satisfied in the practical geometries described earlier (plane-cone viscometer, and cylindrical Couette viscometer).

4.6.2 One-dimensional flow between fixed walls

Let us go back to the two-dimensional steady-state flow between two horizontal walls $y = 0$ and $y = a$, as discussed in Section 4.5.3 for Newtonian fluids. Projecting the equation of motion 4.25 in the x-direction of the flow, we obtain:

$$\frac{\partial \sigma'_{xy}}{\partial y} = \frac{\partial p}{\partial x}. \tag{4.113}$$

Along the plane of symmetry, $y = a/2$, of the flow, we must have $\sigma'_{xy} = 0$, because there is no reason for momentum transport to occur preferentially in one direction rather than the opposite. By integrating Equation 4.113, we then obtain:

$$\sigma'_{xy} = \left[y - \frac{a}{2}\right]\frac{\partial p}{\partial x}. \tag{4.114}$$

Then, on the walls at $y = 0$ and $y = a$, one has respectively:

$$\sigma'_{wall}(y = 0) = -\frac{a}{2}\frac{\partial p}{\partial x} \quad (4.115a) \qquad \text{and:} \qquad \sigma'_{wall}(y = a) = \frac{a}{2}\frac{\partial p}{\partial x}. \quad (4.115b)$$

We observe that the force per unit area has the same absolute value on each of the walls. As we pointed out in Section 4.5.1 for Newtonian fluids, Equation 4.114 expresses the equilibrium between the x-components of the viscous stresses and of the pressure on volume elements of the fluid. We also note that both the velocity gradients and the viscous stresses vanish in the plane $y = a/2$, and reach a maximum in absolute value on the walls (this is a result of Equation 4.114, and of the fact that the function f increases monotonically).

If $\dot{\gamma}_{xy} = \partial v_x/\partial y = f(\sigma'_{xy})$, we obtain, by using Equation 4.114 for σ'_{xy}, and by integration from y to a, the velocity $v_x(y)$ at position y:

$$v_x(y) = -\int_y^a f\left(u\frac{\partial p}{\partial x}\right) du. \tag{4.116}$$

The flow rate Q between the two planes (per unit depth in the third, z-direction) is obtained by further integration in the y-direction:

$$Q = -2\int_{a/2}^{a}\left[\int_y^a f\left(u\frac{\partial p}{\partial x}\right) du\right] dy \tag{4.117}$$

(We integrate over half the interval, and double the result). Integrating then by parts, changes the integral into the even simpler form:

$$Q = -2 \int_{a/2}^{a} y\, f\left(y \frac{\partial p}{\partial x}\right) dy. \qquad (4.118)$$

For a Newtonian fluid, the function $f(u)$ simplifies to $f(u) = u/\eta$, and we recover the value $Q = -(a^3/12\eta)(\partial p/\partial x)$ (Equation 4.68). Note that the use of the above equations requires only that the function f be defined, monotonic and integrable. Specifically, we would have no problem in treating cases where the function f has a different value on the two sides of a threshold value of the stress.

We now calculate the shear rate at the wall, $\dot{\gamma}_{\text{wall}} = \partial v_x/\partial y(y = a)$, using the variation of the flow rate Q as a function of the pressure gradient, without using the function f. Combining Equations 4.114 and 4.115, we find at first:

$$y = \frac{a}{2} \frac{\sigma'_{xy}}{\sigma'_{\text{wall}}}. \qquad (4.119)$$

Substituting the right hand of Equation 4.119 for y, and $\dot{\gamma}_{xy}$ for the function f, Equation 4.118 becomes:

$$Q \sigma'^2_{\text{wall}} = -\frac{a^2}{2} \int_0^{\sigma'_{\text{wall}}} \sigma'_{xy} \dot{\gamma}_{xy}\, d\sigma'_{xy}. \qquad (4.120)$$

Taking the derivative of the above equation with respect to σ'_{wall} and observing that Q is a function of $\partial p/\partial x$, and thus of σ'_{wall}, we obtain:

$$2Q + \sigma'_{\text{wall}} \frac{dQ}{d\sigma'_{\text{wall}}} = -\frac{a^2}{2} \dot{\gamma}_{\text{wall}} \quad (4.121) \qquad \text{i.e.:} \qquad \dot{\gamma}_{\text{wall}} = -\frac{2Q}{a^2}\left(2 + \frac{d \operatorname{Log} Q}{d \operatorname{Log} \Delta p}\right). \quad (4.122)$$

We note that Equation 4.122, known as the Mooney–Rabinovitch equation, involves only macroscopic parameters which are relatively easy to measure (flow rate, pressure difference, etc.). Moreover, the wall stress σ'_{wall} results immediately from the equation $\sigma'_{\text{wall}} = (a/2)(\partial p/\partial x)$. The local rheological characteristic $\dot{\gamma}_{xy} = f\left(\sigma'_{xy}\right)$ can therefore be obtained from the global relationship between the flow rate and the pressure gradient along the flow, but only if the function f does not depend on time.

Corresponding results for cylindrical capillaries

Starting with Equation 4.25, and taking into account that the equation $\partial p/\partial r = 0$ still holds, we obtain:

$$\frac{\partial}{\partial r}(\sigma'_{zr} r) = r \frac{\partial p}{\partial z} \quad (4.123) \qquad \text{whence:} \qquad \sigma'_{zr} = \frac{r}{2}\frac{\partial p}{\partial z}. \quad (4.124)$$

We again assume in going from Equation 4.123 to Equation 4.124, that $\sigma'_{zr} = 0$ along the axis $r = 0$. Combining Equation 4.124 and the relationship between the stress and rate of strain $\partial v_z/\partial r = f(\sigma'_{zr})$, we obtain:

$$v_z(r) = -\int_r^R f\left(\frac{u}{2}\frac{\partial p}{\partial z}\right) du. \qquad (4.125)$$

We then calculate the flow rate Q by integrating $2\pi\, r\, v_z(r)$ along r. Integrating by parts, we finally obtain the equivalent of Equation 4.118:

$$Q = -\pi \int_0^R r^2 f\left(\frac{r}{2}\frac{\partial p}{\partial z}\right) dr. \qquad (4.126)$$

Replacing, just as in the case of the two parallel planes, r and $(\partial p/\partial z)$ by σ'_{zr} and the wall stress σ'_{wall}, we obtain:

$$Q\sigma'^{3}_{\mathrm{wall}} = -\pi R^3 \int_0^{\sigma'_{\mathrm{wall}}} \sigma'^{2}_{zr}\, \dot{\gamma}_{zr}\, \mathrm{d}\sigma'_{zr}. \tag{4.127}$$

Taking the derivative of this equation with respect to σ'_{wall} and then dividing by $\sigma'^{2}_{\mathrm{wall}}$, we finally obtain:

$$\dot{\gamma}_{\mathrm{wall}} = -\frac{Q}{\pi R^3}\left(3 + \frac{\mathrm{d\,Log}\, Q}{\mathrm{d\,Log}\,\Delta p}\right). \tag{4.128}$$

Here again, the above equation in combination with the relation between $\partial p/\partial z$, σ'_{wall} and the radius of the tube allows us to determine the rheological characteristics of the fluid from the relationship between the flow rate and the pressure difference. In practical terms, we carry out several measurements on tubes of the same diameter, but of differing lengths, in order to estimate and correct for the influence of tube end effects at the entrance. Also, comparing the results obtained along tubes of differing diameters allows us to detect possible effects of slip at the wall.

4.6.3 Velocity profiles for simple rheological behavior

Power-law fluids

Let us assume that, just as in Equation 4.40, $\sigma'_{xy} = D\,(\partial v_x/\partial y)^{(1-\alpha)}$. When $\alpha = 0$, this equation simplifies to that for Newtonian fluids with $D = \eta$. For $\alpha < 0$, the apparent viscosity $\sigma'_{xy}/(\partial v_x/\partial y)$ increases with the shear rate (shear thickening fluid). For $\alpha > 0$, the apparent viscosity decreases as a function of the shear rate (shear thinning fluid). We then have: $f(u) = D^{-1/(1-\alpha)}u^{1/(1-\alpha)}$. By substituting this expression into Equations 4.116 and 4.125, we obtain for the flow between two planes separated by a distance a and symmetric with respect to the plane $y = a/2$, the profile:

$$v_x(y) = \frac{1-\alpha}{2-\alpha}\left[\frac{1}{D}\left|\frac{\partial p}{\partial x}\right|\right]^{\frac{1}{1-\alpha}}\left(\left(\frac{a}{2}\right)^{\frac{2-\alpha}{1-\alpha}} - \left(\left|y-\frac{a}{2}\right|\right)^{\frac{2-\alpha}{1-\alpha}}\right) \tag{4.129a}$$

and, inside a cylindrical tube:

$$v_z(r) = \frac{1-\alpha}{2-\alpha}\left[\frac{1}{2D}\left|\frac{\partial p}{\partial z}\right|\right]^{\frac{1}{1-\alpha}}\left(R^{\frac{2-\alpha}{1-\alpha}} - r^{\frac{2-\alpha}{1-\alpha}}\right) \tag{4.129b}$$

[We multiply these expressions by –1 if $\partial p/\partial x > 0$ (for the plane geometry), or $\partial p/\partial z > 0$ (for the cylindrical case)]. The corresponding respective values for the flow rate are:

$$Q = \frac{2-2\alpha}{3-2\alpha}\left[\frac{1}{D}\left|\frac{\partial p}{\partial x}\right|\right]^{\frac{1}{1-\alpha}}\left(\frac{a}{2}\right)^{\frac{3-2\alpha}{1-\alpha}} \tag{4.130a}$$

and: $$Q = \frac{\pi(1-\alpha)}{4-3\alpha}\left[\frac{1}{2D}\left|\frac{\partial p}{\partial z}\right|\right]^{\frac{1}{1-\alpha}} R^{\frac{4-3\alpha}{1-\alpha}}. \tag{4.130b}$$

We see that, for $\alpha > 0$, the exponents for the law relating the change in flow rate to the apertures a or R of the channels, at a given pressure gradient, are higher than those (respectively

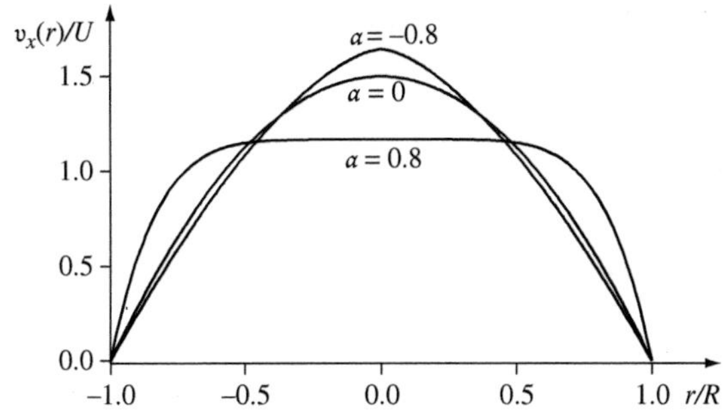

Figure 4.25 *Profile along a diameter of the ratio of the local velocity to the average velocity in a cylindrical tube for fluids displaying a power-law rheological behavior of the type* $\sigma'_{xy} = D\,(\partial v_x/\partial y)^{(1-\alpha)}$*; upper curve:* $\alpha = -0.8$ *(shear thickening fluid); middle curve:* $\alpha = 0$ *(Newtonian fluid); flattest curve:* $\alpha = 0.8$ *(shear thinning fluid)*

3 and 4) given by the Poiseuille laws in the case of Newtonian fluids; we can, in fact, rewrite the exponent for a as $3 + \alpha/(1-\alpha)$, and that for R as $4 + \alpha/(1-\alpha)$. The ratio between the rates of flow in two channels of different cross-sections, under a given pressure gradient is therefore increased by the shear thinning characteristics of the fluid. This difference is, on the contrary, reduced for a shear thickening fluid.

Let $U = Q/a$ and $U = Q/(\pi R^2)$ be the respective average values of the velocity in each of the two cases; we can rewrite the velocity profiles in the form:

$$\frac{v_x}{U} = \frac{3-2\alpha}{2-\alpha}\left[1-\left(2\left|\frac{y}{a}-\frac{1}{2}\right|\right)^{\frac{2-\alpha}{1-\alpha}}\right] \quad (4.131a) \qquad \text{and:} \qquad \frac{v_z}{U} = \frac{4-3\alpha}{2-\alpha}\left[1-\left(\frac{r}{R}\right)^{\frac{2-\alpha}{1-\alpha}}\right]. \quad (4.131b)$$

As we can see in Figure 4.25, the velocity profiles are sharper for the values $\alpha < 0$ (shear thickening fluids) than for a Newtonian fluid ($\alpha = 0$); moreover, the curvature is effectively infinite at $y = a/2$ or where $r = 0$. The velocity profiles obtained for $\alpha > 0$ (shear thinning fluids) are, on the other hand, increasingly flattened as the value of α increases.

Qualitatively, the shape of the profile increases the velocity gradient in the neighborhood of the wall and, thus, allows it to take advantage of an even lower effective viscosity; the effect is opposite in the case of a shear thickening fluid.

Bingham fluid between two parallel planes

As we have previously discussed, Bingham fluids are threshold fluids for which there exists a linear relationship between the stress and the strain beyond a threshold σ_c. We assume that:

$$\frac{\partial v_x}{\partial y} = \frac{1}{D}(\sigma'_{xy} - \text{sign}(\sigma'_{xy})\,\sigma_c) \quad \text{if} \quad |\sigma'_{xy}| > \sigma_c > 0 \qquad (4.132a)$$

and:

$$\frac{\partial v_x}{\partial y} = 0 \quad \text{if} \quad |\sigma'_{xy}| \le \sigma_c. \qquad (4.132b)$$

In this case, the function $f(u)$ is of the form $(1/D)(u - \text{sign}(u)\,\sigma_c)$ for $|u| > \sigma_c$ and $f(u) = 0$ for $|u| \le \sigma_c$. Hereafter, we assume that $\partial p/\partial x < 0$ and $v_x \ge 0$. The equation $\sigma'_{xy} = (y - a/2)(\partial p/\partial x)$ remains valid, and the stress σ'_{xy} vanishes for $y = 0$. We therefore have a region near the plane of symmetry where $\sigma'_{xy} < \sigma_c$ and where $\partial v_x/\partial y$ is thus zero, such that the velocity has a constant value V_{max}. The boundaries y_1 and y_2 of this region are symmetric with respect to the plane $y = a/2$ and satisfy the condition:

$$y_1 - \frac{a}{2} = -\left[y_2 - \frac{a}{2}\right] = \frac{\sigma_c}{\left|\frac{\partial p}{\partial x}\right|}. \qquad (4.133)$$

If $y_1 > a$, the velocity is everywhere zero. If $y_1 < a$, the velocity profile near the wall $y = a$ is obtained by simple integration of Equation 4.132a:

$$v_x(y) = \frac{1}{D}\int_y^a \left[u\left|\frac{\partial p}{\partial x}\right| - \sigma_c\right] du \tag{4.134}$$

and we have a symmetric profile near the wall $y = 0$. We then obtain, by using Equations 4.132a and 4.132b, the velocity profiles:

$$v_x(y) = \frac{1}{2D}\left|\frac{\partial p}{\partial x}\right| [a - y]\ [a + y - 2y_1] \tag{4.135}$$

and:

$$v_x(y) = V_{max} = \frac{1}{2D}\left|\frac{\partial p}{\partial x}\right| [a - y_1]^2, \tag{4.136}$$

respectively for $y > y_1$ and $0 < y \leq y_1$ (we have a symmetric velocity profile in the region $y < a/2$). Qualitatively, we observe the shapes of velocity profiles shown in Figure 4.26.

Where the stress σ'_{xy} is everywhere less than the threshold stress, the fluid is at rest (top graph). As the applied pressure gradient increases, we observe velocity profiles with a flat region in the middle (region of *plug flow*) whose extent becomes smaller as the applied pressure gradient increases (velocity profiles in the center and bottom graphs). More generally, we observe, looking at Figures 4.25 and 4.26 that the existence of a non-zero yield stress or of a viscosity which decreases with the shear stress (which plays a similar role) leads in both cases to a flattening of the velocity profile.

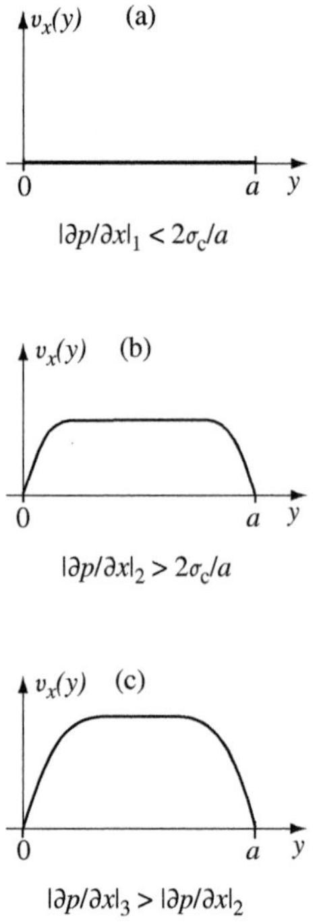

Figure 4.26 *Velocity profiles observed for the case of a Bingham fluid, for different values of the longitudinal pressure gradient. (Note that the slope is continuous at the points where the flat region connects to the regions on each of its sides)*

Bingham fluid in a cylindrical tube

The problem of a Bingham fluid in a cylindrical tube is solved like that of the flow between two parallel planes by using Equation 4.124: $\sigma'_{zr} = (r/2)\ (\partial p/\partial z)$. In the central area of the profile, the velocity gradient $\partial v_z/\partial r$ is zero all the way to the radius $r_0 = 2\sigma_c/\ |\partial p/\partial x|$. In this region we have a constant velocity V_{max} which is zero if $r_0 > R$. If $r_0 < R$, the velocity profile near the wall satisfies the equation:

$$v_z(r) = \frac{1}{D}\int_r^R \left[\frac{u}{2}\left|\frac{\partial p}{\partial z}\right| - \sigma_c\right] du. \tag{4.137}$$

We then obtain the respective velocity profiles for $r > r_0$ and $r < r_0$:

$$v_z(r) = \frac{1}{4D}\left|\frac{\partial p}{\partial z}\right| [R - r]\ [R + r - 2r_0] \tag{4.138a}$$

and:

$$v_z(r) = V_{max} = \frac{1}{4D}\left|\frac{\partial p}{\partial z}\right| [R - r_0]^2. \tag{4.138b}$$

These profiles are shown qualitatively in Figure 4.26.

4.6.4 Flow of a viscoelastic fluid near an oscillating plane

We have discussed, in Section 4.5.4., the flow of a Newtonian fluid near a solid plane oscillating parallel to itself, with a frequency ω. Let us now assume that the fluid is a viscoelastic Maxwell fluid with rheological characteristics described by Equation 4.51b. We shall follow in this case the same procedure as in the previous one, by assuming that the velocity

v_x parallel to the plane satisfies $v_x(y, t) = f(y)\cos(\omega t + \varphi) = \Re e(f(y)\, e^{i\omega t})$(Figure 4.22). In Equation 4.85, we replace the viscosity ν by the complex viscosity $\bar{\nu} = \bar{\eta}/\rho$ which has the form (see Equation 4.51b):

$$\bar{\nu} = \frac{\nu}{1 + i\omega\tau}. \tag{4.139}$$

The characteristic response time of the fluid is related to the elastic modulus G and the dynamic viscosity, $\eta = \nu/\rho$, by $\tau = \eta/G$. Solving Equation 4.85 and using Equation 4.139 for the viscosity, we find solutions of the form:

$$f(y) = C_1 e^{-(\beta + ik)y} + C_2 e^{(\beta + ik)y}, \tag{4.140}$$

where, in contrast to Newtonian case (Equation 4.86), we do not necessarily have $\beta = k$, and where β and k satisfy the equation:

$$i\omega = \nu \frac{1 - i\omega\tau}{1 + \omega^2\tau^2}(\beta^2 - k^2 + 2i\beta k). \tag{4.141}$$

If we consider only flows in the semi-infinite space $y > 0$, we can assume (as for the case of Newtonian fluid) $C_2 = 0$ with $\beta > 0$. The real and imaginary parts of Equation 4.141 lead us to the two relations:

$$\left(\beta^2 - k^2\right) + 2\,\beta k\,\omega\tau = 0 \tag{4.142a}$$

and:

$$\omega\left(1 + \omega^2\tau^2\right) = 2\beta\nu\,k - \nu\omega\tau\left(\beta^2 - k^2\right). \tag{4.142b}$$

Substituting the value of $\beta^2 - k^2$ from one equation into the other one, we find the dependence:

$$\beta\,k = \frac{\omega}{2\nu}. \tag{4.143}$$

This expression thus requires that $k > 0$ if $\beta > 0$. Substituting further in Equation 4.142a, and taking the positive square root of the result obtained, we finally obtain the equation:

$$\frac{k^2}{\omega^2} = \frac{1}{2}\left[\frac{\tau}{\nu} + \frac{\tau}{\nu}\sqrt{1 + \frac{1}{\omega^2\tau^2}}\right]. \tag{4.144}$$

From this, we can determine the value for β from Equation 4.141. At low frequencies, such that $\omega\tau \ll 1$, Equation 4.144 reduces to $k^2 = \omega/2\nu$ and, from Equation 4.143, β has the same value. This variation is identical to that for a Newtonian fluid (Equation 4.86): this was to be expected since the complex viscosity $\bar{\nu}$, according to Equation 4.139, becomes essentially equal to ν. In the opposite limit, where $\omega\tau \gg 1$ (with the period of sinusoidal excitation short relative to the characteristic response time τ of the viscoelastic fluid), Equation 4.144 becomes:

$$\frac{k^2}{\omega^2} = \frac{\tau}{\nu} = \frac{\rho}{G} = \frac{1}{c_s^2}. \tag{4.145}$$

The ratio ω/k is then equal to the velocity of propagation c_s of transverse shear waves in a solid with shear modulus G and density ρ: for high excitation frequencies, the medium behaves as an elastic solid (the complex modulus $\bar{G}$ in Equation 4.51 then simplifies to the real value G) in which such waves can propagate, in contrast to a Newtonian fluid in which they attenuate very rapidly.

The distance $1/\beta$ is the characteristic attenuation length for velocity oscillations induced by the motion of the plane; the ratio k/β thus represents the order of magnitude of the number of wavelengths over which attenuation occurs. By means of Equations 4.143 and 4.145, and using the definition (Equation 4.89) of the penetration depth of oscillations $\delta(\omega)$ in a Newtonian fluid, we obtain:

$$2\,k^2\,\delta^2(\omega) = 4\omega\tau \tag{4.146}$$

and:

$$\frac{k}{\beta} = \frac{k^2}{\omega/2\nu} = k^2\delta^2(\omega). \tag{4.147}$$

The condition $\omega\tau \gg 1$ (elastic behavior of the fluid) thus implies both $k\delta(\omega) \gg 1$ (wavelength much shorter than the viscous penetration depth) and $k/\beta \gg 1$ (large number of wavelengths over the attenuation distance of the velocity oscillations). Since Equation 4.143 can be rewritten in the form: $\beta\,\delta(\omega) = 1/(k\,\delta(\omega)) \ll 1$, we can also infer that the attenuation distance $1/\beta$ in the elastic regime is much greater than the penetration depth $\delta(\omega)$ which would result only from viscosity effects. We note, however, that, for real fluids such as those discussed in Figure 4.14, the deviations from the Maxwell relation observed at high frequencies will in practice reduce the distance over which the waves propagate.

4A Appendix - Representation of the equations of fluid mechanics in different systems of coordinates

4A.1 Representation of the stress-tensor, the equation of conservation of mass and the Navier–Stokes equations in Cartesian coordinates (x, y, z)

Stress Tensor

$$\sigma_{xx} = -p + 2\eta \frac{\partial v_x}{\partial x} \qquad \sigma_{yz} = \eta \left(\frac{\partial v_y}{\partial z} + \frac{\partial v_z}{\partial y} \right)$$

$$\sigma_{yy} = -p + 2\eta \frac{\partial v_y}{\partial y} \qquad \sigma_{zx} = \eta \left(\frac{\partial v_z}{\partial x} + \frac{\partial v_x}{\partial z} \right)$$

$$\sigma_{zz} = -p + 2\eta \frac{\partial v_z}{\partial z} \qquad \sigma_{xy} = \eta \left(\frac{\partial v_x}{\partial y} + \frac{\partial v_y}{\partial x} \right)$$

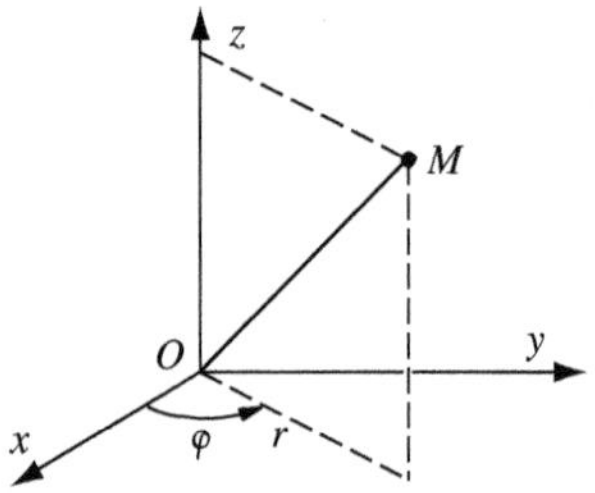

Figure 4A.1 *Definition and orientation of cylindrical coordinates*

Equation of conservation of mass for an incompressible fluid ($\nabla \cdot \mathbf{v} = 0$)

$$\frac{\partial v_x}{\partial x} + \frac{\partial v_y}{\partial y} + \frac{\partial v_z}{\partial z} = 0$$

Navier–Stokes equations

$$\frac{\partial v_x}{\partial t} + v_x \frac{\partial v_x}{\partial x} + v_y \frac{\partial v_x}{\partial y} + v_z \frac{\partial v_x}{\partial z} = -\frac{1}{\rho} \frac{\partial p}{\partial x} + \nu \left(\frac{\partial^2 v_x}{\partial x^2} + \frac{\partial^2 v_x}{\partial y^2} + \frac{\partial^2 v_x}{\partial z^2} \right) + f_x$$

$$\frac{\partial v_y}{\partial t} + v_x \frac{\partial v_y}{\partial x} + v_y \frac{\partial v_y}{\partial y} + v_z \frac{\partial v_y}{\partial z} = -\frac{1}{\rho} \frac{\partial p}{\partial y} + \nu \left(\frac{\partial^2 v_y}{\partial x^2} + \frac{\partial^2 v_y}{\partial y^2} + \frac{\partial^2 v_y}{\partial z^2} \right) + f_y$$

$$\frac{\partial v_z}{\partial t} + v_x \frac{\partial v_z}{\partial x} + v_y \frac{\partial v_z}{\partial y} + v_z \frac{\partial v_z}{\partial z} = -\frac{1}{\rho} \frac{\partial p}{\partial z} + \nu \left(\frac{\partial^2 v_z}{\partial x^2} + \frac{\partial^2 v_z}{\partial y^2} + \frac{\partial^2 v_z}{\partial z^2} \right) + f_z$$

4A.2 Representation of the stress-tensor, the equation of conservation of mass, and the Navier–Stokes equations in cylindrical coordinates (r, φ, z) (see Figure 4A.1 for the definition and orientation of the cylindrical coordinates)

Stress Tensor

$$\sigma_{rr} = -p + 2\eta \frac{\partial v_r}{\partial r} \qquad \sigma_{\varphi z} = \eta \left(\frac{\partial v_\varphi}{\partial z} + \frac{1}{r} \frac{\partial v_z}{\partial \varphi} \right)$$

$$\sigma_{\varphi\varphi} = -p + 2\eta \left(\frac{1}{r} \frac{\partial v_\varphi}{\partial \varphi} + \frac{v_r}{r} \right) \qquad \sigma_{zr} = \eta \left(\frac{\partial v_z}{\partial r} + \frac{\partial v_r}{\partial z} \right)$$

$$\sigma_{zz} = -p + 2\eta \frac{\partial v_z}{\partial z} \qquad \sigma_{r\varphi} = \eta \left(\frac{1}{r} \frac{\partial v_r}{\partial \varphi} + \frac{\partial v_\varphi}{\partial r} - \frac{v_\varphi}{r} \right)$$

Equation of conservation of mass for an incompressible fluid ($\nabla \cdot \mathbf{v} = 0$)

$$\frac{1}{r}\frac{\partial(r v_r)}{\partial r} + \frac{1}{r}\frac{\partial v_\varphi}{\partial \varphi} + \frac{\partial v_z}{\partial z} = 0$$

Navier–Stokes equations

$$\frac{\partial v_r}{\partial t} + v_r\frac{\partial v_r}{\partial r} + \frac{v_\varphi}{r}\frac{\partial v_r}{\partial \varphi} + v_z\frac{\partial v_r}{\partial z} - \frac{v_\varphi^2}{r} =$$

$$-\frac{1}{\rho}\frac{\partial p}{\partial r} + \nu\left(\frac{\partial^2 v_r}{\partial r^2} + \frac{1}{r^2}\frac{\partial^2 v_r}{\partial \varphi^2} + \frac{\partial^2 v_r}{\partial z^2} + \frac{1}{r}\frac{\partial v_r}{\partial r} - \frac{2}{r^2}\frac{\partial v_\varphi}{\partial \varphi} - \frac{v_r}{r^2}\right) + f_r \qquad \text{[}r\text{-component]}$$

$$\frac{\partial v_\varphi}{\partial t} + v_r\frac{\partial v_\varphi}{\partial r} + \frac{v_\varphi}{r}\frac{\partial v_\varphi}{\partial \varphi} + v_z\frac{\partial v_\varphi}{\partial z} + \frac{v_r v_\varphi}{r} =$$

$$-\frac{1}{\rho}\left(\frac{1}{r}\frac{\partial p}{\partial \varphi}\right) + \nu\left(\frac{\partial^2 v_\varphi}{\partial r^2} + \frac{1}{r^2}\frac{\partial^2 v_\varphi}{\partial \varphi^2} + \frac{\partial^2 v_\varphi}{\partial z^2} + \frac{1}{r}\frac{\partial v_\varphi}{\partial r} + \frac{2}{r^2}\frac{\partial v_r}{\partial \varphi} - \frac{v_\varphi}{r^2}\right) + f_\varphi \qquad \text{[}\varphi\text{-component]}$$

$$\frac{\partial v_z}{\partial t} + v_r\frac{\partial v_z}{\partial r} + \frac{v_\varphi}{r}\frac{\partial v_z}{\partial \varphi} + v_z\frac{\partial v_z}{\partial z} =$$

$$-\frac{1}{\rho}\frac{\partial p}{\partial z} + \nu\left(\frac{\partial^2 v_z}{\partial r^2} + \frac{1}{r^2}\frac{\partial^2 v_z}{\partial \varphi^2} + \frac{\partial^2 v_z}{\partial z^2} + \frac{1}{r}\frac{\partial v_z}{\partial r}\right) + f_z \qquad \text{[}z\text{-component]}$$

4A.3 Representation of the stress-tensor, the equation of conservation of mass, and the Navier–Stokes equations in spherical polar coordinates (r, θ, φ)

Stress Tensor

$$\sigma_{rr} = -p + 2\eta\frac{\partial v_r}{\partial r} \qquad \sigma_{\theta\varphi} = \eta\left(\frac{1}{r\sin\theta}\frac{\partial v_\theta}{\partial \varphi} + \frac{1}{r}\frac{\partial v_\varphi}{\partial \theta} - \frac{v_\varphi \cot\theta}{r}\right)$$

$$\sigma_{\theta\theta} = -p + 2\eta\left(\frac{1}{r}\frac{\partial v_\theta}{\partial \theta} + \frac{v_r}{r}\right) \qquad \sigma_{\varphi r} = \eta\left(\frac{\partial v_\varphi}{\partial r} - \frac{v_\varphi}{r} + \frac{1}{r\sin\theta}\frac{\partial v_r}{\partial \varphi}\right)$$

$$\sigma_{\varphi\varphi} = -p + 2\eta\left(\frac{1}{r\sin\theta}\frac{\partial v_\varphi}{\partial \varphi} + \frac{v_r}{r} + \frac{v_\theta \cot\theta}{r}\right) \qquad \sigma_{r\theta} = \eta\left(\frac{1}{r}\frac{\partial v_r}{\partial \theta} + \frac{\partial v_\theta}{\partial r} - \frac{v_\theta}{r}\right)$$

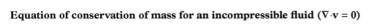

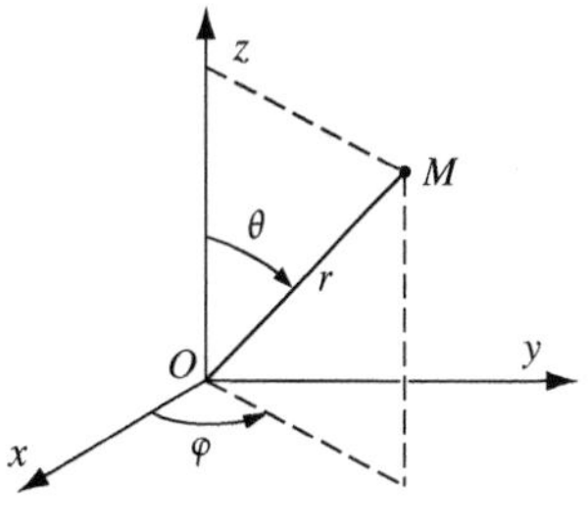

Figure 4A.2 *Definition and orientation of spherical polar coordinates*

Equation of conservation of mass for an incompressible fluid ($\nabla \cdot \mathbf{v} = 0$)

$$\frac{1}{r^2}\frac{\partial(r^2 v_r)}{\partial r} + \frac{1}{r\sin\theta}\frac{\partial(\sin\theta\ v_\theta)}{\partial \theta} + \frac{1}{r\sin\theta}\frac{\partial v_\varphi}{\partial \varphi} = \frac{\partial v_r}{\partial r} + 2\frac{v_r}{r} + \frac{1}{r}\frac{\partial v_\theta}{\partial \theta} + \frac{\cot\theta\ v_\theta}{r} + \frac{1}{r\sin\theta}\frac{\partial v_\varphi}{\partial \varphi} = 0$$

Navier–Stokes equations

$$\frac{\partial v_r}{\partial t} + v_r \frac{\partial v_r}{\partial r} + \frac{v_\theta}{r}\frac{\partial v_r}{\partial \theta} + \frac{v_\varphi}{r\sin\theta}\frac{\partial v_r}{\partial \varphi} - \frac{v_\theta^2 + v_\varphi^2}{r} =$$

$$-\frac{1}{\rho}\frac{\partial p}{\partial r} + \nu\left(\frac{1}{r}\frac{\partial^2 (r v_r)}{\partial r^2} + \frac{1}{r^2}\frac{\partial^2 v_r}{\partial \theta^2} + \frac{1}{r^2 \sin^2\theta}\frac{\partial^2 v_r}{\partial \varphi^2} + \frac{\cot\theta}{r^2}\frac{\partial v_r}{\partial \theta}\right.$$

$$\left. -\frac{2}{r^2}\frac{\partial v_\theta}{\partial \theta} - \frac{2}{r^2\sin\theta}\frac{\partial v_\varphi}{\partial \varphi} - \frac{2 v_r}{r^2} - \frac{2\cot\theta}{r^2} v_\theta\right) + f_r \qquad \text{[}r\text{-component]}$$

$$\frac{\partial v_\theta}{\partial t} + v_r \frac{\partial v_\theta}{\partial r} + \frac{v_\theta}{r}\frac{\partial v_\theta}{\partial \theta} + \frac{v_\varphi}{r\sin\theta}\frac{\partial v_\theta}{\partial \varphi} + \frac{v_r v_\theta}{r} - \frac{v_\varphi^2 \cot\theta}{r} =$$

$$-\frac{1}{\rho}\left(\frac{1}{r}\frac{\partial p}{\partial \theta}\right) + \nu\left(\frac{1}{r}\frac{\partial^2 (r v_\theta)}{\partial r^2} + \frac{1}{r^2}\frac{\partial^2 v_\theta}{\partial \theta^2} + \frac{1}{r^2 \sin^2\theta}\frac{\partial^2 v_\theta}{\partial \varphi^2} + \frac{\cot\theta}{r^2}\frac{\partial v_\theta}{\partial \theta}\right.$$

$$\left. -\frac{2\cos\theta}{r^2\sin^2\theta}\frac{\partial v_\varphi}{\partial \varphi} + \frac{2}{r^2}\frac{\partial v_r}{\partial \theta} - \frac{v_\theta}{r^2\sin^2\theta}\right) + f_\theta \qquad \text{[}\theta\text{-component]}$$

$$\frac{\partial v_\varphi}{\partial t} + v_r \frac{\partial v_\varphi}{\partial r} + \frac{v_\theta}{r}\frac{\partial v_\varphi}{\partial \theta} + \frac{v_\varphi}{r\sin\theta}\frac{\partial v_\varphi}{\partial \varphi} + \frac{v_r v_\varphi}{r} + \frac{v_\theta v_\varphi \cot\theta}{r} =$$

$$-\frac{1}{\rho}\left(\frac{1}{r\sin\theta}\frac{\partial p}{\partial \varphi}\right) + \nu\left(\frac{1}{r}\frac{\partial^2 (r v_\varphi)}{\partial r^2} + \frac{1}{r^2}\frac{\partial^2 v_\varphi}{\partial \theta^2} + \frac{1}{r^2 \sin^2\theta}\frac{\partial^2 v_\varphi}{\partial \varphi^2} + \frac{\cot\theta}{r^2}\frac{\partial v_\varphi}{\partial \theta}\right.$$

$$\left. +\frac{2}{r^2\sin\theta}\frac{\partial v_r}{\partial \varphi} + \frac{2\cos\theta}{r^2\sin^2\theta}\frac{\partial v_\theta}{\partial \varphi} - \frac{v_\varphi}{r^2\sin^2\theta}\right) + f_\varphi \qquad \text{[}\varphi\text{-component]}$$

EXERCISES

1) **Laminar Poiseuille flow between two coaxial tubes.**
A Newtonian fluid (viscosity η) fills the space between two coaxial cylindrical tubes with axis z, constant radii R_1 and $R_2 > R_1$ and length L. A pressure difference Δp is applied between the ends of the tubes and we neglect the entry effects. Assuming a parallel flow $v_z(r)$ and using Equations 4.73, determine the velocity profile and show that the flow rate Q is: $Q = \frac{\pi\,\Delta p\left(R_2^2 - R_1^2\right)}{8\eta L}\left[R_2^2 + R_1^2 - \frac{R_2^2 - R_1^2}{\mathrm{Log}(R_2/R_1)}\right]$. Apply this result in the two limits $R_1 \to 0$ and $R_1 \to R_2$ and indicate the corresponding situations.

2) **Flow of a Bingham fluid on a tilted plane**
We consider the flow of a layer of Bingham fluid (described in Section 4.6.3) of constant thickness h deposited upon a plane surface (x,z). The z-axis is horizontal and the x-axis is tilted at an angle α (> 0) with respect to the horizontal. The free surface is in contact with air at a pressure p_0, and we assume that the normal stresses are isotropic. The flow (if any) is parallel to x with a velocity $v_x(y)$ $\left(v_y = v_z = 0\right)$. Write the x- and y-components of the equation of motion (Equation 4.25) and the boundary conditions.

What is the pressure variation in the fluid volume? Solve the equations of motion and study the change of the profile $v_x(y)$ as the angle α increases from 0.

3) **Dynamics of a meniscus rising in a cylindrical tube.**
The bottom end of an open vertical cylindrical tube is put in contact at $t = 0$ with the horizontal surface of large area of a liquid of density ρ, viscosity η and surface tension γ. The wetting angle is θ, and the diameter d of the tube is small compared to the capillary length l_c. What is the variation with time of the height $h(t)$ of the air–liquid interface in the tube (the variation of the velocity is slow enough so that acceleration terms are negligible)? Write explicitly the variation $h(t)$ at short and long times.

4) **Poiseuille flow in an elastic tube.**
A stationary Poiseuille flow of flow rate Q of a fluid of viscosity η is established in an horizontal tube of circular cross-section $A(x)$ and of length L parallel to the direction x; the walls of the tube are elastic (e.g. as those of blood vessels) and $A(x)$ satisfies: $A(x)=A_0(1+K(p(x)-p_0))$ where K is a constant related to the elasticity of the tube. The pressure has the same value p_0 outside the wall and at the outlet $x = L$; $p(x)$ is the pressure inside the tube at the distance x from the inlet. The Poiseuille relation 4.77 is assumed to remain valid in a given section x=*cst*. Use it to write the equation satisfied by the variation of the area $A(x)$ as a function of x and show that: $A = A_0\left(1 + (24\pi\,\eta\,Q\,K/A_0^2)(L-x)\right)^{1/3}$. What is then the variation of the pressure drop $p(0)$-p_0 as a function of the flow rate Q?

Conservation Laws

This chapter deals with various conservation laws for a moving fluid: conservation of mass, of momentum and of energy. The conservation of circulation of the velocity (angular momentum) will be treated in detail in Chapter 7. The question of the conservation of mass, already discussed in Chapter 3, is only briefly restated here (Section 5.1).

The equations of motion for real fluids have been derived in the previous chapter. Together with the equation of conservation of mass, they lead to the derivation of the equation of conservation of momentum (Section 5.2). By applying this equation to a suitably chosen volume within the fluid, called the control volume, we are able to analyze the exchange of momentum in simple flows.

We thereupon discuss the equation of conservation of energy, stated in the form of Bernoulli's equation (Section 5.3), applying it then to a number of classical examples (Pitot tube, Venturi gauge, etc.).

Finally, we conclude (Section 5.4) by dealing with a few more complex problems. These latter illustrate how we can analyze quantitatively a number of flows by means of the conservation laws, without requiring a complete determination of the velocity field in the fluid.

5.1 Equation of conservation of mass

The equation of conservation of mass was derived earlier in Section 3.3.1. We recall here the two equivalent ways of expressing it, corresponding to the two different reference frames, i.e. the Eulerian and Lagrangian frames:

Eulerian viewpoint (Equation 3.25): $\dfrac{\partial \rho}{\partial t} + \mathbf{\nabla}.\,(\rho \mathbf{v}) = \dfrac{\partial \rho}{\partial t} + \dfrac{\partial \left(\rho v_j \right)}{\partial x_j} = 0.$ (5.1)

Lagrangian viewpoint (Equation 3.27): $\dfrac{\mathrm{d} \rho}{\mathrm{d} t} + \rho \, \mathbf{\nabla} \cdot \mathbf{v} = \dfrac{\mathrm{d} \rho}{\mathrm{d} t} + \rho \dfrac{\partial v_j}{\partial x_j} = 0.$ (5.2)

The Lagrangian (or convective) derivative $\mathrm{d}\rho/\mathrm{d}t$ refers to the change in density of a particle of fluid that we follow along in its motion (just as in the previous chapters, we retain the Einstein convention of summing over repeated indices).

In situations where there is a source term for the fluid, e.g., a reactive flow in which a certain chemical species A of partial density ρ_A can be formed, Equation 5.1 then becomes

$$\frac{\partial \rho_A}{\partial t} = -\mathbf{\nabla}\cdot(\rho_A \mathbf{v}) + q_A. \qquad (5.3)$$

Physical Hydrodynamics. Second Edition. Etienne Guyon *et al.*
© Oxford University Press 2015. Published in 2015 by Oxford University Press.

Here, q_A indicates the rate at which the density of species A is being produced per unit time. Equation 5.3 expresses the rate of change of density $\partial \rho_A / \partial t$ in terms of the negative divergence, or influx, of $\rho_A \mathbf{v}$ together with a volume source term q_A.

The presence of a spatial gradient of concentration of species A causes a diffusive flux $\mathbf{J}_{DA}$ proportional to this gradient ($\mathbf{J}_{DA} = -D_A \nabla C_A$). We will need to add $\mathbf{J}_{DA}$ to ρ_A $\mathbf{v}$ in the first term of the right-hand side of Equation 5.3. We will see that the expression thus obtained corresponds to the most general form of conservation equations.

5.2 Conservation of momentum

5.2.1 The local equation

The momentum per unit volume of a fluid is equal to $\rho \mathbf{v}$. The time derivative of its ith component ($i = x, y, z$) is then:

$$\frac{\partial (\rho v_i)}{\partial t} = v_i \frac{\partial \rho}{\partial t} + \rho \frac{\partial v_i}{\partial t}. \tag{5.4}$$

We then combine Equations 5.4 and 5.1 (after having multiplied the latter by v_i), together with the general form of the equation of motion 4.25 for a fluid (restated below) along the x_i-axis:

$$\rho \frac{\partial v_i}{\partial t} = -\rho v_j \frac{\partial v_i}{\partial x_j} - \frac{\partial p}{\partial x_i} + \frac{\partial \sigma'_{ij}}{\partial x_j} + \rho f_i, \tag{5.5}$$

where ρf_i is a force per unit volume, and σ'_{ij} is the viscous stress tensor. We then obtain:

$$\frac{\partial}{\partial t}(\rho v_i) = -v_i \frac{\partial (\rho v_j)}{\partial x_j} - \rho v_j \frac{\partial v_i}{\partial x_j} - \frac{\partial p}{\partial x_i} + \frac{\partial \sigma'_{ij}}{\partial x_j} + \rho f_i,$$

i.e.:

$$\frac{\partial}{\partial t}(\rho v_i) = -\frac{\partial}{\partial x_j}\left(\rho v_i v_j + p\delta_{ij} - \sigma'_{ij}\right) + \rho f_i. \tag{5.6}$$

In the above derivation, we have included the pressure term $\partial p / \partial x_i$ in the parenthesis as $\partial (p\delta_{ij}) / \partial x_j$. In this very general form, Equation 5.6 is valid for all fluids, Newtonian or not, compressible or incompressible. We find here a term $\partial (\rho v_i) / \partial t$ for the rate of change of momentum, the divergence of a flux ($\rho v_i v_j + p\delta_{ij} - \sigma'_{ij}$) and a source term ρf_i. In order to understand more precisely the physical significance of these various terms, we look for a global expression of Equation 5.6, by integrating over a macroscopic volume of the fluid.

5.2.2 The integral expression of the law of conservation of momentum

Integral of the equation of conservation of momentum

Going back to the equation above, and integrating over a volume $\mathcal{V}$ fixed in space (such that particles of fluid can cross its outer boundaries), we obtain:

$$\iiint_{\mathcal{V}} \frac{\partial (\rho v_i)}{\partial t} \mathrm{d}V = -\iiint_{\mathcal{V}} \frac{\partial}{\partial x_j}\left(\rho v_i v_j + p\,\delta_{ij} - \sigma'_{ij}\right) \mathrm{d}V + \iiint_{\mathcal{V}} \rho f_i \,\mathrm{d}V. \tag{5.7}$$

Transforming now the first integral on the right-hand side by means of the divergence theorem, we have further:

$$\iiint_{\mathcal{V}} \frac{\partial (\rho v_i)}{\partial t} \mathrm{d}V = -\iint_{S}\left(\rho v_i v_j + p\,\delta_{ij} - \sigma'_{ij}\right) n_j \,\mathrm{dS} + \iiint_{\mathcal{V}} \rho f_i \,\mathrm{d}V, \tag{5.8}$$

$\mathcal{S}$ represents the surface bounding the volume of integration $\mathcal{V}$. Using the fact that this volume $\mathcal{V}$ is fixed, we can write for the time-derivative of the ith component of the total momentum within $\mathcal{V}$:

$$\frac{\mathrm{d}}{\mathrm{d}t}\left(\iiint_{\mathcal{V}} \rho\, v_i\, \mathrm{d}V\right) = -\iint_{\mathcal{S}} \left(\rho\, v_i v_j + p\,\delta_{ij} - \sigma'_{ij}\right) n_j\, \mathrm{d}S + \iiint_{\mathcal{V}} \rho f_i\, \mathrm{d}V. \qquad (5.9)$$

Or, in vector form, with **n** being the outward normal to the volume V:

$$\frac{\mathrm{d}}{\mathrm{d}t}\left(\iiint_{\mathcal{V}} \rho\,\mathbf{v}\,\mathrm{d}V\right) = -\iint_{\mathcal{S}} (\rho\,\mathbf{v}\,(\mathbf{v}\cdot\mathbf{n}) + p\,\mathbf{n} - [\boldsymbol{\sigma}']\cdot\mathbf{n})\mathrm{d}\,S + \iiint_{\mathcal{V}} \rho\,\mathbf{f}\,\mathrm{d}V. \qquad (5.10)$$

We can write Equation 5.10 in the following alternative form (where we take $\mathbf{f} = \mathbf{g}$):

$$\frac{\mathrm{d}}{\mathrm{d}t}\left(\iiint_{\mathcal{V}} \rho\,\mathbf{v}\,\mathrm{d}V\right) = -\iint_{\mathcal{S}} (\rho\,\mathbf{v}(\mathbf{v}\cdot\mathbf{n}) - [\boldsymbol{\sigma}']\cdot\mathbf{n})\,\mathrm{d}S + \iiint_{\mathcal{V}} (\rho\,\mathbf{g} - \nabla p)\,\mathrm{d}V. \qquad (5.14)$$

This alternate expression makes it more obvious that it is the difference between the pressure gradient and the hydrostatic pressure gradient (equal to $-\rho\,\mathbf{g}$), and not the pressure itself, that generates momentum. The change of momentum within a volume of fluid is then the difference between this contribution and the fluxes due to viscous diffusion and convection through the boundaries. For instance, in some numerical simulations, one applies a volume force rather than a pressure gradient in order to create flow.

The first integral on the right-hand side of the above equation describes the contribution to the change of momentum of the flux through surface $\mathcal{S}$. Momentum being a vector quantity, the corresponding flux is a second-rank tensor having the most general form:

$$\Pi_{ij} = \rho\, v_i\, v_j + p\,\delta_{ij} - \sigma'_{ij}, \qquad (5.11)$$

Π_{ij} is called the *momentum flux tensor* per unit area: it is the flux in the j-direction of the ith component of the momentum. It consists of three terms:

- $\rho\, v_i v_j$, the transport of the ith component of the momentum, ρv_i, by particles moving in the j-direction;
- $p\,\delta_{ij}$, the momentum transport associated with pressure forces;
- $-\sigma'_{ij}$, the transport of momentum associated with the viscous friction forces.

The integral $-\iint_{\mathcal{S}} p\,\mathbf{n}\,\mathrm{d}S$ is the resultant of the pressure forces exerted normal to the surface $\mathcal{S}$.

The integral $\iint_{\mathcal{S}} [\boldsymbol{\sigma}']\cdot\mathbf{n}\,\mathbf{dS}$ is equal to the tangential component of the viscous friction force exerted on the surface $\mathcal{S}$.

The second integral on the right-hand side represents the rate of generation of momentum, due to the external force field **f**, within the volume under consideration.

By using the notation $[\boldsymbol{\Pi}]$ for the momentum flux tensor, we obtain the more compact form for Equation 5.10:

$$\frac{\mathrm{d}}{\mathrm{d}t}\left(\iiint_{\mathcal{V}} \rho\,\mathbf{v}\,\mathrm{dV}\right) = -\iint_{\mathcal{S}} [\boldsymbol{\Pi}]\cdot\mathbf{n}\,\mathrm{d}S + \iiint_{\mathcal{V}} \rho\,\mathbf{f}\mathrm{dV}. \qquad (5.12)$$

Equations 5.10 and 5.12 appear in the classical form characteristic of any conservation law: the time derivative of a certain physical quantity (here, the momentum $\rho\,\mathbf{v}$ of the fluid located within the volume of integration) is equal to the sum of a *flux* term and a *source* term. The importance of this equation is that it allows us, in certain cases, to determine the parameters of a flow without needing to know all its details within the volume $\mathcal{V}$. It is sufficient to know what the flow is on the boundaries of the volume. Indeed, if we are dealing with a stationary flow (velocity and pressure fields independent of time), the term on the left-hand side of the equation of conservation 5.10, the only one involving a volume integral of the velocity field of the fluid, is then identically zero. The equation then takes on a much simpler form:

$$\iint_S \rho \mathbf{v}(\mathbf{v}\cdot\mathbf{n})\,dS + \iint_S p\,\mathbf{n}\,dS - \iint_S [\boldsymbol{\sigma}']\cdot\mathbf{n}\,dS - \iiint_{\mathcal{V}} \rho\,\mathbf{f}\,dV = 0. \qquad (5.13)$$

By an appropriate choice of the volume of integration, usually called the *control volume* (bounded by the walls of a channel within which the fluid flows, or by surfaces which either coincide with the flow tubes or are normal to them), we can then easily determine the force exerted on the boundary of this volume by the moving fluid. We illustrate the use of this property by means of a few specific examples in Section 5.4.

The case of an incompressible Newtonian fluid

Throughout the remainder of this section, we confine ourselves to the study of *incompressible Newtonian fluids*. In such a case, we have the simple relationship $[\boldsymbol{\sigma}'] = 2\eta[\mathbf{e}]$ between the stress tensor $[\boldsymbol{\sigma}']$ and the rate-of-strain (or deformation) tensor $[\mathbf{e}]$ (Equation 4.16). Equation 5.10 then becomes:

$$\frac{d}{dt}\left(\iiint_{\mathcal{V}} \rho\,\mathbf{v}\,dV\right) = -\iint_S (\rho\mathbf{v}\,(\mathbf{v}\cdot\mathbf{n}) + p\,\mathbf{n} - 2\eta\,[\mathbf{e}]\cdot\mathbf{n})\,dS + \iiint_{\mathcal{V}} \rho\,\mathbf{f}\,dV. \qquad (5.15)$$

We now discuss a few examples of simple flows to which we can apply local and integral momentum conservation equations.

Application of the momentum conservation law to simple flows

In Section 4.5 we used the equation of motion for a viscous fluid and applied it to simple flows between parallel planes, such as Poiseuille flow or Couette shear flow. Here we analyze these flows in terms of the conservation of momentum, rather than in terms of the equilibrium of forces.

Consider first the case of a stationary, simple shear (Couette) flow of an incompressible Newtonian fluid located between two planes $y = 0$ and $y = a$, the lower plane being fixed while the upper one moves parallel to itself at the constant velocity V_0 in the x-direction (Figure 5.1a). As shown in Chapter 4, the velocity field $\mathbf{v}$ has components $(V_0 y/a, 0, 0)$. We assume that gravity effects are insignificant. Let us consider the conservation of momentum for a volume $d\mathcal{V}$, of unit depth in the z-direction, and with dimensions δx and δy along the two other directions. Since the velocity field is everywhere in the x-direction, the only non-zero component of momentum $\rho\, v_x$ is along the x-axis. In this direction, the momentum flux entering through the faces located at x and $x + \delta x$ is, according to Equation 5.11:

$$\Pi_{xx}\, n_x\, \delta y = -\left(\rho\, v_x^2 + p\right) n_x\, \delta y$$

(the unit vector $\mathbf{n}$ is oriented toward the outside of the volume and perpendicular to the boundary). But the velocity v_x depends only on the y-coordinate and the pressure p is constant throughout the flow. The algebraic value of the flux exiting through the face located at $x + \delta x$ (where $n_x = 1$) is thus opposite to the flux through the face located at x (for which $n_x = -1$). As a result, there is simply passive transport of momentum in the x-direction across the element $\delta\mathcal{V}$.

In the y-direction, the algebraic flux of the momentum ρv_x through the faces located at y and $y + \delta y$ can be written:

$$\Pi_{xy}\, n_y\, \delta x = -\sigma'_{xy}\, n_y\, \delta x = -\eta\,(\partial v_x/\partial y)\, n_y\, \delta x. \qquad (5.16)$$

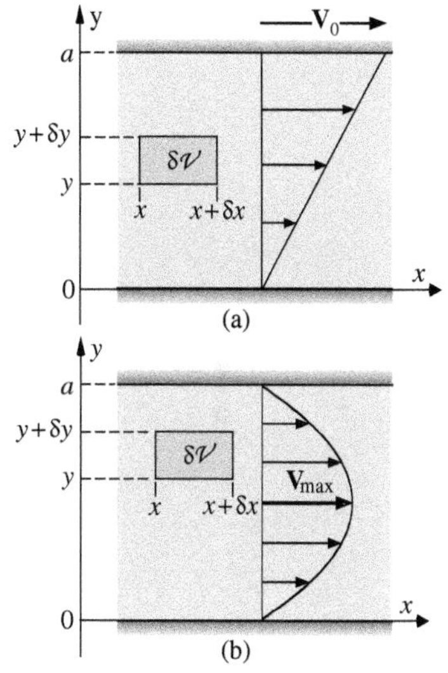

Figure 5.1 *Definition of the volume element $\delta\mathcal{V}$ for the calculation of the conservation of momentum (a) for simple shear flow (Couette flow) between two parallel planes without any external pressure gradient; (b) for Poiseuille flow between two planes $y = 0$ and $y = a$, resulting from a horizontal pressure gradient $\partial p/\partial x$*

For a velocity gradient as shown in Figure 5.1a, this momentum transfer is directed downward because the layers of fluid located near the plane moving with the greater velocity V_0 drag along the others. Here again, since the velocity gradient $\partial v_x/\partial y$ does not depend on y, the algebraic flux across the faces located at y and $y+\delta y$ are in opposite directions ($n_y = -1$ on the face at y and $n_y = +1$ on the other) We have simply a passive transfer of momentum in the y-direction (here again, for the specific case of Couette flow).

We obtain a very different result in the case of stationary Poiseuille flow, again in the absence of gravity, resulting from a constant pressure gradient $\partial p/\partial x$ between the two fixed parallel planes $y = 0$ and $y = a$ (Figure 5.1b). In this case, the velocity component v_y is zero, and v_x has a parabolic profile such that, following Equations 4.66 and 4.67:

$$v_x(y) = -\frac{a^2}{2\eta}\frac{\partial p}{\partial x}\frac{y(a-y)}{a^2}. \tag{5.17}$$

As in the previous case, the pressure gradient in the y-direction remains zero ($p = p(x)$), and only the momentum component $\rho\, v_x$ in the x-direction is non-vanishing. Its flux across the faces perpendicular to the x-direction of the volume element $\delta\mathcal{V}$ is equal to:

$$\Pi_{xx} n_x \delta y = -\left(\rho\, v_x^2\,(y) + p\,(x)\right) n_x\,\delta y.$$

In this case, the flux varies both in the x-direction (because of the pressure gradient $\partial p/\partial x$) and in the y-direction (because of the change in $\rho\, v_x^2(y)$). In the same way, the momentum flux through the faces located at y and $y+\delta y$, again given by Equation 5.16, depends on y. In contrast, the time-rate of change of the momentum P_x within the volume element $\delta\mathcal{V}$ must still be zero because the flow is stationary. We thus obtain, applying Equation 5.12 and expanding the result for small values of δx and δy, the equation:

$$\frac{\partial P_x}{\partial t} = \frac{\partial}{\partial t}(\rho\, v_x)\,\delta x\,\delta y = -\left[\frac{\partial \Pi_{xx}}{\partial x} + \frac{\partial \Pi_{xy}}{\partial y}\right]\delta x\,\delta y = 0, \tag{5.18}$$

which can be rewritten in the form:

$$\left[-\eta\frac{\partial^2 v_x(y)}{\partial y^2} + \frac{\partial p}{\partial x}\right]\delta x\,\delta y = 0,$$

which is identical to the Navier-Stokes equation 4.59 in parallel flow.

Writing the equation of conservation of momentum gives us then simply a different point of view of the equation of motion. Equation 5.18 indicates that the contribution of the pressure gradient ($\partial\Pi_{xx}/\partial x$ term) to the momentum is compensated by the change $\partial\Pi_{xy}/\partial y$ of the viscous flux of momentum $\Pi_{xy} = -\eta\,\partial v_x/\partial y$ in the transverse y-direction. For Poiseuille flow we find, using Equation 5.17, that this latter flux has the value $\Pi_{xy} = [(a/2)-y]\;\partial p/\partial x$; it is therefore zero at $y = a/2$ (half-way between the two planes) and has a maximal absolute value along the walls. For $\partial p/\partial x < 0$, this flux is directed toward the nearest wall; at a given transverse position y, it represents the momentum added per unit time by the pressure gradient between the planes $a/2$ and y.

5.3 The conservation of kinetic energy; Bernoulli's Equation

We first evaluate the rate of change in the kinetic energy per unit volume for a moving fluid. For the specific case of an ideal fluid this leads us to *Bernoulli's equation*, one of the forms which express the conservation of energy. Thereafter, we discuss applications of Bernoulli's equation to a few real flow situations where viscosity may be neglected.

5.3.1 The conservation of energy for a flowing incompressible fluid with or without viscosity

Derivation of the conservation equation

We consider the case of incompressible fluid flow. The kinetic energy per unit volume can be written:

$$e_c = \frac{dE_c}{dV} = \frac{\rho v^2}{2}.$$

Calculating the Eulerian derivative of e_c with respect to time at a fixed point, we obtain the following equation:

$$\frac{\partial}{\partial t}\left(\frac{\rho v^2}{2}\right) = -\nabla\cdot\left[\mathbf{v}\left(\frac{\rho v^2}{2}+p\right) - ([\sigma']\cdot\mathbf{v})\right] + \rho\,\mathbf{v}\cdot\mathbf{f} - \sigma'_{ij}\frac{\partial v_i}{\partial x_j}. \tag{5.19}$$

Derivation of Equation 5.19. We can write:

$$\frac{\partial}{\partial t}\left(\frac{\rho v^2}{2}\right) = \rho\,\mathbf{v}\cdot\frac{\partial \mathbf{v}}{\partial t} = \rho\, v_i \frac{\partial v_i}{\partial t}. \tag{5.20}$$

Replacing $\partial \mathbf{v}/\partial t$ by its value from the equation of motion of the fluid (Equation 4.25):

$$\rho\left(\frac{\partial v_i}{\partial t} + v_j\frac{\partial v_i}{\partial x_j}\right) = -\frac{\partial p}{\partial x_i} + \frac{\partial \sigma'_{ij}}{\partial x_j} + \rho f_i, \tag{5.21}$$

we obtain:

$$\begin{aligned}\frac{\partial}{\partial t}\left(\frac{\rho v^2}{2}\right) &= -\rho\, v_i v_j\frac{\partial v_i}{\partial x_j} - v_i\frac{\partial p}{\partial x_i} + v_i\frac{\partial \sigma'_{ij}}{\partial x_j} + \rho\, v_i f_i \\ &= -v_j\frac{\partial}{\partial x_j}\left(\frac{\rho v^2}{2}\right) - v_i\frac{\partial p}{\partial x_i} + \frac{\partial}{\partial x_j}\left(v_i\sigma'_{ij}\right) - \sigma'_{ij}\frac{\partial v_i}{\partial x_j} + \rho\, v_i f_i,\end{aligned} \tag{5.22}$$

i.e.:

$$\frac{\partial}{\partial t}\left(\frac{\rho v^2}{2}\right) = (-\mathbf{v}\cdot\nabla)\left(\rho\frac{v^2}{2}+p\right) + \nabla\cdot([\sigma']\cdot\mathbf{v}) - \sigma'_{ij}\frac{\partial v_i}{\partial x_j} + \rho\,\mathbf{v}\cdot\mathbf{f}. \tag{5.23}$$

Applying the general result from vector calculus, $\nabla\cdot(\alpha\,\mathbf{A}) = \mathbf{A}\cdot\nabla\alpha + \alpha\,\nabla\cdot\mathbf{A}$, we find:

$$\nabla\cdot\left[\mathbf{v}\left(\frac{\rho v^2}{2}+p\right)\right] = \left(\frac{\rho v^2}{2}+p\right)(\nabla\cdot\mathbf{v}) + (\mathbf{v}\cdot\nabla)\left(\frac{\rho v^2}{2}+p\right). \tag{5.24}$$

By then combining the condition for incompressibility $\nabla\cdot\mathbf{v} = 0$ with Equations 5.23 and 5.24, Equation 5.19 follows in a straightforward manner.

In order to understand more easily the physical meaning of the terms in the *local* Equation 5.19, we carry out an integration leading to the *integral* form of the same equation. We choose a volume of integration $\mathcal{V}$, fixed in space and bounded by the surface $\mathcal{S}$, and apply Gauss' divergence theorem to introduce the corresponding flux vectors:

$$\frac{\mathrm{d}}{\mathrm{d}t}\left(\iiint_{\mathcal{V}} \rho\frac{v^2}{2}\,\mathrm{d}V\right) = -\iint_S \rho\frac{v^2}{2}\,\mathbf{v}\cdot\mathbf{n}\,\mathrm{d}S$$
$$-\iint_S p\mathbf{v}\cdot\mathbf{n}\,\mathrm{d}S + \iint_S ([\boldsymbol{\sigma}']\cdot\mathbf{n})\cdot\mathbf{v}\,\mathrm{d}S + \iiint_{\mathcal{V}} \rho\,\mathbf{f}\cdot\mathbf{v}\,\mathrm{d}V \qquad (5.25)$$
$$-\iiint_{\mathcal{V}} \sigma'_{ij}\frac{\partial v_i}{\partial x_j}\,\mathrm{d}V.$$

- The first term on the right-hand side of Equation 5.25 represents the overall flux of kinetic energy $\rho v^2/2$ convected by the fluid across the surface S.
- The next three terms of the equation represent the work done by the combination of forces acting on the volume $\mathcal{V}$ under consideration:
  The first two of these give respectively the work done by the pressure forces and by the components of the viscous stresses which act *normal* to the surface S, enclosing the volume $\mathcal{V}$.
  The third term corresponds to the increase in energy resulting from external forces (equal to $\rho\,\mathbf{f}$ per unit volume). This term is positive when $\rho\,\mathbf{f}$ and $\mathbf{v}$ are in the same direction, as would be the case for a fluid flowing downward within the earth's gravitational field ($\mathbf{f} = \mathbf{g}$), leading to an increase in the kinetic energy. If the force field $\mathbf{f}$ is the derivative of a potential, we can consider that the work resulting from it indicates an exchange of kinetic and potential energy.
- Finally, we will see that the term $\left(\iiint_{\mathcal{V}} \sigma'_{ij}\,(\partial v_i/\partial x_j)\,\mathrm{d}V\right)$ represents the irreversible transformation of kinetic energy, by viscous dissipation, into internal energy of the fluid in the form of heat.

We now discuss, in greater detail, the respective contribution of the two terms of Equation 5.25 which involve the viscous shear stresses $[\boldsymbol{\sigma}']$.

Kinetic energy dissipation through viscosity in a parallel plane flow

Consider a volume element $\delta\mathcal{V}$ of sides δx and δy along the x- and y-directions respectively, and of unit depth in the z-direction in a stationary plane parallel flow with velocities $v_x(y)$ such as those indicated in Figure (5.1a–b). Let us assume that only the components $\sigma'_{xy} = \sigma'_{yx}$ of the viscous stress tensor are non-zero (as will be the case for Newtonian fluid) and that the component in the x-direction of the volume force is zero. The work done per unit time by these forces on the upper and lower faces of the volume element, i.e., the third term on the right-hand side of Equation 5.25, has the value:

$$\left(\sigma'_{xy}(y+\delta y)\,v_x(y+\delta y) - \sigma'_{xy}(y)\,v_x(y)\right)\delta x = \frac{\partial}{\partial y}\left(\sigma'_{xy}(y)\,v_x(y)\right)\delta x\,\delta y.$$

The difference in sign results from the fact that the normal $\mathbf{n}$ is oriented toward positive y for the upper face, and toward negative y for the lower face. The derivative of the product of the second term can be split into two terms:

- The first is equal to $v_x(\partial\sigma'_{xy}(y)/\partial y)\;\delta x\;\delta y$. It represents the work done per unit time by the resultant $\sigma'_{xy}(y+\delta y) - \sigma'_{xy}(y)$ of the viscous stresses on the lower and upper faces of the volume element. This work would result in a change of the global kinetic energy of the element but, in the present case, according to Equation 4.113, it is opposite to the work done by the pressure forces on the faces of height δy of the element. This latter work has the value $(p(x) - p(x + \delta x))\, v_x\, \delta y = -(\partial p/\partial x)\, v_x\, \delta x\, \delta y$. The kinetic energy thus remains constant, as expected since the flow is stationary.

- The second term resulting from taking the derivative of the product has the value $\sigma'_{xy}\, (\partial v_x(y)/\partial y)\, \delta x\, \delta y$ and corresponds, in contrast, solely to work due to deformation. In fact, in a reference frame moving at the average velocity $v_x(y)$, it represents the work done per unit time by the force $\sigma'_{xy}\, \delta x$ exerted on the upper face of the volume element: the point at which this force acts moves at velocity $(\partial v_x(y)/\partial y)\delta y$ relative to the lower face. If the element was solid, this work would be stored in the form of potential energy of elastic deformation; for a fluid, it is dissipated in the form of thermal energy and transformed into internal energy.

Kinetic energy dissipation through viscosity in an arbitrary flow geometry

Equation 5.25 represents simply the generalization of the previous decomposition to a flow with arbitrary geometry. It expresses the result that the work done by forces which act on a volume element and the increase of kinetic energy by convection contribute to the change in the kinetic energy of the volume on the one hand, and on the other hand, are transformed into thermal energy. The rate of viscous dissipation of energy in the volume $\mathcal{V}$ of the fluid is thus:

$$\delta E_c / \delta t = \iiint_{\mathcal{V}} \sigma'_{ij} \left(\partial v_i / \partial x_j\right) \mathrm{d}V = \iiint_{\mathcal{V}} \sigma'_{ij}\, e_{ij}\, \mathrm{d}V.$$

Proof

Consider the integral: $\delta E_c/\delta t = \iiint_{\mathcal{V}} \sigma'_{ij} \left(\partial v_i / \partial x_j\right) \mathrm{d}V$. Making use of the symmetry of the tensor $[\boldsymbol{\sigma}']$ $(\sigma'_{ij}{=}\sigma'_{ji})$, we can write: $\sigma'_{ij}\, \partial v_i/\partial x_j = \frac{1}{2}\sigma'_{ij}\, \partial v_i/\partial x_j + \frac{1}{2}\sigma'_{ji}\, \partial v_i/\partial x_j$. In order to calculate $\delta E_c/\delta t$, we sum over the two indices i and j and interchange these two indices in the second term of the previous sum in order to obtain, after taking out the factor σ'_{ij}:

$$\frac{\delta E_c}{\delta t} = \iiint_{\mathcal{V}} \left[\frac{1}{2}\sigma'_{ij}\frac{\partial v_i}{\partial x_j} + \frac{1}{2}\sigma'_{ji}\frac{\partial v_i}{\partial x_j}\right] \mathrm{d}V = \iint_{\mathcal{V}} \sigma'_{ij}\frac{1}{2}\left[\frac{\partial v_i}{\partial x_j} + \frac{\partial v_j}{\partial x_i}\right] \mathrm{d}V.$$

We recognize the component e_{ij} of the deformation tensor in the factor following σ'_{ij}.

Kinetic energy dissipation in a Newtonian fluid

For the particular case of an incompressible Newtonian fluid, the above equation for the rate of viscous dissipation can be combined with Equation 4.16 to obtain:

$$\frac{\delta E_c}{\delta t} = \frac{\eta}{2} \iiint_{\mathcal{V}} \left(\frac{\partial v_i}{\partial x_j} + \frac{\partial v_j}{\partial x_i}\right)^2 \mathrm{d}V = 2\,\eta \iiint_{\mathcal{V}} e_{ij}^2\, \mathrm{d}V, \qquad (5.26)$$

where e_{ij} is the tensor of the rate of deformation. Since the term $\delta E_c/\delta t$ corresponds to an irreversible energy dissipation and the quadratic form within the integral is positive definite, this equation agrees with the fact that the viscosity coefficient is necessarily positive (see Section 4.1.3).

5.3.2 Bernoulli's equation and its applications

Bernoulli's equation expresses the conservation of energy for *ideal, incompressible* fluids when *the volume forces* **f** result from a potential φ with $\mathbf{f} = \nabla\varphi$ (one has $\varphi = gy$ when the volume force is gravity and the vertical coordinate y is positive upwards). We discuss first the case of stationary flows, then that of non-stationary ones and, finally, we will illustrate the underlying physics with a few applications of this equation.

Bernoulli's equation for stationary flow

For an ideal fluid in stationary flow, one can neglect the losses of energy due to viscosity and Equation 5.19 becomes:

$$\nabla \cdot \left[\mathbf{v}\left(\frac{\rho v^2}{2} + p\right)\right] - \rho\,\mathbf{v}\cdot\mathbf{f} = \nabla \cdot \left[\mathbf{v}\left(\frac{\rho\, v^2}{2} + p\right)\right] + \rho\,(\mathbf{v}\cdot\nabla)\,\varphi = 0. \tag{5.27}$$

Using again the general vector result, $\nabla\cdot(\alpha\ \mathbf{A}) = \mathbf{A}\cdot\nabla\alpha + \alpha\nabla\cdot\mathbf{A}$ and the incompressibility equation $\nabla\cdot\mathbf{v} = 0$, together with its consequence that the fluid density ρ is constant, Equation 5.27 can be rewritten in the form:

$$(\mathbf{v}\cdot\nabla)\left(\frac{\rho\, v^2}{2} + p + \rho\,\varphi\right) = 0. \tag{5.28}$$

The scalar product in the above equation represents the time rate of change $d\mathcal{P}/dt$ of the quantity $\mathcal{P} = \rho(v^2/2) + p + \rho\,\varphi$ during the course of a displacement along a streamline (tangent at all points to the velocity vector). In fact, because the flow is stationary, the Eulerian derivative $\partial\mathcal{P}/\partial t$ is zero, so that:

$$\frac{d\mathcal{P}}{dt} = (\mathbf{v}\cdot\nabla)\,\mathcal{P} = 0. \tag{5.29}$$

We hence infer a first representation of Bernoulli's equation:

$$\rho\frac{v^2}{2} + p + \rho\,\varphi = \text{constant } \textit{along a streamline}. \tag{5.30}$$

The quantity $\rho v^2/2$, having the dimensions of a pressure, is called the *dynamic pressure*, and the expression $p + (\rho\, v^2)/2$ is the *total* or *stagnation pressure*.

Let us analyze this result by considering a uniform horizontal flow with velocity **U** which impinges perpendicularly on an obstacle (Figure 5.2). There appears a stagnation point S, where both the tangential and the normal velocity components are zero even for the case of an ideal fluid.

Figure 5.2 *Flow in the neighborhood of a stagnation point S for a fluid incident on an obstacle. A measurement of the difference between the stagnation pressure p_S and the pressure p_O far from the obstacle leads to a measurement of the fluid velocity U*

Let us denote by p_O the pressure at a point O located sufficiently far from the obstacle on the same horizontal streamline as S (φ is then constant), and where the velocity has the magnitude U. We have, between p_O and the pressure p_s at S, the relationship:

$$p_s = p_O + \rho U^2/2. \tag{5.31}$$

The fact that the fluid velocity is zero at S accounts for the name for p_s, the *stagnation pressure*, which, by extension is given to the sum $p_s = p + \rho\, U^2/2$. This expression is also referred to as the *total pressure*.

Equation 5.30 predicts that an increase in the velocity along a streamline is associated to a decrease of the pressure along the same path. This effect explains *cavitation* phenomena:

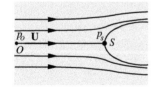

if the velocity increases sufficiently so that the pressure drops down to the saturation vapor pressure of the fluid at the corresponding temperature, the fluid displays local boiling. The resulting vapor bubbles formed within the fluid implode afterwards on solid surfaces (turbine blades, propellers, . . .) causing, in time, undesirable pitting of the surfaces. The noise associated with the emission of such bubbles is also and important factor in the detection of military submarines. Moreover, some species of shrimp can close their claws at a large enough velocity (~20 m/s) that a cavitation bubble is emitted which can paralyze its nearby prey; as a result of the multitude of such shellfish, the resultant noise can even, on occasions, be loud enough to interfere with sonar detectors!

Bernoulli's equation for potential flow

Let us now consider the case of potential flow (discussed in detail in Chapter 6). The velocity field $\mathbf{v}$ is then derivable from a potential Φ such that:

$$\mathbf{v} = \nabla\Phi \tag{5.32}$$

We continue to assume that the flow is incompressible, with constant density ρ, and acted upon by volume forces $\mathbf{f}$ derivable from a potential φ. However, we no longer assume that the velocity field is stationary. We now derive Bernoulli's equation directly from Euler's equation 4.31; we could also, of course, have used this same method to derive Bernoulli's equation in the most general case:

$$\rho \frac{\partial \nabla\Phi}{\partial t} = \rho \nabla \left(\frac{\partial \Phi}{\partial t} \right) = -\rho \, (\mathbf{v} \cdot \nabla) \, \mathbf{v} - \nabla p - \nabla (\rho \, \varphi). \tag{5.33}$$

We make use of the following identity from vector calculus:

$$(\mathbf{v} \cdot \nabla)\mathbf{v} = \nabla \frac{v^2}{2} - \mathbf{v} \times (\nabla \times \mathbf{v}). \tag{5.34}$$

Moreover, due to the fact that the flow is curl free (since the velocity field is derivable from a potential function and, consequently, $\nabla \times \mathbf{v} = 0$), Equation 5.33 becomes:

$$\nabla \left(\rho \frac{\partial \Phi}{\partial t} \right) = -\nabla \left(\rho \frac{v^2}{2} \right) - \nabla p - \nabla (\rho \, \varphi), \tag{5.35}$$

so that, after integrating:

$$\rho \frac{\partial \Phi}{\partial t} + \rho \frac{v^2}{2} + p + \rho \, \varphi = \text{constant}. \tag{5.36}$$

This equation is another representation of Bernoulli's equation. If the flow is stationary, such that $\rho(\partial\Phi/\partial t) = 0$, this equation seems to revert to Equation 5.30. The crucial difference here is in that the quantity $(\rho v^2/2 + p + \rho\,\varphi)$ is constant *throughout the volume of the flow*, instead of only along a streamline as in the previous discussion. However, this more general result cannot be derived merely from the law of conservation of energy.

5.3.3 Applications of Bernoulli's equation

Bernoulli's equation leads to the understanding of a large number of effects in which a change in the velocity along the flow (e.g., due to the narrowing of a tube) results in an opposite change in the pressure. The examples below give a few illustrations or applications of these effects. Moreover, we will see how Bernoulli's equation can be applied, at least to a good approximation, to the flow of real fluids.

Preliminary reminder: an important property of one-dimensional flows

Before taking up these examples, we should recall a key result already discussed for Couette and Poiseuille viscous flows in Chapter 4. Consider a volume of flowing fluid within which the velocity is always in the same x-direction. We demonstrate that, in this case, the transverse change in pressure along a cross-section normal to the flow reduces to the gradient of the hydrostatic pressure: there is no additional term due to the flow. Indeed, because the condition of incompressibility, $\nabla \cdot \mathbf{v} = 0$, together with the fact that $v_y = v_z = 0$, leads to $\partial v_x / \partial x = 0$, the term $(\mathbf{v} \cdot \nabla)\mathbf{v}$ is identically zero. Moreover, the viscous friction forces are in the x-direction. The equation of motion thus reduces to:

$$\frac{\partial p}{\partial y} + \rho g = 0 \qquad \text{in the vertical direction (normal to the flow)} \qquad (5.37a)$$

and:

$$\frac{\partial p}{\partial z} = 0 \qquad \text{in the other direction normal to the flow.} \qquad (5.37b)$$

In the absence of gravitational effects, the pressure is thus constant throughout the cross-section normal to the flow. This is a very important result in a large number of problems involving the flow of real fluids.

Moreover, in high-velocity flow, real fluids display ideal fluid behavior, except within a very thin boundary layer near the solid walls. It is in this *boundary layer*, discussed in detail in Chapter 10 that the transition occurs between the zero tangential velocity condition at the wall, and the bulk flow velocity expected for an ideal fluid. Since the flow in this boundary layer is essentially locally parallel to the wall, we have, as a consequence of Equation 5.37b, continuity of the pressure between the wall and the region just outside the boundary layer.

The Pitot tube

The *Pitot tube* involves a direct application of Bernoulli's equation, allowing for a determination of the velocity of a fluid by means of a pressure measurement. In the example discussed below, it consists of a fixed obstacle placed within a fluid in motion. In quite a number of practical applications, however, the Pitot tube is attached to a moving object (airplane, boat, etc.), and used to measure its velocity relative to the fluid. The device consists of two concentric tubes (Figure 5.3): the inner tube has an opening S at its apex, oriented normal to the incoming flow, while the second tube is perforated with a series of tiny openings A spaced uniformly along the circumference of a circle, on the outer surface of the device. A differential manometer connected to each of the two tubes measures the pressure difference Δp between points S and A.

If we neglect viscous effects, assuming that they are only significant within a very thin boundary layer near the walls of the tubes, we can apply Equation 5.30 along the streamline OS, coinciding with the axis of the tubes:

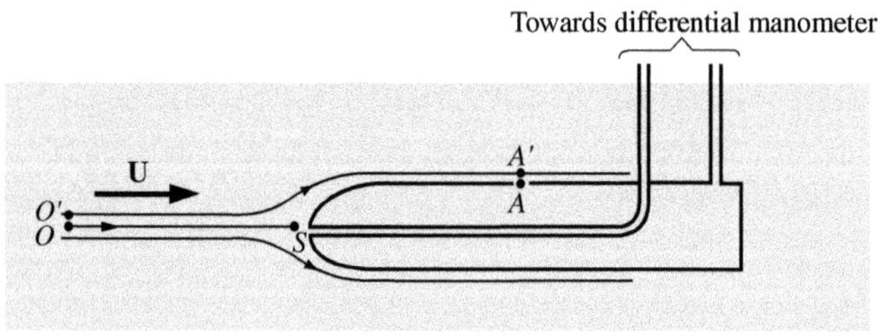

Figure 5.3 *Principle of the Pitot tube. Bernoulli's equation is applied along the two streamlines, from O to S, and from O' to A'*

$$p_O + \rho\frac{U^2}{2} = p_s. \tag{5.38}$$

If we apply now Equation 5.30 along the streamline $O'A'$ (the point A' is on the same vertical axis as the pressure inlet A but outside the boundary layer):

$$p_{O'} + \rho\frac{U^2}{2} = p_{A'} + \rho\frac{v_{A'}^2}{2} = p_A + \rho\frac{v_{A'}^2}{2} = p_A + \rho\frac{U^2}{2}. \tag{5.39}$$

Indeed, as we have just seen, the pressure remains constant as we cross the quasi one-dimensional flow in the boundary layer normal to this flow, and we thus have $p_A = p_{A'}$. On the other hand, the velocity at A' is effectively equal to U if A' is sufficiently far downstream of S and if the cross-section of the Pitot tube is small relative to the diameter of the flow channel. Finally, the pressures at points O and O' infinitesimally close to each other and located far upstream of the obstacle have the same values. We then obtain by combining Equations 5.38 and 5.39:

$$\Delta p = p_S - p_A = \rho\frac{U^2}{2} \tag{5.40}$$

The magnitude of the flow velocity is thus directly obtainable from a measurement of the pressure difference Δp.

The Venturi gauge

The *Venturi gauge* (Figure 5.4) uses Bernoulli's equation to evaluate the decrease in pressure at a constriction in a tube. It is frequently used in practical applications (to draw in the air-gas mixture in an automobile carburettor, in vacuum nozzles, in flow meters . . .).

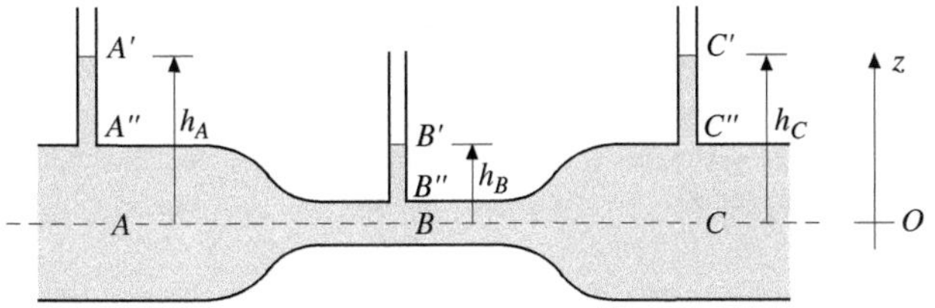

Figure 5.4 *Lowering of pressure in the higher fluid-velocity region of a Venturi gauge*

When flow is initiated in this system, there appears a height difference, proportional to the square of the flow rate, between the levels h_A and h_B in the manometer tubes A and B. On the other hand, levels h_A and h_C are practically equal (h_C being slightly lower than h_A, if the flow is directed from A to C, as a result of the head loss in the tube due to viscosity, which we neglect in the present discussion). At the surface of the liquid in the three manometer tubes, we have a pressure equal to the atmospheric pressure p_0:

$$p_{A'} = p_{B'} = p_{C'} = p_0. \tag{5.41}$$

Provided the manometer tubes are sufficiently narrow in diameter, they cause little disturbance in the flow. The latter then remains parallel in the sections of the tube where points A, B and C are located (assuming the pressure detection points are sufficiently far away from the regions where the cross-section of the tube is varying, so that the velocity can be assumed uniform at their location). Equation 5.37 therefore holds. The pressure gradients between A and A'', B and B'', C and C'' then reduce to the hydrostatic pressure gradient.

Throughout the discussion on the left, we neglect viscous forces; it might thus seem that this model corresponds to an approximation of an ideal fluid. On the contrary, the phenomenon of the Venturi gauge would not be observed in an ideal fluid in potential flow, because the boundary condition for zero velocity at the walls plays an essential part! The reason for this apparent paradox is the fact that the key hypothesis of our discussion (i.e., the fact that the transverse pressure gradient reduces to the hydrostatic pressure) is no longer obeyed. Let us write Bernoulli's equation 5.36 assuming that the only volume force is gravity and that the flow is *potential throughout the fluid*:

$$\frac{v^2}{2} + \frac{p}{\rho} + g\,z = \text{constant}. \tag{5.47}$$

If the manometer tubes are sufficiently long, the effect of main flow decreases sufficiently as we go up in the tubes, so that we can write:

$$v_{A'} = v_{B'} = v_{C'} = 0. \tag{5.48}$$

Thus, applying Equation 5.47 to points A', B' and C', where the pressure p is equal to p_0, we find:

$$gh_A = gh_B = gh_C, \tag{5.49}$$

i.e.,

$$h_A = h_B = h_C. \tag{5.50}$$

There would therefore not be any difference between the levels in the three tubes. We would still have a pressure difference between the points A and B but this difference would be exactly compensated in the opposite direction due to the velocity changes as we go up inside the manometric tubes toward A'' and B''. Indeed, as we no longer have the condition of zero tangential velocity at the walls, the flow penetrates a finite distance into the tubes. The condition that the velocity field be parallel at the points A'' and B'' where the pressures are being detected is thus no longer satisfied. In a viscous flow, on the other hand, the velocity is zero or very small both at the walls and within the manometer tubes: we have, therefore, a virtually parallel flow as soon as we are a small distance away from the walls.

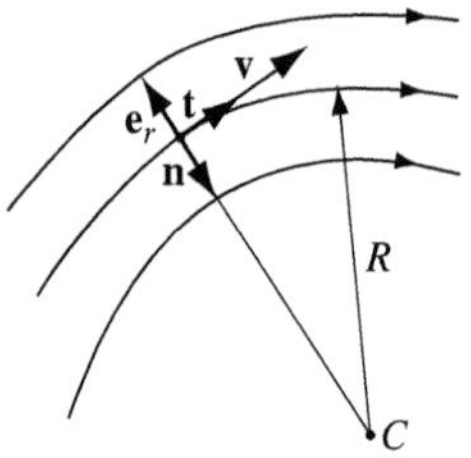

Figure 5.5 *Variation of the pressure in a flow in which the streamlines are curved*

The flow hardly penetrates into the manometer tubes, and the pressure gradient there is also hydrostatic only. We can therefore write:

$$p_A = p_{A'} + \rho\,g\,h_A = p_0 + \rho\,g\,h_A. \tag{5.42}$$

Similarly: $p_B = p_0 + \rho\,g\,h_B$ (5.43) and: $p_C = p_0 + \rho\,g\,h_C.$ (5.44)

Let us now assume that the flow is uniform in each of the sections with respective velocities v_A, v_B and v_C, except very close to the solid walls. The transition to the condition of zero tangential velocity at the solid wall occurs over a very narrow boundary layer without any transverse pressure gradient, as we have discussed above. If the energy losses through viscous friction are sufficiently small relative to the kinetic energy of the fluid, we can apply Bernoulli's equation 5.30 along the horizontal streamline ABC with:

$$p_A + \frac{1}{2}\rho\,v_A^2 = p_B + \frac{1}{2}\rho\,v_B^2 = p_C + \frac{1}{2}\rho\,v_C^2, \tag{5.45}$$

i.e., substituting for p_A, p_B and p_C their values given by Equations 5.42–5.44, and dividing the result by ρg, we obtain:

$$h_A + \frac{1}{2}\frac{v_A^2}{g} = h_B + \frac{1}{2}\frac{v_B^2}{g} = h_C + \frac{1}{2}\frac{v_C^2}{g}. \tag{5.46}$$

This predicts indeed a lower level in the manometer tube B placed in the region of highest velocity and a difference in level between A and B proportional to $(v_B{}^2 - v_A{}^2)$. In order to make use of Equation 5.45, we have assumed that the velocity is uniform throughout each of the cross-sections A, B and C. In actual experimental conditions, this assumption is not strictly valid, because of the boundary conditions of zero velocity at the wall; we must therefore introduce an experimental correction factor dependent on the velocity profile.

Application of Bernoulli's equation to flow along a curve

Consider a flow with curved streamlines of radius of curvature R (Figure 5.5). Assuming also that we can neglect viscous friction and volume forces, the equilibrium condition between the pressure gradient and the acceleration of a particle of incompressible fluid can be written:

$$\rho\frac{d\mathbf{v}}{dt} = \left(\rho\frac{dv}{dt}\right)\mathbf{t} + \left(\rho\frac{v^2}{R}\right)\mathbf{n} = -\nabla p. \tag{5.51}$$

where $\mathbf{t}$ and $\mathbf{n}$ are unit vectors tangent and normal to the streamlines, and $d\mathbf{v}/dt$ is the Lagrangian acceleration. By taking the scalar product of Equation 5.51 with $\mathbf{t}$, and denoting by s the coordinate along the length of the streamline, we obtain:

$$\rho\,\mathbf{v}\cdot\frac{\partial \mathbf{v}}{\partial s} = -\frac{\partial p}{\partial s}, \tag{5.52}$$

which is a local form of the equation of motion (recall that $v = ds/dt$). Similarly, by taking the scalar product of the same equation with $\mathbf{n}$, we have:

$$\rho\frac{v^2}{R} = -\mathbf{n}\cdot\nabla p = -\mathbf{n}\cdot\frac{\partial p}{\partial r}\,\mathbf{e}_r = \frac{\partial p}{\partial r} \tag{5.53}$$

($\mathbf{e}_r$ is the radial unit vector opposite in direction to $\mathbf{n}$). The pressure therefore increases as we go away from the center of curvature C of the streamline.

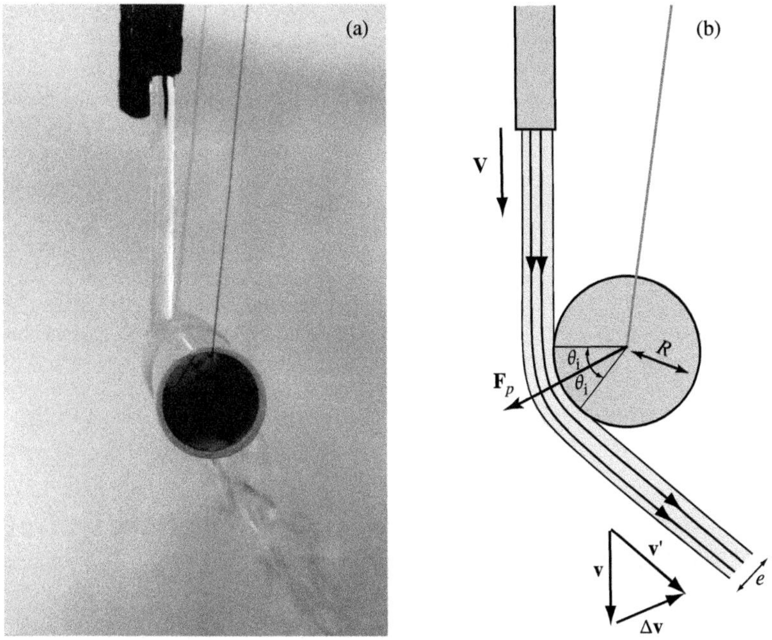

Figure 5.6C *Deviation of a jet of fluid along the curved surface of a cylinder. (a) Deflection of the jet and attraction force on the cylinder. (b) schematic diagram displaying the balance of the forces acting (document Belin publishers)*

The Coanda effect

The above result underlies the explanation of the *Coanda effect*, which can be observed by placing a cylindrical object under a jet of liquid, normal to the jet, but slightly off-axis (Figure 5.6a): the jet tends to stick to the obstacle and undergoes a deflection. At the same time, the fact that the wires holding up the cylinder are inclined to the vertical indicates that the cylinder is attracted by the jet. According to Equation 5.53, the curvature of the streamlines creates a pressure gradient $\partial p/\partial r > 0$: the pressure at the level of the cylinder is lower than the atmospheric pressure outside the jet, providing an explanation for the mutual attraction of the jet and the cylinder.

Proof

In order to calculate the corresponding force of attraction, let us assume that we have, instead of the jet, a sheet of liquid of thickness e in a plane parallel to the axis of the cylinder (Figure 5.6b). By assuming that the atmospheric pressure p_{at} acts at the outer surface of the sheet, it should be equal, according to Equation 5.53, to $p_{at} - \rho v^2 e/R$ on the inner surface (if $e \ll R$). Then the force $\mathbf{F}_p$ exerted by the jet on the cylinder, per unit depth parallel to the cylinder axis, satisfies:

$$F_p = \rho \frac{v^2}{R} e \int_{-\theta_i}^{\theta_i} R \cos\theta \, \mathrm{d}\theta = 2\rho\, e v^2 \sin\theta_i.$$

The force $\mathbf{F}_p$ is perpendicular to the bisector of the angle between the flow directions upstream and downstream of the contact points and is oriented away from the axis of the cylinder.

We come to the same result by analyzing the change in the momentum of the jet between the point where it comes into contact with the cylinder, and the point where it leaves. Let us take the component in the y-direction of Equation (5.13): the deflection of the jet corresponds to a change in velocity $\Delta \mathbf{v} = (\mathbf{v}' - \mathbf{v})$ and to a difference between the entering and exit fluxes of the component of momentum in the y-direction equal to $2\,\rho\,e v^2 \sin\theta_i$. If we neglect viscosity forces, in order to have conservation of momentum this difference must be compensated by a resultant pressure force, which is in fact the one which we have just calculated.

This same mechanism allows us to explain how a very light ball can be *levitated* within an air jet impinging at a small angle to the vertical, slightly above the ball. Just as in the case of the cylinder, the compensating force results from the curvature of the streamlines which lowers the pressure, and not because of the impact of the jet of air. A comparable phenomenon is the *teapot effect*: the liquid stream which flows from the spout of a teapot seems to be attached to the spout's surface after having followed the curve of the spout, instead of flowing directly into a cup.

The Coanda effect is not the only one which acts in the teapot effect; recent studies have indicated that the liquid film remains much less in contact with the surface of the spout if the latter is non-wetting (e.g., coated with teflon).

5.4 Applications of the laws of conservation of energy and momentum

We describe in this section a number of flows which can be analyzed by the application of the laws of conservation of momentum, of mass and of energy. In this manner we avoid a complete determination of the velocity field of the fluid, a task frequently impossible in a real situation. Also, in this section, we assume that one deals with an ideal fluid so that viscous effects can be neglected.

5.4.1 Jet incident onto a plane

This problem is closely related to simple observations of the impingement on obstacles of different shapes of a stream of water. Consider a two-dimensional rectangular jet of liquid of width h and unit depth in the z-direction, perpendicular to the plane of the figure, impinging upon a plane wall at an angle α to the normal (Figure 5.7). Assume that the velocity U of

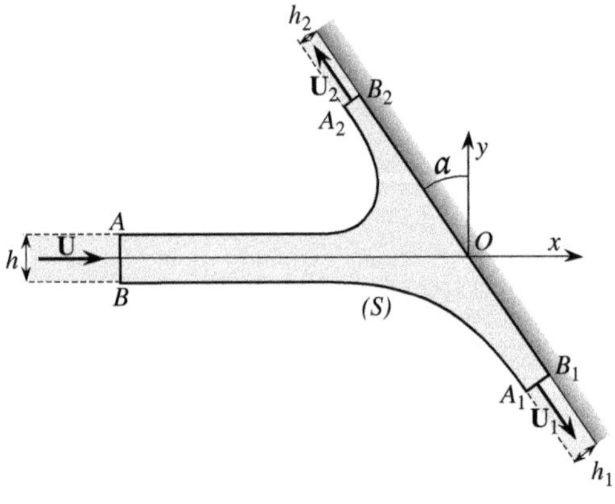

Figure 5.7 *Schematic view of the splitting of the fluid flow and of the control volume for a two-dimensional jet incident on a flat surface*

the fluid is uniform throughout its cross-section. Upon impact with the wall, the jet splits up into two sheets of thicknesses h_1 and h_2, and respective velocities U_1 and U_2. We will show that:

$$h_1 = (h/2)\,(1 + \sin\alpha) \qquad (5.54a) \qquad\qquad h_2 = (h/2)\,(1 - \sin\alpha). \qquad (5.54b)$$

On the other hand, the component of force normal to the plane and per unit depth has the value:

$$F_\perp = F_x \cos\alpha + F_y \sin\alpha = \rho\, U^2 h \cos\alpha. \qquad (5.55)$$

This equation displays the standard dependence of inertial resistance forces in the absence of viscosity, i.e., a force proportional to the square of the velocity, and to the density of the fluid.

Proof

Assuming, as mentioned above, an ideal fluid in potential flow, we neglect viscosity effects. We equally neglect the effect of gravity, an assumption justified by the fact that the experimentally observed sheets of fluid are quite thin. The balance of the rates of flow, required by mass conservation (assuming unit depth in the z-direction) implies:

$$hU = h_1 U_1 + h_2 U_2. \qquad (5.56)$$

If the flow is potential, Bernoulli's Equation 5.36 applies throughout the volume of fluid, so that:

$$p + \frac{1}{2}\rho\, U^2 = p_1 + \frac{1}{2}\rho\, U_1^2 = p_2 + \frac{1}{2}\rho\, U_2^2. \qquad (5.57)$$

But, since the streamlines in each of the three cross-sections AB, A_1B_1 and A_2B_2 of the sheets of fluid are all parallel, the pressure p does not vary across them so that we have:

$$p = p_1 = p_2 = p_0 = (\text{atmospheric pressure}). \qquad (5.58)$$

Combining Equations 5.57 and 5.58, we conclude that the velocity is identical throughout the liquid sheet:

$$U = U_1 = U_2. \qquad (5.59)$$

Inserting this into Equation 5.56, the conservation of mass for fluid flow, we obtain a simple relation for the thicknesses of the fluid sheets:

$$h = h_1 + h_2. \qquad (5.60)$$

The conservation of momentum (Equation 5.13) now allows us to calculate the force of the jet on the plane: selecting a control volume bounded by the boldface lines in Figure 5.7 and of unit depth in the z-direction, and taking components of this vector equation along the x- and y-axes, while still neglecting gravity and viscous shear forces, we obtain:

$$\rho\left(U_1^2 h_1 \sin\alpha - U_2^2 h_2 \sin\alpha - U^2 h\right) + \iint_{\text{plane}} (\delta p\; n_x)\, \mathrm{d}S = 0 \qquad (5.61a)$$

and:
$$\rho\left(-U_1^2 h_1 \cos\alpha + U_2^2 h_2 \cos\alpha\right) + \iint_{\text{plane}} \left(\delta p\, n_y\right) \mathrm{d}S = 0. \qquad (5.61b)$$

The two surface integrals are evaluated on the part of the solid plane located between the cross-sections A_1B_1 and A_2B_2. They represent the components Fx and Fy of the total pressure force **F** on the plane in the x- and y-directions. The atmospheric pressure p_0 acts through integrals of the form $\iint_S (p_0\, n_i)\, \mathrm{d}S$ (where $i = x$ or y and S is the entire surface bounding the control volume), integrals which in this case are identically zero. Recalling that the velocities U, U_1, and U_2 are all equal, Equations 5.61a and 5.61b reduce to:

$$F_x = \rho\, U^2 \left(h - (h_1 - h_2)\sin\alpha\right) \quad (5.62a) \qquad \text{and:} \qquad F_y = \rho\, U^2 (h_1 - h_2)\cos\alpha. \quad (5.62b)$$

We now determine the thicknesses h_1 and h_2 of the two sheets of liquid from the fact that the component of the force **F** parallel to the plane is zero (on the continued assumption that viscous friction forces are zero, and that only pressure forces act normal to the plane). This leads to the condition:

$$F_\parallel = F_x \sin\alpha - F_y \cos\alpha = 0. \qquad (5.63)$$

Combining Equations 5.62a and 5.62b, we find:

$$\rho\, U^2 \left(h - (h_1 - h_2)\sin\alpha\right)\sin\alpha - \rho\, U^2 (h_1 - h_2)\cos^2\alpha = 0,$$

whence:
$$h_1 - h_2 = h\,\sin\alpha. \qquad (5.64)$$

Combining this last result with Equation 5.60, we finally obtain Equations 5.54 a and b for h_1 and h_2. We also obtain F_x and F_y using Equations 5.62a, 5.62b and 5.64 and, then, $F_\perp$ using the first equality in Equation 5.55.

5.4.2 Exit jet from an opening in a reservoir

We consider here a container emptying through a small circular opening located in its lower region (Figure 5.8a). If the cross-sectional area S_0 of the opening is not too small, viscous losses can again be neglected here. As usual, we denote the atmospheric pressure by p_0, and we observe that it acts on the free surface of the container, and on the outside of the exit jet.

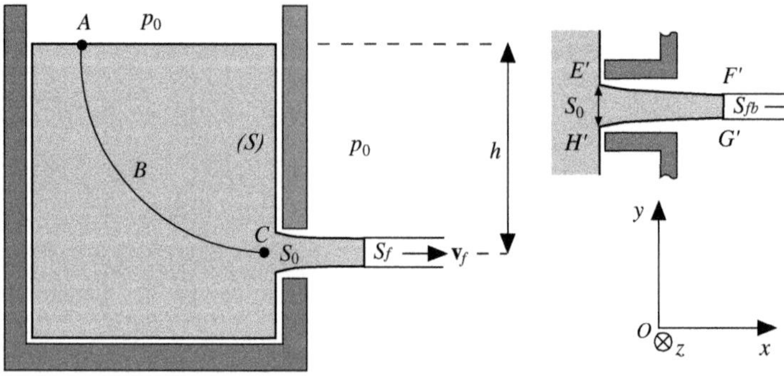

Figure 5.8 *(a) Jet emerging from a reservoir through a circular orifice; (b) reentrant tube, known as Borda's mouthpiece, for which it can be shown that the final cross-sectional area of the jet, S_{fb}, is equal to half the cross-section of the original exit orifice S_0*

Determination of the velocity of the exit jet

Experimental observations indicate that, as the jet exits from the opening, it contracts down to a minimum cross-section, the *vena contracta*, which we denote by S_f. When this minimum value has been attained, the streamlines have become parallel. The pressure has therefore, in accordance with Equations 5.37a and 5.37b, the same value p_0 in the jet as outside it (we neglect the gradient of hydrostatic pressure over the dimension of the jet). Also, if the area of the container is large compared to that of the section S_f, the velocity at which the level of the liquid inside the large container drops is negligible relative to the velocity of the jet (since the flow rate must be conserved). Let y_0 be the vertical coordinate of the fluid surface in the container and h be the difference in elevation between this fluid surface and the orifice. Writing Bernoulli's equation along a streamline, such as ABC on Figure 5.8, leads us then to:

$$p_0 + \rho g y_0 = p_0 + \frac{1}{2}\rho v_f^2 + \rho g (y_0 - h). \tag{5.65}$$

The velocity v_f at the minimal section is then:

$$v_f = \sqrt{2gh}. \tag{5.66}$$

We observe that v_f has the same value as for a body dropping from a height h under the action of gravity in the absence of friction.

***Calculation of the* vena contracta**

The minimum cross-section S_f can only be calculated exactly for the particular case of a tube re-entrant toward the interior of the container (*Borda's mouthpiece*, shown in Figure 5.8b). We show below that one has then:

$$S_{fb} = S_0/2. \tag{5.67}$$

Experimentally, for a circular orifice drilled directly in the wall, it is observed that $S_{fb}/S_0 \approx 0.6$. This ratio, larger than predicted by Equation 5.67, results from the fact that the pressure is lower along the wall in the neighborhood of point C (Figure 5.8a) than for *Borda's mouthpiece*: the reaction force is therefore greater than in this latter case, so that the jet, whose rate of flow must provide the balancing momentum, but which exits with the same velocity, must have a larger cross-section. More generally, for various configurations of the opening mouthpiece, one finds experimentally values of the contraction coefficient S_f/S_0, ranging usually between 0.5 and 1.

Proof

Let us apply the conservation of momentum (Equation 5.13) to a control volume bounded by a fixed surface S, consisting of the free surface at the top of the reservoir, of the walls of the container "wet" by the fluid, of the surface of the jet up to its minimum cross-section S_{fb}, and finally of S_{fb} itself. The x-component of the vector equation 5.13 is then:

$$\iint_S \rho\, v_x (v_j n_j)\, dS + \iint_S p\, n_x\, dS = 0. \tag{5.68}$$

The first integral reduces simply to $\rho\, S_{fb}\, v_{fb}^2$. The second one can be calculated on the basis of the hypothesis of the reentrant mouthpiece. The pressure p is everywhere equal to the hydrostatic pressure except on the lateral surface of the jet, and across its cross-section S_f where it is equal to p_0. If p were everywhere equal to the hydrostatic pressure:

$$p_{\text{hydro}} = p_0 + \rho g\,(y_0 - y),$$

we would have found an integral $\iint_S p_{\text{hydro}}\, \mathbf{n}\, dS$ equal to the weight $\mathbf{W}$ of the fluid contained within S (y is the vertical coordinate of the point in question). Thus, the x-component of this integral is zero.

$$\iint_S p_{\text{hydro}}\, n_x\, dS = 0. \tag{5.69}$$

Since p differs from p_{hydro} only over the region $E'F'G'H'$ of the surface of integration, where p is equal to p_0, we have:

$$\iint_S p\, n_x\, dS = \iint_S p_{\text{hydro}} n_x dS + \iint_{(E'F'G'H')} (p_0 - p_{\text{hydro}})\, n_x dS,$$

or, using the equation for p_{hydro}:

$$\iint_S p\, n_x\, dS = p_0 S_0 - (p_0 + \rho\, g\,(y_0 - y))S_0 = p_0 S_0 - (p_0 + \rho\, g\, h)S_0 = -\rho\, g\, h\, S_0. \tag{5.70}$$

The cross-section S_0 of the opening is indeed equal to the projected area of the surface $E'F'G'H'$ onto a plane perpendicular to the x-axis. We have also assumed that the relative variation of $\rho\, g\,(y_0\text{-}y)$ across the section S_0 is negligible. Inserting Equation 5.70 into Equation 5.68, we obtain:

$$\rho\, S_f\, v_{fb}^2 = \rho\, g\, h\, S_0. \tag{5.71}$$

Equation 5.67 then results when we insert the value of the velocity from Equation 5.66.

Force exerted by the fluid on the container

The force $\mathbf{F}$ exerted by the fluid on the entire container is equal to the integral of the pressure forces on the surface *wetted* by the liquid. In order to evaluate its component F_x in the x-direction, we use Equation 5.70, subtracting from the integral the horizontal component of the pressure forces on the section $E'F'G'H'$ of S corresponding to the free jet.

$$F_x = \iint_{(\text{walls})} p\, n_x\, dS = \iint_S p\, n_x\, dS - \iint_{(E'F'G'H')} p_0\, n_x\, dS, \tag{5.72}$$

i.e.:

$$F_x = -(\rho S_{fb}\, v_{fb}^2 + p_0\, S_0) = -(\rho\, g\, h + p_0)S_0. \tag{5.73}$$

It is not surprising that the pressure p_0 appears in this expression: indeed, the pressure on the free surface of the reservoir must necessarily play a role. On the other hand, this external pressure does not appear in the summation of all the forces exerted on the walls of the container, since the external fluid (in this case, air) exerts its counter-pressure p_0 on the *outer* face of the container.

5.4.3 Force on the walls of an axially symmetric conduit of varying cross-section

Assume that we have a conduit in the shape of a surface of revolution around the x-axis, with a section in which the channel diverges smoothly and that, on both sides of this section, we have regions of uniform cross-section S_1 and S_2 in which the flow velocities U_1 and U_2 are parallel to the x-axis (Figure 5.9). We will show that the component F_x in the x-direction of the forces exerted on the walls of the conduit has the value:

$$F_x = p_1 (S_1 - S_2) + \frac{1}{2}\rho U_1^2 S_1 \left(2 - \frac{S_2}{S_1} - \frac{S_1}{S_2}\right) = p_1 (S_1 - S_2) - \frac{1}{2}\rho U_1^2 S_1 \left(\sqrt{\frac{S_1}{S_2}} - \sqrt{\frac{S_2}{S_1}}\right)^2, \quad (5.74)$$

where p_1 is the pressure at the inlet. This equation allows us, by the use of the equations of the conservation of mass, momentum and energy, to express the force F_x simply in terms of the variables p_1 and U_1, without the need for a complete calculation of the velocity field of the flow. As in the previous case, this equation neglects any viscosity effects and is therefore only an approximation for viscous fluids.

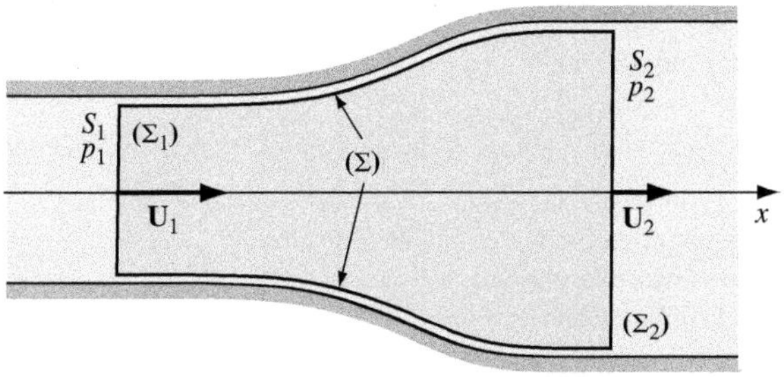

Figure 5.9 *Determination of the force exerted on the walls of a conduit in the shape of a surface of revolution, based on the laws of conservation of momentum and of mass over a control volume bounded by the surfaces Σ, Σ_1 and Σ_2*

Proof

Let us choose a fixed control volume bounded by a surface of revolution S consisting of the two sections normal to the x-axis (Σ_1 and Σ_2) and of the region of the tube wall located between these (Σ). Assuming that the flow is stationary, we can therefore use Equation 5.13 to express the conservation of momentum. By taking components in the x-direction, we obtain:

$$\iint_S \rho\, v_x (v_j n_j)\, \mathrm{d}S + \iint_S p\, n_x \mathrm{d}S - \iint_S \sigma'_{xj}\, n_j\, \mathrm{d}S = 0. \quad (5.75)$$

Gravity forces have no component in the horizontal x-direction and, accordingly, do not appear in this expression. The integral containing σ'_{xj} involves only the contribution along the lateral wall Σ, but we neglect hereafter this contribution due to viscous effects. In the term $\iint_S p\, n_x\, \mathrm{d}S$, let us now separate the contributions of the conduit walls (surface Σ) from those of the cross-sections (Σ_1 and Σ_2); we note that the contribution of surface Σ to the first integral of Equation 5.75 is zero. Denoting by $F_{px\Sigma}$ the x-component of the pressure forces on Σ, we obtain, by grouping the terms involving p and ρv_x^2 integrated over the surfaces Σ_1 and Σ_2:

$$F_{px\Sigma} = \iint_{\Sigma_1} \left(p + \rho v_x^2\right) \mathrm{d}S - \iint_{\Sigma_2} \left(p + \rho v_x^2\right) \mathrm{d}S. \quad (5.76)$$

In the above equations, we observe that there appear, in addition to the pressure terms p, terms of the form $(\rho\, v_x^2)$ which correspond to normal stresses due to the motion of the fluid. They represent a flux in the x-direction of the component of momentum ρv_x, i.e., $v_x(\rho v_x)$. In order to obtain a value for $F_{px\Sigma}$, it is enough then to know the distribution of pressures and velocities in sections Σ_1 and Σ_2.

In the example of the conduit illustrated in Figure 5.9, the cross-section slowly increases from a value S_1 to the value S_2. We can then apply Equations 5.37a, 5.37b and 5.38 in each of the regions where the areas of sections Σ_1 and Σ_2 are constant and equal to S_1 and S_2. The pressure gradient in both sections Σ_1 and Σ_2 then reduces to the hydrostatic pressure gradient. This does not affect the motion of the fluid, and can therefore be neglected in the equation above. We also assume that the velocities are constant over each of the cross-sections and respectively equal to U_1 and U_2, as would be the case for a fluid with negligible viscosity. Equation 5.76 then becomes:

$$F_{px\Sigma} = \left[p_1 S_1 + \rho U_1^2 S_1\right] - \left[p_2 S_2 + \rho U_2^2 S_2\right]. \quad (5.77)$$

As mentioned above, there is, in addition to the pressure forces, a term related to the momentum convected by the flowing fluid. If we now write down the condition for energy conservation, by applying Bernoulli's equations along a streamline, we obtain:

$$p_2 - p_1 = \frac{1}{2}\rho\left(U_1^2 - U_2^2\right). \quad (5.78)$$

We then obtain Equation 5.74 by combining Equations 5.77 and 5.78 with the law of conservation of mass, which in this case is given by:

$$U_2 = U_1\ S_1/S_2. \quad (5.79)$$

5.4.4 Liquid sheets of varying thickness: the hydraulic jump

Qualitative properties of hydraulic jumps

Anyone who has looked at water flowing down from a tap into a sink has probably noticed, on the surface on which the water impinges, a circular bulge in the water flow centered around the jet (Figure 5.10). This bulge, known as a *hydraulic jump*, occurs at the border between a central region where the fluid is shallow, and an outer one where the fluid is considerably deeper. The boundary between these two regions corresponds to a transition of the flow velocity $U(x)$ from a value higher than the local velocity $c(h)$ of the surface waves in the central region, to a subcritical value further out. We will see in Section 6.4 that $c = \sqrt{gh}$ represents the local velocity of these *gravity waves* along the surface of a layer of fluid of thickness h. As a result, there appears a rather abrupt transition: when the fluid becomes deeper, the fluid velocity decreases while the wave velocity increases.

A breaking wave represents a moving version of the phenomenon of the hydraulic jump: here, the jump moves along at the velocity of the breaking wave's crest. In that case, if we wish to apply the theoretical analysis developed below, we need to do so in a reference frame

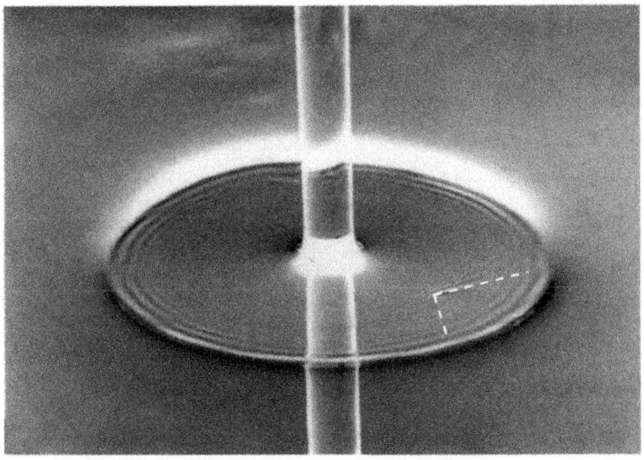

Figure 5.10 *Formation of a bulge, which can be considered as an hydraulic jump, when a jet of water impinges on a solid flat horizontal surface. The velocity of the fluid in the central region exceeds that of the surface waves in that region, and then become smaller than the latter in the region beyond the bulge and all the way to the edges. We observe in the right center region a "V"-shaped ripple opening outward (dotted lines), due to an imperfection on the surface, which corresponds to a kind of shock wave, somewhat similar to the supersonic "bang" from an airplane breaking the sound barrier (document S. Middleman)*

Figure 5.11C *Tidal bore heading up (to the right) the Petitcodiac river, near its estuary in the Bay of Fundy, in the Canadian province of New Brunswick (document C. L. Gresley)*

fixed relative to this moving region. Another particularly spectacular moving hydraulic jump is the *tidal bore* which can appear where the flow of a river meets a rising tide (Figure 5.11).

The *tidal bore* is only observed in a few regions of the world. It results from the combination of large amplitude tides, a funnel-shaped estuary much wider at its mouth, a gentle slope and shallow water in the corresponding river.

The hydraulic jump is quantitatively characterized by the ratio of the velocity of the fluid and of the surface waves, known as the *Froude number:*

$$Fr = \frac{U(x)}{\sqrt{g\,h(x)}}. \tag{5.80}$$

In the experiment illustrated in Figure 5.10, the Froude number changes from a value greater than 1 near the center to a value less than 1 in the outer regions.

This phenomenon is the analog of the shock wave which is formed near a supersonic airplane and which corresponds to the change of the velocity of the air from a value greater than to a value less than that of sound. The Mach number, $M = v/c$ (where c indicates the velocity of sound), is the *analog* of the Froude number, Fr. Moreover, we often observe behind an obstacle the formation of a characteristic "V" (Figure 5.10), at an angle which depends, just as for compressible flows, on the ratio between the velocity of the fluid and that of the surface waves.

Liquid flow over a weir (an underwater obstacle)

Hydraulic jumps can appear in flowing liquid layers downstream of an obstacle (Figure 5.12). Le us first discuss the flow over the obstacle: we denote by h the initial level of the fluid, and by $h(x)$ the change with distance of the thickness of the fluid layer, $e_0(x)$ the elevation of the bottom and let us assume that the velocity $U(x)$ is uniform in every vertical cross-section of the flow; the relationship between $U(x)$ and $e_0(x)$ is given for a stationary flow by:

$$\frac{1}{U(x)}\frac{dU(x)}{dx}\left(-gh(x)+U^2(x)\right)+g\frac{de_0(x)}{dx}=0. \tag{5.81}$$

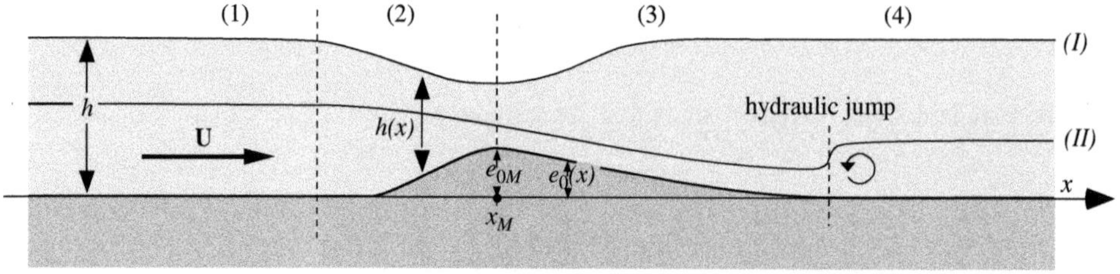

Figure 5.12 *Flowing liquid passing over a weir (underwater obstacle) for the two cases of the calculation. Case I: Froude number, Fr < 1; case II: Froude number, Fr > 1. In case (II), a low fluid velocity region (1) is followed by a second region (2) of acceleration, then by a high-velocity region (3) upstream of the hydraulic jump, and finally by a low velocity region downstream of the jump*

Proof

The two equations below express the conservation of mass, and Bernoulli's equation along a streamline just under the surface above which the atmospheric pressure has the constant value p_0:

$$U\,h=U(x)\,h(x) \qquad \text{and:} \qquad p_0+\frac{1}{2}\rho\,U^2+\rho g h=p_0+\frac{1}{2}\rho\,U^2(x)+\rho\,g\,(h(x)+e_0(x)).$$

Taking the derivative of these two equations with respect to x, we obtain:

$$U(x)\frac{dh(x)}{\partial x}+h(x)\frac{dU(x)}{\partial x}=0 \qquad \text{and} \qquad \rho\,U(x)\frac{dU(x)}{dx}+\rho\,g\frac{dh(x)}{dx}+\rho\,g\frac{de_0(x)}{dx}=0.$$

Equation 5.81 follows from combining these last equations in order to eliminate the term dh/dx.

Let us assume that the flow is initially sufficiently slow, and the depth sufficiently large so that:

$$U^2-g\,h<0\ (\text{i.e., } Fr<1).$$

We see the appearance of two types of behavior, at the moment when the flow passes over the position x_M where the crest of the weir, of height e_{0M}, is located, and where $(de_0/dx)=0$. Equation 5.81 can, in fact, be satisfied in two ways:

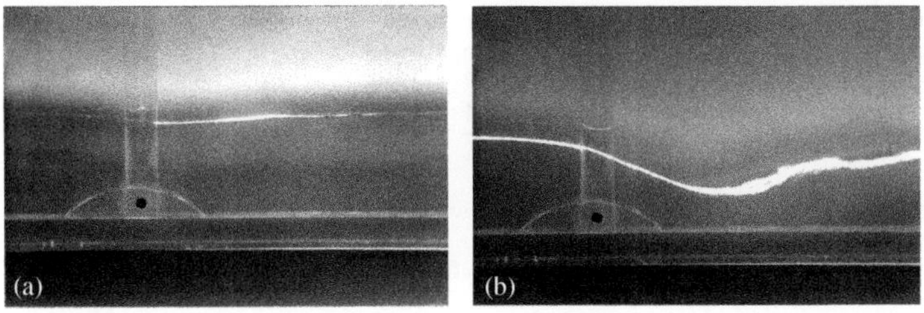

Figure 5.13 *Flow of a fluid over a weir (an underwater obstacle) in a rectangular channel; (a) situation where the Froude number is everywhere smaller than unity; (b) case where the flow is supercritical everywhere upstream of the region where the weir's height is a maximum (plates courtesy of M. Devillers, ENSTA)*

(i) $\mathrm{d}U(x)/\mathrm{d}x = 0$. In this case, we have also, from the equation of conservation of mass, $\mathrm{d}h(x)/\mathrm{d}x = 0$. (This corresponds to case I in Figure 5.12 and to plate (a) in Figure 5.13). After passing over the crest of the weir, the thickness of the fluid layer increases again, and the velocity returns to its original value U.

(ii) $U^2(x) = g\,h(x)$. In this second case, $\mathrm{d}U(x)/\mathrm{d}x$ no longer changes sign. Thus, the velocity continues to increase, and the thickness of the fluid to diminish, after passing over the point x_M. Equation 5.82 can still be satisfied because $U^2(x) - gh(x)$ becomes positive, and $(\mathrm{d}e_0/\mathrm{d}x)$ also changes sign, as it passes over the point $x = x_M$ (case II in Figure 5.12 and plate (b) in Figure 5.13).

We can see the key role that the Froude number plays in these phenomena. In case (i), the Froude number is everywhere less than one. In case (ii), it increases and goes through the value 1 *exactly* at the point x_M, then becoming greater than one (supercritical flow). The fluid then returns to a state of smooth flow with a large depth in a very abrupt manner, by means of a hydraulic jump, as illustrated in Figures 5.12 and 5.13b (we discuss this phenomenon further on). We can observe the transition from one behavior to the other by gradually increasing the velocity of the fluid at constant depth until the value $Fr = 1$ is reached at the point $x = x_M$. We observe hydraulic jumps of the kind displayed in Figure 5.13b in spillways downstream of dams.

The discussion above indicates that there can be two different values of the velocity, and of the depth of the fluid, for the same height $e_0(x)$ of the weir at a fixed flow rate: one corresponding to a Froude number Fr greater than unity, the other to a number Fr' less than one. For the specific case $e_0 = 0$ (region far downstream of the obstacle, where U' is the velocity and h' the depth of the fluid), we find a second solution $Fr' \neq Fr$ such that:

$$Fr^{2/3}\,Fr'^{2/3}\left(Fr^{2/3} + Fr'^{2/3}\right) = 2. \tag{5.82}$$

If Fr is less than one, this equation can effectively only be satisfied if Fr' is greater than one. Indeed, if this were not the case, the sum in Equation 5.82 would be less than two, and each of the factors in the product would be smaller than unity.

Proof

Let us rewrite Equation 5.81, eliminating the height $h(x)$ by means of the flow rate condition $Q = U(x)\,h(x)$. We then obtain:

$$gQ\frac{\partial}{\partial x}\left(\frac{1}{U(x)}\right) + \frac{1}{2}\frac{\partial\,[U(x)]^2}{\partial x} + g\frac{\partial e_0(x)}{\partial x} = 0,$$

so that, after integrating between two points located respectively far upstream and far downstream of the obstacle, where the flow velocities are U and U':

$$gQ\left(\frac{1}{U}-\frac{1}{U'}\right)+\frac{1}{2}\left(U^2-U'^2\right)=0.$$

This equation is satisfied either by $U = U'$, or when:

$$gQ = UU'\frac{U+U'}{2}. \qquad (5.83)$$

Rewriting this last equation by means of the Froude numbers written in the form:

$$Fr = \frac{U}{\sqrt{gh}} = \frac{U^{3/2}}{\sqrt{gUh}}, \qquad Fr' = \frac{U'}{\sqrt{gh'}} = \frac{U'^{3/2}}{\sqrt{gU'h'}}$$

and using the equations $Q = Uh = U'h'$, we obtain Equation 5.82 from Equation 5.83, after cancelling out $g\,Q$.

Analogy with compressible flow and shock waves

At the beginning of Section 5.4.4, we have already underscored the close analogy between these problems and those of compressible flows. The system equivalent to the problem we have just discussed is the convergent-divergent nozzle (known as the *de Laval nozzle*) illustrated in Figure 5.14. The analog of the Froude number is the Mach number, M.

For low flow rates, the flow velocity is everywhere lower than the speed of sound and reaches a maximum at the point where the cross-section is a minimum: *the throat* (flow corresponding to case (a) in Figure 5.13). But, when the velocity reaches the speed of sound at the throat ($M = 1$), it continues to increase with distance and becomes supersonic in the downstream region. This increase in velocity comes about as a result of the continuous decrease of the gas pressure required to maintain the conservation of the mass. Further downstream, a shock wave appears which forms a transition toward the higher pressure regions beyond the nozzle exit, where the Mach number is again less than unity. This shock front is the analog of the hydraulic jump illustrated in Figures 5.12 and 5.13, which we now proceed to discuss in greater detail.

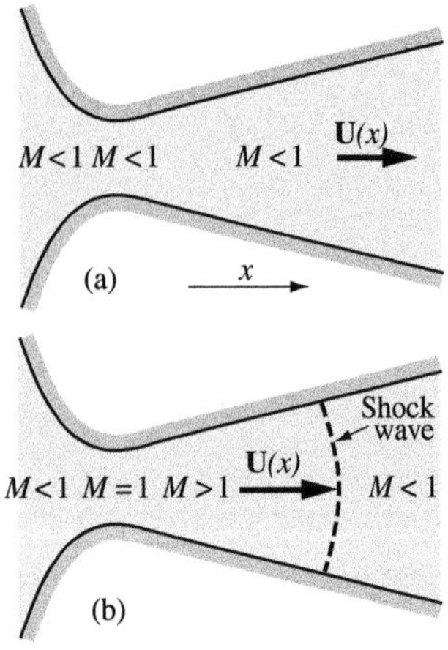

Figure 5.14 *Flow of a compressible fluid within a convergent-divergent nozzle. (a) Case where the flow remains subsonic throughout the nozzle; (b) case where the flow reaches the speed of sound at the throat of the nozzle, resulting in a shock wave indicated by the line of dashes*

Hydraulic jump: conservation equations

As we have mentioned above, a hydraulic jump corresponds to a sudden transition from a flow with Froude number Fr greater than one to a value less than one. As opposed to the previous discussion as to how $h(x)$ varies above an underwater obstacle, we can no longer apply Bernoulli's equation, because there is significant energy dissipation in the region of the jump. We then rely simply on the equations of conservation of mass and of momentum.

In Figure 5.15 (next page), we show schematically a hydraulic jump with two regions where the flow is parallel and uniform, with velocities U and U' respectively upstream and downstream. Using the equation of conservation of momentum, we can obtain the following relationships between U, U' and the levels h and h' upstream and downstream of the jump:

$$U' = \sqrt{g\frac{h}{h'}\frac{(h+h')}{2}}, \quad (5.84a) \qquad U = \sqrt{g\frac{h'}{h}\frac{(h+h')}{2}}. \quad (5.84b)$$

By squaring and then subtracting, we find that U and U' satisfy:

$$U' < \sqrt{gh'} \quad (5.85a) \qquad \text{and} \qquad U > \sqrt{gh}. \quad (5.85b)$$

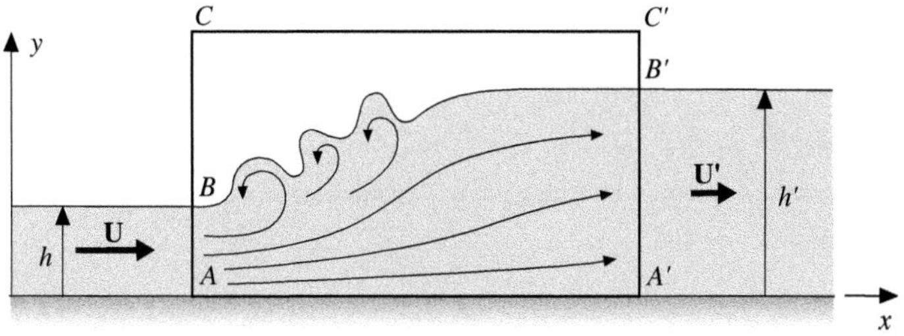

Figure 5.15 *Hydraulic jump. The upstream fluid velocity is greater than the speed $\sqrt{gh}$ of surface waves, while the downstream fluid velocity is smaller (than $\sqrt{gh}$). We display the control volume ABCC'B'A'A through which we calculate the conservation of momentum and of mass*

given that h is smaller than h'. The flow is thus supercritical (velocity greater than that of the surface waves) in the region before the jump, and sub-critical in the other. From a physical point of view, this result indicates that the jump will have a stable structure. The ripples which might escape upstream in the super-critical region, taking away a corresponding part of the energy, will be brought back by the flow which moves more rapidly than they do. Those which would be driven downstream by the flow beyond the jump move fast enough to go upstream against the flow and return toward the jump.

Proof

As in the preceding case, we use Equation 5.13 which expresses the conservation of momentum by assuming that the flow is stationary and neglecting the viscosity along the boundaries of the control volume. We choose a control volume bounded by the surface (S) denoted by $ABCB'A'A$ in the plane of the figure, and of unit depth in the perpendicular direction. Note that part of the control volume is in air, and accordingly subjected to atmospheric pressure p_0. Let us assume that the flow velocity is uniform and equal to U and U' in sections AB and $A'B'$ of the control volume, which are the only ones through which there is non-zero mass transport; we then obtain:

$$\iint_S \rho\, v_x (v_j n_j) \mathrm{d}S = \rho\, U'^2 h' - \rho\, U^2 h. \tag{5.86}$$

As before, we determine the pressure p on AB and $A'B'$ from Equations 5.37:

where $\quad p = p_0 + \rho\, g\,(h - y) \quad$ and $\quad p' = p_0 + \rho\, g\,(h' - y).$

By adding a term $-p_0\ (h' - h)$, corresponding to the integral of p_0 along the section BC we obtain:

$$\iint_s p\, n_x \mathrm{d}S = p_0\left(h - h'\right) - \int_0^h (p_0 + \rho g\,(h - y))\,\mathrm{d}y + \int_0^{h'} (p_0 + \rho g\,(h' - y))\,\mathrm{d}y = \frac{\rho g}{2}\left(h'^2 - h^2\right). \tag{5.87}$$

Equation 5.13 can then be written, after taking into consideration Equations 5.86 and 5.87, and neglecting any viscosity:

$$\left(U'^2 h' - U^2 h\right) + \frac{g}{2}\left(h'^2 - h^2\right) = 0. \tag{5.88}$$

The conservation of mass from upstream to downstream of the jump then requires:

$$U'\, h' = U\, h. \tag{5.89}$$

Using Equation 5.89, in order to eliminate U' (or U) from Equation 5.88, we then obtain Equations 5.84a and 5.84b.

Ratio between the fluid levels and the velocities on either side of the jump

In order to determine the ratio h'/h, we rewrite Equation 5.84b in the form:

$$gh'^2 + ghh' - 2U^2h = 0. \tag{5.90}$$

Taking the positive root of this quadratic equation in h', we find:

$$\frac{h'}{h} = \frac{U}{U'} = \frac{-gh + \sqrt{(gh)^2 + 8\,U^2gh}}{-gh} = \frac{-1 + \sqrt{1 + 8\,Fr^2}}{2}. \tag{5.91}$$

Here, Fr is the Froude number of the upstream flow. This again leads to the result that, if h'/h is greater than unity, this implies that $Fr > 1$, and that the flow is super-critical upstream of the jump. For $Fr = 1$, we find that $h' = h$, corresponding to the limiting case of a jump of infinitesimal amplitude where both velocities are very close to the value $\sqrt{gh}$.

EXERCISES

1) **Momentum conservation using a moving control volume**
Assume that the control volume used in Equation 5.10 is moving at a constant velocity $\mathbf{w}$ while $\mathbf{v}$ is the velocity in the fixed reference frame. Show that we must replace in this equation one of the velocity vectors $\mathbf{v}$ by the relative velocity $\mathbf{v}$-$\mathbf{w}$. Write the mass-conservation equation for this same control volume. These results can be used in Exercises (2) and (4).

2) **Principle of a rotating lawn sprinkler**
This device consists of a horizontal, straight tube of length $2R$, rotating around a vertical axis in the middle and with a vertical injection of water along this axis into the tube (total flow rate $2q$). A nozzle is attached perpendicular to the tube at each of its ends, creating a horizontal jet of cross-section S and velocity $\boldsymbol{U}$ relative to the nozzle. If the sprinkler rotates at an angular velocity Ω, compute the tangential force component F_t

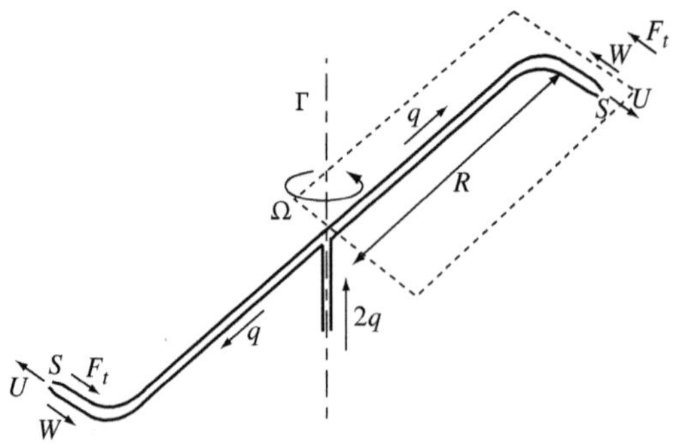

on the nozzle due to the flow. Write the equation of motion if $\Gamma(\Omega)$ is the torque due to friction opposing the rotation. What is the value of Ω if $\Gamma = 0$? What is the torque which must be applied in order to keep $\Omega = 0$? Considering the system as an engine, when is the greatest power available (use the solution from Exercise (1)).

3) **Impact of a circular jet on a plane and a cone**
We would like to extend the results obtained for a two-dimensional sheet of liquid in Section 5.4.1 to a circular laminar jet of perfect fluid in potential flow. We start with a large plate upon which the jet impinges at an angle of incidence α which creates on the plate layer of liquid flowing radially outward. The radius of the jet is R and the fluid velocity inside it U. What is the velocity $U(r)$ in the radial flow on the plane? In the particular case $\alpha = 0$, is the thickness $e(r)$ constant as a function of r and/or what is its variation? What is the total component of the force normal to the plane and its variation with α?

We then consider a conical surface of tip O, half-angle β and with the same x-axis as the jet. What is the variation of the thickness $e(h)$ with the distance h from the tip O along the x-axis? What is the force $\mathbf{F}$ on the cone?

4) **Principle of a turbine flow-meter**
Some flow-meters use a helix of geometry similar to that of a boat propeller, located inside a circular channel with a very small clearance between the end of the blades and the channel wall (see Figure (a) in the margin). The flow velocity $\mathbf{U}$ upstream of the blades is measured from their angular velocity Ω around the x-axis of the helix. In a simple model (see Figure (b) in the margin), the blades are represented as twisted plates, tilted at an angle $\alpha(r)$ with respect to x at a distance r from this axis. We consider the relative velocity $\mathbf{V}_{rel}(r)$ in the moving reference frame of the twisted blade. What are the values of the tangential components of $\mathbf{V}_{rel}$ and of the velocity of the blade $v_\varphi(r)$? Assuming now that there is no momentum loss by solid or viscous friction, and using again the results of Exercise (1), what is the angular velocity Ω at which the local exchange of tangential momentum between the blade and the fluid is zero? What must the variation of $\alpha(r)$ be so that this relation is satisfied for all values of r? In this case, what is the flow velocity at the level of the blades? What happens if the energy loss during the rotation is non-zero?

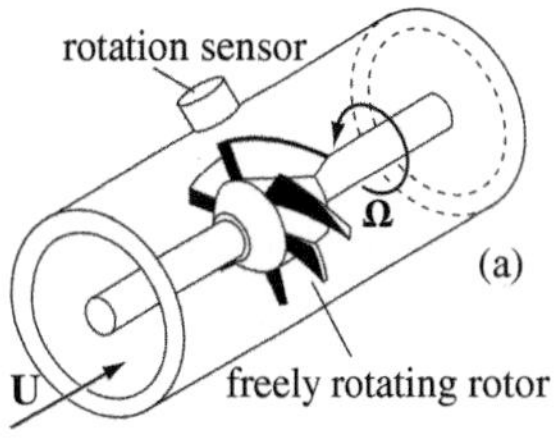

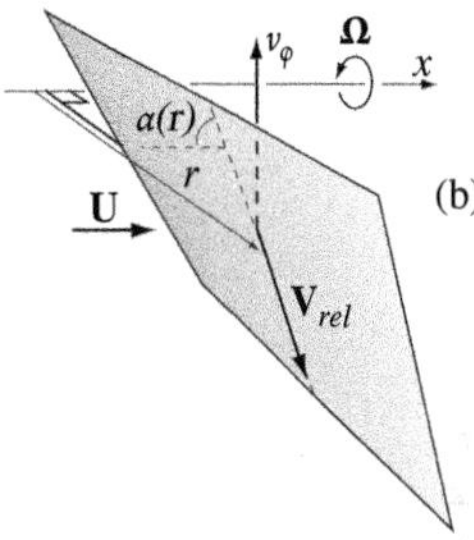

6

Potential Flow

The study of an ideal (perfect) fluid or "dry water," as Richard P. Feynman called it in the *Feynman Lectures in Physics,* is an important part of fluid dynamics. In the mechanics of solids, the laws of motion can be expressed, in the absence of friction, in terms of conservation laws. In a similar way, assuming a zero viscosity led to simpler energy and momentum conservation laws in Chapter 5 and allowed us to solve a number of problems without requiring a detailed knowledge of the local motion. This lack of viscosity also leads, indirectly, to a persistence of the irrotational nature of the velocity field: we derive this result in Section 6.1 of this chapter, where we also list the cases for which the theory of potential flow applies. We then introduce the velocity potential and its properties (Section 6.2), together with a number of illustrations. Next, in Section 6.3, we deal more generally with the case of potential flow around an obstacle of arbitrary shape. We then continue with a discussion of linear waves along the surface of a fluid, a specific example of potential flow (Section 6.4). The following section (Section 6.5) discusses the analogy between potential flow and electromagnetic theory. Then, in Section 6.6, we introduce the concept of the complex potential, providing a number of illustrative examples, concluding finally with a description, and explicit examples, of the use of conformal mapping in the solution of flow problems.

6.1 Introduction

First and foremost, it is ideal (non-viscous) fluids which lead to potential flows. As seen in Section 4.2.3, their motion is described by Euler's equation:

$$\rho \frac{\partial \mathbf{v}(\mathbf{r},t)}{\partial t} + \rho\,(\mathbf{v}\cdot\nabla)\,\mathbf{v}\,(\mathbf{r},t) = \rho\,\mathbf{f} - \nabla p. \qquad (6.1)$$

The *potential flow* of an ideal fluid is such that its velocity field $\mathbf{v}\,(\mathbf{r},\,t)$ can be derived from a *velocity potential* $\Phi(\mathbf{r},\,t)$, with:

$$\mathbf{v}\,(\mathbf{r},t) = \nabla\Phi\,(\mathbf{r},t). \qquad (6.2)$$

The flow velocity field $\mathbf{v}\,(\mathbf{r},\,t)$ must therefore satisfy both Euler's equation and the curl-free condition $\nabla\times\mathbf{v}=0$, a consequence of Equation 6.2. We will show in Section 7.2.1, as a result of Euler's equation, that, if the flow of an ideal fluid is curl-free at a given instant of time (e.g., if the fluid is initially at rest with $\mathbf{v(r)}\equiv 0$), it will remain so at any subsequent time provided that any external volume forces are also the derivatives of a potential, and that the fluid density is constant, or only a function of the pressure. In these circumstances, potential flow of an ideal fluid can continue indefinitely, and we can, moreover, apply the formulation (Equation 5.36) of Bernoulli's theorem as derived in Section 5.3.2:

$$\rho\frac{\partial\Phi}{\partial t} + \rho\frac{v^2}{2} + p + \rho\,\varphi = \text{constant within the volume of the flow},$$

where φ is the potential from which the volume forces derive. This equation allows us to determine, by integration, the velocity potential at any arbitrary time from its value at a specific time: this is in accordance with the fact that such flows remain permanently potential. Let us note that the mathematical determination of the velocity field $\mathbf{v}(\mathbf{r})$ is quite analogous to that of the electrostatic field $\mathbf{E}(\mathbf{r})$, or even of a time-dependent field at sufficiently low frequencies such that $\nabla \times \mathbf{E}(\mathbf{r}) = 0$. Equation 6.2 is thus analogous, within a minus sign, to the relationship between the electrostatic field $\mathbf{E}(\mathbf{r})$ and the corresponding potential $V(\mathbf{r})$.

From an experimental point of view, we have mentioned, in Section 4.2.3, that the only fluid with properties approaching those of an ideal fluid is superfluid liquid helium. We will discuss in the appendix of Chapter 7 the characteristic features of superfluid flow which satisfies the curl-free condition except along singular lines (quantum vortices), where the vorticity is concentrated.

On the other hand, viscous fluids can also lead to approximately potential flow, at least in a large part of their volume. This is the case when the transit time of a particle of viscous fluid past a solid body is short: then, the perturbation brought to the flow by the zero velocity boundary condition at the surface can only propagate by viscous diffusion over a small transverse distance. In this way, *non-turbulent* flows at large Reynolds numbers frequently display potential-flow properties except for a thin boundary layer near walls and a narrow wake downstream from any obstacles. We will discuss such flows in Chapter 10.

Similar results are obtained for high-frequency sinusoidal flow ($\omega = 2\pi/T$) near an object of characteristic length L. In this case, the resultant flow will be approximately potential if the diffusion distance of the velocity gradients resulting from the zero velocity boundary condition at the walls is small compared to L (this diffusion length is proportional to $\sqrt{\nu T}$ as we have seen in Section 2.1.2). We will see the same effect in a transient phase of duration T, which shows similar behavior for flow around an object that is set into sudden motion.

The existence of free surfaces can also lead to the appearance of almost potential flow: in fact, as we have seen in the case of an ideal fluid in Section 4.3.2, we do not have a requirement of zero tangential velocity at such free surfaces. We can therefore set fluid in motion without the appearance of a velocity gradient (and thus of a non-zero curl) near the wall. An example of such flow is that due to *Taylor bubbles* (large bubbles rising in tubes of a diameter comparable to that of the bubble, which we will discuss in Section 6.4.4).

6.2 Definitions, properties and examples of potential flow

6.2.1 Characteristics and examples of velocity potentials

If we assume incompressible flow (Equation 3.28), Equation 6.2 becomes:

$$\nabla \cdot \mathbf{v} = \nabla \cdot [\nabla \Phi (\mathbf{r})] = \nabla^2 \Phi (\mathbf{r}) = 0. \tag{6.3}$$

In electrostatics, this case corresponds to the field obtained in the absence of free charges. The resulting vector velocity field reduces to the scalar solution of Laplace's equation for the potential. Here, we have the benefit of the entire arsenal of methods developed for electrostatics (we have already mentioned this equivalence in Section 3.3.3).

The *boundary conditions* are that the normal component of the velocity $v_\mathbf{n}$ is zero relative to that of solid walls. The impossibility of fluid flow through a solid wall S leads to the condition:

$$[v_\mathbf{n}]_S = \left[\frac{\partial \Phi}{\partial n}\right]_S = 0, \tag{6.4}$$

in which $[v_\mathbf{n}]_S$ is the velocity component of the wall normal to its surface. At the interface (S) between two ideal fluids 1 and 2, the boundary condition $v_{\mathbf{n}1} = v_{\mathbf{n}2}$ is expressed by:

$$\left[\frac{\partial \Phi_1}{\partial n}\right]_S = \left[\frac{\partial \Phi_2}{\partial n}\right]_S. \tag{6.5}$$

There is no boundary condition applying to the tangential component of the velocity because we have no viscous forces: in fact, viscosity requires that these tangential components be equal in real fluids, as we have seen in Section 4.3.1.

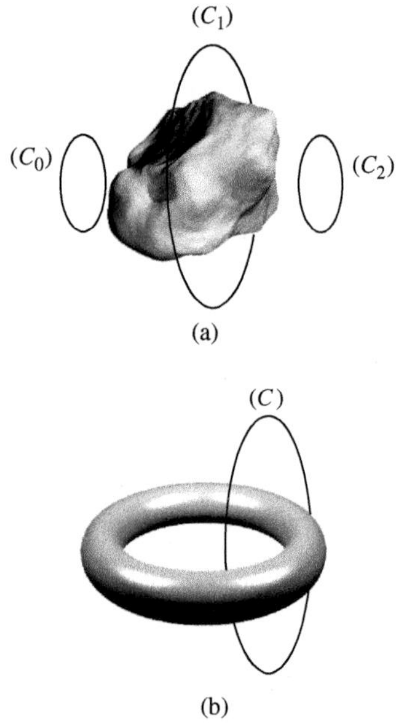

Figure 6.1 *(a) Example of a simply connected volume of fluid: the area of the curves (C) can be reduced to zero by being continuously deformed without intersecting the surface of the solid. (b) Doubly connected volume of fluid: the area of the curve (C) cannot be reduced to zero by a continuous deformation*

6.2.2 Uniqueness of the velocity potential

Here again, we will call upon the classical proofs from electrostatics. Within a *simply connected* volume of fluid (Figure 6.1a), there exists a unique velocity field, potential and incompressible, which corresponds to the normal velocity components at the solid walls and to a given velocity at infinity.

Proof

Assume that we have two velocity fields $\mathbf{v}_1 = \nabla\Phi_1$ and $\mathbf{v}_2 = \nabla\Phi_2$ corresponding to the same boundary conditions. We show that the integral $\iiint (\mathbf{v}_1 - \mathbf{v}_2)^2 \, d\tau$ evaluated over the entire volume is zero. This indicates that the velocity fields $\mathbf{v}_1$ and $\mathbf{v}_2$ are identical. Let us write $\mathbf{v} = \mathbf{v}_1 - \mathbf{v}_2$ and $\Phi = \Phi_1 - \Phi_2$. We then obtain:

$$\iiint_{\mathcal{V}} (\mathbf{v}_1 - \mathbf{v}_2)^2 d\tau = \iiint_{\mathcal{V}} \mathbf{v}\cdot(\nabla\Phi)\, d\tau = \iiint_{\mathcal{V}} \nabla\cdot(\mathbf{v}\,\Phi)\, d\tau - \iiint_{\mathcal{V}} \Phi(\nabla\cdot\mathbf{v})\, d\tau.$$

The integral on the extreme right-hand side is zero because $\nabla\cdot\mathbf{v} = 0$. The adjacent integral can be transformed into a surface integral on the solid walls and at infinity:

$$\iiint_{\mathcal{V}} \nabla\cdot(\mathbf{v}\,\Phi)\, d\tau = \iint_{S} \Phi\,\mathbf{v}\cdot\mathbf{n}\, dS.$$

This integral is zero on the solid walls because of the boundary condition on the normal velocity component, which requires that $\mathbf{v}\cdot\mathbf{n} = 0$. The surface integral also vanishes toward infinity because, as we will see in Section 6.2.4, the effect on the velocity field of the existence of any obstacle decreases as $1/r^3$ with the distance r. As a result, the integral of any such terms on a spherical surface with area proportional to r^2 tends to zero at large distances.

A method similar to the one used to demonstrate the uniqueness allows us to show that the velocity field of this flow is that which minimizes the total kinetic energy. This is true for all fields which satisfy simultaneously $\nabla\cdot\mathbf{v} = 0$ and the boundary conditions along the walls and at infinity.

A volume of fluid will be *multiply connected* for solid walls of which at least one dimension is infinite (e.g., an infinitely long cylinder), or for walls with a toroidal geometry (Figures 6.1b and 6.2). In both these cases, we can have a closed curve *(C) entirely contained within the fluid*, for which the area cannot be reduced to zero by continuous deformation within the fluid. In this instance, we cannot define an unambiguous value of the velocity potential Φ because the circulation $\int_C \mathbf{v}\cdot d\mathbf{l} = \int_C (\nabla\Phi)\cdot d\mathbf{l}$ of the velocity along the contour (C) can take on an arbitrary finite value Γ.

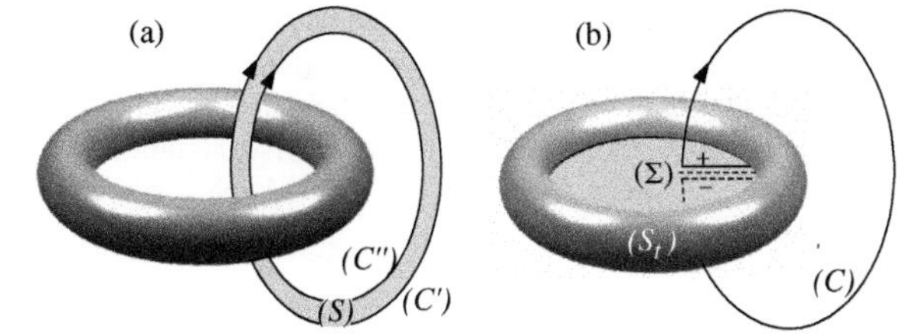

Figure 6.2 *Flow geometry used in the proof of the uniqueness of the solutions of Laplace's equation, for the case of a doubly connected volume of fluid. (a) The surface (S) rests on the two curves (C′) and (C″) along which the circulations are equal. (b) The evaluation of the integral of $(\mathbf{v}_1 - \mathbf{v}_2)^2$ over the volume of flow, excluding the infinitesimal volume between the two surfaces Σ^+ et Σ^-, that are infinitely close but on different sides of the surface (Σ). For clarity, only the intersections of these two surfaces with a surface bounded by the contour (C) have been shown.*

Proof

Consider a geometry, such as that illustrated in Figure 6.2, in which the fluid volume is bounded by a solid toroidal surface (S_t). We calculate the potential Φ starting with a line integral of the velocity along a closed curve (C) (Figure 6.2b).

We first show that, though the nature of the flow is potential and the fluid velocity $\mathbf{v}$ uniquely defined at every point, the resulting circulation $\Gamma = \int_C \mathbf{v} \cdot d\mathbf{l}$ along the curve (C) no longer necessarily vanishes. Applying Stokes' theorem to the circulation, we have:

$$\Gamma = \int_C \mathbf{v} \cdot d\mathbf{l} = \iint_S (\nabla \times \mathbf{v}) \cdot \mathbf{n} \, dS.$$

This equation gives the dependence of the circulation of the velocity $\mathbf{v}$ along the curve (C) on the flux of $\nabla \times \mathbf{v}$ through a surface (S) anchored on this curve. For (S) to be completely enclosed in fluid, either the path of integration (C) must not encircle the toroidal surface (S_t), or the surface (S) must be enclosed by the two curves (C') and (C'') both of which encircle the surface (S_t) (it is the combination of these two curves which then makes up the contour (C)). It is this latter case which is displayed in Figure 6.2a for surface (S). We then have:

$$\iint_S (\nabla \times \mathbf{v}) \cdot \mathbf{n} \, dS = \int_{C''} \mathbf{v} \cdot d\mathbf{l} - \int_{C'} \mathbf{v} \cdot d\mathbf{l} = 0.$$

Thus the circulation $\Gamma = \int_C \mathbf{v} \cdot d\mathbf{l}$ is not uniquely defined; we have merely demonstrated that it has the same value for any given space curve, such as (C') and (C''), which encircles the torus (S_t) once only, and in the same direction.

If Γ does not vanish, a single-valued potential function Φ cannot be defined. In fact, the integral $\Gamma = \int_C \mathbf{v}.d\mathbf{l}$ satisfies:

$$\int_C \mathbf{v} \cdot d\mathbf{l} = \int_C (\nabla \Phi) \cdot d\mathbf{l} = \Delta \Phi = \Gamma,$$

where $\Delta\Phi$ is the total increment in Φ for a complete loop along the curve (C'). Should $\Delta\Phi$ be non-vanishing, the function Φ is correspondingly multivalued, and defined only within an integer multiple of Γ: this problem can be dealt with by postulating along (C) a point where the function Φ is discontinuous, and changes by an amount $+\Gamma$ or $-\Gamma$ depending on the direction of travel. Since the position of this point is arbitrary, we can postulate the existence inside the torus (S_t) of a cut surface (Σ), the locus of all such points of discontinuity (the trace of (Σ) is shown in light gray in Figure 6.2b). This is quite analogous to the approach used to define the value of an angle in polar coordinates.

We now demonstrate that, with this modification, there does exist a unique potential velocity field corresponding to a given Γ. For this purpose we evaluate the integral $\iiint_{\mathcal{V}} (\mathbf{v}_1 - \mathbf{v}_2)^2 d\tau$ over the volume $(\mathcal{V})$ (Figure 6.2b), bounded by the toroidal surface (S_t) and by two additional surfaces of integration (Σ^+) and (Σ^-); we assume that (Σ^+) and (Σ^-) are infinitesimaly close to (Σ) and on each side of it. The volume $\mathcal{V}$ is simply connected so that we can proceed now as in this latter case and obtain:

$$\iiint_{\mathcal{V}} (\mathbf{v}_1 - \mathbf{v}_2)^2 d\tau = \iint_S (\Phi_1 - \Phi_2)\,(\mathbf{v}_1 - \mathbf{v}_2)\cdot\mathbf{n}\, dS.$$

The surface of integration bounding the volume $(\mathcal{V})$ consists of the surfaces (S_t), (Σ^+) and (Σ^-). Since $\mathbf{v}_1$ and $\mathbf{v}_2$ have a zero normal component on (S_t), there are contributions only from the integrals over (Σ^+) and (Σ^-). $\mathbf{v}_1$ and $\mathbf{v}_2$ are continuous at the surface (Σ) and thus each have the same value on all three surfaces (Σ), (Σ^+) and (Σ^-). Thus:

$$\iiint_{\mathcal{V}} (\mathbf{v}_1 - \mathbf{v}_2)^2 d\tau = \iint_\Sigma (\Phi_{1+} - \Phi_{1-} - \Phi_{2+} + \Phi_{2-})\,(\mathbf{v}_1 - \mathbf{v}_2)\cdot\mathbf{n}\, dS.$$

Moreover the difference $\Phi_+ - \Phi_-$ between the values of Φ, at two corresponding points of the surfaces (Σ^+) and (Σ^-), equals Γ. We thus have:

$$\iiint_{\mathcal{V}} (\mathbf{v}_1 - \mathbf{v}_2)^2 d\tau = \iint_\Sigma (\Gamma_1 - \Gamma_2)\,(\mathbf{v}_1 - \mathbf{v}_2)\cdot\mathbf{n}\, dS = (\Gamma_1 - \Gamma_2)\,(Q_1 - Q_2),$$

where Q_1 and Q_2 are the respective fluxes of the two velocity fields across the surface Σ. For the existence of a unique solution, it is therefore sufficient that the circulation has a given fixed value Γ, in addition to the equality of the normal components of the velocities at the walls. Indeed, if $\Gamma_1 = \Gamma_2$, we have $\iiint_{\mathcal{V}}(\mathbf{v}_1 - \mathbf{v}_2)^2\, d\tau = 0$, so that $\mathbf{v}_1 = \mathbf{v}_2$.

The problem of multiply connected geometries occurs also in the study of magnetic fields associated with currents. The circulation of the magnetic field around a closed contour which encloses an infinitely long wire, or a wire loop, carrying a current is, according to Ampere's law, independent of the contour (for a given number of turns about the wire, or of crossings of the loop, in the same direction).

6.2.3 Velocity potentials for simple flows and combinations of potential functions

In this section, we discuss four elementary flows: uniform, source, vortex and dipole flows. We then analyze how the velocity fields for these flows can be superimposed to solve more complicated problems. Finally, in Section 6.2.4, we illustrate these ideas with a few examples of the velocity fields around objects of simple shape. Since we are discussing incompressible fluids, we also determine the stream function (defined in Section 3.4) for these flows. A table of the velocity potentials and stream functions for the most common flows is included at the end of this chapter.

Uniform parallel flow

Consider uniform flow with velocity $\mathbf{U}$ in the x-direction, with corresponding velocity components:

$$v_x = U = \textit{constant}; \qquad v_y = 0 \qquad \text{for a } \textit{two-dimensional flow}$$

and:

$$v_x = U = \textit{constant}; \qquad v_y = v_z = 0 \qquad \text{for a } \textit{three-dimensional flow}.$$

In *two dimensions*, from Equations 3.38, we have:

$$\frac{\partial \Phi}{\partial x} = \frac{\partial \Psi}{\partial y} = v_x = U \quad \text{and} \quad \frac{\partial \Phi}{\partial y} = -\frac{\partial \Psi}{\partial x} = v_y = 0,$$

whence: $\Phi = U x$ (6.6a) and: $\Psi = U y.$ (6.6b)

The *streamlines* (Ψ = constant) are thus straight lines in the x-direction (and consequently, as required, parallel to the velocity $\mathbf{U}$). The *equipotentials* (Φ = constant) are parallel straight lines in the y-direction (and perpendicular to the streamlines).

In three dimensions, since the flow is axially symmetric, we can obtain a similar result by the use of the Stokes stream function Ψ which we have defined in Section 3.4. Here, we assume that the flow is in the z-direction.

- In cylindrical coordinates (r, φ, z), we obtain by using Equations 3.50:

$$\frac{\partial \Phi}{\partial z} = -\frac{1}{r}\frac{\partial \Psi}{\partial r} = v_z = U \quad \text{and} \quad \frac{\partial \Phi}{\partial r} = \frac{1}{r}\frac{\partial \Psi}{\partial z} = v_r = 0,$$

whence: $\Phi = U z$ (6.7a) and $\Psi = -(U r^2)/2.$ (6.7b)

- In spherical polar coordinates (r, θ, φ), we have similarly, in accordance with Equations 3.52:

$$\frac{\partial \Phi}{\partial r} = \frac{1}{r^2 \sin\theta}\frac{\partial \Psi}{\partial \theta} = v_r = U\cos\theta, \quad \frac{1}{r}\frac{\partial \Phi}{\partial \theta} = -\frac{1}{r\sin\theta}\frac{\partial \Psi}{\partial r} = v_\theta = -U\sin\theta,$$

whence: $\Phi = U r \cos\theta$ (6.8a) and $\Psi = (U r^2 \sin^2\theta)/2.$ (6.8b)

We should recall that the lines (or surfaces) Ψ = constant are streamlines (or surfaces) along which the current flows. They obey the equations r = constant and $r \sin\theta$ = constant, in their respective cylindrical and spherical polar coordinates, and they are parallel to the direction of the velocity U. The equipotential lines (or, in three dimensions, surfaces) are respectively the curves (or surfaces) normal to the flow.

Vortex flow

Two-dimensional vortex flow is flow around an axis perpendicular to the x-y plane and passing through the origin O. The velocity field is azimuthal (i.e., normal to the plane formed by the radius vector and the axis), with components v_r and v_φ obeying, in polar coordinates:

$$v_r = 0, \qquad v_\varphi = \frac{\Gamma}{2\pi r},$$

where Γ is a constant. Equivalently, according to Equations 3.39a and 3.39b:

$$\frac{1}{r}\frac{\partial \Phi}{\partial \varphi} = -\frac{\partial \Psi}{\partial r} = v_\varphi = \frac{\Gamma}{2\pi r}, \qquad \frac{\partial \Phi}{\partial r} = \frac{1}{r}\frac{\partial \Psi}{\partial \varphi} = v_r = 0.$$

If we evaluate the circulation of the velocity along a circle (C), of radius r centered at O, we find:

$$\int_C \mathbf{v} \cdot \mathbf{dl} = \int_0^{2\pi} \frac{\Gamma}{2\pi r} r \, d\varphi = \Gamma,$$

Γ is therefore the circulation along any curve looping once around the origin. We obtain:

$$\Phi = \frac{\Gamma \varphi}{2\pi} \qquad (6.9a) \qquad \text{and:} \qquad \Psi = -\frac{\Gamma}{2\pi}\mathrm{Log}\frac{r}{r_0}, \qquad (6.9b)$$

where r_0 is an arbitrary constant which maintains the dimensionless character of the argument of the logarithm (Ψ and Φ are always defined within an arbitrary additive constant, because ultimately only their derivative has physical significance).

It should be noted that this is a case of a doubly connected flow. The singular line $r = 0$ (extending infinitely far along the z-axis) takes on the role of the solid surface S_t shown in Figure 6.2 and around which we evaluated the circulation of velocity (in this instance, the radius of curvature of the torus is infinite). The velocity potential Φ is not uniquely defined since it contains the angle φ. As a result, the circulation of velocity along the contour looping n times in the positive direction around the line $r = 0$ equals $n\Gamma$. This situation is analogous to that of a magnetic field resulting from an infinitely long straight wire, of very small diameter, carrying an electric current.

Sources and sinks

Elementary potential flows streaming at a rate of flow Q away from, or toward, a point are known, respectively, as sources and sinks ($Q > 0$ for a source, while $Q < 0$ for a sink).

In *two dimensions*, the flow from a source is given in cylindrical coordinates by (Figure 6.3b):

$$v_r(r) = \frac{Q}{2\pi r}, \qquad v_\varphi = 0.$$

The value of the velocity flux across a circle of radius r about the origin (evidently equal to the rate of flow, Q), is:

$$\int_C \mathbf{v} \cdot \mathbf{n}\, dl = \int_0^{2\pi} r\, v_r\, d\varphi = Q.$$

Using Equations 3.19a and 3.19b, we obtain for this flow:

$$\frac{\partial \Phi}{\partial r} = \frac{1}{r}\frac{\partial \Psi}{\partial \varphi} = \frac{Q}{2\pi r}, \qquad \frac{1}{r}\frac{\partial \Phi}{\partial \varphi} = -\frac{\partial \Psi}{\partial r} = 0,$$

whence: $$\Phi = \frac{Q}{2\pi}\mathrm{Log}\left(\frac{r}{r_0}\right), \qquad (6.10a) \qquad \Psi = \frac{Q}{2\pi}\varphi. \qquad (6.10b)$$

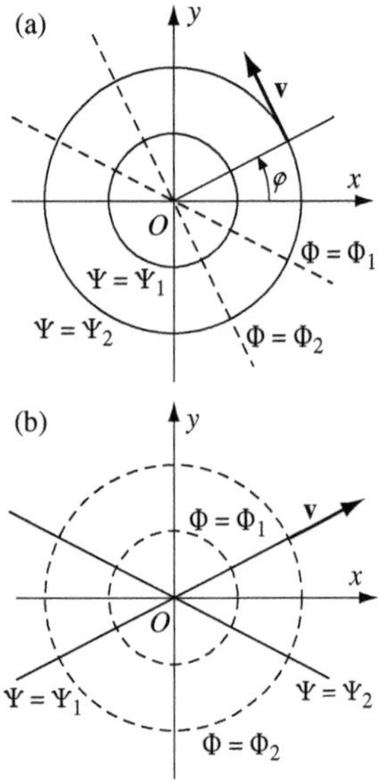

Figure 6.3 *Streamlines and equipotential lines (a) for a plane vortex flow around the z-axis and (b) for flow resulting from a source located at the origin of the coordinate axes*

As in the previous case, Φ and Ψ are defined within an arbitrary constant. We note the close correspondence, in *two dimensions*, between the solutions obtained here for a source, and those just derived for a vortex. The functions Φ and Ψ are merely interchanged between one flow and the other, but their mathematical dependence is the same. For the source, the radial streamlines are identical to the equipotentials for vortex flow. The streamlines of the vortex, concentric circles about the origin, correspond to the equipotentials in the presence of a source. This correspondence is more closely analyzed in Section 6.6, where we introduce the concept of a complex velocity potential.

In *three dimensions*, the velocity field resulting from a point source with outflow at rate Q, can be written in spherical coordinates:

$$v_r = \frac{Q}{4\pi r^2}, \qquad v_\theta = v_\varphi = 0.$$

The flow must effectively be radial, and such that the flux through a sphere of arbitrary radius r always equals Q. We then have, using Equations (3.52a) and (3.52b):

$$\frac{\partial \Phi}{\partial r} = \frac{1}{r^2 \sin\theta}\frac{\partial \Psi}{\partial \theta} = v_r = \frac{Q}{4\pi r^2}, \qquad \frac{1}{r}\frac{\partial \Phi}{\partial \theta} = -\frac{1}{r \sin\theta}\frac{\partial \Psi}{\partial r} = v_\theta = 0.$$

The velocity potential and the stream function can then be written

$$\Phi = -\frac{Q}{4\pi r}, \quad (6.11a) \qquad \Psi = -\frac{Q}{4\pi}\cos\theta. \quad (6.11b)$$

Dipole flow

Consider a point sink at S_1 and a source at S_2, separated by a distance d, with the same absolute value of flow rate, Q. Letting their separation $d \to 0$ while keeping the product $p = Q\,|d|$ constant, we obtain a dipole flow, with vector moment $\mathbf{p} = Q\,(\mathbf{S}_1\mathbf{S}_2) = Q\mathbf{d}$, oriented from the sink to the source (Figure 6.4).

In *two dimensions*, the velocity potentials induced at the point P ($\mathbf{OP} = \mathbf{r}$) by the source S_2, or the sink S_1, respectively located at $\mathbf{OS}_2 = \mathbf{r}_2$ and $\mathbf{OS}_1 = \mathbf{r}_1$, can be expressed in cylindrical coordinates by:

$$\Phi_2 = \frac{Q}{2\pi}\mathrm{Log}\frac{|\mathbf{r}-\mathbf{r}_2|}{r_0}, \qquad \Phi_1 = -\frac{Q}{2\pi}\mathrm{Log}\frac{|\mathbf{r}-\mathbf{r}_1|}{r_0}.$$

We have therefore, for the source and sink pair:

$$\Phi = \Phi_1 + \Phi_2 = \frac{Q}{2\pi}\left(\mathrm{Log}\,|\mathbf{r}-\mathbf{r}_2| - \mathrm{Log}\,|\mathbf{r}-\mathbf{r}_1|\right).$$

Let us now carry out a Taylor series expansion of Log $|\mathbf{r}-\mathbf{r}_2|$ to the lowest non-vanishing order near $|\mathbf{r}| = |\mathbf{OP}|$. In the limit in which we are interested (i.e. $d \to 0$ while the product $p = Q\,d$ remains constant), $|\mathbf{r}|$ is much greater than $|\mathbf{r}_1| = |\mathbf{OS}_1|$ and $|\mathbf{r}_2| = |\mathbf{OS}_2|$ (and thus also than d), we thus obtain:

$$\Phi_2 = \frac{Q}{2\pi}\left(\frac{\partial\,(\mathrm{Log}\ r)}{\partial r}\left(|\mathbf{r}-\mathbf{r}_2| - |\mathbf{r}|\right) + \cdots\right)$$

and:

$$\Phi_1 = -\frac{Q}{2\pi}\left(\frac{\partial\,(\mathrm{Log}\ r)}{\partial r}\left(|\mathbf{r}-\mathbf{r}_1| - |\mathbf{r}|\right) + \cdots\right).$$

Whence, in the limit where $d \to 0$:

$$\Phi = \frac{Q}{2\pi}\frac{1}{r}\left(|\mathbf{r}-\mathbf{r}_2| - |\mathbf{r}-\mathbf{r}_1|\right) = -\frac{Q d}{2\pi}\frac{\cos\varphi}{r} = -\frac{p}{2\pi}\frac{\cos\varphi}{r}, \quad (6.12)$$

i.e.:

$$\Phi = -\frac{\mathbf{p}\cdot\mathbf{r}}{2\pi\, r^2}, \quad (6.13)$$

where $\mathbf{p} = Q\,(\mathbf{S}_1\mathbf{S}_2)$ ($p = |\mathbf{p}| = Q\,d$) and φ is the angle between $\mathbf{S}_1\mathbf{S}_2$ and the radius vector $\mathbf{r}$ of magnitude r.

From the gradient of the potential Φ, we obtain the components of the velocity:

$$v_r = \frac{\partial \Phi}{\partial r} = \frac{p}{2\pi}\frac{\cos\varphi}{r^2}, \quad (6.14a) \qquad v_\varphi = \frac{1}{r}\frac{\partial \Phi}{\partial \varphi} = \frac{p}{2\pi}\frac{\sin\varphi}{r^2}. \quad (6.14b)$$

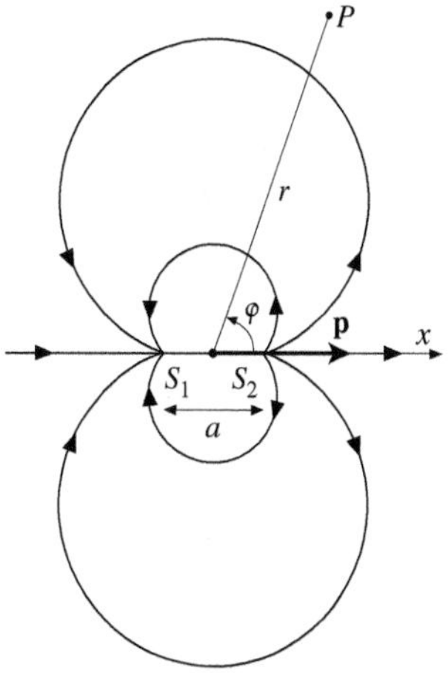

Figure 6.4 *Two-dimensional flow resulting from the combination of a sink S_1 and a source S_2, having the same absolute value Q of the flow rate*

At distances r large compared to d, the potential and the velocity field resulting from this dipole flow are seen therefore to have the same mathematical dependence as the potential and the electric field due to an electric dipole with dipole moment $\mathbf{p}$. The stream function Ψ is once again obtained by integration of the velocity components:

$$\Psi = \frac{p}{2\pi}\frac{\sin\varphi}{r} = \frac{\mathbf{p}\times\mathbf{r}}{2\pi r^2}. \qquad (6.14c)$$

Going through a similar set of steps for a *three-dimensional* flow and a moment $\mathbf{p}$ parallel to the z-axis ($\theta = 0$), we obtain the potential for dipole flow in spherical polar coordinates:

$$\Phi = -\frac{p\,\cos\theta}{4\pi r^2} = -\frac{\mathbf{p}\cdot\mathbf{r}}{4\pi r^3}. \qquad (6.15a)$$

with corresponding components v_r and v_θ for the velocity field:

$$v_r = \frac{p\,\cos\theta}{2\pi r^3}, \qquad (6.15b) \qquad \text{and:} \qquad v_\theta = \frac{p\,\sin\theta}{4\pi r^3} \qquad (6.15c)$$

and a stream function Ψ obtained by use of Equations 3.52, satisfies:

$$\Psi = \frac{p\,\sin^2\theta}{4\pi r}. \qquad (6.15d)$$

Solutions of Laplace's equation: superposition and separation of variables

As a result of the linearity of Laplace's equation, linear combinations of solutions are themselves also solutions. We can therefore construct the velocity field for a potential problem by superimposing simple solutions in such a way that the resulting function obeys the boundary conditions. Also, just as for the case of the electrostatic potential resulting from a distribution of electric charges, we can write the velocity potential for a flow as a *multipole expansion*. This corresponds to a sum of elementary potentials related to distributions of more and more complicated sources of fluid (simple source, dipole, quadrupole, etc.). We will see that the velocity potential of a dipole can describe simple velocity fields such as those corresponding to flow around a sphere or a cylinder.

Another approach consists of looking for particular solutions of Laplace's equation written as a product of separate functions of the individual variables: in that case, we must use coordinate systems which reflect the symmetry of the problem. Problems with cylindrical symmetry lead to solutions involving Bessel functions. Those with spherical symmetry involve the same Legendre polynomials that are also used in the solution of the Schrödinger equation encountered in quantum mechanics for the description of atomic orbitals.

6.2.4 Examples of simple potential flows

The examples which follow involve the elementary flows studied above. We discuss, in order, uniform flow around a circular cylinder, three-dimensional flow around a sphere located in an otherwise uniform flow, flow around a Rankine solid and, finally, the superposition of flows resulting from a sink and a vortex.

Flow around a circular cylinder

We consider uniform flow at velocity $\mathbf{U}$, perturbed by the presence of a circular cylinder of radius R, with its axis normal to the direction of the velocity. Given that this problem is invariant for translations along the axis of the cylinder, we can treat it as two-dimensional.

We examine first the case where (i) there is no circulation of the velocity around the cylinder, and, then (ii), we include the effects of an existing circulation, thus observing the effect of lift. We discuss this effect in a more general approach, in Section 6.3.1, where we treat the case of a two-dimensional obstacle of arbitrary cross-section.

(i) Circular cylinder with no circulation Let us consider, in polar coordinate notation, the velocity potential Φ resulting from a superposition of the potential corresponding to uniform flow at velocity $\mathbf{U}$ in the direction $\varphi = 0$ (Equation 6.6a) and of a dipole of moment $\mathbf{p}$ pointing in the same direction (Equation 6.12). We then have:

$$\Phi = \Phi_{\text{uniform flow}} + \Phi_{\text{dipole}} = U\,r\cos\varphi - \frac{p\,\cos\varphi}{2\pi\,r} = \left(U\,r - \frac{p}{2\pi\,r}\right)\cos\varphi. \qquad (6.16)$$

This equation provides a first logical step: the dipole potential is in fact the first non-zero term of a multipole expansion since, by assumption, there is no source of fluid within the cylinder. Given the uniqueness of potential solutions with a given circulation, Equation 6.16 is the solution of our problem if it satisfies the boundary conditions at the walls. If this is not the case, we will have to add in higher order terms of the multipole expansion. From Equation 6.16 for the potential, we obtain the velocity components:

$$v_r = \frac{\partial \Phi}{\partial r} = \left(U + \frac{p}{2\pi\,r^2}\right)\cos\varphi, \qquad v_\varphi = \frac{1}{r}\frac{\partial \Phi}{\partial \varphi} = -\left(U - \frac{p}{2\pi r^2}\right)\sin\varphi.$$

We now look for a value of p such that the velocity field obeys the boundary conditions: $\mathbf{v} = \mathbf{U}$ at infinity, and $v_r\,(r{=}R) = 0$ (normal component of the velocity zero at the surface $r = R$ of the cylinder).

The first condition is obeyed trivially, since the dipole contribution vanishes as $1/r^2$ at large distances. The second implies:

$$\frac{p}{2\pi R^2} = -U, \qquad \text{whence:} \qquad \Phi = U\,r\cos\varphi\left(1 + \frac{R^2}{r^2}\right),$$

thus yielding the required value of the dipole moment p as a function of the velocity U. Accordingly, the velocity field has the form:

$$v_r = U\left(1 - \frac{R^2}{r^2}\right)\cos\varphi, \qquad (6.17a) \qquad\qquad v_\varphi = -U\left(1 + \frac{R^2}{r^2}\right)\sin\varphi. \qquad (6.17b)$$

Because of the uniqueness of the solutions of Laplace's equation, this velocity field, obeying the required boundary conditions at infinity and along the surface of the cylinder, must therefore be the correct solution for our problem.

The stream function Ψ for this flow can equally be constructed from those for uniform flow and for the dipole. However, we can obtain it more easily from the velocity field derived above, by direct integration of Equations (6.17a,b) and from the expression for Ψ in cylindrical coordinates. We thus obtain:

$$\Psi = U\,r\,\sin\varphi\left(1 - \frac{R^2}{r^2}\right).$$

The streamlines for this flow are shown in Figure 6.5 (following page). We observe a particular streamline corresponding to $\Psi = 0$. It consists of two semi-infinite lines starting from the *stagnation points* $r = R$ and $\varphi = 0$ or π (the points along the circumference of the cylinder where the velocity is zero), and of the circumference of the cylinder itself.

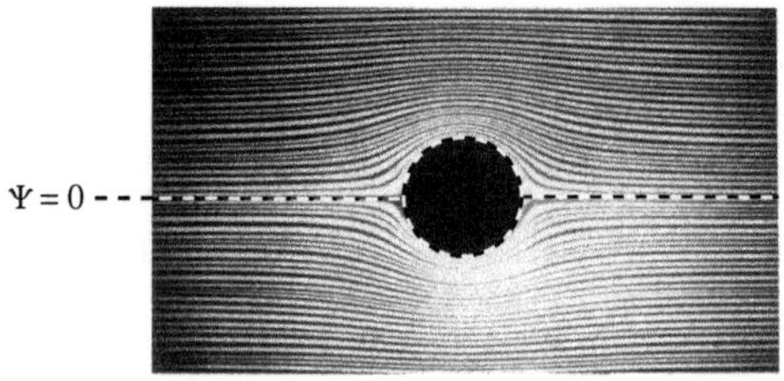

Figure 6.5 *Shape of the streamlines around a circular cylinder placed within a flow which is uniform at infinity, for the case where there is no circulation of the velocity around the cylinder. This figure was obtained experimentally using a Hele-Shaw cell which allows us to simulate a two-dimensional potential flow from that around an obstacle placed between two parallel plates with very small separation. This technique is discussed in Chapter 9, Figure 9.23 (plate courtesy of H. Peregrine, "An Album of fluid motion")*

(ii) Flow around a circular cylinder in the presence of circulation In this case, we obtain the velocity potential by adding to the previous potential (Equation 6.16) the potential of a vortex of circulation Γ (Equation 6.9a). As a matter of fact, the velocity field of the vortex is tangent to the circles r = constant, and vanishes at infinity. It therefore automatically satisfies the two boundary conditions. Since there is only a single potential flow corresponding to a given value of the circulation Γ, the sum of these two velocity potentials must be the correct solution to the problem. Thus, if we write U as the magnitude of the velocity parallel to the x-axis:

$$\Phi = \left(U\,r - \frac{p}{2\pi\, r}\right)\,\cos\varphi + \frac{\Gamma}{2\pi}\varphi.$$

In the same manner, we obtain the velocity components by superposition of the solutions:

$$\mathbf{v} = \mathbf{v}_{\text{cylinder}} + \mathbf{v}_{\text{vortex}}$$

because each of these velocity fields satisfies independently the boundary conditions. We therefore conclude:

$$v_r = U\left(1 - \frac{R^2}{r^2}\right)\,\cos\varphi, \qquad (6.18a) \qquad v_\varphi = -U\left(1 + \frac{R^2}{r^2}\right)\,\sin\varphi + \frac{\Gamma}{2\pi\, r}. \qquad (6.18b)$$

Let us now determine if there still exist stagnation points on the surface of the cylinder. They must be such that:

$$v_\varphi\,(r = R) = -U\left(1 + \frac{R^2}{R^2}\right)\,\sin\varphi + \frac{\Gamma}{2\pi R} = 0, \qquad \text{i.e.,:} \qquad \sin\varphi = \frac{\Gamma}{4\pi R U}. \qquad (6.19)$$

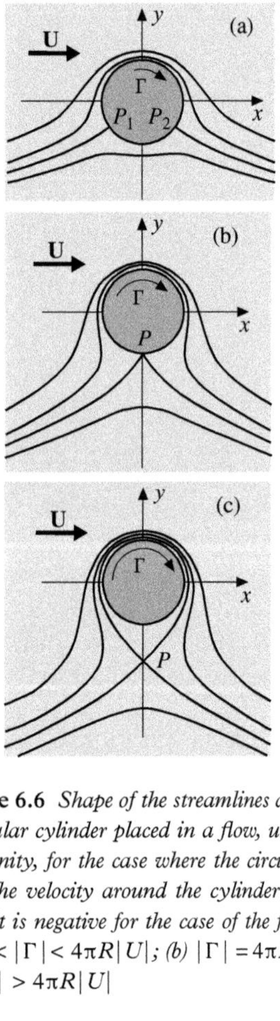

Figure 6.6 *Shape of the streamlines around a circular cylinder placed in a flow, uniform at infinity, for the case where the circulation Γ of the velocity around the cylinder is not zero (it is negative for the case of the figure); (a): $0 < |\Gamma| < 4\pi R|U|$; (b) $|\Gamma| = 4\pi R|U|$; (c) $|\Gamma| > 4\pi R|U|$*

Equation 6.19 leads us to a distinction between two different regimes according to the relative values of the respective magnitudes of circulation and velocity, $|\Gamma|$ and $|U|$:

- for $0 < |\Gamma| < 4\pi\ R\ |U|$, there are two stagnation points P_1 and P_2, symmetrically located relative to the y-axis (Figure 6.6a). Their position is determined by the angles φ, solutions of Equation 6.19. As the circulation increases, P_1 and P_2 move closer to each other, starting from their diametrically opposite positions for $\Gamma = 0$ (the case previously discussed). Ultimately they merge into a unique stagnation point P on the surface of the cylinder when $|\Gamma| = 4\pi\ R\ |U|$ (Figure 6.6b).
- for $|\Gamma| > 4\pi\ R\ |U|$, there are no longer any stagnation points along the surface of the cylinder. Immediately next to the cylinder, we observe closed streamlines, and open ones further away (Figure 6.6c). On a trajectory intermediate between these

two cases we find a single stagnation point P external to the cylinder. Its coordinates φ and r satisfy $\sin\varphi = \pm 1$ (depending on whether Γ and U have the same sign) and:

$$-|U|\left(1+\frac{R^2}{r^2}\right)+\frac{|\Gamma|}{2\pi r}=0, \tag{6.20a}$$

whence:

$$r = R\left(\frac{|\Gamma|}{4\pi R|U|}+\sqrt{\left(\frac{|\Gamma|}{4\pi R|U|}\right)^2-1}\right) \tag{6.20b}$$

(in fact, the second solution of Equation 6.20a corresponds to a value smaller than R).

The force exerted by the fluid on the cylinder is normal to the axis and has two components: one, along the direction of the velocity $\mathbf{U}$ is called the *drag force*; the other, in the perpendicular direction, is the *lift*. To evaluate these components of the force, we determine the resultant of the pressure on the cylinder from the pressure field $p(r = R, \varphi)$ along the surface. The pressure obeys Bernoulli's Equation 5.36, applicable throughout the fluid, since we are dealing with potential flow. Taking as a reference a point infinitely distant radially (where the pressure is p_0 and the velocity $\mathbf{U}$), we obtain:

$$p(r=R,\,\varphi)+\frac{1}{2}\rho\, v_\varphi^2(r=R,\,\varphi)=p_0+\frac{1}{2}\rho U^2,$$

whence

$$p = p_0+\frac{1}{2}\rho U^2\left(1-\left[-2\sin\varphi+\frac{\Gamma}{2\pi RU}\right]^2\right).$$

The lift, per unit length of cylinder, is then the total resultant component F_L of the pressure in the y-direction:

$$F_L = -\int_{\text{cylinder surface}} p\,\sin\varphi\, R\,\mathrm{d}\varphi.$$

The only non-zero term in the integral comes from the term involving $\sin\varphi$, in the equation we have just derived for the pressure, so that:

$$F_L = -\int_0^{2\pi}\frac{\rho U\Gamma}{\pi}\sin^2\varphi\,\mathrm{d}\varphi = -\rho\, U\,\Gamma; \tag{6.21}$$

here, for the cases illustrated in Figure 6.6, F_L is directed upward. This result for the lift, also known as the *Magnus force*, is derived more generally in Section 6.3.1 (Equations 6.43 and 6.44).

Moreover, the x-component of the pressure forces, the drag force F_D, is zero. This follows from the fact that the magnitude of the velocity at points on the cylinder symmetrically located relative to the y-axis is the same, and hence we have the same pressure. The x-component of the global pressure force cancels because of this symmetry. This result may be shown to be valid for all stationary flows of an ideal fluid around an obstacle. In every such case, the vanishing of the drag force is consistent with the absence of a viscous dissipation mechanism.

Sphere in a flow with uniform velocity at infinity

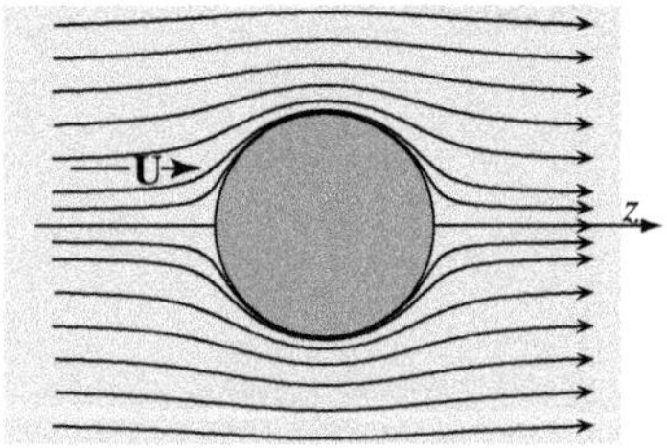

Figure 6.7 *Streamlines around a motionless sphere placed within a uniform potential flow*

We now discuss the case of uniform flow at velocity **U** perturbed by a stationary sphere of radius R located at the origin (Figure 6.7). In keeping with the analogy with the electrostatic problem, and following the procedure we used for the case of the cylinder, we pick as a trial function for the velocity potential the superposition of the potentials for uniform flow (Equation 6.8a) and for a dipole (Equation 6.15a). We thus obtain, in spherical polar coordinates:

$$\Phi = U r \cos\theta - \frac{p \cos\theta}{4\pi r^2} = \left(U r - \frac{p}{4\pi r^2}\right) \cos\theta. \tag{6.22}$$

The components of the velocity can then be written:

$$v_r = \frac{\partial \Phi}{\partial r} = \left(U + \frac{p}{2\pi r^3}\right)\cos\theta, \qquad v_\theta = \frac{1}{r}\frac{\partial \Phi}{\partial \theta} = -\left(U - \frac{p}{4\pi r^3}\right)\sin\theta \quad v_\varphi = 0. \tag{6.23}$$

The dipole moment p is determined, as before, from the boundary condition at the surface of the sphere:

$$v_r(r = R) = \left(U + \frac{p}{2\pi R^3}\right)\cos\theta = 0,$$

so that: $\quad \mathbf{p} = -2\pi\, \mathbf{U} R^3 \quad$ (6.24a) $\quad$ and: $\quad \Phi = U r \cos\theta \left(1 + \dfrac{R^3}{2r^3}\right).$ $\quad$ (6.24b)

The condition that the velocity equals **U** at infinity is satisfied by the form chosen for the potential Φ (the dipole contribution vanishes toward infinity). We obtain for the velocity field:

$$v_r = U\left(1 - \frac{R^3}{r^3}\right)\cos\theta, \qquad v_\theta = -U\left(1 + \frac{R^3}{2\, r^3}\right)\sin\theta, \qquad v_\varphi = 0. \tag{6.25}$$

The streamlines are obtained from the stream function, itself derived from the velocity potential by integration of the two equations below (see Equations 3.52):

$$\frac{\partial \Psi}{\partial \theta} = \left(r^2 \sin\theta\right) v_r = U\left(r^2 - \frac{R^3}{r}\right)\sin\theta \cos\theta, \tag{6.26a}$$

$$\frac{\partial \Psi}{\partial r} = -(r \sin\theta)\, v_\theta = U\left(r + \frac{R^3}{2r^2}\right)\sin^2\theta. \tag{6.26b}$$

Thus, within a constant of integration:

$$\Psi = \frac{U}{2}\left(r^2 - \frac{R^3}{r}\right)\sin^2\theta. \tag{6.27}$$

Figure 6.7 displays the distribution of the corresponding streamlines. The surface $\Psi = 0$ is made up of the sphere ($r = R$) and of the axis of symmetry ($\theta = 0$ and $\theta = \pi$). We should note that the magnitude of the deviation of the velocity from **U** decreases as $1/r^3$ for large values of r.

We will discuss in Section 9.4.1 the flow of a fluid around a sphere at low Reynolds numbers, when viscosity forces dominate (contrary to the present case); in that case, the magnitude of the velocity decreases as $1/r$, i.e. much more slowly than for the potential flow discussed here.

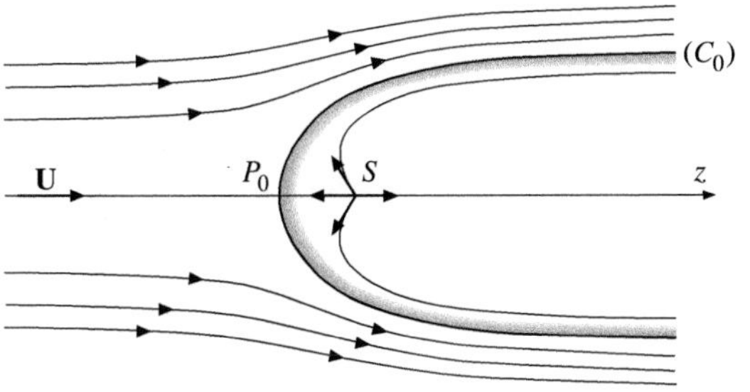

Figure 6.8 *Flow around an axially symmetric solid, known as a Rankine solid. This flow is equivalent to the superposition of a uniform flow and of the flow resulting from a source S*

The Rankine solid

This example is the case of a source placed in a uniform flow (Figure 6.8). We show that the resultant flow field is identical to that of the flow around an obstacle having a specific shape, called a *Rankine solid*, which has axial symmetry around the direction of the unperturbed flow.

We start with a velocity potential and a stream function corresponding to the superposition of a uniform flow (Equation 6.8) and that of a source located at the origin (Equation 6.11):

$$\Phi = \Phi_{\text{uniform flow}} + \Phi_{\text{source}}, \qquad \Psi = \Psi_{\text{uniform flow}} + \Psi_{\text{source}},$$

so that, in spherical coordinates:

$$\Phi = U r \cos\theta - \frac{Q}{4\pi r} \quad (6.28a) \qquad \text{and:} \quad \Psi = U\frac{r^2}{2}\sin^2\theta - \frac{Q}{4\pi}\cos\theta. \quad (6.28b)$$

We obtain the velocity components from Equation 6.28a:

$$v_r = \frac{\partial\Phi}{\partial r} = U\cos\theta + \frac{Q}{4\pi r^2} \qquad v_\theta = \frac{1}{r}\frac{\partial\Phi}{\partial\theta} = -U\sin\theta \qquad v_\varphi = 0. \quad (6.29)$$

There is a stagnation (zero-velocity) point P_0 on the x-axis ($\theta = 0$ or π depending on the sign of Q relative to that of U) when:

$$r = r_0 = \sqrt{\frac{|Q|}{4\pi\,|U|}}. \quad (6.30)$$

The streamlines are surfaces of revolution around the z-axis. More precisely, each streamline is the intersection of one of these surfaces of revolution with a plane bounded by the (polar) z-axis and extending radially outward at a given angle of azimuth φ, with equation:

$$U\frac{r^2}{2}\sin^2\theta - \frac{Q}{4\pi}\cos\theta = \Psi = \text{constant}. \quad (6.31)$$

A set of such streamlines is illustrated in Figure 6.8.

The value Ψ_0 of the stream function along the streamline (C_0) in the plane of the figure and passing through the stagnation point P_0, determined above, ($r = r_0$, $\theta = \pi$) can be written:

$$\Psi_0 = \Psi\,(r = r_0, \theta = \pi) = \frac{Q}{4\pi}.$$

The obstacle described by Equation 6.32 belongs to the more general family of Rankine ovoidal solids: these are obtained by superimposing current sheets corresponding to a uniform flow together with a source and a sink of equal flow rate. The parameter which changes the shape of the solid is the distance between the source and the sink. When this distance tends to zero, we get back to the flow around a sphere which we have studied in the preceding example. For the case where this distance becomes infinitely large (e.g., if the sink goes to infinity), we have the particular case of the semi-infinite solid discussed here: the sink then becomes only an additional component of the uniform flow.

Substituting the above value in Equation 6.31, we obtain the equation for (C_0):

$$r^2 = \frac{Q}{2\pi U}\frac{1+\cos\theta}{\sin^2\theta}. \qquad (6.32)$$

This streamline is made up, on one hand, of the z-axis ($\theta = 0, \pi$) and, on the other hand, of the curve which separates the fluid space into two regions in which the respective fluid streams belong to each of the two basic flows (uniform flow and flow resulting from the source). We can replace this flow tube by a solid obstacle without altering the remaining flow.

Sink and vortex

We now treat the superposition of the two-dimensional flow resulting from the simultaneous presence of a sink, with flow rate $-Q$ ($Q > 0$), and a vortex of circulation Γ, both centered at the origin. The resulting flow approximates the case of a cylindrical container being emptied through a central hole (sink), while simultaneously receiving a peripheral inflow so as to maintain the vortex motion. The only feature of such a flow not represented here, because this is a two-dimensional model, is a depression of the free surface which is generally observed in the center of the container. The resultant velocity potential, and stream function, can be written as:

$$\Phi = \Phi_{\text{sink}} + \Phi_{\text{vortex}} \quad \text{and} \quad \Psi = \Psi_{\text{sink}} + \Psi_{\text{vortex}},$$

i.e., in cylindrical coordinates (Equations 6.9 and 6.10):

$$\Phi = -\frac{Q}{2\pi}\text{Log}\frac{r}{r_0} + \frac{\Gamma}{2\pi}\varphi \quad (6.33) \quad \text{and:} \quad \Psi = -\frac{Q}{2\pi}\varphi - \frac{\Gamma}{2\pi}\text{Log}\frac{r}{r_0}. \quad (6.34)$$

from which we obtain the velocity components:

$$v_r = \frac{\partial\Phi}{\partial r} = -\frac{Q}{2\pi r} \quad (6.35a) \quad \text{and:} \quad v_\varphi = \frac{1}{r}\frac{\partial\Phi}{\partial\varphi} = -\frac{\Gamma}{2\pi r}. \quad (6.35b)$$

The equation of the streamlines is then, still in cylindrical coordinates:

$$\Psi = \text{constant} = K - \frac{Q}{2\pi}\varphi - \frac{\Gamma}{2\pi}\text{Log}\frac{r}{r_0} \quad \text{whence} \quad r = r_1\, e^{-\left(\frac{Q}{\Gamma}\right)\varphi}. \quad (6.36)$$

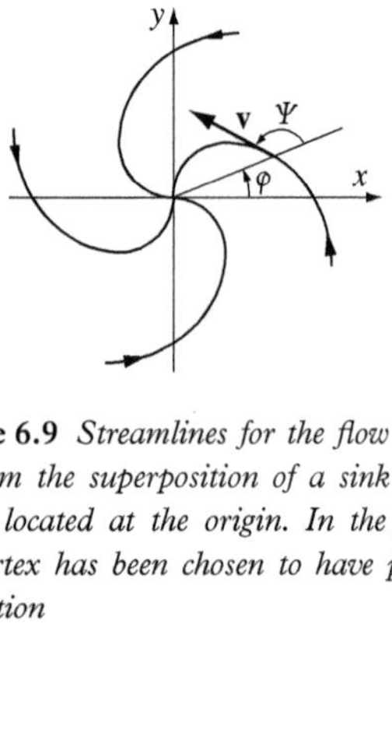

Figure 6.9 *Streamlines for the flow resulting from the superposition of a sink and a vortex located at the origin. In the figure, the vortex has been chosen to have positive circulation*

They are therefore logarithmic spirals (Figure 6.9), with the factor r_1 as the parameter identifying a specific streamline. These correspond to the trajectory of a particle which goes around the sink, gradually approaching it. The velocity vector $\mathbf{v}$ makes a constant angle ψ with the radius vector r such that:

$$\tan\psi = \frac{v_\varphi}{v_r} = -\frac{\Gamma}{Q}.$$

This angle is zero (radial streamlines) if $\Gamma = 0$ (sink only), or $\pi/2$ if $Q = 0$ (vortex only).

6.3 Forces acting on an obstacle in potential flow

In this section, we discuss the problem of the forces exerted by a fluid in potential flow on a solid obstacle of arbitrary shape placed within it. The treatment is based on the fact, already discussed above (Section 6.2.3), that we can use a multipole expansion of the velocity potential (the terms of the expansion must be solutions of Laplace's equation). We must also comply with the boundary conditions for the normal component of the velocity at the surface of the object (zero for an object at rest). For calculating the field, at large distances r from the object, we take into account only the first non-vanishing contribution, to lowest order in $1/r$, i.e., the one which approaches zero most slowly is the one which we must use in the calculation of the velocity field.

By means of the equations of conservation of momentum, the force exerted can be determined from the velocity field, evaluated far enough from the body that the preceding dipole approximation holds. In two dimensions, we must also take into account the possible existence of a circulation, related to the lack of uniqueness of the solutions for the potential, as discussed in Section 6.2.2.

6.3.1 Two-dimensional flows

Velocity potential

A circular cylinder, or the cross-section of an infinitely long wing, are classic examples of two-dimensional obstacles. We assume these objects to be fixed, as would be the case in a wind tunnel in which the fluid has velocity $\mathbf{U}$, uniform at large distances from the obstacle, and normal to the z-axis of the cylinder. We also assume, *a priori*, that a finite circulation Γ exists around the obstacle, without worrying about the manner in which this circulation might be induced (this point is addressed in Section 7.5.2).

At a distance r from the obstacle, large compared to its dimensions in the x-y plane, we can write the velocity in the form:

$$\mathbf{v}(\mathbf{r}) = \mathbf{U} + \nabla\Phi_1(\mathbf{r}) + \nabla\Phi_2(\mathbf{r}) = \mathbf{U} + \mathbf{v}_1 + \mathbf{v}_2. \tag{6.37}$$

- The term $\nabla\Phi_1(\mathbf{r})$, where $\Phi_1(\mathbf{r}) = (\Gamma/2\pi)\ \varphi$ + constant (Equation 6.9a), expresses the effect on the velocity field of the circulation Γ around the obstacle. The sum of the first two terms represents the combination of a uniform flow with velocity $\mathbf{U}$ and of a vortex.
- The potential $\Phi_2(\mathbf{r})$ expresses changes in the velocity potential due to the shape and non-zero transverse dimension of the obstacle. We write Φ_2 as a multipole expansion (simple source, dipole, quadrupole, etc....). Normally, the flow does not contain sources, so that the first non-vanishing term is the dipole and is, using (Equation 6.13):

$$\Phi_2 = \frac{\mathbf{A}\cdot\mathbf{r}}{r^2}, \tag{6.38}$$

  where $\mathbf{A}$ is a constant vector characteristic of the dipole. If the obstacle is a circular cylinder, this potential is the exact solution of the problem (Section 6.2.4, example (i)). For an obstacle of arbitrary shape, it merely represents the correction term which dominates at large distances. This correction to the value of the potential thus decreases as $1/r$ with distance, which leads to a decrease in $1/r^2$ of the corresponding contribution to the velocity. We now demonstrate that this expansion is sufficient to determine the forces which act on the obstacle.

Drag and lift forces on a two-dimensional obstacle

We calculate the *lift* ($\mathbf{F}_L$) and *drag* ($\mathbf{F}_D$) forces exerted on the body in the directions respectively normal and parallel to the flow. For this purpose, we evaluate the components along the x- and y-directions of the momentum conservation equation (Equation 5.10), for a cylindrical slice located around the obstacle, of radius r large relative to its dimensions in the x-y plane and of unit length along the z–axis (Figure 6.10a): we thus avoid the integration of the pressure field over the entire surface of the obstacle. We shall see that the

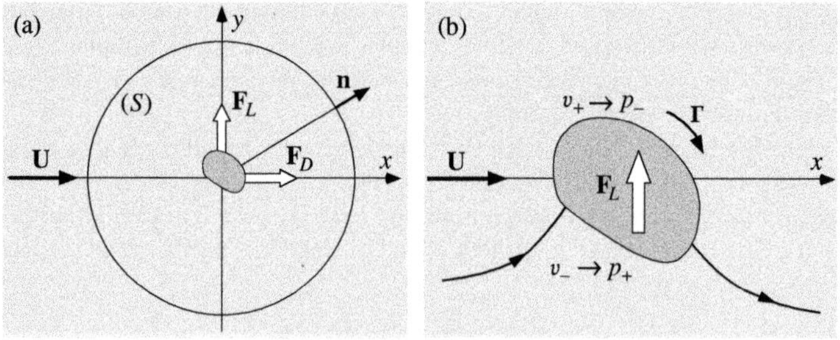

Figure 6.10 *The evaluation of the lift force $\mathbf{F}_L$, and of the drag force $\mathbf{F}_D$, on a cylindrical obstacle with its axis transverse to a uniform flow when there is also a finite circulation Γ of the velocity around the obstacle. (a) (S) represents the surface of the circular cylinder within which we carry out the momentum balance. (b) Determination of the direction of $\mathbf{F}_L$ as a function of the direction of the circulation Γ from the application of Bernoulli's Equation*

exact form of the cross-section of the obstacle in the plane of the figure is not relevant to the calculation of the lift force $\mathbf{F}_L$. This results from the fact that only the dominant term Φ_1 of the multipole expansion of the velocity potential is required for the determination of $\mathbf{F}_L$.

Let us apply the momentum balance equation (Equation 5.10), derived in the preceding chapter, to a cylindrical element of radius r surface area (S), and of unit depth in the direction normal to the plane of the Figure 6.10a. Since we have ideal, potential flow, we can ignore any effects of the viscosity stress tensor σ'_{ij}. We then obtain:

$$-\iint_S (\rho\, v_x(\mathbf{v}\cdot\mathbf{n}) + p\, n_x)\, \mathrm{d}S + (-F_D) = 0, \tag{6.39a}$$

$$-\iint_S \left(\rho\, v_y\, (\mathbf{v}\cdot\mathbf{n}) + p n_y\right) \mathrm{d}S + (-F_L) = 0. \tag{6.39b}$$

In these equations, $\mathbf{n} = [\cos\varphi, \sin\varphi, 0]$ is the unit vector normal to the surface element of the cylinder $\mathrm{d}S = r\,\mathrm{d}\varphi$; F_D and F_L, representing the respective drag and lift forces per unit length along the z-axis, are considered to be a special kind of fictitious volume forces: they are the forces which must be applied by an observer, or an external mechanism, in order to maintain the obstacle in its place by balancing out the effect of the lift and the drag exerted by the fluid. Separating out the pressure and inertial terms, we obtain:

$$F_D = -\int_0^{2\pi} \rho\left(v_x^2 \cos\varphi + v_x v_y \sin\varphi\right) r\, \mathrm{d}\varphi - \int_0^{2\pi} (p\cos\varphi) r\, \mathrm{d}\varphi, \tag{6.40a}$$

$$F_L = -\int_0^{2\pi} \rho\left(v_x v_y \cos\varphi + v_y^2 \sin\varphi\right) r\, \mathrm{d}\varphi - \int_0^{2\pi} (p\sin\varphi) r\, \mathrm{d}\varphi. \tag{6.40b}$$

Using the expansion of the velocity $\mathbf{v}(\mathbf{r})$ from Equation 6.37, and removing the terms which become negligible after the pressure has been evaluated, Equations 6.40a and 6.40b become:

$$F_D = \int_0^{2\pi} \rho\left[-U v_{1x} \cos\varphi - U v_{1y} \sin\varphi\right] r\, \mathrm{d}\varphi, \tag{6.41a}$$

$$F_L = \int_0^{2\pi} \rho\left[-U v_{1y} \cos\varphi + U v_{1x} \sin\varphi\right] r\, \mathrm{d}\varphi. \tag{6.41b}$$

Proof

We first replace $\mathbf{v}$ by $\mathbf{U} + \mathbf{v}_1 + \mathbf{v}_2$ in Equations 6.40a and 6.40b. Each of the terms containing components of $\mathbf{v}_2$ effectively decreases as $1/r^2$ (or even more rapidly if the terms involve their product or their square). After having multiplied by r, we then have a contribution in $1/r$ to the integral which approaches zero at large distances. Moreover, the pressure field is given by Bernoulli's Equation, which is valid throughout the medium, since the flow is potential:

$$p + \frac{1}{2}\rho v^2 = p_0 + \frac{1}{2}\rho U^2.$$

where p_0 is the uniform pressure at a distance from the obstacle, sufficiently large so that $|\mathbf{v}| = |U|$. By using the expansion of $\mathbf{v}(\mathbf{r})$ and, as previously, neglecting the terms resulting from the components of $\mathbf{v}_2$, we have:

$$p = p_0 + \frac{1}{2}\rho U^2 - \frac{1}{2}\rho(\mathbf{U} + \mathbf{v}_1 + \mathbf{v}_2)^2 = p_0 - \rho U v_{1x} - \frac{1}{2}\rho v_1^2.$$

We can therefore rewrite Equations 6.40a and 6.40b in the form:

$$F_D = -\int_0^{2\pi} \rho\Big[(U + v_{1x})^2 \cos\varphi + (U + v_{1x})\, v_{1y} \sin\varphi\Big]\, r\, d\varphi + \int_0^{2\pi} \rho\Big[U v_{1x} + \frac{1}{2}v_1^2\Big] r \cos\varphi\, d\varphi,$$

$$F_L = -\int_0^{2\pi} \rho\left[(U + v_{1x})\, v_{1y} \cos\varphi + v_{1y}^2 \sin\varphi\right] r\, d\varphi + \int_0^{2\pi} \rho\left[U v_{1x} + \frac{1}{2}v_1^2\right] r \sin\varphi\, d\varphi.$$

In fact, the term with p_0 constant vanishes since the integrals of $\cos\varphi$ and $\sin\varphi$ between zero and 2π vanish. We can also neglect second-order terms of the kind $v_{1x}\, v_{1y}$, v_{1y}^2 or v_{1x}^2: the corresponding integrals will indeed decrease as $1/r$ or even more rapidly since $\mathbf{v}_1$ varies as $1/r$ and these terms are multiplied by r. Only the terms $-\rho U v_{1x}$ and $-\rho U v_{1y}$ which have a contribution constant with r at large distances will contribute to the integrals. We then recover easily Equations 6.41a and 6.41b.

Equation 6.41a for the drag force can be rewritten in the form:

$$F_D = -\rho\, U \int_0^{2\pi} (\mathbf{v}_1 \cdot \mathbf{n})\; r\, d\varphi = -\rho\, U \int_0^{2\pi} (\mathbf{v}_1 \cdot \mathbf{n})\, dS = 0, \tag{6.42}$$

where $\mathbf{n}$ has components $\cos\varphi$ and $\sin\varphi$, and represents the unit vector normal to the surface of integration. F_D is then proportional to the flux of the velocity field $\mathbf{v}_l$ through the surface (S) and is equal to zero (i.e., it is a vortex-type velocity field and not a source-like term).

We will find again in Section 6.3.2 this zero value of the drag force for three-dimensional bodies: it represents the absence of energy dissipation in potential flows of ideal fluids under stationary conditions.

Regarding the lift force $\mathbf{F}_L$, we obtain, denoting by (C) the curve which represents the intersection of (S) with the plane of the figure:

$$F_L = -\rho U \int_C (v_{1x}\, dx + v_{1y}\, dy) = -\rho U \int_C \mathbf{v} \cdot \mathbf{dl} = -\rho U\, \Gamma. \tag{6.43}$$

Thus, only the term $\Phi_1 = \Gamma\varphi/2\pi$ resulting from the presence of a circulation of the velocity around the obstacle contributes to the lift force $\mathbf{F}_L$, also known as the *Magnus force*. This can be expressed in vector form by using an axial circulation vector $\boldsymbol{\Gamma}$ parallel to the z-direction. $\boldsymbol{\Gamma}$ will be oriented toward the back of the plane of the figure, for a rotation in the same direction as in Figure 6.10b, and toward the front in the opposite case (this representation

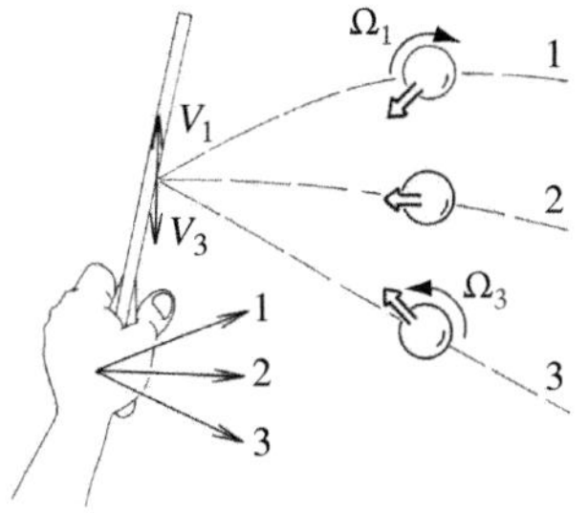

Figure 6.11 *The aerodynamics of ping-pong ball strokes. When the ball is hit, the player influences the curvature of the ultimate trajectory of the ball by three different techniques. (1) Top spin: If the hand moves upward at the instant of impact, the edge of the ball starts to move with velocity V_1 in the direction tangent to the impact, since it does not slip relative to the racket. A rotation $\Omega_1 \approx V_1/R$ (where R is the radius of the ball) results and as a consequence the ball experiences a lift force proportional to Ω_1. The trajectory will curve downwards leading to a higher initial velocity, without overreaching the boundaries of the table. (2) Straight stroke: This is the normal stroke of the beginner pushing the ball. The player does not affect in any way the curvature of the ball's path. (3) Slice: This is the converse of top spin with a tangential velocity V_3 which creates an upward lift, and a much flatter, almost horizontal trajectory*

will become more obvious in the next chapter, where the circulation will be related to the vorticity vector, also pointing in the z-direction). The lift force then satisfies:

$$\mathbf{F}_L = \rho\, \mathbf{U} \times \mathbf{\Gamma}. \tag{6.44}$$

Interpretations such as that given on figure 6.11 dealing with a real fluid are not rigourous for two reasons. The objects we discuss are in fact not two-dimensional but, in contrast to the assumptions made above, have three finite dimensions. We can then find that, for whatever integration contour (C) we choose, there is a surface (S) anchored on (C) which is totally enclosed in the fluid. The circulation of the velocity $\mathbf{v}$ along the contour (C) is equal to the flux of $\nabla\times\mathbf{v}$ through the surface (S), and would then vanish identically (just as with the Magnus force) for potential flow. In the case of balls spinning in a real viscous fluid, the drag of the latter by viscous forces near the walls results, on the contrary, in the appearance of the circulation and of a non-vanishing Magnus lift force. In the same way, the circulation Γ around a wing of finite length would vanish for a purely ideal potential flow: we will see in Section 7.5.2 (Figures 7.27 to 7.29) that, in order to have a non-vanishing Γ, we must have a vortex of circulation equal to Γ originating from the tip of the wing.

We also find the sign of the force, by applying Bernoulli's Equation at two opposite points located above and below the obstacle (Figure 6.10b). With the direction of the circulation which we have chosen around the obstacle, the absolute value of the velocity v_+ at a point above the obstacle is higher than the velocity v_- below. The pressure p_- above the obstacle is thus less than the pressure p_+ below, leading to a lift force directed upward.

Magnus force for real flows

The Magnus force is responsible for the fact that airplanes can fly through the air. In fact, there is a circulation of the velocity field around a cross-section of the wings (see Section 6.6.3). This force is also the basis of the mechanisms for propulsion and support resulting from the action of a propeller (in ships, airplanes, helicopters, etc.). Finally, we can use it to explain a pitcher's curve ball in baseball (or the bowler's in cricket), the *topspin* and *slices* in tennis or ping-pong strokes (Figure 6.11 on previous page), and the "curving" kick in soccer.

6.3.2 Added mass effects for a three-dimensional body undergoing acceleration in an ideal fluid

For a body with three finite dimensions, in relative motion at a constant velocity in an infinite volume of an ideal fluid, the drag force vanishes just as for a two-dimensional body (Equation 6.42); like in the previous case, this reflects the lack of viscous dissipation of the energy. For purely potential flow, this is also true for the lift force: as we saw above, this is the result of the absence of circulation of the velocity around the body.

On the other hand, when a body is accelerated we must take into account the drag due to inertial effects in the incompressible fluid displaced by the solid: i.e., when the solid object is accelerated, the fluid it displaces must also undergo acceleration. It is this phenomenon of the solid being *dressed*† by the surrounding fluid which causes an increase in the inertial effects, and leads to a fictitious mass called the *added mass*.

†Physicists with some exposure to quantum electrodynamics will recognize the analogy with the difference between a *bare* particle's mass or charge and those of one *dressed* by interactions.

We calculate now in the following order, the kinetic energy of the fluid set in motion by the object, the impulse which it senses as a result of the acceleration of the object and, finally, the force exerted on it.

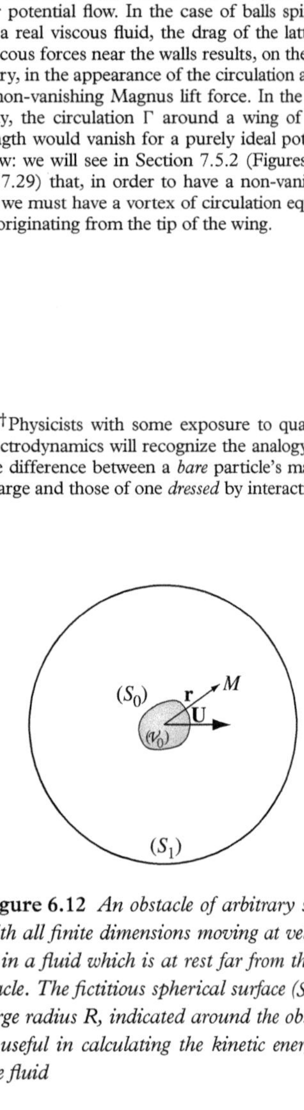

Figure 6.12 *An obstacle of arbitrary shape with all finite dimensions moving at velocity U in a fluid which is at rest far from the obstacle. The fictitious spherical surface (S_1), of large radius R, indicated around the obstacle is useful in calculating the kinetic energy of the fluid*

Derivation of the velocity potential and of the pressure field around a finite three-dimensional obstacle

We consider a solid body of arbitrary shape and volume V_0 bounded by a surface S_0, moving at velocity $\mathbf{U}$ in a fluid which is at rest far from the body (Figure 6.12).

Just as in the two-dimensional case discussed above, we can expand the velocity potential at a point M, located at a distance r from the object, in powers of $1/r$, where r is large compared to the greatest dimension of the object. Here again we encounter the analog of the multipole expansion of a distribution of charges in electrostatics: the terms in $1/r$ correspond to free charges, those in $1/r^2$ to dipoles, etc.

$$\Phi(\mathbf{r}) = \frac{A_1}{r} + \frac{\mathbf{p}}{4\pi}\cdot\nabla\frac{1}{r} + O\left(\frac{1}{r^3}\right). \tag{6.45}$$

The first term is a source term such that the flux of the velocity field through a sphere (S_1) of large radius R, surrounding the object, satisfies:

$$\iint_{S_1} \mathbf{v} \cdot \mathbf{n}\, \mathrm{d}S = \frac{A_1}{R^2} 4\pi R^2 = 4\pi A_1 .$$

In the present example, we have no source of fluid so that $A_1 = 0$. The second term of Equation 6.45 is the expression for the potential due to a dipole which we derived as Equation 6.15 in Section 6.2.3. Keeping in mind the principle of superposition, the components of the dipole moment $\mathbf{p}$ and those of the velocity $\mathbf{U}$ obey a linear relationship, which can be written:

$$p_i = \alpha_{ij} U_j \qquad \text{with summation implied over the repeated index } j, \qquad (6.46)$$

where α_{ij} is an element of a tensor $[\boldsymbol{\alpha}]$ characteristic of the shape of the solid object. For the particular case of the sphere, Equation 6.46 takes the form

$$\mathbf{p} = 2\pi\, \mathbf{U} R^3, \qquad (6.47\text{a}) \qquad \text{i.e.:} \qquad \alpha_{ij} = 2\pi R \delta_{ij}. \qquad (6.47\text{b})$$

Proof

This relationship is deduced from Equation 6.24a, by changing the sign of the dipole moment $\mathbf{p}$: the flow around a sphere moving at velocity $\mathbf{U}$ is indeed, in the reference frame of the sphere, equal and opposite to that for a motionless sphere within a flow of velocity $-\mathbf{U}$.

Kinetic energy of the fluid

We calculate the kinetic energy E_k for a large volume of fluid $(\mathcal{V}_1)$, surrounding the object, and bounded by the sphere (S_1) of radius R. For this purpose, we use the following expansion allowing one to take into account the volume V_0 of the object:

$$E_k = \frac{\rho}{2} \iiint_{\mathcal{V}_1} \mathbf{v}^2 \mathrm{d}V = \frac{\rho}{2} \iiint_{\mathcal{V}_1} \mathbf{U}^2 \mathrm{d}V + \frac{\rho}{2} \iiint_{\mathcal{V}_1} (\mathbf{v} - \mathbf{U})\, (\mathbf{v} + \mathbf{U})\, \mathrm{d}V = \frac{\rho}{2} (I_1 + I_2). \qquad (6.48)$$

From this we infer (see below):

$$E_k = \frac{\rho}{2} \left[\mathbf{p} \cdot \mathbf{U} - V_0\, U^2 \right]. \qquad (6.49)$$

The total kinetic energy of the fluid E_k has a square law dependence on the velocity U, because $\mathbf{p}$ varies linearly with U following Equation 6.46. Thus:

$$E_k = \frac{\rho}{2} \left(\alpha_{ij} U_i U_j - V_0\, U_i U_i \right). \qquad (6.50)$$

Proof

We have first:

$$I_1 = U^2 \left[\frac{4}{3} \pi R^3 - V_0 \right]. \qquad (6.51)$$

In order to evaluate I_2, we begin with the vector identity

$$\boldsymbol{\nabla} \cdot (f\, \mathbf{v}) = (\mathbf{v} \cdot \boldsymbol{\nabla}) f + f\, \boldsymbol{\nabla} \cdot \mathbf{v}.$$

Writing $f = \Phi + \mathbf{U} \cdot \mathbf{r}$, and making use of the condition of incompressibility $\nabla \cdot \mathbf{v} = 0$, we then have:

$$(\mathbf{v} - \mathbf{U})\,(\mathbf{v} + \mathbf{U}) = \nabla \cdot [(\mathbf{v} - \mathbf{U})\,(\Phi + \mathbf{U} \cdot \mathbf{r})].$$

The volume integral I_2 is therefore equal to the sum of two surface integrals, one over the surface (S_0) of the object, and the other over the surface (S_1) of the sphere of radius R much larger than the size of the object. We thus obtain:

$$I_2 = \iint_{S_0} [(\mathbf{v} - \mathbf{U})\,(\Phi + \mathbf{U} \cdot \mathbf{r})] \cdot \mathbf{n}\,\mathrm{d}S + \iint_{S_1} [(\mathbf{v} - \mathbf{U})\,(\Phi + \mathbf{U} \cdot \mathbf{r})] \cdot \mathbf{n}\,\mathrm{d}S.$$

The first integral on the right-hand side is identically zero because, at the surface of the object, $(\mathbf{v} - \mathbf{U}) \cdot \mathbf{n} = 0$. Expanding the product in the second integral, we find:

$$I_2 = \iint_{S_1} [\Phi\,\mathbf{v} - \Phi\,\mathbf{U} + (\mathbf{U} \cdot \mathbf{r})\,\mathbf{v} - (\mathbf{U} \cdot \mathbf{r})\,\mathbf{U}] \cdot \mathbf{n}\,\mathrm{d}S. \tag{6.52}$$

The first term in the integral is of order $(1/R^5)$, so that its surface integral vanishes as R tends to infinity. The velocity field resulting from the potential $\Phi = -(\mathbf{p} \cdot \mathbf{n}) / (4\pi r^2)$ of the dipole (Equation 6.15a) can be written:

$$\mathbf{v} = -\frac{\mathbf{p}}{4\pi r^3} + 3\,(\frac{\mathbf{p} \cdot \mathbf{n}}{4\pi r^3})\,\mathbf{n}, \quad \text{with:} \quad \mathbf{n} = \frac{\mathbf{r}}{r}.$$

We substitute these two expressions for $\mathbf{v}$ and $\mathbf{r}$ so that we can evaluate the last three terms of Equation 6.52. We can then write I_2 in the form:

$$I_2 = \iint_{S_1} \left[(\mathbf{U} \cdot \mathbf{n})\,(3\,\mathbf{p} \cdot \mathbf{n}) - 4\pi\,(\mathbf{U} \cdot \mathbf{n})^2 R^3\right] \frac{\mathrm{d}\Omega}{4\pi},$$

where $\mathrm{d}\Omega$ is the element of the solid angle subtended by the surface element $\mathrm{d}S$ at the center of the sphere (S_1) with radius R ($\mathrm{d}S = R^2\,\mathrm{d}\Omega$). In order to compute I_2, we use the following general vector identity, applicable for any constant vectors $\mathbf{A}$ and $\mathbf{B}$:

$$\int (\mathbf{A} \cdot \mathbf{n})(\mathbf{B} \cdot \mathbf{n})\mathrm{d}\Omega = A_i B_j \int n_i n_j \mathrm{d}\Omega = \frac{4\pi}{3} A_i B_j \delta_{ij} = \frac{4\pi}{3} \mathbf{A} \cdot \mathbf{B}.$$

We then obtain:

$$I_2 = \mathbf{p} \cdot \mathbf{U} - \frac{4\pi}{3} U^2 R^3 \tag{6.53}$$

Substituting Equations 6.51 and 6.53 into Equation 6.48, we obtain Equation 6.49.

Impulse

Let us denote by $\mathbf{P}$ the momentum of the fluid associated with the motion resulting from the displacement of the solid body ($\mathbf{P}$ should not be confused with the dipole moment $\mathbf{p}$). For a change $\delta\mathbf{U}$ in the velocity of the object, the resulting variation in the kinetic energy of the fluid set in motion is related to $\mathbf{P}$ by:

$$\delta E_k = \mathbf{P} \cdot \delta\mathbf{U}. \tag{6.54}$$

If we now use Equation 6.50 for the kinetic energy and assume that $\alpha_{ij} = \alpha_{ji}$ (which can always be achieved by symmetrizing the expression), we can then write:

$$\delta E_k = \frac{\rho}{2}\left(\alpha_{ij} 2\, U_j \delta U_i - 2 V_0 U_i \delta U_i\right) = \rho\left[\alpha_{ij} U_j - V_0 U_i\right] \delta U_i .$$

Comparing this with Equation 6.54 and using Equation 6.46, we find:

$$\mathbf{P} = \rho\,\mathbf{p} - \rho\, V_0\, \mathbf{U}. \tag{6.55}$$

The significance of these two terms is the following: assume that we eliminate the object and replace it by the (pure) dipole moment $\mathbf{p}$. The velocity field resulting at the surface (S_1) will be the same in both cases. The term $\rho\,\mathbf{p}$ represents therefore the momentum associated with the dipole itself, while $\rho V_0\, \mathbf{U}$ is then the added effect of the finite solid object.

Force on the solid object

From the expression for the impulse, we deduce the force that the fluid exerts on the object:

$$F = -\frac{\mathrm{d}\mathbf{P}}{\mathrm{d}t} = -\rho\frac{\mathrm{d}\mathbf{p}}{\mathrm{d}t} + \rho\, V_0 \frac{\mathrm{d}\mathbf{U}}{\mathrm{d}t}. \tag{6.56}$$

For the special case where the object moves with constant velocity $\mathbf{U}$, $\mathbf{p}$ is also a constant, and the resultant force $\mathbf{F}$ (lift + drag) vanishes as predicted.

Special case for a spherical object

For the case where the object is a sphere of radius R, the dipole moment $\mathbf{p}$ can be expressed as a function of $\mathbf{U}$ (Equation 6.47a). Using Equation 6.49, we then obtain for the kinetic energy of the fluid:

$$E_k = \frac{\pi}{3}\rho R^3 U^2. \tag{6.57a}$$

In the same manner, we deduce a value for the force $\mathbf{F}$ from Equations 6.47 and 6.56:

$$\mathbf{F} = -\frac{\rho\, V_0}{2}\frac{\mathrm{d}\mathbf{U}}{\mathrm{d}t}. \tag{6.57b}$$

The force which must be applied to the sphere so as to obtain the acceleration $\mathrm{d}\mathbf{U}/\mathrm{d}t$ is thus increased by a value $-\mathbf{F}$, the same as if we had added to the mass of the sphere half the mass of the fluid which it displaces (this represents the *added mass*).

6.4 Linear surface waves on an ideal fluid

An important class of problems, which can be described in terms of the properties of an ideal fluid, so long as we are willing to neglect attenuation effects which arise from viscosity, is that of surface waves. These problems involve the coupling between deformations of the surface and the bulk flows which result from them. The mechanisms which tend, as the wave propagates, to restore the free surface to its equilibrium configuration are gravity, which counteracts deviations of the surface from the horizontal, and surface tension, which opposes any curvature of the interface and tends to minimize its area.

6.4.1 Swell, ripples and breaking waves

We begin by listing the various wave regimes which can exist at the surface of a fluid. Figure 6.13 below indicates the dependence of the propagation velocity c of a wave on the wave

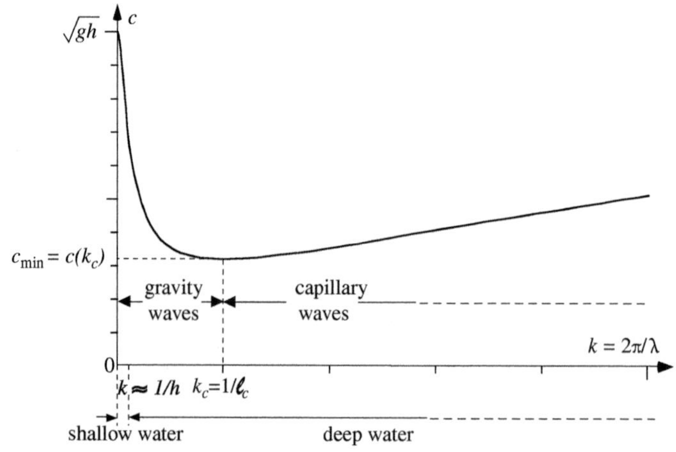

Figure 6.13 *Variation of the phase velocity of a surface wave as a function of the wave vector k for a liquid layer of thickness h large compared to the capillary length ℓ_c. As k increases, there is a transition from the regime of shallow water to that of deep water waves for $k \approx 1/h$ and then from gravitational toward capillary waves governed by surface tension for $k \approx k_c = 1/l_c$*

vector $k = 2\pi/\lambda$. We see that the variation is not monotonic and, specifically, that there exists a minimum value c_{min} for a certain value k_c of the wave vector k (for water c_{min} is 0.23 m/s). We see below that k_c equals the reciprocal of the capillary length of the fluid $\ell_c = \sqrt{\gamma/\rho g}$, which we defined by Equation 1.65 (Section 1.4.4). For a wave of wave number k (corresponding to an angular frequency ω), propagating along the surface of a layer of fluid of thickness h, density ρ, and surface tension γ, the velocity c is given by the equation:

$$c^2 = \frac{g}{k} \tanh(kh) \left(1 + \frac{\gamma k^2}{\rho g}\right). \tag{6.58}$$

This result is derived at the end of this section. It is easily generalized to the case of waves at the interface between two immiscible liquids. We discuss this particular problem in the framework of interfacial instabilities in Section 11.4.1.

Let us evaluate the relative importance of the different terms in Equation 6.58. We can do this by introducing explicitly the capillary length ℓ_c:

$$c^2 = \frac{g}{k} \tanh(kh) \left(1 + k^2 \ell_c^2\right). \tag{6.59}$$

In the discussion below, we consider the case where the thickness h of the fluid layer is significantly larger than the capillary length ℓ_c: for water this length is about 3 mm, amply justifying this assumption.

In Equation 6.59, the factor tanh (kh) is of the order of unity for those surface waves, called *deep water waves*, for which h is large relative to the wavelength $\lambda = 2\pi/k$. For that case, Equation 6.59 simplifies to:

$$c^2 \approx \frac{g}{k} \left(1 + k^2 \ell_c^2\right). \tag{6.60}$$

- The first term on the right-hand side, dominant for wavelengths large relative to the capillary length ℓ_c ($k\ell_c << 1$), corresponds to *gravity waves*. Their phase velocity is:

$$c = \sqrt{\frac{g}{k}}, \tag{6.61}$$

which decreases as the wave vector increases. This wave corresponds to ocean *swells*.

- In the other limit, for short wavelengths ($k \ell_c \ll 1$), the second term dominates. The phase velocity for *capillary waves* then becomes:

$$c \approx \sqrt{g k} \, \ell_c \approx \sqrt{\frac{\gamma k}{\rho}}, \qquad (6.62)$$

The minimum in the wave velocity corresponds to the case where the respective contributions due to capillarity and gravity are of the same order of magnitude, i.e. where the corresponding wavelength λ_c is comparable to ℓ_c. The capillary length is thus the boundary between the domains where gravitational and capillarity effects each dominate.

$$\lambda_c = 2\pi \, \ell_c = 2\pi \sqrt{\frac{\gamma}{\rho g}}, \qquad (6.63)$$

In the case where the wavelength is large compared to both the depth h of fluid and the capillary length ℓ_c (i.e., $kh \ll 1$ and $k\ell_c \ll 1$), Equation 6.59 takes on the approximate form:

$$c = \sqrt{gh}. \qquad (6.64)$$

In this case, known as that of *shallow water gravity waves*, it is the thickness of the fluid layer which determines the wave velocity. It provides the explanation for the phenomenon of *breaking waves*, in a situation such as that at a steep beach where the depth changes rapidly. The crest of the wave, where the water is significantly deeper than in the trough just ahead of it, propagates faster and actually overtakes the leading trough.

Insects and other small animals living on the surface of the water often take advantage of surface waves in different ways. First of all, the generation of such waves and their associated momentum contribute in a similar manner to other mechanisms (vortex emission, friction on the liquid, etc.) by which some of these insects move on the surface. The emission of waves, however, both requires energy and dissipates it: other small insects avoid this emission and the equivalent frictional resistance which results from that by moving at a velocity less than the minimum $c_m = 0.23$ m/s of the wave velocity. Finally, other insects find further applications for this: whirligig beetles spin and thus emit spiral waves which allow them to avoid obstacles thanks to the reflection of the waves from these obstacles.

Derivation of Equation 6.58 for the dispersion of surface waves

Consider a layer of liquid of average depth h (Figure 6.14) bounded below by the plane $y = 0$. We seek solutions for two-dimensional waves characterized by the velocity potential $\Phi(x, y, t)$. On one hand, we assume that the amplitude of the wave is sufficiently small that terms of the type $(\mathbf{v}.\nabla)\mathbf{v}$ are negligible, and, on the other hand, that the curvature of the interface remains always small.

This is equivalent to assuming $\partial y_0(x, t)/\partial x \ll 1$ where $y_0(x,t)$ represents the instantaneous position of the surface. Bernoulli's Equation (Equation 5.36) for a non-stationary problem of this type takes then the form (writing $\varphi = g \, y$ and neglecting the $v^2/2$ term):

$$\frac{\partial \Phi}{\partial t} + \frac{p}{\rho} + g \, y = \text{constant}. \qquad (6.65)$$

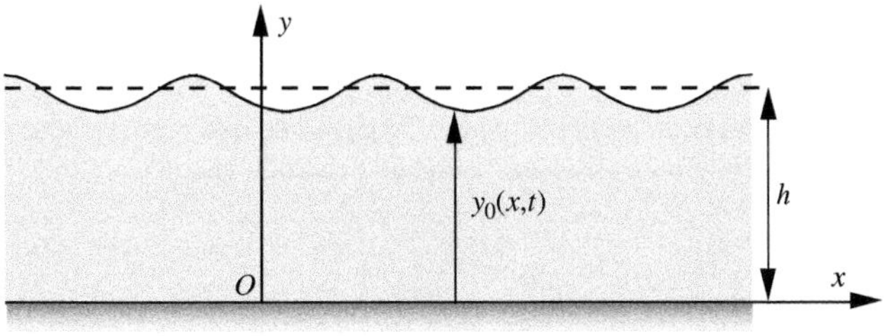

Figure 6.14 *Geometry of a liquid layer for the analysis of the propagation of surface waves*

The pressure at the surface of the liquid is given by the Young-Laplace law (Equation 1.58 with $R' \to \infty$):

$$p = p_0 - \frac{\gamma}{R}, \tag{6.66}$$

where p_0 is the external pressure above the interface, γ the coefficient of surface tension and $R \simeq (\partial^2 y/\partial x^2)^{-1}$ is the instantaneous local curvature of the interface. At the bottom of the container (at $y = 0$), the velocity normal to the surface is zero, so that:

$$\left(\frac{\partial \Phi}{\partial y}\right)_{y=0} = 0. \tag{6.67}$$

At the interface, we require the vertical component of the velocity of the fluid to equal that of the interface, so that:

$$v_y(y = y_0) = \left(\frac{\partial \Phi}{\partial y}\right)_{y=y_0} = \frac{\partial y_0}{\partial t}. \tag{6.68}$$

The equality on the right is, in fact, correct only to first order. There is a higher order term, whose significance we shall discuss in Section 11.4.1, which takes into account the effect of the horizontal convective motion on the vertical displacement of the interface when the interface is not horizontal.

The potential function we seek here is a solution of Laplace's equation $\nabla^2 \Phi = 0$, satisfying the boundary conditions stated above. We obtain a solution for Φ by the method of separation of variables. We therefore assume a solution in the form of the product of a function $f(u)$, where $u = x - ct$, representing a wave traveling at speed c in the positive x-direction, with a function $q(y)$, dependent on the vertical y coordinate, i.e.:

$$\Phi(x, y, t) = f(u)\, q(y). \tag{6.69}$$

Substitution into Laplace's equation leads to:

$$\frac{\partial^2 f}{\partial u^2} q + f \frac{\partial^2 q}{\partial y^2} = 0 \qquad \text{or} \qquad \frac{1}{f}\frac{\partial^2 f}{\partial u^2} = -\frac{1}{q}\frac{\partial^2 q}{\partial y^2}.$$

Since each side of the latter equation is respectively a function of only one of the *independent* variables u and y, each side must necessarily be a constant, so that:

$$\frac{1}{f}\frac{\partial^2 f}{\partial u^2} = -\frac{1}{q}\frac{\partial^2 g}{\partial y^2} = \text{constant} = -k^2. \tag{6.70}$$

We have chosen the constant to be negative for the x-dependent left-hand side, because we want sinusoidally propagating solutions of the form $e^{i(kx-\omega t)}$ in the x-direction. Taking into account the boundary condition (Equation 6.67), we then obtain for the potential function a solution of the form:

$$\Phi(x, y, t) = f(x - ct)\, q(y) = A\, e^{i(kx-\omega t)} \cosh(ky), \tag{6.71}$$

where A is a constant proportional to the amplitude of the wave. The y-dependence corresponds to an attenuation of the wave with depth. Taking the derivative of Equation 6.65 with respect to time, and using once again the boundary condition (Equation 6.68), we obtain the equation:

$$\left[\frac{\partial^2 \Phi}{\partial t^2} + g\frac{\partial \Phi}{\partial y} - \frac{\gamma}{\rho}\frac{\partial^3 \Phi}{\partial x^2 \partial y}\right]_{y=y_0} = 0. \tag{6.72}$$

Finally, substituting in this last equation the explicit form of the solution for the potential (Equation 6.71), and replacing $y_0(x, t)$ by its average value, h, we find:

$$\omega^2 = \left(g k + \frac{\gamma k^3}{\rho}\right) \tanh(kh), \tag{6.73}$$

which reduces to Equation 6.58, the result stated at the beginning of this section, since the phase velocity is $c = \omega/k$.

6.4.2 Trajectories of fluid particles during the passage of a wave

If we use the solution for the velocity potential (Equation 6.71):

$$\Phi(x, y, t) = A\,\mathrm{e}^{\mathrm{i}(kx-\omega t)} \cosh(ky)$$

derived in the previous section, the corresponding velocity fields obtained from the equation $\mathbf{v}(x, y, t) = \nabla\Phi$, are:

$$v_x(t) = \frac{\partial\Phi}{\partial x} = A\ \mathrm{i}\ k\,\mathrm{e}^{\mathrm{i}(kx-\omega t)} \cosh(ky), \qquad v_y(t) = \frac{\partial\Phi}{\partial y} = A\,k\,\mathrm{e}^{\mathrm{i}(kx-\omega t)} \sinh(ky).$$

We then integrate the above fields with respect to the time t. Because this appears only in the exponentials, integration with respect to time is equivalent to dividing by $-\mathrm{i}\omega$, – i.e., multiplying by i/ω. We set the constant of integration to zero, which amounts to looking only at the oscillating component of the corresponding coordinate. We obtain:

$$\Delta x(t) = -A\frac{k}{\omega}\,\mathrm{e}^{\mathrm{i}(kx-\omega t)}\cosh(ky), \qquad \Delta y(t) = \mathrm{i}A\frac{k}{\omega}\,\mathrm{e}^{\mathrm{i}(kx-\omega t)}\sinh(ky).$$

We then take the real parts of the above results, in order to see the displacements in the real space, using Euler's identity $\mathrm{e}^{\mathrm{i}u} = \cos u + \mathrm{i}\sin u$. This then leads to the equation (with time as a parameter) for the trajectories of the particles of fluid in the (x, y)-plane:

$$\Delta x(t) = -A\frac{k}{\omega}\cos(kx - \omega t)\cosh(ky), \qquad \Delta y(t) = -A\frac{k}{\omega}\sin(kx - \omega t)\sinh(ky).$$

Here, x and y are the mean position coordinates for each particle and $\Delta x(t)$ and $\Delta y(t)$ are the components of the displacement of each particle at time t relative to this position. We now eliminate the expressions involving time in these equations, obtaining the equation of an ellipse:

$$\frac{(\Delta x)^2}{\cosh^2(ky)} + \frac{(\Delta y)^2}{\sinh^2(ky)} = \frac{A^2 k^2}{\omega^2}. \tag{6.74}$$

- In *deep water*, far from the bottom, $k|y| \gg 1$. Therefore:

$$\cosh^2(ky) = \sinh^2(ky) = \frac{\mathrm{e}^{2ky}}{4}.$$

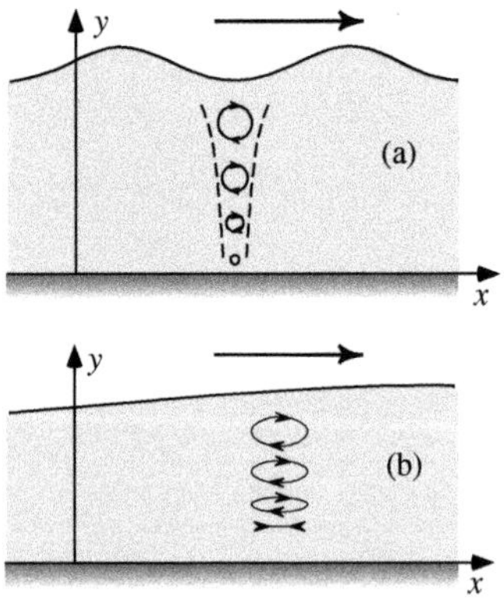

Figure 6.15 *The trajectories of fluid particles as a wave passes by (a) for deep water and (b) for shallow water. The horizontal arrow indicates the direction of propagation of the wave*

The trajectories are then circular, with a radius $R_t(y) = R_t(h)\ e^{k(y-h)}$ ($R_t(h)$ being the value at the surface). The amplitude of the displacement decreases exponentially with the depth and becomes negligible as soon as the latter exceeds a few wavelengths (Figure 6.15a).

- In *shallow water*, we have $k|y| \ll 1$, for all y. The trajectories are therefore ellipses with major axis $\{2A(k/\omega)\}$ in the x-direction, and minor axis $\{2A(k/\omega)\ ky\}$ vertically: they are highly elongated near the surface, and gradually becoming flat straight-line segments near the bottom. The ratio of the axes of the ellipse is of the order of ky. The amplitude of the displacement of the particles in the (horizontal) x–direction remains basically constant (Figure 6.15b), while the vertical one is a factor ky smaller. Thus, the shallow-water approximation is in fact equivalent to neglecting the velocity component v_y for all values of y, and to assuming that the velocity component v_x is independent of y.

6.4.3 Solitons

Here, we enter the rich and complex domain of non-linear waves, for the effects of which surface waves provide one of the simplest illustrations. We present here a semi-quantitative discussion of some of their properties.

We consider the propagation of a perturbation of the free surface of a fluid layer of thickness h, having amplitude A, *not* negligible in comparison to h, and localized in a region of width Δ (Figure 6.16). We show that, by a compensating effect, such a perturbation can propagate without deformation, resulting in what is known as a *soliton*, or *solitary wave*.

Two effects compete to modify the profile of the disturbance as it propagates:

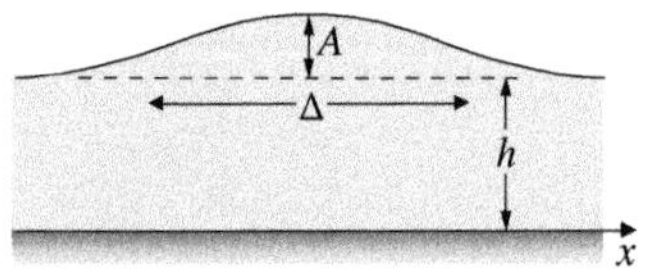

Figure 6.16 *A soliton, with amplitude A that is non-negligable relative to the thickness h of the fluid layer, can propagate along the surface of the liquid without undergoing any deformation*

- A spreading due to dispersion: the perturbation, of width Δ, can be considered as a wave packet, or superposition of sinusoidal waves of nearly equal frequencies. Assuming that kh is not too large, we expand to lowest order Equation 6.58 predicting the velocity of the gravity waves. If we also neglect the term corresponding to capillarity, we obtain:

$$c(k) \approx \sqrt{g\,h}\left(1 - \frac{k^2h^2}{6}\right). \tag{6.75}$$

We note that components of the wave with the shorter wavelengths (larger k values) propagate more slowly. If the spatial extent of the wave is Δ, the spectrum of its wave vectors extends approximately over $0 < k < k_{max} \approx 1/\Delta$. The velocity difference δc between these extremes has order of magnitude:

$$|\delta c| \approx \sqrt{g\,h}\,\frac{h^2}{\Delta^2}. \tag{6.76}$$

This velocity difference between the various components of the wave spectrum tends to cause spreading of the wave packet, a result typical of the classical phenomenon known as dispersion.

- The steepening of the wave front results from non-linear effects; since the wave velocity increases with increasing depth h, the crest and the trough of the deformation tend to propagate with respective velocities:

$$c' = \sqrt{g\,(h+A)} \qquad \text{and} \qquad c = \sqrt{g\,h}.$$

The crest thus propagates more rapidly than the trough, tending to steepen the wave, and often causing it to break. The corresponding difference in velocity is of order:

$$c' - c \approx \sqrt{\frac{g}{h}\frac{A}{2}}. \tag{6.77}$$

The dispersion and non-linearity effects cancel out when the two above velocity differences (Equations 6.76 and 6.77) are of the same order of magnitude, i.e., when:

$$\sqrt{gh}\,\frac{h^2}{\Delta^2} \approx \sqrt{\frac{g}{h}}\,A, \qquad \text{so that:} \qquad A \approx \frac{h^3}{\Delta^2}. \tag{6.78}$$

The wave propagates without any deformation only if the cancellation is exact everywhere along the profile of the wave. It can be shown by a more elaborate calculation that this can be achieved for a precisely determined shape, having the mathematical form:

$$y = h + \frac{A}{\cosh^2(x/\Delta)}, \quad (6.79a) \qquad \text{with:} \qquad A\,\Delta^2 = 4\,h^3/3, \quad (6.79b)$$

and a velocity of propagation:

$$c = \sqrt{g\,(h + A)}. \tag{6.79c}$$

The propagation of such solitary waves can be observed in canals over distances of several miles. We recall the story of Scott Russell, who in 1838 discovered solitons as he followed on horseback the wave resulting from the sudden stopping of barge on a canal. Also, some of the tidal bores discussed in Section 5.4.4 (the slower ones) are composed a leading shock-like wave followed by a train of solitons.

6.4.4 Another example of potential flow in the presence of an interface: the Taylor bubble

Big bubbles of air rising in a large diameter water-filled tube, close to that of the bubble (10 cm or more in diameter), frequently display in the forward direction a stable quasi-spherical shape (Figure 6.17a). On the other hand, the wake which they leave behind is generally turbulent. In 1950, G.I. Taylor showed that the velocity U_b at which the bubble rises is well predicted by a simple model, which assumes a potential flow in the region near the tip of the bubble, neglecting wall effects. This is why we discuss here this phenomenon, although it involves global movement of a water-air interface rather than the propagation of a wave.

Given the assumption of potential flow, the rising velocity, for a bubble of diameter much greater than the capillary length l_c (Section 1.4.4, Equation 1.65), and comparable to that of the tube, is given by:

$$U_b \approx \left(\frac{2}{3}\right)\sqrt{gR} \approx 0.47\sqrt{2gR}. \tag{6.80a}$$

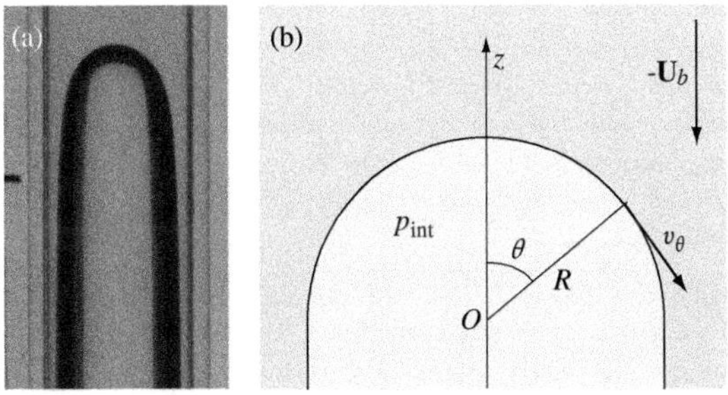

Figure 6.17 *(a) View of a Taylor bubble rising in a tube full of water with a diameter* $d = 8$ cm *Document courtesy of S. Madani, O. Caballina, M. Souhar). (b) Schematic diagram of the flow around the leading edge of the spherical shape of the Taylor bubble (the reference frame of the spherical coordinates is assumed to be stationary relative to the bubble).*

Experimentally, measurements carried out in tubes of diameter d (slightly greater than $2R$) yield, in the limit of large diameters and low viscosities:

$$U_b \approx 0.35\sqrt{gd}. \tag{6.80b}$$

For smaller diameters and/or more viscous fluids, the velocity will be lower, due to surface tension and viscosity effects.

Proof

In a reference frame where the bubble is stationary under the given assumptions, this problem is equivalent to that of a uniform stationary flow with velocity $-U_{\text{bubble}}$ around a sphere. The tangential velocity at the surface of the bubble is then $v_\theta = -(3/2)\, U \sin\theta$, according to Equation 6.25. Capillary effects will be negligible and the pressure in the liquid at the surface of the bubble with be equal to the pressure p_{int} of the gas in the bubble, a pressure which is effectively constant. If we apply Bernoulli's Equation at the surface of the bubble between the stagnation point at the tip of the bubble ($\theta = 0$, $z = R$) and a point on the leading face ($\theta < \pi/2$, $z = R\cos\theta$):

$$p_{\text{int}} + \frac{9}{8}\rho\, U_b^2 \sin^2\theta + \rho\, g\, R\cos\theta = p_{\text{int}} + \rho\, g\, R. \tag{6.81}$$

Assuming that one remains in the vicinity of the tip with $\theta \ll 1$, one can take $(1-\cos\theta) = 2\sin^2(\theta/2) \cong \theta^2/2$ and $\sin^2\theta \cong \theta^2$, so that Equation 6.80a results from Equation 6.81. For water in a tube of diameter $2R = 0.1$ m, the velocity of an air bubble of diameter nearly equal to $2R$ will be 0.35 m/s, corresponding to a Reynolds number $Re = 2RU_b/\nu$ around 35,000.

As Taylor writes in his original paper, the result of this demonstration is valid because of the effective absence of viscous stress at the surface of the bubble given the small value of the viscosity of air within the bubble. For a solid body of the same geometry as the bubble, a boundary layer would appear at the surface so as to satisfy the condition of zero relative velocity at the wall.

6.5 Electrical analog for two-dimensional potential flows

Consider a two-dimensional, potential, incompressible fluid flow. As indicated in Equations 3.38 and 6.2, known as the Cauchy-Riemann conditions, the velocity field can be derived from the velocity potential $\Phi(x, y)$, or from the stream function $\Psi(x,y)$ with:

$$v_x = \frac{\partial \Phi}{\partial x} = \frac{\partial \Psi}{\partial y} \quad (6.82a) \qquad \text{and:} \qquad v_y = \frac{\partial \Phi}{\partial y} = -\frac{\partial \Psi}{\partial x}. \quad (6.82b)$$

Since $\nabla \cdot \mathbf{v} = 0$ and $\nabla \times \mathbf{v} = 0$, we conclude that each of the functions $\Phi(x, y)$ and $\Psi(x, y)$ is harmonic, i.e., obeys Laplace's equation:

$$\nabla^2\Phi = 0, \quad (6.83a) \qquad \nabla^2\Psi = 0. \quad (6.83b)$$

Similarly, the electrical potential V resulting, in vacuum, from a static or a quasi-static distribution of charges obeys the identical equation:

$$\nabla^2 V = 0. \tag{6.84}$$

By comparing Equations 6.83a and 6.84, with due care in the use of appropriate boundary conditions, it is therefore possible to establish a correspondence between the electrical potential V and, either the velocity potential Φ (*direct analog*) or the stream function Ψ

(*inverse analog*). In the latter case, we see that it is sufficient to prescribe an electrical equipotential function with the geometry of the solid walls; all other equipotentials then describe the streamlines. This correspondence is used in analog models of the flow of ideal fluids in two dimensions. These models are constructed by passing an electric current through an electrolytic tank, or through a sheet of paper or plastic covered by a weakly electrically conducting layer (e.g., graphite).

6.5.1 Direct analog

In this case, the velocity potential $\Phi(x, y)$ corresponds to the electrical potential $V(x, y)$, and the velocity field $\mathbf{v} = \nabla\Phi$, to the current density $\mathbf{j} = \sigma\mathbf{E} = -\sigma\nabla V$, σ being the electrical conductivity of the medium. The streamlines of the hydrodynamic problem, being tangent to a solid obstacle, require this object to be replaced by an insulator of identical geometry. The equipotentials of the electrical problem (easily determined experimentally) correspond then to velocity equipotentials. The streamlines for the hydrodynamic problem could then be obtained by drawing the system of curves orthogonal to these equipotentials (e.g., by using a resistive sheet with a circular hole to represent the flow around the cylinder shown in Figure 6.5). The resulting method is somewhat impractical; it is much preferable to use the *inverse analog*.

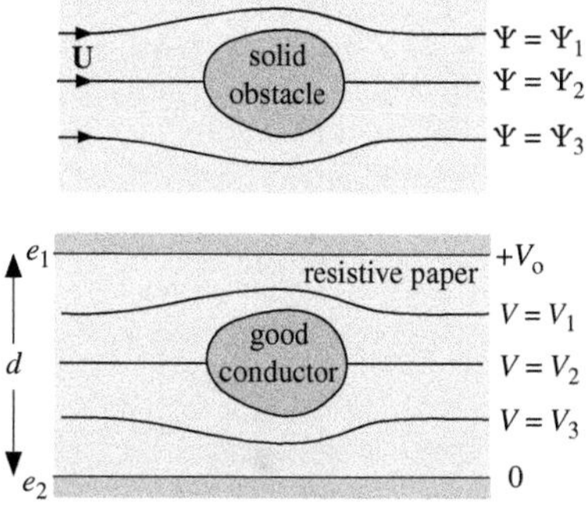

Figure 6.18 *(a) A representation of uniform flow at velocity U between two solid parallel planes, as it is perturbed by a rigid obstacle. (b) The inverse electrical analog of the flow in (a) using a sheet of paper of low electrical conductance (clear shade), two lateral electrodes e_1 and e_2, and a highly conducting central region (grey) in the shape of the obstacle*

6.5.2 Inverse analog

In this case, the electrical equipotentials correspond directly to the streamlines of the fluid. Here, the hydrodynamic obstacle is replaced by a *perfect conductor* acting as an equipotential surface: the particular streamline, which forms the boundary of the object in the hydrodynamic problem, corresponds to the equipotential boundary for the similar object in the electrical model.

In practice, one uses a weakly conducting piece of paper on which is painted, with a more conducting paint, a surface of the same shape as that of the obstacle in hydrodynamic flow. The flow conditions, far from the obstacle, are simulated by applying an electrical potential between a pair of appropriately located, well-separated, lateral electrodes (Figure 6.18) e_1 and e_2. The streamlines are obtained point-by-point, by moving an electrode along the surface of the paper, in such a way as to follow the electrical equipotential lines.

Instead of the electrically conducting paper, one can use a shallow layer of electrolytic solution.

Finally, we demonstrate that, in the inverse analog, the existence of non-zero circulation ($\Gamma = \int_{(c)} \mathbf{v}\cdot\mathbf{d}\ell$) around the obstacle can be simulated by injecting (or withdrawing, depending on the sign of Γ) an electrical current of magnitude I from the conducting obstacle toward the conducting paper. The total amount I of current emitted satisfies:

$$I \propto \Gamma. \tag{6.85}$$

Proof

The strength I of the current emitted can be written:

$$I = \int \mathbf{j}\cdot\mathbf{n}\,(h\,\mathrm{d}\ell) = \sigma\,h \int \mathbf{E}\cdot\mathbf{n}\,\mathrm{d}\ell, \tag{6.86}$$

where h is the thickness of the conducting sheet, σ its electrical conductivity, and $\mathbf{j}$ the current density; $\mathbf{n}$ is the unit vector normal to the contour (C), along which the circulation

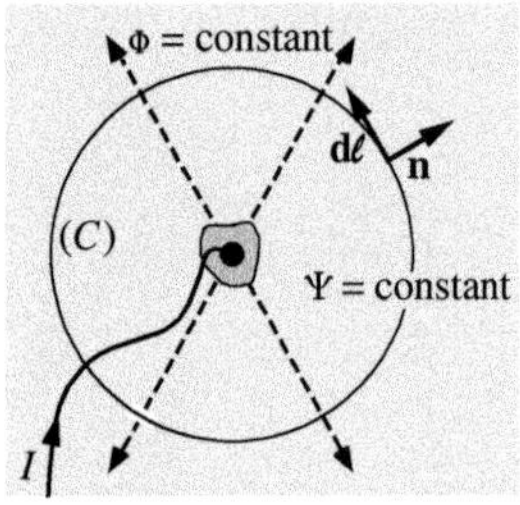

Figure 6.19 *Simulation of the circulation of the velocity around an obstacle by injection of an electrical courent through the conducting electrode into the resistive circuit. The equipotentials then have the same geometry as the streamlines around the obstacle*

is calculated (Figure 6.19), and $d\ell$ is the element of length along this contour. However, in the inverse analog, the electrical field lines correspond to the velocity equipotentials and must therefore be orthogonal to the streamlines in the fluid. We therefore have:

$$\mathbf{E} = \alpha \nabla \Psi, \tag{6.87}$$

where the constant α stands for the ratio between the units of the electrical case and the hydrodynamic one. We then recover Equation 6.85 by rewriting Equation 6.86, using Equation 6.82a and Equation 6.82b and noting that $\mathbf{n}\, d\ell = (-dy, dx, 0)$:

$$I = \sigma h \int_{(C)} (\nabla \Psi \cdot \mathbf{n})\, d\ell = \alpha \sigma h \left[\int_{(C)} (-v_y)(-dy) + (v_x)(dx) \right] = \alpha \sigma h \int_{(C)} \mathbf{v} \cdot \mathbf{d\ell} = \alpha\, \sigma\, h\, \Gamma .$$

The proportionality factor corresponds to the product of σ h by the ratio between the electrical and velocity potentials.

6.6 Complex velocity potential

Conformal mappings which result from the use of a complex potential allow us to set up a correspondence between the two-dimensional velocity fields described above and problems involving more complex geometries, such as the profile of a two-dimensional wing.

6.6.1 Definition of the complex potential

Equations 6.82 satisfied by the conjugate harmonic functions Φ and Ψ represent the Cauchy-Riemann conditions: they are both necessary and sufficient for the *complex potential function*:

$$f(z) = \Phi(x, y) + \mathrm{i}\, \Psi(x, y) \tag{6.88}$$

to be an analytic function of the complex variable $z = x + \mathrm{i}y$. The derivative of $f(z)$ is then called the *complex velocity:*

$$\frac{\mathrm{d} f(z)}{\mathrm{d}z} = w(z) = v_x - \mathrm{i}\, v_y. \tag{6.89}$$

Proof
Choosing, for instance, the increment $\mathrm{d}z$ along the real axis (= $\mathrm{d}x$ + i 0), we find:

$$w(z) = \frac{\partial(\Phi + \mathrm{i}\Psi)}{\partial x} = \frac{\partial \Phi}{\partial x} + \mathrm{i}\frac{\partial \Psi}{\partial x} = v_x - \mathrm{i}\, v_y, \tag{6.90a}$$

while an independently chosen displacement in the imaginary direction ($\mathrm{d}z = 0 + \mathrm{i}\, \mathrm{d}y$) leads to the identical conclusion:

$$\frac{\partial(\Phi + \mathrm{i}\,\Psi)}{\mathrm{i}\ \partial y} = -\mathrm{i}\frac{\partial \Phi}{\partial y} + \frac{\partial \Psi}{\partial y} = -\mathrm{i}\, v_y + v_x = w(z). \tag{6.90b}$$

Thus, the result is completely general, applying for any arbitrary increment $\mathrm{d}z = \mathrm{d}x + \mathrm{i}\, \mathrm{d}y$.

We can also determine the physical meaning, over a closed contour (C), of the complex circulation $\overline{\Gamma}(z)$ defined by:

$$\overline{\Gamma}(z) = \int_{(C)} w\,(z)\mathrm{d}z = \int_{(C)} (v_x - \mathrm{i}\, v_y)\,(\mathrm{d}x + \mathrm{i}\,\mathrm{d}y) = \int_{(C)} (v_x\mathrm{d}x + v_y\mathrm{d}y) + \mathrm{i}\int_{(C)} (v_x\mathrm{d}y - v_y\mathrm{d}x),$$

i.e.:

$$\overline{\Gamma}(z) = \int_{(C)} \mathbf{v}\cdot \mathbf{dl} + \mathrm{i}\int_{(C)} \mathbf{v}\cdot \mathbf{n}\,\mathbf{dl} = \Gamma + \mathrm{i}\,Q \tag{6.91}$$

where $\mathbf{n}$ represents a unit vector normal to the line element $\mathbf{dl}$. The real part of $\overline{\Gamma}(z)$, Γ, is thus the circulation of the fluid, and the imaginary part, Q, the rate of flow of the fluid (per unit length normal to the plane of the flow), resulting from sources inside the contour (C). For a source-free flow within a simply connected region $\overline{\Gamma}(z) = 0$. Along a closed contour, the function $f(z)$ is uniquely defined at each point.

6.6.2 Complex velocity potential for several types of flow

The following complex potentials can be shown to correspond to the flows discussed in Sections 6.2.3 and 6.2.4:

(i) ***Uniform parallel flow***

$$f(z) = U\,z.$$

The potential, where $\overline{U} = U_x - \mathrm{i}\,U_y$ is a complex constant, represents then a uniform flow in an arbitrary direction in the x-y plane.

(ii) ***Source and vortex***

We have seen above, in Section 6.2.3, the correspondence between the streamlines of one of these flows and the equipotentials of the other one. These two flows can be described together by the complex potential:

$$f(z) = a_0 \operatorname{Log} z, \tag{6.92}$$

where the coefficient a_0 is complex.

The complex velocity is, as always, the derivative of the complex potential. Since $f(z) = a_0$ Log z, we have then:

$$w(z) = v_x - \mathrm{i}\,v_y = \frac{a_0}{z} = \frac{a_0}{r}e^{-\mathrm{i}\,\varphi}$$

or, in polar coordinates:

$$v_r - \mathrm{i}v_\varphi = a_0/r,$$

because we can easily show by expanding and identifying corresponding terms that:

$$v_r - \mathrm{i}\,v_\varphi = (v_x - \mathrm{i}\,v_y)\,e^{\mathrm{i}\,\varphi}.$$

- For real a_0 ($a_0 = Q/2\pi$) then:

  $$v_r = Q/(2\pi\,r), \qquad v_\varphi = 0,$$

  which we recognize as the velocity field of a source with flow rate Q.

- If a_0 is a pure imaginary ($a_0 = -\mathrm{i}\,\Gamma/2\pi$), then:

$$v_r = 0, \qquad v_\varphi = \Gamma/(2\pi\, r),$$

which represents the velocity field of a vortex with circulation Γ.

(iii) ***Dipole flow***

The complex potential of a two-dimensional dipole and its corresponding velocity field can be respectively written:

$$f(z) = -p/2\pi z \qquad (6.93) \qquad \text{and:} \qquad w(z) = p/2\pi\, z^2,$$

whence:

$$v_r = p \cos\varphi/2\pi\, r^2, \qquad v_\varphi = p \sin\varphi/2\pi\, r^2.$$

These equations are identical to Equations 6.14a and 6.14b.

(iv) ***Flow around a corner or near a stagnation point***

As a final example of the complex potential method, we now discuss a class of flows which have significant practical applications. These flows are described by complex potentials of the form:

$$f(z) = C\, z^{m+1}, \tag{6.94}$$

corresponding to two-dimensional flow in the presence of a wedge-shaped corner formed by the intersection of two planes intersecting at the origin. In polar coordinates, the velocity potential and the stream function may be respectively written:

$$\Phi = C\, r^{m+1} \cos(m+1)\varphi, \quad \Psi = C\, r^{m+1} \sin(m+1)\varphi.$$

The components of the velocity are obtained by taking the derivatives of Φ:

$$v_r = \frac{\partial \Phi}{\partial r} = (m+1)\, C\, r^m \cos(m+1)\,\varphi, \qquad v_\varphi = \frac{1}{r}\frac{\partial \Phi}{\partial \varphi} = -(m+1)\, C\, r^m \sin(m+1)\,\varphi. \tag{6.95}$$

The straight lines with equations $\varphi = 0$ and $\varphi = n\pi/(m+1)$, with n a positive or negative integer, are therefore streamlines for which $\Psi = 0$ for all r. They represent the intersection of plane rigid walls with the plane of the flow. We now discuss different shapes of these corners determined by different values of the parameter m (Figure 6.20 below).

- The case $m > 1$ represents the flow inside an acute-angled re-entrant corner (Figure 6.20a). In this instance, the magnitude of the velocity at the origin approaches zero as r^m.
- For $m = 1$, $f(z) = C z^2$, we have flow inside a right angle or in the neighborhood of a stagnation point at a plane wall (provided that, in this second case, we add in the symmetric flow relative to the y-axis) (Figure 6.20b). We see, in this situation, a streamline perpendicular to the wall which ends at the stagnation point on the wall, where the velocity is zero.

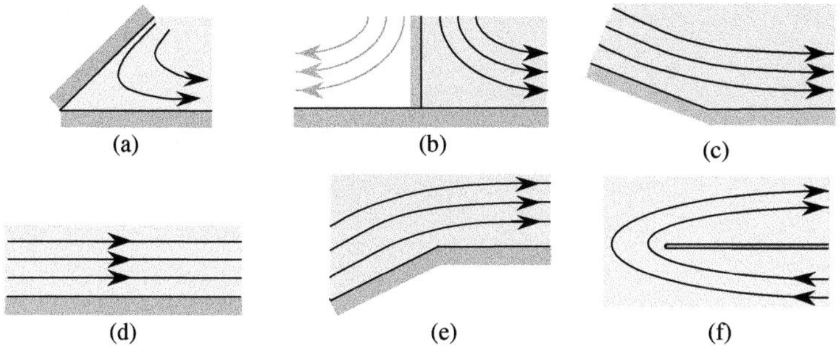

Figure 6.20 *Flows for which the velocity potential has the form: $f(z) = C\ z^{m+1}$. (a) $m > 1$; (b) $m = 1$; (c) $0 < m < 1$; (d) $m = 0$; (e) $-1/2 < m < 0$; (f) $m = -1/2$*

- When $0 < m < 1$, the function $f(z)$ represents flow inside an obtuse-angled corner with opening $\alpha = \pi/(m + 1)$, bounded by the solid walls $\varphi = 0$ and $\varphi = \pi/(m+1)$, corresponding respectively to $n = 0$ and $n = 1$ (Figure 6.20c).
- If $m = 0$, we have the simple case of flow parallel to a plane (Figure 6.20d).
- For $-1/2 < m < 0$, we have this time, still with values $n = 0$ and $n = 1$ at the solid walls, flow around the outside of a corner, obtuse-angled (if $m > -1/3$) or acute-angled (when $m < -1/3$), as shown in Figure 6.20e.
- Finally, $m = -1/2$ corresponds to flow around the edge of a semi-infinite flat plate (Figure 6.20f).

For negative m, with the flow outside the corner, the magnitude of the velocity at the origin *diverges* as r^m, in contrast with the situation when m was positive. Such divergence is, of course, unphysical; we shall see, in Section 10.5.2, that we must reconsider this particular result, taking into account the viscosity of this fluid. The existence of viscosity requires that the velocity approach zero continously at stationary, solid walls. The solutions that we have just discussed apply at sufficiently large distances from such walls; they are found to connect to the condition of zero velocity, by means of a transition region known as a boundary layer.

6.6.3 Conformal mapping

The conformal mapping method

Consider the pair of complex variables $z = x + iy$ and $Z = X + iY$, each represented by a point in its own (complex) plane, and such that Z is an *analytic* function of z:

$$Z = g(z). \tag{6.96}$$

Equation 6.96 is then called a *conformal* transformation, *mapping* a given point (x,y) in the z–plane, into a specific point (X,Y) in the Z-plane. Consequently, it maps the family of streamlines $\Psi(x,y) = \text{constant}$ (and the corresponding equipotentials $\Phi(x,y) = \text{constant}$) into respective families in the (X, Y), i.e., Z-plane (Figure 6.21).

One of the fundamental properties of conformal mappings is that angles between intersecting curves are preserved, as shown directly below. Thus, while two arcs of curves (C_1) and (C_2), intersecting at a point z_0, mapping into Z_0 in the Z-plane, are macroscopically distorted and rotated, the angles between their tangents are preserved (Figure 6.21 below). More specifically, streamlines and equipotentials, mutually orthogonal in the x-y plane,

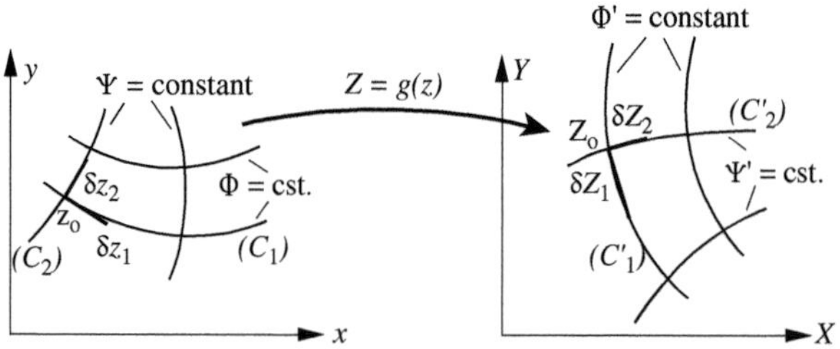

Figure 6.21 *Mapping of the families of streamlines and of the corresponding equipotentials by means of the conformal mapping $Z = g(z)$. The angles of intersections between the curves are unchanged preserving the orthogonality of the curves whenever $g(z)$ is analytic at the point of intersection z_0 and its derivative $g(z_0) \neq 0$*

remain so in their representations as mapped onto the X-Y plane. The corresponding curve increments can be written as:

$$\delta Z_1 = g'(z_0)\,\delta z_1 \qquad \text{and} \qquad \delta Z_2 = g'(z_0)\,\delta z_2.$$

So long as $g'(z_0)$ remains finite, we can conclude:

$$\frac{\delta Z_2}{\delta Z_1} = \frac{|\delta Z_2|}{|\delta Z_1|} e^{i(\arg\delta Z_2 - \arg\delta Z_1)} = \frac{|\delta z_2|}{|\delta z_1|} e^{i(\arg\delta z_2 - \arg\delta z_1)},$$

i.e.,:

$$\frac{|\delta Z_2|}{|\delta Z_1|} = \frac{|\delta z_2|}{|\delta z_1|} \qquad \text{and} \qquad \arg(\delta Z_2) - \arg(\delta Z_1) = \arg(\delta z_2) - \arg(\delta z_1).$$

The singular points of the conformal mapping correspond to points z_0 where $g(z)$ is zero or infinity. At these points, angles are not conserved, and we will show that the ratio of the initial angles and their images equals the order of the first non-vanishing derivative of the function $g(z)$ at the point z_0. Let us call this order n; expanding the function $g(z)$ to order n around z_0 we find:

$$\delta Z \approx \frac{(\delta z)^n}{n!}\left(\frac{\partial^n g(z)}{\partial z^n}\right)_{z=z_0}.$$

The ratios of the two increments δz_1 et δz_2 and of their corresponding transforms δZ_1 et δZ_2 thus satisfy:

$$\frac{\delta Z_2}{\delta Z_1} = \frac{|\delta Z_2|}{|\delta Z_1|} e^{i(\arg\delta Z_2 - \arg\delta Z_1)}$$

$$= \frac{|\delta z_2|^n}{|\delta z_1|^n} = \frac{|\delta z_2|^n}{|\delta z_1|^n} e^{in(\arg\delta z_2 - \arg\delta z_1)}.$$

So that we have:
$\arg\delta Z_2 - \arg\delta Z_1 = n(\arg\delta z_2 - \arg\delta z_1)$.

Thus the scaling factor for the length of the segments, which results from the transformation, does not depend on their direction. Moreover, the angle between the curves (C_1) and (C_2) is equal to the angle between their images, except at the points with coordinate z_0 such that $g'(z_0)$ is either equal to zero or to infinity.

As a result of this property that angles are preserved, we conclude that the images of the equipotentials $\Phi(x, y) = $ constant and $\Psi(x, y) = $ constant, are orthogonal to each other just as they were in the original reference frame.

A second important property is that, if a function $f(x, y)$ is harmonic i.e. (if it obeys Laplace's equation in the x-y plane), the function $F(X, Y)$ resulting from the mapping is also harmonic with respect to the X-Y variables. As a result, the streamlines and equipotentials, not only preserve their mutual orthogonality through the transformation, but become the streamlines and equipotentials of the mapped flow. Thus, if we determine these curves in a certain flow geometry, we can deduce from them, in general, *the solutions corresponding to any conformal mapping of the original geometry*!

Let us assume that the complex potential $f(z)$ characterizes a flow in a domain of the plane (x, y) and that $h(Z)$ represents the inverse transformation of the conformal mapping $Z = g(z)$ defined by Equation 6.96. The function:

$$f(h(Z)) = F(Z) \tag{6.97}$$

describes the flow in the (X, Y) plane for which the equipotentials and the streamlines are respectively the images obtained by transforming equipotentials and streamlines located in the (x, y)-plane. Specifically, the mapping law transforms the obstacles in the *object* plane into the obstacles in the *image plane*. It is therefore possible to obtain directly the velocity and the complex potential by means of the $g(z)$ transform.

Transformation of a plane into a corner

In Section 6.6.2 (iv), we studied the flow of fluid inside or outside a corner, by the method of complex potentials. An equivalent method for the study of this kind of flow is to carry out a conformal mapping of flow parallel to a plane by means of the following transform:

$$z = h(Z) = Z^{m+1}. \tag{6.98}$$

In using this transform, the complex potential $f(z) = U\, z$ of the initial flow is transformed into a potential:

$$F(Z) = U\, Z^{m+1}, \tag{6.99}$$

which precisely describes the kind of flow we wish to examine, as we showed in Section 6.6.2 (iv). Note that the fact that the plane is transformed into a corner is due to the singularity of the inverse function of $h(Z)$ at the origin; at this point, angles are not conserved, and the zero angle in the object plane is transformed into an acute, or obtuse angle, depending on the value of the parameter m.

The Joukowski transformation — Modeling an airplane wing in potential flow

This transformation is the first of a series of conformal mappings which allow us to change from flow around a circular cylinder to flow around a two-dimensional profile of an airplane wing. As a result, we can determine the velocity field of the flow around an element of the flat plane, located in a flow which is uniform at infinity and inclined at an angle with respect to the plane.

(i) ***Definition***

The Joukowski transformation is defined by the equation:

$$Z = g\,(z) = z + \frac{R^2}{z}, \tag{6.100}$$

where R is a real number. It transforms a circle, centered at the origin and of radius r in the (x, y) plane, into an ellipse in the (X, Y)-plane. Effectively, for $z = r\,e^{i\varphi}$ (polar equation of a circle), we have:

$$Z = X + iY = re^{i\varphi} + \frac{R^2}{r}e^{-i\varphi} = \left(r + \frac{R^2}{r}\right)\cos\varphi + i\left(r - \frac{R^2}{r}\right)\sin\varphi.$$

By eliminating the angle φ between the equations for X and Y, we obtain:

$$\frac{X^2}{\left(r + \frac{R^2}{r}\right)^2} + \frac{Y^2}{\left(r - \frac{R^2}{r}\right)^2} = 1,$$

which is the equation of an ellipse for which the foci P_1 et P_2 are located on the X-axis, at the points $X = \pm\, 2R$ (Figure 6.22).

When $r = R$ (the radius of the circle equals the parameter of the Joukowski transformation), we have:

$$Z = 2\,R\cos\,\varphi.$$

The ellipse then becomes the straight line segment (Σ), bounded by the points with coordinates $Y = 0$ and $X = \pm\, 2R$. In this last case, when an object point makes one turn

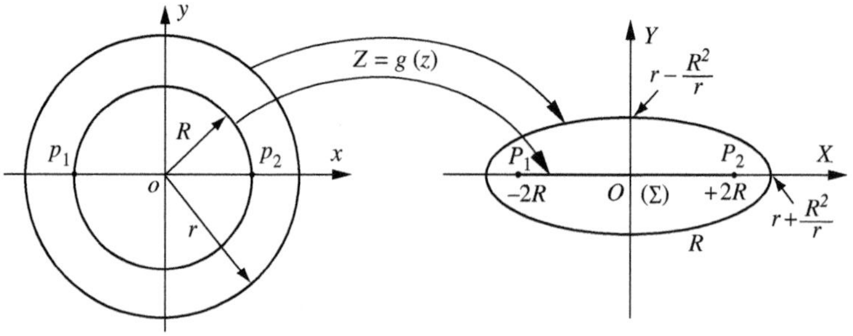

Figure 6.22 *Mapping of a circle under the Joukowski transformation. The resulting ellipse collapses into a straight-line segment (Σ) [−2R, +2R] when the radius of the circle being mapped equals the parameter R in the equation for the transformation*

around the circle, its image under the transformation covers the segment in both directions; first it covers the "upper" part of the line segment (for θ varying between 0 and π), and next it covers the "lower" part (for θ varying between π and 2π). We observe that angles at the points P_1 and P_2 which are the transforms of the points p_1 and p_2 located at $z = \pm R$, are not conserved. The derivative $g'(z) = 1 - R^2/z^2$ of the function $g(z)$ which defines the conformal mapping, vanishes indeed at points p_1 and p_2.

(ii) ***Inverse transform of a flow parallel to a segment of a flat plane***

Let us now consider in the *image* (*X*, *Y*) plane, a uniform flow parallel to a plane (II), which is traced in the plane of the figure precisely by the segment (Σ) discussed above (Figure 6.23). The complex potential for this flow is:

$$F(Z) = U\,Z.$$

In the *object* (*x*,*y*)-plane, we have a corresponding flow around a circle (C) of radius R centered at the origin, with uniform velocity U far from the circle (Figure 6.23b). The corresponding complex potential can be written directly:

$$f(z) = F\,[g(z)] = U(z + R^2/z). \tag{6.101}$$

Taking the real part of $f(z)$, we recover the velocity potential Φ, already derived in Section 6.2.4.

In the neighborhood of the stagnation point p_1 of the circle (C), we have locally the same kind of flow as one normal to a plane (Figure 6.20b). In the transformation, this corresponds to flow at the leading edge of the plate (Σ) and parallel to it (Figure 6.20f).

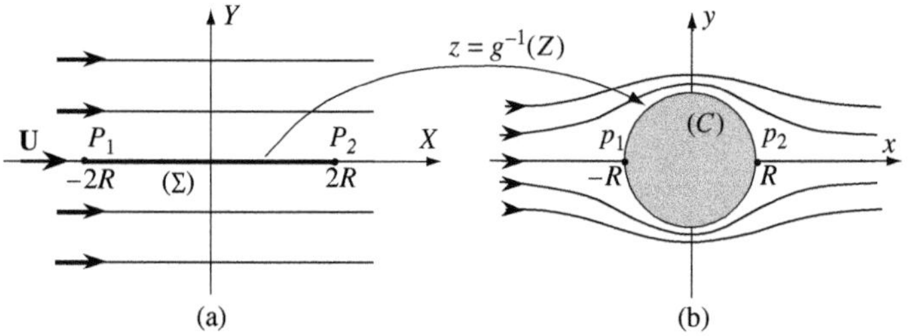

Figure 6.23 *(a) Flow around a plate at zero angle of incidence. (b) Its image under the Joukowski transformation (Equation 6.100)*

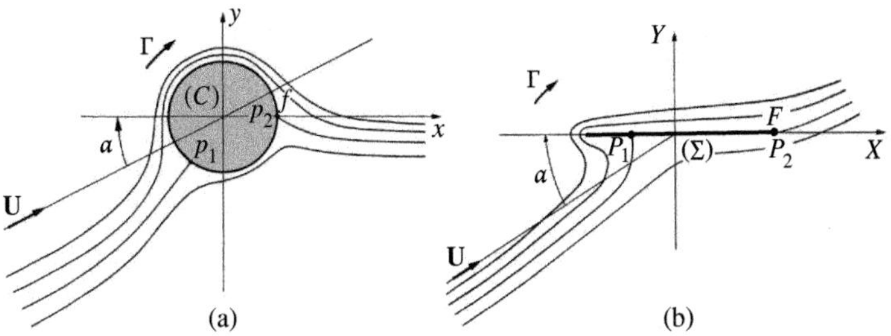

Figure 6.24 *(a) Flow around a circular cylinder at an angle of incidence α and with circulation Γ. (b) mapping of this flow by means of the Joukowski transformation*

(iii) ***Complex potential for a flow incident at an angle onto a plate segment***

Consider now, in the (X,Y) plane, the flow around a plate segment (Σ) when the direction of (Σ) is at an angle of incidence α to the direction of the velocity U, and where, moreover, there is a circulation Γ of the velocity around (Σ). This problem is a simplified model of the flow around a wing. We begin by deriving the complex potential in the (x,y)-plane. This is easily done by rotating the axes through an angle α (Figure 6.24a): z is then mapped into the point $z_1 = z\,\mathrm{e}^{-\mathrm{i}\alpha}$ in the potential described by Equation 6.101. We then add to this the complex potential corresponding to flow with circulation Γ around the cylinder (Equation 6.92). This leads to:

$$f_2(z) = U\left(z\,e^{-\mathrm{i}\alpha} + \frac{R^2}{z\,\mathrm{e}^{-\mathrm{i}\alpha}}\right) - \mathrm{i}\frac{\Gamma}{2\pi}\mathrm{Log}\left(\frac{z\,\mathrm{e}^{-\mathrm{i}\alpha}}{R}\right). \tag{6.102}$$

Conformal mapping preserves the angles at all points where the derivative does not vanish. The angle of incidence in the (X,Y) plane is then also equal to α. It can also be shown that the conformal mapping also preserves the circulation. The complex potential:

$$F_2(Z) = f_2\left(g^{-1}(z)\right)$$

obtained from Equation 6.102 by the transform inverse to $g(z)$, then represents the potential for the flow around the element of the plate (Σ) (Figure 6.24b).

As a matter of fact, the transform inverse to $g(z)$ is not an analytic function and, consequently, the corresponding potential F_2 cannot be written directly. However, we are here only interested in the complex velocity which is easily determined.

(iv) ***Complex velocity at the solid plane***

We have:

$$W(z) = \frac{\mathrm{d}F_2}{\mathrm{d}Z} = \frac{\mathrm{d}f_2}{\mathrm{d}z}\frac{\mathrm{d}z}{\mathrm{d}Z} = \frac{\mathrm{d}f_2}{\mathrm{d}z}\frac{1}{g'(z)}, \tag{6.103a}$$

i.e.,:

$$W(z) = \frac{z^2}{z^2 - R^2}\left(U\left(\mathrm{e}^{-\mathrm{i}\alpha} - \frac{R^2}{z^2\mathrm{e}^{-\mathrm{i}\alpha}}\right) - \mathrm{i}\frac{\Gamma}{2\pi\,z}\right). \tag{6.103b}$$

We shall confine our analysis to the calculation of the complex velocity at the surface of the plate segment (Σ). Since this is the image of a circle (C) we need only substitute $z = R\,\mathrm{e}^{\mathrm{i}\varphi}$ in the above equation, so that:

$$W(z) = \frac{1}{1-e^{-2i\varphi}}\left(U\left(e^{-i\alpha} - e^{-i(\alpha-2\varphi)}\right) - i\frac{\Gamma}{2\pi R}e^{-i\varphi}\right).$$

We note that the velocity W at P_1 in the *image plane* has been expressed as a function of the radius R of the circle and of the polar angle φ in the *original plane* of Figure 6.24a. We should recall that $Z = X + iY$ and z are related by the mapping formula (6.100). On the circle $r = R$, the relationship is $X = 2R\cos\varphi$ and $Y = 0$. Multiplying both the numerator and the denominator by $e^{i\varphi/2}$, we obtain for the variation of W along the surface of the plate (Σ):

$$W(\varphi) = \frac{U\sin(\varphi-\alpha) - \Gamma/(4\pi R)}{\sin\varphi} \tag{6.104}$$

As it might have been predicted, the expression we have obtained is real, since the component of the velocity normal to the plane segment must vanish. We see in Figures 6.24a and 6.24b, respectively, the outline of the streamlines in the (x,y)- and (X,Y)-planes. We should pay particular attention to the two stagnation points P_1 and P_2 on (Σ). These two points correspond to values of φ such that:

$$\sin(\theta-\alpha) = \frac{\Gamma}{4\pi RU}$$

which are located asymmetrically in the presence of a circulation. These points are the places where $W = 0$, and exist only if the magnitude of the circulation Γ is smaller than $4\pi RU$.

(v) ***The Kutta condition***

We now seek the particular value of the circulation which results in the point P_2, the image of the stagnation point p_2, being located precisely at the sharp, trailing (downstream) edge F of the plate, as illustrated in Figure 6.24b. It can be shown that this corresponds to the stable configuration of flow around the cross-section of a wing at a finite angle of incidence (angle of attack) α: the value of the circulation around the wing adjusts itself to that necessary to satisfy this condition (paradoxically, this is the condition, for flow around a wing of a *viscous* fluid, that allows it to be treated as though it were non-viscous). The point p_2 then coincides with the point of the contour (C) whose image is the trailing edge F in the $X-Y$ plane. This point corresponds to $\varphi = 0$, and we must therefore have:

$$W(\varphi = 0) = 0, \qquad \text{i.e.,:} \qquad 0 = -U\sin\alpha - \Gamma/4\pi R.$$

Whence:

$$\Gamma = -4\pi R U \sin\alpha. \tag{6.105}$$

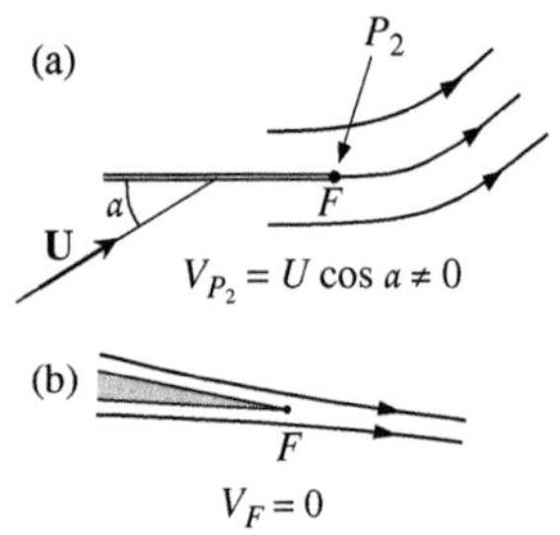

Figure 6.25 *Comparison of the velocities at the trailing edge at the corner of an angle of intersection. (a) Angle* 0 *and (b) finite angle*

The above equation constitutes the *Kutta condition.* We should nonetheless observe that at the point P_2 (the image of the stagnation point p_2), the velocity is not zero since the denominator of the complex velocity W also vanishes (Figure 6.25a). The trailing edge is a singular point of the transformation; we find the velocity at that point by taking into account Equation 6.105, and expanding Equation 6.104 in the neighborhood of $\varphi = 0$:

$$W(\varphi = 0) = U\cos\alpha = V_{p_2} = V_F.$$

This last result is only valid if the angle between the upper and lower bearing surfaces is zero, as shown in Figures 6.24 and 6.25a. On the other hand, for real profiles, this angle is not zero, as shown in Figure 6.25b. In this case, the velocity at the point F becomes zero: we have, in fact, seen in Section 6.6.2 (iv), by analyzing flows in the neighborhood of a dihedral corner, that the velocity at the corner vanishes if the angle at the corner is smaller than 180° (as is the case in Figures 6.20a, 6.20b and 6.20c).

(vi) ***Estimate of the lift force on a wing***

The expression obtained above for the circulation allows us to evaluate the lift force on an airplane wing. It can, in fact, be expressed, according to Equation 6.44, in the form:

$$\mathbf{F}_L = \rho \mathbf{U} \times \mathbf{\Gamma}. \qquad (6.105)$$

Taking into account Equation 6.105, we find, for the order of magnitude of the lift force per unit length in the direction normal to the plane of the flow:

$$F_L = \pi \rho U^2 (4\,R) \sin \alpha$$

If we call L the wing-span of the wing and $\ell = 4R$, its width, known as the *chord*, we finally obtain, for the lift force on the entire wing:

$$F_L = \pi \rho U^2 (L\,\ell) \sin \alpha \qquad (6.106)$$

The orders of magnitude for three different types of airplanes are as follows:

Aircraft type	**Takeoff speed U**	**Chord ℓ**	**Wingspan**	**Angle of attack (degrees)**	**Lift force F_L**	**Gross takeoff weight**
Boeing 747 airliner	300 km/h	9 m	60 m	10°	3×10^7 N (6.7×10^6 lb-f)	300,000 kg 660,000 lb
Cessna private aircraft	100 km/h	1.70 m	9 m	13°	10^4 N (2,200 lb-f)	900 kg 2,000 lb
Mirage F1 fighter airplane	350 km/h	5 m	9 m	20°	6×10^5 N (130,000 lb-f)	16,000 kg 35,000 lb

We note that an airliner takes off thanks to its large wing area, but with a small attack angle, for reasons of passenger comfort. The small private plane requires much smaller lift, because of its small weight. Finally, the fighter plane takes off at very high angles of attack and at high velocity, to compensate for its much smaller wing area. We will come back in more general terms to the idea of lift in Section 7.5.2.

The finite lifting plate segment we have just discussed represents only a crude approximation to the cross-section of a real airplane wing. It illustrates, however, the major physical principles behind the mechanisms which generate lift, by the combined effect of flow at an angle of attack, and of the presence of circulation of the velocity around the section. Moreover, if the circle to which we apply the Joukowski transformation (Figure 6.23) is gradually shifted so that its center is no longer at the origin, we obtain images, known as *Joukowski airfoils*, which bear a much closer resemblance to the cross-sections of real wings.

In closing, we remark that these two-dimensional models of flow around a wing neglect important effects related to its three-dimensional nature. Specifically, in real situations, there appears a vortex structure which trails off from the wing tips (see Section 7.5.2).

6A Appendix: Velocity potentials and stream functions

6A.1 Velocity potentials and stream functions for two-dimensional flows

Type of flow	Velocity potential	Stream function
uniform flow in two dimensions	$\Phi = Ux$	$\Psi = Uy$
uniform flow in three dimensions at velocity U (cylindrical coordinates)	$\Phi = Uz$	$\Psi = (-Ur^2)/2$
uniform flow in three dimensions at velocity U (spherical polar coordinates)	$\Phi = Ur\cos\theta$	$\Psi = (Ur^2\sin^2\theta)/2$
Vortex (cylindrical coordinates)	$\Phi = \dfrac{\Gamma\varphi}{2\pi}$	$\Psi = -\dfrac{\Gamma}{2\pi}\mathrm{Log}\dfrac{r}{r_0}$
2-D point source (cylindrical coordinates)	$\Phi = \dfrac{Q}{2\pi}\mathrm{Log}\left(\dfrac{r}{r_0}\right)$	$\Psi = \dfrac{Q}{2\pi}\varphi$
3-D point source (spherical polar coordinates)	$\Phi = -\dfrac{Q}{4\pi\,\mathrm{r}}$	$\Psi = -\dfrac{Q}{4\pi}\cos\theta$
2-D dipole flow (cylindrical coordinates)	$\Phi = -\dfrac{\mathbf{p}\cdot\mathbf{r}}{2\pi\, r^2}$	$\Psi = -\dfrac{p}{2\pi}\dfrac{\sin\varphi}{r}$
3-D dipole flow (spherical polar coordinates)	$\Phi = -\dfrac{\mathbf{p}\cdot\mathbf{r}}{4\pi\, r^3}$	$\Psi = \dfrac{p\sin^2\theta}{4\pi\, r}$
Flow around a circular cylinder (cylindrical coordinates)	$\Phi = U\,r\cos\varphi\left(1+\dfrac{R^2}{r^2}\right)$	$\Psi = U\,r\sin\varphi\left(1-\dfrac{R^2}{r^2}\right)$
Flow around a sphere (spherical polar coordinates)	$\Phi = U\,r\left(1+\dfrac{R^3}{2r^3}\right)\cos\theta$	$\Psi = \dfrac{U}{2}\left(r^2-\dfrac{R^3}{r}\right)\sin^2\theta$
Corner with (dihedral) angle $\alpha = \pi/(m+1)$ (cylindrical coordinates)	$\Phi = C\,r^{m+1}\cos(m+1)\varphi$	$\Psi = C\,r^{m+1}\sin(m+1)\varphi$

6A.2 Derivation of the velocity components from the stream function

2-D flow (Cartesian coordinates)	$v_x = \dfrac{\partial \Psi}{\partial y}$	$v_y = -\dfrac{\partial \Psi}{\partial x}$
2-D flow (polar coordinates)	$v_r = \dfrac{1}{r}\dfrac{\partial \Psi}{\partial \varphi}$	$v_\varphi = -\dfrac{\partial \Psi}{\partial r}$
Axially symmetric flow (cylindrical coordinates)	$v_r = \dfrac{1}{r}\dfrac{\partial \Psi}{\partial z}$	$v_z = -\dfrac{1}{r}\dfrac{\partial \Psi}{\partial r}$
Axially symmetric flow (spherical polar coordinates)	$v_r = \dfrac{1}{r^2 \sin\theta}\dfrac{\partial \Psi}{\partial \theta}$	$v_\theta = -\dfrac{1}{r\sin\theta}\dfrac{\partial \Psi}{\partial r}$

6A.3 Derivation of the velocity components from the velocity potential function

The equations displayed in this table are merely explicit representations of the general relationship $\mathbf{v} = \nabla\Phi$ in the respective Cartesian, polar, cylindrical and spherical polar coordinates.

2-D flow (Cartesian coordinates)	$v_x = \dfrac{\partial \Phi}{\partial x}$	$v_y = \dfrac{\partial \Phi}{\partial y}$	
2-D flow (polar coordinates)	$v_r = \dfrac{\partial \Phi}{\partial r}$	$v_\varphi = \dfrac{1}{r}\dfrac{\partial \Phi}{\partial \varphi}$	
3-D flow (Cartesian coordinates)	$v_x = \dfrac{\partial \Phi}{\partial x}$	$v_y = \dfrac{\partial \Phi}{\partial y}$	$v_z = \dfrac{\partial \Phi}{\partial z}$
3-D flow (cylindrical coordinates)	$v_r = \dfrac{\partial \Phi}{\partial r}$	$v_\varphi = \dfrac{1}{r}\dfrac{\partial \Phi}{\partial \varphi}$	$v_z = \dfrac{\partial \Phi}{\partial z}$
3-D flow (spherical polar coordinates)	$v_r = \dfrac{\partial \Phi}{\partial r}$	$v_\theta = \dfrac{1}{r}\dfrac{\partial \Phi}{\partial \theta}$	$v_\varphi = \dfrac{1}{r\,\sin\theta}\dfrac{\partial \Phi}{\partial \varphi}$

EXERCISES

1) **The method of images: line source in front of a solid plane surface**
Consider an infinite line parallel to the z-axis with $x = 0$ and $y = a$. A perfect fluid is emitted radially at a flow rate q per unit length of the line. In an infinite fluid, the flow field is potential and invariant along z with $v_z = 0$. It corresponds in a plane z = constant to the two-dimensional flow from a point source (known as a *line source*).

An infinite solid wall is put in the plane $y = 0$. Show that the flow remains the same in the half-space $y > 0$ if we replace this wall by a second parallel line source. Using Equation 6.92, which corresponds to a two-dimensional point source in an infinite fluid, show that the corresponding complex potential is $q/2\pi \,\mathrm{Log}\,(z^2 + a^2)$. Compute the flow velocity on the plane surface and show that the deviation of the pressure from its value far upstream (under zero gravity) is: $\delta p(x,\ 0) = -\dfrac{\rho}{2}\left(\dfrac{q}{\pi a}\right)^2 \dfrac{x^2/a^2}{\left(1 + x^2/a^2\right)^2}$. Compute then the corresponding force F_p on the plane per unit distance along z (using the result: $\displaystyle\int_{u_1}^{u_2} \frac{u^2}{\left(u^2+1\right)^2}\,du = \left[\frac{1}{2}\left(\mathrm{atan}(u) - \frac{u}{u^2+1}\right)\right]_{u_1}^{u_2}$). In what direction is the force on the plane oriented? Give a physical interpretation of this apparently counter-intuitive result. Does the direction of the force change if q is replaced by $-q$?

2) **Force on a half-cylinder placed in a uniform flow**

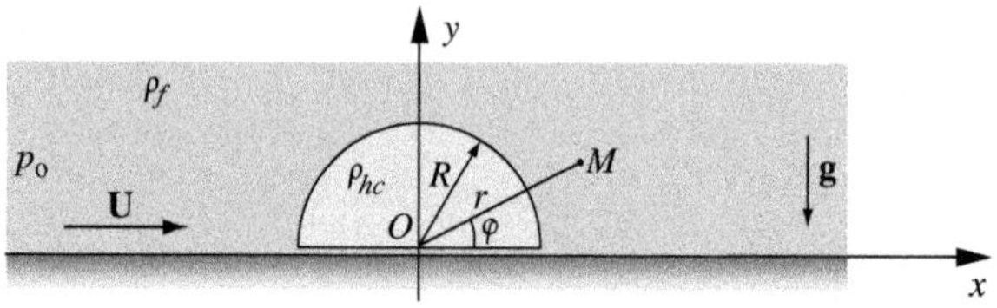

A half-cylinder of density ρ_{hc} and radius R is placed upon a solid plane $y = 0$ at the bottom of a large, deep bath of a perfect fluid of density $\rho_f < \rho_{hc}$. The axis of the cylinder is parallel to the z-axis, and the flow is potential and two-dimensional in the (x, y)-plane with a velocity U parallel to x at large distances from the obstacle, where the pressure has a constant value p_0. Using the results of Section 6.2.4(i), show that the pressure distribution $p(\varphi)$ at the surface of the cylinder is: $p(\varphi) = p_0 + \rho_f U^2 (2\cos^2\varphi - 3/2)$. Compute the total pressure force per unit length of the cylinder (the pressure in the thin space between the cylinder and the plane is assumed to be equal to that on the surface of the cylinder at $\varphi = 0$ and $\varphi = \pi$). What is the velocity U_M above which the cylinder does not remain at the bottom of the bath?

3) **Gravity waves at the interface between fluids of different densities**

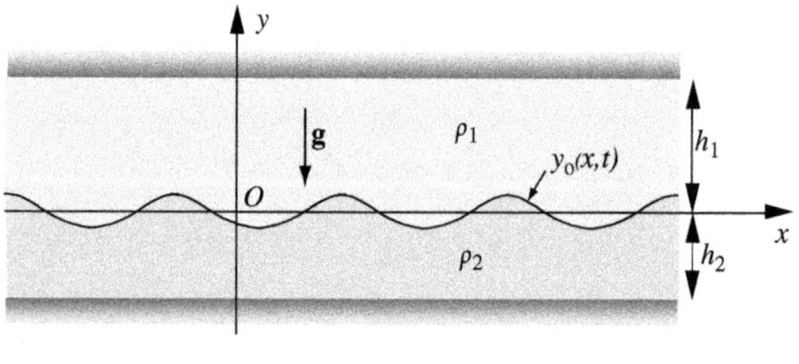

Two layers with thicknesses h_1 and h_2 of ideal fluids of different densities ρ_1 and ρ_2 fill the space between two horizontal solid walls $y = h_1$ and $y = -h_2$. We neglect the influence of surface tension and study the gravity waves distorting the interface between the fluids initially located at $y = 0$ and represented by a function $y_0(x, t)$. We follow the same approach as in Section 6.4.1 and assume that the velocities induced by the

waves in each fluid correspond to potential functions of the type $\Phi_j = f_j(x, t)\, g_j(y)$ with $j = 1, 2$. What are the boundary conditions which need to be satisfied at the walls and at the interface (in particular that resulting from the Bernoulli equations). Assuming that $f_j(x, t) = A_j\, e^{i(kx-\omega t)}$, what is the equation satisfied by $g_j(y)$ in order to comply with Laplace's equation, and what are its solutions satisfying the boundary conditions at the walls? Using these latter results, write Bernoulli's Equation and the relation expressing the continuity of the vertical velocity at the interface. Show then that: $\omega^2(\rho_1 \coth kh_1 + \rho_2 \coth kh_2) = (\rho_2 - \rho_1)gk$. Write the variation of the phase velocity c of the wave as a function of k in the two limiting cases $kh_2, kh_1 \gg 1$ and $kh_2, kh_1 \ll 1$.
Compare these results to those obtained for one liquid with a free surface.

4) **Reflection of a gravity wave due to a variation of depth**

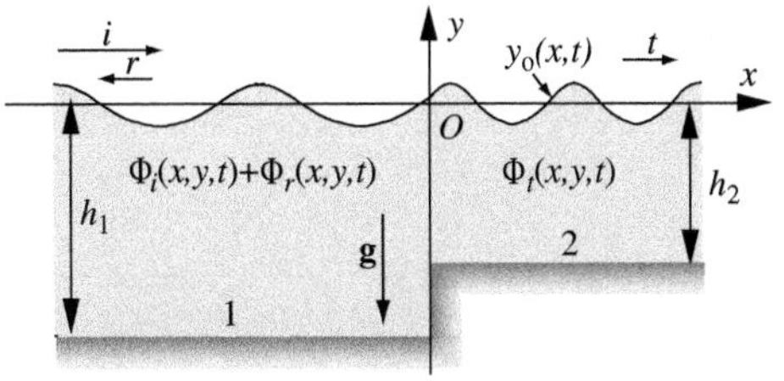

A plane gravity wave propagates in the direction $x > 0$ at the free surface of a layer of perfect fluid with zero surface tension, and, at $x = 0$, reaches a point at which the depth decreases from h_1 to h_2. The two layers have an infinite length in the x-direction. We assume that the incident (i) wave generates both a reflected (r) and a transmitted (t) wave at the step: the velocity potentials are respectively Φ_i, Φ_r and Φ_t and are assumed (as in the previous exercise) to be of the form: $\Phi_j = f_j(x, t)\, g_j(y) = A_j\, e^{i(\pm kx-\omega t)}$ $\cosh k_j(y+h_j)$ (the index j corresponds to (i), (t) or (r)). As in the previous exercise, we use the approach described in Section 6.4.1. What are the phase velocities of the three different waves in the shallow water limit $kh \ll 1$? What are the relations between the different potentials at the step ($x = 0$)? Write the corresponding equations and take their limit when $kh \ll 1$ (shallow water limit). Obtain then the values of A_t and A_r as a function of A_i.

7 Vorticity, Vortex Dynamics and Rotating Flows

In the discussion of deformations of an element of fluid in Section 3.2, we have seen the appearance of a term involving the local rotation, i.e. the antisymmetric part of the velocity gradient tensor. The corresponding local rotation vector equals half the vorticity, defined as the curl of the velocity field. Thus, vorticity is the tool which enables us to characterize the local rotation within a fluid. In some cases, it can be localized in space, as in vortices; in others, it is continuously distributed, as in a fluid in uniform rotation.

We begin this chapter by reviewing the definition of vorticity, illustrating it by using the example of vortex lines. This will allow us to examine in detail the mathematical parallel between the velocity and a magnetic field, and between the vorticity and the electric current which generates that field (Section 7.1). We then discuss the transport of vorticity in an ideal fluid: on the one hand, from the point of view of the dynamics of the circulation (Kelvin's theorem, Section 7.2), and on the other hand, by deriving directly an equation for the evolution of the vorticity (Section 7.3). Kelvin's theorem allows us to justify, a posteriori, the discussion of potential flows that we have carried out in Chapter 6. We demonstrate that, for an ideal fluid under a few specific restrictions, flow which was initially potential remains potential at subsequent times. We describe the dynamics of a system of vortices, corresponding to flows within which all the vorticity is concentrated along singular lines (Section 7.4). We then consider, in Section 7.5, propulsion mechanisms in water and air involving vorticity. Section 7.6 studies rotation flows where the effects of solid body rotation are superimposed on the vorticity existing in the rotating reference frame. Such a discussion plays an important part in our understanding of atmospheric and oceanographic phenomena. Finally, Section 7.7 discusses secondary flows which are frequent near walls and/or when centrifugal forces are present.

7.1 Vorticity: its definition, and an example of straight vortex filaments

7.1.1 The concept of vorticity

Vorticity and the velocity field

In Section 3.2.3, we have defined the *vorticity* pseudovector $\boldsymbol{\omega}(\mathbf{r})$ at a point $\mathbf{r}$, by:

$$\boldsymbol{\omega}(\mathbf{r}) = \nabla \times \mathbf{v}(\mathbf{r}), \tag{7.1}$$

Physical Hydrodynamics. Second Edition. Etienne Guyon *et al.*

where $\mathbf{v(r)}$ is the velocity field of the flow. We have shown that $\boldsymbol{\omega}$ is equal to twice the local rotation vector $\boldsymbol{\Omega}$ of the fluid (in some outdated references, this latter vector $\boldsymbol{\Omega}$ is called *vorticity vector* instead of rotation vector).

Just as in the case of streamlines, we define *vorticity lines* which are tangent at every point to the vorticity pseudovector. Similarly, *vorticity tubes* represent that portion of space bounded by a surface made up of vorticity lines anchored on a closed space curve.

The components ω_k of the vorticity $\boldsymbol{\omega}(\mathbf{r})$ are related to the anti-symmetric part $\omega_{ij} = 1/2(\partial v_i/\partial x_j - \partial v_j/\partial x_i)$ of the velocity gradient tensor $G_{ij} = \partial v_i/\partial x_j$ by $\omega_k = -\varepsilon_{ijk}\omega_{ij}$; ε_{ijk} has value zero if any two indices are equal, +1 if the permutation $i \to j \to k$ is cyclic, and -1 if it is inverted.

Vorticity for various types of flow

We find vorticity at any time that the flow is not ideal and potential and, therefore, for viscous fluids. It plays a particularly important role in turbulent flows, which can often be considered as a superposition of an average translational movement and local rotational motions, on a set of highly variable scales. In such flows, vorticity is distributed throughout the volume of the fluid.

In other circumstances, however, the flow might be almost everywhere potential, except for a line (or *core*) of radius ξ, small compared to the overall scale of the flow: all the vorticity is then localized in this core, around which the fluid rotates. We encounter such vortex flows in tornadoes and hurricanes (Figure 7.1a) and in atmospheric waterspouts (Figure 7.1b), in the whirlpools formed when a bathroom sink or tub empties or, on a completely different scale, in the vortices in superfluid helium ($\xi \approx 10^{-10}$m), which we will study in the appendix to this chapter.

In other instances, the vorticity is less clearly localized, but rotational flow structures remain easily identifiable. This is the case for shear layers between fluids moving at different velocities, which we discuss in Section 7.4.2. Finally, structures of this type are clearly visible on a scale of several hundreds of kilometers on satellite photos of hurricanes or cyclones, or in the atmosphere of Jupiter on scales of several tens of thousands of kilometers. In Sections 2.4.2 and 2.4.3, we also came across the periodic formation of vortex-type structures in the wakes behind an obstacle.

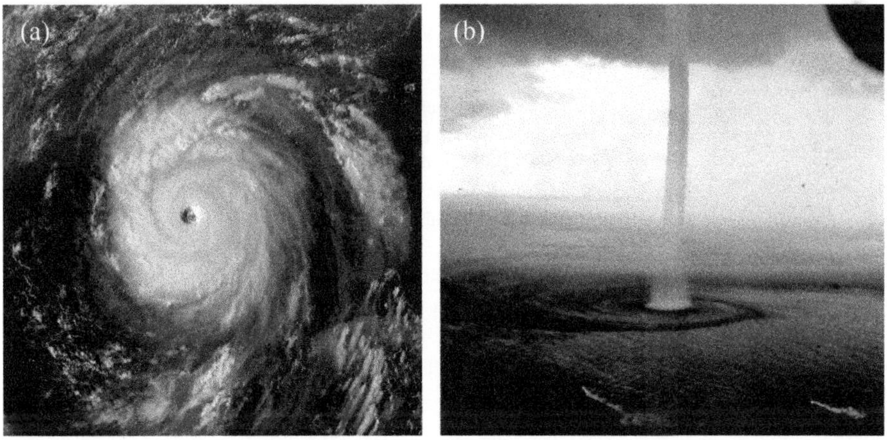

Figure 7.1C *(a) View, from a satellite above the Gulf of Mexico, of Hurricane Katrina, which devastated New Orleans in August 2005 (document NOAA). (b) Waterspout on the ocean offshore from the southern Florida islands. Note the spiral secondary flow at the surface of the water (document V. Golden, NOAA)*

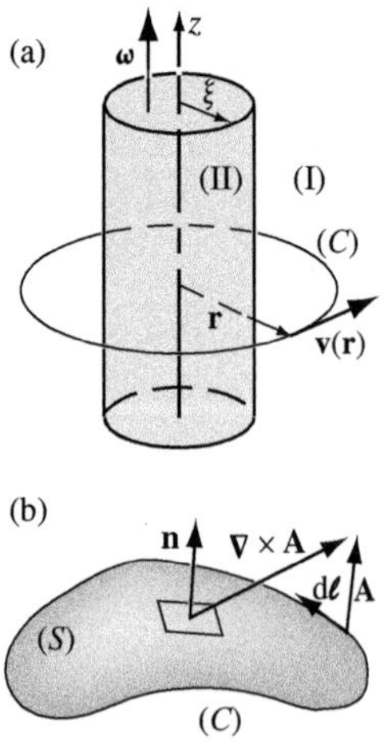

Figure 7.2 *(a) Velocity field generated by a linear vorticity tube. (b) Calculation of the circulation of a vector field* **A** *along a curve (C), from the flux of its curl through a surface (S) anchored on the curve*

7.1.2 A simple model of a line vortex: the Rankine vortex

The Rankine vortex is a model representation of a linear filament where the vorticity is concentrated in a small diameter region, the *core* of the vortex (Figure 7.2a): in this model, we assume that there is a uniform distribution of vorticity $\omega_z = \omega_z^0$ localized within a circular cylinder of radius ξ and infinitely long in the direction z (region II). We assume that the other components ω_x and ω_y vanish everywhere, that the flow field has a rotational symmetry of axis z and that it is invariant along z. For $r > \xi$ (region I), $\omega_z = 0$.

Velocity field of a Rankine vortex

We calculate the corresponding velocity field by using *Stokes' theorem*, which states that the circulation of a vector field **A** along a closed contour (C) is equal to the flux of the curl of **A** through a surface (S) anchored on this contour (Figure 7.2b):

$$\int_C \mathbf{A} \cdot \mathrm{d}\mathbf{l} = \iint_S (\nabla \times \mathbf{A}) \cdot \mathbf{n}\, \mathrm{d}S. \qquad (7.2)$$

Region I ($r > \xi$): In order to calculate the flow velocity component v_φ orthogonal to the radial vector **r**, let us first apply Equation 7.2 along a circle of radius $r > \xi$ centered on the axis of the tube and in a plane perpendicular to the z-axis. Assuming that the flow is azimuthal along the axis and, as mentioned above, independent of z, and according to the assumptions made about ω_z, we find that the circulation Γ of the velocity vector around the circle has the value:

$$\Gamma = 2\,\pi r\, v_\varphi(r) = \int_0^\xi 2\pi\, \omega_z(r)\, r\ \mathrm{d}r = \omega_z^0\, \pi\, \xi^2. \qquad (7.3)$$

The circulation Γ is therefore independent of r and equal to $\omega_z^0\, \pi\, \xi^2$ so long as r is greater than ξ: it characterizes the strength of the vortex. The component v_φ varies thus as $1/r$ with:

$$v_\varphi(r) = \frac{\Gamma}{2\pi r}. \qquad (7.4)$$

The decrease of v_φ as $1/r$ is characteristic of any azimuthal field with zero curl. Let us assume that only the v_φ component does not vanish, and that it depends only on r. The condition $\nabla \times \mathbf{v} = 0$ becomes:

$$(\nabla \times \mathbf{v})_z = \frac{1}{r}\frac{\partial}{\partial r}(r\, v_\varphi) = 0,$$

so that v_φ is proportional to $1/r$.

More generally, the circulation has the same value for every arbitrary closed contour which goes once around the cylinder. We can check this by applying Stokes' theorem to any such contour. These results remain valid in the limit when the vortex core shrinks down to a line with an infinite density of vorticity: provided that the integral of this singular distribution remains finite, it will again give the value of the circulation of the velocity around the core of the vortex.

Region II ($r < \xi$): the circulation of the velocity has the value:

$$2\pi\ r\ v_\varphi(r) = \int_0^r 2\pi\ \omega_z r\ \mathrm{d}r\ =\ \omega_z^0\ \pi r^2,$$
$$v_\varphi(r) = \frac{\omega_z^0 r}{2} = \frac{\Gamma\, r}{2\pi\, \xi^2}. \qquad (7.5)$$

We thus have a solid body rotation in the core of the vortex with $v_\varphi(r) \propto r$. We observe that the angular velocity of rotation equals $\omega_z^0/2$.

Physically, the flow in region I thus appears as a potential flow in a multiply connected geometry, which we have already seen in Section 6.2; this flow is superimposed on a solid body rotation in the *core* on the axis of the vortex (region II), which then plays the role of the obstacle around which the circulation of the fluid appears.

In the flow of a viscous fluid, the effect of viscosity increases as r decreases and as the velocity gradient $\partial v_\varphi/\partial r$ increases. The model of the Rankine vortex assumes that, below a radius ξ, which is the boundary of the vortex core, these effects are such that only solid-body rotation of angular velocity Ω is possible: in this region, $v_\varphi = \Omega\, r$.

Pressure field within a Rankine vortex

- **Region I** ($r > \xi$). In the curl-free region of the velocity field of the vortex, the variation of the pressure p is obtained by applying Bernoulli's equation (Equation 5.36). The pressure p decreases as we approach the core of the vortex in such a way that (if we neglect gravitational effects):

$$p(r) + \frac{1}{2}\rho\, v_\varphi^2 = p(r) + \frac{1}{2}\rho \frac{\Gamma^2}{4\pi^2 r^2} = p_\infty, \tag{7.6}$$

  in which p_∞ is the limiting value of the pressure $p(r)$ as $r \to \infty$. If $p(r)$ decreases sufficiently as r decreases, we can have boiling of the rotating fluid. We have discussed this *cavitation* phenomenon in Section 5.3.2. A further consequence of this decrease in pressure in the central region is that a solid object of finite size placed close to the axis of a vortex line is subjected to a radial pressure directed toward the center of the core; in fact, according to Equation 7.6, the pressure is higher on the side of the object which is farther from the vortex core. Thus solid particles, or gas bubbles collected by a vortex filament, can be swept into the central region and write (this observation has been used to visualize vortex structures). In the same way, vortex lines can be trapped by the asperities on a solid wall.

- **Region II**: ($r < \xi$). In the solid body rotation region, the liquid is at rest in the reference frame which rotates with angular velocity $\omega_z^0/2$. We will derive in Section 7.6.1 the equations of motion of a fluid in a rotating reference frame. Still neglecting gravity, we assume here that the radial pressure gradient compensates the centrifugal force associated to the rotation, so that:

$$\frac{\partial p}{\partial r} = \rho\left(\frac{\omega_z^0}{2}\right)^2 r = \rho\left(\frac{\Gamma}{2\pi\,\xi^2}\right)^2 r. \tag{7.7}$$

The pressure increases quadratically with r, and, for $r < \xi$:

$$p(r) = p(0) + \rho\left(\frac{\omega_z^0}{2}\right)^2 \frac{r^2}{2} = p(0) + \rho\left(\frac{\Gamma}{2\pi\,\xi^2}\right)^2 \frac{r^2}{2}. \tag{7.8}$$

By requiring that Equations 7.6 and 7.8 be simultaneously satisfied at $r = \xi$, we can calculate $p(0)$ as a function of the pressure p_∞ at large distances.

Kinetic energy per unit length of a Rankine vortex

The kinetic energy e_k of a Rankine vortex has the value, per unit length:

$$e_k = \rho\frac{\Gamma^2}{4\pi}\left(\mathrm{Log}\frac{L}{\xi} + \frac{1}{4}\right). \tag{7.9}$$

Proof

Let us consider, first of all, the case where the radius of the core is zero. We then obtain:

$$e_k = \frac{1}{2}\iint \rho\, v_\varphi^2\, dS = \frac{\rho}{2}\int \left(\frac{\Gamma}{2\pi r}\right)^2 2\pi r\, dr. \quad (7.10)$$

This integral of the form $\int dr/r = \int d(\mathrm{Log}\, r)$ diverges, as r approaches both infinity and zero. The first divergence disappears if we limit r to the size L of the container ($L = R$ in the previous example). The second divergence disappears when the radius ξ of the core is finite. The additional contribution of the core to the kinetic energy is finite and equal to $\mathcal{J}_\xi\, \Omega_\xi^2/2$, where $\mathcal{J}_\xi = \pi\,\rho\,\xi^4/2$ is its moment of inertia, and $\Omega_\xi = |\boldsymbol{\omega}(\xi)|/2 = \Gamma/(2\,\pi\,\xi^2)$, the angular velocity of its rotation. We then obtain Equation 7.9 for the kinetic energy by adding this contribution to the kinetic energy of the external part.

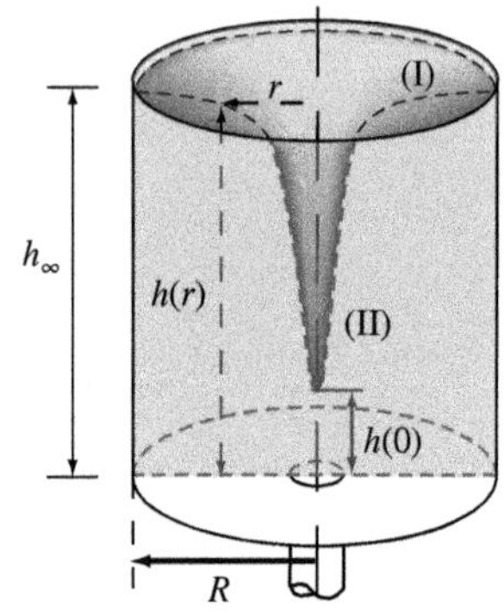

(a)

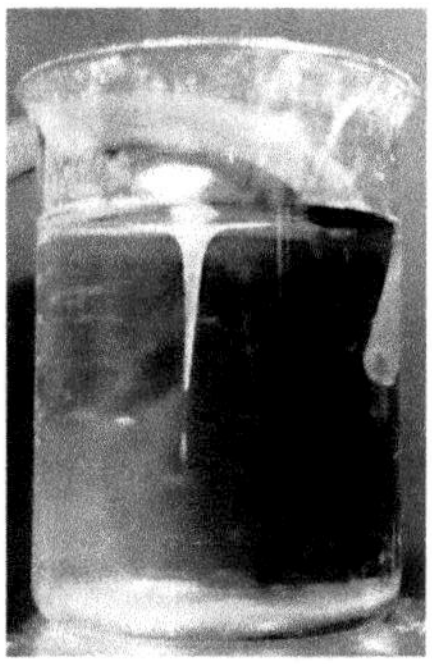

(b)

Figure 7.3 *(a) Vortex generated by the emptying of a container through a circular orifice. (b) Experimental set-up: continuous filling through the tube shown, allows the vortex to be examined under stationary conditions*

An example of a line vortex: the Rankine vortex at a free surface

As an experimental model for the *Rankine vortex*, we can use a cylindrical container where water is maintained at a constant level: water is injected tangentially near the outer walls at a flow rate Q while simultaneously emptying through a central, circular opening at the bottom. The observation of the free surface (Figure 7.3) allows one to distinguish an external, convex region (I) merging into an interior, parabolic one (II). In order to explain this observation, we represent the flow as a Rankine vortex where region (I) corresponds to a curl-free flow and region (II) is in solid-body rotation. The preceding derivations thus remain fundamentally valid: we will need, however, to take into account the hydrostatic pressure-gradient terms which we have so far neglected. Their equilibrium with the terms involving the centrifugal force determines the shape of the free surface.

Region (I) ($r \geq \xi$)

The free surface has a typically hyperbolic profile obeying the equation:

$$h(r) = -\frac{\Gamma^2}{8\pi^2 g\, r^2} + h_\infty. \quad (7.11)$$

Proof

We can use again Equation 7.6 by adding a hydrostatic pressure term $\rho\, g\, z$ just as we saw in Section 5.3.2. Moreover, at a large distance r from the axis, the term in $1/r^2$ becomes negligible, and the fluid surface approaches a constant level $z = h_\infty$. Since the pressure above the surface equals the atmospheric pressure p_0, Equation 7.6 allows us to write:

$$p(r,z) + \rho\, g\, z + \frac{1}{2}\rho\frac{\Gamma^2}{4\pi^2 r^2} = p_0 + \rho\, g\, h_\infty. \quad (7.12)$$

Moreover, if we neglect capillary effects due to the curvature of the interface, the level $z = h(r)$ of the free surface, still at a distance $r > \xi$, must be such that $p(r, h(r)) = p_0$. By replacing p and z by their values in Equation 7.12, we obtain Equation 7.11.

***Region* (II) $(r \leq \xi)$**

We have here an interface of parabolic shape:

$$h(r) = h(0) + \left(\frac{\Gamma}{2\pi\,\xi^2}\right)^2 \frac{r^2}{2g}. \tag{7.13}$$

In order to obtain the level $h(0)$ on the axis as a function of h_∞, we need only write that Equations 7.11 and 7.13 give the same value for $r = \xi$.

Proof

Equation 7.7 for the radial pressure gradient is still valid, but we must take into account the gradient of the hydrostatic pressure: $\partial p/\partial z = -\rho\, g$. By integrating with respect to r and z, and using the condition $p = p_0$ for $z = h(0)$ and $r = 0$, we obtain:

$$p(r, z) = p_0 + \rho\left(\frac{\Gamma}{2\pi\,\xi^2}\right)^2 \frac{r^2}{2} + \rho\, g\,(h(0) - z).$$

Equation 7.13 for the surface follows then from the fact that $p = p_0$ along it.

Many ideas which we will encounter in more general distributions of vorticity follow from this example. Vorticity is in fact often concentrated in the form of filaments, in localized regions of space. We will see in Section 7.3.2 how such a concentration can appear in a longitudinal rotational flow (which represents, in a simplified manner, the velocity field for an emptying container).

7.1.3 Electromagnetic analogies

The Helmholtz analogy

Principle of the analogy Equation 7.1, relating the velocity field $\mathbf{v}(\mathbf{r})$, and the vorticity $\boldsymbol{\omega}(\mathbf{r})$, is formally identical to the electromagnetic equation, $\mathbf{j}(\mathbf{r}) = (\nabla \times \mathbf{B}(\mathbf{r}))/\mu_0$, which gives the dependence, under stationary or quasi-stationary conditions, of the current density $\mathbf{j}(\mathbf{r})$ on the magnetic field $\mathbf{B}(\mathbf{r})$ in a vacuum.

This equation is generally used in the quasi-stationary regime of Maxwell's equations where one can neglect the term $\varepsilon_0\, \partial A/\partial t$ and which corresponds to what we normally call *magnetostatics*.

We now elaborate on this analogy. First of all, for an incompressible fluid, the field $\mathbf{v}$ is source free, as is the magnetic field $\mathbf{B}$, since they satisfy respectively:

$$\nabla \cdot \mathbf{H} = \nabla \cdot (\mathbf{B}/\mu_0) = 0 \quad (7.14a) \qquad \text{and} \qquad \nabla \cdot \mathbf{v} = 0. \quad (7.14b)$$

Similarly, we recall Ampere's law which states:

$$I = \int_C \mathbf{H} \cdot \mathrm{d}\mathbf{l} = \iint_S \mathbf{j} \cdot \mathrm{d}\mathbf{S}. \tag{7.15}$$

It relates the circulation of the magnetic excitation field $\mathbf{H}(\mathbf{r})$ along the space curve (C), which goes around a conductor carrying an electric current, and the flux I of the current density $\mathbf{j}(\mathbf{r})$ through a surface (S) anchored on this contour. This equation is the analog of Equation 7.2, which gives the value of the circulation Γ around a vortex line.

The velocity field resulting from a vortex filament can thus be shown to correspond to the field $\mathbf{H}$ resulting from an electric current in a conducting wire. The circulation Γ of the velocity vector is then the analog of the circulation of the field $\mathbf{H}$, and thus of the electric current I.

A mathematical correspondence is thus seen to exist between the magnetic field $\mathbf{H}$ and the fluid velocity $\mathbf{v}$, *provided that the boundary conditions are also identical*. In hydrodynamics, the normal component of the velocity is zero at all static solid walls ($v_\perp = 0$): this condition, discussed in Section 4.3.1, leads to streamlines which are parallel to solid boundaries. We find similar boundary conditions in electromagnetic problems only in the very special case of superconductors which, due to a property known as the Meissner effect, expel magnetic fields from their interior. In the discussion below, we therefore use the analogy between the velocity and the magnetic field only when the latter results from a system of current-carrying wires in free space. This is equivalent to the case of a fluid in an infinitely large container, where we need not be concerned with boundary conditions at the walls.

We now study this correspondence explicitly for a straight vortex tube, and an infinitely long, straight wire.

Correspondence fields resulting from vortex filaments and straight conducting wires The straight vortex filaments which we have discussed above are thus analogous to straight conducting wires. The velocity field of such a vortex filament (assumed of infinite length) in the geometry of Figure 7.4a is given by Equation 7.4; for a conducting wire (Figure 7.4b), we have an equivalent expression:

$$H_\varphi(r) = \frac{I}{2\pi r}, \tag{7.16}$$

where $H_\varphi(r)$ is the tangential component of the field $\mathbf{H}$ at a distance r from the wire. In both cases, only the tangential component of the velocity (or, respectively, of the magnetic field) is non-zero.

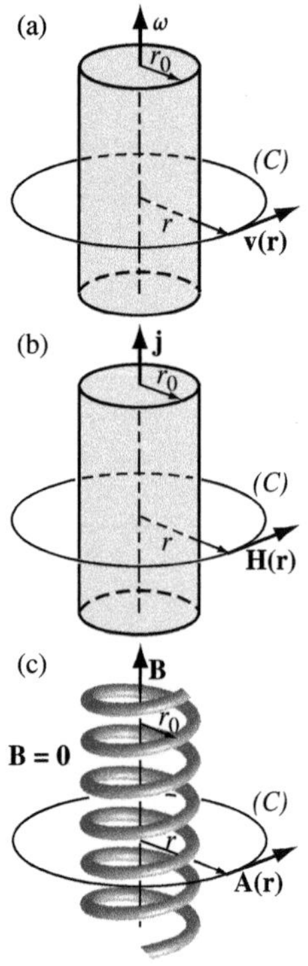

Figure 7.4 *(a) Velocity field of a straight vortex filament (b) Magnetic field induced by a current density in a tube (Helmholtz analog) (c) Magnetic field and vector potential created by a solenoid (Maxwell analog)*

Calculation of the velocity field from the vorticity field In practical terms, the main function of the Helmholtz analogy is that it provides us with a method to calculate everywhere the velocity field of a fluid from its vorticity field. In electromagnetics, the field $\mathrm{d}\mathbf{H}$ created at a point O (which we choose as the origin) by a line element $\mathbf{dl}$ of electrical conductor located *in empty space* at a point M obeys the law of Biot and Savart:

$$\mathrm{d}\mathbf{H} = -\frac{1}{4\pi} I \frac{\mathbf{dl} \times \mathbf{r}}{r^3}, \tag{7.17}$$

in this equation, I is the electrical current in the conductor, $\mathbf{r}$ is the vector $\mathbf{OM}$ and r its modulus. Similarly, the field resulting from a distribution $\mathbf{j(r)}$ of current density within a volume $(\mathcal{V})$ satisfies:

$$\mathbf{H} = \iiint_{\mathcal{V}} \frac{1}{4\pi} \frac{\mathbf{j(r)} \times \mathbf{r}}{r^3} \mathrm{d}V. \tag{7.18}$$

We find identical equations for the velocity field $\mathbf{v}$, induced by an element $\mathbf{dl}$ of a vortex filament with circulation Γ, or by a distribution $\boldsymbol{\omega}(\mathbf{r})$ of vorticity in a volume $(\mathcal{V})$ (Figure 7.5). For this, one uses the equivalences $\mathbf{B}/\mu_0 \leftrightarrow \mathbf{v}$ and $I \leftrightarrow \Gamma$. We thus obtain the equations which correspond to Equations 7.17 and 7.18 for the velocity induced at the origin $\mathbf{r} = 0$:

$$\mathrm{d}\mathbf{v} = -\frac{1}{4\pi} \Gamma \frac{\mathbf{dl} \times \mathbf{r}}{r^3} \quad (7.19a) \qquad \text{and} \qquad \mathbf{v} = \iiint_{\mathcal{V}} \frac{1}{4\pi} \frac{\boldsymbol{\omega}(\mathbf{r}) \times \mathbf{r}}{r^3} \mathrm{d}V. \quad (7.19b)$$

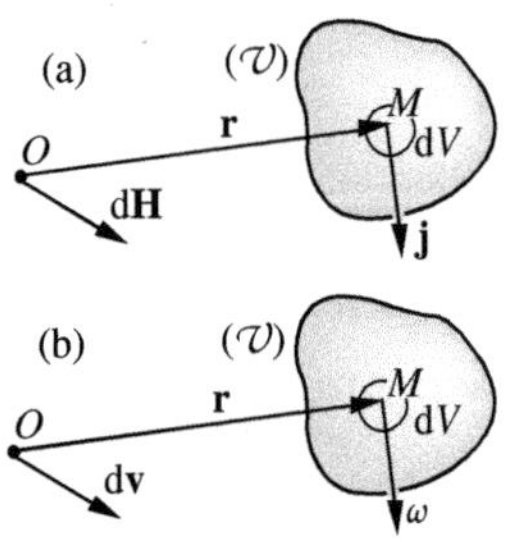

Figure 7.5 *(a) Magnetic field induced by a current density* $\mathbf{j(r)}$; *(b) velocity field resulting from a distribution of vorticity* $\boldsymbol{\omega}(\mathbf{r})$

Application: self-induced velocity field on a curved vortex filament We show below that the order of magnitude of the velocity $\mathbf{u}_l$ self-induced by a curved vortex at a point O (Figure 7.6 on next page) is:

$$|\mathbf{u}_l| = -\frac{|\Gamma|}{4\pi R} \mathrm{Log}\left(\frac{R}{\xi}\right), \tag{7.20}$$

where R is the radius of curvature of the vortex line, Γ its circulation, and ξ the radius of its core. The velocity vector $\mathbf{u}_l$, perpendicular to the plane of the vortex filament, results mainly from elements of the vortex very close to the point of interest. It is this component

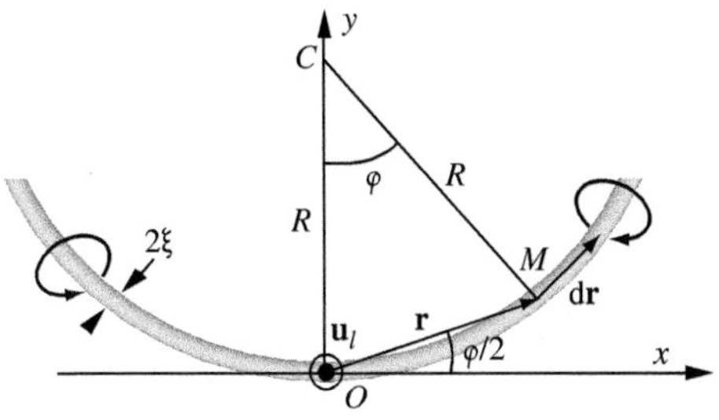

Figure 7.6 *Self-induced velocity field on a curved vortex filament: the velocity vector* $\mathbf{u}_l$ *induced at point O is pointing out of the picture, for the direction of the circulation shown on the figure*

$\mathbf{u}_l$ which accounts for the motion of curved vortices even in the absence of external flow. Indeed, we show in Section 7.2.1 that an element of vortex filament in an ideal fluid moves at a velocity equal to the local velocity of the fluid (Kelvin's theorem). Such a velocity is the sum of the external velocity field and of the velocities induced by the other elements of the vortex.

Proof

The self-induced velocity field is calculated at a point O on the curved vortex line (where we assume that the radius of curvature R is very large compared to the radius ξ of the vortex core). We evaluate only the local contribution $\mathbf{u}_l$, due to the neighboring points M, obtained by carrying out a line integral over the region of the vortex from M' to M, where M' is the mirror image point to M on the vortex, relative to the y-z plane. From Equation (7.19a), the velocity $\mathbf{u}_l$, directed along the z-axis, is:

$$\mathbf{u}_l = -\frac{\Gamma}{4\pi}\int \frac{\mathbf{dr}\times\mathbf{r}}{r^3}.$$

In the reference frame with origin O, the components of $\mathbf{r}$ are: $[R\sin\varphi, R\,(1-\cos\varphi), 0]$. So that:

$$\mathbf{dr} = [R\cos\varphi\,\mathrm{d}\varphi, R\sin\varphi\,\mathrm{d}\varphi, 0].$$

Taking into account the symmetry of the problem relative to the y-z plane, we find:

$$|\mathbf{u}_l| = \frac{\Gamma}{4\pi}\left|\int \frac{\mathbf{dr}\times\mathbf{r}}{r^3}\right| = 2\left(\frac{\Gamma}{4\pi}\right)\int_{\varphi_{min}}^{\varphi_{max}} \frac{2R^2\sin^2(\varphi/2)\,\mathrm{d}\varphi}{8R^3\sin^3(\varphi/2)} = \frac{\Gamma}{8\pi R}\int_{\varphi_{min}}^{\varphi_{max}} \frac{\mathrm{d}\varphi}{\sin(\varphi/2)}.$$

The velocity induced at each point of the vortex accordingly varies as the inverse of the radius of curvature of the vortex ring. Specifically, we note that the velocity thus vanishes for a straight, vortex filament ($R \to \infty$), as we would have expected from symmetry considerations.

For small values of φ, the integral diverges, as $\int 2\,\mathrm{d}\varphi/\varphi$, i.e., as 2 Log φ. As before, this is due to the fact that the above calculation treats the vortex as an infinitesimally thin singularity, neglecting any finite radial extent of the core; it cannot, therefore, be applicable below values of φ_{min} so small that distances r become comparable to the core radius ξ. The existence of this divergence indicates, also, that the largest contributions to the velocity field induced at O result from those elements of the vortex line nearest that point. An exact treatment of the problem would need to take into account the fact that the vorticity density is distributed over the radius of the core, ξ, and an even more precise approximation would consider ξ as a lower limit for the variable r. For our present treatment then, it is sufficient to obtain an order of magnitude estimate by taking $\varphi_{min} \approx \xi/R$, as the lower limit of the integral. These assumptions then lead to Equation 7.20: the effect of the upper limit $\varphi_{\max}$ is merely an additive constant depending weakly on φ_{max} which can be, for instance, taken equal to $\pi/2$.

The Maxwell analogy

We have just seen the usefulness of the Helmholtz analogy which allowed us to evaluate the velocity field from a distribution of vorticity. However, this analogy associates an *axial pseudovector* (for which the direction depends on the choice of reference frame), the vorticity, to a *polar vector*, the current density. In a similar manner, it creates a correspondence between the velocity field (polar vector) and the magnetic field (axial vector). A second analogy, which follows directly from Maxwell's equations, creates a direct correspondence between the vorticity $\boldsymbol{\omega}$ and the magnetic field $\mathbf{B}$, on the one hand, and the hydrodynamic velocity field $\mathbf{v}$ and the magnetic potential vector $\mathbf{A}$, on the other hand. This analogy associates a polar vector to a polar vector, and an axial vector to an axial vector. The correspondence between the equations is then the following:

In classical electromagnetic theory, the vector potential $\mathbf{A}(\mathbf{r})$ is generally considered as only a convenient intermediate step in the calculation of magnetic fields, just like the scalar potential for an electric field. Its physical significance, which Maxwell emphasized, is subject to controversy. The difficulty arises from the fact that if the vector potential $\mathbf{A}$ is defined from its relationship to the magnetic field by $\mathbf{B} = \nabla \times \mathbf{A}$, it is only determined within an arbitrary gradient of a scalar function.

Using the vector potential $\mathbf{A}$ allows one to express the equations of fluid mechanics in completely identical terms to Maxwell's equations for electromagnetic theory: this is of particular interest in situations where the two fields (magnetic and hydrodynamic) are coexistent. We will see in the appendix of this chapter (Section 7A.4), that, for quantum phenomena, such a vector potential has a clear physical significance.

$$\mathbf{B} = \nabla \times \mathbf{A} \quad (7.21a) \qquad \text{and:} \qquad \boldsymbol{\omega} = \nabla \times \mathbf{v}. \quad (7.21b)$$

$$\nabla \cdot \mathbf{A} = 0 \quad (7.22a) \qquad \text{and:} \qquad \nabla \cdot \mathbf{v} = 0. \quad (7.22b)$$

In this analogy, the equivalent of a straight line vortex is an infinitely long solenoid with a radius equal to that of the vortex core (Figure 7.4c): the field $\mathbf{B}$ differs from zero only within the solenoid (just as $\boldsymbol{\omega}$ is non-zero only within the core). It is the vector potential $\mathbf{A}$ which, just as the velocity due to the vortex, decreases at $1/r$ outside the solenoid; only its tangential component A_φ is non-zero and it is related to the field B_i within the solenoid and to its radius ξ by $A_\varphi = B_i\,\xi^2/2r$.

7.2 Dynamics of the circulation of the flow velocity

Having introduced the concepts of vortex lines and of continuous distributions of vorticity, we discuss, in this section, the dynamics of the vorticity by considering the variation of the circulation along an arbitrary closed contour "drawn" within the fluid and dragged along by it. In the section that follows, we use yet another approach by deriving directly, from the Navier-Stokes equation, the law of evolution of vorticity with time. The results derived in these two sections are applicable both to continuous distributions of vorticity and to those cases where the vorticity is localized along singularities.

7.2.1 Kelvin's theorem: conservation of the circulation

Derivation of Kelvin's theorem

Kelvin's theorem expresses the fact that, for a closed contour in which each point moves with the velocity of the fluid at that location, the circulation is constant, provided the following conditions are satisfied:

- the fluid is inviscid (i.e., an ideal fluid for which $\eta = 0$),
- the external forces are derivable from a potential function φ: $\mathbf{f} = -\nabla\varphi$,
- the fluid density ρ is constant or, more generally, only a function of pressure: $\rho = f(p)$.

These assumptions are identical to those made in Section 5.3.2, in deriving Bernoulli's theorem. Kelvin's theorem is expressed by the equation:

$$\frac{\mathrm{d}}{\mathrm{d}t}\left[\int_C \mathbf{v}\cdot\delta\boldsymbol{\ell}\right] = 0, \tag{7.23}$$

where, as defined earlier, $\mathrm{d}/\mathrm{d}t$ is the convective (Lagrangian) derivative calculated following the motion of fluid particles. The integral is evaluated along the closed contour (C); we have used the notation $\delta\boldsymbol{\ell}$ in order to differentiate this element of length from variations of $\boldsymbol{\ell}$ with time. The space curve along which the integration is carried out moves along with the fluid, as indicated in Figure 7.7.

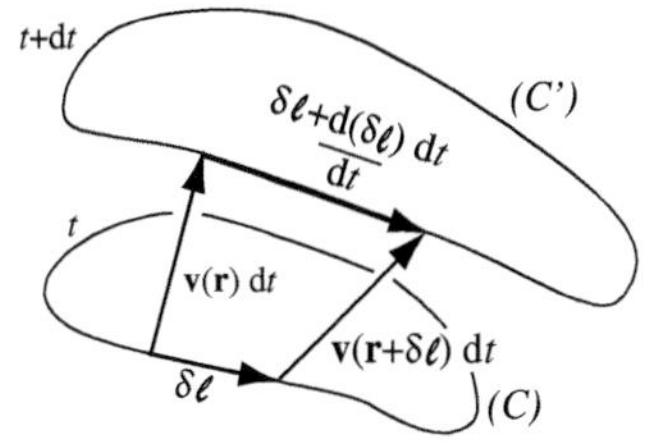

Figure 7.7 *Change in an element of a material curve (C) which follows the motion of the fluid*

Proof

Changes in the circulation are the sum of two contributions: one due to the variation with time of the velocity at the points of the contour of integration, and the other to the fact that the contour (C) is itself deformed as it undergoes displacements. These two effects can be separated out by dividing the integral (Equation 7.23) into the two terms:

$$\frac{\mathrm{d}}{\mathrm{d}t}\left[\int_C \mathbf{v}\cdot\delta\boldsymbol{\ell}\right] = \int_C \frac{\mathrm{d}\mathbf{v}}{\mathrm{d}t}\cdot\delta\boldsymbol{\ell} + \int_C \mathbf{v}\cdot\frac{\mathrm{d}\,(\delta\boldsymbol{\ell})}{\mathrm{d}t}. \tag{7.24}$$

The first integral is evaluated by the use of Euler's equation which can be written:

$$\frac{\mathrm{d}\mathbf{v}}{\mathrm{d}t} = -\nabla\varphi - \frac{1}{\rho}\nabla p. \tag{7.25}$$

If the fluid density ρ depends only on the pressure ($\rho = f(p)$), the second term of Equation 7.25 can also be expressed as the gradient of a function $g(p) = \int \mathrm{d}p/f(p)$. In accordance with the fundamental property of gradients, the circulation of these two gradients and, consequently, that of the vector $\mathrm{d}\mathbf{v}/\mathrm{d}t$ along the closed curve (C), is then zero. The second integral of Equation 7.24 obeys the following sequence of equalities:

$$\int_C \mathbf{v}\cdot\frac{\mathrm{d}\,(\delta\boldsymbol{\ell})}{\mathrm{d}t} = \int_C v_i\cdot\mathrm{d}\frac{\delta\ell_i}{\mathrm{d}t} = \int_C v_i\left(\frac{\partial v_i}{\partial x_j}\delta\ell_j\right) = \int_C \frac{\partial}{\partial x_j}\left(\frac{\mathbf{v}^2}{2}\right)\cdot\delta\ell_j = \int_C \nabla\left(\frac{\mathbf{v}^2}{2}\right)\cdot\delta\boldsymbol{\ell} = 0.$$

Indeed, the variation with time of a line element $\delta\boldsymbol{\ell}$ of the contour is due to the difference $\delta\mathbf{v}$ between the velocities of the two points located at the two ends of this element (Figure 7.7), so that each component $\mathrm{d}(\delta\ell_i)/\mathrm{d}t$ is equal to the product $(\partial v_i/\partial x_j)\,\delta\ell_j$. Combining these two results leads then to Equation 7.23.

By using Equation 7.2, Kelvin's theorem can be rewritten in the form:

$$\frac{\mathrm{d}}{\mathrm{d}t}\left(\iint_S \nabla\times\mathbf{v}\cdot\delta\mathbf{S}\right) = \frac{\mathrm{d}}{\mathrm{d}t}\left(\iint_S \boldsymbol{\omega}\cdot\delta\mathbf{S}\right) = 0. \tag{7.26}$$

Thus, the flux of the vorticity vector $\boldsymbol{\omega}$ through any surface anchored on the space curve (C) moving with the fluid, i.e., the total vorticity in (C), remains constant during the flow.

Physical meaning and consequences of Kelvin's theorem

The result just derived above expresses the conservation of angular momentum in an ideal fluid, thus completing our presentation of the conservation laws in Chapter 5.

Justification in the case of an ideal fluid

Consider an elementary tube of vorticity of length δL resting on a circle of radius r perpendicular to $\boldsymbol{\omega}$ (Figure 7.8). We have seen previously that the local angular velocity vector, $\boldsymbol{\Omega}$, of an element of fluid equals $\boldsymbol{\omega}/2$. Since the circulation Γ of the velocity along the circle (C) of radius r is equal to the total flux of the vorticity through the circle, Γ is related to the magnitude ω of the vorticity by:

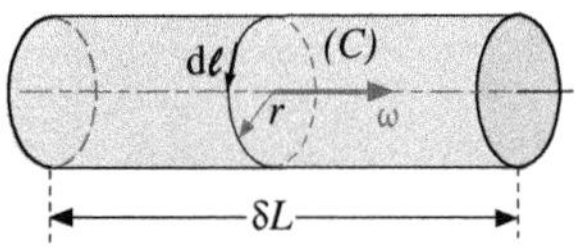

Figure 7.8 *Calculation of the angular momentum for an element of a vortex tube*

$$\Gamma = \int \mathbf{v} \cdot d\boldsymbol{\ell} = \pi r^2 \omega. \tag{7.27}$$

The product $\pi r^2 \omega$ can be rewritten so as to include explicitly the magnitude $\Omega(=\omega/2)$ of the local rotational velocity, as well as the moment of inertia $\mathcal{J}$ of the cylinder of length δL:

$$\pi r^2 \omega = \frac{\delta m\, r^2}{2} \frac{\omega}{2} \frac{4\pi}{\delta m} = K \mathcal{J} \Omega. \tag{7.28}$$

Here $\delta m = \rho \pi r^2 \delta L$ is the mass of fluid, with density ρ, contained in the element of the cylinder, $\mathcal{J} = \delta m\, r^2/2$ is the moment of inertia associated with this fluid and $K = 4\pi/\delta m$. Thus, since the mass δm of fluid in the element of vorticity tube is necessarily constant, conservation of the circulation Γ with time is equivalent, according to Equations 7.27 and 7.28, to the conservation of the angular momentum $\mathcal{J}\Omega$ of the fluid in the vortex-tube element.

We can draw a number of physical conclusions from the circulation theorem:

A specific application of this property is the appearance of an incipient vortex downstream of the trailing edge of an airplane wing when the latter is set in motion (Figure 7.27b).

(i) If, at an initial time, the circulation around any closed contour is zero, it remains zero subsequently. Specifically, an inviscid fluid (i.e., for which $\eta = 0$) set in motion from a state of rest will continue, at any subsequent time, in irrotational flow (i.e., such that the vorticity vector $\boldsymbol{\omega}(\mathbf{r})$ is identically zero everywhere). In fact, the circulation along any closed curve drawn within the fluid being zero initially, it remains so for all time. Since such curves can be drawn arbitrarily small, it follows, by the application of Stokes' theorem, that $\boldsymbol{\omega}(\mathbf{r})$ is everywhere zero. This result, originally stated in Section 6.1, allowed us to make the connection between the study of potential flows and that of ideal fluids.

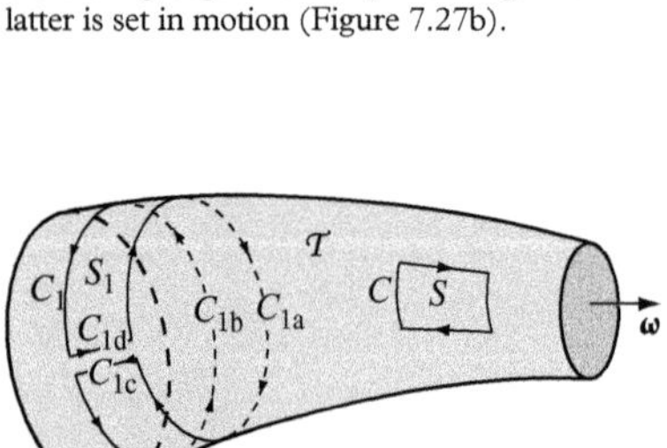

Figure 7.9 *Variation of the flux of vorticity through material contours drawn on the wall of a vortex tube*

(ii) In a flow where the vorticity vector $\boldsymbol{\omega}(\mathbf{r})$ is no longer identically zero, vortex lines (and tubes) follow precisely the motion of the curves (or surfaces) made up of fluid particles which are called *material* lines. Let us draw such a closed material contour (C) on the wall of a vortex tube $(\mathcal{T})$. In Figure 7.9, (C) does not make a loop around the tube: thus, there exists a surface (S) which is entirely part of the tube wall and rests on curve (C). The vorticity $\boldsymbol{\omega}$ is tangent to the wall so that its flux through the surface (S) is zero: it will remain so at all later times while the material contour (C) is convected by the flow. This is true for all contours like (C)

and, therefore, the volume of matter made of fluid particles belonging initially to $(\mathcal{T})$ and convected by the flow is also a vortex tube.

(iii) Consider now a contour (C_1) drawn on the walls of the tube $(\mathcal{T})$ and made of two loops (C_{1a}) and (C_{1b}) drawn around the tube and connected by two parallel paths (C_{1c}) and (C_{1d}) (Figure 7.9). The total circulation of the velocity along (C_1) remains zero, like for (C), because there exists a surface (S_1) resting on (C_1) and belonging entirely to the tube wall; in contrast, the circulation on either (C_{1a}) or (C_{1b}) is non-zero because any surface resting on one of these curves must extend across the tube section. The paths (C_{1c}) and (C_{1d}) can be brought infinitely close to each other and are followed in opposite directions so that their contributions cancel out. The sum of the circulations along (C_{1a}) and (C_{1b}) then becomes equal to that along (C_1) and, therefore, to zero: these circulations have a different sign because they are followed in opposite directions and their absolute values are equal. The flux of vorticity (equal to its circulation) is therefore the same through all curves surrounding once the vortex tube: this is just like the flux of the velocity vector in a flow tube, in the absence of sources.

(iv) In Section 7.4, we discuss the motion of vortex filaments for which the vorticity is concentrated along singular lines (*vortex cores*). In that case, the results we have just derived show that the cores of these vortices always move at the local velocity of the fluid (a velocity resultant from the external flow velocity and the velocity induced by other vortex filaments, as shown in Section 7.1.3). Indeed, if a little material contour, be it very small, completely surrounds the core of the vortex at a given instant of time, it will always loop around it, after being displaced by the fluid: the circulation of the velocity along this contour thus remains constant, as well as the flux of the vorticity in the core of the vortex which is equal to it.

Moreover, the cores of these vortex filaments must either close on themselves (case of vortex rings), or they must terminate on a solid wall or liquid interface. They cannot have a dangling, loose, end in an ideal fluid. If, in fact, we assume that such a free end exists, and calculate the circulation Γ of the velocity of a fluid along a contour (C) surrounding the vortex, we find that Γ equals the flux of the vorticity $\boldsymbol{\omega}$ across a surface (S) anchored on (C). Depending on the position of (S) relative to the free end, (S) might or might not intersect the core of the vortex, so that the value of the integral would be different depending on the position of (S). Such a result would then be contrary to the existence of a well defined value of circulation of the velocity along (C).

7.2.2 Sources of circulation

Let us recall the proof of the circulation theorem starting with Equation 7.24. When the three conditions stated at the beginning of Section 7.2.1 are not satisfied, the circulation of the velocity around a particular contour of matter is no longer constant. The second term of that equation remains zero, and we need to evaluate the first term by expressing $d\mathbf{v}/dt$ starting from the Navier-Stokes equation (Equation 4.30) instead of Euler's equation (Equation 4.31) which we used in Section 7.2.1:

$$\frac{d\mathbf{v}}{dt} = \mathbf{f} - \frac{1}{\rho}\nabla p + \nu\nabla^2\mathbf{v}. \tag{7.29}$$

Equation 7.24 then becomes:

$$\frac{\mathrm{d}}{\mathrm{d}t}\left[\int_C \mathbf{v}\cdot\delta\boldsymbol{\ell}\right] = \int_C \frac{\mathrm{d}\mathbf{v}}{\mathrm{d}t}\cdot\delta\boldsymbol{\ell} = \underbrace{\int_C \mathbf{f}\cdot\delta\boldsymbol{\ell}}_{\mathbf{I}} - \underbrace{\int_C \frac{1}{\rho}(\nabla p)\cdot\delta\boldsymbol{\ell}}_{\mathbf{II}} + \underbrace{\int_C \nu\,\nabla^2\mathbf{v}\cdot\delta\boldsymbol{\ell}}_{\mathbf{III}}. \tag{7.30}$$

We now examine, term by term, the physical meaning of each of the three terms **I, II** and **III**.

Non-conservative volume forces (term I in Equation 7.30)

Any forces **f** (per unit volume) which are not derivable from a potential, and which are therefore such that their circulation along a closed contour is not zero, are capable of creating circulation. We find, in hydrodynamic problems, two important examples of such forces:

Coriolis (fictitious) forces They appear as a term $(-2\boldsymbol{\Omega}\times\mathbf{v})$, when we write the equation of motion of a fluid in a rotating reference frame with a rotation vector $\boldsymbol{\Omega}$ (the term "fictitious" indicates that these forces are not really physical forces but merely result from the change of reference frame, since the velocity **v** is measured in the rotating frame). For example, atmospheric and ocean current flows occur in a reference frame which rotates at $\boldsymbol{\Omega}$, the local rotation vector of the earth, for which the direction (just as that of $\boldsymbol{\Omega}$) depends on the hemisphere of the earth considered (Section 7.6.1).

More generally, we will see in Section 7.6 that Coriolis forces have a key influence on large-scale flows in the atmosphere.

In a laboratory setting, the effect of Coriolis forces may be observed by allowing for a large diameter cylindrical container (of the order of 2 meters in diameter) to empty through a hole located in its center. The presence of a radial component of the velocity, directed toward the hole through which the fluid empties, results in a Coriolis force which depends on the local value of $\boldsymbol{\Omega}$ and leads to rotation of the fluid. This result is quite similar to that of the flow around a region of low atmospheric pressure. It is however essential that the experimental conditions be very carefully controlled: more precisely, the fluid must be allowed to rest for an extended time, in order for any residual vorticity to be completely dissipated. In the emptying of bathroom sinks or bathtubs, it is the effects of the residual vorticity (see Section 7.3.2), amplified as the container empties, which ultimately result in the appearance of the drain whirlpool: the final direction of rotation is often random and, in any case, unrelated to the direction of $\boldsymbol{\Omega}$.

Magneto-hydrodynamic forces Vorticity can also be created by magneto-hydrodynamic forces which are produced by the action of a magnetic field **B** on an electrically conducting fluid. Consider a conducting fluid with zero net electrical charge subjected to an electric field **E** and a magnetic field **B** (**E** and **B** are defined in the laboratory reference frame). In the presence of an electric current density $\mathbf{j}(\mathbf{r}, t)$ in the fluid, there appears a *Laplace force*: $\mathbf{F}_{\text{Laplace}} = \mathbf{j}\times\mathbf{B}$ per unit volume.

Proof

The Lorentz force on an individual particle (electron, ion, etc.) with charge q_i and velocity $\mathbf{v}_{pi}$ in the reference frame of the laboratory is equal to $q_i(\mathbf{E} + \mathbf{v}_{pi}\times\mathbf{B})$. The total velocity, $\mathbf{v}_{pi}$, is the sum of the velocity of the particle relative to the fluid and the velocity **v** of the fluid itself. Let us now calculate the total force per unit volume $\mathcal{V}_1$ equal to: $\Sigma_{i\in\mathcal{V}_1} q_i(\mathbf{E}+\mathbf{v}_{pi}\times\mathbf{B})$. The component due to the electric field vanishes because the fluid is charge neutral with $\Sigma_{i\in\mathcal{V}_1} q_i = 0$. Moreover, the sum $\Sigma_{i\in\mathcal{V}_1} q_i\mathbf{v}_{pi}$ is the electrical current density **j** in the fluid, which then leads to a *Laplace force* per unit volume: $\mathbf{F} = \Sigma_{i\in\mathcal{V}_1} q_i\,\mathbf{v}_i\times\mathbf{B} = \mathbf{j}\times\mathbf{B}$. We observe that, again for a neutrally charged fluid, **j** is uniquely determined by the relative motion of the charges with respect to the fluid (in opposite directions according to the sign of the charge). In fact, the velocity **v** of the flow gives a vanishing contribution $\Sigma_{i\in\mathcal{V}_1} q_i\mathbf{v}$ (for the same reason, **j** does not depend on the reference frame selected).

Including this Laplace force in the equation of motion of the fluid, the latter becomes:

$$\frac{\mathrm{d}\mathbf{v}}{\mathrm{d}t} = \mathbf{f} - \frac{1}{\rho}\nabla p + \frac{\mathbf{j}\times\mathbf{B}}{\rho} + \nu\,\nabla^2\mathbf{v}. \tag{7.31}$$

We now assume that the quasistatic approximation (previously used in Section 7.1.3) is valid with sufficiently slow changes and correspondingly low velocities such that we can neglect the displacement currents $\partial\mathbf{D}/\partial t$ and use Maxwell's equation in the form $\nabla\times\mathbf{H}=\nabla\times\mathbf{B}/\mu=\mathrm{j}$. We will also assume that $\mathbf{B}$ and $\mathbf{H}$ are proportional to each other, with the magnetic permeability $\mu=\mathbf{B/H}$ possibly being spatially inhomogeneous. In rewriting the equation for the Laplace force using these equations and the vector identity $(\mathbf{B}\cdot\nabla)\mathbf{B}/\mu=\nabla(\mathbf{B}^2/2\mu)+\{\nabla\times(\mathbf{B}/\mu)\}\times\mathbf{B}$, Equation 7.31 for the motion becomes:

$$\frac{d\mathbf{v}}{dt}=\mathbf{f}-\frac{1}{\rho}\nabla\left(p+\frac{\mathbf{B}^2}{2\mu}\right)+\frac{1}{\rho}(\mathbf{B}\cdot\nabla)\frac{\mathbf{B}}{\mu}+\nu\nabla^2\mathbf{v}. \tag{7.32}$$

The term $\mathbf{B}^2/2\mu$, added to the pressure in the expression for the gradient, cannot contribute as a source of vorticity (if ρ is constant): we frequently refer to this term as a *magnetic pressure*. In contrast, the term $(\mathbf{B}\cdot\nabla)(\mathbf{B}/\mu)$ is not a gradient of a potential function and thus can be a source for the vorticity.

If the magnetic field can create a vorticity in the fluid, the flow of a conducting fluid can, reciprocally, generate a magnetic field. The magnetic field $\mathbf{B}$ satisfies indeed the equation:

$$\frac{\partial\mathbf{B}}{\partial t}=\nabla\times(\mathbf{v}\times\mathbf{B})+\nu_m\nabla^2\mathbf{B}, \tag{7.33a}$$

where $\nu_m=(\mu\sigma_{el})^{-1}$ is the magnetic diffusivity (σ_{el} is the electrical conductivity of the fluid and μ, its magnetic permeability). By using the identity $\nabla\times(\mathbf{v}\times\mathbf{B})=-(\mathbf{v}\cdot\nabla)\mathbf{B}+(\mathbf{B}\cdot\nabla)\mathbf{v}$, Equation 7.33a can be rewritten in the form:

$$\frac{d\mathbf{B}}{dt}=\frac{\partial\mathbf{B}}{\partial t}+(\mathbf{v}\cdot\nabla)\mathbf{B}=(\mathbf{B}\cdot\nabla)\mathbf{v}+\nu_m\nabla^2\mathbf{B}. \tag{7.33b}$$

We see on the right hand side of the first equality the sum of a term due to non-stationary flow and of a convective term. The first term on the right side of the second equality corresponds to the effects of stretching and tipping over of magnetic field tubes. We discuss such effects in Section 7.3.1 in relation to Equation 7.41: the latter determines the evolution of the vorticity and has a structure identical to that of Equation 7.33b. We note that, in the case of stretching (e.g., a term $v_z\,\partial B_z/\partial z>0$ for a field along the z-direction) there is an amplification of the magnetic field. Finally, the last term on the right hand side of the second equation expresses the diffusion of the field $\mathbf{B}$, which always tends, in contrast, to attenuate it.

Proof

In a fluid at rest, the electrical current is related to the electric field $\mathbf{E}$ in the laboratory reference frame by Ohm's law $\mathbf{j}=\sigma_{el}\mathbf{E}$. When the fluid is in motion at a velocity $\mathbf{v}$ relative to the laboratory reference frame, this equation is still valid if $\mathbf{E}$ is replaced by the field $\mathbf{E}'$, taken this time in the reference frame of the fluid. $\mathbf{E}'$ is related to the fields $\mathbf{E}$ and $\mathbf{B}$ by $\mathbf{E}'=\mathbf{E}+\mathbf{v}\times\mathbf{B}$. In the laboratory frame, the equation $\mathbf{j}=\sigma_{el}\mathbf{E}'$ for Ohm's law thus becomes a function of the fields $\mathbf{E}$ and $\mathbf{B}$ with: $\mathbf{j}=\sigma_{el}(\mathbf{E}+\mathbf{v}\times\mathbf{B})$.

In calculating the curl of Maxwell's equation, $\nabla\times(\mathbf{B}/\mu)=\mathbf{j}$, and using the above derived equation, we obtain:

$$\nabla\times(\nabla\times(\mathbf{B}/\mu))=\sigma_{el}(\nabla\times\mathbf{E}+\nabla\times(\mathbf{v}\times\mathbf{B})). \tag{7.34}$$

Using the vector identity, Equation 7.45, to calculate $\nabla\times(\nabla\times\mathbf{B})$ as well as the other Maxwell equations $\nabla\cdot\mathbf{B}=0$ and $\nabla\times\mathbf{E}=\partial\mathbf{B}/\partial t$, we retrieve Equation 7.33a with $\nu_m=(\mu\,\sigma_{el})^{-1}$ in the case where the magnetic permeability μ and the electrical conductivity

This dimensionless number $Re_m = UL/\nu_m$ (where U and L are the characteristic velocity and dimension of the flow) is called the *magnetic Reynolds number*.

σ_{el} are constant throughout the fluid. As in the case of other transport phenomena, we introduce here a dimensionless number Re_m to characterize the relative importance of the convective and diffusing terms in Equations 7.33a and 7.33b.

A particularly important example in geophysics of the application of Equations 7.32, 7.33a and 7.33b is the *dynamo effect*, which is the origin of the earth's magnetic field; the conductive fluid is, in that case, the liquid core of the interior of the earth, and the magnetic field is the result of the convective flow within this core. This involves a complex coupled system, where an initial fluctuation of the magnetic field will result in forces and, consequently, in an electric current in the fluid, which, in turn, induces a magnetic field amplifying the initial fluctuation.

This dynamo effect has been reproduced in the laboratory in a cylindrical container filled with liquid sodium (another type of conducting fluid) within a turbulent flow resulting from two coaxial turbines located at the two ends of the container and rotating in opposite directions. A magnetic field then spontaneously appears above a threshold angular velocity, and random reversals of this field, comparable to those of the earth's magnetic field, have been observed. A key factor for the appearance of the dynamo effect is the value of the magnetic Reynolds number Re_m defined above.

There are numerous applications of magnetohydrodynamics: a few examples are the *Tokamak* machines for thermonuclear fusion, where a conductive plasma (ionised gas) is subjected to a strong magnetic field, propulsion systems for satellites or some manufacturing processes involving liquid metals. Plasmas submitted to magnetic fields are also widespread in astrophysics (e.g., in stars) or in the atmosphere of the earth, particularly at very high altitudes.

Non-barotropic fluids (term II in Equation 7.30)

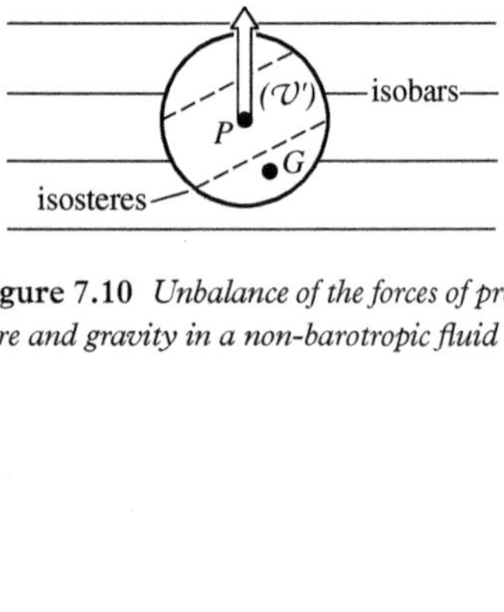

Figure 7.10 *Unbalance of the forces of pressure and gravity in a non-barotropic fluid*

A *barotropic* fluid is one for which $\rho = f(p)$, i.e., such that *isobars* (surfaces for which p = constant) coincide with the surfaces of constant density (*isosteres*). If this is not the case, the term $(-\nabla p)/\rho$ cannot be expressed as the gradient of a scalar potential function: then, the corresponding integral in Equation 7.30 does not vanish. In this case, if we consider an isolated element of fluid $(\mathcal{V}')$, its center of gravity G no longer necessarily coincides with the center of buoyancy P, determined (according to Archimedes' principle) by the isobars of the external fluid (Figure 7.10). Consequently a torque results, tending to cause local rotation of the fluid and to create a circulation of the velocity.

Proof

A necessary and sufficient condition for $(\nabla p)/\rho$ to be the gradient of a potential function is that $\nabla \times ((\nabla p)/\rho) = 0$. Recalling the general vector identity:

$$\nabla \times (\alpha \mathbf{A}) = \alpha \nabla \times \mathbf{A} + (\nabla \alpha) \times \mathbf{A}, \tag{7.35}$$

we must have:
$$\nabla \times \left(\frac{\nabla p}{\rho}\right) = \frac{1}{\rho} \nabla \times \nabla p - \frac{1}{\rho^2} \nabla \rho \times \nabla p = -\frac{1}{\rho^2} \nabla \rho \times \nabla p = 0. \tag{7.36}$$

Accordingly, a potential function exists for $(\nabla p)/\rho$ if, and only if, the vectors ∇p and $\nabla \rho$ are everywhere parallel, i.e., if the surfaces of constant pressure and constant density, normal to their respective gradients, coincide.

Quantitatively, term II of Equation 7.30 can be rewritten by using Stokes' theorem, relating the flux to the circulation:

$$-\int_C \frac{1}{\rho}\nabla p \cdot d\mathbf{l} = -\iint_S \nabla \times \left(\frac{1}{\rho}\nabla p\right) \cdot d\mathbf{S}. \qquad (7.37)$$

The integral of the right-hand term is evaluated over a surface (S) anchored on the closed curve (C) located inside the fluid, the vector $\mathbf{dS}$ being normal to the surface (S). In combining Equation 7.37 with Equation 7.36, we find the contribution of the non-barotropic effect to the change in the circulation:

$$\frac{d}{dt}\left[\int_C \mathbf{v} \cdot d\mathbf{l}\right] = \int_C \frac{1}{\rho}\nabla p \cdot d\mathbf{l} = -\iint_S \frac{1}{\rho^2}\nabla \rho \times \nabla p \cdot d\mathbf{S}. \qquad (7.38)$$

Examples of non-barotropic fluids A first example of a non-barotropic fluid is the case of the fluid placed between two vertical plates at different temperatures, resulting in a horizontal temperature gradient. We have studied this situation in Section 4.5.5: the changes in the density with temperature result, in this configuration, in the appearance of a horizontal gradient of the density ρ. The latter is thus perpendicular to the pressure gradient, which coincides with the hydrostatic pressure gradient, vertical if the fluid is in an equilibrium state. We thus have a creation of vorticity such that a thermal convection motion appears.

A second example is the case where density variations are associated with variations in the concentration of a solution. Such is the case of a solution of a liquid whose concentration varies with depth, obtained, for example, by carefully filling a container with a sugar solution of concentration C decreasing with the vertical distance from the bottom of the container (Figure 7.11). The solution is in stable equilibrium, since the density *decreases* with height. The concentration C is then constant along a given horizontal plane. Since the fluid is at rest, the only possible mass transport occurs through molecular diffusion, with a flux, $\mathbf{j} = -D_m \nabla C$, which is, in this case, directed vertically upward.

If a flat plate is inserted at an angle, at a fixed position in this solution, there cannot be any mass diffusion across the solid surface; thus, as indicated in the preceding equation, we must have $(\nabla C)_\mathbf{n} = 0$ at the solid wall (where $\mathbf{n}$ is the unit vector normal to the plate). Lines of constant density no longer remain everywhere horizontal because they must intersect the inclined plate at a right angle. Therefore, there appears a horizontal component of the gradient of concentration and, as a result of this, a gradient in density analogous to that of the preceding example. Along a given horizontal line, the density is higher near the inclined solid surface. Consequently, there is an imbalance in the hydrostatic pressure, which creates a convective movement. Its direction is such that it tends to decrease the gradients of concentration. This in turn leads to the creation of circulation of the velocity vector.

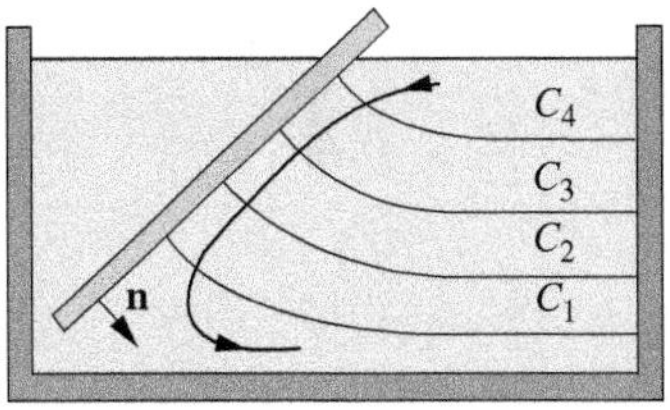

Figure 7.11 *Convective motion of a fluid, in the presence of an oblique solid surface, for which there exists a vertical gradient in the density ($C_1 > C_2 > C_3 > C_4$)*

Viscosity effects (term III in Equation 7.30)

Viscosity leads to the existence of velocity gradients in the neighborhood of walls, here again creating circulation. Starting, for example, from a fluid at rest, the generation of such a flow is often accompanied by the appearance of vorticity, and hence of circulation. The integral along the closed contour in the term **III** of Equation 7.30, involving dissipative viscosity forces, does not then vanish during this transitional phase. We have already encountered this effect in Section 2.1.1, in the example discussing the setting into rotational motion of a cylinder full of fluid. Initially, the fluid was at rest, while the stationary, time-independent, final condition of the fluid was solid body rotation, corresponding to a uniform density of vorticity. It is the diffusive transport of vorticity, by means of viscous forces, which results in that distribution.

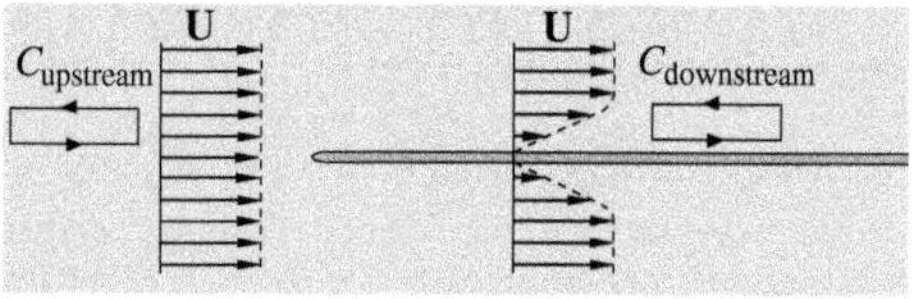

Figure 7.12 *Vorticity creation associated to viscous forces in a fluid incident on a flat plate*

Let us, for example, analyze a *Blasius flow* of fluid (see Sections 10.2 and 10.4.1) which is near the edge of a semi-infinite plate parallel to the average flow (Figure 7.12). Circulation is created in the immediate neighborhood of the leading edge of the plate. Upstream of the edge, the velocity field **U** is uniform, and the net circulation of the velocity is zero along a closed contour of the type ($C_{upstream}$) shown on Figure 7.12. On the other hand, downstream of the leading edge of the plate, a velocity gradient appears because the velocity must be zero at the wall, while again approaching **U** sufficiently far from the plate. The circulation along the closed contour ($C_{downstream}$), bounded on one side by the plate, is thus non-zero downstream of the sharp edge. The creation and growth of the *boundary layer*, the region near the plate where velocity gradients exist, is discussed in Chapter 10. An even more striking example of the generation of vorticity due to viscosity effects is the downstream shedding of vortices from a cylinder placed normal to a potential upstream flow (Figures 2.9 and 2.10).

7.3 Dynamics of vorticity

7.3.1 Transport equation for vorticity, and its consequences

The Helmholtz equation for an incompressible fluid

We now generalize the results of the preceding section by considering directly the evolution of the vector field of vorticity $\boldsymbol{\omega}$. We start by writing the Navier-Stokes equation in the form:

$$\frac{\partial \mathbf{v}}{\partial t} - \mathbf{v} \times \boldsymbol{\omega} + \nabla\left(\frac{\mathbf{v}^2}{2}\right) = \mathbf{f} - \frac{1}{\rho}\nabla p + \nu \nabla^2 \mathbf{v}, \tag{7.39}$$

where, from Equation 5.34, the $(\mathbf{v}.\nabla)\mathbf{v}$ term has been replaced by its equivalent expression: $-\mathbf{v} \times \boldsymbol{\omega} + \nabla(\mathbf{v}^2/2)$. Taking the curl of this equation, we obtain:

$$\frac{\partial}{\partial t}(\nabla \times \mathbf{v}) - \nabla \times (\mathbf{v} \times \boldsymbol{\omega}) = \nabla \times \left(\mathbf{f} - \frac{1}{\rho}\nabla p\right) + \nu \nabla \times \left(\nabla^2 \mathbf{v}\right). \tag{7.40}$$

Assuming now that the volume forces **f** are conservative (i.e., equal to the gradient of a potential), that the density ρ is constant, and that the kinematic viscosity is finite(assumptions equivalent to neglecting type **I** and **II** terms in Equation 7.30, the evolution equation for the vorticity), Equation 7.40 becomes

$$\frac{\partial \boldsymbol{\omega}}{\partial t} + (\mathbf{v} \cdot \nabla)\,\boldsymbol{\omega} = (\boldsymbol{\omega} \cdot \nabla)\,\mathbf{v} + \nu\, \nabla^2 \boldsymbol{\omega}. \tag{7.41}$$

The above equation plays, for $\boldsymbol{\omega}(\mathbf{r}, t)$, a role similar to the Navier-Stokes equation for $\mathbf{v}(\mathbf{r}, t)$. Recognizing that the terms on the left hand side are none other than the Lagrangian derivative, we can equally write:

$$\frac{d\boldsymbol{\omega}}{dt} = (\boldsymbol{\omega} \cdot \nabla)\,\mathbf{v} + \nu\nabla^2\boldsymbol{\omega}. \qquad (7.42)$$

This transport equation holds for all kinds of flows, whether laminar or turbulent. Describing a flow in terms of its vorticity field is thus always a legitimate alternative to a description in terms of the velocity field. The choice of either approach is governed by practical considerations related to the specific configuration of the flow.

Proof of Equation 7.41: In Equation 7.40, we split up $\nabla \times (\mathbf{v}\times\boldsymbol{\omega})$ by using the following general vector identity, applicable to arbitrary vector fields **A** and **B**:

$$\nabla \times (\mathbf{A} \times \mathbf{B}) = (\mathbf{B} \cdot \nabla)\,\mathbf{A} - (\mathbf{A} \cdot \nabla)\,\mathbf{B} - \mathbf{B}\,(\nabla \cdot \mathbf{A}) + \mathbf{A}\,(\nabla \cdot \mathbf{B})\,. \qquad (7.43)$$

Assuming the fluid to be incompressible, so that $\nabla \cdot \mathbf{v} = 0$, and given that $\nabla \cdot \boldsymbol{\omega} = \nabla \cdot (\nabla \times \mathbf{v}) \equiv 0$ for any vector field, we have:

$$\nabla \times (\mathbf{v} \times \boldsymbol{\omega}) = (\boldsymbol{\omega} \cdot \nabla)\,\mathbf{v} - (\mathbf{v} \cdot \nabla)\,\boldsymbol{\omega}. \qquad (7.44)$$

To evaluate the last term, $\nu\nabla^2\boldsymbol{\omega}$, we further recall the vector identity:

$$\nabla \times (\nabla \times \mathbf{A}) = \nabla\,(\nabla \cdot \mathbf{A}) - \nabla^2\mathbf{A}. \qquad (7.45)$$

Letting $\mathbf{A} = \mathbf{v}$, we obtain:

$$\nabla^2\mathbf{v} = \nabla\,(\nabla \cdot \mathbf{v}) - \nabla \times (\nabla \times \mathbf{v}) = -\nabla \times \boldsymbol{\omega}. \qquad (7.46)$$

Taking the curl of both terms, this equation becomes:

$$\nabla \times \left(\nabla^2\mathbf{v}\right) = -\nabla \times (\nabla \times \boldsymbol{\omega}).$$

So that, again using Equation 7.45:

$$\nabla \times \left(\nabla^2\mathbf{v}\right) = \nabla^2\boldsymbol{\omega} - \nabla\,(\nabla \cdot \boldsymbol{\omega}) = \nabla^2\boldsymbol{\omega}. \qquad (7.47)$$

We thus show that the last two terms in Equations 7.40 and 7.41 are identical.

In Equation 7.41, the first two terms describe the effect of the non-stationary flow and of the convection of the vorticity; the last term gives the decay of vorticity due to viscous effects. Were it not for the term $(\boldsymbol{\omega} \cdot \nabla)\mathbf{v}$, this equation would be very similar to the heat transport and mass diffusion equations (Equations 1.17 and 1.26). Here, we see the kinematic viscosity ν as the diffusion coefficient for the vector vorticity $\boldsymbol{\omega}$.

An important consequence of this equation is the persistence of the irrotational state in an ideal fluid ($\eta = 0$) initially at rest. Indeed, if $\boldsymbol{\omega}(\mathbf{r}, t = 0)$ vanishes identically at the start, we have:

$$\frac{d\boldsymbol{\omega}}{dt} = (\boldsymbol{\omega} \cdot \nabla)\,\mathbf{v} = 0. \qquad (7.48)$$

So that, at all times t, $\boldsymbol{\omega}(\mathbf{r}, t)$ remains identically zero. We have already stated this fundamental property during our discussion of potential flow at the beginning of Chapter 6, and also derived it in the preceding section, on the basis of Kelvin's theorem (Section 7.2.1(i)).

Elongation and tilting of vortex tubes

The additional term $(\boldsymbol{\omega}\cdot\nabla)\mathbf{v}$ in Equation 7.41 involves spatial variations of the velocity vector $\mathbf{v}$ for displacements in the direction of the vorticity vector $\boldsymbol{\omega}$. This term is present even in the case of ideal fluids, in situations where an initial, non-zero, vorticity $\boldsymbol{\omega}$ exists. Throughout the present section, in fact, viscosity effects will be neglected; their consequences are treated in the section that follows.

Here, we calculate the change in $\boldsymbol{\omega}$ from the deformations of an element of a vorticity tube of length $\delta\ell$ parallel to $\boldsymbol{\omega}$ and perpendicular cross-section S (Fig. 7.13). We have already seen in Section 7.2.1 that tubes of vorticity are convected by the fluid just as a set of material lines. For the purposes of our analysis, we split the term $(\boldsymbol{\omega}\cdot\nabla)\mathbf{v}$ into a component parallel to the vector $\boldsymbol{\omega}$ (assumed, in the following, to be parallel to the z-axis) and a component perpendicular to $\boldsymbol{\omega}$ (labelled "$\perp$").With this choice of coordinates, Equation 7.41 can then be written in the form:

$$\frac{\mathrm{d}\boldsymbol{\omega}}{\mathrm{d}t} = \omega_z \frac{\partial v_z}{\partial z}\mathbf{e}_z + \omega_z \frac{\partial v_\perp}{\partial z}\mathbf{e}_\perp, \tag{7.49}$$

where $\mathbf{e}_z$ and $\mathbf{e}_\perp$ are the respective unit vectors along the z-axis, and normal to it.

- The term $\omega_z\ \partial v_z/\partial z$ represents the effect of the elongation of the element. As $\delta\ell$ increases (Figure 7.13a), the cross-section S decreases, and the magnitude of the vorticity $\boldsymbol{\omega}$ increases.
- The term $\omega_z\ \partial v_\perp/\partial_z$ corresponds to the tilt of the vortex tube, with no change in its length, nor in the magnitude of $\boldsymbol{\omega}$ induced by the variation along z of $v_\perp$ (Figure 7.13b).

Physically, the elongation term is the direct consequence of the conservation of the angular momentum, already mentioned in Section 7.2.1, associated with the conservation of circulation around the vortex tube. Indeed, any elongation of an element of stream tube in the fluid must be accompanied by a decrease in its cross-section, since the mass $\rho S\,\delta\ell$ of the element remains constant. As a result, the moment of inertia $\mathcal{J}$ of the tube decreases, since $\mathcal{J}$ is proportional to $S^2\delta\ell$. The angular velocity Ω, and thus the vorticity, must therefore also increase, in order that the angular momentum $\mathcal{J}\,\Omega$ be conserved. By reasoning in terms of the conservation of circulation, rather than that of angular momentum, we write that the quantity $\Gamma = \omega\ S$ is constant (Γ is effectively, according to Stokes' theorem, the circulation of the velocity around the vortex tube). Since $S\ \delta\ell$ is constant, we conclude that ω (and hence Ω) is proportional to $\delta\ell$ and that the ratio $\omega\ /\delta\ell$ must remain constant.

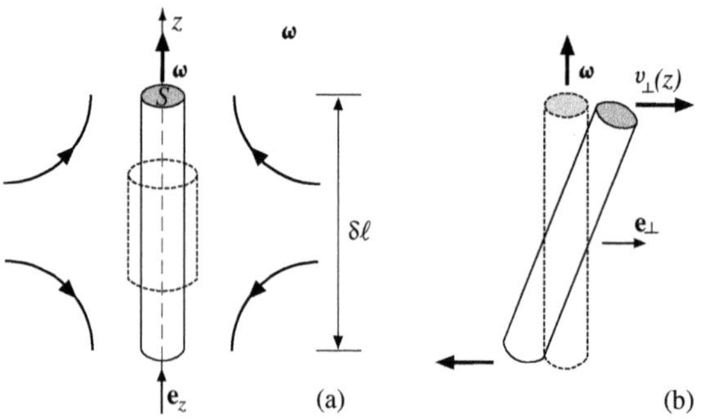

Figure 7.13 *(a) Change in the vorticity* $\boldsymbol{\omega}$ *associated to deformations of a vortex-tube by elongation; (b) change in* $\boldsymbol{\omega}$ *as the vortex-tube tilts*

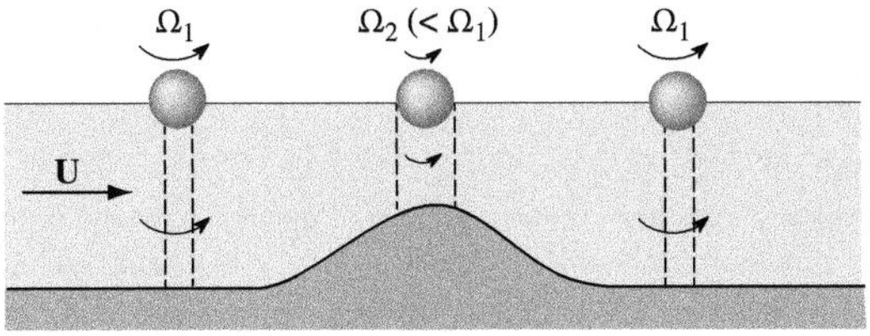

Figure 7.14 *The change in vorticity, resulting from a change in the depth of a fluid, is displayed by the change in the rotational velocity of a ball placed on the surface of the liquid, flowing with velocity **U** (from the movie "Vorticity" of NCFMF)*

This effect can be elegantly demonstrated by generating a vertical vortex line in a horizontally flowing layer of liquid. The vorticity of the vortex can be visualized by floating a small ball on the surface of the liquid, and observing its resultant rotation (Figure 7.14). The presence of a bump along the bottom of the channel leads to a decrease in the length of a vortex tube, and hence to a slowing of the rotation of the ball (ω and $\delta\ell$ decrease proportionately in order to maintain their constant ratio).

We will see in Section 7.6, which discusses rotating flows, the significance of these elongational effects for atmospheric phenomena in the presence of elevation differences in the ground level.

An example of the application of the conservation of vorticity: Hill's spherical vortex

Hill's vortex represents a limiting case, where vorticity is distributed throughout the volume of a sphere of radius R; the other limit corresponds to the distributions of vortex lines discussed previously. The components of $\boldsymbol{\omega}(\mathbf{r})$ can be written in cylindrical coordinates as:

$$\omega_\varphi = A\,r,\ \omega_r = \omega_z = 0 \text{ (inside the sphere)}, \quad (7.50a) \qquad \boldsymbol{\omega} = 0 \text{ outside the sphere.} \quad (7.50b)$$

The vortex lines are therefore circles centered on the axis of symmetry, the z-axis and normal to it (Figure 7.15). We assume that the fluid is ideal, and incompressible, and that the vorticity was generated as part of the initial conditions.

The significance of the shape of this distribution can be understood by analyzing the evolution of a vorticity tube, consisting of a toroid of average radius r, with cross-sectional radius a (Figure 7.15(a)); the circulation around this toroid has magnitude $\pi a^2 \omega_\varphi(\mathrm{r})$. If the vortex tube is deformed while being dragged along by the fluid motion, its volume $2\pi^2 r(t) a^2(t)$ must remain constant because of the incompressibility of the fluid. Conservation of the circulation in the tube results therefore in the quantity $a^2 \omega_\varphi(r) \propto \omega_\varphi(r)/r(t)$ remaining constant as the tube moves along with the fluid (Lagrangian

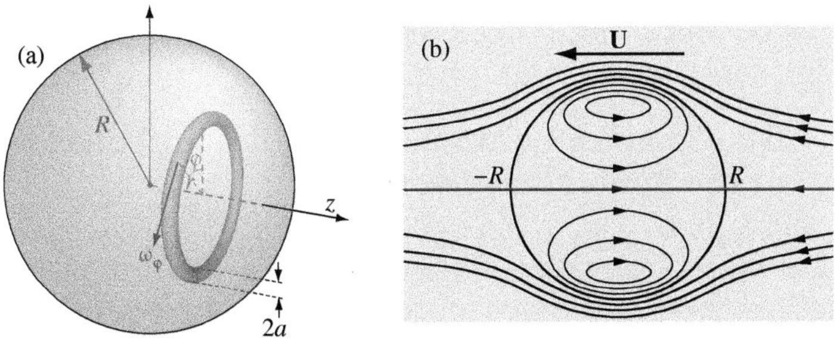

Figure 7.15 *Hill's spherical vortex; (a) schematic diagram of a toroidal vortex tube inside the vortex; (b) shape of the streamlines, both in the interior and around this vortex, in a reference frame moving with the velocity of the entire vortex*

point of view). The choice $\omega_\varphi = Ar$ therefore corresponds to a distribution of vorticity which obeys the above condition at all times.

Quantitative proof
We write the azimuthal component of the equation of evolution for the vorticity (Equation 7.42), denoting by $\mathbf{e}_r$ and $\mathbf{e}_\varphi$ the respective unit vectors in the radial and azimuthal directions. In cylindrical coordinates, this equation has only one non-zero component, the one in the φ-direction. Expanding the term $(\boldsymbol{\omega} \cdot \nabla)\mathbf{v}$ in cylindrical coordinates, while using the relationship $\partial \mathbf{e}_r / \partial \varphi = \mathbf{e}_\varphi$, and the rotational symmetry of the problem around the z-axis (which implies that $\partial v_r / \partial \varphi = 0$), we find:

$$\frac{\mathrm{d}\omega_\varphi}{\mathrm{d}t}\mathbf{e}_\varphi = \omega_\varphi \frac{1}{r}\frac{\partial}{\partial \varphi}(v_r\, \mathbf{e}_r) = \frac{\omega_\varphi v_r}{r}\mathbf{e}_\varphi.$$

This, in turn, leads to:

$$\frac{\mathrm{d}}{\mathrm{d}t}\left(\frac{\omega_\varphi}{r}\right) = \frac{1}{r}\frac{\mathrm{d}\omega_\varphi}{\mathrm{d}t} + \omega_\varphi \frac{\mathrm{d}}{\mathrm{d}t}\left(\frac{1}{r}\right) = \frac{1}{r}\frac{\omega_\varphi v_r}{r} - \frac{\omega_\varphi}{r^2}v_r = 0. \tag{7.51}$$

We have thus shown that the relation $\omega_\varphi / r = \text{constant}$ remains satisfied at all times by the vorticity in Hill's spherical vortex since it allows one to satisfy the evolution equation. Moreover, one can show, too, that, *in a reference frame moving along with the average displacement of the vortex*, the distributions of vorticity and velocity are stationary: fluid particles located inside the sphere of radius R move along the closed trajectories shown on Figure 7.15b. We can calculate the resulting velocity field by applying the law of Biot and Savart (Equation 7.19b) to this distribution of vorticity.

Stream function for Hill's spherical vortex In a reference frame moving at the mean velocity $\mathbf{U}$ of the vortex, it can be shown that the stream function Ψ, defined, in cylindrical coordinates, by Equation 3.52 is, within the vortex:

$$\Psi = -\frac{A}{10}r^2\left(R^2 - z^2 - r^2\right). \tag{7.52a}$$

The surface $z^2 + r^2 = R^2$ and the axis $r = 0$ are therefore streamlines. Moreover, outside the vortex, the vorticity vanishes: the flow is then potential and identical to that around a sphere of radius R. The macroscopic displacement velocity of the vortex, in a reference frame where the fluid at infinity is at rest, is therefore related to the constant A by:

$$U = \frac{2}{15}AR^2 \tag{7.52b}$$

Finally, the tangential component of the velocity, at the vortex surface $r = R$, equals $-(1/5)A\, r\, R$.

7.3.2 Equilibrium between elongation and diffusion in the dynamics of vorticity

An important feature of the equation of transport of the vorticity (Equation 7.42) is the coexistence of a viscous diffusion term which tends to spread out the distribution of vorticity, and of an elongation term which, in contrast, tends to concentrate the vorticity and to increase its magnitude. The equilibrium between these two terms is a fundamental aspect of the dynamics of the vorticity in turbulent flows, which will be discussed in Chapter 12.

Evolution of the vorticity in an axially symmetric, elongational flow

This model provides an approximate representation of the dynamics of the exchanges of vorticity for ordinary flow near an orifice at the bottom of a cylindrical container (emptying of a container). We have already discussed (Section 7.1.2) the Rankine vortex, one model for the velocity field of a vortex due to such a flow. We need to understand now why, in that case, the vorticity remains localized within a small-diameter core instead of spreading out uniformly throughout the fluid.

Consider an axially symmetric, incompressible, irrotational flow (Figure 7.16) with velocity components:

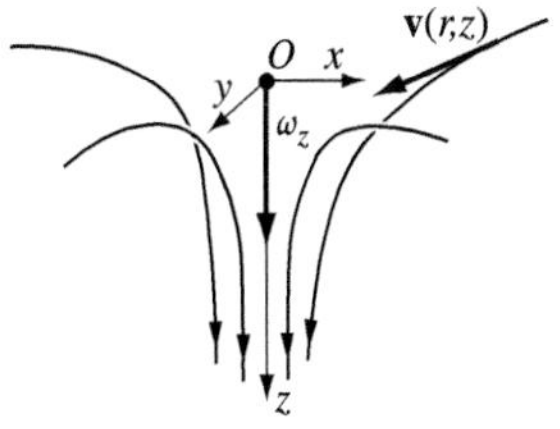

Figure 7.16 *Model of an axially-symmetric elongational flow (see Figure 7.3)*

$$v_r = -(a/2)\,r \qquad \text{and} \qquad v_z = a\,z \qquad \text{with } (a > 0). \tag{7.53}$$

This flow corresponds to an elongation along the z-axis, compensated by the radial flow needed to obey mass conservation; it is reasonably representative of the velocity field near an orifice at the bottom of a container.

Assume now that this flow is *perturbed* by the introduction of a small amplitude distribution of vorticity $\omega_z(\mathbf{r}, t)$ (the emptying whirlpool). Let us now write, for this distribution of vorticity, the equation of transport (Equation 7.41) in cylindrical coordinates:

$$\frac{\partial \omega_z}{\partial t} = \frac{a}{2r}\frac{\partial}{\partial r}\left(\omega_z r^2\right) + \frac{\nu}{r}\frac{\partial}{\partial r}\left(r\frac{\partial \omega_z}{\partial r}\right). \tag{7.54}$$

The first term on the right-hand side corresponds to the sum of the elongation and convection terms in Equation 7.41. Under stationary conditions, the left-hand side is zero $(\partial \omega_z/\partial t) = 0$, so that Equation 7.54, after integration with respect to r, becomes:

$$\frac{a}{2}\omega_z r^2 + \nu\, r\frac{\partial \omega_z}{\partial r} = \text{constant}. \tag{7.55}$$

The constant in the above equation must be necessarily zero; otherwise, further integration of Equation 7.55, in the region of small r, leads to a divergence of the vorticity ω_z. Furthermore, at large values of the radius r, it would decrease as $1/r^2$, so that the total vorticity, integrated over the radial cross section of the flow, i.e. the circulation, would also diverge. From physical considerations, however, it must remain finite and constant, because in this situation there exists no mechanism for creating vorticity. The distribution of vorticity must therefore obey the equation:

$$\omega_z = \omega_1\, e^{-ar^2/4\nu}. \tag{7.56}$$

This result corresponds to a balance, over a characteristic distance $\delta_D \approx \sqrt{\nu/a}$, between the effects of elongation of the vorticity under the action of the elongational field $\mathbf{v}$, and those of a spreading due to diffusion. In the case of flow through a hole of diameter d, with a characteristic velocity U, we have:

$$a \approx \frac{U}{d} \qquad \text{and} \qquad \frac{\delta_D}{d} \approx \sqrt{\frac{\nu}{U\,d}} \approx \sqrt{\frac{1}{Re}}, \tag{7.57}$$

where Re is the Reynolds number ($Re = Ud/\nu$). Thus, the larger the Reynolds number characteristic of the flow, the more concentrated will be the vorticity inside a small diameter core. The model of the Rankine vortex (Section 7.1.2) represents only an approximation to the structure of such vortices (there is no definite boundary to the core), but it describes correctly the fact that most of the vorticity of the flow is concentrated within a very small radius inside the core. The variation of this radius as $1/\sqrt{Re}$ is indicative of the balance between convection and diffusion. We will come across an identical dependence in Chapter 10 as we study boundary layers, where equilibria of the same type are observed.

Creation and annihilation of vorticity in turbulent flow

The above result suggests an analogy with the volume distributed turbulence, which we study in Section 12.6. In this model, we assume that the energy transfer is carried out by convective, non-dissipative mechanisms, from large-scale turbulent structures to smaller and smaller vortices in which vorticity is concentrated (*Kolmogorov energy cascade*). The elongation and bending of vortex tubes plays an essential role in this process. On the other hand, viscosity comes in only at the scale of the smallest vortices and, just as in the problem we have discussed, it is mainly significant over distances on the scale of the vortex core.

In contrast, in two-dimensional flows where the velocity field $\mathbf{v}$ is independent of one of the coordinates (e.g., the vertical), the $(\boldsymbol{\omega}\cdot\nabla)\mathbf{v}$ term is identically zero. This property is characteristic, on large scales, of the turbulence in atmospheric and oceanographic flows. It is due to the finite vertical extent of these flows (corresponding to the depth of oceans, or the thickness of the atmosphere), and especially also to the influence of the angular velocity $\boldsymbol{\Omega}$ of the earth: this rotation tends to decouple the components of motion perpendicular and parallel to the surface of the earth, as we will see in Section 7.6, where we discuss rotating flows. For the same reason, two-dimensional turbulence, where stretching of vortices cannot take place, has very different characteristics from that in three dimensions.

We have previously discussed the dynamics of vortex lines by considering them isolated, and assuming that they are subjected to an external velocity field. In both two- and three-dimensional flows, we must also take into account the interaction between different vortex tubes. Finally, for curved vortices, we must also consider the interaction with the velocity field of the vortex itself (Section 7.1.3). We study these effects in the following section, dedicated to the dynamics of filamentary vortices.

7.4 A few examples of distributions of vorticity concentrated along singularities

7.4.1 Vorticity concentrated along specific lines

At the beginning of this chapter (Section 7.1.1) we have mentioned examples of flows where the vorticity is concentrated along filaments. We might also recall the formation of straight vortex filaments in alternate rows, downstream of a cylindrical obstacle placed normal to a flow. In that case, they form an alternating alley (the *Bénard-von Karman vortex street*), seen previously in Section 2.4.2. A similar, but single, row of vortices is observed at the *mixing layer* between two parallel flows moving at different velocities, whether of the same fluid or of different fluids (see Section 11.4.1). In the latter cases, the ratio of the size of the vortex core to the overall dimensions of the structure is larger than for the vortex filaments, but the rotation around a central zone is clearly visible.

In other cases, vortex lines form closed loops (vortex rings). Specifically, we might mention smoke rings emitted from the apertures of pipes or other circular openings: for example, from the mouth of a cigarette smoker or, on a larger scale, from active volcanic craters (Figure 7.17). Another illustration, this time microscopic, is that of vortex rings smaller than a micron in size, which have been identified in superfluid helium, and whose dynamics have been extensively studied experimentally (see Section 7A.5 at the end of this chapter).

Figure 7.17C *Observation of a vortex ring rising above the volcano Etna, as seen from the "Torre del Filosofo". One can see, as the ring moves, the displacement of its shadow on the outer slope of the crater (document J. Alean)*

In the remainder of this section, we model various flows by means of *vortex filaments* with very small core radii. We treat first the case of straight-line vortices, then that of vortex rings.

7.4.2 Dynamics of a system of parallel-line vortices

We discuss distributions of parallel-axis, straight-line vortex filaments corresponding to several of the examples cited above. In the absence of viscosity, every element of the vortex core moves at the local fluid velocity at that point. This velocity is the sum of the external velocity field and of the velocity resulting from all other vortices, since a straight-line vortex does not affect itself. As seen in Section 7.2.1, this result is a direct consequence of Kelvin's theorem as it applies to ideal fluids.

Parallel, line-vortex pairs

We begin with the simplest example of two parallel vortex filaments with respective circulations Γ_1 and Γ_2 (Figure 7.18a). The fluid velocity at the cores O_1 and O_2 of each filament is simply equal to the velocity induced by the other vortex. This latter velocity is normal to the line segment $\mathbf{O}_1\mathbf{O}_2$ and has respective magnitudes: $\Gamma_2/(2\pi\, d)$ and $\Gamma_1/2\pi\, d$ where d is the separation between the vortex cores.

Two cases are particularly important:

- $\Gamma_1 + \Gamma_2 = 0$ (the vortex pair shown in Figure 7.18b). The pair of vortices moves at constant velocity in the direction normal to the line joining them. In this case, this linear velocity has magnitude $V_p = \Gamma/(2\,\pi d)$ where Γ is the magnitude of each circulation. We find a similar result in the case of vortex rings. There, the circulations of the fluid velocity around the core at two diametrically opposite points of the ring are opposite.
- $\Gamma_1 = \Gamma_2$ (Figure 7.18c). The pair of lines rotates around the center C of the line segment $\mathbf{O}_1\mathbf{O}_2$, with angular velocity $\Gamma/\pi d^2$ where Γ again is the magnitude of each circulation.

More generally, for an arbitrary value of the ratio Γ_1/Γ_2, the vortex lines rotate around their "center of gravity" (we assign to them respective coefficients Γ_1 and Γ_2) with an angular

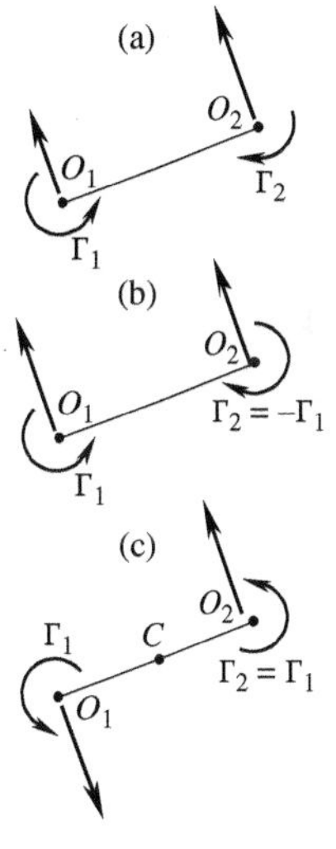

Figure 7.18 *Velocity field associated with two parallel line vortices with respective circulations Γ_1 et Γ_2; (a) the most general case; (b) $\Gamma_1 + \Gamma_2 = 0$, the line $\mathbf{O}_1\mathbf{O}_2$ moves parallel to itself; (c) $\Gamma_1 = \Gamma_2$, the line $\mathbf{O}_1\mathbf{O}_2$ rotates about its center C*

velocity $(\Gamma_1 + \Gamma_2)/(2\pi d^2)$. In our example of the pair of opposite vortices, the center of gravity is at infinity.

The results just discussed for the case of vortex pairs are an indirect consequence of the laws of conservation of momentum and angular momentum, for the ideal fluid within which the lines are moving. Thus, in the more general case of arbitrary numbers of lines, the following results hold:

- the total circulation $\Sigma\Gamma_i$ remains constant during the motion;
- the center of gravity G of the system of vortices, defined by the equation $\Sigma\Gamma_i\, \mathbf{GO}_i = \mathbf{0}$, remains fixed.

We now continue by discussing a number of specific distributions of vorticity, which are good approximations of real flows.

Continuous and discrete vortex sheets

A tangential-velocity discontinuity (a free shear layer) can result from the superposition of two layers (whether or not of the same fluid) initially separated by a very thin wall and in tangential contact with each other at two different velocities, constant within each layer:

$$v_x = U_1 (y > 0), \qquad v_x = U_2 (y < 0)$$

Corresponding to this flow, there is an infinitely thin vortex sheet, continuous along the plane $y = 0$ and with a uniform density of vorticity γ_1 per unit length along the z-axis with:

$$\gamma_1 = \lim_{\varepsilon \to 0} \left[\int_{-\varepsilon}^{\varepsilon} \omega_z \, \mathrm{d}y \right]; \tag{7.58}$$

Figure 7.19 *Circulation along a contour (C), associated to a flow displaying a discontinuity of the tangential velocity*

γ_1 can be evaluated by applying Ampere's law to the contour (C) in Figure 7.19:

$$(U_1 - U_2)\, d = \gamma_1 d, \quad (7.59) \qquad \text{whence:} \qquad \gamma_1 = (U_1 - U_2).$$

It should be noted that, in these equations, γ_1 has the dimensions of a velocity, since it represents a circulation per unit length. In a real experimental situation, such a vortex sheet is unstable; the vorticity concentrates in vortex cores periodically located along a line parallel to the x-axis, along the direction of the flow. This instability is discussed in more detail in Section 11.4.1.

Simple vortex street

We now consider a large number of parallel straight line vortices evenly spaced along a straight line (simple vortex street). The velocity field at an arbitrary point can be easily calculated by superposition of the velocity fields resulting from each individual vortex. We have seen in Section 6.6.2(ii) that the complex velocity potential $w(z) = v_x - \mathrm{i}\, v_y$ at a coordinate point $z = x + \mathrm{i}\, y$ is, for an isolated vortex located at the point with coordinates x_i and y_i $(z_i = x_i + \mathrm{i}\, y_i)$ and with an algebraic value Γ of the circulation:

$$f(z) = -\frac{\mathrm{i}\,\Gamma}{2\pi} \mathrm{Log}\,(z - z_i). \tag{7.60}$$

Corresponding to this potential, there exists a tangential velocity field $v_\varphi = \Gamma/(2\pi\,|z - z_i|)$ due to the vortex located at z_i.

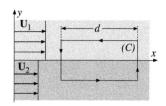

For an infinite system of parallel vortex lines periodically located along the real axis between $-\infty$ and $+\infty$ at the points $z_m = m\,a$ (where m is any integer), we have a complex velocity:

$$w(z) = \frac{dF}{dz} = -\frac{i\,\Gamma}{2a}\cot\left(\frac{\pi\,z}{a}\right). \tag{7.61}$$

In order to obtain the complex velocity $w_m(z)$ for each vortex, located at $z_m = m\,a$, we must subtract from $w(z)$ the contribution of that particular vortex so that we obtain:

$$w_m(z) = -\frac{i\,\Gamma}{2a}\left[\cot\left(\frac{\pi\,z}{a}\right) - \frac{a}{\pi(z - ma)}\right]. \tag{7.62}$$

It can be shown, by expanding the cotangent function in the neighborhood of each point $z_m = m\,a$, that $w_m(z)$ is zero for $z = ma$, and that, consequently, this velocity is zero at the center of each vortex. The vortex row remains motionless, a fact which might have easily been predicted from symmetry arguments.

Proof

We begin by calculating the complex potential of the vortex street from the velocity potential (Equation 7.60) for each vortex:

$$F(z) = -\frac{i\,\Gamma}{2\pi}\left(\sum_{m=-\infty}^{m=\infty} \mathrm{Log}\,[z - z_m]\right) = -\frac{i\,\Gamma}{2\pi}\mathrm{Log}\left(\prod_{m=-\infty}^{m=\infty} [z - ma]\right). \tag{7.63}$$

Rearranging, in the equation above, terms corresponding to the same absolute value of m, in the form:

$$\mathrm{Log}\,(z - ma) + \mathrm{Log}\,(z + ma) = \mathrm{Log}\left[z^2 - ma)^2\right] = \mathrm{Log}(ma)^2 + \mathrm{Log}\left[\frac{z^2}{m^2a^2} - 1\right],$$

we obtain: $$F(z) = -\frac{i\,\Gamma}{2\pi}\mathrm{Log}\left(z\prod_{m=1}^{\infty}\left(1 - \frac{z^2}{m^2a^2}\right)\right) - \frac{i\,\Gamma}{2\pi}\mathrm{Log}\left(\prod_{m=1}^{\infty}(-1)^n(ma)^2\right).$$

Denoting by F_0 the second term (independent of z, which appears in this equation), and using the identity:

$$\sin x = x\prod_{m=1}^{\infty}\left(1 - \frac{x^2}{m^2\pi^2}\right),$$ we obtain: $$F(z) = -\frac{i\,\Gamma}{2\pi}\mathrm{Log}\left[\sin\left(\frac{\pi\,z}{a}\right)\right] - \frac{i\,\Gamma}{2\pi}\mathrm{Log}\frac{a}{\pi} + F_0.$$

The complex velocity $w(z)$ is then obtained by taking the derivative of $F(z)$ with respect to z.

Benard-von Karman vortex street

The above calculation scheme may also be applied to the case of the double, alternating, vortex street. We have seen in Section 2.4.2 that, above Reynolds numbers of the order of 50, alternating vortices are periodically shed downstream of cylindrical obstacles. This system of vortices is known as the *Bénard-von Karman (BvK) vortex street* (Figure 2.9). We now calculate the velocity field along the vortex street, which we represent as two, parallel, single rows of the type just treated (Figure 7.20), each shifted relative to the other by

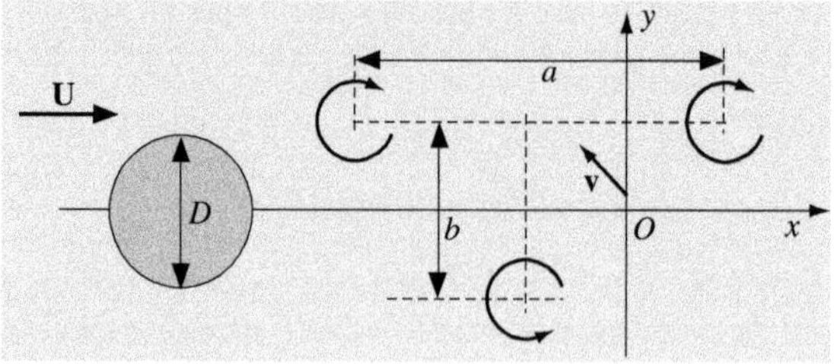

Figure 7.20 *Schematic diagram of the Bénard-von Karman double vortex street generated in a uniform velocity stream passing across a cylinder (see also Figure 2.9c)*

half the spacing between vortices. The sign of the circulation is identical for every vortex in the same row, but opposite for the ones in the facing row.

The double vortex street moves lengthwise, parallel to itself, as a result of both the overall flow velocity **U** and the velocity induced on a given vortex by all those of the other row (we have indeed just proved that velocity contributions due to vortices in the same row cancel, by symmetry). Moreover, considering the effects of pairs of vortices in the facing row symmetric with respect to the vortex of interest shows that there can be only an x-component to the velocity induced on it; since the velocity is identical at each vortex, the system moves then with no deformation. The detailed calculation of this velocity is found below. It can be shown, by a linear instability analysis, that the only solution for an infinitely long vortex street, stable to first order, is one for which the vortices are located in alternate positions in the two rows, with a ratio $b/a \approx 0.3$ independent of the velocity.

Calculation of the velocity for a double vortex street Consider the case of two rows of vortices, each infinitely long, like those of the simple vortex street; let us denote by W the complex velocity induced by the vortices of the lower row on a vortex of the upper row, located at the point $(0, b/2)$. Taking into account the fact that the upper row induces a zero resultant velocity on this vortex, and using Equations 7.61 and 7.62 derived for the case of the simple vortex street, as well as the identity $\tan(\mathrm{i}x) = \mathrm{i}\tanh(x)$, we obtain:

$$w(z) = -\frac{\mathrm{i}\,\Gamma}{2a}\cot\left[\frac{\pi}{a}\left(z + \frac{a}{2} + \mathrm{i}\frac{b}{2}\right)\right],$$

i.e.:

$$W = w\left(\mathrm{i}\frac{b}{2}\right) = \frac{\mathrm{i}\,\Gamma}{2a}\tan\left(\frac{\mathrm{i}\,\pi b}{a}\right) = -\frac{\Gamma}{2a}\tanh\left(\frac{\pi b}{a}\right). \tag{7.64}$$

We show in the same manner that the velocity induced by the vortices in the upper row on those of the lower row has the same value. This demonstrates that the entire double vortex street moves globally at the velocity W given by Equation 7.64.

Profile of the flow velocity between the double row of vortices Up to this point, we have been mainly interested in the velocity induced on each vortex by all the others. It is equally important to determine the velocity field in the remainder of the fluid, and specifically between the rows of vortices. Moreover, perturbations of the velocity field by the vortex street allow us to estimate the momentum which it carries along and the force that would result on any obstacle. For this purpose we need to calculate the velocity profile $w(y)$ along the y-axis (Figure 7.20), perpendicular to the axis of the double vortex street (chosen as the x-axis) and at the same distance from the two nearest vortices (this avoids perturbations of the velocity field in the region near each vortex). We then obtain:

$$v_x(y) = -\frac{\Gamma}{a}\frac{\sinh(\pi\, b/a)\,\cosh(\pi\, b/a)}{\sinh^2(\pi\, b/a) + \cosh^2(2\pi\, y/a)}, \tag{7.65a}$$

and

$$v_y(y) = \frac{\Gamma}{a}\frac{\cosh(\pi\, b/a)\,\cosh(2\pi\, y/a)}{\sinh^2(\pi\, b/a) + \cosh^2(2\pi\, y/a)}. \tag{7.65b}$$

The two velocity components v_x and v_y are even functions of y, and both change sign according to Γ. In the configuration of Figure 7.20, v_y is positive while v_x is negative. The maximum absolute value of these two components is obtained at $y = 0$ where respectively $v_x(0) = -(\Gamma/a)\tanh(\pi b/a)$ and $v_y(0) = \Gamma/(a\cosh(\pi b/a))$. Finally, at a large positive transverse distance y, the two components v_x and v_y decrease respectively exponentially as $e^{-4\pi y/a}$ and $e^{-2\pi y/a}$.

Proof

In order to calculate the velocity profile $w(y)$, we use Equation 7.61, derived for a vortex row, by replacing z by $z - z_0$ (with a circulation $-\Gamma$) for the upper row, and z by $z + z_0$ (with circulation $+\Gamma$) for the lower row. The coordinates $z_0 = ib/2 + a/4$ and $-z_0$ correspond to the cores of the nearest vortices and the affix of the points where we measure the velocity profile is iy. Summing the contributions of both rows, we obtain the complex velocity field:

$$w(z) = -\frac{i\,\Gamma}{2a}\left[\cot\left(\frac{\pi\,(i\,y + z_0)}{a}\right) - \cot\left(\frac{\pi\,(i\,y - z_0)}{a}\right)\right].$$

Using the trigonometric equations: $\cot(p) - \cot(q) = \sin(q-p)/(\sin p\,\sin q)$ and $\sin(p)\sin(q) = (\cos(p-q) - \cos(p+q))/2$, we obtain, in succession:

$$w(y) = \frac{i\,\Gamma}{2a}\frac{\sin(2\pi\, z_0/a)}{\sin\{\pi\,(i\,y + z_0)/a\}\,\sin\{\pi\,(i\,y - z_0)/a\}} = \frac{i\Gamma}{a}\frac{\sin(2\pi z_0/a)}{\cos(2\pi\, z_0/a) - \cos(2\pi\, iy/a)}.$$

Substituting for z_0 its value and using the equations $\cos(ix) = \cosh(x)$ and $\sin(ix) = i\sinh(x)$, the previous equation becomes for z_0 on the y-axis:

$$w(y) = \left(v_x(y) - i\, v_y(y)\right) = \frac{\Gamma}{a}\frac{\cosh(\pi\, b/a)}{-\sinh(\pi\, b/a) + i\,\cosh(2\pi y/a)},$$

from whence we derive Equations 7.65.

Frequency of vortex shedding in the double vortex street Let us now estimate the vortex shedding frequency f corresponding to the example in Figure 7.20. The velocity of the vortex street relative to the cylindrical obstacle is proportional to U, the incident flow velocity on the cylinder. Indeed, the velocity W, which we have just calculated, is itself proportional to the external flow velocity. On the other hand, the transverse spacing a can be assumed to be proportional to the diameter D of the obstacle. The vortex shedding frequency f thus satisfies:

$$f \approx \frac{U + W}{a} \approx \alpha\frac{U}{D}, \tag{7.66}$$

whence:

$$Sr = \frac{f}{(U/D)} \approx \alpha. \tag{7.67}$$

The dimensionless number Sr, previously defined in Section 2.4.2, is known as the *Strouhal number*. Under the assumptions that we have made, it is independent of the velocity and

These characteristics have led to the design of flow meters, measuring the fluid velocity by means of the frequency at which the vortices are formed. Sharp-edged obstacles are used, since the vortex shedding is then more stable and Sr is less dependent on the Reynolds number. The formation of vortices can be detected by measuring oscillations in the pressure difference between the two faces of the obstacle, parallel to the flow.

The formation of vortices behind structures can also have undesirable consequences, like in the example of the collapse of the Tacoma bridge discussed at the beginning of Section 2.4. More generally, the interaction between fluids and solid bodies makes up a mixed discipline with numerous applications which we will not discuss further here.

the nature of the fluid for a given obstacle and thus constitutes a dimensionless parameter characterizing the frequency of vortex shedding.

In accordance with Equations 7.66 and 7.67, Sr should be less than 1 since we have $a > D$, as seen in Figure 7.20; moreover, $U + W$ is smaller than U because the velocities U and W are in opposite directions. Experimentally, it is observed that Sr is of the order of 0.2 for a circular cylinder, and that it has little dependence on the nature of the fluid or on the Reynolds number, so long as the latter is sufficiently large (over a few thousands, if we take the size of the object as the typical length scale).

Momentum of a set of parallel vortex filaments

Let us begin with the case of two parallel counter-rotating parallel vortices with circulation vectors $\boldsymbol{\Gamma}_1$ and $\boldsymbol{\Gamma}_2 = -\boldsymbol{\Gamma}_1$ (Figure 7.18b). As in the preceding discussion, we will consider this situation in a plane perpendicular to the two vortices (their intersections are the points O_1 et O_2). In order to compute the momentum, we start with a configuration where the two vortex cores coincide at the origin O (the velocity fields in that case cancel out exactly), and we apply to each of them a force $\mathbf{f}_i$ per unit length (i = 1,2); due to this force, each vortex acquires a velocity component $\mathrm{d}\mathbf{r}_i/\mathrm{d}t$ perpendicular to $\mathbf{f}_i$ and to the core of the vortex (with $\mathbf{r}_i = \mathbf{OO}_i$). This velocity is such that the corresponding Magnus force per unit length $\mathbf{F}_{Mi} = \rho(\mathrm{d}\mathbf{r}_i/\mathrm{d}t) \times \boldsymbol{\Gamma}_i$ is equal to $-\mathbf{f}_i$ (see Section 6.3.1): this allows one to keep the global force on each vortex filament equal to zero. Taking $\mathbf{f}_1 = \mathbf{f}_2$ and because $\boldsymbol{\Gamma}_2 = -\boldsymbol{\Gamma}_1$, the velocities $\mathrm{d}\mathbf{r}_i/\mathrm{d}t$ of the two vortices induced by the forces are opposite so that they move apart ($\mathbf{f}_1$ and $\mathbf{f}_2$ must be chosen perpendicular to the orientation of $\mathbf{O}_1\mathbf{O}_2$ desired); at the same time, the work done by the total force $\mathbf{f}_1 + \mathbf{f}_2$ corresponds to the increase in the energy of the vortex pair. We obtain the momentum per unit length of the pair by integrating the total force with respect to time from $t = 0$ to the moment when the filaments reach the distance $|\mathbf{r}_2 - \mathbf{r}_1| = d$ of interest:

$$\mathbf{P} = \int_0^t (\mathbf{f}_1 + \mathbf{f}_2)\mathrm{d}t = \int_0^t \rho\left(\frac{\mathrm{d}\mathbf{r}_1}{\mathrm{d}t} \times \boldsymbol{\Gamma}_1 + \frac{\mathrm{d}\mathbf{r}_2}{\mathrm{d}t} \times \boldsymbol{\Gamma}_2\right)\mathrm{d}t,$$

whence $$\mathbf{P} = \rho\,(\mathbf{r}_1 \times \boldsymbol{\Gamma}_1 + \mathbf{r}_2 \times \boldsymbol{\Gamma}_2) = -\rho\,(\mathbf{O}_1\mathbf{O}_2 \times \boldsymbol{\Gamma}_1).$$

When $|\mathbf{O}_1\mathbf{O}_2| = d$, the modulus of $\mathbf{P}$ is then $P = \rho\, d\, \Gamma$ where Γ is the magnitude of the circulations. One notes that the result is independent both on the choice of the origin O and on the magnitude of the forces $\mathbf{f}_i$.

Let us now generalize this result to an arbitrary number of parallel vortices. We proceed as before by having them all start at the same origin O; we then subject them to a force $\mathbf{f}_i$ per unit length, which results in a velocity $\mathrm{d}\mathbf{r}_i/\mathrm{d}t$ such that the Magnus force $\mathbf{F}_{Mi} = -\rho(\mathrm{d}\mathbf{r}_i/\mathrm{d}t) \times \boldsymbol{\Gamma}_i$ equals $-\mathbf{f}_i$ (we then have a zero resultant force). It is enough to choose $\mathbf{f}_i$ such that $\mathrm{d}\mathbf{r}_i/\mathrm{d}t$ is directed along $\mathbf{OO}_i$. By adding up the various contributions, as in the preceding equation, we find:

In order for this result to be independent of the origin O the sum of all the circulations must be zero, which also results in a zero circulation of the velocity at infinity.

$$\mathbf{P} = \rho \sum_i (\mathbf{r}_i \times \boldsymbol{\Gamma}_i). \quad (7.68a)$$

For the case of a continuous distribution of vorticity, still two-dimensional, we can adjust this equation by applying it to elementary vortices of cross-section $\mathrm{d}S$ perpendicular to their length and then integrating in the same plane.

A rigorous, but delicate, proof of the preceding results, as well as their generalization to three dimensions, has been carried out by P.G. Saffman.

$$\mathbf{P} = \rho \iint (\mathbf{r} \times \boldsymbol{\omega}(\mathbf{r}))\mathrm{d}S. \quad (7.68b)$$

7.4.3 Vortex rings

A vortex ring can be imagined as a small diameter vortex tube closed on itself, very similar to an infinitesimally thin doughnut (Figure 7.21). The circulation Γ is constant along any contour looping once around the vortex core. A vortex ring is a very stable vorticity structure, frequently observed in hydrodynamics whenever obstacles or orifices with circular symmetry are present (Figure 7.17).

We can generate experimentally vortex rings by the means of a cylindrical can in which one of the end plates has been replaced by a thin stretched elastic membrane, while a circular hole has been carved in the other end plate. After having introduced smoke into the can, a smoke ring can be emitted at the hole by flicking the membrane; we can then study its trajectory, or how it is deviated in the presence of an obstacle.

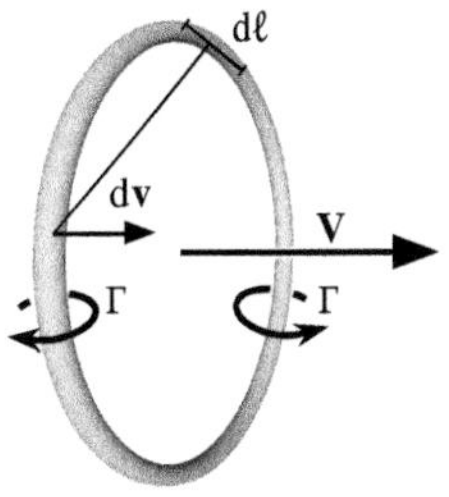

Figure 7.21 *Schematic diagram of a vortex ring with circulation Γ moving at uniform velocity*

Velocity of a vortex ring

We study below the case of a plane circular vortex ring, of radius R, moving in an ideal fluid as a result of the velocity induced at each point M by the other elements $\mathbf{dl}$ of the vortex core (we assume a zero external velocity field). We have seen in Section 7.1.3 that the corresponding velocity component $\mathbf{dv}$ is normal both to the line element and to the vector connecting M to $\mathbf{dl}$; $\mathbf{dv}$ is therefore perpendicular to the plane of the vortex ring. The total velocity induced at a given point is then also normal to the plane of the ring and, because of the symmetry of the problem, it is the same at every point of the vortex core. The ring therefore moves without deformation, parallel to its axis, at a velocity $\mathbf{V}$.

The order of magnitude of $\mathbf{V}$ can be calculated by means of Equation 7.20, which gives the velocity induced by an arc of the vortex line at each of its points:

$$V \approx \frac{\Gamma}{4\pi R} \operatorname{Log}\left(\frac{R}{\xi}\right),$$

where ξ is a length of the order of magnitude of the vortex core. When $\xi \ll R$, the displacement velocity results mainly from the divergent contribution of elements of the vortex line closest to the point in question. This contribution is here greater than that due to the velocity induced by the diametrically opposite vortex ring elements by a factor Log (R/ξ). An exact calculation, assuming that the vortex has a cylindrical core in uniform rotation, gives a closely similar value:

$$V = \frac{\Gamma}{4\pi R}\left(\operatorname{Log}\frac{8R}{\xi} - \frac{1}{2}\right). \tag{7.69}$$

For comparable values of circulation Γ, the larger the radius of the vortex ring, the slower it moves. This result is very reminiscent of that obtained above for two parallel line vortices of opposite circulation Γ, separated by a distance d, for which the displacement velocity, equal to $(\Gamma/(2\pi\, d))$, also varies as $1/d$.

Moreover, a vortex ring has kinetic energy and momentum, corresponding to the kinetic energy and momentum of the fluid it causes to move because of its structure: this is the reason why an impulse can be detected when a smoke ring impacts onto a solid surface. We now estimate directly the magnitude of these quantities.

Kinetic energy of a vortex ring

We start with the expression (Equation 7.9) of the kinetic energy e_k per unit length of a straight vortex filament. If the radius of curvature R of the ring is large relative to the core radius ξ, the total kinetic energy E_k of the ring is approximately $(2\pi\, R\, e_k)$, so that, neglecting the additive term due to the core:

$$E_k \approx \rho\, \Gamma^2 \frac{R}{2} \operatorname{Ln}\frac{R}{\xi}.$$

In the above equation, we have neglected any distortion of the velocity field by distant elements of the vortex line and we assume $L = R$ as the upper length scale of the radius r in the integral. In fact, at larger distances, the decrease of the velocity field is much faster (as $1/r^3$, as for a magnetic dipole) so that its contribution to E_k becomes negligible. The exact value of E_k, obtained by a full integration, is very close to our estimate:

$$E_k = \frac{\rho\,\Gamma^2 R}{2}\left(\mathrm{Log}\frac{8R}{\xi} - \frac{3}{2}\right). \tag{7.70}$$

We see therefore that in spite of the fact that the displacement velocity *decreases* in vortex rings, as their radius becomes larger, the overall kinetic energy *increases*!

Momentum of a vortex ring

Using the same approach as for straight vortex filaments, we calculate the momentum associated with a vortex ring by assuming that the vortex core is subjected to a constant force **f**, directed along the velocity **V**: the vortex responds to this force by an increase in its radius, which allows it to store the work done by the force **f** as additional kinetic energy. This change in radius leads to a Magnus force per unit length which balances exactly the force **f** with: $\mathbf{f}_M = \rho(\mathrm{d}\mathbf{r}/\mathrm{d}t) \times \mathbf{\Gamma} = -\mathbf{f}$. The magnitude $F = 2\pi\, rf$ of the total force on the ring then satisfies:

$$F = 2\pi\rho\,\Gamma\, r\frac{\mathrm{d}r}{\mathrm{d}t}.$$

The magnitude of the momentum P of a vortex of radius R is equal to the integral $\int F\mathrm{d}t = \int 2\pi\,\rho\,\Gamma\, r\,\mathrm{d}r$ evaluated between the radii zero and R. We therefore obtain:

$$P = \pi\,\rho\,\Gamma\, R^2. \tag{7.71}$$

Using Equations 7.70 and 7.71 for the energy E_k and momentum $\mathbf{P}$ of the ring, we can determine the group velocity $V_g(R)$ of the ring from the classical definition:

$$V_g\,(R) = \frac{\mathrm{d}E_k}{\mathrm{d}P} = \frac{\mathrm{d}E_k/\mathrm{d}R}{\mathrm{d}P/\mathrm{d}R} = \frac{\rho\,\Gamma^2}{2}\left(\frac{\mathrm{Log}\,(8R/\xi) - 3/2 + 1}{2\pi\rho\,\Gamma\, R}\right) = \frac{\Gamma}{4\pi R}\,(\mathrm{Log}\,(8R/\xi) - 1/2) = V. \tag{7.72}$$

V_g is therefore equal to V, the velocity at which the vortex ring moves, which is also that for energy transport.

This behavior is very different from that of usual material systems, where an increase in the kinetic energy and momentum is usually accompanied by a simultaneous increase in the velocity. In contrast, in the case of vortex rings, the velocity decreases while the radius, kinetic energy and momentum all increase. The laws governing the dynamics of vortex rings that we have just described have been verified with great precision in the case of superfluid helium (see the appendix to this chapter).

Interactions between vortex rings, or between a ring and a solid wall

The dynamics of a system of vortex rings can be described, like that for parallel vortex lines, by considering the field induced at one vortex by all the others.

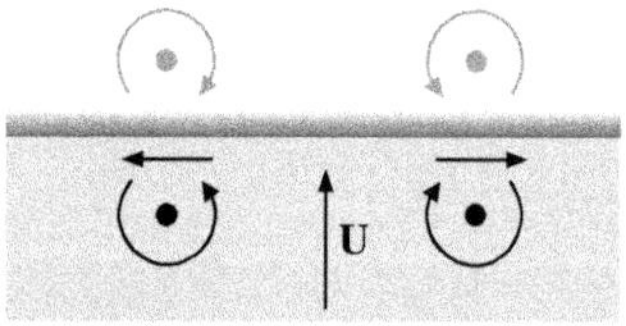

Figure 7.22 *A vortex ring impacting onto a solid plane*

Impact of a vortex ring on a solid plane The behavior of a ring approaching a flat plane can be described by replacing the plane with an image ring symmetric with respect to the plane (Figure 7.22). The presence of the ring image ensures that the normal component of the velocity at the plane is zero; the tangential component need not vanish, because we are discussing the case of a non-viscous fluid. As a result of its interaction with the image, the radius of the ring increases indefinitely while it moves slower and slower towards the plane (the ring does not bounce away either!). The image ring induces indeed on the original one an outward, radial,velocity component which increases as the ring gets closer to the plane. Moreover, the velocity components normal to the plane induced on a given element by the other parts of the ring is largely cancelled by those induced by the image ring.

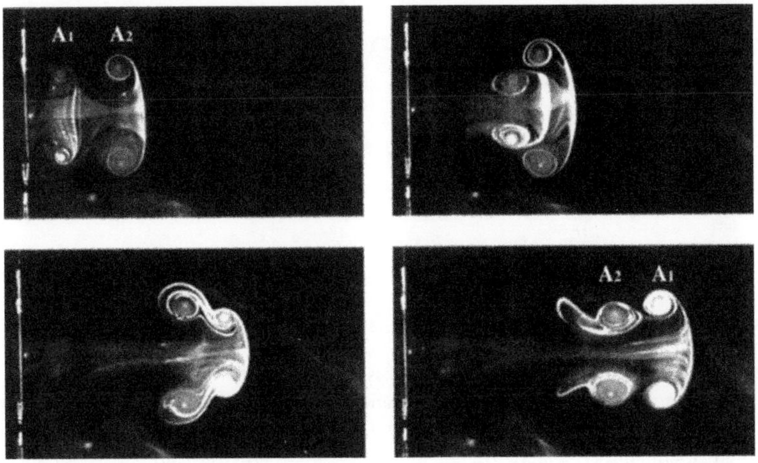

Figure 7.23 *Relative "leapfrog"-like motion of two coaxial vortex rings is described by this sequence of four images. (An Album of Fluid Motion; M. Van Dyke)*

Coaxial vortex rings with identical circulation The interaction of two vortex rings, such as might be created at the exit of a circular jet, leads to an amusing leapfrog phenomenon (Figure 7.23). Each ring passes through the inside of the ring just ahead, while its radius decreases. It is then itself passed by the other ring which is now behind it, and so on. The velocity field, induced on ring A_1 by the ring A_2 ahead of it, includes a radial component of velocity V_{r_1}, at A_1, directed towards the axis: this decreases the radius of A_1 and consequently, increases its velocity. Simultaneously, ring A_1 induces on A_2 an outward radial velocity component V_{r_2} which, conversely, increases its radius, and causes it to slow down. The process goes on until A_1 has caught up with A_2, and passed through it, a sequence which repeats indefinitely.

7.5 Vortices, vorticity and movement in air and water

Vortices and, more generally, the circulation of velocity frequently play a crucial role in the movement of fluids by inducing thrust (or drag) forces which lead to (or retard) the motion, but also create lift. These effects are significant, whether the fluid is air or water, for vehicles or animals moving at a Reynolds number large enough to generate the vortices. The opposite case of displacements governed by viscosity, such as for microorganisms, will be discussed in Chapter 9.

7.5.1 Thrust due to an emission of vortices

Many kinds of animal species, such as fish and birds, propel themselves by emitting vortex streets from the beating of their tails or wings. Figure 7.24 illustrates this phenomenon by

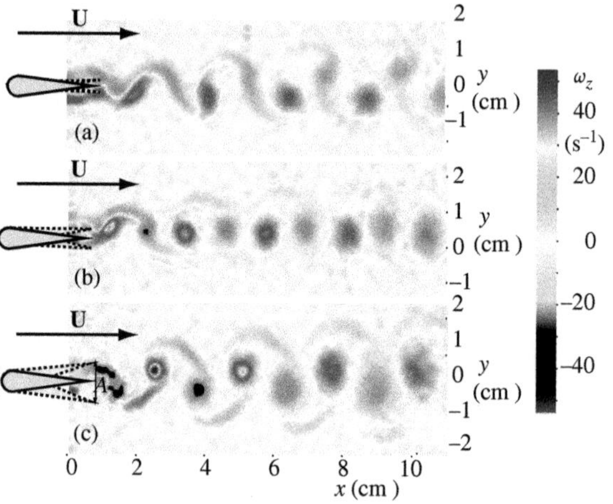

Figure 7.24C *Vorticity field behind an oscillating wing of width d = 0.5 cm within a flow of constant Reynolds number Re = 255 for several amplitudes A of the oscillations of the wing characterized by normalized values $A_d = A/d$ = 0.36 (a), 0.71 (b), 1.07 (c) (the respective Strouhal numbers are $Sr_A = fA/U$: 0.08, 0.16, 0.24). The color scale corresponds to the local vorticity (in s^{-1}) (document courtesy R. Godoy-Diana, J.L. Aider and J.E. Wesfreid)*

displaying the velocity fields in the fluid, as measured downstream of a simple model using a wing which oscillates at constant frequency f and an amplitude A. The wing has a thickness d and it is placed in a flow of constant velocity $\mathbf{U}$ (with Reynolds number $Re = Ud/\nu$).

At small amplitudes (Figure 7.24a), we observe a Bénard-von Karman (BvK) vortex street of the same nature as observed behind a fixed obstacle placed within a flow (see Sections 2.4.2 and 7.4.2). As the amplitude increases, this vortex street is replaced by another one where the vortices rotate in the direction opposite to that in the previous case (Figure 7.24c). At the boundary between these two regimes, we observe an intermediate structure, i.e. a linear region of alternating vortices (Figure 7.24b).

The inversion of the direction of the vortices results in the appearance of a thrust force opposed to the average flow, whereas there is a drag force for the BvK vortex street. The component $\mathbf{F}_D$ of these forces in the direction of $\mathbf{U}$ is, per unit length of the vortex:

$$\mathbf{F}_D = \varepsilon \rho \Gamma \frac{b}{a} (\mathbf{U} + \mathbf{W}), \tag{7.73}$$

where $\varepsilon = 1$ for the BvK vortex street and $\varepsilon = -1$ for the inverse one. The induced velocity $\mathbf{W}$ given by Equation 7.64 is of opposite sign to $\mathbf{U}$ in the first case and of the same sign in the second case (but we always have $|W| < |U|$).

Proof

According to Equation 7.68, the total momentum of a vortex street equals: $\mathbf{P} = \rho\, \Sigma_i (\mathbf{r}_i \times \mathbf{\Gamma}_i)$. The generation of an additional vortex pair leads to a variation in the component of the momentum in the direction of the average flow $\mathbf{U}$ with a value:

$$\Delta P_U = \rho\, (\mathbf{OO}_+ \times \mathbf{\Gamma} + \mathbf{OO}_- \times -\mathbf{\Gamma}) \cdot \mathbf{e}_U = \rho\, (\mathbf{O}_-\mathbf{O}_+ \times \mathbf{\Gamma}) \cdot \mathbf{e}_U; \tag{7.74}$$

$\mathbf{e}_U$ is the unit vector in the direction of $\mathbf{U}$ and $\mathbf{O}_-\mathbf{O}_+$ is the vector joining the intersections with a normal plane of the cores of the two vortices which have been created. We observe that the variation ΔP_U is independent of the choice of the vortex pair selected (and even of that of the neighbors with which a vortex is associated). The mixed product $\rho\,(\mathbf{O_{-}O_{+}} \wedge \mathbf{\Gamma})\cdot \mathbf{e}_U$ is always equal to Γb (Figure 7.20): the drag force is then obtained by multiplying $\Delta P_U \mathbf{e}_U$ by the number of vortex pairs generated per unit time, which is the frequency f (Equation 7.66). Finally, the sign of ΔP_U depends on the orientation of the component transverse to $\mathbf{U}$ of the vector $\mathbf{O_{-}O_{+}}$, which allows us to justify Equation 7.73.

This latter equation is only an approximation, and an exact calculation needs to take into account the velocity variations in the potential region of the flow, such as those given by Equation 7.65. This introduces some correction terms, but preserves the dominant term proportional to $\rho\,\Gamma b\,U/a$.

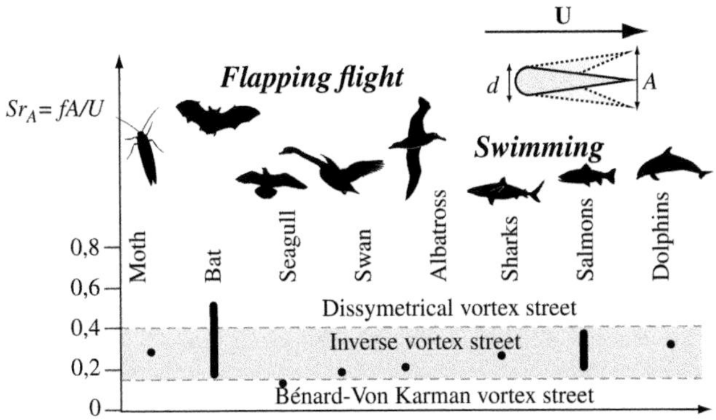

Figure 7.25 *Observation domains of the various flow regimes behind an oscillating wing, as a function of the normalized Strouhal number $Sr_A = fA/U$; (•) values of Sr_A corresponding to the propulsion mechanisms for different animals (from R.Godoy-Diana, J.L. Aider and J.E. Wesfreid)*

Figure 7.25 shows the observation regions for various regimes resulting from experiments carried out at different values of the amplitude and of the frequency of oscillation. These measurements indicate that the type of regime observed depends principally on the Strouhal number $Sr_A = fA/U$. The transition between the BvK vortex street and the inverse one is observed for a value of Sr_A of the order of 0.15. For values of Sr_A above 0.4, the vortex street loses its periodicity and its average direction is deviated relative to that of the velocity $\mathbf{U}$.

The Strouhal number Sr_A is defined here by the use of the oscillation amplitude A as a characteristic length.

The most effective region of thrust is observed in the regime of the inverse vortex street ($0.15 \le Sr_A \le 0.4$): we observe in Figure 7.25 that these values of Sr_A correspond well to those observed for a broad variety of birds, insects and fish.

In addition to the effect of the vortex emission that we have just described, flight and swim strategies frequently rely on deformations of the wing or the fin at the frequency of beating, which can be created deliberately or merely result from the flexibility of the wing.

7.5.2 The effects of lift

As defined in Section 6.3.1 and Section 6.6.3, lift is the component of the force perpendicular to the relative velocity of an object and a fluid. The lift on airplane wings clearly plays a crucial role in their flight, as it does in the flight of birds. Fish (and submarines) also use the effects of lift (in addition to Archimedes' buoyancy) to allow them to move at a fixed depth. We will first discuss the lift on airplane wings.

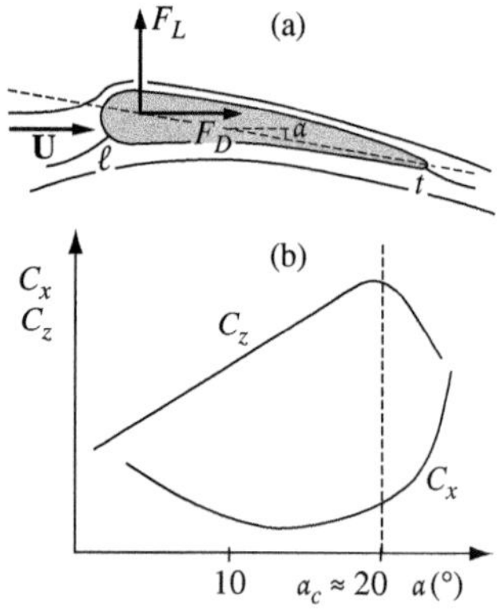

Figure 7.26 *(a) Schematic diagram of the flow around a wing, and of the lift and drag forces, ℓ and t are the leading and trailing edges (see discussion of Figures 6.22 et seq.); (b) Dependence of the lift and drag coefficients, respectively C_z and C_x, on the angle of attack α*

Lift and drag on an airplane wing

Figure 7.26a reminds us of the definition of the lift and drag forces on an airplane wing that we will at first consider as a two-dimensional object (with geometry not dependent on the direction normal to the plane of the figure).

The *lift force* $\mathbf{F}_L$ is perpendicular to the velocity **U**, and balances the weight of the airplane: one seeks to maximize this lift. The *drag force* $\mathbf{F}_D$ is parallel and opposite to the velocity: one needs therefore to minimize it. Practically, one characterizes the lift and the drag by two dimensionless coefficients (respectively C_z and C_x) which depend uniquely on the geometry and the angle of attack α of the wing; C_z and C_x are defined in terms of the lift, the drag, the area S of the wing, and the density ρ of the air by:

$$F_L = \frac{1}{2}\rho U^2 S\, C_z(\alpha) \quad (7.75a) \qquad \text{and} \qquad F_D = \frac{1}{2}\rho U^2 S\, C_x(\alpha). \quad (7.75b)$$

The lift is due to the circulation of the velocity of the fluid around the wing: this is in fact a manifestation of the Magnus force which has been described in Section 6.3.1 (Equation 6.44). This circulation results from the shape of the cross-section of the wing, in such a way that the stagnation point of the flow on the upper surface of the wing is located at the trailing edge t: this condition, known as the *Kutta condition* has been analyzed in Section 6.6.3 (Equation 6.105). The circulation Γ and the lift force $\mathbf{F}_D$ increase as the velocity **U** of the wing relative to the fluid increases (Equation 6.106); the airplane can take off when **U** is large enough so that $\mathbf{F}_D$ exceeds its weight.

Figure 7.26b indicates the variation of the lift and drag coefficients as a function of the angle of incidence α of the wing relative to the average velocity **U**. One observes that C_z increases nearly linearly with α up to a value α_c: then, as α increases (while remaining below α_c) the speed of the airplane for which the lift force compensates for the weight of the airplane decreases. Above the critical angle α_c, the lift coefficient decreases rapidly (*stalling phenomenon*): this process is related to the *boundary-layer separation* discussed in Section 10.6.1.

Lift on an airplane wing

In Section 2.1 of this chapter, we have shown that, for an ideal fluid, if the initial circulation of the velocity **v** of the fluid along a curve (C) is equal to zero, it remains equal to zero at all subsequent times around the curves $(C'(t))$, which correspond to particles of the fluid initially located on (C) and carried along by the flow (Kelvin's theorem). In contrast, in the case of a boundary layer, the presence of a vorticity layer very close to the wall allows for the appearance of a circulation around the cross-section of the wing and, consequently, of a lift force. We can however apply Kelvin's theorem to a curve (C) which surrounds the

Figure 7.27 *(a, b) Appearance of circulation around an airplane wing as it begins to accelerate: (a) initially, the circulation of the fluid velocity along the contour (C) around the static wing is zero; (b) the creation of circulation around the wing as it begins to move is compensated by the appearance of a vortex which is left behind; (c) three-dimensional structure of the distribution of vorticity around an airplane: the vortex line which includes the wing and the starting vortex is closed by two vortices emitted from the tip of each wing*

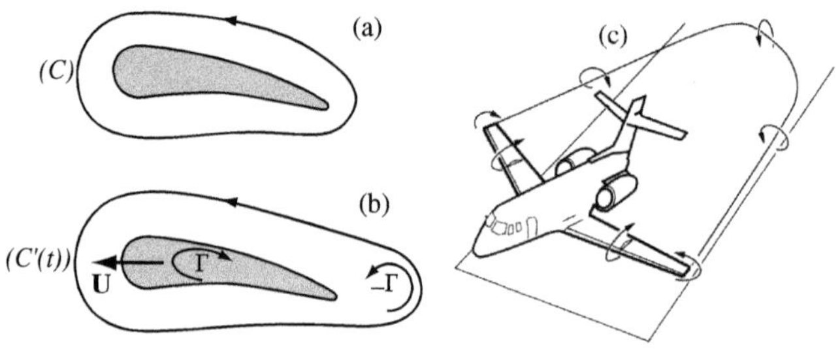

wing outside the boundary layer (Figure 7.27a) and far enough so that the flow can be considered as ideal everywhere along this curve. Consider a wing set in motion at an initial time: we observe a vortex appearing on the trailing edge of the wing (Figure 7.27b). The circulation $-\Gamma$ of the velocity of the fluid around this *start-up vortex* must be opposite to the value Γ around the wing in order to maintain zero circulation along $(C'(t))$: the initial circulation along (C) is indeed zero because the fluid is still at rest. This phenomenon can also be observed at the trailing edge of a spoon displaced in a coffee cup while remaining parallel to itself. The start-up vortex is left behind when the motion starts; moreover, in a real fluid, the distribution of the vorticity spreads out by viscous diffusion.

In Section 7.2.1, we have also shown that, in the case of an ideal fluid, a vortex line must close up on itself or needs to have its two ends localized on a solid wall or a liquid interface; given that the length of the wings is finite, the vorticity circuit which includes the wing of the airplane and the take off vortex must be closed by two vortices with axes parallel to the velocity $\mathbf{U}$ of the airplane and emitted at the ends of the wings (Figures 7.27c, 7.28a and 7.28b).

Behind a large airplane, these *trailing edge vortices* can be sufficiently large so as to unbalance another airplane that follows it too closely; this is especially the case when the velocity of the first airplane is low and the values of C_z, of α and thus of the circulation Γ must, accordingly, be quite high, so that the corresponding lift compensates for the weight of the airplane (Equation 6.106).

These vortices are also a source of needless energy dissipation: winglets, placed at the extreme ends of the wing, allow one to limit this effect in modern airplanes. Other strategies

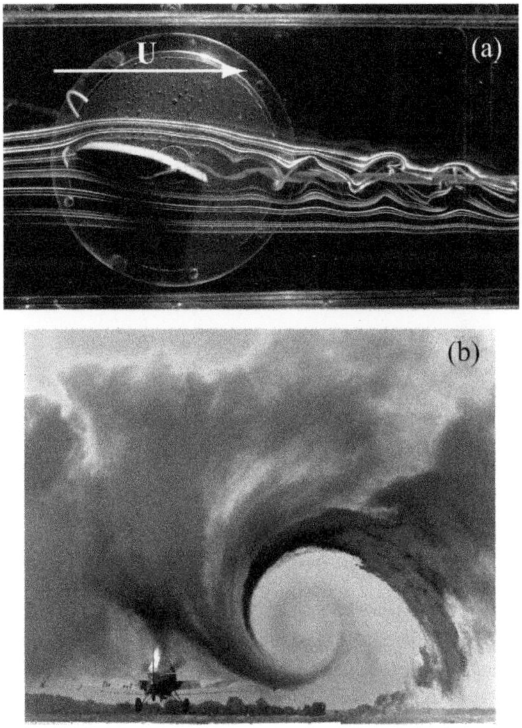

Figure 7.28C *(a) Visualization of a flow around a wing with the appearance of a vortex at the tip of the wing (plate courtesy O. Cadot and T. Pichon, ENSTA). (b) Visualization, by injection of smoke at ground level, of the vortex from the end of the wing of a plane spraying agricultural products (NASA document)*

can also be considered to reduce the wake e.g., by the emission of secondary vortices which are superimposed on the wake and destabilize and destroy the original vortices (Crow's instability).

7.5.3 Lift and propulsion

Propulsion by the turning blades of a propeller

The propeller blades of a boat and those of an airplane also induce lift: they are indeed inclined at a small angle relative to the plane in which they rotate in such a way that the lift is parallel to the axis of rotation and thus corresponds to a force of propulsion. Moreover, just like for an airplane wing, the drag force (which is located in the plane of rotation) will be small and result in a minimal energy dissipation.

This differs completely from instances such as a watermill where the speed at which the vanes move is perpendicular to their surface: in this case, only the *drag force* on the vanes has an effect, and the lift plays no role. The motion of the vanes communicates indeed a rearward momentum to the water while pushing it backward: however, there is a significant energy dissipation due to the work of the drag force resulting from the difference between the speed of the vanes and that of the water. The efficiency is therefore very poor. The oars of a rowing boat are yet another means of propulsion which uses drag.

The blades of a helicopter, aside from their very long shape, function essentially just as those of a propeller but they serve directly to support the helicopter since the axis of rotation is vertical (in this case we talk about lift from rotation).

Figure 7.29C *Diagram of the flow around the sails of a yacht designed for the America's cup: vortices (twisted coloured streamlines) are generated at the top of the mast, at the upper point where the jib is fastened, and below the boom bordering the lower side of the mainsail. The yacht sails upwind and the angle between its axis and the wind at the top of the mast is 22° (document courtesy C. Pashias, South African America's Cup Challenge)*

Propulsion of sailboats

We can consider the sail of a boat as similar to an airplane wing mounted vertically with its width decreasing as one moves upward. When a boat *sails upwind*, the angle that the wind makes with the sail is relatively small and its direction is such that it *fills* the sail. The force resulting on the sail is then actually a lift force perpendicular to the sail. The component of this force perpendicular to the axis of the sailboat is compensated by a *keel* or a *centerboard*; these act as vertical plates located under the sailboat parallel to its axis and induce a very large resistance to the drift of the boat perpendicular to its sailing direction. The component of the lift force in the direction of the axis of the boat is then the force which allows for the propulsion.

We have, just as for an airplane wing, a resultant circulation around the sail with, this time, a circulation vector oriented vertically: the vortices at the wingtips of an aeroplane are here replaced by a vortex emitted horizontally from the tip of the mast and another one emitted between the bottom side of the sail and the deck of the boat (Figure 7.29).

Lift, drag and propulsion for animals (and humans)

While swimming, a fish (or a swimmer) must permanently adjust the angle of attack of its fin (or hand) relative to the ambient fluid in order to have maximal efficiency. Too large an angle results in a behavior similar to the vanes of a windmill, with a significant dissipation of energy in their wake. If the hand or the fin is parallel to the flow, the drag force will be small, but there will be no significant lift force usable for propulsion: one therefore needs a sufficiently large angle of incidence. It is the sensitivity of the swimmer (or fish) to the surrounding flow which allows for finding the optimal orientation.

7.6 Rotating fluids

The study of rotating fluids is particularly important in the understanding of atmospheric and oceanic flows. Figure 7.30 indicates the effect of the rotation of the earth, which has an angular velocity $\Omega_0 \simeq 10^{-4}\mathrm{s}^{-1}$. Atmospheric and oceanic flows observed with reference to the surface of the earth take place in a reference frame which rotates at the local angular velocity $\Omega = \Omega_0 \sin\alpha$ (α is the latitude): this is the same velocity as in the *Foucault's pendulum* experiment for which the rotational period relative to an observer on the earth is $T = 2\pi/(\Omega_0 \sin\alpha)$. Geophysicists call the quantity 2Ω: *planetary vorticity.*

Laboratory experiments on rotating plates are often used to model the effect of the terrestrial rotation on flows in the atmosphere or the ocean, and result in spectacular and unexpected effects.

Rotating flows also occur in numerous industrial processes. The structure of flows in rotating containers is strongly influenced by the rotation, particularly in the boundary layers.

We will be interested principally in the case, most important in geophysics, where the deviations of the velocity of the fluid with respect to the global rotation are small.

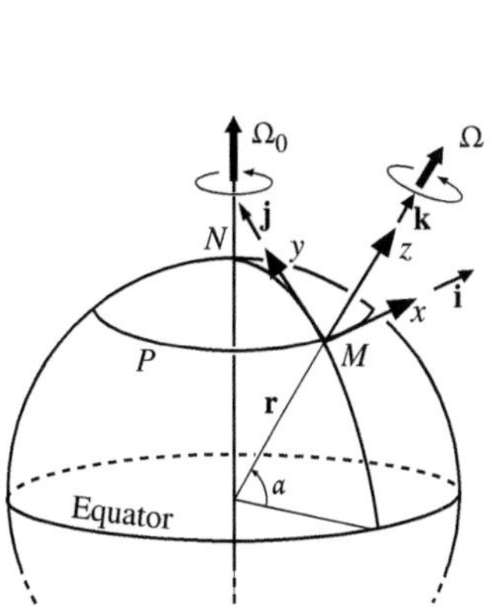

Figure 7.30 *Schematic diagram displaying the definition of the local coordinate system used to describe flow in a rotating reference frame*

7.6.1 Motion of a fluid in a rotating reference frame

Equation of motion of the fluid in a rotating reference frame

We start from the classical equation of motion in a fixed reference frame (the Navier-Stokes equation) in order to derive the corresponding equation in a reference frame rotating at angular velocity $\mathbf{\Omega}$ ($\mathbf{\Omega}$ being parallel to the axis of rotation). We will denote by $\mathbf{v}_a$ the velocity of the fluid in the "absolute" (fixed) reference frame and the "relative" velocity $\mathbf{v}_r$ in the rotating reference frame. The velocities are measured at a point M with $\mathbf{OM} = \mathbf{r}$ (O is the origin of the coordinate axes assumed to be located on the axis of rotation).

The variations with time of an arbitrary vector field $\mathbf{A(r)}$ in the absolute reference frame and in the rotating reference frame, are related by the equation:

$$\left(\frac{\mathrm{d}\mathbf{A}}{\mathrm{d}t}\right)_a = \left(\frac{\mathrm{d}\mathbf{A}}{\mathrm{d}t}\right)_r + \mathbf{\Omega} \times \mathbf{A}, \tag{7.76}$$

where $\mathrm{d}\mathbf{A}/\mathrm{d}t$ represents a Lagrangian derivative. For example, if we have a constant vector $\mathbf{A}$ in the rotating fluid, its rate of variation in the absolute reference frame is $\mathrm{d}\mathbf{A}/\mathrm{d}t = \mathbf{\Omega} \times \mathbf{A}$. Using Equation 7.76 to rewrite the terms of the equation of motion in the rotating reference frame, we obtain:

$$\left(\frac{\mathrm{d}\mathbf{v}_r}{\mathrm{d}t}\right)_r = \frac{\partial \mathbf{v}_r}{\partial t} + (\mathbf{v}_r \cdot \nabla)\,\mathbf{v}_r = -\nabla\left(\frac{p}{\rho} + \varphi - \frac{1}{2}(\mathbf{\Omega} \times \mathbf{r})^2\right) + \nu\nabla^2\mathbf{v}_r - 2\,\mathbf{\Omega} \times \mathbf{v}_r. \tag{7.77}$$

The term $(1/2)\nabla(\mathbf{\Omega} \times \mathbf{r})^2$ is associated with the *centrifugal force* and appears as a volume force. The component $-2\mathbf{\Omega} \times \mathbf{v}_r$, the *Coriolis force*, is normal to the velocity of the particles in the fluid: it is this particular force which explains the rotation of Foucault's pendulum and the eastward derivation of bodies in free fall. We will see that it is also responsible for important atmospheric phenomena.

Proof

Let us first apply Equation 7.76 to the radius vector $\mathbf{r}$ for which the time derivatives are the absolute velocity $\mathbf{v}_a$ and the rotating frame velocity $\mathbf{v}_r$. We then obtain:

$$\mathbf{v}_a = \mathbf{v}_r + \mathbf{\Omega} \times \mathbf{r}. \tag{7.78}$$

Applying again Equation 7.76 to the preceding expression for $\mathbf{v}_a$, we obtain:

$$\left(\frac{\mathrm{d}\mathbf{v}_a}{\mathrm{d}t}\right)_a = \left(\frac{\mathrm{d}\mathbf{v}_a}{\mathrm{d}t}\right)_r + \boldsymbol{\Omega} \times \mathbf{v}_a = \frac{\mathrm{d}\mathbf{v}_r}{\mathrm{d}t} + 2\boldsymbol{\Omega} \times \mathbf{v}_r + \boldsymbol{\Omega} \times (\boldsymbol{\Omega} \times \mathbf{r}). \quad (7.79)$$

The term: $\boldsymbol{\Omega} \times (\boldsymbol{\Omega} \times \mathbf{r})$ is equal to $-\nabla(\boldsymbol{\Omega} \times \mathbf{r})^2/2$ since only the component $\mathbf{r}_\perp$ of $\mathbf{r}$, normal to $\boldsymbol{\Omega}$, is effective, and that both expressions are then equal to $-\Omega^2\mathbf{r}_\perp$. Let us start from the Navier-Stokes equation for which the usual form in an absolute reference frame is:

$$\left(\frac{\mathrm{d}\mathbf{v}_a}{\mathrm{d}t}\right)_a = \frac{\partial \mathbf{v}_a}{\partial t} + (\mathbf{v}_a \cdot \nabla)\,\mathbf{v}_a = -\frac{1}{\rho}\nabla p + \nu\nabla^2\mathbf{v}_a - \nabla\varphi \quad (7.80)$$

(φ represents the potential of the volume forces per unit mass, such that $\varphi = gz$ for gravity). Expressing $\mathbf{v}_a$ as a function of $\mathbf{v}_r$ by means of Equations 7.78 and 7.79, Equation 7.80 becomes:

$$\left(\frac{\mathrm{d}\mathbf{v}_r}{\mathrm{d}t}\right)_r = \frac{\partial \mathbf{v}_r}{\partial t} + (\mathbf{v}_r \cdot \nabla)\,\mathbf{v}_r = -\frac{1}{\rho}\nabla p + \nu\,\nabla^2\mathbf{v}_r - \nabla\varphi - 2\,\boldsymbol{\Omega} \times \mathbf{v}_r + \frac{1}{2}\nabla(\boldsymbol{\Omega} \times \mathbf{r})^2. \quad (7.81)$$

When $\nabla^2\mathbf{v}_a$ is replaced by its expression as a function of $\nabla^2\mathbf{v}_r$ obtained from Equation 7.78, these two terms differ indeed only by $\nabla^2(\boldsymbol{\Omega} \times \mathbf{r})$ which is identically zero since $\boldsymbol{\Omega} \times \mathbf{r}$ is a linear function of the components of $\mathbf{r}$. Grouping the gradient terms, Equation 7.81 gives back Equation 7.77 for the particular case of a constant density fluid (ρ = constant).

Throughout the remainder of this chapter, we drop the subscript r and we use the components of the velocity measured in the rotating reference frame.

Effects of the centrifugal force in a rotating fluid

The centrifugal force can induce flow: an important application is represented by centrifugal pumps: the fluid is injected along an axis, set in motion by vanes, and finally forced to exit by the gradient of the pressure due to the rotation. Another application is the ability to obtain very thin fluid layers of uniform thickness by depositing a droplet onto a rotating disk (spin coating). The effect of the centrifugal force is equally crucial in the formation of tornadoes and water spouts. We will see in Section 7.7 that these forces can result in secondary flows, which become superimposed onto the main flow in some flow geometries (e.g., tubes with curved walls).

In the remainder of this chapter, we will be interested in the opposite case, seen frequently, where the centrifugal forces do not generate flows. Let us assume, for example, a container rotating at uniform velocity. After a sufficiently long time has elapsed, the fluid takes on the angular velocity of the walls of the container if no mechanism creates flow relative to the walls (see Section 2.1.1). In this case, we obtain, taking $\mathbf{v}_r \equiv 0$ in Equation 7.77:

$$\left(\frac{p}{\rho} + \varphi - \frac{1}{2}(\boldsymbol{\Omega} \times \mathbf{r})^2\right) = \frac{p}{\rho} + gz - \frac{1}{2}(\boldsymbol{\Omega} \times \mathbf{r})^2 = \frac{p}{\rho} + gz - \frac{\Omega^2 r_\perp^2}{2} = \text{constant} \quad (7.82)$$

($r_\perp = r\sin\theta$ is the distance to the axis of rotation and Ω is the magnitude of $\boldsymbol{\Omega}$). The centrifugal force creates an additional pressure gradient, which depends only on the distance $r_\perp$ to the axis of rotation: in the remainder of this chapter, we will not take this into account but use implicitly a value of the pressure corrected from this effect by adding the term $-\rho\,\Omega^2(r_\perp)^2/2$.

The centrifugal force has, however, a very obvious effect on the shape of a free surface of a fluid. The pressure there is indeed constant and equal to the atmospheric pressure (from Section 1.4.4, we can neglect the effects of surface tension if the size of the container is larger than a few millimeters). The level $z(r_\perp)$ of the free surface then satisfies: $gz(r_\perp) - \Omega^2 (r_\perp)^2 = $ constant. We find here again the well-known parabolic shape of the free surface, which has a minimum on the axis of rotation ($r_\perp = 0$). The corresponding centrifugal pressure term due to rotation is compensated here by an additional hydrostatic one due to the deformation of the free surface.

Equation of motion of the vorticity in a rotating reference frame

Like for deriving the equation of motion of the vorticity in the absolute reference frame (Section 7.3), we take here the curl of Equation 7.77 which eliminates the gradient terms. We then obtain the equation of transport of the vorticity in the rotating reference frame:

$$\frac{d\boldsymbol{\omega}}{dt} = \frac{\partial \boldsymbol{\omega}}{\partial t} + (\mathbf{v} \cdot \nabla)\boldsymbol{\omega} = ((\boldsymbol{\omega} + 2\boldsymbol{\Omega}) \cdot \nabla)\mathbf{v} + \nu\, \nabla^2 \boldsymbol{\omega} \tag{7.83}$$

(where $\boldsymbol{\omega} = \nabla \times \mathbf{v}_r$). The only difference between this equation and that obtained in the absence of any rotation is the presence of the term of *planetary vorticity* $2\boldsymbol{\Omega}$ in addition to the vorticity $\boldsymbol{\omega}$. The sum $\boldsymbol{\omega} + 2\boldsymbol{\Omega}$ of these two terms is known as the *absolute vorticity*. The term $(\boldsymbol{\omega} \cdot \nabla)\mathbf{v}$ represents the variations of vorticity which accompany changes in the length or orientation of the vorticity tubes (Section 7.3.1). Rotation adds an additional contribution which can be dominant at high rotational velocities.

Order of magnitude of the different terms of the equation of motion; Rossby and Ekman numbers

Let us call L the characteristic length of the flow (such as the size of a channel) and U its characteristic velocity (e.g., the maximum velocity or an average velocity). Multiplying Equation 7.77 by L/U^2 so as to create dimensionless variables, one obtains:

$$\frac{\partial \mathbf{v}'}{\partial t'} + (\mathbf{v}' \cdot \nabla')\mathbf{v}' = -\nabla' p' + \frac{\nu}{UL}\nabla'^2 \mathbf{v}' - 2\frac{|\Omega|\, L}{U}\,\frac{\boldsymbol{\Omega}}{|\Omega|} \times \mathbf{v}', \tag{7.84}$$

where $\mathbf{v}' = \mathbf{v}/U$, $p' = (p - p_0)/(\rho U^2)$; p_0 is the pressure in the absence of flow but in the presence of rotation and of the volume forces with $\nabla(p_0/\rho) = -\nabla(\varphi - (\boldsymbol{\Omega} \times \mathbf{r})^2/2)$. The operator ∇' corresponds to derivatives with respect to the components of the dimensionless radius vector $\mathbf{r}' = \mathbf{r}/L$; $\boldsymbol{\Omega}/|\Omega|$ is the unit vector in the direction of the axis of rotation.

Equation 7.84 only introduces, aside from the dimensionless variables $\mathbf{r}'$, $\mathbf{v}'$ and p', the dimensionless combinations $\nu/(UL)$ (the inverse of the Reynolds number) and $|\Omega|L/U$ which is the inverse of the number $Ro = U/(|\Omega|L)$, which is called the *Rossby number*. The solutions of Equation 7.84 are thus of the form:

$$\frac{\mathbf{v}}{U} = \mathbf{v}' = \mathbf{f}\left(\frac{x}{L}, \frac{y}{L}, \frac{z}{L}, Re, Ro\right), \tag{7.85a}$$

$$\frac{p - p_0}{\rho U^2} = p' = g\left(\frac{x}{L}, \frac{y}{L}, \frac{z}{L}, Re, Ro\right). \tag{7.85b}$$

Two flows with identical geometry, but having different values of U and L, then correspond to similar velocity and pressure fields if the Reynolds and Rossby numbers are the same.

The Rossby number corresponds physically to the ratio between the orders of magnitude of a convective transport term of the equation of motion and of the Coriolis force term. We have:

$$|(\mathbf{v} \cdot \nabla)\, \mathbf{v}| \approx \frac{U^2}{L} \quad \text{and} \quad 2\, |\boldsymbol{\Omega} \times \mathbf{v}| \approx |\Omega|\, U.$$

Thus: $$Ro = \frac{U}{|\Omega|\,L} \approx \frac{\text{convective transport term}}{\text{Coriolis force term}}.$$

In flows at small Rossby number, the effect of rotation and of the Coriolis force dominate. The Rossby number becomes small, for normal flow velocities, only for motion on a very large scale. Thus for an atmospheric flow, with $|\Omega| = 10^{-4}\,\text{s}^{-1}$; $U \approx 10\,\text{m/s}$, $L \approx 10^3\,\text{km}$, we find $Ro = 10^{-1}$. This explains why the rotation of the earth affects the direction of the cyclonic movements which are opposite in the northern and southern hemispheres. This does not apply, however, to tornadoes with velocities of 100 m/s on a scale of kilometers, where $Ro = 10^3$ (nor does it apply to the vortex that forms in a draining bathtub!).

In the same way, we can evaluate the ratio between the viscosity terms and the Coriolis force in the equation of motion where:

$$\nu\,\nabla^2\mathbf{v} \approx \frac{\nu U}{L^2} \quad \text{and} \quad 2\,|(\boldsymbol{\Omega} \times \mathbf{v})| \approx |\Omega|\,U.$$

The ratio of these terms, knows as the *Ekman number*, satisfies:

$$\text{Ekman number} = Ek = \frac{\nu}{|\Omega|\,L^2} \approx \frac{\text{viscous force}}{\text{Coriolis force}}.$$

We will use the Ekman number in the study of flows near the bottom of a container or adjacent to a rotating wall (Sections 7.6.4 and 7.7.2).

Potential vorticity

We can understand the meaning of this parameter by observing a vortex tube passing over a two-dimensional obstacle transverse to a flow in the absence of viscosity.

A related problem is the change in direction of a dominant wind which passes over a long mountain range (in the y-direction in Figure 7.31) and with relatively constant height in this direction. This is partly similar to the effect of a change of velocity on vertical vorticity tubes such as those in Figure 7.14 but, here, it is the absolute vorticity $\boldsymbol{\omega} + 2\boldsymbol{\Omega}$ which affects the flow as opposed to $\boldsymbol{\omega}$, the vorticity of the flow by itself in this first case.

We assume that we have a thin layer of fluid perpendicular to the axis of rotation $\boldsymbol{\Omega}$ in the z-direction and flowing initially with a velocity $\mathbf{V}^0\,(v_x^0,\, v_y^0)$ which is uniform and constant (Figure 7.31). The fluid layer passes over an obstacle infinitely long in the y-direction and with height $h_0 - h(x)$ (h_0 being the initial thickness of the fluid layer); the upper surface, however, remains horizontal. We are interested here in the variation of the ω_z component of the vorticity: we retain therefore all the terms which contain ω_z (with the exception of the terms involving the viscosity, which we neglect throughout this paragraph). Moreover, we assume that the horizontal velocity components are uniform throughout the thickness of the layer, so that $\partial v_x/\partial z = \partial v_y/\partial z = 0$. By taking the derivative relative to z of the condition

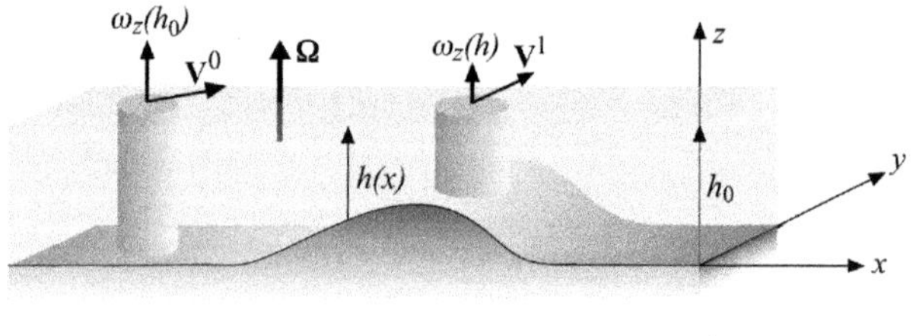

Figure 7.31 *Flow of a layer of fuid of thickness h(x) over an obstacle of infinite length in the y-direction (the entire system rotates with angular velocity* $\boldsymbol{\Omega}$*).* $\mathbf{V}^0$ *is the initial velocity of the fluid upstream of the obstacle, while* $\mathbf{V}^1$ *is that downstream: as the fluid passes over the obstacle, the* v_y*-component undergoes a negative change* $v_y^1 - v_y^0$ *which signals an anticyclone-type deviation (in the northern hemisphere). The local vorticity* $\omega_z(h)$ *decreases with the thickness h of the fluid layer, in accordance with Equation. 7.90*

of incompressibility $\nabla \cdot \mathbf{v} = 0$, we find therefore $\partial^2 v_z/\partial z^2 = 0$. The derivative $\partial v_z/\partial z$ is then a constant with z such that:

$$\frac{\partial v_z}{\partial z} = \frac{1}{h}\frac{\mathrm{d}h}{\mathrm{d}t} \qquad (7.86)$$

since $v_z = 0$ for $z = h_0$ and $v_z = -\mathrm{d}h/\mathrm{d}t$ for $z = h_0 - h$.

Taking into account these approximations, we use the transport equation for vorticity (Equation 7.83) in the form:

$$\frac{\mathrm{d}\boldsymbol{\omega}}{\mathrm{d}t} = ((\boldsymbol{\omega} + 2\boldsymbol{\Omega}) \cdot \nabla)\, \mathbf{v} \qquad (7.87) \qquad \text{whence:} \qquad \frac{\mathrm{d}\omega_z}{\mathrm{d}t} = (\omega_z + 2\,\Omega)\,\frac{\partial v_z}{\partial z} \qquad (7.88)$$

$\mathrm{d}\omega_z/\mathrm{d}t$ is a Lagrangian derivative, which represents the change in the vorticity in an element of fluid which we follow along its motion and Ω is the component of $\boldsymbol{\Omega}$ on the z-axis. We note that, if the slope of the bottom of the fluid layer is small, such that $|\partial h/\partial x|$ and $|\partial h/\partial y| \ll 1$, the vertical component v_z of the velocity of the fluid will also be small relative to v_x and v_y. In contrast, the derivative $\partial v_z/\partial z$ which has a multiplying factor $1/h$ is of the same order of magnitude as the derivatives of v_x and v_y with respect to x and y; this explains the important role which they have in this particular problem. Combining Equations 7.86 and 7.88, we find:

$$\frac{1}{\omega_z + 2\Omega}\frac{\mathrm{d}\omega_z}{\mathrm{d}t} = \frac{1}{h}\frac{\mathrm{d}h}{\mathrm{d}t}, \qquad (7.89a) \qquad \text{whence:} \qquad \frac{\mathrm{d}}{\mathrm{d}t}\mathrm{Log}\left(\frac{\omega_z + 2\Omega}{h}\right) = 0. \qquad (7.89b)$$

We thus obtain, finally:

$$\frac{\omega_z + 2\Omega}{h} = \text{constant}; \qquad (7.90)$$

this conserved quantity is known as the *potential vorticity*. Equation 7.90 plays a fundamental role in geophysics: it generalizes the equation of conservation of angular momentum for an element of matter, expressed by Kelvin's theorem in a non-rotating reference frame ($\Omega = 0$). In this latter case, Equation 7.90 reduces to the condition $\omega_z/h = $ constant, which we have already discussed in Section 7.3.1 (h was then replaced by $\delta\ell$).

Let us now take a cylinder of liquid with initial cross section S_0 before it arrives on the obstacle ($h = h_0$). When the liquid cylinder arrives in a region where the thickness of the liquid layer is h, its cross section $S(h)$ obeys the equation $S(h)h = S_0 h_0$ (conservation of the volume of fluid, which we assume to be incompressible). Equation 7.90 then becomes:

$$(\omega_z(h_0) + 2\Omega)S_0 = (\omega_z(h) + 2\,\Omega)\,S(h). \qquad (7.91)$$

According to Stokes' theorem, $\omega_z(h_0)S_0$ and $\omega_z(h)S(h)$ represent the circulation of the velocity around a cylinder of liquid before and during its passing over an obstacle: Equation 7.91 is thus a generalization of Kelvin's theorem, with two additional terms in 2Ω which correspond to the rotation.

We can also use Equation 7.90 to determine the variation of the horizontal velocity of the fluid layer when it passes over the obstacle. For this purpose, let us calculate the velocity components of the fluid velocity downstream of the obstacle. Taking into account the symmetry of the obstacle, the velocity field is independent of the coordinate y, and $\partial v_x/\partial y = \partial v_y/\partial y = 0$. The component ω_z of the vorticity then reduces to $\omega_z = \partial v_y/\partial x$; ω_z is initially zero throughout the region where $h = h_0$. Equation 7.90 can then be rewritten in the form:

$$\frac{\partial v_y}{\partial x} = -\frac{2\Omega\,(h_0 - h)}{h_0}, \qquad (7.92)$$

so that, after integration with respect to x:

$$v_y^1 = v_y^0 - \frac{2\,\Omega}{h_0}\,A, \tag{7.93}$$

where $A(>0)$ is the area $\int (h_0 - h)\mathrm{d}x$ of the horizontal cross-section of the obstacle (Figure 7.31). Moreover, since the quantity of fluid flowing through the planes x = constant is conserved, we must have $v_x^1 = v_x^0$. Thus, in the northern hemisphere, the wind deviates in a clockwise direction (anti-cyclonic). In nature, this effect will only be observed if a mountain range is sufficiently long; otherwise, and especially if $Ro \ll 1$, the wind will just go around as in the experiments to be discussed in the next section.

7.6.2 Flows at small Rossby numbers

Geostrophic flows

Let us consider again Equation 7.77 assuming, as we did in the discussion above, that the effect of the centrifugal force and of the volume forces result merely in a gradient of the hydrostatic pressure, which we will subtract from the total pressure. Let us assume that the Rossby number Ro is $\ll 1$, which will allow us to neglect the convective transport term $(\mathbf{v}\cdot\nabla)\mathbf{v}$ (as well as the terms $(\mathbf{v}\cdot\nabla)\boldsymbol{\omega}$ and $(\boldsymbol{\omega}\cdot\nabla)\mathbf{v}$ of the equation of conservation of the vorticity $\boldsymbol{\omega}$ which must also obey $\omega \ll \Omega$). Finally, let us assume that the viscosity forces are negligible on the scale of the flows at which we are looking (Ekman number $\ll 1$). Equations 7.77 and 7.83 then reduce to:

$$\frac{\partial \mathbf{v}}{\partial t} = -\nabla\left(\frac{p}{\rho}\right) - (2\,\boldsymbol{\Omega} \times \mathbf{v}) \quad (7.94) \qquad \text{and} \qquad \frac{\partial \boldsymbol{\omega}}{\partial t} = 2\,(\boldsymbol{\Omega}\cdot\nabla)\,\mathbf{v}. \quad (7.95)$$

Let us further assume that the flow is quasi stationary, which will allow us to neglect the terms $\partial\mathbf{v}/\partial t$ and $\partial\boldsymbol{\omega}/\partial t$. If τ is the characteristic time for changes in the flow, this amounts to assuming that the product $\Omega\tau$ is $\gg 1$. A flow which obeys all these conditions is known as a *geostrophic flow*.

Let us assume again that the axis of rotation is the z-axis. Equation 7.94 becomes:

$$\frac{\partial p}{\partial x} = 2\rho\,\Omega\,v_y \quad (7.96a) \qquad \frac{\partial p}{\partial y} = -2\rho\,\Omega\,v_x \quad (7.96b) \qquad \text{and} \qquad \frac{\partial p}{\partial z} = 0. \quad (7.96c)$$

Taking the derivative of the first two equations with respect to z, we obtain:

$$\frac{\partial v_x}{\partial z} = \frac{\partial v_y}{\partial z} = 0. \tag{7.97}$$

Under the same conditions, Equation 7.95 reduces to:

$$\Omega\frac{\partial v_z}{\partial z} = 0. \tag{7.98}$$

Combining this result with the incompressibility condition $\nabla\cdot\mathbf{v} = 0$, we then obtain:

$$-\frac{\partial v_z}{\partial z} = \frac{\partial v_x}{\partial x} + \frac{\partial v_y}{\partial y} = 0. \tag{7.99}$$

We then observe that a geostrophic flow has the following three characteristics:

- This flow is the result of a two-dimensional flow $\mathbf{v}(x, y)$ and of a translation parallel to the axis of rotation with a velocity independent of the coordinate z along this axis. If the boundary conditions (walls, free surface) require $v_z = 0$ at one of the points along the lines parallel to the axis of rotation (as this will frequently be the situation), we then have $v_z = 0$ everywhere. The motion is then two-dimensional with velocity components $v_x(x, y)$ and $v_y(x, y)$.
- The flow is normal to the pressure gradient: the streamlines coincide with the isobars. The fact that the flow is normal to the gradient of the pressure (and, as a result, to the force applied on an element of volume) reminds us of a Magnus force (force perpendicular to the relative velocity of a vortex and of a flow, as well as its vorticity). We can apply this result to help us read meteorological charts which display isobar curves: the tangents to these latter curves represent the direction of the average wind in the region considered, at an altitude high enough so that the influence of the boundary layer near the earth is not significant.
- If the fluid is incompressible, we have $\partial v_x/\partial x + \partial v_y/\partial y = 0$. The cross-section of a plane curve normal to $\boldsymbol{\Omega}$ which moves along with the fluid remains thus constant: particularly, according to Kelvin's theorem, the cross-section of the vorticity tubes also remains constant.

The combination of Equations 7.97 to 7.99 represents the *Taylor-Proudman theorem*, of which we will see spectacular applications.

Displacement of a solid perpendicular to $\boldsymbol{\Omega}$

Let us assume that the fluid has an almost horizontal free surface, and that we displace a solid body of finite dimensions (such as a sphere), at a velocity $\mathbf{U}$ parallel to the surface (Figure 7.32a). If the magnitude of the velocity $\mathbf{U}$ is small enough and constant, the conditions for applying the Taylor-Proudman theorem will be obeyed. Due to the boundary conditions, the local velocity of the fluid must first be equal to $\mathbf{U}$ over the whole outer surface of the solid. From Equations 7.97 and 7.98, this velocity is also equal to $\mathbf{U}$ everywhere in the vertical cylinder of fluid, resting on the outer contour of the solid and rising up to the upper free surface. Below the solid, the resulting flow is more complicated because the condition $\partial\mathbf{v}/\partial z = 0$ cannot be obeyed due to the boundary condition at the solid bottom of the container.

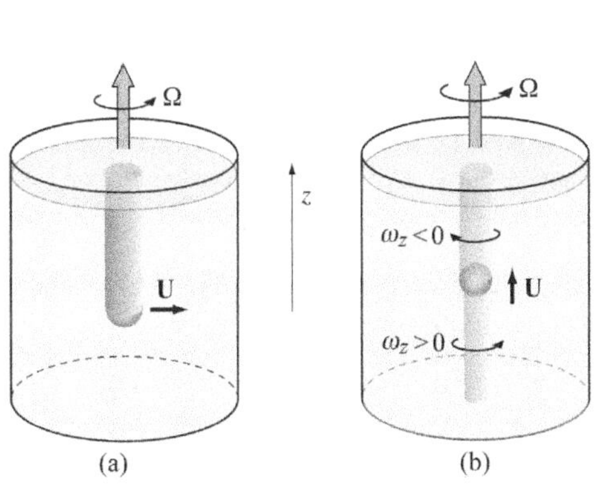

Figure 7.32 *Motion of a solid body within a rotating fluid. (a) The velocity of the displacement* $\mathbf{U}$ *is perpendicular to the axis of rotation. Dye injected into the region of fluid just above the object moves with it; (b) changes in the local vorticity resulting from the motion of the solid object parallel to the axis of rotation, in a rotating container*

In such an experiment, the flow is only approximately geostrophic. The conditions for applying the Taylor-Proudman theorem are not, for example, obeyed at the boundary between the inside and the outside of the virtual cylinder; a small exchange of fluid takes therefore place at this boundary.

As a first approximation, everything thus occurs as if the interior of the upper fluid cylinder remained separated from its outside, while moving along at the uniform velocity $\mathbf{U}$. Outside the cylinder of fluid, the flow is the same as that of an ideal fluid around a vertical solid cylinder. The fluid cylinder represents a *Taylor column*. In fact, we observe experimentally (Figure 7.32a) that dye injected into it remains there for a very long time during the flow; in contrast, any dye injected on the outside avoids this cylinder.

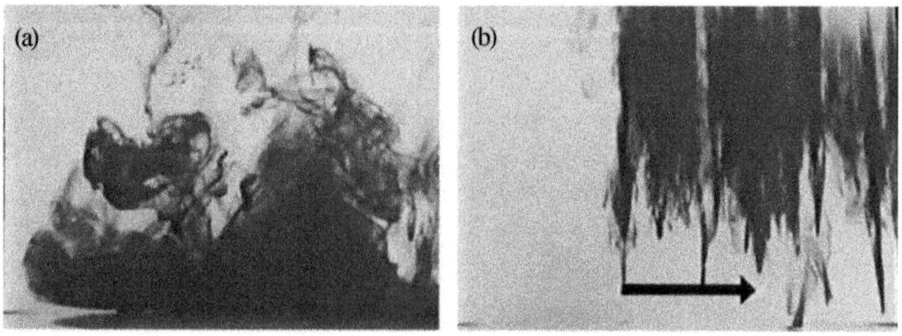

Figure 7.33 *Injection of colored water from above in a container filled with water: (a) stationary, (b) in rotation. The isotropic structure of the mixure is replaced by a filamentary structure resulting from the rotation (copied from Illustrated Fluid Mechanics, NCFMF, MIT Press)*

A visual demonstration of two-dimensional behavior of flows at Ro≪1

Let us inject colored water at the surface of a container full of clear water which can be set in rotation at an angular velocity high enough to obtain small Rossby numbers. If the container is stationary, the colored water is dispersed; then, it spreads out along the bottom without any definite structural form (Figure 7.33a). If the container is now rapidly rotated, the solution takes on the form of vertical streaks parallel to the axis of rotation, thus illustrating that the flow is invariant in the direction of this axis (Figure 7.33b).

We can set up this type of experiment with a turntable upon which we fasten a cylindrical glass container, carefully centered, and filled with water. We can then establish a simplified version of the experiment in Figure 7.32a by placing at the bottom of the container a fixed obstacle away from the axis of rotation (e.g., a cylinder small in height): by making small changes in the rotational velocity we create, because of the inertia of the liquid, a flow of the liquid around the obstacle. Just as in Figure 7.32a, this flow extends all the way to the surface: dye injected above the obstacle will remain localized in this Taylor column where the fluid is at rest relative to the obstacle.

Motion of a solid parallel to the axis of rotation

Let us now take a solid object rising vertically with velocity $\mathbf{U}$ in a container similar to that of Figure 7.32b (the velocity $\mathbf{U}$ is, here, parallel to $\boldsymbol{\Omega}$). Here, again, one can no longer obey rigorously the conditions of the Taylor–Proudman theorem and we do not have a purely geostrophic flow: the vertical component of the fluid velocity cannot indeed be identically zero all along a vertical line rising from the solid surface. We first observe that the force opposing the motion of the solid is much stronger than in the absence of rotation. We also note that the effect of the vertical motion of the solid propagates over large distances above and below the solid (there is, however, a transition zone near the lower and upper boundaries of the fluid volume): if there were no rotation, the flow perturbation induced by the motion of the object would decay over a distance of the order of its size. Moreover, we observe that the motion of the solid induces a rotation of fluid in opposite directions above and below the solid: these rotations can be visualized by locally injecting colored fluid.

In order to evaluate quantitatively these different effects, let us first take the divergence of Equation 7.94. We obtain, by the use of the vector identity:

$$\nabla \cdot (\mathbf{A} \times \mathbf{B}) = -\mathbf{A} \cdot (\nabla \times \mathbf{B}) + \mathbf{B} \cdot (\nabla \times \mathbf{A}):$$

$$2\,\boldsymbol{\Omega} \cdot \boldsymbol{\omega} = \frac{\nabla^2 p}{\rho} \qquad (7.100a) \qquad \text{or} \qquad 2\,\Omega\,\omega_z = \frac{\nabla^2 p}{\rho}. \qquad (7.100b)$$

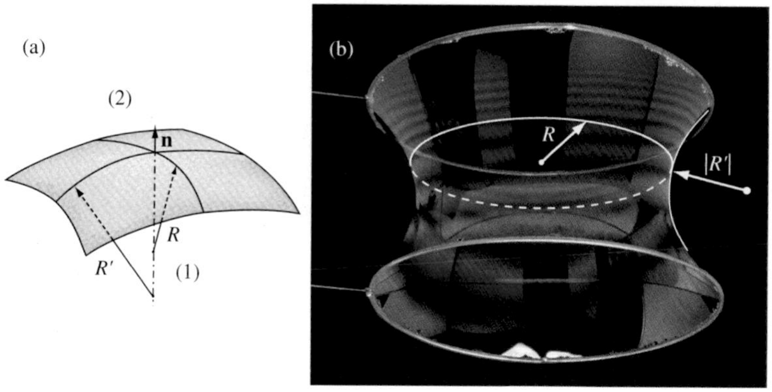

Figure 1.16 *(a) Geometry of an arbitrary boundary surface between two fluids (1) and (2), illustrating the corresponding definition of the principal radii of curvature R and R′; (b) the surface of a soap film stretched between two circular metal rings (at the top and bottom of the figure) is open and, thus, the pressure on the two sides of the film is the same. The local mean curvature* $C = [(1/R) + (1/R')]$ *is thus zero at every point. The surface generated in this way, a catenoid, achieves a minimum in the surface area of the film, while taking into account the boundary conditions imposed by the two rings. (after a photograph by S. Schwartzenberg, © Exploratorium, www.exploratorium.edu)*

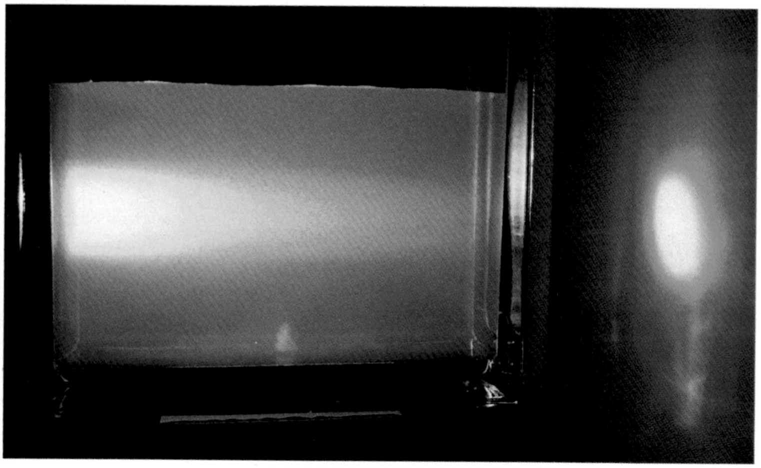

Figure 1.30 *Rayleigh scattering from a beam of white light (coming from the left) incident on a container filled with water to which a tiny amount of milk has been added. The light scattered at right angles to the beam has a bluish tint while the light transmitted to the screen is reddish. (document courtesy of B. Valeur)*

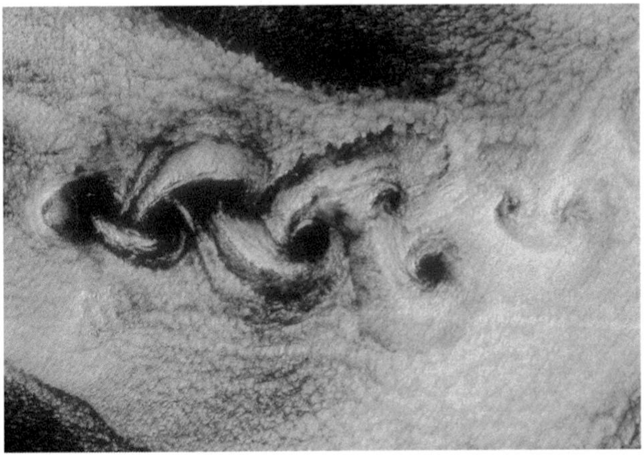

Figure 2.10 *Satellite view of cloud formations in the vortex street emitted behind the volcanic island of Rishiri-to off the northern tip of Hokkaidō in the Sea of Japan (document NASA, mission STS100)*

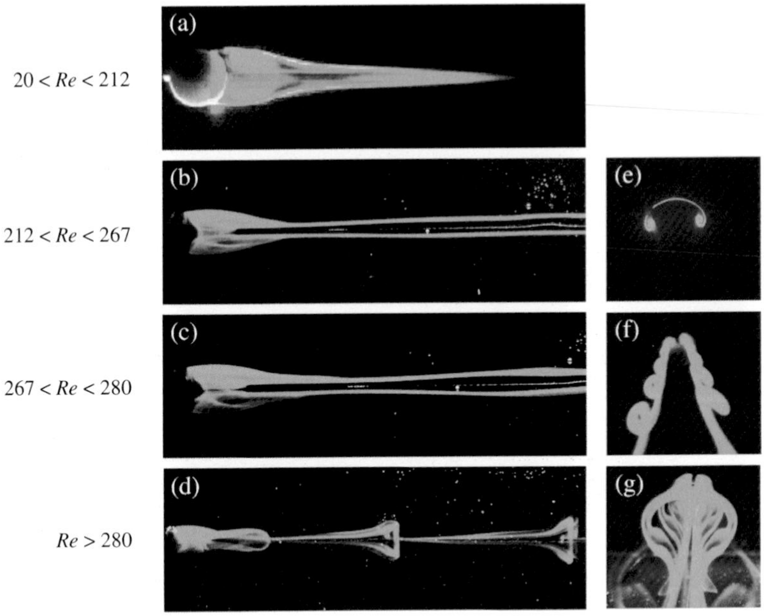

Figure 2.11 *Views from the top (a,b,c,d) and back (e,f,g) of the flow behind a sphere (at the left) visualized by injection of a fluorescent dye. For increasing Reynolds numbers Re, evaluated on the basis of the upstream velocity and the diameter of the sphere, one observes in succession (a) a toroïdal vortex attached on the downstream side of the sphere ($20 < Re < 212$); (b,e) two fixed vortices with their axis parallel to the main flow ($212 < Re < 267$); (c,f) two oscillating vortices ($267 < Re < 280$); (d,g) "hairpin"-like structures ($Re > 280$) (documents A. Przadka and S. Goujon Durand, PMMH-ESPCI)*

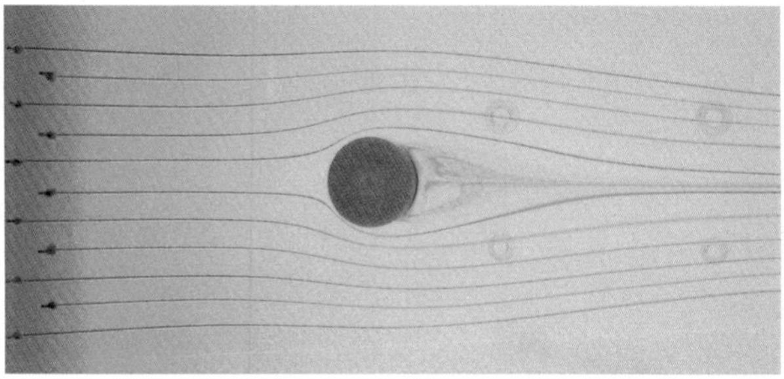

Figure 3.14 *Visualization, by means of streaklines of dye, of the flow around a cylindrical obstacle with its axis perpendicular to the plane of the figure. The flow is from left to right and the dye is injected at 11 discrete points (at the left of the figure). We note the recirculation zone just downstream of the cylinder. (document courtesy of L. Auffray and P. Jenffer, Université Paris-Sud)*

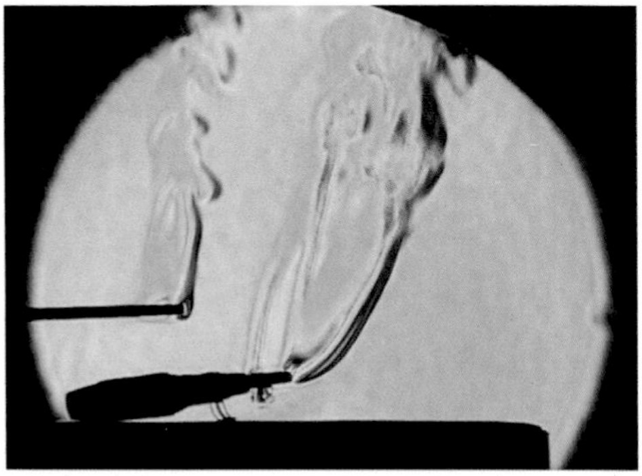

Figure 3.19 *Visualization by the Schlieren method of the variations in the index of refraction of air, resulting from temperature changes near a butane torch (lower plume) and a heated glass rod (upper plume) (document courtesy of I. Smith)*

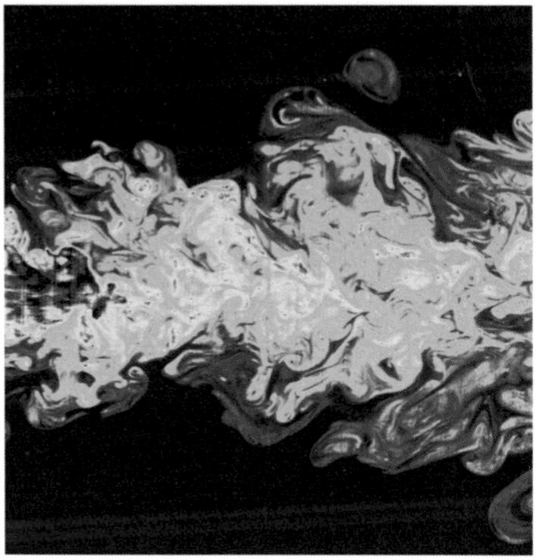

Figure 3.21 *Display of a turbulent jet of liquid injected (from the left) into an identical fluid at rest. A fluorescent dye has been added to the jet, illuminated by a light-sheet in the plane of the figure and coinciding with the axis of the jet. The colours correspond to the local concentration of the dye in the plane of the image (document courtesy of C. Fukushima & J. Westerweel)*

Figure 3.25 *Measurement by means of the PIV technique of (a) the velocity field and (b) the vorticity field of a vortex ring resulting from the propulsion of a squid (Lolliguncula Brevis), as it emits a series of pulsed jets (documents courtesy of I.K. Bartol, P.S. Krueger, W. J. Stewart et J. T. Thompson)*

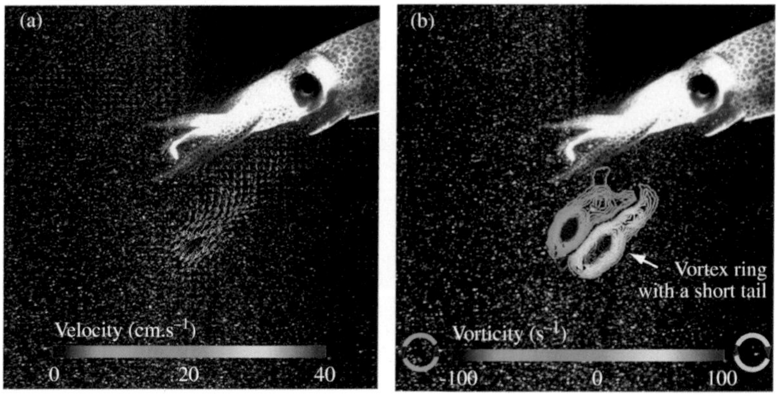

Figure 3.26 *Flow toward a plane (at the bottom of the figure) where there is a zero-velocity stagnation point in the center of the plane: (a) basic image: long exposure visualization of the streamlines by the use of particles carried along by the flow; the colours correspond to the vorticity field calculated from the velocity field of the image (b). Red and dark blue correspond to zones of high vorticities of opposite signs. - (b) velocity field obtained by the micro-velocimetry technique from the observation of the same particles (documents courtesy of C. Pirat, G. Bolognesi and C. Cottin-Bizonne, LPMCN-Lyon)*

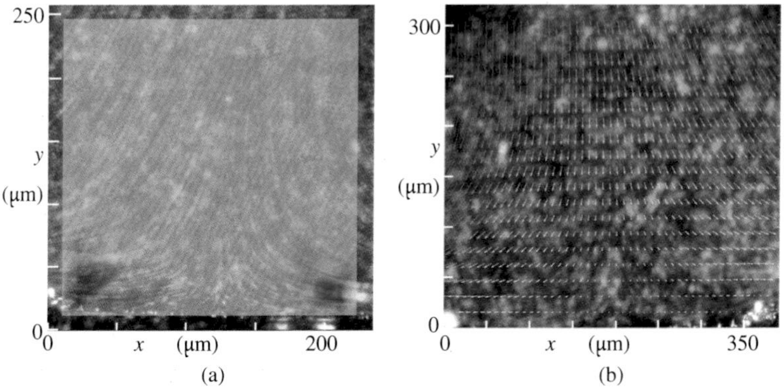

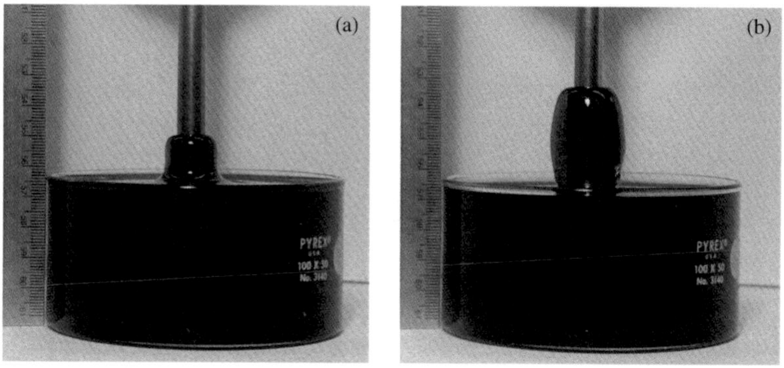

Figure 4.16 *Rise at two successive times, around a rotating axis, of a viscoelastic fluid (polystyrene dissolved in an organic solvent): this phenomenon demonstrates the Weissenberg effect (documents courtesy of J. Bico, G. Mc Kinley, MIT)*

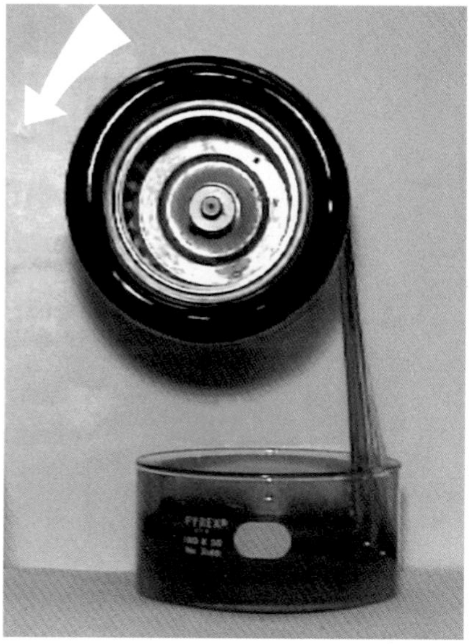

Figure 4.17 *The winding on a rotating spool of a free layer of the viscoelastic fluid shown in Figure 4.16 demonstrates the effect of its high elongational viscosity (documents courtesy of J. Bico, G. Mc Kinley, MIT)*

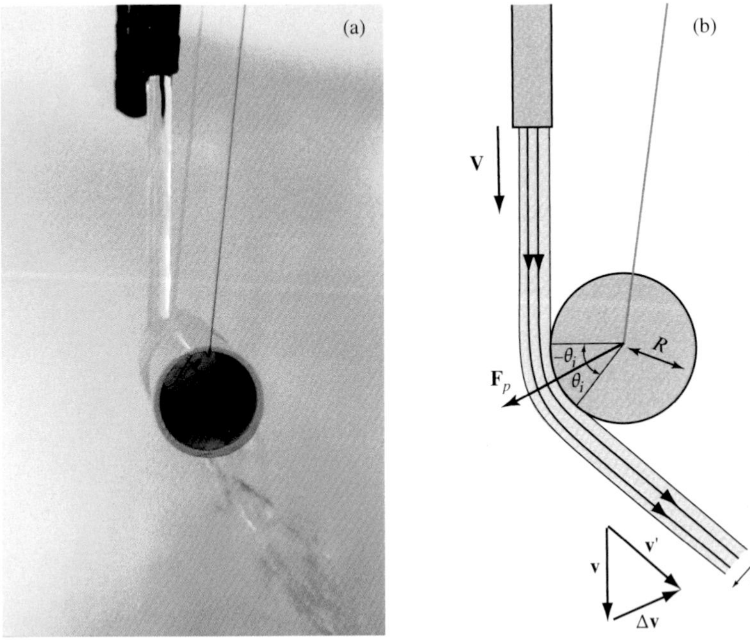

Figure 5.6 *Deviation of a jet of fluid along the curved surface of a cylinder. (a) Deflection of the jet and attraction force on the cylinder. (b) schematic diagram displaying the balance of the forces acting (document Belin publishers)*

Figure 5.11 *Tidal bore heading up (to the right) the Petitcodiac river, near its estuary in the Bay of Fundy, in the Canadian province of New Brunswick (document C. L. Gresley)*

Figure 7.1 *(a) View, from a satellite above the Gulf of Mexico, of Hurricane Katrina, which devastated New Orleans in August 2005 (document NOAA). (b) Waterspout on the ocean offshore from the southern Florida islands. Note the spiral secondary flow at the surface of the water (document V. Golden, NOAA)*

Figure 7.17 *Observation of a vortex ring rising above the volcano Etna, as seen from the "Torre del filosofo". On can see, as the ring moves, the displacement of its shadow on the outer slope of the crater (document J. Alean)*

Figure 7.24 *Vorticity field behind an oscillating wing of width d = 0.5 cm within a flow of constant Reynolds number $Re = 255$ for several amplitudes A of the oscillations of the wing characterized by normalized values $A_d = A/d = 0.36$ (a), 0.71(b), 1.07 (c) (the respective Strouhal numbers are $Sr_A = fA/U$: 0.08, 0.16, 0.24). The color scale corresponds to the local vorticity (in s^{-1}) (document courtesy R. Godoy-Diana, J.L. Aider and J.E. Wesfreid)*

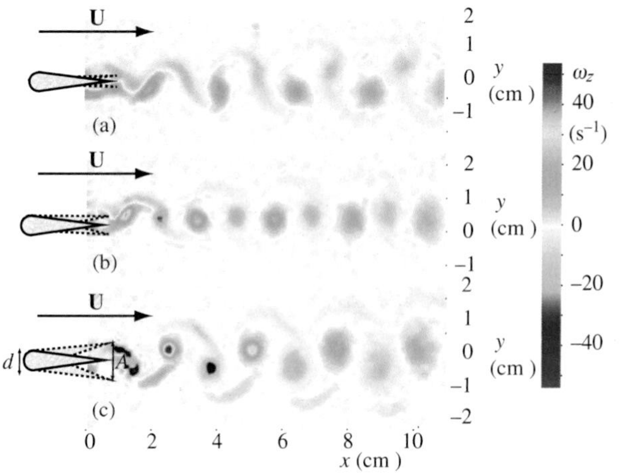

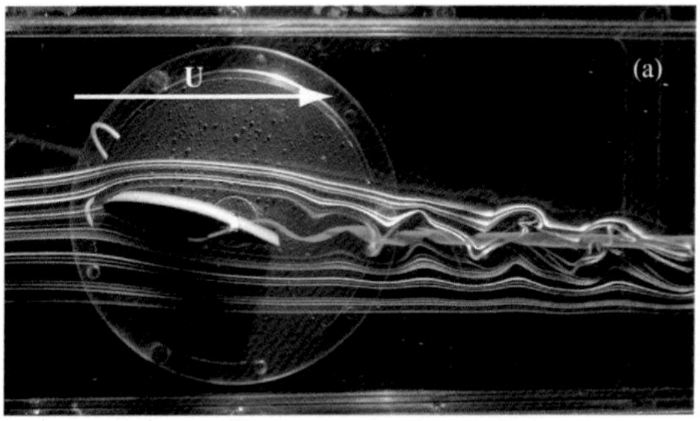

Figure 7.28 *(a) Visualization of a flow around a wing with the appearance of a vortex (in blue) at the tip of the wing (plate courtesy O. Cadot and T. Pichon, ENSTA). (b) Visualization, by injection of smoke at ground level, of the vortex from the end of the wing of a plane spraying agricultural products (NASA document)*

Figure 7.29 *Diagram of the flow around the sails of a yacht designed for the America's cup: vortices (twisted coloured streamlines) are generated at the top of the mast, at the upper point where the jib is fastened, and below the boom bordering the lower side of the mainsail. The yacht sails upwind and the angle between its axis and the wind at the top of the mast is 22° (document courtesy C. Pashias, South African America's Cup Challenge)*

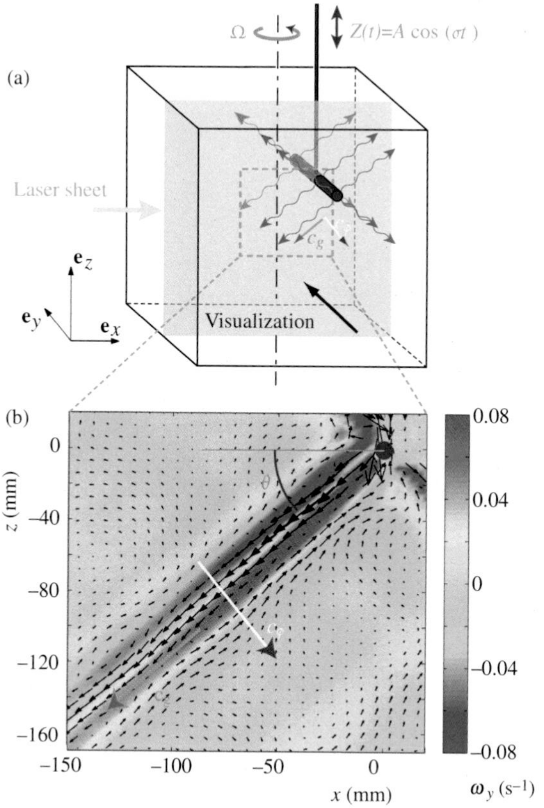

Figure 7.35 *(a) Generation of inertial waves by a horizontal cylinder (in red) oscillating vertically in a fluid rotating around a vertical axis. (b) View of the lower half of one of the beams emitted by the oscillating cylinder. The velocity field is visualized by the Particle Image Velocimetry (PIV) technique in a plane perpendicular to the axis of the cylinder in its middle point. The color scale corresponds to the gradient of the fluid velocity in the direction of the phase velocity* c_φ. *(documents courtesy P.P. Cortet, C. Lamriben and F. Moisy)*

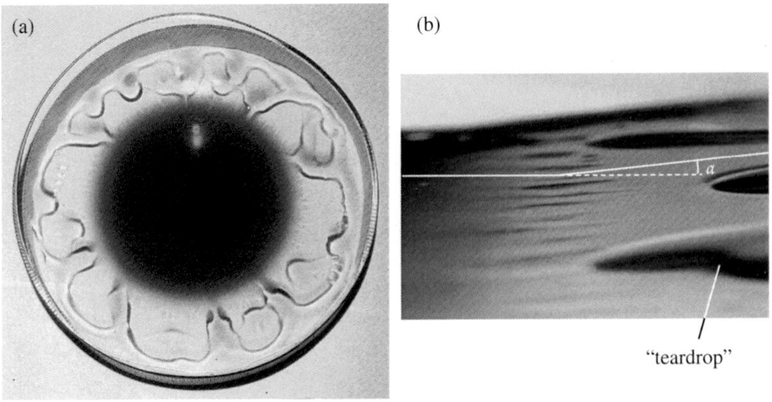

Figure 8.17 *(a) Top view of the rise of a liquid along the walls of a glass due to the Marangoni effect, and of the fall of the liquid with a lower alcohol concentration in the form of "teardrops" or "wine legs." (b) Close-up side view of a model experiment which displays this phenomenon. One pulls a plate inclined at an angle α relative to the horizontal from the water-alcohol mixture: this creates on the plate a fluid film which drops back as legs or teardrops. The narrower transverse ridges correspond to a different flow instability (documents courtesy of J. Bush and P. Hosoi, MIT)*

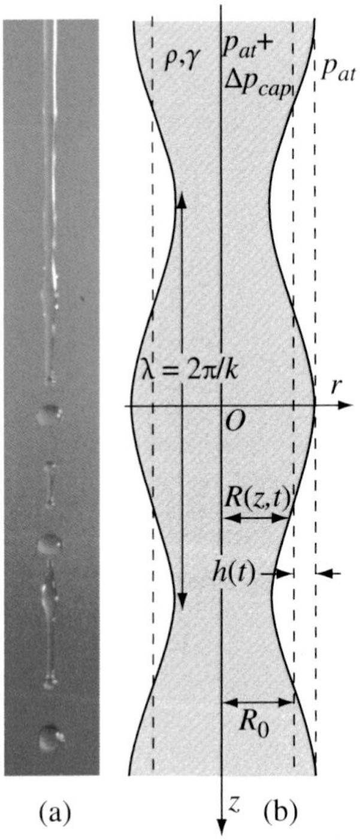

Figure 8.19 *Destabilization, by the Rayleigh-Plateau mechanism, of a vertical cylindrical liquid jet. a) experimental observation of the development of the instability, followed by the formation of drops (document courtesy of J. Aristoff and J. Bush, MIT); b) Schematic diagram for the development of the instability*

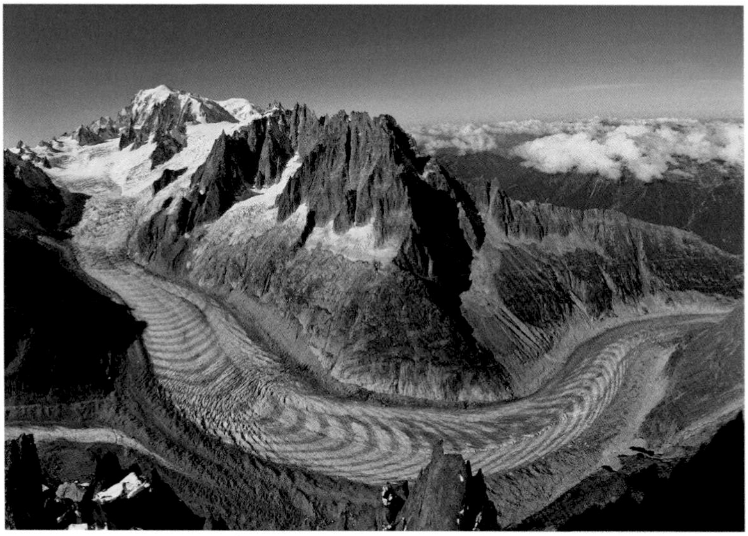

Figure 9.1 *Illustration of Forbes' bands on the "Mer de Glace" glacier (the top of Mont Blanc is visible at the back). The shape of these bands as well as the downstream increase of the amplitude of their deformation indicate the transverse profile of the flow-velocity of the glacier (from left to right on the image), which is considerably smaller near the edges. The separation between the bands corresponds to the displacement over a whole year: they are formed downstream of "seracs" (fracture zones high up on the glacier) and indicate the different level of penetration of solids and dust particles into the ice, depending on the snow coverage (document, courtesy of J-F Hagenmuller (lumieresdaltitude.com)*

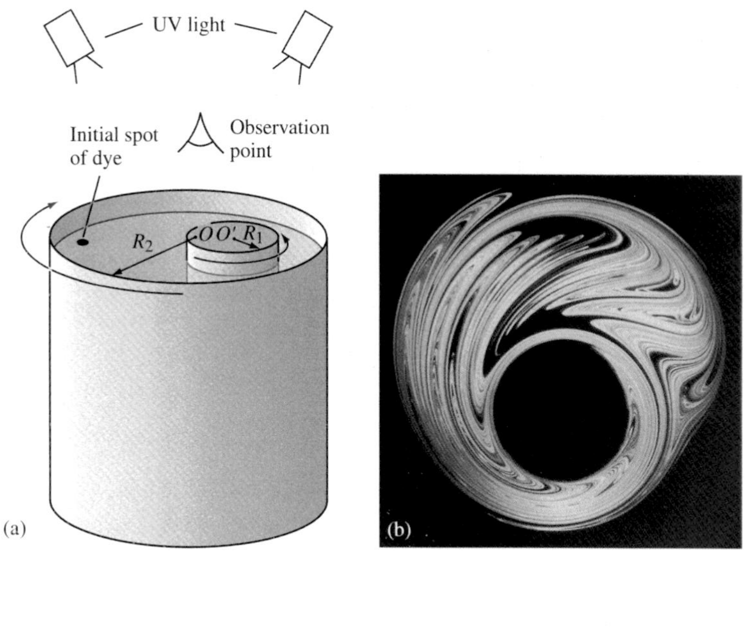

Figure 9.3 *Experiment displaying the chaotic dispersion of a drop of dye initially located between two cylinders with different axes (distance $OO' = 0.3\ R_2$) turning in alternating directions and through different angles. (a) schematic diagram of the experiment; (b) distribution of the dye after ten sequences of alternate rotations through angles $\theta_0 = 270°$ (exterior cylinder) and $-3\,\theta_0$ (interior cylinder) (image courtesy J.M. Ottino, Northwestern University)*

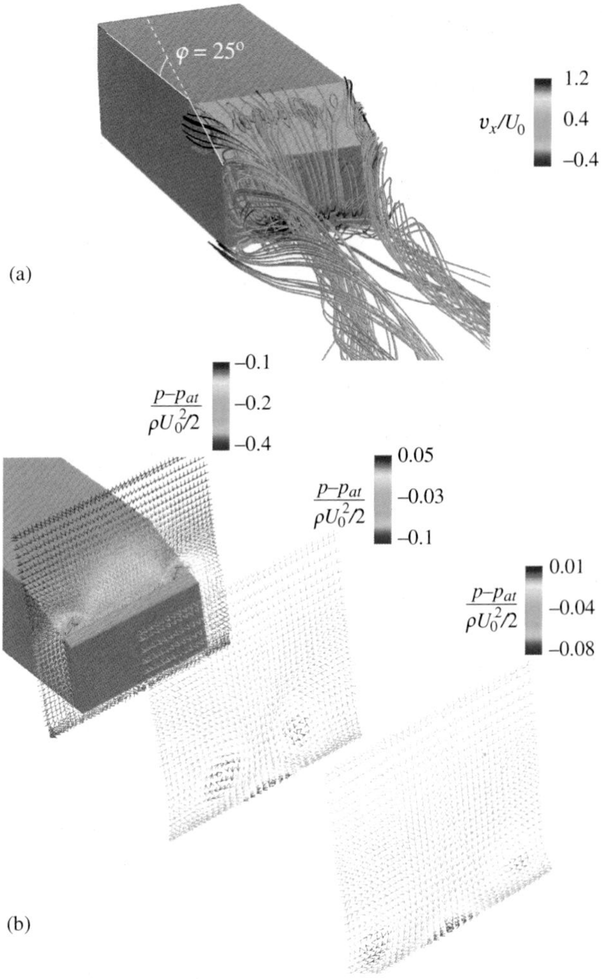

Figure 10.15 *Large-scale simulation (see Section 12.3.4) of the flow behind an Ahmed body with angle* $\varphi = 25°$ *in a flow of average velocity* $U_0 = 40$ m/s *upstream of the body (the flow is from left to right). Two longitudinal vortices are visible on the simulation. (a) Streamlines: the color corresponds to the normalized component* v_x/U_0 *of the velocity in the direction of the average flow. (b) Pressure change* $(p - p_{at})/(\rho U_0^2/2)$ *relative to the normalized atmospheric pressure (colors) and velocity field (vectors). The vortices are displayed as regions of low pressure (in blue on Figure (b)). They represent an important contribution to the drag force (courtesy M. Minguez, R. Pasquetti, and E. Serre)*

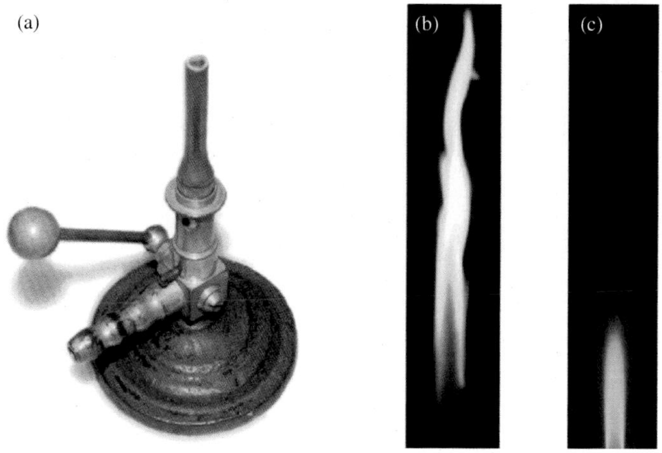

Figure 10.24 *(a) Bunsen burner (collection: Powerhouse Museum, Sidney); (b) diffusion-type flame (closed ferrule); (c) premixed flame (open ferrule); ((b-c): excerpts from a document by A. J. Fijałkowski)*

Figure 10.28 *Visualization of a premixed flame illuminated by a plane light sheet. The white region indicates the boundary between the fresh gases (blue-green) and the burnt ones (red). The fresh gases have been seeded with small refractory grains of caesium oxide so as to visualize the streamlines: these are deflected as they cross the flame front (plate courtesy J. Quinard and G. Searby, IRPHE)*

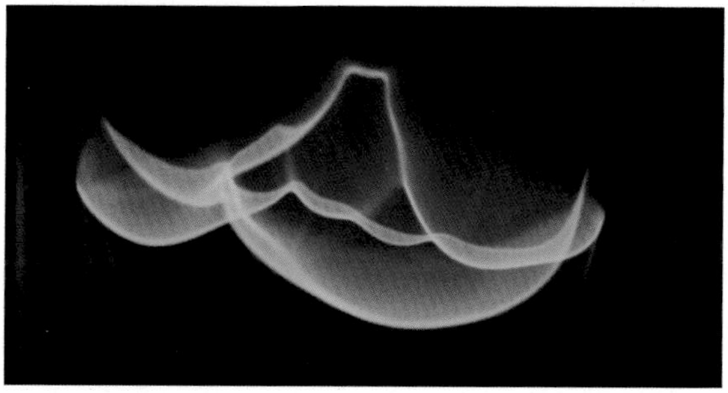

Figure 10.30 *Landau-Darrieus instability of a premixed methane-air flame-front, propagating freely from top to bottom within a tube of diameter* 14 cm, *when no acoustic resonance occur (plate courtesy of J. Quinard, G. Searby, B. Denel, and J. Graña-Otero, IRPHE)*

Figure 11.6 *Temperature-inversion layer and accumulation of "smog" above downtown Los Angeles (plate courtesy M. Luethi)*

Figure 11.16 *(a) Experimental visualization of an instability at the interface between two immiscible liquids in a parallelepipedic cell initially horizontal and tilted downward to the right. (plate courtesy O. Pouliquen). (b) Kelvin–Helmholtz instability in a layer of clouds above the bay of Jervis, New South Wales, Australia (plate courtesy G. Goloy)*

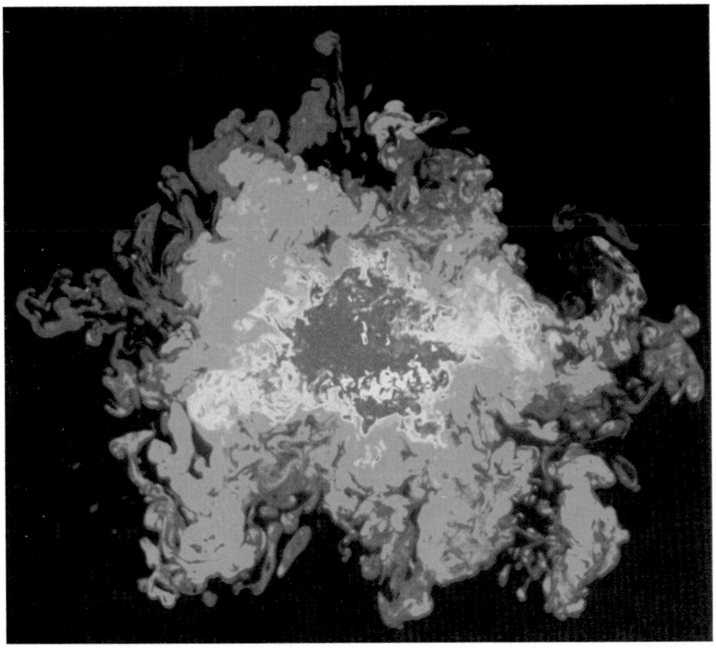

Figure 12.2 *Cross-sectional view of a turbulent jet lighted by a plane of light perpendicular to its axis and injected into a volume of the same fluid. The jet contains a fluorescent dye while the external fluid is pure. The colours indicate the concentration of the dye, increasing from blue to red (plate courtesy H.E. Catrakis et P.E. Dimotakis)*

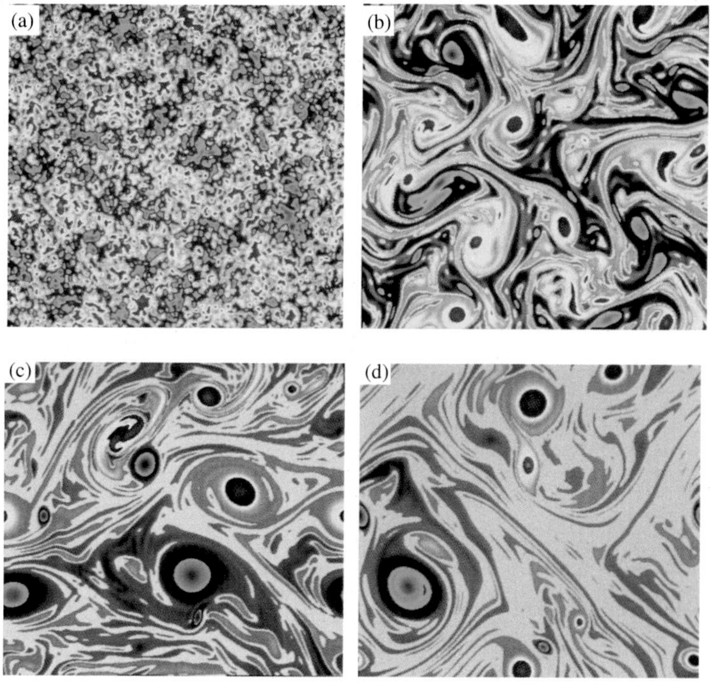

Figure 12.19 *Numerical simulation of the evolution, in time, of a distribution of vorticity, initially random within a two-dimensional turbulent flow, and evolving without further addition of external energy. The colour code is used to represent the vorticity field: red for the strongly positive values, blue for the strongly negative values, grey for low values and yellow for the value zero. The images (a), (b), (c) and (d) correspond respectively to normalized times* $t = 0$, $t = 1$, $t = 3$, $t = 5$ *(plates courtesy M. Farge and J-F. Colonna)*

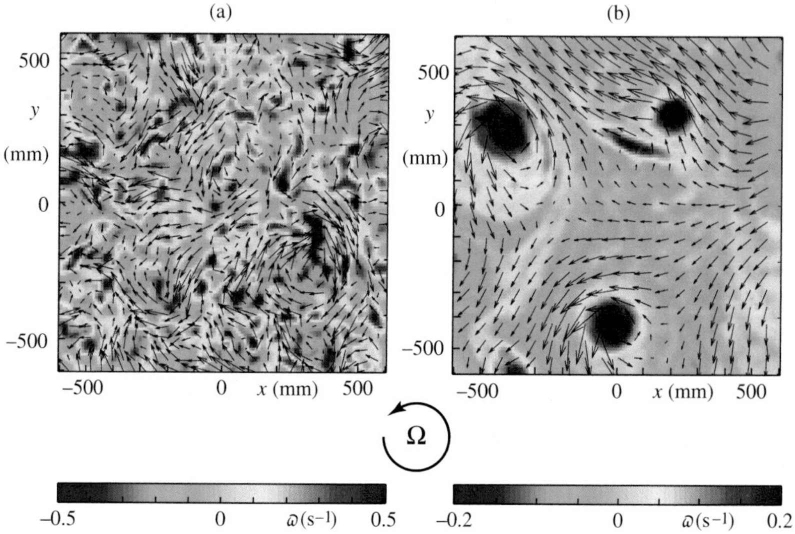

Figure 12.20 *Vorticity distribution in a* 13 m *diameter tank rotating around a vertical axis normal to the plane of the figure (rotation period* = 30 s*): (a) after 2 periods (b) after 10 periods. The turbulent flow has been created by the motion, in the plane of the figure, of a grid perpendicular to it. The component of the vorticity along the axis is shown by the colours and the velocity field, by the vectors. The cyclonic vorticity zones (in red) corresponds to a local rotation of same orientation as the applied rotation; the anticyclonic zones of opposite sign are in blue (plates courtesy F. Moisy, C. Morize, M. Rabaud and J. Sommeria)*

We can combine this result with the component of the equation for vorticity in the z-direction (Equation 7.95):

$$\frac{\partial \omega_z}{\partial t} = 2\,\Omega\,\frac{\partial v_z}{\partial z}, \qquad (7.101) \qquad \text{whence:} \qquad 4\rho\,\boldsymbol{\Omega}^2\frac{\partial v_z}{\partial z} = \frac{\partial}{\partial t}\left(\nabla^2 p\right). \qquad (7.102)$$

If we displace the solid in the direction parallel to the axis of rotation, the conditions for zero velocity at the free surface and at the bottom of the container require that $\partial v_z/\partial z$ will not vanish (<0 above the solid and >0 below, for $U > 0$). There appears therefore, in accordance with Equation 7.101, a vorticity ω_z of amplitude initially proportional to the time and to Ω, with opposite signs above and below the solid.

According to Equation 7.102, the Laplacian $\nabla^2 p$ of the pressure increases proportionately to the time and to Ω^2, also with opposite signs above and below the solid. Let us assume that the solid is axially symmetric around an axis parallel to the rotation vector $\boldsymbol{\Omega}$, and that $\partial p/\partial r = 0$ for $r = 0$ (r being the distance to the axis of symmetry); using the fact that we must match the external pressure, which is not perturbed at large distances from the axis, we find that, for $U > 0$, there is an excess of pressure above the solid and a pressure reduction below, both proportional to $\boldsymbol{\Omega}^2$. Moreover, for small amplitudes, this excess pressure is linearly proportional to the displacement: if we release the object after having displaced it rapidly parallel to $\boldsymbol{\Omega}$, it moves slightly backwards and can even display a few oscillations.

Everything occurs as if the rotating fluid had a certain elasticity: this elasticity is not related to the compressibility of the fluid in question, but results from the quasi two-dimensional character of the flow (Equation 7.99) due to the Coriolis force. Let us assume, for example, that one induces a radial flow perturbation of velocity v_r so as to increase the area of a circle (C) in a plane normal to $\boldsymbol{\Omega}$ (Figure 7.34). The Coriolis force associated to v_r results in a tangential velocity component v_t; this component is, in turn, associated to a Coriolis force inducing a radial velocity component v_{r2} of direction opposite to v_r which tends to bring back the curve to its original area.

A related phenomenon is the behavior, in the absence of a pressure gradient, of a fluid particle to which is applied a small initial velocity $\delta\mathbf{v}$ in a fluid initially at rest in the rotating reference frame. The resulting change in velocity obeys the simplified form of Equation 7.94:

$$\frac{\partial \mathbf{v}}{\partial t} = -2\,(\boldsymbol{\Omega} \times \mathbf{v}). \qquad (7.103)$$

The velocity components are thus in the form:

$$v_x = A\cos\left(-2\,\Omega\,t + \varphi\right), \qquad (7.104a)$$

$$v_y = A\sin\left(-2\,\Omega\,t + \varphi\right), \qquad (7.104b)$$

$$v_z = \text{constant}. \qquad (7.104c)$$

The particle does not move along a straight line as it would in the absence of rotation; on the contrary, it follows, in the plane perpendicular to $\boldsymbol{\Omega}$, circular paths with an angular velocity $-2\,\Omega$ (twice the angular velocity of rotation of a Foucault pendulum); it returns therefore periodically to its initial location. 2Ω thus represents a characteristic frequency of the rotating fluid, which is actually observed in oceanographic systems. In the atmosphere, such motion can be superimposed onto a geostrophic flow. This motion is analogous to that of a charged particle of charge q and mass m in a magnetic field $\mathbf{B}$: the motion of the charged particle in the plane perpendicular to $\mathbf{B}$ is also circular and its angular frequency is the Larmor frequency qB/m. The analog of the Coriolis force $-2(\boldsymbol{\Omega} \times \mathbf{v})$ is, in this case, the Laplace force $q(\mathbf{v} \times \mathbf{B})$.

7.6.3 Waves within rotating fluids

We introduce here several mechanisms for wave propagation in rotating fluids, all of these associated with deviations relative to the basic geostrophic flows (or appearing in a fluid at rest); such waves do not involve the compressibility of the fluid, but are associated to the Coriolis force. These waves will be modelled using Equations 7.94 and 7.95 which include non-stationary terms; second-order terms in $\mathbf{v}$ and $\boldsymbol{\omega}$ and centrifugal forces are neglected, as well as the effects of viscosity and gravity. The fact that these equations of motion, valid at small Rossby numbers, are linear, allows for a simple study of these waves, which play a very important role in the atmosphere and oceans.

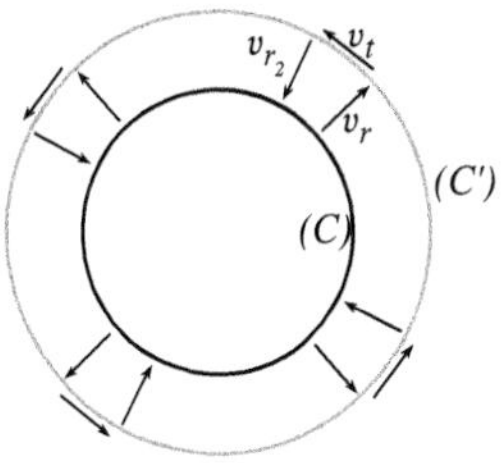

Figure 7.34 *A perturbation of the velocity v_r which tends to increase the area of a circular material curve (C) to (C') leads, due to the effect of the Coriolis forces, to fluctuations v_t and then v_{r2} which tend to bring it back to the initial radius*

Inertial waves

The "elasticity" of rotating fluids, already discussed in Section 7.6.2 can be sufficient to allow for the propagation of waves known as *inertial waves*.

We can generate these waves by creating an oscillation, parallel to the axis of rotation, of a solid object (a cylinder in Figure 7.35a) with a small amplitude $z_0(t) = A\,exp(i\,\sigma\,t)$. The frequency σ needs to be sufficiently large so that the non-stationary terms $\partial\mathbf{v}/\partial t$ and $\partial\boldsymbol{\omega}/\partial t$ are not negligible. In order to achieve this, the ratio σ/Ω, of σ to the rotational angular velocity Ω, must not be much less than 1 (otherwise, as we will see further down, we are back in the case of geostrophic flows).

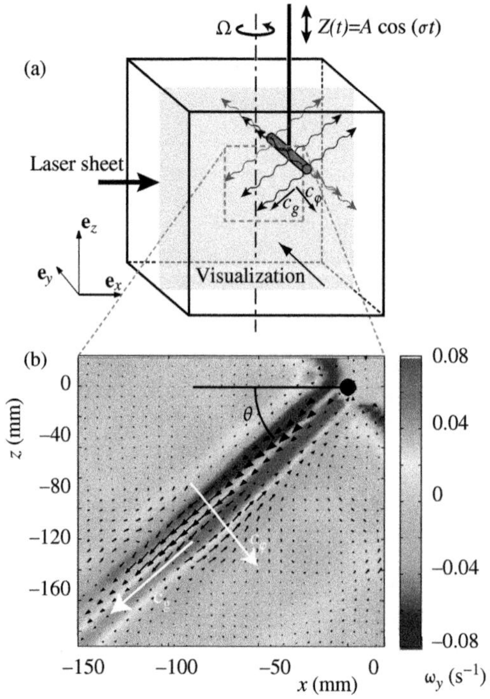

Figure 7.35C *(a) Generation of inertial waves by a horizontal cylinder oscillating vertically in a fluid rotating around a vertical axis. (b) View of the lower half of one of the beams emitted by the oscillating cylinder. The velocity field is visualized by the Particle Image Velocimetry (PIV) technique in a plane perpendicular to the axis of the cylinder in its middle point. The color scale corresponds to the gradient of the fluid velocity in the direction of the phase velocity c_φ. (documents courtesy P.P. Cortet, C. Lamriben and F. Moisy)*

During a period $2\pi/\sigma$ of the oscillation, the fluid particles follow circular trajectories in the plane of the beams: we see in Figure 7.35b that the orientation of the velocity of the fluid at a given time is constant in the direction parallel to this plane but varies in the perpendicular direction. Since this orientation is determined by the phase of the motion along the circular trajectories, the planes which are parallel to that of the beam correspond to a constant phase: the wave vector $\mathbf{k}$ and the phase velocity $\mathbf{c}_\varphi$ are therefore normal to them. In contrast, the group velocity $\mathbf{c}_g$ must be oriented in the direction of the beam, since it corresponds to propagation of energy: it is thus perpendicular to the phase velocity $\mathbf{c}_\varphi$ indicating an unusual kind of anisotropy for a propagating wave.

Experimentally, we observe that the motion of the fluid resulting from this oscillation is concentrated in two thin plane beams which cross at the cylinder and make equal angles θ with the horizontal (Figure 7.35).

Assuming that the wave is planar and that its wave vector $\mathbf{k}$ makes an angle θ with the angular velocity vector $\boldsymbol{\Omega}$, we find that the dispersion relation between σ and the wave vector $\mathbf{k}$ is:

$$\sigma = 2\,\Omega\cos\theta. \tag{7.105}$$

The frequency σ does not depend on the magnitude of the wave vector, but solely on its orientation. The frequency of these inertial waves has a maximum value 2Ω for $\theta = 0$, which is the characteristic frequency of the particles of fluid which we have already discussed in the preceding section ($\mathbf{k}$ has then the direction of the axis of the rotation).

One does indeed find in this case circular trajectories of the particles in a plane perpendicular to $\boldsymbol{\Omega}$ (Equation 7.104).

In the low-frequency limit $\sigma \to 0$ (the quasi-stationary case), we find, in contrast, geostrophic flows with $\theta = \pi/2$ which corresponds to a phase of the wave invariant in the z-direction, in agreement with the two-dimensional nature of these flows.

Proof

Let us consider the z-direction component of Equation 7.94, taking into account the non-stationary term:

$$\frac{\partial v_z}{\partial t} = -\frac{1}{\rho}\frac{\partial p}{\partial z} \tag{7.106}$$

and let us combine this with Equation 7.102. We obtain the propagation equation for pressure fluctuations:

$$4\,\Omega^2\,\frac{\partial^2 p}{\partial z^2} = -\,\frac{\partial^2}{\partial t^2}(\nabla^2 p). \qquad (7.107)$$

From this, we obtain Equation 7.105, by assuming a pressure variation of the type $p = p_0\,e^{i(\sigma t - \mathbf{k}.\mathbf{r})}$, corresponding to a plane wave, and using this expression in Equation 7.107.

The phase velocity $c_\varphi = \sigma/k$ of the wave has a value $(2\Omega\cos\theta)/k$: it is non-zero and directed along **k**. For the group velocity c_g, we must use a vector definition: the velocity component $c_{g//}$ along **k**, given by the usual expression $c_g = d\sigma/dk$ and using Equation 7.105 is indeed zero. For the perpendicular component, which does not vanish, we have $c_{g\perp} = (1/k)\,d\sigma/d\theta = -(2\Omega\sin\theta)/k$.

Kelvin waves

In contrast to the preceding case, we consider here surface waves of the same nature as those which we have studied in Section 6.4, but altered by the effect of the Coriolis force. The hydrostatic pressure gradients resulting from the passage of the wave are, indeed, partially balanced by this force. We study here the case where the waves propagate along a vertical wall ($x = 0$) which corresponds to the situation of a wave propagating along a coast line; it is only in this case that we observe plane waves, where the velocity of the fluid is everywhere parallel to the direction of propagation.

We use here the geometry displayed in Figure 7.36a, assuming that we are in a regime of *shallow water waves,* for a fluid with negligible viscosity: the height h of the fluid at rest is assumed to be constant (horizontal bottom), and small compared to the characteristic distances for the variation of the level of the surface in the x and y directions.

Let us denote by $z_0(x, y, t)$ the instantaneous thickness of the fluid layer and let us assume that the waves have a small amplitude; the component v_y of the velocity of the fluid, assumed to be independent of the vertical coordinate z, can then be written (see proof below):

$$v_y(x, y, z, t) = v_{y1}(y + ct)\,e^{-x/R} + v_{y2}(y - ct)\,e^{x/R}, \qquad (7.108a)$$

with:

$$R = \frac{\sqrt{gh}}{2\,\Omega} = \frac{c}{2\,\Omega}. \qquad (7.108b)$$

One then finds that the depth z_0 can be expressed in the form:

$$z_0(x, y, t) = h - \sqrt{\frac{h}{g}}\,\big(v_{y1}(x, y + ct) - v_{y2}(x, y - ct)\big). \qquad (7.108c)$$

The components v_{y1} and v_{y2} vary exponentially with x and with opposite arguments but, in the geometry of Figure 7.36a (with $\Omega > 0$), only the solution varying as $e^{-x/R}$ has physical significance: as we go away from the wall, the other solution would in fact correspond to an exponential growth of the deformation. The wave is therefore confined along the vertical wall $x = 0$ and its amplitude $z_0 - h$ decreases exponentially over a distance R, known as the *deformation distance.* The value of R is smaller, the greater the angular velocity of rotation: for an ocean depth of the order of 100 meters with $\Omega = 10^{-4}\,s^{-1}$, we find a value of R of the order of 150 km. This length R appears also in other geophysical problems which involve vertical motion of fluids. Finally, one has a single possible direction of propagation of the wave (that associated to the decreasing exponential); it changes, according to Equation 7.108b, with the direction of the rotation.

Such waves can be associated with tides, or with the action of winds parallel to the shore line (Figure 7.36b). For two shores close enough to each other, these waves can result in a significant difference of the amplitude of the tides, as is the case between the English and French coasts of the English Channel. A tide that travels from west to east can, indeed, be

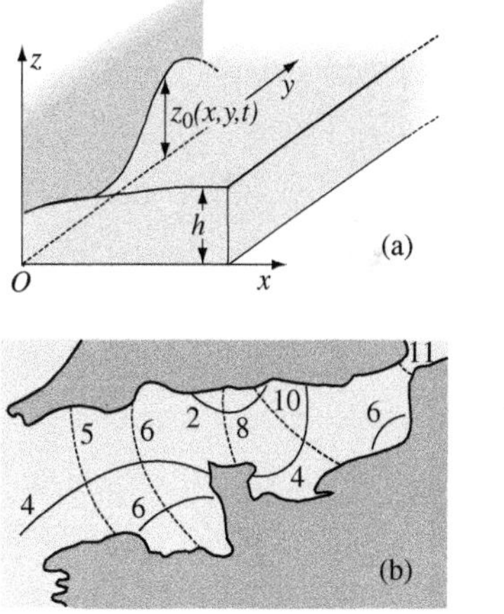

Figure 7.36 *(a) Schematic diagram for the propagation of a Kelvin wave along a vertical wall ($x = 0$); (b) propagation of tides in the English Channel with time indicated in hours (dashed lines) and lines of equal amplitude of the tides in metres (solid lines), indicating a maximum value of the tide near the French coast (after J. Proudman (1953) and A. E. Gill (1982))*

considered as a Kelvin wave propagating along the French coast: the latter plays the role of the wall $x = 0$, and the resulting tides are consequently higher. This is equally the case for rising and falling tides which represent opposite directions of the sea-level but which always propagate in the same direction (the direction of the currents is, of course, opposite for the two cases).

Waves analogous to Kelvin waves can propagate along the equator, still from west to east, without the presence of a coastline being needed. In fact, since the direction of the local component of the rotation of the earth becomes inverted as we pass through the equator, we can satisfy the boundary conditions for the equations with components decreasing exponentially both northward and southward, while still propagating in the same direction. Such waves appear, just as Rossby waves which we will now describe, during the phenomenon known as "El Niño" which perturbs significantly the meteorologic conditions in the Pacific Ocean west of the coast of North and South America. In this case, we deal with waves propagating near the American coastlines as well as near the equator. Such waves have been detected by satellite observations (from Topex-Poseidon and also Jason 1 and 2) able to detect changes of the order of a centimeter of the average level of the ocean.

Proof

The assumptions of a shallow depth of the fluid layer, and of a small amplitude for the wave, allow us to write a simpler form of the equation of motion than for the general case discussed in Chapter 6. In the absence of viscosity, the pressure gradient in the x- and y-directions is related to the changes in the level of the surface by:

$$\frac{\partial p}{\partial x} = \rho g \frac{\partial z_0}{\partial x} \quad (7.109a) \qquad \text{and} \qquad \frac{\partial p}{\partial y} = \rho g \frac{\partial z_0}{\partial y} \quad (7.109b)$$

(the pressure near the surface is indeed equal to the atmospheric pressure and the vertical-pressure gradient is purely of hydrostatic origin). It follows then from Equation 7.94:

$$\frac{\partial v_x}{\partial t} - 2\,\Omega\, v_y = -g\frac{\partial z_0}{\partial x} \quad (7.110a) \qquad \text{and} \qquad \frac{\partial v_y}{\partial t} + 2\,\Omega\, v_x = -g\frac{\partial z_0}{\partial y}. \quad (7.110b)$$

Moreover, the property of incompressibility $\nabla\cdot\mathbf{v} = 0$ can be written, evaluating $\partial v_z/\partial z$ just as we did in deriving Equation 7.86:

$$\frac{1}{h}\frac{\partial z_0}{\partial t} + \frac{\partial v_x}{\partial x} + \frac{\partial v_y}{\partial y} = 0 \quad (7.111)$$

(in a first-order approximation, we have substitued h for z_0 in the first factor on the left hand side). The component v_x of the velocity normal to the wall, must necessarily vanish at the wall ($x = 0$): we will then assume throughout the remainder of this section that $v_x = 0$ throughout the volume of the fluid, and show that we are able to solve the equations of motion under this assumption. Equations 7.110 and 7.111 thus become:

$$-2\,\Omega\, v_y = -g\frac{\partial z_0}{\partial x}, \quad (7.112a) \qquad \frac{\partial v_y}{\partial t} = -g\frac{\partial z_0}{\partial y} \quad (7.112b)$$

and

$$\frac{1}{h}\frac{\partial z_0}{\partial t} + \frac{\partial v_y}{\partial y} = 0. \quad (7.112c)$$

We observe that, according to Equation 7.112a, the slope of the surface in the x-direction is a function both of the direction of the rotation and of the instantaneous velocity. If we eliminate z_0 from Equations 7.112b and 7.112c after taking derivatives with respect to t and y, we obtain the wave equation:

$$\frac{\partial^2 v_y}{\partial t^2} = g\,h\,\frac{\partial^2 v_y}{\partial y^2} = c^2\frac{\partial^2 v_y}{\partial y^2}, \quad (7.113)$$

where $c = \sqrt{gh}$ is the wave velocity: this velocity is identical to that of gravity waves in shallow water, in the absence of rotation (Equation 6.64) and it is independent of Ω. We will now see, however, that Ω plays a key role in determining the changes in z_0 as a function of the distance from the vertical wall (as indicated by Equation 7.112a). Let us now break up the velocity field v_y and the depth z_0 into two components traveling in opposite directions, so that:

$$v_y(x, y, t) = v_{y1}(x, y + ct) + v_{y2}(x, y - ct) \quad (7.114a)$$

and

$$z_0(x, y, t) = \delta z_{01}(x, y + ct) + \delta z_{02}(x, y - ct) + h, \quad (7.114b)$$

where $c = \sqrt{gh}$. By substituting these two equations into Equation 7.112b and then integrating, we prove Equation 7.108c. By combining this last equation with Equations 7.112a, 7.114a and 7.114b, we obtain:

$$\frac{\partial}{\partial x} v_{y1}(x, y + ct) - \frac{\partial}{\partial x} v_{y2}(x, y - ct) = -\frac{2\Omega}{\sqrt{gh}} \left(v_{y1}(x, y + ct) + v_{y2}(x, y - ct) \right). \quad (7.115)$$

We then obtain Equation 7.108a by integration with respect to x.

Rossby waves due to changes in the local velocity of the earth

The mechanism underlying these waves is quite different, and based on the variation with the latitude of the local velocity of the rotation of the earth. These are also called *planetary waves*, because their wavelength is of the same order of magnitude as the scale factors characteristic of the entire earth. They are characterized by sinusoidal oscillations which propagate from east to west in mid-latitude regions. The Rossby waves are closely related to the global atmospheric circulation, and to the occurence of cyclonic and anti-cyclonic zones alternating over distances of several thousand kilometers. Just like Kelvin waves, they play an important role in the climatic variations related to major ocean currents (e.g., such as the "El Niño" phenomenon).

The local component Ω of the rotation of the earth (projection of the vector of the rotation of the earth Ω_0 onto the local perpendicular to the surface of the earth) obeys the equation:

$$\Omega = \Omega_0 \sin\alpha, \quad (7.116)$$

in which α is the latitude (Figure 7.30). In local coordinates, with x oriented eastward and y northward, a low-order expansion of Equation 7.116 around the reference latitude $y = 0$ yields:

$$\Omega(y) = \Omega_{y=0} + \frac{\beta y}{2} \quad (7.117)$$

where $\beta = (2\Omega \cos\alpha)/R$ (R is the radius of the earth). Geophysicists speak of the *β-plane model* to take into account the changes with latitude of the value of Ω in Equation 7.90 for the potential vorticity. We analyze below the propagation of Rossby waves with and without an average flow of component V_0 in the x-direction (Figure 7.37 on the next page). We further assume that the depth h of the layer of fluid remains constant.

Let us first assume that $V_0 = 0$, and impose an oscillatory motion $\delta y(t, x = 0) = A\, exp(-\mathrm{i}\,\sigma t)$ in the y-direction (South-North) of fluid particles in the plane $x = 0$. Let us further assume that v_x vanishes and that the vorticity ω_z is also zero for $\delta y = 0$. We seek to understand how this oscillation may propagate towards other regions of the fluid ($x \neq 0$). According to Equation 7.90, the total vorticity $\omega_z + 2\Omega$ of the fluid

In the real world, these assumptions are frequently not satisfied; we would then have to take into account the variations of h, and consequently, of the velocity of gravity waves. We might also need to take into account internal waves which deform the interface between fluid layers of different densities (e.g., within the ocean). We will neglect these effects in the model which we consider here, which is a good approximation describing the essential physical phenomenon.

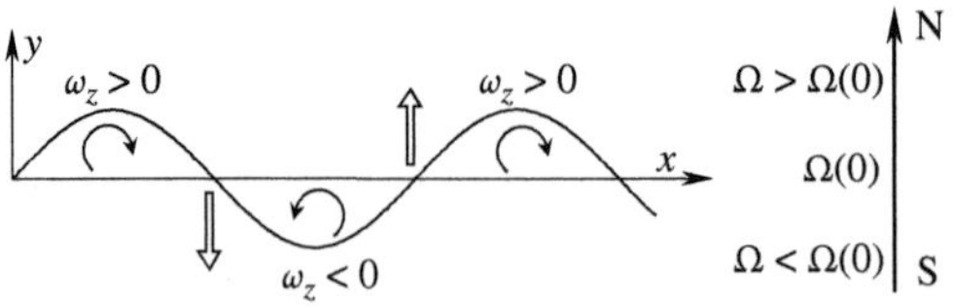

Figure 7.37 *Schematic diagram of the mechanism of the propagation of Rossby waves from east to west (in the negative x-direction)*

particles must remain constant; for a displacement δy, the variation $\beta\,\delta y$ of 2Ω, resulting from Equation 7.117, must be compensated by the appearance of a vorticity component:

$$\omega_z = \frac{\partial v_y}{\partial x} = -\beta\;\delta y. \tag{7.118}$$

The sign of this vorticity component depends on the direction of the displacement. Equation 7.118 indicates that the oscillations may propagate beyond the plane $x = 0$, but that the velocity v_y will depend on the distance. If we assume (as will be shown below) that we generate a plane wave with: $\delta y = A e^{i(kx - \sigma t)}$, Equation 7.118 reduces to $k\sigma = -\beta$ (taking $v_y = d(\delta y)/dt$). The velocity of propagation is then $c = \sigma/k = -\beta/k^2$: the negative sign indicates that the waves propagate in the direction $x < 0$. These *free Rossby waves* observed in the absence of flow ($V_0 = 0$) are tranverse waves propagating *westward* ($x < 0$).

Let us now consider the more general case where we have a global flow V_0 and where we take into account the changes in velocity in the y-direction, assuming that they are characterized by a component ℓ of the wave-vector. Retaining the assumption of a constant depth of the fluid layer, we obtain (as will be proved below) the more general dispersion relation:

$$c = \frac{\sigma}{k} = V_0 - \frac{\beta}{k^2 + \ell^2} \tag{7.119}$$

Without a global flow ($V_0 = 0$) and for $\ell = 0$, we obtain again the equation $c = -\beta/k^2$ found above. The velocity c increases with the wave length and decreases as one moves away from the equator.

At intermediate latitudes, the westerly winds can have a sufficiently high velocity so that we can have $c = 0$ in Equation 7.119. In such *stationary regimes*, a Rossby wave can be initiated, for example, by the deviation of a westerly wind over a long mountain range oriented in the North-South direction (Figure 7.31). Air streams moving eastward display a north-south oscillation with a wavelength of a few hundred kilometers but the velocity field remains invariant as a function of time. Taking $c = 0$ in Equation 7.119, we then obtain:

$$k^2 = \frac{2\Omega_0 \cos\alpha}{R V_0}. \tag{7.120}$$

Substituting for $2\Omega_0 \cos\alpha$ the value $10^{-4}\,\mathrm{s}^{-1}$, and assuming that V_0 is of the order of 10 m/s, we find a wavelength of the order of 5000 km; this is indeed the order of magnitude for the waves observed.

Proof

Let us write Equation 7.118 for the conservation of the potential vorticity in the form:

$$\frac{d\omega_z}{dt} + \beta\, v_y = 0. \tag{7.121}$$

We assume that this motion is a superposition of an constant flow of velocity V_0 in the x-direction, and of a perturbation $\mathbf{v}'$ varying sinusoidally in the x- and y-directions, i.e.:

$$v_x = V_0 + v'_{0x}\,\mathrm{e}^{\mathrm{i}(kx+\ell y-\sigma t)}, \qquad v_y = v'_{0y}\,\mathrm{e}^{\mathrm{i}(kx+\ell y-\sigma t)}, \qquad \omega_z = \omega_0 + \omega'_0\,\mathrm{e}^{\mathrm{i}(kx+\ell y-\sigma t)}.$$

We represent the perturbation in the velocity by means of a stream-function Ψ (as in Chapter 6, Section 6.5) such that:

$$v'_x = -\frac{\partial \Psi}{\partial y}, \quad (7.122a) \qquad v'_y = \frac{\partial \Psi}{\partial x}. \quad (7.122b)$$

Equation 7.121 then becomes:

$$\left(\frac{\partial}{\partial t} + V_0\frac{\partial}{\partial x}\right)\nabla^2\Psi + \beta\,\frac{\partial \Psi}{\partial x} = 0 \qquad (7.123)$$

We look for solutions in the form $\Psi = \Psi_0\,\mathrm{e}^{\mathrm{i}(kx+\ell y-\sigma t)}$. Substituting this into the previous equation, we find:

$$-\left(-\sigma + kV_0\right)\left(k^2 + \ell^2\right) + k\beta = 0 \qquad (7.124)$$

From this, we derive Equation 7.119 for the phase velocity.

As we have mentioned above, the model which leads to this equation is only an approximation, as it does not take into account changes in the depth of the fluid layer, nor is it applicable in the region near the equator where the Rossby approximation becomes invalid. The velocities of propagation of the waves are higher than the ones predicted here, most particularly near the equator.

Rossby waves due to the changes in depth of the fluid layers

The same phenomenon can result, not from the β effect, but because of the progressive change in the depth h of oceans as one moves away from the coast. Then, the change in the potential vorticity is induced by the variation of h (Equation 7.90): such *topographic Rossby waves* are observed for instance in the Gulf of Mexico.

Let us use again the previous approach and assume (Figure 7.38) that the thickness h of the fluid layer is equal to h_0 on the axis $y = 0$ and that it varies linearly with distance so that $h(y) = h_0 - \gamma y$ ($\gamma \ll 1$). On the other hand, h is assumed to be independent of x. The unperturbed flow is uniform, at a velocity $\mathbf{V}_0$ in the x-direction. Let us now assume that we cause a small perturbation by creating a velocity component v_y normal to the mean flow (we

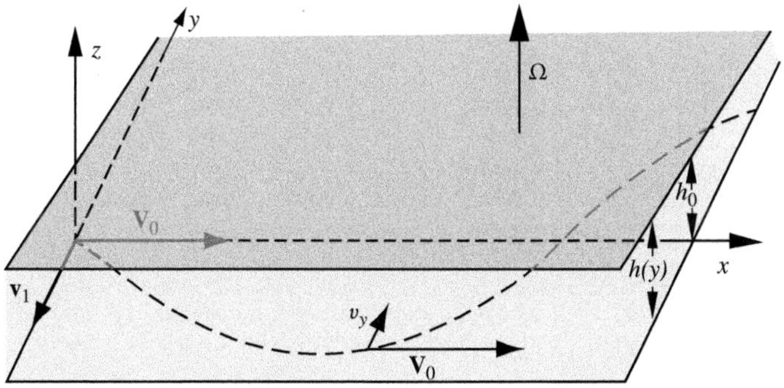

Figure 7.38 *Schematic diagram of the flow of a rotating fluid layer with a depth varying in the direction y normal to the mean flow*

can, for example, introduce a tiny obstacle at $x = 0$, transverse to the flow): we are interested here in the stationary flow resulting from this perturbation, which we assume to be constant. The initial vorticity $\partial v_x/\partial y$ is assumed to be zero.

Let us follow the trajectory of an element of fluid which starts at the point $x = 0, y = 0$ (so that $h = h_0$), at time $t = 0$, with an initial velocity $(\mathbf{V}_0, \mathbf{v}_1)$ where $\mathbf{v}_1$ is the small perturbation. We apply Equation 7.90 to this small element between the initial time and an arbitrary time, assuming $\omega_z = 0$ for $h = h_0$ in the form:

$$\frac{\partial v_y}{\partial x} = \omega_z = 2\,\Omega\left(\frac{h}{h_0} - 1\right) = -\frac{2\gamma\,\Omega\,y}{h_0}. \tag{7.125}$$

If γ and Ω have the same sign and, if $V_0 > 0$, the component v_y in the y-direction of the velocity of the small element of the fluid decreases as it goes further from the plane $y = 0$, while being carried along in the direction $x > 0$ by the average flow. The component v_y ends up becoming zero and turns negative. The element of fluid then crosses the plane $y = 0$ and oscillates between both sides of this plane, as it is carried along. If, on the other hand, γ and Ω have opposite signs (still with $V_0 > 0$), the fluid element takes on a velocity component which increases as it moves further from the plane $y = 0$: we thus do not have any oscillation of the trajectory. If $V_0 < 0$, the requirements are reversed: for an oscillation to appear, the product $\Omega V_0 \gamma$ must be >0.

For a sinusoidal oscillation, we find that the wavevector k has the value:

$$k = \sqrt{\frac{2\,\gamma\,\Omega}{h_0\,V_0}}. \tag{7.126}$$

Throughout this discussion of the case of a varying depth, we considered spatially periodic oscillations of the velocity and the trajectory of the fluid particles which are stationary in time in the reference frame of the laboratory. Just as in the previous case of the β effect, one can also create similar waves without any average flow, by means of a periodic excitation of angular frequency σ and wave vector k.

The stationary case studied above corresponds to a superposition of a wave of this type together with a flow such that the phase velocity satisfies $c_\varphi = \sigma/k = -V_0$. Combining this expression with Equation 7.126 leads to:

$$\sigma = -2\,\gamma\,\Omega/\,(h_0\,k). \tag{7.129}$$

In this case, σ is the angular frequency of the temporal oscillations of the coordinate y of a given fluid particle as one follows its motion along its trajectory. Taking the derivative of Equation 7.129 shows that the corresponding group velocity $c_g = \mathrm{d}\sigma/\mathrm{d}k$ is equal to $-c_\varphi = V_0$. For the transport of energy, the global velocity is therefore $2\,V_0$ because c_g must be added to the velocity of the mean flow.

In the general case of an arbitrary value of V_0, Equation 7.129 remains valid, except for the sign; it allows one to determine k as a function of σ and to deduce the following expressions for the phase and group velocities:

$$c_\varphi = \frac{\sigma}{k} = -\frac{h_0\,\sigma^2}{2\,\Omega\,\gamma} \tag{7.130a}$$

and

$$c_g = \frac{\mathrm{d}\sigma}{\mathrm{d}k} = \frac{h_0\,\sigma^2}{2\,\Omega\,\gamma}. \tag{7.130b}$$

Proof

Let us assume that the component in the y-direction of the velocity of the particle of fluid varies sinusoidally along its trajectory with:

$$v_y = v_1 \mathrm{e}^{-\mathrm{i}kx}, \tag{7.127a}$$

where k is the wave-vector of the oscillation. We also assume that the particle is located at the origin ($x = 0, y = 0$) at time $t = 0$ and that its position in the x-direction obeys the condition $x = V_0 t$. The distance y of the particle relative to the plane $y = 0$ must then be given by:

$$y = \mathrm{i}\frac{v_1}{kV_0}\,\mathrm{e}^{-\mathrm{i}kx}. \tag{7.127b}$$

We obtain this dependence from Equation 7.127a by substituting, in the integration with respect to time, $\mathrm{d}t$ by $\mathrm{d}x/V_0$. By substituting Equations 7.127a and 7.127b into Equation 7.125, we obtain:

$$-\mathrm{i}\,k\,v_1\,\mathrm{e}^{-\mathrm{i}kx} = \frac{2\mathrm{i}\,\gamma\,\Omega}{h_0}\frac{v_1}{kV_0}\mathrm{e}^{-\mathrm{i}kx}. \tag{7.128}$$

from which Equation 7.126 follows immediately.

We find therefore that we may have a periodic oscillation of the y-coordinate in the trajectory of the particle only if $\Omega V_0\,\gamma > 0$, leading to a real value of k. Equations 7.127 and 7.128 then predict stable and stationary oscillating trajectories in the rotating reference frame.

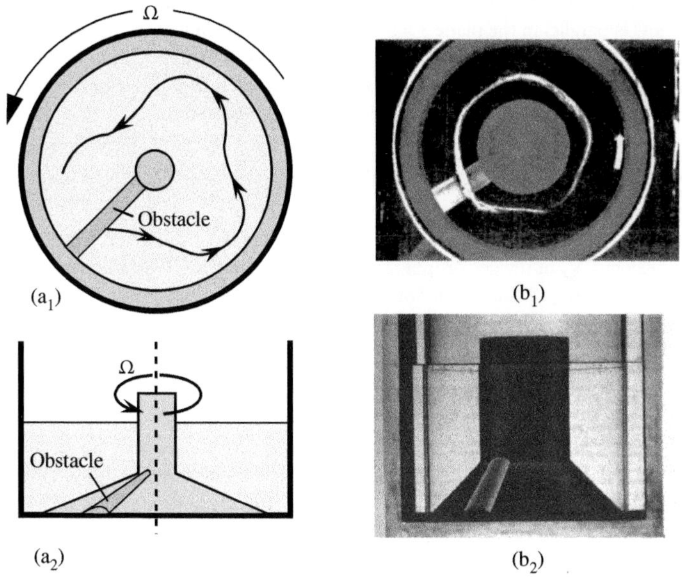

Figure 7.39 *Experimental demonstration of Rossby waves in a layer of rotating fluid of varying depth in the radial direction; (a) principle of the device ((a$_1$) and (a$_2$)); (b) experimental observations (b$_1$) and (b$_2$)) (plates from "Illustrated experiments of fluid Mechanics, NCFMF, MIT Press")*

In order to observe Rossby waves experimentally in the laboratory, the above discussion shows therefore that we can rotate a layer of fluid which is thinner near the axis of rotation than further out (e.g. a cone-shaped bottom such as that in Figure 7.39a). An obstacle placed across the fluid layer can then create a transverse component of the velocity: one visualizes the oscillations by means of solid particles in suspension, after inducing a relative motion of the fluid with respect to the bottom by varying slightly the angular velocity (Figure 7.39b).

7.6.4 The effect of viscosity near the walls: the Ekman layer

In the preceding sections, we have overlooked all the effects of viscosity. We will now study the transition region between a flow affected by viscosity near the walls and a geostrophic flow perpendicular to the pressure gradient away from the walls. This transition occurs over a layer of finite thickness: the *Ekman layer.* While the effects of the rotation of the earth are only observed on a large scale far away from any boundaries (at the surface of the earth or of the ocean), effects associated with the Ekman layer are important on local scales. Within this region, we will make use of Equation 7.77 by assuming that the flow is stationary ($\partial \mathbf{v}/\partial t = 0$), and neglecting the terms associated with gravity or with centrifugal forces:

The Ekman layer was introduced, in 1903, in the doctoral thesis of the Swede V. Ekman, who explained the deviation of the ship of the Arctic explorer F. Nansen. Nansen had observed several years earlier that his ship, caught in icy polar waters, had deviated by some 20 degrees to the right of the direction of the wind.

$$(\mathbf{v} \cdot \nabla)\, \mathbf{v} = -\nabla \left(\frac{p}{\rho}\right) - 2\,(\boldsymbol{\Omega} \times \mathbf{v}) + \nu\, \nabla^2 \mathbf{v}. \qquad (7.131)$$

The effect of a wall perpendicular to the axis of rotation

Let us apply Equation 7.131 to a stationary flow near a solid wall located at the plane $z = 0$ and perpendicular to the rotation vector $\boldsymbol{\Omega}$. The flow results from the application of a pressure gradient $\partial p/\partial y$ = constant at a great distance from the plane ($\partial p/\partial x = 0$). We assume that this flow is parallel to the plane $z = 0$ and that it is uniform with components $v_x(z)$ and $v_y(z)$ in a plane at a given position z. The component in the z-direction of Equation 7.131 is $\partial p/\partial z = 0$: the pressure gradient is thus independent of z and, as a result, constant and equal to $\partial p/\partial y$ throughout the fluid. Equation 7.131 thus becomes:

$$-\frac{1}{\rho}\frac{\partial p}{\partial y} + \nu\frac{\partial^2 v_y}{\partial z^2} - 2\Omega\, v_x = 0 \quad (7.132a) \qquad \text{and} \qquad \nu\frac{\partial^2 v_x}{\partial z^2} + 2\,\Omega\, v_y = 0. \quad (7.132b)$$

Let us assume that the effect of viscosity disappears far from the plane $z = 0$ so that we match the velocity $\mathbf{V}_g$ of the geostrophic flow given by Equations 7.96; we then find that the components v_x and v_y of the velocity of the fluid obey:

$$v_x = V_g\left(1 - e^{-\sqrt{\frac{\Omega}{\nu}}\,z}\cos\sqrt{\frac{\Omega}{\nu}}\,z\right) \quad (7.133a) \qquad \text{and} \qquad v_y = V_g\, e^{-\sqrt{\frac{\Omega}{\nu}}\,z}\sin\sqrt{\frac{\Omega}{\nu}}\,z. \quad (7.133b)$$

As we go farther from the wall, we approach exponentially the geostrophic velocity over a distance of the order of $\sqrt{\nu/\Omega}$. This distance represents the thickness δ_E of the *Ekman layer*; for water, where $\nu = 10^{-6}\,\mathrm{m^2/s}$, δ_E is of the order of 10 cm for $\Omega = 10^{-4}\,\mathrm{s^{-1}}$. Very near the wall, v_x and v_y tend to vanish; they are of the same order of magnitude with:

$$v_x \cong v_y \cong V_g\,\frac{z}{\delta_E}. \quad (7.134)$$

The local velocity of the fluid then makes an angle of 45° with the pressure gradient.

Proof

According to Equations 7.96, the geostrophic velocity $\mathbf{V}_g$ must be in the x-direction (at a right angle with the pressure gradient $\partial p/\partial y$) with:

$$V_g = -\frac{1}{2\,\Omega\,\rho}\,\frac{\partial p}{\partial y}. \quad (7.135)$$

If we now introduce a complex velocity, $w = v_x - \mathrm{i}\, v_y$, we obtain from Equations 7.132:

$$2\,\Omega\,(w - V_g) - \mathrm{i}\,\nu\,\frac{\partial^2 w}{\partial z^2} = 0. \quad (7.136)$$

We then integrate this equation in the form:

$$w - V_g = A\, e^{-(1+\mathrm{i})\sqrt{\frac{\Omega}{\nu}}\,z}, \quad (7.137a) \qquad \text{whence:} \qquad w = V_g\left(1 - e^{-(1+\mathrm{i})\sqrt{\frac{\Omega}{\nu}}\,z}\right) \quad (7.137b)$$

(we have kept only the exponent which tends towards zero as $z \to \infty$ and, in order to determine A, we have used the boundary condition of zero velocity $w = 0$ for $z = 0$); Equations 7.133 then follow.

We note in conclusion that the geostrophic velocity is not the maximum value of the component v_x: this component reaches indeed a value of 1.1 times V_g when the argument z/δ_E, of the cosine becomes of the order of π (the flow has however the same direction as $\mathbf{V}_g$).

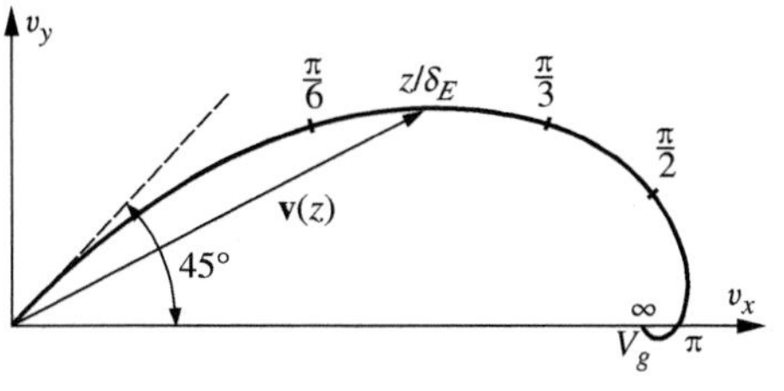

Figure 7.40 *Parametric representation of the variations of the components of the fluid velocity as a function of the distance z to the solid horizontal wall. At a large distance ($z/\delta_E \to \infty$) the velocity becomes equal to the geostrophic velocity $\mathbf{V}_g$ oriented along Ox*

Figure 7.40 displays in a parametric manner the variation of the components v_x and v_y as a function of the angle z/δ_E, for which we indicate a few values (*Ekman spiral*). We obtain similar results for turbulent atmospheric flows, but the value of the angles in the neighborhood of solid surfaces is quite different. One observes indeed that the wind near the ground level blows in a direction slightly different from that at a higher altitude. The directions of the deviations are opposite in the northern and southern hemispheres.

As $\Omega \to 0$, the thickness of the Ekman layer tends toward infinity. In this case, we frequently find that the non-linear terms $(\mathbf{v}\cdot\nabla)\,\mathbf{v}$ in the equation of motion dominate the effects of the Coriolis force, particularly near the surface, and that they balance out the viscous forces. We are then back to the case of a classical boundary layer, with the flow parallel to the pressure gradient.

Motion resulting from surface stresses

An Ekman layer can also appear underneath the surface of oceans in response to a frictional stress $\boldsymbol{\tau}$ resulting from winds which are parallel to the surface and directed in the x-direction (Figure 7.41). We assume that at significant depths, the velocity becomes constant and equal to a value $\mathbf{V}$ (V_x, V_y) independent of the presence of the surface stress ($\mathbf{V}$ then acts only as an additive constant). As in the previous case, the equations of motion appear in the form:

$$\nu \frac{\partial^2 v_y}{\partial z^2} - 2\Omega\,(v_x - V_x) = 0 \quad (7.138a) \qquad \text{and} \qquad \nu \frac{\partial^2 v_x}{\partial z^2} + 2\Omega\,(v_y - V_y) = 0. \quad (7.138b)$$

We then obtain the velocity components:

$$v_x = V_x + \frac{\tau_x}{\rho_0 \sqrt{2\,\nu\,\Omega}}\, \mathrm{e}^{\,z/\delta_E} \cos\left(\frac{z}{\delta_E} - \frac{\pi}{4}\right). \qquad (7.139a)$$

and

$$v_y = V_y + \frac{\tau_x}{\rho_0 \sqrt{2\,\nu\,\Omega}}\, \mathrm{e}^{\,z/\delta_E} \sin\left(\frac{z}{\delta_E} - \frac{\pi}{4}\right) \qquad (7.139b)$$

These equations are very similar to Equations 7.133. In particular, we observe again the appearance of the thickness $\delta_E = \sqrt{\nu/\Omega}$ of the Ekman layer, which indicates here the characteristic thickness of the layer of fluid in which the flow velocity component $\mathbf{v} - \mathbf{V}$ induced

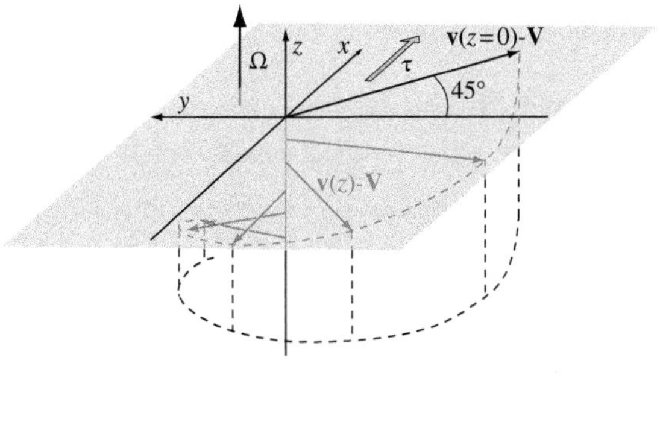

Figure 7.41 *Structure of an Ekman surface layer (the arrows represent the difference $\mathbf{v} - \mathbf{V}$ between the local velocity and its value $\mathbf{V}$ at large depths). The figure displayed corresponds to the northern hemisphere ($\Omega > 0$), and the deviation is observed to the right of the surface stress $\boldsymbol{\tau}$. The deviation is in the opposite direction in the southern hemisphere*

by the wind (independent of the current at greater depths) will be observed. At the surface ($z = 0$), $\mathbf{v} - \mathbf{V}$ is at an angle of 45° with respect to the stress $\boldsymbol{\tau}$ (here parallel to the wind): this deviation is oriented clockwise in the northern hemisphere.

We find again an Ekman spiral to indicate the change in the direction of the flow with increasing depth (Figure 7.41). In order to obtain the total current resulting from the wind, we take the integral of the velocity relative to z over the lower half-space. We find:

$$Q_x = \int_{-\infty}^{0} (v_x - V_x)\, dz = 0 \quad (7.140a) \quad \text{and} \quad Q_y = \int_{-\infty}^{0} (v_y - V_y)\, dz = -\frac{\tau_x}{2\rho_0\, \Omega}. \quad (7.140b)$$

The resulting average flow is then orthogonal to the direction of the wind and this deviation is also clockwise in the northern hemisphere (Figure 7.41).

Proof

If the stress $\boldsymbol{\tau}$ at the surface of the ocean ($z = 0$) is in the x-direction (τ_x, 0), we obtain, assuming that the flow is laminar, the boundary conditions:

$$\rho_0 \nu \frac{\partial v_x}{\partial z} = \tau_x \quad (7.141a) \quad \text{and} \quad \rho_0 \nu \frac{\partial v_y}{\partial z} = 0. \quad (7.141b)$$

Introducing, as in the preceding case, the complex velocities $w = v_x - iv_y$ and $W = V_x - iV_y$, solving these equations and applying the boundary conditions 7.141, we obtain:

$$w = W + A\, e^{(1+i)\sqrt{\frac{\Omega}{\nu}}\, z} \quad (7.142) \quad \text{and} \quad A = \frac{\tau_x (1 - i)}{2\rho_0 \sqrt{\nu\, \Omega}} = \frac{\tau_x\, e^{-i\pi/4}}{\rho_0 \sqrt{2\nu\, \Omega}}. \quad (7.143)$$

We must indeed keep only the solution with an exponential vanishing as $z \to -\infty$. We thus obtain Equations 7.139a-b.

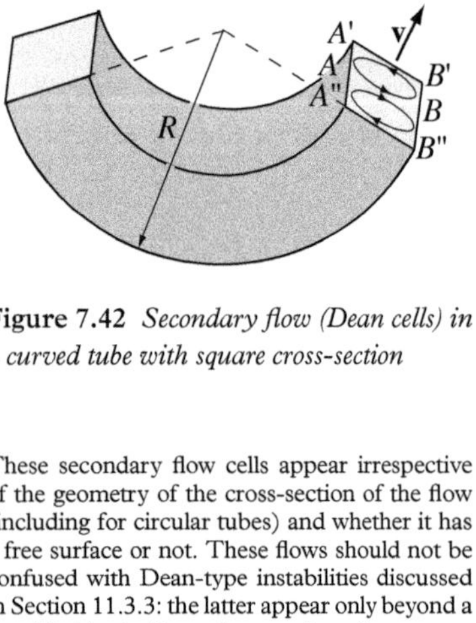

Figure 7.42 *Secondary flow (Dean cells) in a curved tube with square cross-section*

7.7 Vorticity, rotation and secondary flows

Secondary flows are flows which are superimposed on the main flow and, in some cases, can actually replace it: they frequently have a lower velocity and a different direction from the main flow. A brief discussion of such flows does belong in this chapter: a number of such flows are, in fact, associated with the rotation of fluids, with centrifugal forces resulting from a curvature of the streamlines, or with layers of vorticity created by viscous effects in the neighborhood of solid walls. The dynamics of the vorticity and of the angular momentum play, morever, an essential role in these phenomena. Let us now give a few examples of secondary flows and of the mechanisms which can lead to them.

7.7.1 Secondary flows due to the curvature of channels or due to channels with a free surface

In cross-sections of curved channels (Figure 7.42), there appear two recirculation cells (*Dean's recirculation cells*), each symmetric with respect to the plane of the curvature of the channel; the flow is directed respectively opposite to and towards the center of curvature in the plane of symmetry and near the side walls.

These secondary flow cells appear irrespective of the geometry of the cross-section of the flow (including for circular tubes) and whether it has a free surface or not. These flows should not be confused with Dean-type instabilities discussed in Section 11.3.3: the latter appear only beyond a threshhold velocity, and occur along the concave external wall.

In a curved channel with a free surface (e.g., a meandering river) there also appears a recirculation cell, where the flow is in the opposite direction to the center of curvature at the surface, but directed toward this center of the curvature at the bottom (Figure 7.43).

One considers that this mechanism explains, at least partially, the significant asymmetry often observed between the slopes of the two banks of a meandering river; in the cross-section of the stream, the secondary flow has indeed a direction such that it erodes the outer bank and deposits sediments on the inner bank of the meander. The velocity of these secondary flows increases in both these cases with the curvature and with the velocity of the main flow, but they are always present even when the velocities are low.

We will now analyze these phenomena in two different ways: first in terms of the pressure gradients, and then on the basis of the vorticity. We saw in Section 7.1 that it is possible to go from one such description to the other, and that we are discussing the same phenomenon from two different points of view.

The curvature of the streamlines of a flow creates a radial pressure gradient directed opposite to the center of curvature (Equation 5.53). This gradient is of the same nature as that associated with the centrifugal force in Equation 7.77. It balances out the centrifugal force when we are far from a wall, e.g., along a segment AB of Figure 7.42 or close to the surface of the channel in Figure 7.43. Near the side walls (e.g. at the level of the segments $A'B'$ and $A''B''$ of Figure 7.42) we find this same radial pressure gradient (assuming that the flow is locally parallel): however, it is no longer balanced by the centrifugal force because the velocity is small and the term v^2/R in Equation 5.53 is also small. This radial gradient results then in a local flow toward the centre of curvature: it is balanced out by viscous forces associated with this flow. In order to take into account mass conservation, this new flow must be compensated by another one in the opposite direction at the center of the tube (or on the surface of the channel); we recover therefore the direction of the observed circulation.

We describe now the same phenomenon by using vorticity considerations based on the rotation, due to curvature, of the vorticity layers near the walls. Let us go back to a flow with depth h in a channel for which the width is assumed to be large compared to h (Figure 7.43), which describes a shallow river; in this configuration, we have a vorticity component ω_y directed transversely to the average flow, (which we originally assumed to be in the x-direction). ω_y is associated with the velocity gradient $\partial v_x/\partial z$ between the bottom of the river, where the velocity is zero, and the surface, where the velocity is the maximum. The vorticity $\boldsymbol{\omega}$ represents a local rotation of the fluid and, consequently, an angular momentum $\mathbf{J}$. The curvature of the meandering flow causes the vorticity $\boldsymbol{\omega}$ (and thus $\mathbf{J}$), which is always perpendicular to the flow, to rotate by the same angle as the velocity of the fluid: the angular momentum of the main flow varies therefore from $\mathbf{J}_0$ to $\mathbf{J}_1$ before and after the curved zone. If we assume that there is conservation of the global angular momentum, the variation $\mathbf{J}_0 - \mathbf{J}_1 = \mathbf{j}_L$ which is parallel to the average flow reflects the appearance of a component of longitudinal vorticity $\omega_x \propto j_L$. This corresponds to a rotating secondary flow within the section of the main flow. We can apply the same line of reasoning for the curved channel in Figure 7.42.

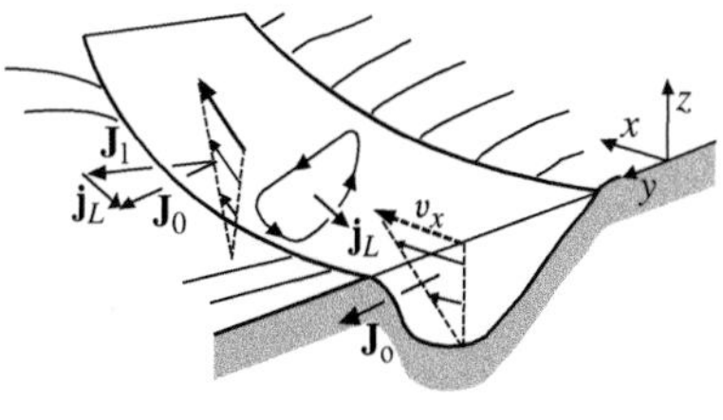

Figure 7.43 *Schematic distribution of the velocity and vorticity in a river meander. The total angular momentum* $\mathbf{J}_0$, *sum of that resulting from the circulation* ($\mathbf{j}_L$) *and that due to the velocity gradient* $\partial v_x/\partial z$ ($\mathbf{J}_1$) *is conserved.*

The Ekman layer, discussed above in Section 7.6.4 can also be considered an example of secondary flow. Indeed, far from the wall, the main geostrophic flow corresponds to a balance between the pressure gradient and the Coriolis flow (the latter replaces the centrifugal force in the example of the meandering river). Near the wall, the pressure gradient is the same because the flow is entirely parallel; but, this time, it is balanced by a viscous stress: this stress results in a secondary flow with an orientation parallel to the pressure gradient and different from the main geostrophic flow. The Ekman spiral describes a transition region between the principal and the secondary flows. We discuss this type of secondary flow in Section 7.7.3.

The appearance of a component ω_x is discussed in the facing text associated with the term $(\boldsymbol{\omega}\cdot\nabla)\mathbf{v}$ in equation 7.41 above governing the variation of the vorticity (this term is needed, moreover, as we have seen, to guarantee the conservation of angular momentum). Just as we have assumed the depth of the fluid layer to be small relative to its transverse dimension, we also assume that the component ω_z of the vorticity is negligible: we discuss in Section 9.7.3 a similar example of flow in very thin cells made up of parallel plates (Hele-Shaw cells) where the average flow parallel to the plates is potential, thus with zero vorticity. The derivatives $\partial v_x/\partial y$ and $\partial v_y/\partial x$ must therefore be equal. The component $\omega_y \partial v_x/\partial y$ of $(\boldsymbol{\omega}\cdot\nabla)\mathbf{v}$ in the x-direction is then of the order of $\omega_y \partial v_y/\partial x \approx \omega_y v_x/R$, where R is the radius of curvature of the meander; the order of magnitude of $\partial v_y/\partial x$ is, indeed: $v_x \partial\theta/\partial x \cong v_x/R$ so long as the angle of deviation θ of the trajectory relative to the x-direction is small. Thus, there appears a longitudinal component ω_x of the vorticity which has a time dependence $\mathrm{d}\omega_x/\mathrm{d}t = (\boldsymbol{\omega}\cdot\nabla)v_x \approx \omega_y v_x/R \approx v_x^2/(Rh)$, if h is the thickness of the fluid layer (i.e. the depth of the river); we have taken here: $\omega_y \cong \partial v_x/\partial z \approx v_x/h$).

7.7.2 Secondary flows in transient motion

In this section we discuss a secondary flow associated with the centrifugal force which results, in turn, in another secondary flow. Let us stir tea in a teacup in which we have kept a few tea leaves as markers. The liquid stops spinning within about 20 seconds, and we notice that, when they stop, the tea leaves have gathered in the center of the cup. This involves a twofold paradox! If the slowing down were due simply to the viscous diffusion of momentum, the order of magnitude of the expected characteristic stopping time should be 10 to 100 times longer ($R^2/\nu = 1{,}600$ s, where $R = 4$ cm is the radius of the cup and $\nu = 10^{-6}\mathrm{m}^2/\mathrm{s}$ for water). Moreover, we might think that the centrifugal force which acts on the tea leaves which are heavier than the fluid, would force them toward the periphery. We explain now this effect by looking at the phenomenon of ***spin-up*** which is similar but better controlled: we will show that it is the secondary flows created near the bottom of the cup which both displace the leaves toward the center, and ensure momentum transfers slowing down the fluid much more rapidly than simple viscous diffusion (Figure 7.44).

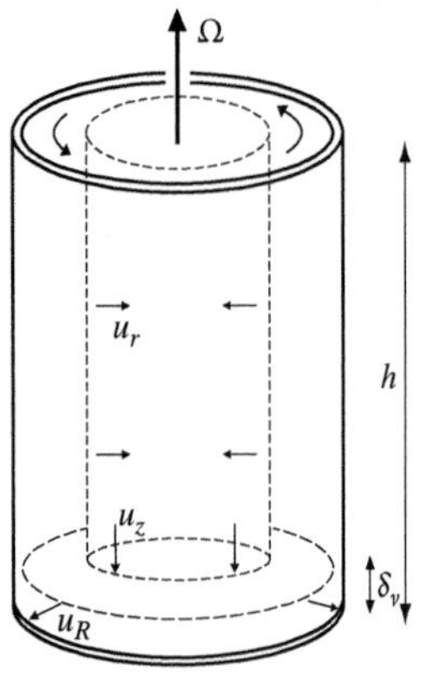

Figure 7.44 *Principle of the "spin-up" effect*

We now examine the evolution of the motion of a fluid within a cylindrical container (with solid upper and lower walls) set into rotation with an angular velocity Ω around a vertical axis, starting at an initial time, $t = 0$. If the cylinder was infinitely long, the onset of the rotation within the liquid would be uniform along the height of the cylinder: as seen in Chapter 1, this latter problem is formally equivalent to that of reaching a uniform temperature in a cylinder by radial thermal diffusion (Section 1.2.1). The viscous diffusion of momentum only plays a dominant role within a very short period of time past the onset of rotation, so long as the liquid near the bottom of the container (or near its top wall) has not yet been set into rotation over a significant depth.

After the onset of rotation, the radial pressure gradient remains, by continuity, almost the same in the layers which have been set into rotation as in the rest of the fluid (since the largest velocity components are those in the plane perpendicular to the axis, the vertical pressure gradient reduces to the hydrostatic pressure and is constant throughout the section). In the central region of the cylinder, where the fluid remains at rest, the radial pressure gradient $\partial p/\partial r$ is negligible and will thus be equally negligible near the upper and lower faces. Therefore, this term $\partial p/\partial r$ cannot compensate in Equation 7.77, the centrifugal force term $1/2\ \partial(\Omega^2 r^2)/\partial r = \Omega^2 r$, which induces therefore a radial flow away from the axis near the bottom and the top of the container. This outflow is compensated by an axial flow toward the nearest horizontal end face, of fluid coming from the central part of the height of the container and located close to its axis. This axial flow generates, in turn, a radial flow towards the axis in regions away from the horizontal faces. This convective movement of the fluid replaces the mechanism of viscous diffusion, and leads much more rapidly than the latter to a uniform distribution of the momentum of the fluid and of its rotation.

By illuminating the cylinder transversely with a plane of laser light, after having injected a small amount of fluorescent dye, we see the vertical boundary between the rotating external liquid and that at rest in the interior move at a relatively constant velocity toward the interior of the cylinder, no longer in a diffusive manner.

Evaluating the effect

Instead of causing a change in the angular velocity from 0 to Ω, let us consider a small, abrupt change in the angular velocity from Ω to $\Omega(1+\varepsilon)$ with $\varepsilon \ll 1$ (a small change is more appropriate for evaluating orders of magnitude by means of first-order mathematical expansions). The increase in the angular velocity at the level of the bottom of the container is of order $\Omega\varepsilon$. The characteristic viscous-diffusion time τ_D over a distance R is of order R^2/ν: it would represent the time required for the change in angular velocity to propagate throughout the volume of the fluid, in the absence of any secondary flow mechanism. A second natural time-constant, $t_0 = 1/\Omega$, is the reciprocal of the angular frequency of the cylinder. The ratio t_0/τ_D of these two characteristic times is the Ekman number $Ek = \nu/(R^2\Omega)$, defined in Section 7.6.1, and which is, in this instance, very much smaller than unity.

Initially then, only the fluid near the upper and lower walls of the cylinder is dragged along with the new rotational velocity by the effect of viscosity. The depth of the layer involved can be taken equal to $\delta_\nu = \sqrt{\nu\, t_0}$, which represents the characteristic distance of viscous diffusion over time t_0 and equals the thickness of the Ekman layer discussed in Section 7.6.4. Since, in this last case, the radial and tangential components of the fluid velocity can be of comparable magnitude, we will then take $u_R \approx \varepsilon\, \Omega\, R$ (only the velocity change relative to the original steady state flow of angular velocity Ω is significant). This radial flow is also confined within layers of thickness δ_ν near the upper and lower walls of the cylinder; it is oriented outward and must be balanced by vertical flows of velocity u_z from the central region of the cylinder, themselves compensated by radial flows with velocity u_r now directed toward the axis, spread over most of the height H of the cylinder over a distance from the axis $r < R$. For a cylinder of approximate radius R, mass conservation may then be written as:

$$2\pi\, r\, H\, u_r \cong 2\pi\, \delta_\nu\, R\, u_R \cong 2\pi\, \varepsilon\, \delta_\nu R^2\, \Omega; \tag{7.144}$$

in this equation, the flow in the Ekman layer has been estimated in a cylinder of radius R. We will now attempt to estimate the characteristic time τ_r needed to attain a new steady-state flow. Let us first evaluate the variation δr, resulting from the change in velocity, of the distance from the axis for fluid particles located in the central region; we estimate then τ_r which must be of order $\delta r/u_r$. In the central region, the viscosity has negligible effect, and we must conserve the angular momemtum of the particles of fluid. If we consider an annular region of mass M located at a distance r from the axis, δr must be such that:

$$M\, r^2 \Omega = M(r - \delta r)^2 \Omega\, (1 + \varepsilon), \tag{7.145a}$$

whence, retaining only the first order terms in δr and ε:

$$\delta r = -\varepsilon\, r/2 \tag{7.145b}$$

From the two equations above, we can estimate the characteristic time for establishing a new flow configuration:

$$\tau_r \cong \frac{\delta r}{u_r} \cong \frac{\varepsilon r}{2} \frac{H\, r}{\Omega\, R^2 \varepsilon\, \delta_\nu} \cong \frac{r^2}{R^2} \frac{H}{R} \frac{1}{\Omega} \sqrt{\frac{R^2 \Omega}{\nu}} \cong \frac{r^2}{R^2} \frac{H}{R} t_0 \sqrt{Ek^{-1}}. \tag{7.146a}$$

If the cylinder is not too stretched out, the two geometrical coefficients r^2/R^2 and H/R are of order unity, we thus have:

$$\tau_r \approx t_0 \sqrt{Ek^{-1}}. \tag{7.146b}$$

The characteristic time τ_r is then independent of ε and much greater than t_0, since Ek is small compared to unity. It remains, however, much smaller than the characteristic time τ_D for viscous diffusion, which is of order $t_0\, Ek^{-1}$. The existence of secondary flows, then allows us to achieve, much more rapidly, a rotational steady-state.

7.7.3 Secondary flows associated with Ekman layer effects

Examples of such oceanographic flows, are the rising (*"upwelling"*) and sinking (*"downwelling"*) of masses of water observed near some coasts, and in equatorial regions. These flows result generally from deviations due to earth's rotation of flows induced by a dominant wind over the ocean (see Section 7.6.4 for effects of this type).

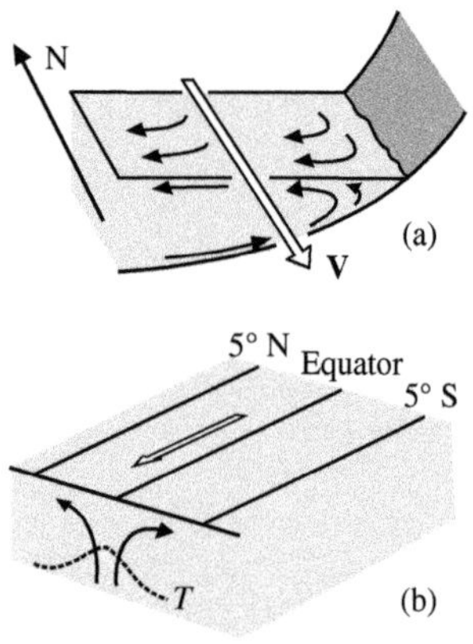

Figure 7.45 *(a) Upwelling effect associated with winds near a coastline; (b) configuration of the surface winds and of the resulting vertical and transverse flow in the ocean near the equator. These deep currents also result in a deformation of the "thermocline" (dashed line)*

Coastal upwelling and downwelling

Such secondary flows are particularly significant near coastlines. For example, in the northern hemisphere (Figure 7.45a), a north wind parallel to a coast induces, because of the Coriolis forces, a flow perpendicular to and away from the coast. In order to insure mass conservation, this current is compensated by a rising current of fish-rich waters (*upwelling*) along the coast. The inverse phenomenon (*downwelling*) is observed if there is a change in the direction of the wind, or in the local component of the earth's rotation (in the Southern hemisphere), or in the orientation of the coastline.

Equatorial upwelling

Another kind of *upwelling* can occur along the equator, where the prevailing *trade winds* are directed westward (Figure 7.45b). On both sides of the equator, the stresses resulting from such winds create flows in opposite directions (because of the change in sign of the local component of the earth's rotation), which push masses of water away from the equator. These currents are again compensated by cold water rising from the depths, here localized near the equator. A further result is the rise of the "thermocline" (the region where a large temperature gradient marks, at a certain depth, the boundary between masses of water at different temperatures). One can also observe significant resulting deformations (of several tens of centimeters) in the sea-level.

7A Appendix - An almost perfect fluid: superfluid helium

7A.1 General considerations

At a transition temperature, $T_\lambda = 2.1720\,\mathrm{K}$ (at a saturated vapor pressure of 37.80 mm of Hg), liquid helium, specifically in its most common isotopic form of atomic mass 4, undergoes a second-order phase transition to a *superfluid* state. At that point, its viscosity vanishes, so that liquid helium at low temperatures represents a model of an ideal fluid.

Because He^4 atoms are bosons (the nucleus has an even number of nucleons), this transition has some similarities with a *Bose-Einstein* condensation where a finite fraction of the atoms are in the same quantum state. We will see that this fluid fraction can be represented by a macroscopic quantum wave-function. The Bose-Einstein model assumes, however, that one deals with an ideal gas which is not the case here because the interactions between the helium atoms are quite significant (just as in an ordinary liquid).

The isotope He^3 undergoes a somewhat similar transition, but at much lower temperatures (of the order of 10^{-3}K). In its particular case, the atoms are fermions: we are then dealing, just as in the case of electron pairs in superconductors, with a condensation of interacting fermion pairs, which behave like quasi-bosons.

Authentic Bose condensations have been observed when the de Broglie wave-functions of a great number (10^6) of atoms (Rb, Na) grouped in close proximity overlap. The hydrodynamic properties of such *Bose-Einstein condensates* actually display a significant analogy with those of superfluid helium.

7A.2 Two-fluid model for superfluid helium

Experimental measurements indicate that at a finite temperature (somewhat lower than T_λ), we can consider that an intimate mixture of two fluids coexists: one of these is a superfluid with zero viscosity and the other one a normal,viscous fluid which can interact with walls (this corresponds to thermal excitations, rotons and phonons, within the fluid). We can define for each of the fluid phases a velocity ($\mathbf{v}_s$ and $\mathbf{v}_n$) and a partial density (ρ_s and ρ_n) such that $\rho_s + \rho_n = \rho$ (the overall density of liquid helium), and $\rho_s\mathbf{v}_s + \rho_n\mathbf{v}_n = \rho\mathbf{v}$, where $\mathbf{v}$ is the global velocity of the liquid. Depending on the nature of the experiments carried out, the influence of one fluid or the other may be dominant.

- When a stack of flat disks separated by a few millimeters oscillates in a helium bath, they drag along only the normal fluid: one can thus measure the relative density of this normal fluid, which changes from 100% of the overall density ρ at T_λ to almost zero for $T < 0.8K$ where the thermal excitations are negligible.
- On the other hand, only the superfluid can pass easily through extremely fine porous filters (with openings ranging from 1 to 10^3 nm), through which the rate of flow of normal fluid is negligible.

7A.3 Experimental evidence for the existence of a superfluid component which flows without any energy dissipation

Superfluid helium films

Just above the surface of the liquid, an extremely thin film (≈ 10nm) of molecules held against the wall by *van der Waals* forces appears at the walls of the container. In the case of a normal fluid, the viscous forces are much too strong, and prevent any flow whatsoever. On the other hand, in superfluid helium, such a film, known as a *Rollin film*, can flow at a sufficiently high rate that the container empties into another one beneath it.

Flow through extremely tiny holes

One can observe superfluid helium flowing through very tiny holes, with zero pressure difference (thus without any dissipation), until a critical velocity is reached; this velocity is frequently greater, the tinier the opening (typically 10μm or less) and can reach several meters per second at low temperatures.

Persistent currents

Within toroidal samples of porous material, one can generate persistent currents of the superfluid which last indefinitely, analogous to those observed in superconductors.

Experimental evidence

Persistent currents were measured experimentally either by measuring the associated angular momentum by means of an experiment based on a gyroscopic effect, or by measuring the Doppler effect in *fourth sound* propagating in a torus shaped porous sample filled with superfluid helium and in which the persistent current is generated. The *fourth sound* is a particular type of sound wave, corresponding to oscillations involving only the superfluid component of the fluid; any motion of the normal component is indeed completely inhibited by the viscous forces, which are strongly dominant within the small pores of the medium. This *fourth sound* propagates along the ring at a different velocity depending on whether the propagation is in the direction of the velocity V_s of the persistent current, or in the opposite direction: comparing the sound velocities in these two directions thus allows us to determine V_S. The velocity V_s decreases logarithmically with time (just like the electric current in superconducting solenoids): thus, after an initial phase of rapid change, we observe a decrease of only a few percent, over several days. This suggests the possibility of a persistent current of indefinite duration (comparable to the age of the universe) . . . provided one could keep the helium in the superfluid state!

7A.4 Superfluid helium: a quantum fluid

Macroscopic phase of the superfluid Phase

We consider that the superfluid fraction defined on the basis of the two-fluid model, can be described by a macroscopic wave-function $\sqrt{\rho_s}\, e^{i\varphi(r)}$ which is common to all the atoms involved. The function $\varphi(x, y, z, t)$ can be interpreted as the phase of this *macroscopic wave-function*. Absent any superfluid flow, we find long-range order characterized by the fact that the phase is the same at all points (just like the direction of the magnetization in a magnetic system). The superfluid velocity $\mathbf{v}_s$ can be derived from the potential $\varphi(x, y, z, t)$, just as in a normal quantum current, by:

$$\mathbf{v}_s = \frac{h}{2\pi\, m} \nabla \varphi \tag{7A.1}$$

(where h is Planck's constant and m is the mass of the helium atom).

Quantisation of the circulation and vortex filaments

The phase $\varphi(x, y, z, t)$ defined above is defined within an integer multiple of 2π. Let us assume that the fluid volume contains an infinitely long one-dimensional solid, or a singularity: the circulation of $\nabla\varphi$ around the solid, or around the singularity, is then not necessarily zero (Section 6.2.2), and equals $2\,n\,\pi$ where n is any integer. The circulation Γ of the superfluid velocity $\mathbf{v}_s$ along a closed curve (C) thus obeys:

$$\Gamma = \int_C \mathbf{v}_s \cdot \mathbf{d}\boldsymbol{\ell} = \int_C \frac{h}{2\pi\, m} \nabla\varphi . \mathbf{d}\boldsymbol{\ell} = \frac{n\,h}{m}. \tag{7A.2}$$

The circulation of the superfluid velocity is thus quantized and a multiple of the elementary circulation $h/m = 10^{-7}\,\mathrm{m}^2/\mathrm{s}$.

The case $n = 0$ corresponds to a potential flow in a simply-connected volume of helium without any singularity of the wave function. We find the value $n \neq 0$ when the volume of helium is multiply connected (e.g., if we have a situation where there is a continuous solid cylinder around which the superfluid is in rotation) or in a system where singular vortex lines are present. This last case is often represented by vortex filaments for which $n = 1$ and the radius ξ of the core is of the order of size of an atom $\xi \cong 0.2\,\mathrm{nm}$; the line vortex discussed in Section 7.1.2 of the present chapter is thus an excellent representation of these objects.

The *kinetic* energy of vortices increases as n^2, while the corresponding circulation increases only as n: it is therefore energetically more favorable to have $n = 1$ for all vortices.

The equivalent of the Aharonov-Bohm effect for the case of superfluid helium

Several experiments have been able to demonstrate the validity of Equation 7A.1. More generally, the quantum nature of helium allows us to think about observing quantum phenomena known in other systems.

Let us consider the example of the *Aharonov-Bohm effect*: this effect corresponds to the quantum interference between a beam of charged particles which splits around a region within which there exists a localized magnetic field, e.g., in a solenoid (Figure 7.4c).

An analogous representation of this effect is the propagation of a plane wave along the surface of a fluid in which we have created a vortex perpendicular to this surface. The hydrodynamic equivalent in the *Maxwell analog* corresponds to a current circulating around a vortex (Figure 7.4a).The plane wave undergoes an unequal phase change on the two sides of the vortex.

The phase difference between the two sides of the electromagnetic wave corresponds to the circulation of the vector potential around the solenoid. Just as we have seen in Section 7.1.3, this potential does not vanish in the region outside the solenoid (unlike the magnetic field) and has a direct influence on the charged particles. In the case of superfluid helium, it is the velocity itself which is the equivalent of the vector potential.

7A.5 Experiments involving superfluid vortices

Generation of rotation within a volume of superfluid helium

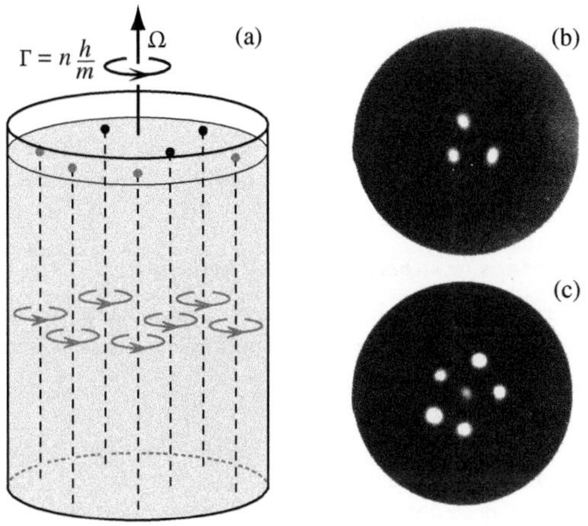

Figure 7A.1 *(a) Schematic diagram of the distribution of vortex filaments in a rotating cylindrical container full of superfluid helium. (b), (c) Top view of superfluid vortices in a cylindrical container full of superfluid helium rotating at angular velocities Ω_b and $\Omega_c > \Omega_b$ (electric charges, trapped in the cores of the vortices, are focused onto a fluorescent screen for the display) (plates courtesy of R. Packard and G. Williams)*

Quantized vortices appear, for example, in a container full of superfluid helium which is set into rotation. If the curl, $\nabla \times \mathbf{v} = 0$, was conserved within the entire volume of fluid, the helium would remain at rest. However, we observe that, above a very small critical velocity, it begins to have an average rotation similar to an ordinary fluid, with its surface taking on the expected parabolic shape. The corresponding average velocity field is generated by an ensemble of vortex lines with an average direction parallel to the axis of rotation (Figure 7A.1a): the density n_s of these lines per unit surface area is uniform and it is such that $n_s h/m$ corresponds to the vorticity of the overall flow. It has been in fact possible to visualize experimentally the positions of the cores of such vortices within cylindrical containers a few millimeters in diameter (Figures 7A.1b and c).

Superconducting analogy: If a plate of certain kinds of superconductors (known as type II superconductors) is perpendicular to a magnetic field **B**, this field penetrates in very localized magnetic field tubes: around these, we have currents due to the superconductors which maintain a zero magnetic field outside these tubes (these objects are known as *superconducting vortices*). We then have a perfect correspondance between these two systems in the representation of the Maxwell analog (Section 7.1.3) since the magnetic field **B** corresponds to the vorticity $\boldsymbol{\omega}$. Just as in the case of the circulation around the vortices, the flux of the magnetic field **B** in these vortices is quantized: the flux quantum has the value $h/2e$. Networks of vortices similar to those which we have just described have also been recently observed in rotating Bose condensates.

Experimental evidence for the quantisation of the circulation in superfluid helium

In this experiment, a container with superfluid helium is set into rotation. A magnetic wire is stretched at the center of the container: electric excitation coils induce transverse vibrations

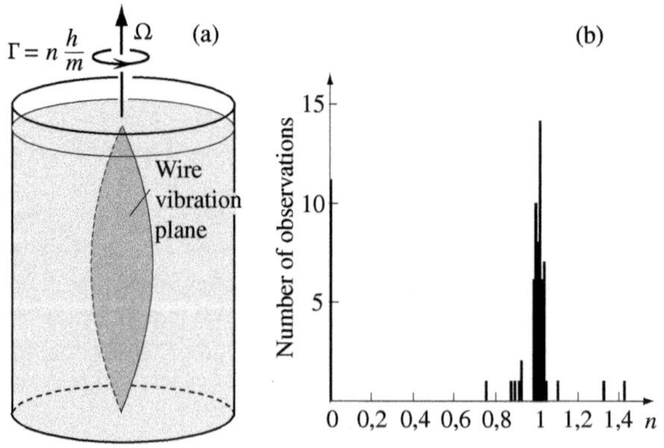

Figure 7A.2 *(a) Schematic diagram of the experiment carried out by H.E. Hall and W.F. Vinen to demonstrate the quantization of the circulation of the velocity in a rotating container of superfluid helium. (b) Distribution statistics of the experimental values of the circulation (normalized by h/m)*

of the wire at its mechanical resonance frequency ω_0 (Figure 7A.2a). For a sufficiently high rotation rate, a vortex line appears, and gets trapped on the wire: we observe there a circulation Γ which must have the value h/m. This circulation results in a Magnus force perpendicular to the plane of the vibration. This leads to a precession at an angular frequency Ω directly related to Γ. The resonant frequency of the wire is then split and becomes $\omega_0 \pm \Omega$. The Doppler effect can be understood if we recall that a linear vibration can be decomposed into two equal and opposite circular motions. The observed experimental results are indeed found to be of order h/m as we see in Figure 7A.2b.

Dynamics of vortex rings in superfluid liquid helium

The dynamics of vortex rings can also be studied by trapping charged particles and applying a uniform electric field; this, in turn, provides a well-defined kinetic energy to each ring (increasing its radius). One can even measure directly the radius of the rings by passing them through calibrated micro-grids.

These experiments must be carried out at very low temperatures ($T \cong 0.5$K) in order to avoid the energy losses associated with the interaction of the vortex cores with thermal excitations at higher temperature.

Thus, one has been able to ascertain directly, within better than a few percent, that the velocity of a vortex ring varies inversely with its energy, as predicted by Equations 7.69 and 7.70. One finds also the expected value h/m for Γ and obtains the value $\xi \cong 0.18$nm for the radius of the vortex core.

EXERCISES

1) **Vortex and plane boundary**
A semi-infinite volume ($y > 0$) of perfect fluid is bounded by the solid plane $y = 0$. A vortex filament of circulation Γ parallel to the z-axis is located initially at the point (x, y). Show that the influence of the plane may be replaced by a second vortex. What must be its characteristics and what will be the motion of the vortex?

2) **Vortex in a corner**
A volume ($x > 0$, $y > 0$) of perfect fluid is bounded by two semi-infinite solid planes ($x = 0$, $y = 0$) intersecting at a right angle. A linear vortex filament of circulation Γ, parallel to the z-axis, has its core initially located at point (x_0, y_0). As in

Exercise 1, show that the planes can be replaced by a set of filaments. Show that the complex potential of the flow velocity induced on the filament may be written as $f(z) = \mathrm{i}(\Gamma/2\pi)\mathrm{Log}\,(xy/(x+\mathrm{i}y))$. Show then that the trajectory of the filament satisfies $\mathrm{d}y/\mathrm{d}x = -y^3/x^3$ and compute this trajectory. What is its limit at long times?

3) **Geostrophic flow in a rotating channel**
A Newtonian fluid fills up the space between two infinite, parallel, solid planes $y = \pm a/2$ and flows with a stationary velocity profile $v_x(y)$ ($v_y = v_z = 0$) in a reference frame rotating around the z-axis at an angular velocity Ω. Centrifugal forces and gravity are neglected and the pressure is p_0 on the z-axis. We denote by Q the volume flow rate of the fluid per unit length along the vertical z-direction. Compute the pressure $p(x, y)$ and the velocity profile $v_x(y)$ after writing the equations of motion for v_x in the rotating reference frame. For which values of Ω does the pressure gradient along y become much larger than that along x? What is then the structure of the flow?

In the two following problems, we consider stationary flows for which the geostrophic approximation discussed in Section 7.6.2 is valid: such flows correspond to Ekman and Rossby numbers $Ek \ll 1$ and $Ro \ll 1$ and result from a balance between the pressure gradients and the Coriolis forces. In both problems, hydrostatic pressure terms must be taken into account by replacing Equation 7.96c by $\partial p/\partial z = -\rho\, g$.

4) **Tilt of the free surface of the oceans between two coasts**
A layer of fluid of finite thickness $h(y)$ ($h(0) = h_0$) flows at a constant velocity V_x between vertical planes $y = \pm L/2$ with $h \ll L$ (Figure(a)). Both the fluid and the walls rotate at an angular velocity Ω around the vertical z-axis. Write the equations of motion and show that the slope is $\partial h/\partial y = \tan\beta$ (as mentioned in Section 7.6.1, the centrifugal force terms are assumed to have been included in the pressure and are not taken into account). Compute the value of β and the difference between the liquid levels near the two walls for $2\Omega = 10^{-4}\ \mathrm{s}^{-1}$, $L = 100$ km, $V_x = 3$ m/s.

In order to model an oceanic current of temperature and composition different from those of the surrounding fluid, we study the stationary counterflow of juxtaposed fluids of densities ρ_1 and $\rho_2 > \rho_1$, and velocities $V_{x1}(y)$ and $V_{x2}(y)$. The interface between fluids is $h_i(y)$ ($y<0$) and those of the fluids with air $h_1(y)$ ($y<0$) and $h_2(y)$ ($y<0$) with $h_i(0) = h_1(0) = h_2(0) = h_0$. Write the equations of motion in the two fluids as well as the

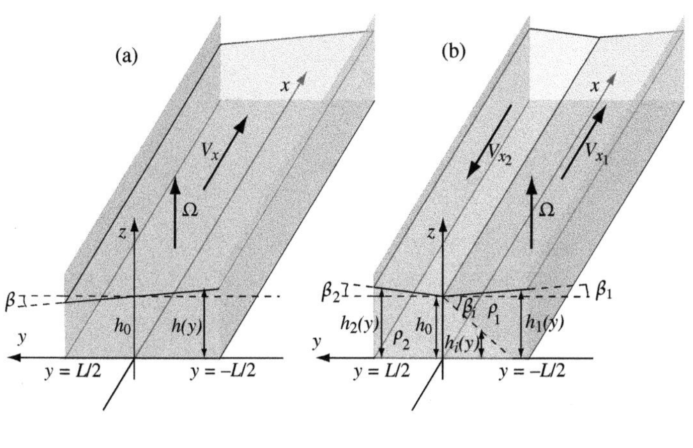

equality between the pressures of the two fluids at their interface. Compute then the angles β_i, β_1 and β_2. What is the relative magnitude of β_i and β_1 or β_2, if $\rho_2 - \rho_1 \ll \rho_1$?

5) **Modelization of thermal winds**
Let us consider winds induced by variations in temperature (and therefore in density) of the air in the direction parallel to the surface of the earth (high-altitude jetstreams are a particular kind of thermal wind). To start, let us consider a layer of incompressible fluid with a temperature $T(x, y)$ independent of the vertical coordinate z. Ω is the projection of the earth's rotation vector on the z-axis. Assuming that the fluid density varies with the temperature as $\rho = \rho_0(1 - \alpha(T - T_0))$, show that an horizontal temperature gradient induces a flow of velocity varying with the elevation as: $\partial \mathbf{v}/\partial z = (\alpha g/2\,|\boldsymbol{\Omega}|^2)\boldsymbol{\Omega} \times \nabla \mathrm{T}$.

In the case of the earth's atmosphere, air is compressible and assumed to be a perfect gas with: $p/\rho = RT/M$ (the molar mass $M = 29 \times 10^{-3}$ kg and $R = 8.32$ J/K). It is then convenient to use the pressure instead of z as the variable characterizing the elevation (this is in line with the practical estimation of the altitude from pressure measurements). Show that $\mathrm{d}z = -(RT/Mg)\ \mathrm{d}(\mathrm{Ln}\ p)$ (at constant x and y) and deduce from it that $(\partial \mathbf{v}/\partial(\mathrm{Ln}\ p))_{x,y} = -(R/2\mathrm{M}|\boldsymbol{\Omega}|^2)\ \boldsymbol{\Omega} \times (\nabla \mathrm{T})_p$ (the notation $(\nabla T)_p$ indicates that the variations of T with x and y are considered for displacements along which p, and not z, is kept constant).

Compute the order of magnitude of the variation of $\mathbf{v}$ for a horizontal temperature gradient of about 1K/1000 km at a latitude such that $\Omega = 10^{-4}\ \mathrm{s}^{-1}$ and for a variation of the pressure by a factor $p_2/p_1 = 4$.

Quasi-Parallel Flows – Lubrication Approximation

8

A significant number of practically important flows (application of coatings or paints, lubrication flows between tight-fitting parts, liquid jets, etc.) have streamlines which are almost parallel. We discuss first (Section 8.1) the so-called lubrication approximation, which allows us to perform a theoretical calculation for these flows, and we give several examples of flows between two solid surfaces. We use then the same method (Section 8.2) to deal with fluid films which have a free surface: we treat specifically problems of wetting, dynamic contact angle and analyze the spreading of liquid layers or droplets. A particularly interesting case is that of Marangoni effects, where the flow results from spatial variations of the surface tension at the interface. Finally, we discuss a similar problem involving the drop of a viscous jet and the Rayleigh-Plateau instabilities which can appear, and which occur because of surface tension (Section 8.3).

8.1 Lubrication approximation

8.1.1 Quasi-parallel flows

We have discussed in Chapter 4 one-dimensional flows for which only one component of the velocity does not vanish. In this case, the non-linear term $(\mathbf{v} \cdot \nabla)\, \mathbf{v}$ of the Navier-Stokes equation vanishes identically, because the gradient of the velocity $\mathbf{v}$ is normal to it. We then end up with a linear equation of motion for the fluid (Equation 4.59) which applies whatever the Reynolds number of the flow might be (so long as instabilities do not appear).

In this chapter, we study flows for which the streamlines are almost parallel, such as, for example, flows between solid walls which are at a very small angle θ to each other, or flows in a thin liquid layer. We can then neglect the non-linear terms, but only if certain specific conditions are satisfied by the Reynolds number: in that case, we speak of the *lubrication approximation.*

Later on, in Chapter 9, we discuss flows with arbitrary geometries: in that case, the condition on the Reynolds number for being able to neglect non-linear terms, is even more stringent ($Re \ll 1$).

We encounter quasi-parallel flows in a number of applications such as the spreading of a fluid film, or the lubrication of rotating machinery. We can thus calculate the dynamics of the spreading of a film in the former case, or the forces between the moving surfaces in the latter, by making the assumption that the flows are effectively parallel to the surfaces of these films.

Physical Hydrodynamics. Second Edition. Etienne Guyon *et al.*

8.1.2 Assumptions of the lubrication approximation

We assume that we have an angle $\theta \ll 1$ at all points between the walls, so that the characteristic distances over which flow parameters change are everywhere much greater than their separation, as shown in Figure 8.1. In the following, we consider a two-dimensional configuration where only the x- and y-directions are involved, but these results can be easily generalized to a three-dimensional case.

Let us find out what becomes of the different terms of the Navier-Stokes equation (Equation 4.30) for two-dimensional flow in the x-y plane. As we have previously indicated (Section 4.2.2), the flow does not result only from the pressure gradient, but from that of the combination $(p - \rho \mathbf{g}\cdot\mathbf{r})$, where $\mathbf{r}$ is the radius-vector for the point in question. In order to simplify the writing, we assume that ∇p, in fact, denotes $\nabla(p - \rho\mathbf{g}\cdot\mathbf{r})$. The components of Equation 4.30 then become:

$$\frac{\partial v_x}{\partial t} + (\mathbf{v}.\nabla)\, v_x = -\frac{1}{\rho}\frac{\partial p}{\partial x} + \nu\left(\frac{\partial^2 v_x}{\partial x^2} + \frac{\partial^2 v_x}{\partial y^2}\right) \tag{8.1a}$$

and:

$$\frac{\partial v_y}{\partial t} + (\mathbf{v}.\nabla)\, v_y = -\frac{1}{\rho}\frac{\partial p}{\partial y} + \nu\left(\frac{\partial^2 v_y}{\partial x^2} + \frac{\partial^2 v_y}{\partial y^2}\right). \tag{8.1b}$$

We must add to this equation the condition for conservation of mass for an incompressible fluid, i.e. $\nabla\cdot\mathbf{v} = 0$:

$$\frac{\partial v_x}{\partial x} + \frac{\partial v_y}{\partial y} = 0. \tag{8.2}$$

We first consider a stationary flow, which allows us to neglect terms such as $\partial/\partial t$, and we carry out the discussion based on an approximate evaluation of the orders of magnitude of the different terms. In the case of a stable and slow *laminar* flow, we observe experimentally that the trajectories of the fluid particles follow the surface of the walls, as shown in Figure 8.1. Therefore, in the middle of the flow, we can assume that the angle between the velocity and the wall $y = 0$ is of the order of the angle θ ($\ll 1$) between the two walls, so that:

$$v_y \approx v_x\,\theta \approx U\theta, \tag{8.3}$$

where U is the characteristic velocity of the flow (e.g., the average velocity or even the maximum velocity at the center of the channel, both of which can be considered to be of the same order of magnitude in terms of the approximation that we are carrying out). Considering that we are assuming a Poiseuille-type velocity profile between the walls, we can consider that the typical distance, over which variations of the velocity in the y-direction occur, is the local thickness $e(x)$, which we will assume to have a typical value e_0 (this amounts to saying that the relative change in the thickness is small along the length L). We thus obtain:

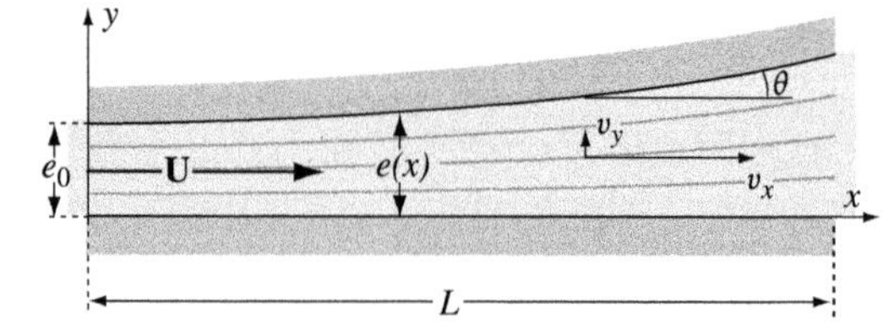

Figure 8.1 *Schematic diagram of the geometry of a lubrication flow*

$$\frac{\partial v_x}{\partial y} \approx \frac{U}{e_0} \quad (8.4a) \qquad \text{and:} \qquad \frac{\partial v_y}{\partial y} \approx \frac{U\theta}{e_0}, \quad (8.4b)$$

from which, using Equations 8.2:

$$\frac{\partial v_x}{\partial x} = -\frac{\partial v_y}{\partial y} \approx \frac{U\,\theta}{e_0} \quad (8.5)$$

(the orders of magnitude are specified in terms of their absolute values). Using the above results, we estimate in the same way the second derivatives:

$$\frac{\partial^2 v_x}{\partial x\,\partial y} \approx \frac{U\theta}{e_0^2}, \quad (8.6a) \qquad \frac{\partial^2 v_x}{\partial y^2} \approx \frac{U}{e_0^2} \quad (8.6b)$$

and:

$$\frac{\partial^2 v_y}{\partial y^2} \approx \frac{U\,\theta}{e_0^2}. \quad (8.6c)$$

We also state an upper limit for the absolute value of the terms $\partial^2 v_x/\partial x^2$ and $\partial^2 v_y/\partial x\partial y$ (which are opposite to each other because of the incompressibility condition) by taking $\partial/\partial x \approx 1/L$; this leads to:

$$\frac{\partial^2 v_x}{\partial x^2} = -\frac{\partial^2 v_y}{\partial x\partial y} \approx \frac{U\,\theta}{e_0\,L} \quad (8.7a) \qquad \text{and, similarly:} \qquad \frac{\partial^2 v_y}{\partial x^2} \approx \frac{U\,\theta}{L^2}. \quad (8.7b)$$

These terms are therefore very small relative to the second derivatives with respect to y; the latter will therefore be the only viscosity terms retained in Equations 8.1a and 8.1b.

Let us now consider the conditions under which the non-linear term of the component of the equation of motion, in the direction of the average flow, is negligible relative to the viscosity term. By using the previous results, we obtain:

$$v_x\frac{\partial v_x}{\partial x} \approx \frac{U^2\,\theta}{e_0}, \quad (8.8a) \qquad v_y\frac{\partial v_x}{\partial y} \approx \frac{U^2\,\theta}{e_0} \quad (8.8b) \qquad \text{and} \quad \nu\frac{\partial^2 v_x}{\partial y^2} \approx \frac{\nu\,U}{e_0^2}. \quad (8.8c)$$

The non-linear terms are therefore negligible if:

$$\frac{U^2\,\theta}{e_0} \ll \nu\frac{U}{e_0^2}, \quad (8.9a) \qquad \text{i.e.:} \qquad Re = \frac{U\,e_0}{\nu} \ll \frac{1}{\theta}. \quad (8.9b)$$

For flows in arbitrary geometry, the condition obeyed by the Reynolds number, in order to be able to neglect the non-linear terms, is $Re \ll 1$. Equations 8.9b is then much less restrictive because in the lubrication geometries the angle θ is small relative to 1. We can thus continue to use a linear equation even for flows which correspond to a Reynolds number Re significantly greater than 1. In the limiting case of a parallel flow, the non-linear term vanishes at all values of the Reynolds number.

Let us calculate the component $\partial p/\partial y$ of the pressure gradient in the direction normal to the average flow on the basis of Equations 8.1b. We estimate, as in the previous case:

$$v_x\frac{\partial v_y}{\partial x} \approx \frac{U^2\,\theta}{L}, \quad (8.10a) \qquad v_y\frac{\partial v_y}{\partial y} \approx \frac{U^2\,\theta^2}{e_0}, \quad (8.10b) \qquad \nu\frac{\partial^2 v_y}{\partial y^2} \approx \nu\frac{U\theta}{e_0^2}. \quad (8.10c)$$

The viscous term resulting from Equations 8.10c is thus smaller by an order of magnitude in θ than that given by Equations 8.8c. If $Re \ll 1/\theta$ and e_0/L is sufficiently small, the non-linear terms in Equations 8.10a and 8.10b will be smaller than the viscous term in Equations 8.10c and, in any case, very small, in comparison with the viscous term of the equation, in the x-direction. We can thus consider that the pressure gradient transverse to the flow is zero (or, more precisely, that it reduces to the hydrostatic pressure) when θ is small, and the Reynolds number sufficiently small.

We should, however, note that this entire discussion assumes that there is a fully developed unperturbed laminar flow. Beyond certain critical flow velocities, even if Equation 8.9b is satisfied, the flow between parallel planes becomes unstable, and significant velocity components, normal to the average flow, appear. The preceding results are then no longer valid.

8.1.3 Non-stationary effects

Let us now go back to Navier-Stokes equation, written as in Equation 8.1, and assume that the flow changes over a characteristic time T, or that it is periodic with angular frequency ω (a parameter better suited for a periodic flow). The term $\partial \mathbf{v}/\partial t$ will then be either of order U/T or, equivalently, of $U\omega$. It will be negligible relative to the viscosity term if $|\partial \mathbf{v}/\partial t| \ll |\nu \nabla^2 \mathbf{v}|$, i.e.:

$$\frac{U}{T} \ll \nu \frac{U}{e_0^2} \qquad \text{or} \qquad Nt_\nu = \frac{e_0^2}{\nu T} \ll 1.$$

The ratio Nt_ν of the characteristic times appeared already in Equation 4.95 during our discussion of alternating flows between two planes (Section 4.5.4). The condition $Nt_\nu \ll 1$ expresses that the characteristic time e_0^2/ν for the diffusion of the velocity gradients (or of the vorticity) over the thickness e_0 of the fluid film is much smaller than the characteristic time T for the evolution of the flow or, equivalently, small compared to $1/\omega$. One notes that e_0^2/ν represents also the time needed to establish a stationary velocity profile: if the above condition is satisfied, the flow can then be considered as quasi-stationary, and the terms of the type $\partial \mathbf{v}/\partial t$ can be neglected.

8.1.4 Equations of motion in the lubrication approximation

In this case, if instabilities do not appear in the flow, and the conditions $\theta \ll 1$ and $Re \ll 1/\theta$ are satisfied, Equations 8.1a and 8.1b simplify, for a steady state flow, to:

$$\frac{1}{\rho}\frac{\partial p}{\partial x} = \nu \frac{\partial^2 v_x}{\partial y^2} \qquad (8.11a) \qquad \text{and:} \qquad \frac{\partial p}{\partial y} = 0, \qquad (8.11b)$$

which represents exactly the same system of equations as for a parallel flow. The velocity v_x will vary in the x-direction if the thickness is no longer a constant, but this variation will be much slower than in the y-direction, for which the characteristic distance for change is e_0. In order to calculate the velocity profile, we then integrate Equation 8.11a relative to y, as though $\partial p/\partial x$ were a constant (the x and y variables become decoupled). In effect, if the requirements for the lubrication approximation are satisfied, $\partial p/\partial x$ is practically independent of y, just as for a parallel flow; on the other hand, it varies, in the x-direction, over distances which are much larger than the thickness e_0 of the fluid layer. Weakly non-stationary flows can be treated in the same manner if the time-dependent change in the velocity is sufficiently slow. We will see, further down, examples where the lack of stationarity plays a significant role.

8.1.5 An example of the application of the equation for lubrication: stationary flow between two moving planes making a small angle to each other

If we let a sheet of paper slide parallel to the horizontal surface of a smooth table, the presence of a film of air between the table and the sheet helps the sliding; on the other hand, if the sheet of paper has a few holes in it, it slides very poorly because there no longer exists a pressure difference between the outside air and this intermediate sheet. The pressure difference which exists in the former case is due to the formation of a wedge toward the back between the sheet and the table, shown schematically in Figure 8.2.

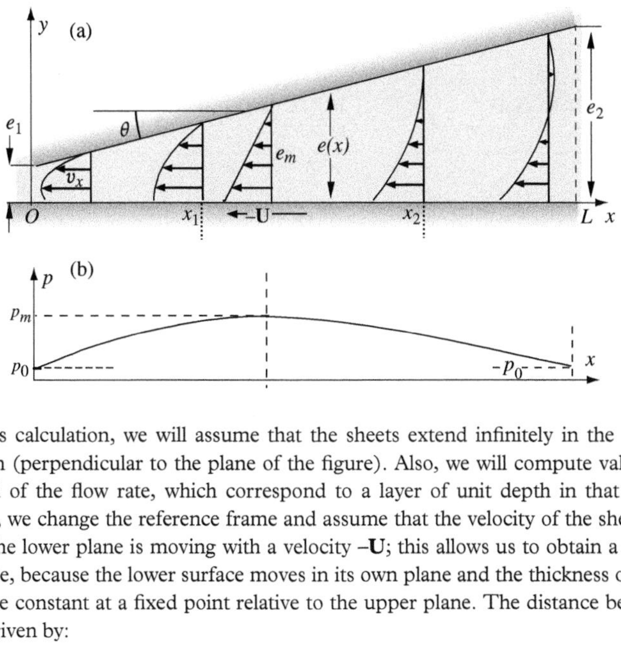

Figure 8.2 *Schematic diagram of the flow resulting from the motion of an inclined plane moving relative to a fixed horizontal plane; (a) the velocity field is shown in the reference frame in which the inclined plane is fixed; (b) variation of the pressure in the region between the two planes*

For this calculation, we will assume that the sheets extend infinitely in the transverse z-direction (perpendicular to the plane of the figure). Also, we will compute values of the force, and of the flow rate, which correspond to a layer of unit depth in that direction. Moreover, we change the reference frame and assume that the velocity of the sheet is zero, and that the lower plane is moving with a velocity $-\mathbf{U}$; this allows us to obtain a stationary flow profile, because the lower surface moves in its own plane and the thickness of the fluid is therefore constant at a fixed point relative to the upper plane. The distance between the planes is given by:

$$e(x) = e_1 + \theta x,$$

where the angle $\theta = (e_2 - e_1)/L$ is assumed to be small. In this case, and if $Re \ll 1/\theta$, Equation 8.11b shows that the pressure p and, therefore, $\partial p/\partial x$ are independent of y.

Integrating Equation 8.11a with respect to y under this assumption and taking into account the boundary conditions $v_x(y=0) = -U$ and $v_x(y=e(x)) = 0$, we obtain then:

$$v_x(x, y) = -\frac{1}{2\eta}\frac{dp}{dx}\, y\,[e(x) - y] - U\,\frac{e(x) - y}{e(x)}. \tag{8.12}$$

Just as in Section 4.5.3, the velocity field corresponds to the superposition of a Poiseuille flow (parabolic term in y associated with the pressure gradient) and of a Couette flow (term linear in y, related to the movement of the plane with velocity $-\mathbf{U}$). In Figure 8.2a, we have shown the parabolic velocity profiles between the two planes corresponding to different distances along x.

We compute now the pressure distribution in the upper plane by integrating Equation 8.12 with respect to y between 0 and $e(x)$; the flow rate Q (per unit depth in the z-direction) is constant with x because the flow is stationary in the reference frame selected so that:

$$Q = \text{constant} = \int_0^{e(x)} v_x \mathrm{d}y = -\frac{1}{\eta}\frac{\mathrm{d}p}{\mathrm{d}x}\frac{e(x)^3}{12} - \frac{U\,e(x)}{2}. \tag{8.13}$$

As we might have expected, this equation coincides with Equation 4.72 when we replace V_0 by $-U$ and a by $e(x)$. We therefore conclude, replacing x by e as the variable:

$$\frac{\mathrm{d}p}{\mathrm{d}x} = \theta\,\frac{\mathrm{d}p}{\mathrm{d}e} = -\frac{12\,\eta\,Q}{e(x)^3} - \frac{6\,\eta U}{e(x)^2}. \tag{8.14}$$

We can obtain the value e_m of the thickness at the point where the pressure has an extreme value ($\mathrm{d}p/\mathrm{d}e = 0$), so that:

$$e_m = -2\,\frac{Q}{U}. \tag{8.15}$$

Finally we obtain the pressure $p(x)$ by integrating Equations 8.14 relative to e with $p\,(x = 0) = p_0$ (atmospheric pressure outside the thin film). We find that:

$$p(x) = p_0 + \frac{6\,\eta\,Q}{\theta}\left[\frac{1}{e(x)^2} - \frac{1}{e_1^2}\right] + \frac{6\,\eta U}{\theta}\left[\frac{1}{e(x)} - \frac{1}{e_1}\right]. \tag{8.16}$$

From this, we determine the value of the flow rate Q by writing that the pressure is also equal to p_0 at the other end of the plane (where $e = e_2$). We then obtain:

$$Q = -\frac{e_1\,e_2}{e_1 + e_2}U; \quad (8.17) \qquad \text{whence:} \qquad e_m = 2\frac{e_1\,e_2}{e_1 + e_2}. \quad (8.18)$$

Equation 8.18 indicates that, when $e_1 \ll e_2$, we have $e_m \cong 2e_1$: the point where the pressure has an extreme value is then very close to the side where the thickness has the value e_1. In all cases, according to Equation 8.17, Q and U are opposite in sign, i.e. the average flow of the fluid has the direction of the motion of the lower plane. Equations 8.16 can then be transformed by replacing Q by its value as a function of U, so that we obtain:

$$p(x) = p_0 + \frac{6\,\eta U}{\theta}\frac{(e_2 - e(x))\,(e(x) - e_1)}{e(x)^2\,(e_1 + e_2)}. \tag{8.19}$$

The difference $p\,(x) - p_0$ has then necessarily the same sign as U/θ, since $e(x)$ is bounded by the values e_1 and e_2. Figure 8.2b gives the corresponding pressure distribution.

The normal force F_N on the lower plane, due to this excess pressure resulting from the flow, is given by the integral:

$$F_N = -\int_0^L (p - p_0)\,\mathrm{d}x = -\frac{1}{\theta}\int_{e_1}^{e_2}(p - p_0)\,\mathrm{d}e = -\frac{6\,\eta U}{\theta^2}\left[\mathrm{Log}\,\frac{e_2}{e_1} - \frac{2\,(e_2 - e_1)}{e_2 + e_1}\right]. \tag{8.20}$$

Let us consider the velocity profiles between the two planes on the basis of Equations 8.12, taking the derivative with respect to y and replacing $\partial p/\partial x$ by $\theta\,(\partial p/\partial e)$, and $\partial p/\partial e$ by combining Equations 8.14 and Equations 8.17. We are specifically interested in the existence of a maximum in the velocity profile. We take the derivative:

$$\frac{\partial v_x}{\partial y} = \frac{2y - e(x)}{2\eta}\theta\frac{\partial p}{\partial e(x)} + \frac{U}{e(x)} = \frac{2U}{e(x)^2}\left[-\frac{3e_1 e_2(2y - e(x))}{e(x)(e_1 + e_2)} + 2e(x) - 3y\right]. \tag{8.21}$$

Near the lower wall ($y = 0$) and the upper wall ($y = e(x)$), we have, respectively:

$$\frac{\partial v_x}{\partial y} = \frac{2\,U}{e(x)^2}\left[-\frac{3\,e_1\,e_2}{(e_1 + e_2)} + 2\,e(x)\right] \tag{8.22a}$$

and

$$\frac{\partial v_x}{\partial y} = \frac{2\,U}{e(x)^2}\left[\frac{3\,e_1\,e_2}{(e_1 + e_2)} - e(x)\right]. \tag{8.22b}$$

The derivatives $\partial v_x/\partial y$ of the velocity profile on the upper and lower planes then vanish, respectively when $e\,(x_1) = 3e_1e_2/\,(2\,(e_1 + e_2)) = 3e_m/4$ and $e\,(x_2) = 3e_1e/\,(e_1 + e_2) = 3e_m/2$, i.e. on both sides of the point where the pressure is a maximum. For distances smaller than x_1, the velocity has a minimum; for distances greater than x_2, the velocity is positive in the neighborhood of the upper wall, and it has a maximum. In the section of thickness e_m corresponding to the pressure maximum, the velocity varies linearly with the distance y: this is easily shown by taking the second derivative $\partial^2 v_x/\partial y^2$ of the velocity profile $v_x(y)$ from Equation 8.21 so as to obtain $\partial^2 v_x/\partial y^2 = (\theta/\eta)(\partial p/\partial e(x))$. The distance x at which the curvature becomes zero corresponds to the maximum in the pressure: this is quite expected since, in the absence of a pressure gradient, the Poiseuille component of the flow becomes zero and only a Couette-type flow remains.

We also determine the tangential frictional force on the lower plane, i.e.:

$$F_T = \int_0^L \eta \frac{\partial v_x}{\partial y} \mathrm{d}x = \frac{1}{\theta} \int_{e_1}^{e_2} \left(\frac{e(x)}{2} \frac{\mathrm{d}p}{\mathrm{d}x} + \frac{\eta U}{e(x)} \right) \mathrm{d}e$$
$$= \frac{2\eta U}{\theta} \left[2\,\mathrm{Log}\frac{e_2}{e_1} - \frac{3(e_2 - e_1)}{(e_2 + e_1)} \right]. \tag{8.23}$$

For $e_2/e_1 = 10$, the prefactors of $\eta U/\theta^2$ and $\eta U/\theta$ respectively in Equations 8.20 and 8.23 are –4 and 4.3.

In the general case, the relative role of the values of θ, e_1 and e_2 is somewhat subtle because, when $\theta \to 0$ and $e_2 \to e_1$, both the numerator and the denominator of the expressions for F_T and F_N approach zero. We will evaluate these two components for the particular case where the thickness e_2 is much greater than the thickness e_1. In that case, we find:

$$F_N \simeq -\left(\mathrm{Log}\frac{e_2}{e_1} - 2 \right) \frac{6\eta U}{\theta^2} \quad (8.24\text{a}) \qquad \text{and} \qquad F_T \simeq \left(4\,\mathrm{Log}\frac{e_2}{e_1} - 6 \right) \frac{\eta U}{\theta}. \quad (8.24\text{b})$$

where $\theta = (e_2 - e_1)/L \simeq e_2/L$ (for a given angle θ, the influence of e_2/e_1 on the values of F_T and F_N is small). When the angle θ is small, F_N can take on large values, while the tangential frictional force F_T will be smaller by an order of magnitude: this is the fundamental result of the lubrication model.

This property is extremely useful in a wide range of applications: axles rotating within a bearing of barely larger diameter, thus able to support much greater normal stresses without excessive friction (*bearings* of rotating machinery, *wheel axles* of vehicles...). In some cases, the normal forces are so large that solid pieces can be deformed in regions where their cross-section is very small; for this reason they are known as *elasto-hydrodynamic forces.*

Lubrication forces can also have rather undesirable effects: if we walk on an oil slick, the normal forces will support our weight while the tangential force is insufficient to allow us to maintain our balance. There is a similar effect when a car skids out of control along a wet road (a phenomenon known as *aquaplaning*): the thin film of water between the tires and the road can support the weight of a car, while the forces that prevent slip are too weak.

The net resultant force on the fluid in the space between the planes must be zero since the pressure difference between the entry and exit of this space is also zero. The x- and y-components of the force on the upper plane must therefore be equal to $-F_T$ and $-F_N$; they can also be computed from the pressure and velocity on that plane. However, the pressure and tangential forces are respectively perpendicular and parallel to the plane and, therefore, at an angle θ respectively to y and x: they must then be projected on these axis in order to retrieve the values $-F_T$ and $-F_N$.

From Equations 8.24a–b, F_T and F_N are approximately proportional to $1/\theta$ and $1/\theta^2$ only, provided $e_2/e_1 \gg 1$ (then, $\theta \approx e_2/L$). If $e_2/e_1 \to 1$, we must use Equations 8.20 and 8.23, leading to $F_N = 0$ when $e_2 = e_1$ (parallel planes).

For $e_2 \gg e_1$, $p(x)$ is highest for $e(x) \approx 2e_1$. More generally, in lubrication flows, the pressure is highest in the regions of small liquid thickness: this allows one to obtain approximate solutions. Take, for instance, two spheres of radius R approaching each other with a minimum separation $e_1(t)$: the force between them may be estimated from the pressure distribution on the spherical cap with radius $\sqrt{R e_1(t)}$ where the local separation between the surfaces of the two spheres is between e_1 and $2e_1$ (see similar problems below).

8.1.6 Flow of a fluid film of arbitrary thickness

In this section, we consider the more general case of a flow of a thin film of fluid between two solid surfaces with varying separation, and with relative motion in arbitrary directions.

Reynolds' equation

We retain the assumption of a film of fluid sufficiently thin relative to the characteristic distances along x over which the velocity and the thickness parallel to the film vary, so that we can still consider the flow as quasi-parallel (Figure 8.3). We also assume that the lower surface is a stationary plane $y = 0$ and that each point of the upper surface located at a height $h(x, t)$ has a velocity with components $U(x, t)$ and $V(x, t)$ in the x- and y-directions. The longitudinal pressure gradient $\partial p/\partial x$ is considered constant over the thickness of the film, but can exhibit a slow variation along this film. As seen in Section 8.1.2, the term $\partial p/\partial x$ needs to be replaced by $\partial p/\partial x - \rho\, g_x$ when the component g_x of gravity in the plane of the film does not vanish. In the present section, the upper surface is solid but not necessarily plane.

As in the preceding case, the local flow is a superposition of a Poiseuille-type and a Couette-type flow. This can result from the relative displacement of the solid walls, which limits the upper and lower surfaces of the film and/or, possibly, by an applied pressure gradient.

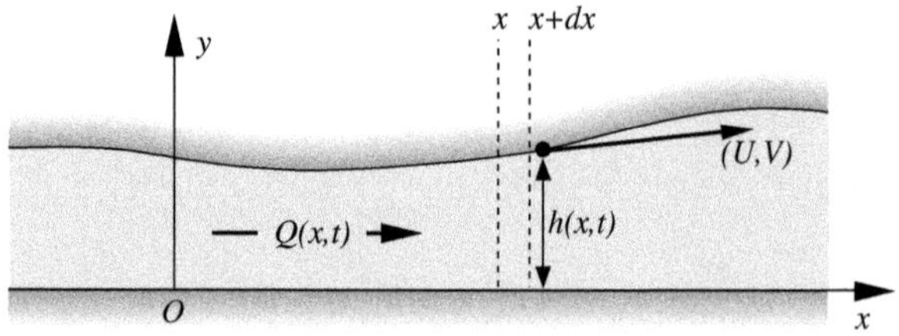

Figure 8.3 *Schematic diagram of a quasi-one-dimensional flow in a thin fluid layer between a plane and a solid upper surface with a distance h(x, t) from the plane, which can be a function of both time and position*

Let us rewrite Equations 8.13 relating the local flow rate $Q(x, t)$ in the film (per unit length in the z-direction), the gradient $\partial p/\partial x$ and the velocity component $U(x, t)$ in the x-direction of the upper surface, with:

$$Q(x,\, t) = -\frac{h^3}{12\,\eta}\,\frac{\partial p}{\partial x} + \frac{U\,h}{2}. \tag{8.25}$$

In contrast with the case of Equations 8.13, Q can depend on x as well as on t. Let us now analyze the conservation of mass in a thin layer of the fluid film over the range $(x,\, x + \mathrm{d}x)$: the change $(\partial h/\partial t)\mathrm{d}x$ of the volume of fluid in this layer, per unit time, is equal to the difference $Q(x) - Q(x + \mathrm{d}x) = -\,(\partial Q/\partial x)\mathrm{d}x$ between the rates of flow of fluid entering and exiting. We thus obtain, by taking the derivative of Equations 8.25 with respect to x:

$$\frac{\partial h}{\partial t} = \frac{1}{12\,\eta}\,\frac{\partial}{\partial x}\left[h^3\frac{\partial p}{\partial x}\right] - \frac{1}{2}\left(h\,\frac{\partial U}{\partial x} + U\,\frac{\partial h}{\partial x}\right). \tag{8.26}$$

This result applies to a two-dimensional flow, translationally invariant in the z-direction. Equation 8.26 is known as *Reynolds' equation*. In practice, we frequently fix the values of the pressure at the two ends of the fluid layer.

Let us now take note of the fact that the vertical V and horizontal U components of the local velocity of the wall are related to the derivative $\partial h/\partial t$ by the geometric condition:

$$\frac{\partial h}{\partial t} = V - U\,\frac{\partial h}{\partial x}. \tag{8.27}$$

A horizontal displacement of the upper wall results indeed in a change in the local depth, if the wall itself is not locally horizontal. This variation is added to that resulting from the vertical velocity V of the wall. We see then, according to Equations 8.27, that h remains constant if the upper wall moves in its own plane ($V/U = \partial h/\partial x$).

In the case of a two-dimensional film, where the thickness $h(x,\, z,\, t)$ varies in the x- and z-directions parallel to the lower plane, we must replace Equations 8.25 by:

$$\mathbf{Q}_{//}(x,\, z,\, t) = -\frac{h^3}{12\,\eta}\,\nabla_{//}\,p + \frac{\mathbf{U}_{//}\,h}{2}, \tag{8.28}$$

where $\mathbf{U}_{//}$ is the projection onto the z-x plane of the velocity $\mathbf{U}$ of the upper surface with respective components W and U in the z- and x-directions. The components Q_z and Q_x of $\mathbf{Q}_{//}$ represent the local rates of flow through unit cross-sections respectively perpendicular to the z- and x-directions. The symbol $\nabla_{//}$ indicates the gradient in the directions parallel to the plane. By writing the equation of conservation of mass in the three-dimensional form $\partial h/\partial t + \nabla\cdot\mathbf{Q}_{//} = 0$, we obtain the generalization of Equations 8.26:

$$\nabla\cdot\left[h^3\,\nabla_{//}p\right] = h^3\,\nabla^2 p + 3\,h^2\,(\nabla_{//}h)\cdot(\nabla_{//}p) = 6\eta\left(h\,\nabla_{//}\cdot\mathbf{U}_{//} + \mathbf{U}_{//}\cdot(\nabla_{//}h) + 2\frac{\partial h}{\partial t}\right). \tag{8.29}$$

The problem in Section 8.1.5 (flow between two planes at an angle θ) can be treated as a particular case of Reynolds' equation (Equation 8.26) with $\partial h/\partial t = -\,U\,\partial h/\partial x$, $h(x) = e(x)$ and $\partial e/\partial t = -\,U\,\partial e/\partial x$. In this former section, we used a reference frame fixed relative to the upper plane, where the flow was stationary, whereas, here, we are in a reference frame fixed with respect to the lower plane and in which $\partial h/\partial t$ is non-zero (the local velocity of the upper plane is then $+U$). Equations 8.26 then becomes:

$$\frac{\partial}{\partial x}\left[h^3\,\frac{\partial p}{\partial x}\right] = -6\,\eta U\,\frac{\partial h}{\partial x}. \tag{8.31}$$

We obtain the same equation by multiplying Equations 8.14 by $e(x)^3$ and taking the derivative with respect to x (Q is constant relative to x in the reference frame used in Section 8.1.5, while, here, it varies).

Similarly, Equations 8.27 becomes:

$$\frac{\partial h}{\partial t} = V - \mathbf{U}_{//} \cdot \nabla_{//} h. \qquad (8.30)$$

Application of Reynolds' equation: a sphere dropping toward a plane, in a viscous fluid

Consider a rigid sphere of radius a dropping vertically toward an horizontal flat plane (Figure 8.4): we denote by $h_0(t)$ the minimum distance between the sphere and the plane, i.e. along the axis of the system. Considering the rotational symmetry (the pressure being solely a function of r and t) and the condition $\mathbf{U}_{//} = 0$, Equations 8.29 becomes:

$$\frac{1}{r}\frac{\partial}{\partial r}\left[r\, h^3 \frac{\partial p}{\partial r}\right] = 12\eta\, \frac{\partial h}{\partial t} \qquad (8.32) \qquad \text{with} \qquad \frac{\partial h}{\partial t} = \frac{\mathrm{d}h_0(t)}{\mathrm{d}t}.$$

This last condition corresponds to the requirement of a uniform vertical velocity $\partial h/\partial t$ for all points of the surface of the sphere. We can calculate the global force $\mathbf{F}$ on the sphere by integrating the pressure:

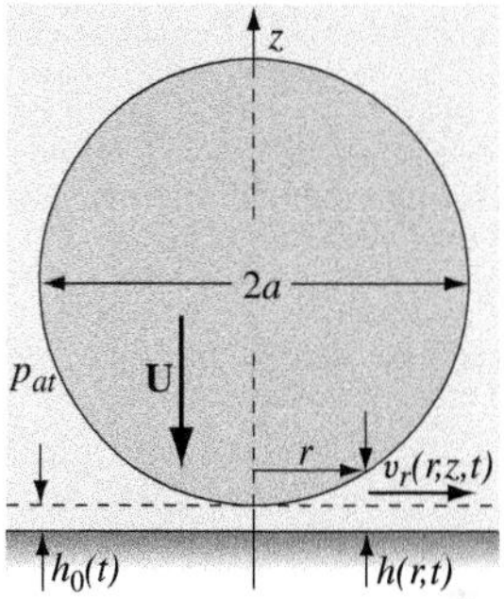

Figure 8.4 *Schematic diagram for a sphere dropping toward an infinite plane*

$$F = \int_0^{r_M} 2\pi\, r\, p(r)\, \mathrm{d}r = \int_{h_0}^{h_M} 2\pi a\, p(h)\, \mathrm{d}h,$$

so that we obtain:

$$F = -\frac{6\,\pi\, a^2}{h_0}\frac{dh_0}{dt}, \qquad (8.33)$$

provided that the thickness h_M as we get away from the region of minimum depth is significantly greater than h_0.

Proof

In order to calculate $p(h)$, we first integrate Equations 8.32 with respect to r, after having multiplied both sides by r. We then obtain:

$$r\, h^3 \frac{\partial p}{\partial r} = 6\,\eta\, r^2 \frac{\partial h}{\partial t}. \qquad (8.34)$$

The constant of integration is zero because none of the derivatives diverge when $r = 0$. In order to calculate the pressure distribution, we substitute h for the variable r, by using the approximate geometrical relationship:

$$h(r,\ t) = h_0(t) + \frac{r^2}{2a}, \qquad (8.35)$$

whence $\mathrm{d}h = r\,\mathrm{d}r/a$ and $\mathrm{d}p/\mathrm{d}r = (r/a)\,\mathrm{d}p/\mathrm{d}h$.

Thus: $\quad \mathrm{d}p = -3\,\eta\, a \dfrac{\mathrm{d}h_0}{\mathrm{d}t}\mathrm{d}\left[\dfrac{1}{h^2}\right] \quad$ and $\quad p(h) = p_0 - \dfrac{3\,\eta\, a}{h^2}\dfrac{\mathrm{d}h_0}{\mathrm{d}t}. \qquad (8.36)$

The pressure must indeed be equal to the atmospheric pressure p_0 when h becomes large.

For a sphere of radius a, falling under its own weight and with respective densities ρ_s and ρ_f of the sphere and the fluid, we find, by equating the weight of the sphere, decreased by the buoyancy due to the Archimedes force, to the viscous force from Equations 8.33 that:

$$\frac{d(\mathrm{Log}\,h_0)}{dt} = -\frac{2\pi}{9}\,\frac{a\,(\rho_s - \rho_f)g}{\eta} \quad (8.37) \qquad \text{i.e.:} \qquad h_0(t) = h_0(0)\,e^{-t/\tau}, \quad (8.38)$$

with $\tau = 9\,\eta/[2\pi a\,(\rho_s - \rho_f)\,g]$. Theoretically, the sphere will take an infinitely long time to touch the plane because its motion is continuously slowing down! Practically, however, the roughness of the surfaces involved will lead to contact when the separation becomes of the order of this micro-roughness, although there still remains a tiny free space for the fluid to be evacuated.

For the case where the sphere is replaced by a flat-bottomed cylinder parallel to the plane, we find that the force varies as $1/h^3$ instead of $1/h$ for a given velocity, because the viscous forces are more uniformly distributed over the bottom of the cylinder, instead of being localized in the region of minimum thickness.

If, on the other hand, we specify a drop velocity $V_z = dh/dt$, the force that needs to be applied for keeping V_z constant diverges, according to Equations 8.36, as $1/h_0$ when the distance h_0 approaches zero, while the pressure diverges as $1/h_0^2$ in the region of minimum thickness h_0: this can result in a local deformation of the surface of the sphere.

8.1.7 Flow between two eccentric cylinders with nearly equal radii

An important industrial application of lubrication is the motion of moving objects with a narrow region between the parts (piston-and-cylinder, axle-and-bearing, etc.) filled with lubricating fluid.

Here we are specifically interested in the flow of the lubricant in the small gap between a rotating axle and its bearing, as well as in the forces which result from it: this system is schematized in Figure 8.5 by two cylinders of nearly equal radii R and $R + \delta R$ ($\delta R/R = \varepsilon \ll 1$), with their axes parallel but offset by a distance $a = \lambda\,\delta R$ ($\lambda \leq 1$). We assume that only the inner cylinder rotates at an angular velocity Ω and that the flow is invariant in the z-direction of the axes of the cylinders. We will take, as the origin of the polar coordinates, the point of intersection O of the axis of the larger cylinder with the plane of the figure; the angle $\theta = 0$ corresponds to the direction of the segment **OO'** (and, also, to the minimum e_0 of the local distance $e(\theta)$ between the cylinders); α is the angle that **OO'** makes with the vertical. The distance $e(\theta)$ satisfies the equation:

$$e(\theta) = \delta R - a\cos\theta = \varepsilon R\,(1 - \lambda\cos\theta). \quad (8.39)$$

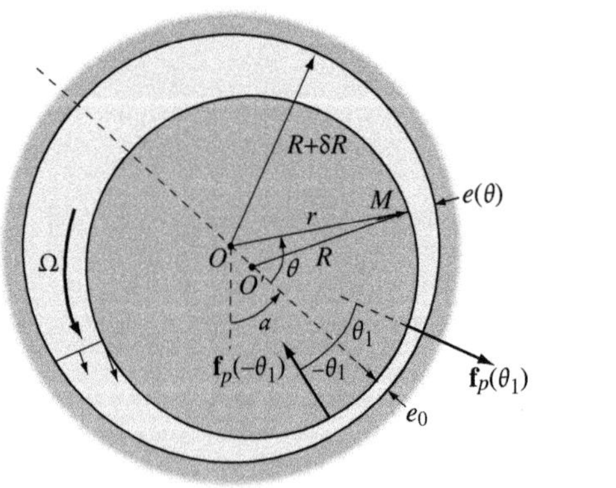

Figure 8.5 *Schematic diagram of the cross-section of an axle rotating within its bearing, the z-axis being perpendicular to the plane of the figure. The points O and O′ correspond to the intersection of the axes of the two cylinders with the plane of the figure*

Proof

The distance $r = |\mathbf{OM}|$ between O and a point M of the inner cylinder such that $\mathbf{OM}$ makes an angle θ with the segment $\mathbf{OO}'$ connecting the axes satisfies, to first-order:

$$r = R + a\cos\theta = R + \lambda\,\delta R\,\cos\theta$$

Equation 8.39 is then obtained by subtracting r from the radius $R+\delta R$ of the outer cylinder and replacing δR *by* εR.

In the region around a given angle θ, we consider that the flow is identical to that between two parallel planes separated by a distance $e(\theta)$, one of which (corresponding to the inner cylinder) moving at a velocity ΩR. This assumption is equivalent to eliminating curvature effects, such as the gradient in the pressure due to the centrifugal force; the latter is actually transverse to the flow and does not affect it. We can then apply Reynolds' equation (Equations 8.26) in the form:

$$\frac{1}{R^2}\frac{\partial}{\partial\theta}\left[e(\theta)^3\frac{\partial p}{\partial\theta}\right] = 6\,\eta\,\Omega\frac{\partial e(\theta)}{\partial\theta} \quad (8.40) \qquad \text{so:} \qquad \frac{\partial p}{\partial\theta} = 6\,\eta\,\Omega\,R^2\frac{1}{e(\theta)^2} + \frac{C}{e(\theta)^3}, \quad (8.41)$$

where C is a constant of integration. In order to go from Equations 8.26 to Equations 8.41, we have merely replaced the derivatives $\partial/\partial x$ by $(1/R)\,\partial/\partial\theta$. The element of length $\mathrm{d}s$ in the direction tangent to the surfaces of the cylinders, corresponding to an angular change $\mathrm{d}\theta$, is indeed equal to $R\,\mathrm{d}\theta$ (still assuming that $\delta R \ll R$); $\mathrm{d}s$ here plays the role of $\mathrm{d}x$ in the problem involving planes. Figure 8.6 displays the changes in pressure as a function of θ calculated for $\lambda = 0.9$, by numerical integration of Equations 8.41, and using Equations 8.39. We have large minima and maxima of the pressure near $x = 0$ because of the large value of the term in $1/e^3$: we can observe here cavitation bubbles in the region of low pressure (see Section 8.3.2).

The characteristics of the curve in Figure 8.6 result from the form of Equations 8.38 and 8.41. The variation of $e(\theta)$ and that of $\partial p/\partial\theta$ are symmetric with respect to $\theta = 0$: as a result, the pressure changes $p(\theta) - p(0)$ obtained by integration are anti-symmetric. We must, moreover, also have the condition that $p(\pi) = p(-\pi)$ since these two values of θ correspond to the same physical location diametrically opposite to the point of minimum thickness. If $p(\theta)$-$p(0)$ is not constant, it must have, at least, two extrema of opposite signs at points which are symmetric relative to $\theta = 0$ so that the constant C must be opposite in sign to Ω (see Equations 8.41). In the neighborhood of $\theta = 0$, the term $1/e^3$ in Equations 8.41 dominates the other term, and the derivative $\partial p/\partial\theta(0)$ is opposite in sign to Ω (< 0 in Figure 8.6). The presence of the term in $1/e^3$ implies that the smaller the minimum thickness $e(0)$ is relative to the maximum value $e(\pi)$, the larger the absolute values of the pressure extrema will be, and closer to the point $\theta = 0$.

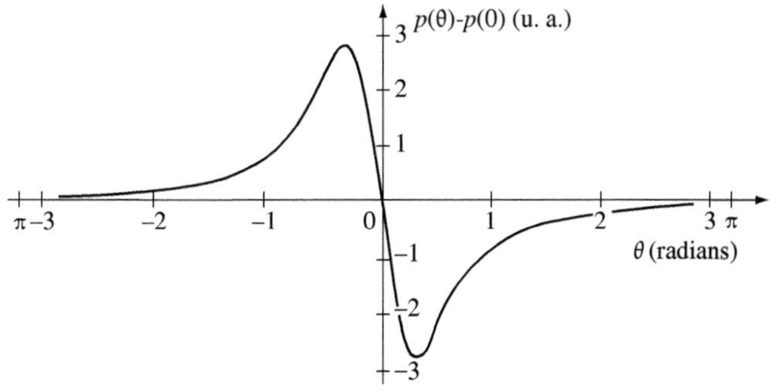

Figure 8.6 *The variation of the pressure in the region between the axle and its bearing in Figure 8.5*

Just as in the case of the two planes which we discussed in Section 8.1.5, we assume that the viscous friction forces, which act tangentially to the cylinders, are negligible relative to the pressure forces in the normal direction. Let us now investigate how these forces can support heavy weights as in the cases of a wheel axle or of the axis of rotating machinery. To this end, let us look at the pressure forces per unit area $\mathbf{f}_p = \mathrm{d}\mathbf{F}_p/\mathrm{d}S$ acting on the inner cylinder at corresponding points θ_1 and $-\theta_1$ (Figure 8.5). We can neglect the effect of the

The form of equation 8.43 may seem counter-intuitive. We might think that the force of gravity would require the point of minimal thickness to be at the lowest point of the bearing in the vertical direction from the axis O: **OO'** would be in that case vertical with $\alpha = 0$. However, the vertical resultant of the pressure forces would vanish, because the contribution of each point at a position θ is cancelled out by that of a point at the angle $-\theta$: the pressure $p(\theta)$ is then opposite to that at $p(-\theta)$ and the vertical component of the normal to the surface is the same. In contrast, if $\alpha = \pi/2$, the pressure force also changes sign when we replace θ by $-\theta$, but so does the vertical component of the normal to the surface. The contributions to the force then have the same sign and they add up. The value of α will adjust to a value intermediate between 0 and $\pi/2$ depending on the weight which must be supported.

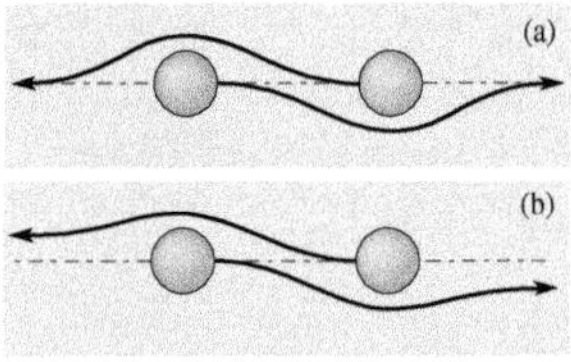

Figure 8.7 *Relative motion of two spheres. (a) ideal case without any interaction other than those due to hydrodynamics; (b) Real case of rough spheres: the two spheres undergo a shift in their trajectories, instead of coming back on the same straight line which they followed initially (an analogous effect is observed in the case where the two spheres display inter-particle interactions such as electrostatic repulsion)*

The reversibility argument assumes that the velocity field takes on, at every instant, the equilibrium configuration which corresponds to the instantaneous distance between the obstacles; we have described a similar problem in Section 8.1.3 which discussed the stationarity of lubrication flows. When the two spheres come together at a distance d, with relative velocity U, the time for establishing the velocity profile is of the order of d^2/ν: this must be small relative to the characteristic time d/U over which the distance between particles evolves. The Reynolds number Ud/ν must then be small. In the opposite case, the dynamics of the motion of the sphere would depend on its previous history. We encounter effects of this type, for which we will not go into detail here, in the case of strongly accelerating particles (specifically, effects known as *Basset forces*).

average pressure $p(0)$ because the integral of the force which results on the entirety of each cylinder is zero. We will therefore assume that $p(\theta_1) = -p(-\theta_1)$. The sum of the vertical components $f_{zp}(\pm\theta_1)$ of the pressure forces per unit area at θ_1 and $-\theta_1$ will be:

$$f_{zp}(\theta_1) + f_{zp}(-\theta_1) = -p(\theta_1)\cos(\alpha+\theta_1) + p(-\theta_1)\cos(\alpha-\theta_1) = 2p(\theta_1)\sin\alpha\sin\theta_1. \quad (8.42)$$

Thus, the vertical resultant of the pressure forces is a maximum when the angle α of **OO'** with the vertical has the value $\pi/2$ (**OO'** is then horizontal). The value $F_{p\pi/2}$ of this force per unit length along the axis is then given by the integral:

$$\mathrm{F}_{p\pi/2} = -\int_{-\theta}^{\theta} \sin(\theta)\, p(\theta) R\, d\theta \text{ which can be shown to equal to: } \mathrm{F}_{p\pi/2} = \frac{6\,\eta A\,\Omega\,R}{\varepsilon^2\,\lambda}. \quad (8.43)$$

Here, A is a constant dependent on λ. This bearing force varies as $1/\varepsilon^2$: when ε is very small, one can generate very significant bearing forces.

Let us now evaluate the viscous frictional force $F_{v\pi/2}$ in the direction of $\theta = \pi/2$. Per unit area, this force is of the order of $\eta\ \Omega\ R/((1-\lambda)\ \varepsilon\ R)$ in the region of minimum thickness; if this is applied along a distance of order R, we obtain then:

$$F_{v\pi/2} \approx \frac{\eta\,\Omega\,R}{(1-\lambda)\,\varepsilon}. \quad (8.44)$$

This viscous frictional force, varying as $1/\varepsilon$, is smaller than the bearing force by an order of magnitude in ε, as we have assumed.

8.1.8 Lubrication and surface roughness

The examples which we have given and, specifically, that of the axle and its bearing, suggest that the distance between the solid surfaces should be as small as possible: this will be limited by the roughness of the surfaces which, in our calculations, have been assumed to be perfectly smooth. If the rough areas of the surfaces actually come into contact, there will appear solid-solid frictional forces which would block any motion. Another example is the interaction between particles in a suspension when they are separated by a small distance. As we have seen in Section 8.1.5, the viscous forces increase when a sphere approaches a wall very closely; when the distance becomes small, we must again take into account the effects of surface roughness. An example is given in Figure 8.7. In the case of two perfectly smooth spheres which come together, the final trajectory coincides with the line of initial displacement (Figure 8.7a): this is the result of the time-reversal invariance in solutions of Stokes' equation of motion at small Reynolds number (see Section 9.2.3). This reversibility can be broken by the effect of the roughness of the particles (Figure 8.7b). Quite a number of problems related to lubrication, to friction and to wear depend on the heterogeneity of surfaces.

8.2 Flow of liquid films having a free surface: hydrodynamics of wetting

Other situations where inertial forces and non-linear terms in the equations of motion of fluids are negligible are encountered in the case of thin fluid films with a free surface: there again, the velocity of the fluid is almost parallel to the surface of the film. Such flows are often present in nature and have, additionally, important applications such as the spreading of coatings, paints or cleaning fluids on various materials; also, some thermal exchangers make use of flowing liquid films. This situation is similar to that of lubrication, but the presence of free surfaces affects the boundary conditions, and additional forces appear due to surface tension. We have discussed this idea in Section 1.4 and have described some static properties of the resulting interfaces. Using the Young-Dupré equation (Equation 1.62), we have also introduced the spreading parameter and the static contact angle.

In this discussion, we are interested in flows of thin films with a free surface resulting from a competition between surface tension, gravity and viscosity. In all cases, we will make broad use of the lubrication approximation.

First of all, we analyze the simple case of the flow of liquid layers uninfluenced by surface tension, such as the falling of a liquid film along a flat isothermal wall. Then, in the situation of complete wetting, we study the dependence of the dynamic contact angle of an interface on the velocity at which the contact line moves. We use this result to predict the spreading of small droplets and compare this to that of larger drops under the influence of gravity. Finally, we look at the case of *Marangoni effects* in which temperature or concentration gradients, in tensioactive compounds, result in surface tension gradients, which, themselves, induce flow.

8.2.1 Dynamics of thin liquid films, neglecting surface-tension effects

The influence of surface tension is negligible for films which have a surface flat or slightly curved because the difference in the Laplace pressure between the two sides of the interface is then either zero or very small. We will discuss situations where the surface tension is constant all over the surface so that Marangoni effects will not come into play.

Flows of liquid films with a free surface have specific characteristics. First of all, if surface tension is not in play, the pressure equals the atmospheric pressure throughout the interface. Just as for other quasi-parallel flows, the pressure gradients normal to the interface (and thus normal to the flow velocity) are reduced to the hydrostatic pressure. Given the small thickness of the films, the pressure is everywhere close to the atmospheric pressure: the pressure gradients parallel to the film (and to the velocity of the fluid) are thus generally much smaller than for the flows between two solid surfaces which we have previously considered. Often, this gradient will be zero (specifically for films of constant thickness) and the driving force of the flow is the component $g_{//}$ of gravity parallel to the surface of the film: we recall that in the Stokes equation (and in that of Navier-Stokes), the driving force of the flow is represented by the sum $\nabla p - \rho\, \mathbf{g}$ and not just by ∇p.

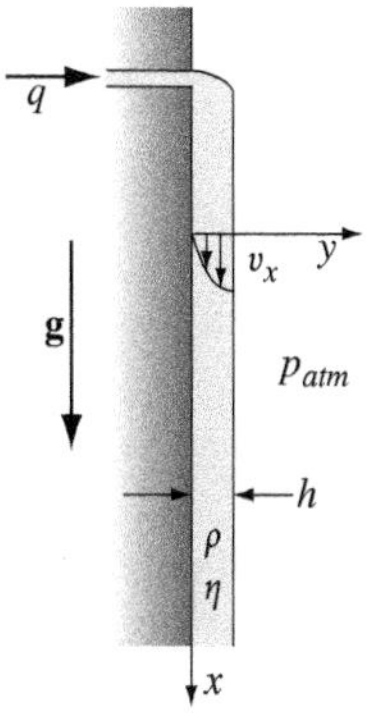

Figure 8.8 *Flow along a vertical wall due to an injection of fluid through a horizontal slit*

If, externally, we have air at rest (or at moderate velocities), we can consider that the stress at the interface vanishes (Section 4.3.2): the derivative of the velocity in the direction normal to the surface thus also vanishes. We have then an extremum of the velocity at the surface and not a zero value as would be the case at a solid wall.

This will no longer be the case when there is a high-velocity flow of air (e.g., high winds above the surface of a lake or river.)

Example: viscous film falling along a vertical wall, at a fixed flow rate
We can create a fluid film (density ρ, viscosity η) along a vertical wall by injecting fluid through a horizontal slit (Figure 8.8). The flow rate q in the slit, per unit width in the z-direction, is constant. We shall now predict the dependence of the thickness of the fluid layer on the flow rate q, in the region where this layer has a constant thickness h. Assuming that the flow is invariant and has zero velocity in the z-direction, the Navier-Stokes equations become:

$$\frac{1}{\rho}\frac{\partial p}{\partial x} - \rho g = \eta \frac{\partial^2 v_x}{\partial y^2} \quad (8.45a) \qquad \text{and} \qquad \frac{\partial p}{\partial y} = 0. \quad (8.45b)$$

For sufficiently large velocities, such flows develop instabilities which cause the thickness of the film to vary and create local curvature at the surface. Then, one needs to take into account both the surface tension and the pressure gradients parallel to the surface.

We obtain similar results for a flow along a plane inclined at an angle θ relative to the vertical. We must then replace $-\rho\, g$ by $-\rho\, g \cos\theta$ in Equations 8.45a and add a term $-\rho g \sin\theta$ into Equation 8.45b. However, since this last term is constant, $\partial p/\partial x$ still vanishes even though the hydrostatic pressure gradient in the y-direction no longer vanishes. Equations 8.46a and 8.46b remain applicable when we replace g by $g \cos\theta$.

Because we have constant external pressure, continuous at the surface and continuous at all horizontal levels, we have $p = p_{atm}$ throughout the fluid film: the pressure gradient parallel to the velocity also vanishes ($\partial p/\partial x = 0$). The flow thus results merely from the vertical component $g > 0$ of gravity. Integrating with the boundary conditions, $v_x(y = 0) = 0$ and $\partial v_x/\partial y(y = h) = 0$, we find:

$$v_x = \frac{\rho\, g}{2\,\eta} y\,(2\,h - y) \qquad (8.46a) \qquad \text{and} \qquad q = \int_0^h v_x\, \mathrm{d}y = \frac{\rho\, g\, h^3}{3\,\eta}. \qquad (8.46b)$$

The thickness of the fluid layer varies with the injected rate of flow as $q^{1/3}$.

8.2.2 Dynamic contact angles

Case of complete wetting: Tanner's law

In contrast to the previous example, surface tension plays a key role in problems involving the contact line between a gas-liquid interface (or a liquid-liquid interface) and a solid surface. Here, we are interested in the changes, of hydrodynamic origin, in the contact angle as a function of the velocity V of the line: it is referred to as a *dynamic contact angle*. We consider here the case of complete wetting for which the initial spreading parameter $S = \gamma_{sg} - \gamma_{sl} - \gamma$ (defined in Equation 1.59) is positive (we use here the notation γ for γ_{lg}). The motion of the interface at a velocity V changes the contact angle from the value $\theta_s = 0$ (*static contact angle*) for $V = 0$ to a finite value $\theta(V)$. In such systems with complete wetting, we always have, upstream of the contact line, a precursor film of submicron thickness: the dynamic contact angle $\theta(V)$ is in fact an *apparent contact angle*, with the wall, of the *macroscopic* region of the meniscus (Figure 8.9).

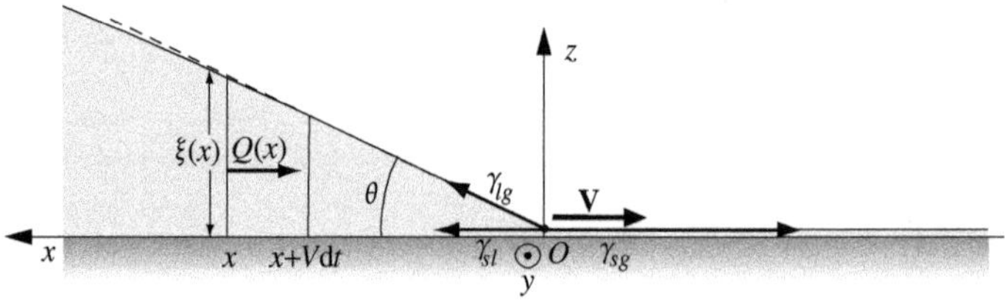

Figure 8.9 *Schematic diagram for the motion of a contact line relative to a solid surface in the presence of the precursor film upstream of the meniscus*

Let us first consider the equilibrium of forces acting on the contact line by assuming a two-dimensional geometry. Following the same steps as those used in Chapter 1 in order to prove the Young-Dupré equation (Equation 1.62), we find a resultant force, per unit length, with horizontal component: $F_r(\theta\,) = \gamma_{sg} - \gamma_{sl} - \gamma \cos\theta$. However, according to this equation, the condition of static equilibrium $F_r = 0$ cannot be satisfied at zero velocity even when $\theta = 0$. The excess force per unit length equals $S(> 0)$, and we assume that its use is to create the precursor film, and that this contribution remains constant when V is not zero. The effective resultant force which causes the motion of the region of macroscopic thickness of the meniscus (to the left of the point O on Figure 8.9) is thus: $F_r(\theta) - S = \gamma\,(1 - \cos\theta)$. The corresponding power (energy per unit time) $\mathrm{d}E_{ts}/\mathrm{d}t$ is then:

$$\frac{\mathrm{d}E_{ts}}{\mathrm{d}t} = \gamma\;(1 - \cos\;\theta)\;V. \qquad (8.47)$$

This energy, provided by the capillary forces, is dissipated through viscosity in the flow of the film resulting from the motion of the interface. We note that $\mathrm{d}E_{ts}/\mathrm{d}t$ should vanish for an angle $\theta = 0$, which would correspond to an almost motionless interface for which all the available energy is dissipated in the residual film.

Let us now analyze the flow in the macroscopic region of the fluid film of thickness $\xi(x)$ at a distance x from the contact line. We assume that we are sufficiently close to this contact line that gravity effects are negligible for the small vertical distances which we are considering. We will also assume that the apparent macroscopic contact angle θ is sufficiently small that we can use everywhere the lubrication approximation. Let us call $Q(x)$ the rate of flow (per unit length normal to the figure) in the cross-section of the film: the change between the times t and $t + dt$ in the volume contained between the cross-section at x and the contact line equals $Q(x)dt$ (this is the volume injected in the same time interval). Let us assume, moreover, that the interface moves without deformation at velocity V: the preceding change in volume must also be equal to that $\xi(x)(V dt)$ of a layer of fluid of thickness $V\, dt$, located in the neighborhood of cross-section x, whence:

$$Q(x) = V\,\xi(x). \tag{8.48}$$

It follows therefore that the average velocity in section x defined by $V_m(x) = Q(x)/\xi(x)$ is constant with x and equal to V. In the lubrication approximation, the tangential stress $\eta\, \partial v_x/\partial z$ on the interface $z = \xi(x)$ vanishes (condition at a free surface); and, moreover, $v_x(0) = 0$. We will then have a velocity profile of the type $v_x(z) = A(x)(\xi(x) - z/2)z$. Integrating between 0 and $\xi(x)$ relative to z so as to calculate $Q(x)$, we find that:

$$v_x(x,\, z) = 3\,\frac{Q(x)}{\xi^3(x)}\left[\xi(x) - \frac{z}{2}\right] z. \tag{8.49}$$

Our approach here, based on an estimate of the driving and dissipation effects, is only approximate. In fact, the assumption of a plane interface implies that there is zero capillary pressure difference between the sides of the interface, and, consequently, a constant pressure within the fluid. This contradicts the prediction (Equations 8.49) for the velocity profile, which requires a pressure gradient $\partial p/\partial x = -3\,\eta Q/\xi^3$. Thus, the interface needs to display a variable radius of curvature near the contact line so that the difference in capillary pressure can balance such gradients.

We note that the velocity $v_x(x,\, \xi)$ of the fluid at the interface equals $3\, Q(x)/(2\, \xi(x))$: it is then higher than the average velocity $V_m(x) = V$ and also independent of x.

Let us now write that the power dE_{ts}/dt given by Equation 8.47 equals the total power dE_η/dt dissipated by viscosity in the flow corresponding to the velocity profile from Equations 8.49. From Equation 5.26, the dissipated power per unit volume is $\eta\, (\partial v_x/\partial z)^2$. We obtain the total dissipation dE_η/dt (again per unit length in the transverse direction) by first integrating this expression between 0 and $\xi(x)$ with respect to z and then with respect to x. Let us now take $\xi(x) = \theta\, x$, which amounts to neglecting the curvature of the interface in the x-plane: assuming that the greatest part of the dissipation occurs very close to the contact line, the resulting error will be quite small. Using Equations 8.48 and 8.49, we then obtain in absolute magnitude:

$$\frac{dE_\eta}{dt} = \int 3\,\eta\,\frac{Q^2(x)}{\xi^3(x)}\, dx = 3\,\eta V^2 \int \frac{1}{\xi(x)}\, dx = 3\,\eta \frac{V^2}{\theta} \int_{x_m}^{x_M} \frac{1}{x}\, dx. \tag{8.50}$$

Here, we needed to introduce an upper limit x_M and lower limit x_m, because of the logarithmic divergence of the integral as x approaches zero and infinity. We can assume that the upper limit x_M corresponds to a length of the order of the size of the droplet in the x-z plane, but we have a much more serious problem from the divergence of the dissipated power at very small distances from the contact line. This issue is, still at this time, the object of significant study and of numerical simulations on a molecular scale, and there is no exact solution presently available. Considering then x_m as an adjustable parameter of this model, we obtain:

$$\frac{dE_{ts}}{dt} = \gamma(1 - \cos\,\theta)V = \frac{dE_\eta}{dt} = 3\,\eta\,\frac{V^2}{\theta}\,\mathrm{Log}\left(\frac{x_M}{x_m}\right). \tag{8.51}$$

In the approximation of small contact angles, where $(1 - \cos\theta) \approx \theta^2/2$, we finally find *Tanner's equation*:

$$\theta^3 = 6\frac{\eta V}{\gamma}\mathrm{Log}\left(\frac{x_M}{x_m}\right) = 6\,Ca\,\mathrm{Log}\left(\frac{x_M}{x_m}\right). \tag{8.52}$$

The dimensionless *capillary number* $Ca = \eta V/\gamma$ indicates the relative importance of the effects due to the viscosity and those due to capillarity. The pressure differences of viscous and capillary origin are indeed respectively of the order of $\eta V/L$ and γ/L, so that their ratio is of the order of Ca, if the gradients of the velocity and the radii of the curvature of the interfaces have the same characteristic length L (if this were not the case, the ratio would still be of the order of Ca, but we would need to introduce a geometrical correction factor). The capillary number can also be considered as the ratio of the velocity V characteristic of the flow to a velocity γ/η characteristic of the fluid (of the order of 10^2 m/s for water). This result does not involve the value of the spreading parameter: this merely indicates the fact that the excess energy corresponding to a positive value of this parameter is assumed to be dissipated in the residual film, without influencing the dynamics of the macroscopic meniscus.

The experimental value of x_m (caption of Figure 8.10) is of the same order of magnitude as that resulting from the small-scale analysis of both the connection between the macroscopic meniscus and the precursor film, and the dissipation of energy in the precursor film. In this model, where the interface can no longer be considered as a sharp corner with angle θ, the transition region starts when the thickness of the film reaches a value smaller than a/θ (a is a length on an atomic scale). The corresponding value of the distance x_m is thus of order a/θ^2; this explains why x_m is large compared to atomic distances and accounts for the experimental value of x_m/x_M.

Figure 8.10 shows the experimental verification of Tanner's law, using measurements carried out by an optical method on a meniscus of silicone oil moving inside a capillary tube. The continuous curve displays the theoretical variation predicted by Tanner's equation (Equations 8.52) but with a multiplying factor equal to 9 instead of 6 as indeed predicted by more complete models. The ratio $\varepsilon = x_m/x_M$, is taken as equal to 10^{-4}, a value which gives the best agreement with the experimental points. These points follow very closely the theoretical prediction up to surprisingly large values of the angle θ_d, of the order of 100°. The corresponding value of x_m is of the order of 100 nm.

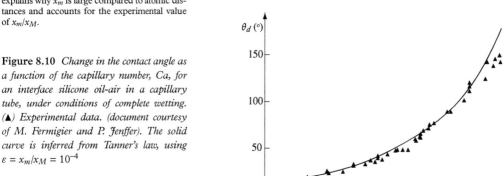

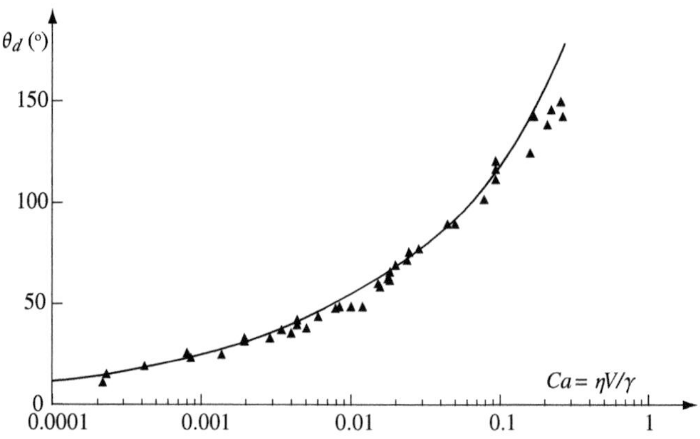

Figure 8.10 *Change in the contact angle as a function of the capillary number, Ca, for an interface silicone oil-air in a capillary tube, under conditions of complete wetting. (▲) Experimental data. (document courtesy of M. Fermigier and P. Jenffer). The solid curve is inferred from Tanner's law, using* $\varepsilon = x_m/x_M = 10^{-4}$

Contact angles under conditions of partial wetting

We can apply a similar approach to the case of partial wetting, so long as the wetting angle is not far from zero. The equations which we then obtain are more poorly obeyed than Tanner's law; we therefore use, in general, a rather empirical approach which generalizes this law. Other models involving molecular adsorption processes have also been suggested.

One of the difficulties involves the definition of the static contact angle, discussed in Section 1.4.3: depending on the direction of the motion of the contact line, the contact angle takes indeed general different values (for forward or backward motion) in the limit of small velocities.

8.2.3 Dynamics of the spread of droplets on a flat surface

Small droplets with complete wetting

Here, we are interested in the changes, as a function of time, of the radius of droplets of a non-volatile liquid which lie on a solid substrate, under conditions of total wetting. We will assume, just as we did in order to derive Tanner's law, that the dynamics of the droplet results from a continuous equilibrium between the viscous dissipation and the work done by capillary forces on the contact line. We assume (as we will discuss later), that it is enough to take into account the viscous dissipation in the region near the contact line (i.e. at distances small relative to the radius of curvature). We will thus apply to this three-dimensional system the equations which we have used above for the two-dimensional case of a straight contact line. More precisely, we assume that they continue to be valid so long as the radius of curvature of the interface is much greater than the thickness of the droplet.

Let us denote by Ω the volume of the droplet and assume that it keeps the shape of a spherical cap of height $h(t)$, with a radius of curvature $R(t)$ and with a contact radius $r_g(t)$ with the plane (Figure 8.11). We then have:

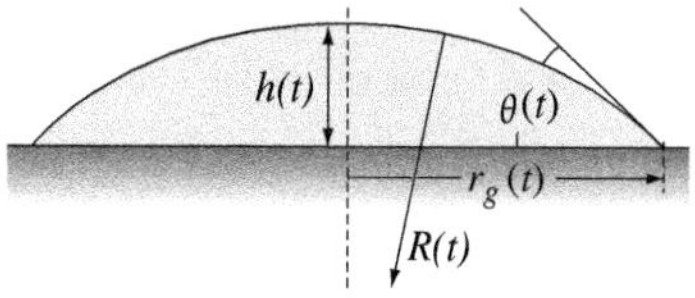

Figure 8.11 *Spreading of a droplet under the action of capillary forces*

$$\Omega = (\pi/4)\, r_g^3\, \theta. \tag{8.53}$$

Proof

We have an exact relationship $(2R-h)\,h = r_g^2$ (Figure 8.11) between the geometrical parameters. We obtain the volume of the spherical cap by integration from zero to $h(t)$ of $\pi\, r^2\, dz = \pi(2R - z)\, z\, dz$ (here r is the radius, expressed in terms of the same equation, of the horizontal cross-section of the cap at an arbitrary value of the height z, taken as positive downward from the top of the cap). We then obtain:

$$\Omega = \pi\left(R\,h^2 - \frac{h^3}{3}\right) \approx \pi\, R(t)\, h^2(t) = \frac{\pi}{2} r_g^2(t)\, h(t). \tag{8.54}$$

Moreover, we also have $h(t) = r_g(t)\,\tan(\theta/2)$ so that we obtain Equation 8.53 by assuming that the angle θ is much smaller than 1.

Combining Equations 8.52 and 8.53 and substituting dr_g/dt for V, we obtain then:

$$\frac{dr_g}{dt} = \frac{\gamma}{\eta}\frac{1}{6\,\mathrm{Log}\left(\frac{x_M}{x_m}\right)}\left[\frac{4\Omega}{\pi r_g^3}\right]^3, \quad (8.55) \quad \text{thus:} \quad [r_g(t)]^{10} = \frac{5}{3}\frac{\gamma}{\eta\,\mathrm{Log}\left(\frac{x_M}{x_m}\right)}\left[\frac{4\,\Omega}{\pi}\right]^3 t. \quad (8.56)$$

We therefore predict a very slow growth, as $t^{1/10}$, of the radius of the droplet as a function of time. Because the product $r_g^3(t)\,\theta(t)$ is constant (Equation 8.53), $\theta(t)$ decreases as $t^{-3/10}$ under these circumstances. The numerical coefficients can be easily calculated by combining Equations 8.54 and 8.56. We note that the solution obtained here can only be approximate because we have assumed that the radius of curvature is constant along the interface, since it is taken as a spherical cap. Also, the pressure just below the free surface is constant along that surface: this in fact contradicts the existence of a spreading flow which leads to pressure gradients resulting from the viscosity (we have already discussed

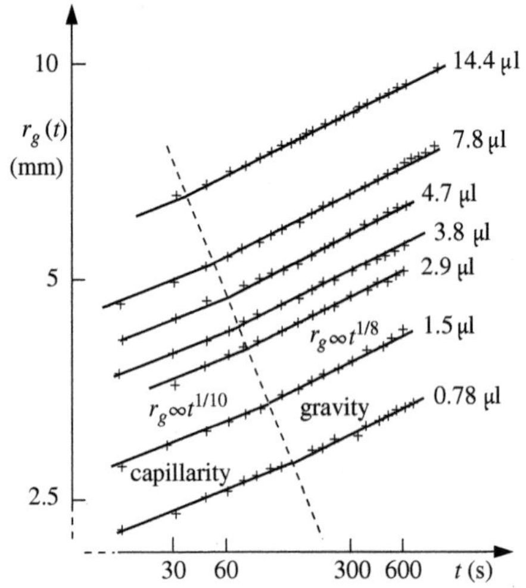

Figure 8.12 *Variation with time (in log-log coordinates) of the radius $r_g(t)$, for a series of droplets of silicone oil of variable volume Ω, ($\eta = 0.02$ Pa.s, γ = 20 mN/m), which are spreading on flat plates of hydrophilic glass. The dashed line indicates the boundary between the regimes dominated by gravity and by capillarity. The solid lines respectively to the right and to the left of the dashed line correspond to the change in the power-law of the exponents, from 1/10 and 1/8. (document courtesy of A.M. Cazabat, and M. Cohen Stuart)*

this concept when we dealt with Tanner's law). The exact solution is quite complicated: it predicts a similar dependence of $r_g(t)$ on the physical parameters, but with somewhat different numerical coefficients.

Figure 8.12 displays the variation as a function of time of the radius of a series of droplets of silicone oil lying on smooth plates of hydrophilic glass. The variation of $r_g(t)$ as $t^{1/10}$ is only observed at short times (to the left of the dashed line) when r_g is small enough so that the capillary effects dominate. At longer times, we enter into a regime dominated by gravity which is applicable to the droplets of larger diameter which we will now describe. We have in fact shown in Section 1.4.4 that it is the relative values of the radius r_g and of the capillary length which determine the relative importance of the capillary and gravity effects.

The gravitational spread of large drops

In this case, gravity plays a dominant role in determining at the same time the geometry of the drop and its dynamics. The drops can then be viewed as flat in the center region, with their curvature localized at the edges (Figure 8.13). We will first compare the energy dissipation per unit time due to the viscosity in the center region ($\mathrm{d}E_{\eta cr}/\mathrm{d}t$) to that near the contact line ($\mathrm{d}E_{\eta cl}/\mathrm{d}t$) which has already been calculated in the preceding case. Their ratio has the value:

$$\left[\frac{\mathrm{d}E_{\eta cr}}{\mathrm{d}t}\right] \Big/ \left[\frac{\mathrm{d}E_{\eta cl}}{\mathrm{d}t}\right] \approx \left[\frac{r_g\,\theta}{4\,h}\right] \Big/ \left[\mathrm{Log}\left[\frac{x_M}{x_m}\right]\right]. \tag{8.57}$$

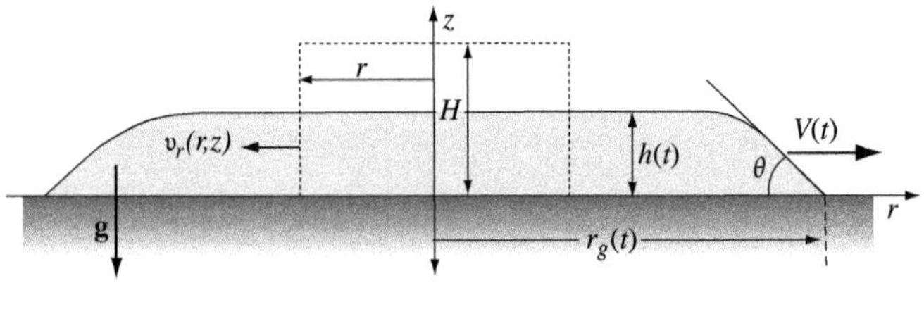

Figure 8.13 *Schematic diagram of a large-radius drop spreading out under the effect of gravity*

Within a logarithmic coefficient, this ratio is of the order of magnitude of the ratio of the radius r_g of the drop to the width ($\approx h/\theta$) of the transition zone between the contact line and the flat center region of the drop. Thus, for drops of outer radius r_g large relative to h/θ, the dissipation by viscosity in the flat region of the drop ends up being the dominant one, in contrast to the small droplets discussed previously for which the dissipation near the contact line is the most significant.

Proof

Let us assume that, in the central region, the drop has a thickness $h(t)$ independent of the distance r to the axis. Let us analyze the conservation of the mass of liquid in a cylinder of radius r, with the same axis as the drop, and of constant height H always larger than the thickness of the drop ($H>h(t)$). The volume rate of flow $Q(r)$ through the walls of this cylinder is opposite to the change per unit time of the amount of fluid inside, so that:

$$Q(r) = 2\pi r h\, V_m(r) = -\pi r^2 \frac{dh}{dt} \quad (8.58a) \qquad \text{or:} \quad \frac{1}{h}\frac{\mathrm{d}h}{\mathrm{d}t} = -2\frac{V_m(r)}{r} = \text{cst with } r, \quad (8.58b)$$

where $V_m(r)$ is the average, over the thickness h of the drop, of the radial velocity $v_r(r, z)$ of the fluid. Moreover, as in Equations 8.48, the average velocity $V_m(r_g)$ equals the spread velocity $V(r_g) = \mathrm{d}r_g/\mathrm{d}t$ at the edge of the drop. Combining this result with Equations 8.58b, applied to the radii r and r_g, we obtain:

$$V_m(r) = V_m(r_g)\frac{r}{r_g} = \frac{dr_g}{dt}\frac{r}{r_g}. \qquad (8.59)$$

Just as in Equation 8.49, the velocity profile $v_r(r, z)$ in the thickness of the drop is a half parabola with a velocity maximum equal to $3V_m(r)/2$ at the interface. We then find that the absolute value of the viscous energy dissipation per unit time $\mathrm{d}E_{\eta cr}/\mathrm{d}t$ in the flat center region of the drop, with radius approximately equal to r_g, satisfies:

$$\frac{\mathrm{d}E_{\eta cr}}{\mathrm{d}t} = \eta \int_0^{r_g} 2\pi r\, \mathrm{d}r \int_0^{h(r)} \left(\frac{\partial v_r}{\partial z}\right)^2 \mathrm{d}z = \frac{3}{2}\pi \eta V^2 \frac{r_g^2}{h} = \frac{3}{2}\pi \eta \left(\frac{dr_g}{dt}\right)^2 \frac{r_g^2}{h}. \qquad (8.60a)$$

Applying Equations 8.51 along the perimeter $2\pi r_g$ of the drop, we find the energy $\mathrm{d}E_{\eta cl}/\mathrm{d}t$ dissipated by the viscosity per unit time, in the area of the contact line:

$$\frac{\mathrm{d}E_{\eta cl}}{\mathrm{d}t} = 6\pi \eta r_g \frac{V^2}{\theta} \mathrm{Log}\left(\frac{x_M}{x_m}\right). \qquad (8.60b)$$

We then recover Equation 8.57 by taking the ratio of these two dissipation terms.

We determine now the spreading equation for drops, for the case where gravity is the dominant force (Figure 8.13). Just as in the previous case, we assume that the liquid in question is non-volatile, in order to avoid evaporation phenomena which, often accompanied by changes in the surface tension, create additional flow by virtue of the Marangoni effect (we will be discussing this in Section 8.2.4, which follows).

We estimate the spreading law by assuming that the thickness of the drop is uniform all over its surface. The dissipation of energy through viscosity must, at every instant of time, correspond to the change in potential energy of the drop which has the value: $(\mathrm{d}/\mathrm{d}t)[(\pi/2)\rho\, g\, r_g^2 h^2]$. Moreover, the volume $\Omega = \pi\, r_g^2 h$ of the drop is constant. We thus have, using Equations 8.60a to evaluate the viscous dissipation, the following energy balance:

$$\frac{\mathrm{d}}{\mathrm{d}t}\left(\frac{\pi}{2}\rho\, g\, r_g^2 h^2\right) = -\frac{3}{2}\eta\, \pi\left(\frac{\mathrm{d}r_g}{\mathrm{d}t}\right)^2 \frac{r_g^2}{h}. \tag{8.61}$$

We can evaluate $r_g(t)$, in the same manner, but with more precise coefficients, by assuming that the thickness $h(r, t)$ of the film is no longer exactly constant as a function of the distance from the axis, but is described by a self-similar profile; more precisely, it must satisfy a relation of the type $h(r, t) = h(0, t)\, f(r/r_g(t))$ in which $r_g^2(t)\, h(0, t)$ is constant with time in order to keep constant the liquid volume, and $f(x) = 0$ for $x > 1$.

Let us now substitute for h its value as a function of Ω and r_g. We then obtain:

$$r_g^7 \frac{\mathrm{d}r_g}{\mathrm{d}t} = \frac{2\,\Omega^3}{3\,\pi^3}\frac{\rho\, g}{\eta}, \tag{8.62}$$

so that:

$$r_g(t) = \left(\frac{\Omega}{\pi}\right)^{\frac{3}{8}}\left(\frac{16}{3}\frac{\rho\, g\, t}{\eta}\right)^{\frac{1}{8}}. \tag{8.63}$$

In this way, we then predict (assuming that the radius of the drop is zero at time $t = 0$), a growth of the drop as $t^{1/8}$, instead of $t^{1/10}$ in the previous case. This result is in agreement with the growth at long times, which we observe on the curves in Figure 8.12.

The spreading behavior as $t^{1/8}$ and $t^{1/10}$ represents two limiting cases, applicable to drops spreading under conditions of complete wetting. Other types of spreading behavior (as $t^{1/4}$) are observed for flow on a rotating plate, or on some surfaces with roughness. Adapting these models to the geometries and properties of more complex fluids has very important applications to the spreading of decorative or protective coatings. Another important practical problem, which we will not discuss here, is the appearance of interfacial instabilities in flowing films, which can be brought on by deformations, often very significant, of the contact line, or by changes in the depth.

8.2.4 Flows resulting from surface-tension gradients: the Marangoni effect

Principle of the Marangoni effect

Gradients in the surface tension due to changes in temperature or in the concentration of solutes (e.g., affecting surface tension) can create surface stresses and cause fluid motion. Fluid flow resulting from such stresses is known as the *Marangoni effect*; this is also known as the *thermo-capillary* effect when it results from temperature gradients.

Thus, if a layer of water covers a surface, and a piece of soap touches a point on that surface, we see this part of the surface "drying out" (Figure 8.14): the surface tension is indeed locally reduced and the forces due to surface tension become unbalanced. We thus have flow toward the neighboring regions where the surface tension is unchanged (this can be visualized by the motion of dust particles initially present on the surface).

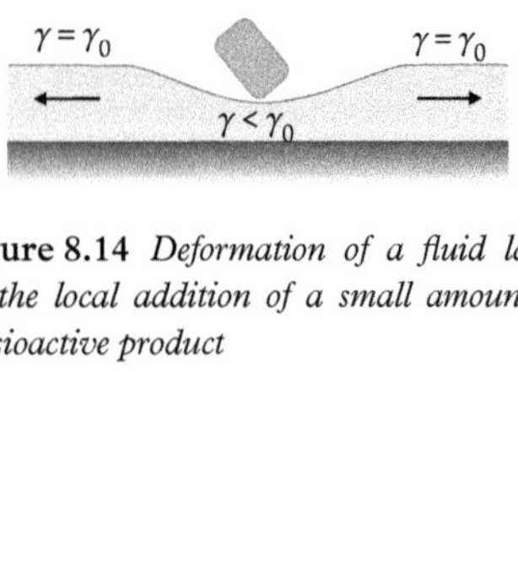

Figure 8.14 *Deformation of a fluid layer by the local addition of a small amount of tensioactive product*

Such motion of the fluid can also be due to variations in the temperature T from one point to another of a liquid-air or liquid-liquid interface. As shown in Section 1.4.1, the surface tension coefficient depends on temperature according to an equation which, for moderate changes in temperature, takes on the linear form:

$$\gamma(T) = \gamma(T_0)\,(1 - b\,(T - T_0)). \tag{8.64}$$

A temperature gradient parallel to the surface of a liquid causes a tangential stress on it (Figure 8.15 below). Along a strip of width δx, the surface tension forces are no longer in balance. The resultant force is directed toward the regions of lower temperature. Corresponding to a temperature gradient $\mathrm{d}T/\mathrm{d}x$, there is a surface-tension gradient equal to:

$$\frac{\mathrm{d}\gamma}{\mathrm{d}x} = \frac{\mathrm{d}\gamma}{\mathrm{d}T}\frac{\mathrm{d}T}{\mathrm{d}x} = -b\,\gamma(T_0)\left(\frac{\mathrm{d}T}{\mathrm{d}x}\right). \tag{8.65}$$

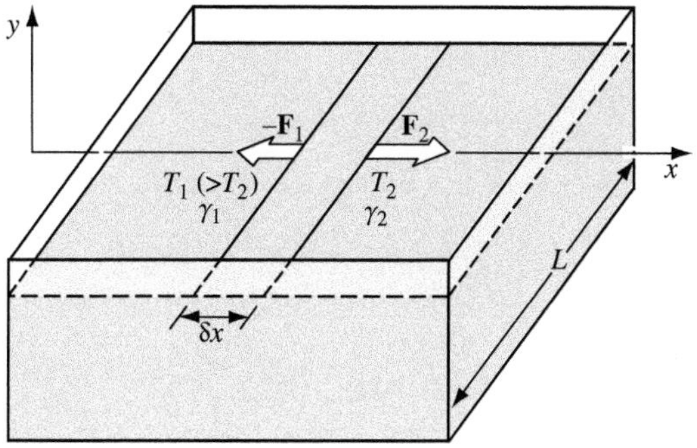

Figure 8.15 *The appearance of tensile stresses at the free surface of a liquid as a result of a horizontal temperature gradient*

This gradient causes a stress $\sigma_{xy}^{(\gamma)}$, in the x-direction, on the surface element $L\,\delta x$, where:

$$\sigma_{xy}^{(\gamma)} = \frac{F_2 - F_1}{L\,\delta x} = \frac{(\gamma_2 - \gamma_1)L}{L\,\delta x} = \frac{\mathrm{d}\gamma}{\mathrm{d}x} = -b\,\gamma(T_0)\left(\frac{\mathrm{d}T}{\mathrm{d}x}\right). \tag{8.66}$$

The negative sign appearing in $\sigma_{xy}^{(\gamma)}$ indicates the fact that the resultant tension (and the flow that results from it) acts in the direction of lower temperature.

Flow created in a horizontal liquid layer by a temperature gradient

The stress $\sigma_{xy}^{(\gamma)}$ on the interface due to the surface-tension gradient will generate a flow of velocity $v_x(y)$ which, in turn, will induce a viscous friction stress $\sigma_{xy}^{(\eta)} = -\eta\,(\partial v_x/\partial y)$ at the interface. For a free, plane, gas-liquid interface, the total tangential stress must vanish so that the stresses $\sigma_{xy}^{(\gamma)}$ and $\sigma_{xy}^{(\eta)}$ must balance each other. We then have:

$$\sigma_{xy}^{(\gamma)} + \sigma_{xy}^{(\eta)} = -b\,\gamma\,(T_0)\,\frac{\mathrm{d}T}{\mathrm{d}x} - \eta\left(\frac{\partial v_x}{\partial y}\right)_{\text{interface}} = 0. \tag{8.67}$$

Let us now calculate the resultant flow-profile in a fluid film bounded from below by a solid horizontal plane $y = 0$, of average thickness a, and extending infinitely in the z-direction (Figure 8.16). We assume, as in the previous sections, that the flow is one-dimensional, the only non-zero component being $v_x(y)$.

The pressure gradient in the y-direction satisfies: $(\partial p/\partial y) = -\rho g$. Let us assume that, at the onset, the upper surface is perfectly horizontal, that the density of the fluid is not

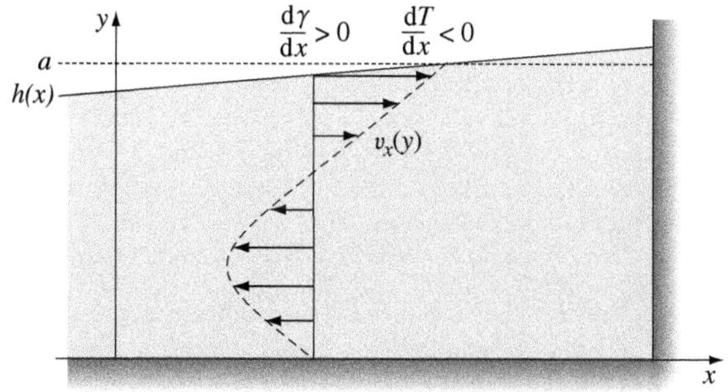

Figure 8.16 *Velocity profile within a finite-length fluid film, in the presence of a horizontal temperature gradient, which causes a recirculating flow due to the Marangoni effect. The pressure gradient resulting from the parabolic velocity profile causes a deformation of the free upper surface*

a function of temperature, and that this density is uniform throughout the depth h. The pressure satisfies everywhere the equation:

$$p = p_{atm} + \rho g (a - y). \tag{8.68}$$

The pressure is then independent of x and the x-component of the equation of motion reduces to:

$$\eta \frac{\partial^2 v_x}{\partial y^2} = 0. \tag{8.69}$$

The velocity v_x then varies linearly with y, just as for simple shear flow. Making use of Equations 8.67, we find:

$$v_x(y) = -\frac{b\ \gamma(T_0)}{\eta} \left(\frac{\mathrm{d}T}{\mathrm{d}x}\right) y. \tag{8.70}$$

In a real situation, the film is of finite length in the x-direction. As a result, the fluid piles upon the side towards which the flow is directed leading to a gradient in the thickness $h(x)$ of the film. Let $\mathrm{d}h/\mathrm{d}x$ be the slope of the free surface ($dh/dx \ll 1$); the flow remains quasi one-dimensional and the vertical pressure gradient is still equal to $-\rho g$. The only effect of the slope in the surface is thus to induce a horizontal pressure gradient $\rho g(\mathrm{d}h/\mathrm{d}x)$. Under stationary conditions, the latter creates a subsurface counterflow which exactly compensates the shear flow near the surface, to give a zero net-flow rate. The profile of the interface then satisfies the condition:

$$h^2(x) - h^2(x_0) = -\frac{3b\,\gamma(T_0)}{\rho g} \left(\frac{\mathrm{d}T}{\mathrm{d}x}\right)(x - x_0). \tag{8.71}$$

The phenomenon we have just described can be easily observed by bringing the tip of a hot soldering iron near a water surface; a dip can be seen in the surface just below the tip.

Proof
The equation of motion corresponding to the overall stationary state can be written:

$$\eta \frac{\partial^2 v_x}{\partial y^2} = \rho g \frac{\mathrm{d}h}{\mathrm{d}x}.$$

Integrating, with the conditions:

$$\int_0^{h(x)} v_x(y)\ \mathrm{d}y = 0 \quad \text{and:} \quad v_x(0) = 0, \qquad \text{we obtain:} \quad v_x = \frac{\rho g}{\eta} \frac{\mathrm{d}h}{\mathrm{d}x} \left(\frac{y^2}{2} - \frac{y\,h}{3}\right). \tag{8.72}$$

This flow is therefore a superposition of a shear flow and a Poiseuille flow. We can find the value of $\mathrm{d}h/\mathrm{d}x$ by using again the condition of Equations 8.67 for the surface stress, thus:

$$h\frac{\mathrm{d}h}{\mathrm{d}x} = -\frac{3}{2} \frac{b\,\gamma(T_0)}{\rho g} \left(\frac{\mathrm{d}T}{\mathrm{d}x}\right). \tag{8.73}$$

Equation 8.71 results from this, by integration.

Surface-tension gradients resulting from temperature differences are the basis of numerous hydrodynamic instabilities. Among the best-known of these is the *Bénard-Marangoni instability* of a horizontal fluid layer with a free surface, heated from below. We will discuss in more detail, in Section 11.3.1, this phenomenon which causes the appearance of a hexagonal lattice of convection cells. We also observe the rise of a liquid film when we dip a vertical plate, heated at the top, into the liquid. Finally, liquid drops begin to move if they are placed on a plate or on a wire where there exists a temperature gradient.

In industrial applications, flow resulting from surface gradients of thermal origin can have significant practical importance. For example, this occurs in the case of very pure single crystals produced by cooling in the presence of temperature-gradients: defects can appear because of this flow. The motion of bubbles resulting from such gradients can also greatly influence thermal transfer for boiling occurring along a heated wall.

Marangoni effect resulting from changes in the chemical composition

We can cause changes in the surface tension of a fluid by adding tensioactive compounds: one example is that of the amphiphilic molecules that we mentioned in Section 1.4.1, which can significantly reduce the surface tension when they are present along an interface. A concentration gradient of such molecules along a gas-liquid or liquid-liquid interface leads to a surface tension gradient which will result in Marangoni flow.

A spectacular example of this is the phenomenon of "*wine legs*" or "*tears of wine*" which can be seen in a glass partly filled with wine having a sufficiently high alcohol content. Swirling the glass creates a liquid film along the walls, above the surface of the wine. We see the liquid film rise, first creating a bulge near the top, followed by the appearance of droplets which fall regularly downward after a certain amount of time.

This phenomenon is the result of the decrease in surface tension of a water-alcohol solution (the wine!), as the concentration of alcohol increases. The evaporation of the alcohol from the liquid film leads to an increase in its surface tension, and causes the rise of the fluid film above the free surface where the surface tension is unchanged. In the bulge, the evaporation diminishes and the gradients in the surface tension can no longer balance the effects of gravity on the droplets, which fall in *"tears of wine"*, also called *"wine legs"* (Figure 8.17).

We have seen at the beginning of this section the example of putting a small amount of soap on a thin, horizontal layer of water: the water *"dessicates"* because the surface tension is highly decreased at this point, and all of the liquid is pulled toward other regions. This phenomenon is found to have practical applications in the drying of certain fragile and very expensive objects, such as the disks (*wafers*) of silicon used in microelectronic applications (in this last case, the process involves blowing alcohol vapor or a similar compound).

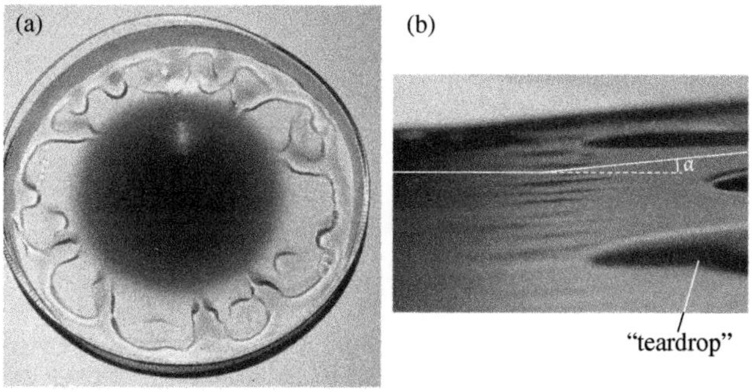

Figure 8.17C *(a) Top view of the rise of a liquid along the walls of a glass due to the Marangoni effect, and of the fall of the liquid with a lower alcohol concentration in the form of "teardrops" or "wine legs." (b) Close-up side view of a model experiment which displays this phenomenon. One pulls a plate inclined at an angle α relative to the horizontal from the water-alcohol mixture: this creates on the plate a fluid film which drops back as legs or teardrops. The narrower transverse ridges correspond to a different flow instability (documents courtesy of J. Bush and P. Hosoi, MIT)*

8.3 Falling liquid cylindrical jet

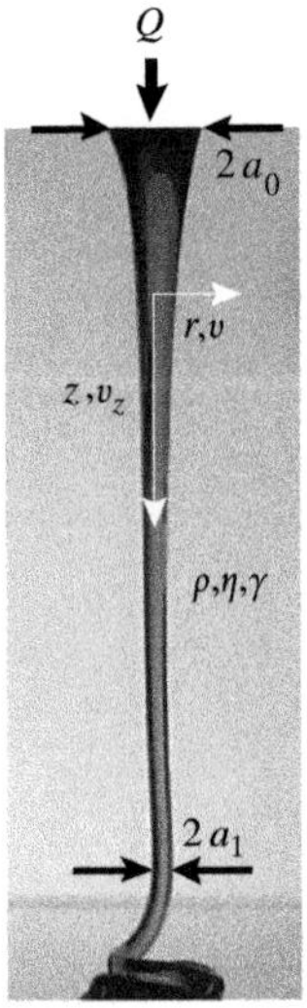

Figure 8.18 *Jet of viscous fluid falling from an orifice. The jet falls onto a plane at the bottom of the figure, which leads to the observed coiling; this coiling is not discussed here, and we will interest ourselves only in the straight portion of the jet (image courtesy of N. Ribe)*

In every example discussed so far in this chapter, the flow occurred in the presence of one or more walls: in the stationary flow regime, forces due to gravity or to the pressure gradients could then be balanced by the viscous stresses on the wall(s).

This is no longer the case when we have a free jet flowing from an opening, as shown in Figure 8.18. The viscous shear stress σ'_{zr} at the free surface of the jet ($r = a(z)$) can, in this case, be considered to be negligible: this leads to the conclusion that the radial gradient of the velocity component v_z also vanishes near the surface (Section 4.3.2, Equation 4.37). The absence of this viscous shear stress component σ'_{zr} which might otherwise cause radial changes in the velocity of the jet leads us to assume that v_z is independent of r.

Even if the cross-section of the jet changes slowly with distance, and the flow is quasi-parallel, the assumptions of the lubrication approximation must be entirely revised: this approximation was indeed based on the existence of equilibrium between the shear stresses and either the pressure gradients or gravity.

Other viscous terms which we have so far neglected relative to σ'_{xr} need to be taken into account: these are, specifically, the diagonal components of the viscous shear tensor which oppose the stretching of the jet as the fluid velocity changes in the x-direction. In the present case, the components which have to be taken into account include: $\sigma'_{zz} = 2\,\eta\,\partial v_z/\partial z$ and $\sigma'_{rr} = 2\,\eta\,\partial v_r/\partial r$ (Chapter 4, appendix 4A.2).

8.3.1 Stable flow regime

Equations of motion

In the one-dimensional model discussed here, the change in the cross-section $A(z) = \pi\,a^2(z)$ and the velocity $v_z(z)$ of the jet satisfy the following equation of motion proven below:

$$\rho g A(z) = \rho A(z)\left(v_z(z)\frac{\partial v_z(z)}{\partial z}\right) + \pi\gamma\,\frac{\partial a(z)}{\partial z} - 3\eta\frac{\partial}{\partial z}\left[A(z)\frac{\partial v_z(z)}{\partial z}\right]. \qquad (8.74)$$

This equation of motion expresses the fact that the force of gravity is balanced by a combination of three terms:

- the inertial term, indicating the acceleration of the fluid along its trajectory;
- the change in the z-direction of the capillary pressure;
- viscous stresses related to the lengthening of the jet in the z-direction, to the decrease of its cross-section and to the resulting change in velocity.

Proof

Let us integrate with respect to r the condition of incompressibility $\nabla\cdot\mathbf{v} = 0$ in cylindrical coordinates (Chapter 4, appendix 4A.2), taking into account the finite velocity v_r at $r = 0$ and the fact that v_z, and thus its gradient $\partial v_z/\partial z$, are not functions of r. We then obtain:

$$v_r = -\frac{r}{2}\frac{\partial v_z}{\partial z}, \qquad (8.75)$$

from which we infer:

$$\sigma'_{rr} = 2\eta\frac{\partial v_r}{\partial r} = -\,\eta\,\frac{\partial v_z}{\partial z} = -\frac{\sigma'_{zz}}{2}. \qquad (8.76)$$

In order to ensure that the jet is in mechanical equilibrium in the transverse direction, we must have $p - \sigma'_{rr}$ independent of r. Since, from Equations 8.76, σ'_{rr} must be independent of

r, the same needs to be true regarding the pressure. On the other hand, if p is the pressure inside the fluid at the level of the interface, we must also have:

$$p - \sigma'_{rr} = p_{atm} + \frac{\gamma}{a(z)}, \qquad \text{i.e.:} \qquad p = p_{atm} + \frac{\gamma}{a(z)} - \eta \frac{\partial v_z}{\partial z}. \tag{8.77}$$

We assume, moreover, that the flow is stationary; the rate of flow Q of the fluid in the jet is:

$$Q = \pi\, a^2(z)\, v_z(z) = A(z)\, v_z(z) \tag{8.78}$$

must be constant as a function of z and of t in order to ensure the conservation of mass ($A(z)$ is the cross-section of the jet at the distance z). Let us now write Equation 5.10 for the conservation of momentum in a slice of the jet located between the cross-sections z and $z + \delta z$:

$$\rho\, g A\, \delta z = \left[\rho\, Q\, v_z(z) + (p(z) - p_{atm})A(z) - \sigma'_{zz}\, A(z)\right]_z^{z+\delta z}. \tag{8.79}$$

Rewriting this equation in differential form, and using Equations 8.75, 8.77 and 8.78, we obtain Equation 8.74.

We assume, first of all, that the change in capillary pressure has a negligible effect, and we will look for the laws involving changes in the two limiting cases, where either the inertial term or the viscous term dominate.

Inertial regime

Equation 8.74 reduces to:

$$g = v_z \frac{\partial v_z}{\partial z},$$

so that, by integration:

$$v_z^2(z) - v_{z0}^2 = 2\, g z. \tag{8.80}$$

We recognize that this is the change in velocity with distance of an object in free fall: this is a logical result since the only frictional forces are of viscous origin, and, here, we assume they are negligible. When $v_z(z) \gg v_{z0}$ the velocity thus increases as the square root of the distance ($v_z(z) = \sqrt{2\, g\, z}$).

Viscous regime

Equation 8.74 becomes:

$$\rho\, g A(z) = -3\, \eta\, \frac{\partial}{\partial z}\left[A(z) \frac{\partial v_z(z)}{\partial z}\right]. \tag{8.81}$$

From this, we infer the following variations of the flow velocity and the cross-section, as functions of z:

$$v_z(z) = \frac{g}{6\nu}\,(z - z_i)^2 \quad \text{(8.82a)} \qquad \text{and} \qquad A(z) = \frac{6\nu\, Q}{g}(z - z_i)^{-2}. \quad \text{(8.82b)}$$

We can then determine z_i (< 0) from the value v_{z0} of the flow velocity at $z = 0$. For $v_z \gg v_{z0}$, the velocity increases as the square of the distance ($v_z = g\, z^2/6\nu$).

For an injection orifice of diameter d, this self-similar power-law solution is only valid far downstream of the orifice, when the cross-section $A(z)$ of the jet is small compared to its initial value $\pi d^2/4$. Combining this condition with Equation 8.82a, we obtain: $z - z_i \gg \sqrt{(24/\pi)\,(\nu\, Q/(d^2 g))}$.

Proof

Gravity is compensated, in the present case, by the viscous stresses resulting from the stretching of the jet. Let us assume that, as in the previous situation, v_z has a power-law dependence on z or, more precisely, on $(z - z_i)$, where z_i is an arbitrary origin so that $v_z = C_\eta\,(z - z_i)^\alpha$: this implies that $A = (Q/C_\eta)\,(z - z_i)^{-\alpha}$. Inserting this into Equations 8.81, we obtain $(z - z_i)^{2-\alpha} = 3\,\nu\,\eta\,C_\eta/g$, which requires $\alpha = 2$ and $C_\eta = g/6\nu$. This leads to Equations 8.82a and 8.82b.

Transition from the viscous to the inertial regime

The viscous regime is found in the start-up phase where the viscous forces maintain a low flow velocity. Subsequently, the inertial effects take over.

Let us evaluate, in the viscous regime and where $v_z \gg v_{z0}$ $(z \gg z_i)$, the change with distance in the inertial and viscous terms in Equation 8.74, by means of Equations 8.82a and 8.82b. We find:

$$\rho\, A\left(v_z \frac{\partial v_z}{\partial z}\right) \simeq \frac{\rho\, Q g z}{3\,\nu} \quad (8.83a) \qquad \text{and:} \qquad -3\,\eta\, \frac{\partial}{\partial z}\left[A \frac{\partial v_z}{\partial z}\right] \simeq \frac{6\,\eta\, Q}{z^2}. \quad (8.83b)$$

The ratio of the inertial to the viscous term is thus of the order of $g\, z^3/(18\,\nu^2)$: the inertial term increases with the distance while the viscous term decreases (because of the decrease of $A(x)$). The distance z_c corresponding to the transition (ratio of the order of unity) is then:

We find Equation 8.84b by writing Equation 8.82a for $z = z_c$, assuming $v_z \gg v_{z0}$, and then multiplying both sides by z_c^2.

$$z_c \cong \left(\frac{18\,\nu^2}{g}\right)^{1/3} \quad (8.84a) \qquad \text{from which we infer:} \qquad v_z(z_c)\, z_c \cong 3\,\nu. \quad (8.84b)$$

In order for the viscous regime to extend throughout the length of the jet, we must have $z_c \geq L$: in the limit $z_c = L$, we then have: $v_L\, L \cong 3\,\nu$. The Reynolds number $Re_L = v_L\, L/\nu$, based on the length of the jet must be at most of the order of unity.

This condition is much stricter than that for the lubrication approximation, for flow in the presence of a wall, which here would be $v_L a_0/\nu < 1/\theta \approx L/a_0$; this is equivalent to $Re_L < (L/a_0)^2$ where $(L/a_0)^2 \gg 1$. This result reflects the much smaller values of the viscous terms in the absence of a wall: these terms then result uniquely from the stretching of the jet, which leads to longitudinal and transverse flow-velocity gradients which are much smaller than for a shear flow in the presence of a wall.

Even for a slow change of radius as a function of the distance (quasi-parallel flow), the viscous terms will only be dominant, in a real situation, if we use very viscous fluids in order, from Equation 8.84a, to satisfy the condition $z_c \geq L$.

In the case of water ($\nu \approx 10^{-6}$ m^2/s), one finds $z_c \approx 0.12$ mm which makes any experimental observation extremely difficult. In order to reach a value of the order of 0.1 m we need a viscosity around 2.5×10^{-2} m^2/s, i.e., about 25,000 times the viscosity of water.

8.3.2 Capillary effects and Rayleigh-Plateau instability of the jet

Until now, we have neglected surface-tension effects in the flow of vertical jets. In fact, for very viscous fluids, the flow described just above remains stable. On the other hand, for liquids with low viscosity, such as water, the effect of surface tension can lead to an instability and to the formation of drops; we can easily observe this by looking at the evolution, as a function of distance, of a jet of water, originally cylindrical (Figures 8.19a,b).

In order to explain this instability, known as the *Rayleigh-Plateau instability*, let us calculate the changes in the capillary pressure resulting from a deformation of the external surface of an originally cylindrical jet of liquid. When no deformation is present, the capillary pressure difference p_{cap} between the inside of the jet and the (constant) atmospheric pressure p_{at} outside satisfies $p^0_{cap} = \gamma/R_0$. Let us assume now that the jet remains axially symmetric, and that its local radius, $R(z)$ is modulated sinusoidally in the z-direction of its axis (Figure 8.19b):

$$R(z, t) = R_0 + h(t) \cos kz. \tag{8.85}$$

Here, we assume that the velocity profile in the jet is uniform and we study the deformations in a reference frame moving at the corresponding (gravity is assumed to have no influence on the instability). We also assume that $h(t) \ll R_0$ and that the slope of the interface remains small ($dR/dz \ll 1$). The variation $\Delta p_{cap}(z, t) = p_{cap}(z, t) - p^0_{cap}$ of the capillary pressure difference due to the deformation is:

$$\Delta p_{cap}(z, t) = \gamma\, h(t) \cos kz \left(k^2 - \frac{1}{R_0^2}\right). \tag{8.86}$$

Justification

The pressure difference $p_{cap}(z)$ is determined from the Young-Laplace law (Equation 1.58), in terms of the sum of the contributions of the two radii of curvature R_1 and R_2 of the jet surface respectively in planes containing and perpendicular to the axis. We then obtain:

$$p_{cap}(z, t) = \gamma \left(\frac{1}{R_1} + \frac{1}{R_2}\right) = \gamma \left(-\frac{\partial^2 R(z, t)}{\partial z^2} + \frac{1}{R(z, t)}\right)$$
$$= \gamma \left(k^2 h(t) \cos kz + \frac{1}{R_0 + h(t) \cos kz}\right), \tag{8.87}$$

Expanding the last term on the right-hand side in terms of h/R_0 and substracting p^0_{cap}, we obtain Equation 8.86.

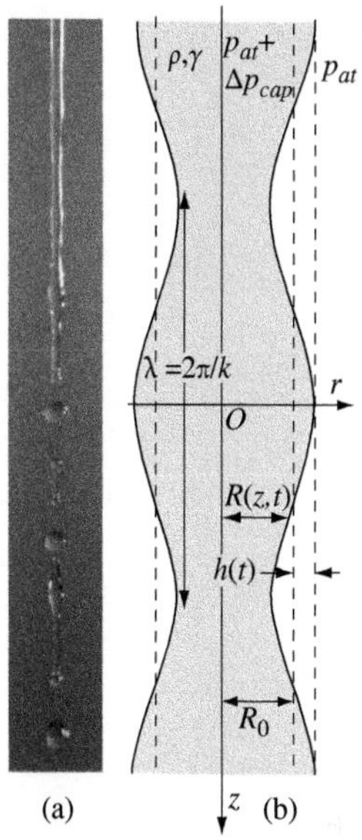

Figure 8.19C *Destabilization, by the Rayleigh-Plateau mechanism, of a vertical cylindrical liquid jet. a) experimental observation of the development of the instability, followed by the formation of drops (document courtesy of J. Aristoff and J. Bush, MIT); b) Schematic diagram for the development of the instability*

When $k^2 - 1/R_0^2 < 0$, Δp_{cap} becomes negative in the regions where $R(x) > R_0$ ($\cos kz > 0$) and, in contrast, positive for $R(z) < R_0$ ($\cos kz < 0$). Since the pressure inside the non-deformed jet is constant (and equal to $p_{at} + \gamma/R_0$), the pressure in the regions where the radius of the jet increases due to the deformation will be larger than in those where it decreases. The flow resulting from these pressure differences thus reinforces the instability. The absolute amplitude of these pressure changes is greatest when $|k| \ll 1/R_0$ and decreases when k approaches $1/R_0$.

This larger value of the changes in capillary pressure as a function of z for long-wavelength deformations might seem paradoxical: capillary effects generally increase indeed with the curvature of the interfaces. But, here, the contribution to the Young-Laplace law of the curvature in planes containing the axis (k^2 term in Equations 8.86) stabilizes the interface; it is the curvature in the planes normal to the axis which is destabilizing (term varying as $-1/R_0^2$). It is thus logical that the destabilising capillary pressure is highest when the first term almost vanishes, for $k \to 0$.

On the other hand, the change as a function of k of the rate of increase σ does not depend only on Δp_{cap} but also on the fact that the time for transfering matter between the regions of smaller and larger radii acts over a distance of the order of a half wavelength: the decrease in this distance as k increases must then, to the contrary, lead to an increase in σ. The approximate estimate of σ (carried out below) for an inertial flow:

$$\sigma^2 \propto k^2 R_0^2 \left(1 - k^2 R_0^2\right) \tag{8.88}$$

combines these two effects. This rate of increase is only positive if $k < 1/R_0$ (unstable case), and has a maximum when $k = 1/(R_0\sqrt{2}) \approx 0.7/R_0$: this value is quite close to that experimentally observed, and corresponds to a wavelength of 1.4 times the perimeter of the jet.

This last result is applicable to numerous instabilities which display a range of unstable wavelengths: the dominant instability is that of the wavelength for which the rate of growth is a maximum. Moreover, the simple model which we discuss here takes no account of the effects of the viscosity or of the average flow rate.

For highly viscous fluids, the Rayleigh-Plateau instability can still be observed, because the changes in the capillary pressure related to the deformations of the interface (Equation 8.86) remain the same. In contrast, Equation 8.88 is no longer valid because the development of the instability then results from the equilibrium between the capillary and the viscous forces.

Justification of Equation 8.88

In the cases for which the instability is best visible (fluids with small viscosity), we are generally in an inertial regime for which the effect of viscosity can be neglected: the acceleration $\partial v_z/\partial t$ of the fluid is then proportional to the pressure gradient in the z-direction. Assuming that this pressure gradient is uniform over the cross-section and equal to the capillary-pressure gradient, and that the velocity is also uniform, we have: $\rho\, \partial v_z/\partial t \approx -\partial \Delta p_{cap}/\partial z$. Using Equation 8.86 to estimate Δp_{cap}, the corresponding flow rate $Q(z,t)$ satisfies:

$$\frac{\partial Q(z,t)}{\partial t} \approx -\frac{\pi R_0^2}{\rho}\frac{\partial \Delta p_{cap}(z)}{\partial z} = \frac{\pi\gamma\, h(t)\, R_0^2}{\rho}\sin kz\; k\left(k^2 - \frac{1}{R_0^2}\right). \tag{8.89}$$

The conservation of the flow rate of fluid can be written locally for a given z in the form:

$$2\pi\, R_0 \cos kz\, \frac{\mathrm{d}h(t)}{\mathrm{d}t} = -\frac{\partial Q}{\partial z}, \tag{8.90}$$

where the left-hand side represents the change with time of the cross-section of the jet (again when $h \ll R_0$). Taking the derivatives of Equations 8.89 and 8.90 respectively with respect to z and to t and setting the results equal, we obtain after dividing both sides by $2\pi\, R_0 h(t) \cos kz$:

$$\frac{1}{h(t)}\frac{d^2h(t)}{dt^2} \approx \frac{\gamma}{2\rho R_0^3}\, k^2R_0^2\left(1-k^2R_0^2\right) \tag{8.91a}$$

or:

$$\sigma^2 \approx \frac{\gamma}{2\rho R_0^3}\, k^2 R_0^2\,(1-k^2R_0^2), \tag{8.91b}$$

assuming that $h(t)$ increases with time as $\exp(\sigma t)$. The exponential growth coefficient σ then satisfies Equations 8.88. If we take into account the pressure and radial velocity gradients, we obtain a similar result, where the factor $k^2R_0^2/2$ is replaced by a function $f(kR_0)$ whose maximum corresponds to $k = 0.697/R_0$, remarkably close to the value obtained from the simple approximation above.

The Rayleigh-Plateau instability can also be observed in the case of some, very soft, solids, such as gels. All solids are characterized by a surface energy: in contrast to liquids, the corresponding forces are, however, generally negligible in comparison with the elastic forces. Their ratio can be characterized by the length $h = \gamma/E$, where γ is the surface energy and E is Young's modulus (the ratio of the stress F/S to the resulting relative deformation $\Delta L/L$). For iron, for example, this length is 3×10^{-13} m and the surface energy has no significant effect on deformations on this scale. This is no longer the case for gels for which Young's modulus has a value of a few Pa (instead of $\approx 2\times10^{11}$ Pa for iron): the length h is then a few mm. Surface-tension forces are then sufficently great to result in a Rayleigh-Plateau instability (but, in contrast, the elastic forces can prevent the breaking up into droplets).

EXERCISES

In the following exercises, the lubrication approximation (see Section 8.1.2) is assumed to be valid and the thickness of the liquid films is assumed to vary slowly with distance parallel to the flow.

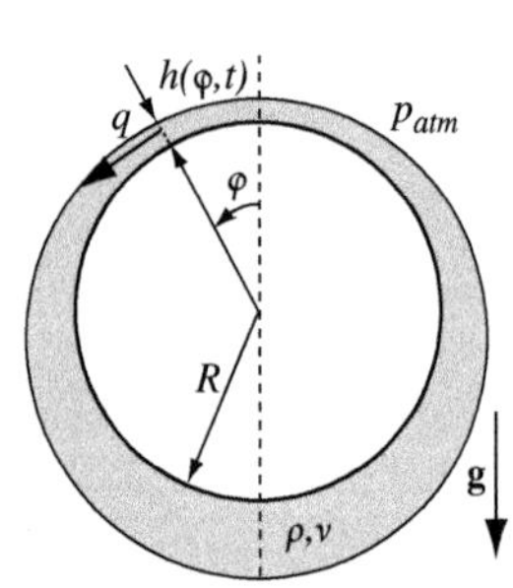

1) **Flow of a fluid layer around a horizontal cylinder**

A horizontal circular cylinder of radius R and axis along z (see in the margin), is coated externally by a layer of a Newtonian fluid of initial thickness $h_0 \ll R$ constant with respect to the coordinates z and φ (using cylindrical coordinates r,φ,z). This represents, for instance, the painting of a tube by means of a brush: we are interested in the subsequent flow of the paint leading to azimuthal variations of its thickness. Show that the influence of the pressure gradients on the flow is negligible compared to that of gravity. Using the procedure of Section 8.2.1, compute the flow rate $q(\varphi,t)$ per unit length along z through a cross-section φ = constant as a function of $h(\varphi,t)$ and the viscosity ν. Both h and q are assumed to remain independent of z. Show that the equation governing the variation of the thickness $h(\varphi, t)$ is: $\partial h/\partial t = -gh^2/(3\nu R)\,(3\partial h/\partial\varphi \sin\varphi + h\cos\varphi)$. Neglecting surface tension effects, compute, for a constant viscosity ν, the variations $h(\varphi, t)/h_0$ at the top and the bottom of the tube ($\varphi = 0$ and $\varphi = \pi$), using the characteristic flow time constant $\tau_f = 3\,\nu\,R/\left(2\,g h_0^2\right)$. Compute then $h(\varphi,t)/h_0$ for $\varphi = 0$ and $\varphi = \pi$ when ν increases exponentially with time as $\nu = \nu_0 \exp(t/\tau_d)$ (in order to simulate the drying of the paint). What is the thickness variation at times short or long compared to τ_d depending on the relative values of τ_d and τ_f?

2) **Spin coating**

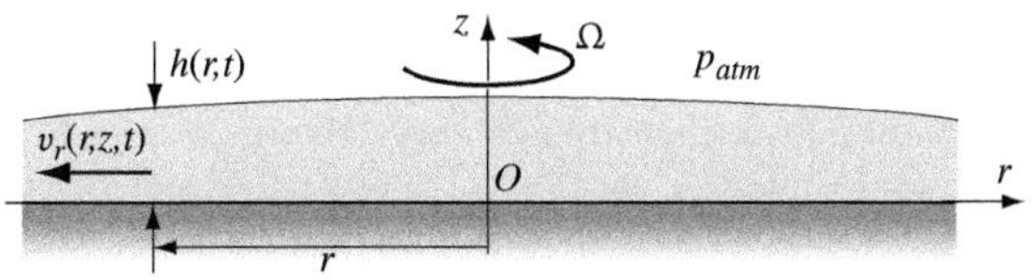

In the "spin coating technique", in order to obtain a thin layer of constant thickness of viscous fluid an initially thicker layer is placed over a horizontal disk rotating at a constant angular velocity Ω around the z-axis. Neglecting surface tension and gravity, show that the flow rate $q(r)$ through a cylinder of radius r and axis Oz, higher than the maximum of the thickness $h(r, t)$ of the layer, is $q(r) = 2\pi\left(\Omega^2 r^2 h^3(r,t)/3\nu\right)$. Write the relation between $\partial h/\partial t$ and q and show that, if $h(r, t)$ is constant with r at a time t, it remains so thereafter. What is then the variation $h(t)$ for an initial thickness h_0? What is the thickness after 10 s and 100 s for $\Omega = 60\,\pi$ rad/s, $\nu = 10^{-5}$ m^2/s and $h_0 = 0.5$ mm?

3) **Rough sphere dropping away from a plane**

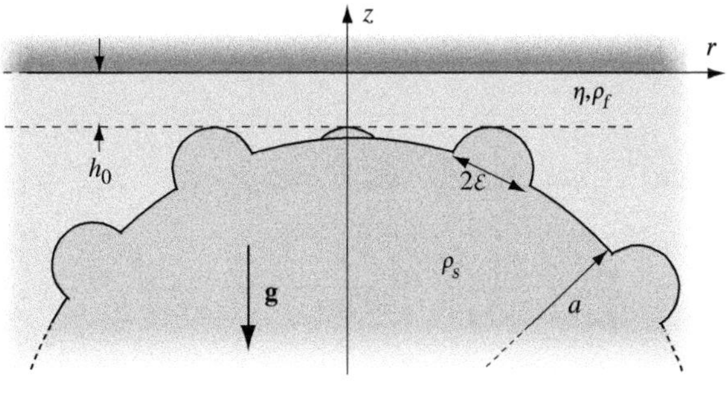

We generalize the analysis of Section 8.1.6 to a rough sphere of radius a. The roughness is represented by a distribution of half spheres of radius $\varepsilon \ll a$. The space between the sphere and the plane is filled by a Newtonian liquid. In contrast to the case of Figure 8.4, the rough sphere is initially below the plane and in contact with it (we assume that three of the small half spheres are at an atomic distance $h_0(0) = h_{0m} \ll \varepsilon$ from the plane); the sphere moves away under its weight thereafter. Due to the reversibility of low Reynolds number flows, the flow velocity field is the opposite, for a same distance from the plane, of that for a sphere falling towards it. In the early phase of the motion during which $h_0 \ll \varepsilon$, how should equation 8.33 be modified? What becomes of the equation when $\varepsilon \ll h_0 \ll a$? At larger distances $h_0 > a$, equations 8.33 must be replaced by the empirical expression $F = -6\pi\,\eta\, a\left(1 + \frac{a}{h_0}\right) \mathrm{d}h_0/\mathrm{d}t$. Show that it is valid for a sphere with smooth walls both for $h_0 \ll a$ and $h_0 \gg a$. Show that the equation of motion of the sphere is then $\frac{4\pi}{3}\, a^3 (\rho_s - \rho_f) g = 6\pi\,\eta\, a\left(1 + \frac{a}{h_0(t)+\varepsilon} + \frac{3\,\varepsilon^2}{a\, h_0}\right) \frac{\mathrm{d}h_0}{\mathrm{d}t}$ and that the distance $h_0(t)$ satisfies the implicit relation:

$$\frac{2a^2}{9\eta}(\rho_s - \rho_f)t = h_0(t) - h_{0m} + a\ \mathrm{Log}\frac{h_0(t) + \varepsilon}{h_{0m} + \varepsilon} + \frac{3\varepsilon^2}{a}\mathrm{Log}\frac{h_0(t)}{h_{0m}}.$$

We want to use this equation in order to determine the relative roughness ε/a of the sphere by measuring the times t_a and t_{2a} for its move away by distances a and $2a$ from the plane (J.R. Smart and D.T. Leighton, 1989). Show that ε/a can be written as a function of solely t_a and t_{2a}.

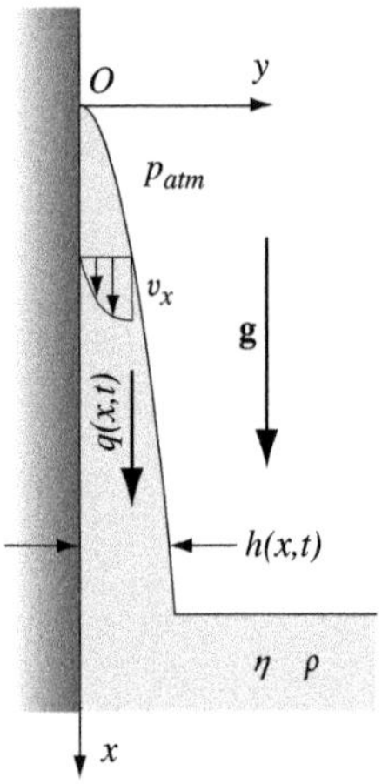

4) **Liquid film draining under gravity along a vertical plate**
A vertical plate ($y = 0$) is dipped into a viscous fluid and then quickly pulled out: it carries with it a layer of fluid of thickness $h(x, t)$ ($x = 0$ at the top of the layer) which drains down due to gravity (surface tension is neglected). Compute the flow rate q per unit length along z through a section x = constant (see Section 8.2.1). Show that, because of the conservation of mass, $h(x, t)$ satisfies the relation $\partial h/\partial t + (g/\nu)\, h^2 (\partial h/\partial x) = 0$. We assume that, after a transitory phase, the variation of h corresponds to a self-similar solution: $h = f(x/t^\alpha)$. Determine α and the function f. What is the corresponding thickness of the film at a distance $x = 0.2$ m for $\nu = 10^{-3}$ m^2/s and $t = 1, 10, 100$ and 10^4 s?

5) **Rayleigh-Plateau instability on a cylindrical fiber**

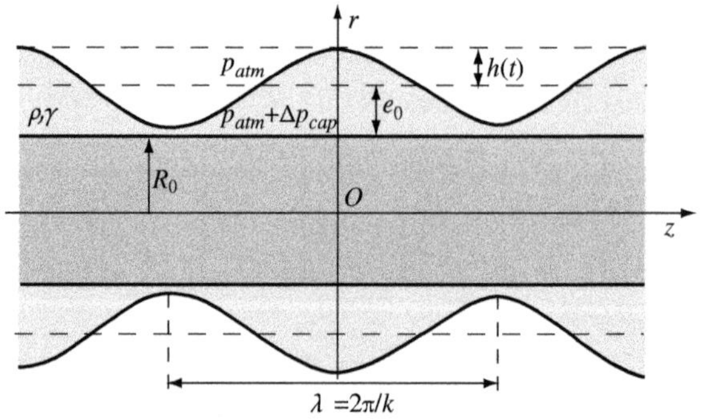

We examine the instability of a viscous film of initially constant thickness e_0 upon a horizontal cylindrical fiber of radius $R_0 \gg e_0$ and axis z. Surface tension and viscosity effects are taken into account while gravity is neglected. The flow along the film is supposed to be quasi-stationary and controlled by viscosity as in Equation 8.25 and not by inertia as in Section 8.3.2. Assuming a sinewave distorsion of the interface ($e(z) = e_0 + h(t) \cos k\, z$) and using a relation similar to Equation 8.86, compute the pressure $p(z)$ and show that the flow rate $Q(z)$ through a cross-section z = constant is $Q(z) = \left[2\pi\,\gamma\,k\,h\,e_0^3/(3\,\eta\,R_0)\right](k^2R_0^2 - 1)\sin kz$. Using this result, write the conservation of mass at a distance z and show that the governing equation for $h(t)$ is $1/\tau(k) = (1/h)\,(dh/dt) = \left[(k^2\,e_0^3\,\gamma)/(3\,\eta\,R_0^2)\right]\,(1 - k^2R_0^2)$. For what value of k does $\tau(k)$ reach its minimum value τ_{min}? Using the results of Exercise 1, what is the condition on τ_{min} which must be satisfied by τ_{min} in order for gravity to be negligible, as assumed above? Compute τ_{min} and the corresponding wavelength for $\eta = 10^{-1}$ Pa.s, $\gamma = 5\times10^{-2}$ N/m, $R_0 = 0.5$ mm and $e_0 = 20\ \mu$m.

6) **Liquid film rising due to the Marangoni effect (Section 8.2.4)**
A vertical plate ($x = 0$) is dipped into a liquid bath of horizontal free surface at $z = 0$. A vertical temperature gradient $\partial T/\partial z < 0$ is established on the plate at the origin time $t = 0$. It creates a vertical gradient $\partial\gamma/\partial z = -b\gamma\ (\partial T/\partial z)$ of the surface tension which causes a liquid film of thickness $h(z, t)$ to move up on the plate. Determine the velocity profile $v_z(x)$ as a function of h and show that the local flow rate q per unit distance along y is $q = \sigma^{(\gamma)}\,(h^2/2\eta) - \rho g\,(h^3/3\eta)$. What is the value of the stress $\sigma^{(\gamma)}$ on the surface? As the film rises, the point where the thickness h has a given value also moves. Show that its velocity V is equal to dq/dh and that it is constant with time and depends only on h (use the equation of conservation of mass to relate the variations of q and h). What is the geometrical transformation leading from the profile $h_1(z)$ at a time t_1 to $h_2(z)$ at the time t_2? Show that the variation $V(h)$ of the velocity with h displays a maximum. What are the corresponding thickness h_{max} and velocity $V(h_{max})$? Assuming that $h > h_{\text{max}}$ for $z = 0$, explain qualitatively why the profile must be cut off (h goes abruptly to zero) for a value of h which must be larger than h_{max}.

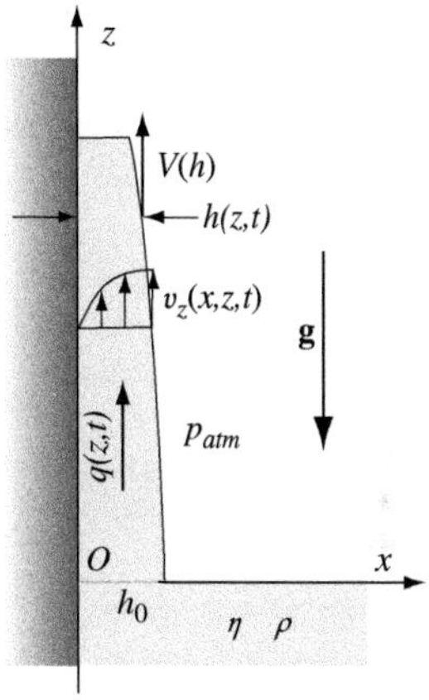

9 Flows at Low Reynolds Number

Flows at low Reynolds number are characterized by the dominance of effects due to the viscosity of the fluid, as opposed to those related to its inertia. The Stokes equation, which governs such flows, is linear because the non-linear convective term $(\mathbf{v}\cdot\nabla)\mathbf{v}$ can be neglected. For parallel flows, which we studied in Chapter 4, this convective term vanishes at all velocities. In Chapter 8, we considered quasi-parallel flows for which viscous terms continue to dominate, for geometrical reasons, up to fairly high Reynolds numbers, Re. In this chapter, we discuss flows of arbitrary geometry where the small value of the inertial contributions comes from the low value of Re.

After several examples of such flows (Section 9.1), we study, in Section 9.2, a certain number of general properties (reversibility, additivity, minimum dissipation) which result from the linear form of the Stokes equation, and which are at the origin of simple solutions; these properties completely distinguish these flows from those at high Reynolds numbers. Flows around small objects (or due to the motion of these objects in a fluid at rest) are an important group of applications which will be discussed in Section 9.3. Flow around a sphere (the Stokes problem) is an important example, although its determination is delicate (Section 9.4). In Section 9.5, we examine the corrections needed at large distances when the value of Re becomes of the order of one. Finally, we will consider the problem of the motion of a group of particles (suspensions, Section 9.6) and that of a fluid flowing through a set of fixed particles (porous medium, Section 9.7).

9.1 Flows at small Reynolds number

9.1.1 Physical meaning of the Reynolds number

We have introduced in Section 2.3.1 the Reynolds number by means of the equation:

$$Re = \frac{\rho UL}{\eta} = \frac{UL}{\nu}, \tag{9.1}$$

where U and L are the respective velocity and characteristic length of the flow and ρ, η and ν are respectively the density, and the dynamic and kinematic viscosities of the fluid. The Reynolds number can have several physical interpretations:

- **The ratio τ_ν/τ_c of the characteristic times** τ_ν for the transport of momentum by viscous diffusion over a distance L ($\tau_\nu = L^2/\nu$) and τ_c for the momentum transport by convection over the distance L ($\tau_c = L/U$).

Physical Hydrodynamics. Second Edition. Etienne Guyon *et al.*

- **The ratio of the inertial to the viscous terms** in the Navier–Stokes equation:

$$\rho \frac{\partial \mathbf{v}}{\partial t} + \rho\, (\mathbf{v} \cdot \nabla)\, \mathbf{v} = -\nabla p + \rho\, \mathbf{f} + \eta\, \nabla^2 \mathbf{v}. \qquad (9.2)$$

- **The ratio of the stresses** associated with the inertia of the fluid (ρU^2) and with the viscous friction ($\eta U/L$).

In this equation, if the various velocity components are of the same order of magnitude, and the length scales in the different directions as well (in contrast to the case of parallel or quasi-parallel flows), the inertial terms $\rho\ (\mathbf{v} \cdot \nabla)\mathbf{v}$ and the viscous terms $\eta \nabla^2 \mathbf{v}$ are respectively of the order of $\rho\ U^2/L$ and $\eta U/L^2$ so that their ratio is of the order of *Re*.

9.1.2 Examples of flows at low Reynolds number

Flows at low Reynolds number ($Re \ll 1$), sometimes known as *creeping flows,* are flows dominated by viscosity, where inertial effects are negligible. Such flows can have quite varied physical origins, since this number is a combination of three different factors:

Small size of the moving objects or of the flow channels

(The small value of *Re* is then directly related to that of L):

- ***The motion of bacteria*** (with lengths of a few microns). For bacteria with a length of 3 μm moving at a velocity of 10 μm/s in water ($\rho = 10^3$ Kg/m^3, $\eta \approx 10^{-3}$ Pa.s), we find $Re \approx 3 \times 10^{-5}$. (This problem will be discussed in Section 9.3.3).
- ***Flows in porous and fractured media*** (Section 9.7) for which the size of the flow channels is of the order of a few microns or less.
- ***Microfluidics*** The technologies of micro-machining and micro-electronics allow one to engrave interconnected networks of micro-channels with smaller and smaller openings (again of a few microns or less). This allows one to create devices for analyzing a very small amount of fluid or for preparing small quantities of pharmaceutical components with high added-value. Mixing and separation play a very important role in these devices and have specific characteristics at very low Reynolds numbers.

For high flow velocities and materials with large-size pores, the condition $Re \ll 1$ will frequently not be satisfied in practical situations.

Very viscous fluids and / or with low velocity

- ***Slow movements of the earth's mantle*** $Re \approx 10^{-20}$. At depths greater than 100 km, the earth's mantle can be considered as a Newtonian fluid of density $\rho \approx 2.1 \times 10^3$ Kg/m^3 and of viscosity $\eta \approx 10^{21}$ Pa.s. With $U \approx 0.05$ m/year (1.5×10^{-9} m/s) and a thickness of the mantle of the order of 2900 km, we get a value of *Re* of the order of 10^{-20}.
- ***Motion of glaciers*** $Re \approx 10^{-17} - 10^{-15}$ In Figure 9.1: on the next page the motion of the tongue of a glacier is marked by bands of rocks (*Forbes' bands*) which have been deformed as a result of the change in velocity between the edges and the center of the glacier. In spite of appearances, the profile of the bands is not parabolic (unlike Poiseuille flow in Section 4.5.3): ice cannot be considered a Newtonian fluid, and, moreover, the geometry of the flow is complex because the thickness of the layer of ice is often much smaller than its width.
- ***Flows of very viscous fluids*** Let us mention, for example, tars, pastes, plastics and honey. Some species of heavy oils have, at room temperature, a viscosity several million times that of water.

Let us assume that $L \approx 1000$ m and $U \approx 0.3$ m/year (10^{-8} m/s). The viscosity of the ice is quite variable within the range $10^{13} < \eta < 10^{15}$ Pa.s with $\rho = 10^3$ Kg/m^3. We then obtain the indicated values of *Re*.

Figure 9.1C *Illustration of Forbes' bands on the "Mer de Glace" glacier (the top of Mont Blanc is visible at the back). The shape of these bands as well as the downstream increase of the amplitude of their deformation indicate the transverse profile of the flow-velocity of the glacier (from left to right on the image), which is considerably smaller near the edges. The separation between the bands corresponds to the displacement over a whole year: they are formed downstream of "seracs" (fracture zones high up on the glacier) and indicate the different level of penetration of solids and dust particles into the ice, depending on the snow coverage (document, courtesy of J-F Hagenmuller /lumieresdaltitude.com)*

9.1.3 Some important characteristic

Stopping distance of a moving object

Let us characterize the effect of inertia by the *stopping distance* d_s of an object moving with an initial velocity U when we suddenly suppress the force which propels it (the opposite of the drag force F_T). At low Reynolds numbers, the stopping distance satisfies $d_s/L \approx (\rho_S/\rho)\, Re$ while, at high Reynolds numbers, we have $d_s/L \approx \rho_S/\rho$ (L is the size of the object, ρ_S its density and ρ and η are the density and viscosity of the fluid). The stopping distance is, therefore, very small if $Re \ll 1$. Thus, when a bacterium a few microns in size stops its propulsion, its velocity goes to zero in a time of the order of a microsecond, and it only travels 10^{-11} m (for the numerical values given in Section 9.1.2)! In contrast, a boat drifting on its own, after having cut off the engine, travels a distance of the order of its length.

Proof

At high Reynolds numbers, we can estimate F_T by the integral $\rho\, U^2 L^2$ of the flux of momentum ρU^2 over the cross-section $S = L^2$ of the object in motion. We will find the same order of magnitude for F_T in Chapter 12 for turbulent flows. The mass m of the moving object is $m = \rho_S L^3$ and its kinetic energy is thus $E_k \approx \rho_S\, U^2 L^3$. The stopping distance d_s is such that the work done by the force F_T zeroes out the kinetic energy E_k and thus satisfies $F_T\, d_s \approx E_k$. We then find: $d_s/L \approx \rho_S/\rho$.

At low Reynolds numbers, the drag force is purely viscous so that: $F_T \approx \eta UL$ (Section 9.2.4) and the stopping distance d_s still satisfies $F_T\, d_s \approx E_k$; we then find $d_s/L \approx (\rho_S/\rho)\, Re$. The corresponding stopping times are then of the order of d_s/U and are respectively equal to $(\rho_S/\rho)\tau_d$ and $(\rho_S/\rho)\tau_c$ for the cases $Re \ll 1$ and $Re \gg 1$. We find here the characteristic viscous diffusion and convection times (Section 9.1.1) but with an extra factor (ρ_S/ρ).

Mixing at low Reynolds number

Flows at low Reynolds numbers have, like all laminar flows, streamlines which only slowly evolve with time: there is thus no mixing comparable to that resulting from velocity fluctuations in turbulent flow. The transfer from one streamline to another of the species in solution to be mixed takes only place through transverse molecular diffusion. In a simple flow with arbitrary geometry, this mechanism is efficient only if the Péclet number, $Pe = UL/D_m$, defined in Section 2.3.2, is smaller than one. This condition is much more stringent than the condition $Re \ll 1$, because, generally, $D_m \ll \nu$ (for water and simple solute ions, $\nu \approx 10^{-6} m^2/s$ and $D_m \approx 10^{-9} m^2/s$). The ratio D_m/ν is smaller for solute molecules of greater molecular mass as well as for more viscous solvents because $D_m \propto \nu^{-1}$ for a given solute and a liquid solvent of viscosity ν (the latter relation applies only to simple liquids). For viscous fluids and large solute molecules, mixing by molecular diffusion is effective only at extremely low velocities.

Let us look at the example of a flow with characteristic length L in all directions and with a characteristic velocity U with $Re = UL/\nu \ll 1$. Let us assume that we want to mix with the remaining flow a tracer (dye, chemical compound, etc.) characterized by a molecular diffusion coefficient D_m and initially localized on a streamline. The time required for the tracer to spread throughout the flow section by diffusion transverse to the streamlines is $\tau_D \approx L^2/D_m$: in order for this mixing to occur during the transit time $\tau_c \approx L/U$ of the tracer along the flow, we must have $\tau_D < \tau_c$, which leads to the condition $Pe < 1$.

We will discuss, in Section 9.2.3, the strategy of *Lagrangian mixing*, also called *chaotic mixing*. It allows one to increase the effectiveness of molecular diffusion by reducing the thickness of the layers of each fluid by a well-chosen sequence of relative motions of the walls of the flow. In Section 10.8.3, we will also discuss the similar problem of Taylor dispersion, this time in a parallel flow, but not necessarily at a low Reynolds number.

9.2 Equation of motion at low Reynolds number

9.2.1 Stokes equation

As we have seen in Section 9.1.1, a first fundamental characteristic of flows where $Re \ll 1$ is the fact that one can neglect the inertial term $\rho(\mathbf{v}\cdot\nabla)\mathbf{v}$ of the Navier–Stokes equation (Equation 9.2) relative to the viscous term $\eta\,\nabla^2\mathbf{v}$.

Another key dimensionless number is the ratio Nt_ν of the non-stationary term $\rho(\partial\mathbf{v}/\partial t)$ to the viscous term:

$$Nt_\nu = \frac{|\rho\,\partial\mathbf{v}/\partial t|}{|\eta\,\nabla^2\mathbf{v}|} = \frac{\rho L^2}{\eta\,T} = \frac{L^2}{\nu T} \qquad (9.3)$$

We have already encountered this ratio in Section 4.5.4, as well as in Section 8.1.3, in the discussion of an alternating flow of a fluid between two planes (in that case, the characteristic time T is the period of the oscillation).

where T is the characteristic time for changes in the velocity.

The restriction of a low Reynolds number implies no assumption about the stationarity of the flow, and thus on the value of Nt_ν. However, throughout most of this chapter, we will confine ourselves to flows for which the velocity profiles are quasi-stationary and where Nt_ν is also very small compared to one. This indicates that, during the time T, the changes in velocity can propagate by viscous diffusion through a distance much larger than the characteristic length L of the flow. Under these conditions, the term $\rho\,\partial\mathbf{v}/\partial t$ of the equation of motion can be considered negligible. By combining the previous assumptions, the Navier–Stokes equation becomes:

$$\nabla p - \rho\,\mathbf{f} = \nabla\,(p - p_0) = \eta\nabla^2\mathbf{v} \qquad (9.4)$$

The equation 9.4 derived from the Navier-Stokes equation, assumes that the flow is incompressible. At high Reynolds numbers, we have seen in Section 3.3.2 that this assumption is valid only if $U \ll c$ (c is the velocity of sound, equal to $1/(\chi\rho)^{1/2}$ where the compressibility χ of the fluid equals $-(1/V)\,(\partial V/\partial p)$). At low Reynolds numbers, we will see in Section 9.2.4 that the pressure differences δp resulting from the flow are of the order of $\eta U/L$ (instead of ρU^2 for $Re \gg 1$). The corresponding changes in volume $\delta V/V$ are thus (in absolute magnitude) of the order of $\chi\eta U/L$ which we can write in the form $\chi\eta U^2/Re = (U^2/c^2)/Re$. The incompressibility condition thus becomes: $U \ll c\sqrt{Re}$. This is much more stringent, and can be difficult to satisfy when the Reynolds numbers are very low, as we have mentioned for the case of the motion of the earth's mantle.

The external force $\mathbf{f}$ per unit mass can be written in the form $\mathbf{f} = (1/\rho)\,(\nabla p_0)$, in the frequent case where it is the derivative of a potential function; when $\mathbf{f}$ corresponds to the acceleration $\mathbf{g}$ due to gravity, ∇p_0 represents the gradient of the hydrostatic pressure.

We do not usually need to consider these volume forces in cases where there is no energy exchange with the flow. The term $\nabla p_0 = \rho\,\mathbf{g}h$ is, in this case, taken into account implicitly in the variable p. Equation 9.4 then takes on the form known as the *Stokes equation*:

$$\nabla p = \eta\nabla^2\mathbf{v} \qquad (9.5)$$

9.2.2 Some equivalent forms of the Stokes equation

The Stokes equation (Equation 9.5) can be written in a more general form which uses the components σ_{ij} of the surface stresses in the fluid. Recalling Equation 4.7:

$$\sigma_{ij} = \sigma'_{ij} - p\,\delta_{ij}$$

where σ'_{ij} is the viscosity stress tensor, such that $\partial\sigma'_{ij}/\partial x_j = \eta\,\nabla^2 \mathbf{v}_i$. Equation 9.5 thus becomes:

$$\frac{\partial \sigma_{ij}}{\partial x_j} = 0 \qquad \textit{first form} \tag{9.6}$$

This equation is applicable to non-Newtonian fluids. We have seen in Section 7.1 that it is possible to describe these flows in terms of their vorticity field $\boldsymbol{\omega} = \nabla \times \mathbf{v}$, rather than of their velocity field $\mathbf{v}$. For flows at low Reynolds number, using the incompressibility condition $\nabla \cdot \mathbf{v} = 0$ and the vector identity:

$$\nabla \times (\nabla \times \mathbf{A}) = \nabla\,(\nabla \cdot \mathbf{A}) - \nabla^2 \mathbf{A} \tag{9.7}$$

which holds for any arbitrary vector field $\mathbf{A}$, Equation 9.5 can then be written in the form:

The equations of motion used in Chapters 4 and 8 to determine parallel, and quasi-parallel, quasi-stationary flows are specific forms of the Stokes equation where the velocity appears only in one of the components and the others reduce to hydrostatic equilibrium. The disappearance of the inertial terms is, in Chapters 4 and 8, related to the geometry of the flow and not to the value of *Re*.

$$\nabla p = -\eta\,(\nabla \times \boldsymbol{\omega}) \qquad \textit{second form} \tag{9.8}$$

This equation implies specifically that:

$$\nabla^2 p = 0 \tag{9.9}$$

By taking the curl of Equation 9.8, the left-hand term vanishes. If we then apply the identities from Equation 9.7 and use the general vector identity $\nabla \cdot (\nabla \times \boldsymbol{\omega}) = 0$, we obtain:

$$\nabla^2 \boldsymbol{\omega} = 0 \qquad \textit{third form} \tag{9.10}$$

This equation is a specific case of the equation of evolution of vorticity derived in Chapter 7 (Equations 7.41 and 7.42) for the case of a stationary flow at low Reynolds number. The transport of the vorticity by viscous diffusion was represented there by the term $\eta\,\nabla^2\boldsymbol{\omega}$. The physical interpretation of Equation 9.10 is thus that, in a stationary flow at low Reynolds number, no diffusion of vorticity occurs: this is related to the fact that the velocity gradients have reached an equilibrium.

9.2.3 Properties of the solutions of the Stokes equation

Uniqueness

In a flow channel, and for given boundary conditions (at infinity or, at finite distances, at the boundary of solid walls), the Stokes equation has a unique solution. This essential property is a consequence of the linearity of the equation. On the other hand, for the flow of a real fluid at a sufficiently high Reynolds number, there exist an infinite number of solutions of the Navier–Stokes equation, and they evolve with time. The non-linear convective terms and the presence of vorticity are at the origin of the multiplicity of such solutions and of their evolution.

Proof

Let us assume that there exist two velocity fields $\mathbf{v}(\mathbf{r})$ and $\mathbf{v}'(\mathbf{r})$ which are solutions of the Stokes equation and which satisfy the same boundary conditions at the walls and at infinity. Let us first show that all the derivatives of the components satisfy everywhere:

$$\frac{\partial v_i}{\partial x_j} = \frac{\partial v_i'}{\partial x_j} \tag{9.11}$$

It is then sufficient to integrate relative to x_j in order to show that $v_i = v_i'$. In order to prove Equation 9.11 for arbitrary i and j, we will establish that:

$$\iiint \left(\frac{\partial v_i}{\partial x_j} - \frac{\partial v_i'}{\partial x_j} \right)^2 \mathrm{d}V = 0 \tag{9.12}$$

(we sum over all i and j and the integral is evaluated over the volume which bounds the flow). We can then write:

$$\iiint \left(\frac{\partial v_i}{\partial x_j} - \frac{\partial v_i'}{\partial x_j} \right)^2 \mathrm{d}V = \iiint \frac{\partial}{\partial x_j} \left[(v_i - v_i') \left(\frac{\partial v_i}{\partial x_j} - \frac{\partial v_i'}{\partial x_j} \right) \right] \mathrm{d}V$$
$$- \iiint (v_i - v_i') \left(\nabla^2 v_i - \nabla^2 v_i' \right) \mathrm{d}V$$

The first term on the right-hand side of this equation can be transformed into a surface integral which vanishes because $v_i = v_i'$ on the walls. By using Stokes' equation, the right-hand term can be written:

$$\frac{1}{\eta} \iiint (v_i - v_i') \left(\frac{\partial p}{\partial x_i} - \frac{\partial p'}{\partial x_i} \right) \mathrm{d}V = \frac{1}{\eta} \iiint \frac{\partial}{\partial x_i} \left[(v_i - v_i')(p - p') \right] \mathrm{d}V$$
$$- \frac{1}{\eta} \iiint (p - p') \frac{\partial}{\partial x_i} (v_i - v_i') \, \mathrm{d}V.$$

The first term on the right-hand side of this equality vanishes because it can also be transformed into a surface integral which is zero because $v_i = v_i'$ on the walls. The second term also vanishes because $\nabla \cdot \mathbf{v} = \nabla \cdot \mathbf{v}' = 0$. The identity (Equation 9.12) and, consequently, the uniqueness of the velocity field are thus shown.

Reversibility

Reversibility is a direct consequence of the linearity of the Stokes equation. Let us assume that we have a known velocity field $\mathbf{v}(x, y, z)$, a solution of the equation that corresponds to a pressure field $p(x, y, z)$; $-\mathbf{v}(x, y, z)$ will also be a solution provided that we change the sign of the pressure gradients, as well as the velocity at every point of the solid walls. Equation 9.5 is still satisfied because its two terms are replaced by their opposites and the boundary conditions are correctly altered. The uniqueness of the solutions thus ensures that we are indeed talking about the same velocity field. It is worth mentioning that, if it is the gradient of the hydrostatic pressure which drives the flow, the direction of the acceleration of gravity $\mathbf{g}$ must also be inverted.

Experimental evidence of the reversibility A classical experiment illustrated by the series of images in Figure 9.2 below uses two coaxial cylinders, with the space between the cylinders filled with a very viscous fluid. A drop of dye of size comparable to the distance

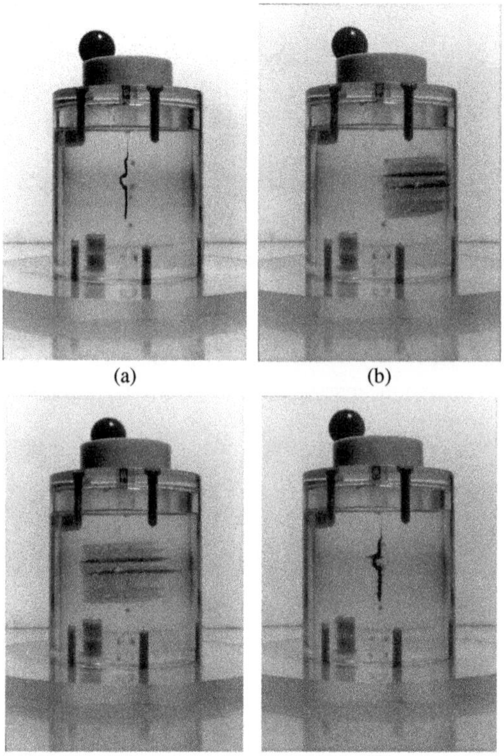

(a) (b) (c) (d)

Figure 9.2 *Experiment displaying the reversibility of flows at low Reynolds number; (a) initial situation; (b) after one turn of the inner cylinder in a clockwise direction; (c) after a turn and a quarter; (d) return to the initial position (images provided by the authors)*

between the cylinders is locally injected into the fluid. The inner cylinder is then rotated very slowly: the particles of dye located near the cylinder follow exactly its motion, while those near the outer cylinder remain almost motionless. The spot of dye spreads along the whole perimeter of the space between the cylinders and, if we move through several turns, becomes virtually invisible because of its spreading. We then stop the rotation and reverse its direction, again at a very low velocity. The velocity at the solid walls, as well as the resulting forces, are thus inverted, so that the velocities of the particles of dye, according to the reversibility principle, become opposite to the previous ones and these particles follow the same trajectory in the opposite direction. After the cylinder has been turned by the same total angle as in the initial maneuver, every particle of dye returns to its inital position and the spot reappears with a geometry close to that at the start ! It is only the molecular diffusion of the dye which is not reversible and, as a result, causes the spot to spread slightly. If we had used a less viscous fluid or had worked with a greater rotational speed, so that the non-linear terms in $(\mathbf{v} \cdot \nabla)\mathbf{v}$ are not negligible, the dye would have been mixed by the effects of the velocity gradients of the vortices which may then appear, and its dispersion would have been irreversible.

Chaos and Lagrangian mixing We do, however, have flows at small Reynolds numbers for which the preceeding reversibility property does not apply. This is the case in the experiment described above, when the two cylinders are not coaxial and when we turn them alternately in opposite directions through different, specific angles. Figure 9.3 shows that a spot of dye initially localized can then be spread over a large section of the surface in the form of multiple sheet-like layers.

In this type of flow, the trajectories followed by the dye display angular points (known as *hyperbolic* points) at the moment when the rotation switches from one cylinder to the other. At the location of these points, the trajectories are very sensitive to infinitesimal changes in the position of the particles. Thus, after a certain number of cycles, two particles of dye that were intially very close end up by following very different trajectories and separating significantly. In such a case we would not, in practical terms, have reversibility of this spreading process, even if we reproduced precisely in the opposite direction the entire sequence of rotations. The least error would indeed be amplified and prevent a return to the initial configuration, even if the instantaneous flow still remains reversible. This result is in contrast to the case of coaxial cylinders, where the trajectories are circular, and thus have no hyperbolic points.

This phenomenon has been called *Lagrangian chaos* because the trajectories in real space have properties similar to those which describe, in the phase space, the chaotic behavior of a dynamical system (in phase space, the coordinates are not those of ordinary space, but are the variables which characterize the state of the system).

Such flows make mixing also much easier, even when $Re < 1$ and $Pe > 1$. Using the same approach as in Section 9.1.3, the time necessary in order to obtain mixing by transverse diffusion between the layers shown in Figure 9.3 will be e^2/D_m where e is the distance between layers: this value is much smaller than the time L^2/D_m corresponding to the diffusion over the global size L of the flow. This type of mixing is frequently called *chaotic mixing*.

Let us now examine some of the consequences of reversibility.

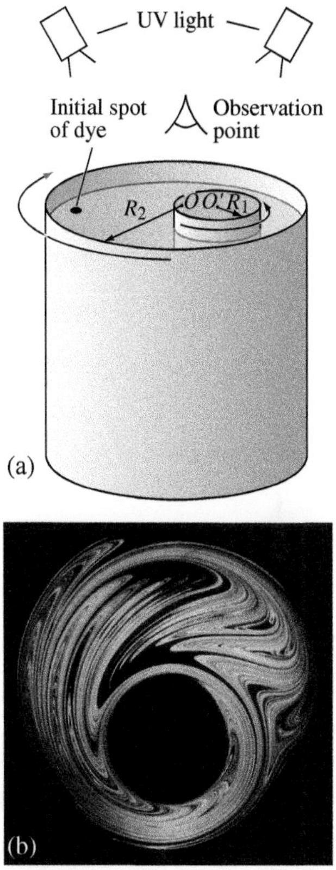

Figure 9.3C *Experiment displaying the chaotic dispersion of a drop of dye initially located between two cylinders with different axes (distance $OO' = 0.3\ R_2$) turning in alternating directions and through different angles. (a) schematic diagram of the experiment; (b) distribution of the dye after ten sequences of alternate rotations through angles $\theta_0 = 270°$ (exterior cylinder) and $-3\,\theta_0$ (interior cylinder) (image courtesy J.M. Ottino, Northwestern University)*

Symmetry of the flow around a body displaying a plane of symmetry We assume that we have a motionless body with three finite dimensions, which displays a plane of symmetry, $x = 0$, and is placed in a flow of velocity $\mathbf{U}$ which, far from the object, is in the x-direction. When the velocity $\mathbf{U}$ far from the body is perpendicular to its symmetry plane $x = 0$, the streamlines upstream of the body are symmetrical to those downstream. This is the case for the flow at Reynolds number $Re = 0.16$ around the cylinder shown in Figure 2.9a in Chapter 2.

Proof

Let us carry out two transformations which both result in a reversal of the velocity $\mathbf{U}$ of the flow far from the body, starting with the same flow $\mathbf{v}^{(0)}(x, y, z)$, and let us then write that, because of the uniqueness property, these flows are identical (Figure 9.4 on the next page). Let us first reverse the direction of the flow: the new velocity field satisfies everywhere the condition $\mathbf{v}^{(1)}(x, y, z) = -\mathbf{v}^{(0)}(x, y, z)$ (reversibility) and, particularly, $\mathbf{U}$ reverses into $-\mathbf{U}$ far from the body. Let us now go back to the initial velocity $\mathbf{v}^{(0)}$ and carry out a symmetry operation relative to the plane $x = 0$. The flow at infinity is again reversed ($\mathbf{U} \to -\mathbf{U}$) and the body transforms into itself because of its symmetry plane.

The flows obtained by these two transformations must then be identical because they correspond to the same velocity at infinity and to the same boundary conditions at the surface of the body. The first transformation (reversal of the direction of the flow) yields the velocity field $\mathbf{v}^{(1)}$ of components:

$$v_x^{(1)}(x, y, z) = -v_x^{(0)}(x, y, z) \quad \text{and} \quad v_{y,z}^{(1)}(x, y, z) = -v_{y,z}^{(0)}(x, y, z).$$

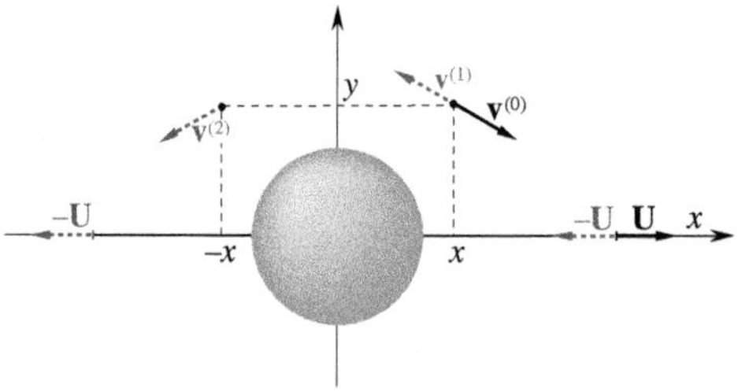

Figure 9.4 *The symmetry of the streamlines of the flow upstream and downstream of an obstacle displaying a symmetry plane (here, the plane x = 0) can be shown by reversing the flow-velocity* **U** *far from the obstacle into* **−U**

The second transformation (symmetry relative to the plane $x = 0$) yields the velocity components:

$$v_x^{(2)}(x, y, z) = -v_x^{(0)}(-x, y, z) \qquad \text{and} \qquad v_{y,z}^{(2)}(x, y, z) = v_{y,z}^{(0)}(-x, y, z).$$

Because the two velocity fields $\mathbf{v}^{(1)}$ and $\mathbf{v}^{(2)}$ must be identical, we thus have:

$$v_x^{(0)}(-x, y, z) = v_x^{(0)}(x, y, z) \qquad \text{and} \qquad v_{y,z}^{(0)}(-x, y, z) = -v_{y,z}^{(0)}(x, y, z).$$

The result is that the streamlines are symmetrical relative to the plane $x = 0$. This result is a very sensitive test for a low value of the Reynolds number. For instance, in the case of the flow around a cylinder (Section 2.4.2), as soon as *Re* becomes of the order of 1, the upstream and downsteam flows are not symmetrical any more. Then, recirculation zones appear downstream for $Re \simeq 5$ (Figure 2.9b) and become unstable for $Re \simeq 60$ with the emission of an alternating vortex street.

If the obstacle does not have a plane of symmetry, the flow at small Reynolds number is not symmetrical between the upstream and downstream sides. However, if one reverses the direction of the flow, the fluid follows the same streamlines in the opposite direction because of the reversibility of the flow at low Reynolds number, independently of the symmetry properties of the obstacles.

In the example which follows, this result is no longer valid as the Reynolds number increases. Imagine a flow obtained by either blowing or sucking through a funnel. At low Reynolds numbers, the streamlines diverge almost radially in the flared section (Figure 9.5a) and are essentially the same whether one blows or sucks; only the direction of the flow is changed. In fact, this is not what happens in a real-life experiment, because one is then under conditions of high Reynolds numbers, and the reversibility of the flow no longer applies: one can always blow out a candle by blowing through the narrow end of a funnel, but never by suction! This results from the fact that, by blowing, one forms an air jet which is confined to a small portion of the section of the funnel (Figure 9.5b) because of the *boundary layer separation* of the flow in its flared section. We will discuss this phenomenon in Section 10.5.4. On the other hand, in the case of suction, the flow is distributed throughout the cross-section of the funnel and, for the same pressure difference, the maximum of the velocity in this section is much smaller.

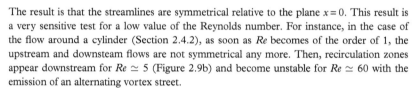

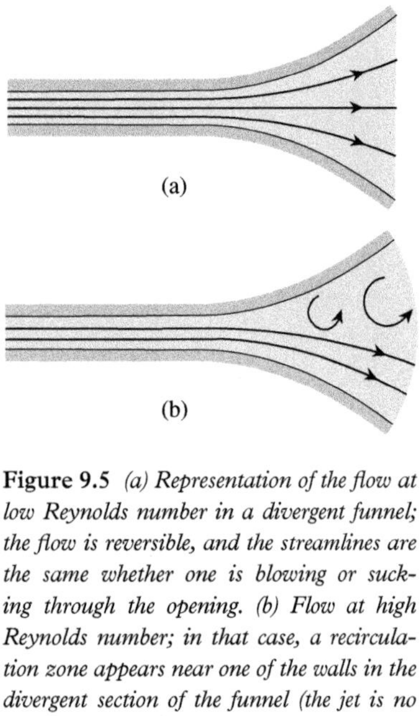

Figure 9.5 *(a) Representation of the flow at low Reynolds number in a divergent funnel; the flow is reversible, and the streamlines are the same whether one is blowing or sucking through the opening. (b) Flow at high Reynolds number; in that case, a recirculation zone appears near one of the walls in the divergent section of the funnel (the jet is no longer symmetric relative to the axis)*

Dropping a sphere along a rigid wall Let us imagine that we have a sphere falling gravitationally (with its weight corrected by the Archimedes buoyancy) in a liquid, near a vertical wall, after being dropped with a zero initial velocity. The reversibility property leads to the result that the sedimentation velocity **U** of the sphere is permanently in the vertical direction. Similar reasons explain why a cylinder sedimenting vertically in a viscous fluid under the action of gravity remains parallel to its original orientation (Section 9.3.2). In the same way, two identical spheres, in close proximity, fall in a liquid with no relative motion to each other (Section 9.4.3).

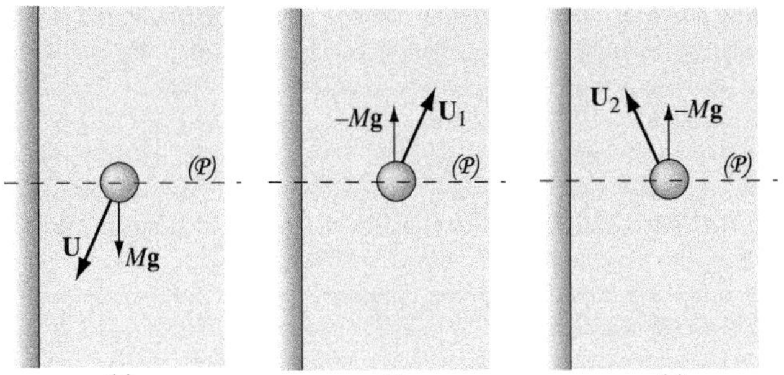

Figure 9.6 *(a) A falling sphere in a viscous fluid near a vertical wall; (b) motion resulting from time reversal; (c) motion after reversal through the plane of symmetry for solution (a)*

Proof

Let us assume that the velocity $\mathbf{U}$ is at an oblique angle, and oriented in such a way that the particle moves toward the wall (Figure 9.6a). One considers that the weight of the sphere is balanced by the combination of hydrodynamic forces on its surface. Let us first reverse everywhere the *local* velocity $\mathbf{v}(\mathbf{r})$ of the fluid flow ($\mathbf{v}'(\mathbf{r}) = -\mathbf{v}(\mathbf{r})$), which implies that the direction of the gravitational acceleration $\mathbf{g}$ is also inverted; because of the boundary conditions at the solid walls (Figure 9.6b), the velocity $\mathbf{U}$ of the sphere is replaced by its opposite $\mathbf{U}_1 = -\mathbf{U}$. Such a flow satisfies the equation of motion of the fluid because the components of the stresses on the sphere are also everywhere replaced by their opposite and, consequently, they balance out the new value $-M\mathbf{g}$ of the weight of the object. Let us come back to the initial configuration *(a)* and carry out a symmetry relative to plane ($\mathcal{P}$), the horizontal diametral plane of the sphere which is also perpendicular to the solid wall (case (c)); one obtains then, if the acceleration due to gravity is also inverted, a velocity $\mathbf{U}_2$. The velocity field of the fluid is then symmetrical to its original direction relative to the plane ($\mathcal{P}$): the components of the velocity in this plane are unchanged while only the vertical component is inverted. The two transformed velocity fields then correspond to the motion of the same body under the effect of the same acceleration, and must therefore be identical so that $\mathbf{U}_1 = \mathbf{U}_2$. This implies that $\mathbf{U}$ should be vertical. One notes that these results are only *exactly* applicable in the limit of a Reynolds number approaching zero; at a finite Reynolds number, even very low, the cumulative effects of the non-linear convection terms end up destroying the reversibility property and its consequences.

Invariance of the streamlines to changes in the flow rate (Re ≪ 1)

This is a further consequence of the linearity of the Stokes equation. Indeed, if $\mathbf{v}$ is a solution, so is $\lambda\mathbf{v}$ (λ real), provided that the velocities at the solid walls, and the external forces, are changed by the same factor λ. The new solution is then obtained by multiplying everywhere the initial solution $\mathbf{v}(x, y, z, t)$ by the factor λ by which the flow rate has been changed (reversibility is a special case of this feature, for which $\lambda = -1$). Since the direction of the velocity remains the same at all points, the shape of the streamlines, which are tangent to it, remains unchanged, and only the magnitude of the velocity varies. Thus, in a flow where the boundaries are at rest, one has the same flow profile for any velocity, so long as the condition of low Reynolds number is obeyed.

Superposition of solutions of the Stokes equation

This property is again an immediate consequence of the linearity of the Stokes equation: if $\mathbf{v}_1(x, y, z, t)$ and $\mathbf{v}_2(x, y, z, t)$ are two solutions of this equation, then $\lambda_1\mathbf{v}_1 + \lambda_2\mathbf{v}_2$ is also a solution with a corresponding pressure gradient which can be written: $\nabla p = \lambda_1 \nabla p_1 + \lambda_2 \nabla p_2$. The velocity at the walls is the linear combination of the velocities for solutions 1 and 2 with the same coefficients λ_1 and λ_2. Since there exists a unique velocity field corresponding to given boundary conditions, it is the solution $\lambda_1\mathbf{v}_1 + \lambda_2\mathbf{v}_2$ which is observed experimentally. One can therefore superimpose linearly velocity fields corresponding to different flows, in channels with identical geometry, provided that one combines linearly, with the same coefficients, the values of the velocities at the walls.

We have already encountered this property in dealing with the superposition of Couette and Poiseuille flows (Section 4.5.3).

Minimum in the energy dissipation

For given boundary conditions at the walls and at infinity, a flow which obeys the Stokes' equation $\nabla p = \eta \nabla^2 \mathbf{v}$ corresponds to a minimum in the rate of dissipation of energy ε. Equation 5.26 gives the following expression of ε:

$$\varepsilon = \iiint \sigma'_{ij} \frac{\partial v_i}{\partial x_j}\, \mathrm{d}V = \frac{\eta}{2} \iiint \left(\frac{\partial v_i}{\partial x_j} + \frac{\partial v_j}{\partial x_i} \right)^2 \mathrm{d}V = 2\eta \iiint e_{ij}^2\, \mathrm{d}V. \tag{9.13}$$

(we have here, according to our usual convention, an implicit summation over the indices i and j).

Proof

Let e_{ij} be the rate-of-strain tensor corresponding to a velocity field $\mathbf{v}$, which is a solution of the Stokes' equation, and assume that another tensor e'_{ij} exists, corresponding to another velocity field $\mathbf{v}'$ which obeys the same two boundary conditions as well as the incompressibility condition, $\nabla \cdot \mathbf{v}' = 0$, but which is not a solution of the Stokes' equation. We can then write:

$$2\eta \iiint e_{ij}'^2\, \mathrm{d}V = \iiint e_{ij}^2\, \mathrm{d}V + 2\eta \iiint \left(e'_{ij} - e_{ij}\right)^2 \mathrm{d}V + 4\eta \iiint \left(e'_{ij} - e_{ij}\right) e_{ij}\, \mathrm{d}V. \tag{9.14}$$

We will prove that the last integral is zero. It can be written in the form:

$$\iiint (e'_{ij} - e_{ij})\, e_{ij}\, \mathrm{d}V = \frac{1}{2} \iiint \left(\frac{\partial v'_i}{\partial x_j} - \frac{\partial v_i}{\partial x_j} \right) \frac{\partial v_i}{\partial x_j}\, \mathrm{d}V$$
$$+ \frac{1}{2} \iiint \left(\frac{\partial v'_i}{\partial x_j} - \frac{\partial v_i}{\partial x_j} \right) \frac{\partial v_j}{\partial x_i}\, \mathrm{d}V = \frac{1}{2}(I_1 + I_2)$$

(recalling again that we use the usual summation convention for repeated indices; these are therefore dummy indices which can be permuted). We have then, for the first integral:

$$I_1 = \iiint \left(\frac{\partial v'_i}{\partial x_j} - \frac{\partial v_i}{\partial x_j} \right) \frac{\partial v_i}{\partial x_j} \mathrm{d}V = \iiint \frac{\partial}{\partial x_j} \left[(v'_i - v_i) \frac{\partial v_i}{\partial x_j} \right] \mathrm{d}V - \iiint (v'_i - v_i) \frac{\partial^2 v_i}{\partial x_j^2}\, \mathrm{d}V.$$

The first term of the right-hand side can be changed into a surface integral, for which the integrand vanishes at the walls. The second term can be written, as a result of the Stokes equation:

$$\iiint (v_i' - v_i) \frac{\partial^2 v_i}{\partial x_j^2} \mathrm{d}V = \frac{1}{\eta} \iiint (v_i' - v_i) \frac{\partial p}{\partial x_i} \mathrm{d}V$$

$$= \frac{1}{\eta} \iiint \frac{\partial}{\partial x_i} \left[p (v_i' - v_i) \right] \mathrm{d}V - \frac{1}{\eta} \iiint p \frac{\partial (v_i' - v_i)}{\partial x_i} \mathrm{d}V.$$

Here again, the first term on the right-hand side can be transformed into a surface integral, which is zero since, by assumption, the two velocities are the same at the walls. The second term vanishes also, because incompressibility requires $\nabla \cdot \mathbf{v} = 0 = \nabla \cdot \mathbf{v}'$.
Let us now evaluate the integral I_2. It can be written:

$$I_2 = \iiint \left(\frac{\partial v_i'}{\partial x_j} - \frac{\partial v_i}{\partial x_j} \right) \frac{\partial v_j}{\partial x_i} dV = \iiint \frac{\partial}{\partial x_j} \left[(v_i' - v_i) \frac{\partial v_j}{\partial x_i} \right] \mathrm{d}V - \iiint (v_i' - v_i) \frac{\partial^2 v_j}{\partial x_i \partial x_j} \mathrm{d}V$$

The first term on the right-hand side can once again be transformed into a surface integral, which vanishes because the velocity vanishes at the walls; the second term also vanishes since $\nabla \cdot \mathbf{v} = 0$.

Thus, the integral $\iiint (e_{ij}' - e_{ij})\, e_{ij}\, \mathrm{d}V$ is zero. Since the second term on the right-hand side of Equation 9.14 is positive, the energy dissipation associated with the tensor e_{ij}' exceeds indeed that corresponding to the velocity field which satisfies the Stokes equation.

The property, just stated, that the dissipated energy is minimized, applies only if the Reynolds number is low: at high Reynolds numbers, turbulent solutions of the Navier–Stokes equation dissipate a greater amount of energy than laminar ones for identical boundary conditions.

The principle of energy minimization applies only when we compare two instances with the same location and geometry for the walls. It does not, in any way, imply that, if the solid walls are free to move, the result will be a configuration minimizing the energy dissipation. For instance, a small rod free to rotate near the stagnation point of a longitudinal flow, assumes, in contrast, an orientation wich maximizes the dissipation of energy.

9.2.4 Dimensional arguments for low Reynolds number flows

The velocity and pressure fields for a flow of given geometry with respective characteristic velocity and length U and L satisfy, if $Re \ll 1$, the equations:

$$\mathbf{v}(x, y, z) = U\, \mathbf{F}\left(\frac{x}{L}, \frac{y}{L}, \frac{z}{L}\right), \qquad (9.15a)$$

$$p(x, y, z) - p_0(x, y, z) = \frac{\eta U}{L} G\left(\frac{x}{L}, \frac{y}{L}, \frac{z}{L}\right). \qquad (9.15b)$$

This result extends the property of proportionality of the whole velocity field $\mathbf{v}(x, y, z)$ to U, as shown in Section 9.2.3.

where $\mathbf{F}$ and G are respectively dimensionless vector and scalar functions which depend only on the geometry of the flow. Thus, the velocity and pressure distributions in two flows with the same geometry but with different sizes, characteristic velocities and viscosities can be inferred one from the other thanks to these equations. Since the solutions are unique, these will be, moreover, the only possible flows.

Proof

We have seen in Section 4.2.4 that the Navier–Stokes equation (Equation 9.2) could be written in dimensionless form by dividing each variable by an appropriate characteristic value.

We will proceed in the same way here, but will make use of the following dimensionless variables which reflect better the dominant role of the viscosity:

$$\mathbf{r}' = \frac{\mathbf{r}}{L}, \ \mathbf{v}' = \frac{\mathbf{v}}{U}, \ t' = \frac{t\,\nu}{L^2}, \quad p' = \frac{(p - p_0)\,L}{\eta\,U}. \qquad (9.16)$$

U and L are again the characteristic velocity and length of the flow; p_0 is the hydrostatic pressure in the absence of flow, while $\tau_\nu = L^2/\nu$ represents the viscous diffusion time over the characteristic length L of the flow (instead of the convection time L/U) and $\eta U/L$ is the viscous stress for a velocity gradient of the order of U/L (in lieu of the pressure ρU^2). We then rewrite the Navier–Stokes equation in the dimensionless form:

$$\frac{\partial \mathbf{v}'}{\partial t'} + Re\left(\mathbf{v}'.\nabla'\right)\mathbf{v}' = -\nabla' p' + \nabla'^2 \mathbf{v}'. \qquad (9.17a)$$

In accordance with the discussion in Section 9.2.1, the non-stationary term is negligible if the flow evolves slowly over a time t long compared to τ_ν (thus over a time $t' \gg 1$). In such a case, the solutions of the equation of motion can be written in the form:

$$\mathbf{v}'(x, y, z) = \mathbf{F}\left(\frac{x}{L}, \frac{y}{L}, \frac{z}{L}, Re\right) \quad (9.17b) \quad \text{and:} \quad p'(x, y, z) = G\left(\frac{x}{L}, \frac{y}{L}, \frac{z}{L}, Re\right), \quad (9.17c)$$

where $\mathbf{F}$ and G are respectively dimensionless vector and scalar functions. When $Re \ll 1$, the inertial term on the left-hand side of Equation 9.17a can be considered zero and, again for a stationary or quasi-stationary flow, one obtains the dimensionless form of the Stokes equation:

$$\nabla' p' = \nabla'^2 \mathbf{v}', \qquad (9.18a)$$

but, this time, with solutions that are independent of Re:

$$\mathbf{v}'(x, y, z) = \mathbf{F}\left(\frac{x}{L}, \frac{y}{L}, \frac{z}{L}\right) \quad (9.18b) \quad \text{and} \quad p'(x, y, z) = G\left(\frac{x}{L}, \frac{y}{L}, \frac{z}{L}\right). \quad (9.18c)$$

From this, one infers Equations 9.15. One notes that the inertial term could only be neglected because it does not display any singularity when $Re \to 0$. It has indeed been shown above that the velocity approaches zero in proportion to U whenever $Re \ll 1$. This is not the case for flows at high Reynolds number with the different normalization introduced in Section 4.2.4. In that case, the viscous term does not vanish as $Re \to \infty$, in spite of the factor $1/Re$, because the velocity gradients diverge.

As a result, one can make similar predictions for the forces exerted on the walls or on moving solids. The local viscous stresses have components of the form $\eta(\partial v_i/\partial x_j)$; they are thus proportional to the product of $\eta U/L$ by dimensionless functions of the scaled coordinates x/L, y/L and z/L. The global viscous drag force is obtained by integration of these stresses over the entire area of the walls, from which one obtains an additional factor L^2 multiplied by a constant vector $\mathbf{A}$. One obtains then the dimensional equation:

$$\text{Viscous drag force} \approx \mathbf{A}\,(\eta\,U/L)\,L^2 \approx \mathbf{A}\,(\eta\,U\,L). \qquad (9.19a)$$

A similar argument may be used for the difference $p - p_0$ between the pressure and the hydrostatic pressure (Equation 9.15b). By integrating this term over the surfaces of the walls, an additional factor L^2 appears again. The total force $\mathbf{F}$ acting on the walls has therefore the following form similar to Equation 9.19a:

$$\mathbf{F} = -\mathbf{B}\,\eta\, U\, L, \quad (9.19b)$$

where $\mathbf{B}$ is a vector dependent only on the orientation and geometry of the flow and of the solid walls. The minus sign indicates that the force $\mathbf{F}$ opposes the motion.

This force is frequently normalized by $(1/2)\ \rho U^2 S$: $(1/2)\ \rho U^2$ is the order of magnitude of the dynamic pressure defined in Section 5.3.2, and $S \approx L^2$ is an area characteristic of the walls. For flows at low Reynolds number, we thus obtain a *coefficient of friction*:

$$C_d \approx \frac{|F|}{\frac{1}{2}\rho\, U^2 L^2} \approx \frac{\eta}{\rho\, U\, L} \approx \frac{1}{Re}. \quad (9.20)$$

This proportionality of the coefficient of friction on the inverse of the Reynolds number is characteristic of flows at small Reynolds numbers. We have already observed this property in the case of laminar flows discussed in Section 4.5, for which the transport of momentum by convection is zero for reasons of the geometry. We note however, that this proportionality to the inverse of the Reynolds number is merely a reflection of the definition of C_d: this definition is actually only physically meaningful for flows with a high Reynolds number, dominated by inertial effects and for which C_d varies little with Re. In the case of pure viscous drag at a low Reynolds number, this definition is less appropriate.

9.3 Forces and torques acting on a moving solid body

As we have just seen, one of the consequences of the linearity of the Stokes equation is the proportionality between the forces on the solid walls and the characteristic velocity of the fluid. We will define this property (Section 9.3.1) and look at its application in two examples of solids with quite different symmetries (Section 9.3.2): a rod and a corkscrew.

9.3.1 Linearity of the equations governing the velocity of the solid body and the forces acting on it

Any motion of the solid body can be represented at a given time as the combination of an overall translation at velocity $\mathbf{U}(t)$ and of a rotation at an angular velocity $\mathbf{\Omega}(t)$ (Figure 9.7). The velocity at any point of the solid can be written as:

$$\mathbf{v} = \mathbf{U} + \boldsymbol{\omega} \times \mathbf{r}, \quad (9.21)$$

where $\mathbf{r}$ is the radius vector relative to the origin of the axes (a change of this origin is equivalent to a change in $\mathbf{U}$).

Let us designate as F_i and G_i the components of the force and torque acting on the solid (the torque is calculated relative the origin O of the coordinate system). These components are the integrals of the viscous stresses and of the local pressure on the surface of the solid. Thus, they satisfy the equations:

$$F_i = -\eta\,(A_{ij}\, U_j + B_{ij}\, \Omega_j) \quad (9.22) \qquad \text{and} \qquad G_i = -\eta\,\left(C_{ij}\, U_j + D_{ij}\, \Omega_j\right). \quad (9.23)$$

The coefficients A_{ij} or, respectively, D_{ij} connect the forces (or, respectively, the torques) to the translations (or, respectively, the rotations). The coefficients B_{ij} and C_{ij} represent, in contrast, the respective cross-effects of the forces resulting from a rotation and of the torques created by a translational motion. Thus, a helical object turning at slow speed about its axis in a very viscous fluid is subjected to a force parallel to this axis. This "*corkscrew*"

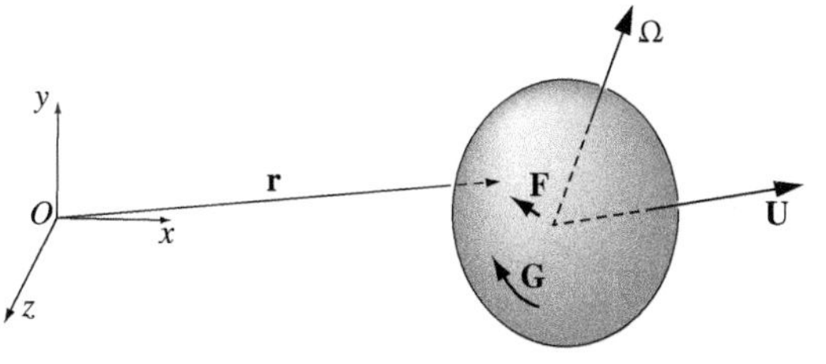

Figure 9.7 *The force and torque acting on an object moving at velocity* $\mathbf{U}$ *and rotating with an angular velocity* $\mathbf{\Omega}$ *in a viscous fluid*

effect will be discussed further in Section 9.3.2. This force results from the absence of a plane of symmetry perpendicular to the axis of rotation.

Equations 9.22 and 9.23 can be obtained by linear combination, with the same coefficients, of two velocity fields which are solutions of the Stokes equation and of the corresponding pressure fields (provided one combines in a similar way the boundary conditions for the velocity along the walls). One also notes that the coefficients A_{ij} have the dimensions of a length, B_{ij} and C_{ij}, that of an area, and D_{ij}, of a volume.

The tensors with components A_{ij}, B_{ij}, C_{ij} and D_{ij} have general symmetry properties independent of the shape of the body; these properties can be derived by using the fact that the stress tensor σ_{ij}, is symmetrical, and they are represented by the equations:

$$A_{ij} = A_{ji}, \quad (9.24a) \qquad D_{ij} = D_{ji}, \quad (9.24b) \qquad B_{ij} = B_{ji}. \quad (9.24c)$$

The equivalence of the tensors B_{ij} and C_{ji} takes into account the reciprocal relationship between the force on a rotating particle and the torque on a particle in translation.

The matrix A_{ij} connecting F_i and U_j, which is symmetric, can be diagonalized. There exists then, regardless of the shape of the body, a Cartesian coordinate system in which each component of the force is proportional to the corresponding component of the velocity:

$$F_i = -\eta\, \lambda_i\, U_i \qquad (9.25)$$

(in contrast to our usual convention, there is no summation here on the repeated index i). The scalar product $\mathbf{F} \cdot \mathbf{U}$ represents the loss of energy by viscous dissipation. It must therefore be negative, regardless of $\mathbf{U}$: this implies that each of the eigenvalues λ_i is positive. From a geometrical point of view, the condition $\mathbf{F} \cdot \mathbf{U} < 0$ indicates that the angle (usually not zero) between the force $\mathbf{F}$ and the direction of motion must always be obtuse.

For solid bodies having specific symmetries (planes or axes), additional relations appear between the coefficients A_{ij}, B_{ij}, C_{ij} and D_{ij}, as we will see for three different examples.

9.3.2 The effect of the symmetry properties of solid bodies on the applied forces and torques

Solid body having a plane of symmetry

Take a solid having $x_1 = 0$ as a plane of symmetry; the coefficients A_{ij} obey:

$$A_{12} = A_{21} = A_{13} = A_{31} = 0. \qquad (9.26)$$

Physically, this indicates that the force corresponding to a motion perpendicular to a plane of symmetry is, itself, normal to this plane. Conversely, a body having a horizontal plane of symmetry which falls under its own weight in a viscous fluid will have no horizontal velocity component. Moreover:

$$C_{11} = C_{22} = C_{33} = C_{32} = C_{23} = 0 \quad (9.27) \qquad \text{and} \qquad B_{11} = B_{22} = B_{33} = B_{32} = B_{23} = 0. \quad (9.28)$$

This shows that a corkscrew effect (the appearance of a force parallel to the axis of rotation x_1 of a solid body) can exist only if the plane perpendicular to the axis of rotation is not a plane of symmetry of the body. Finally:

$$D_{12} = D_{21} = D_{13} = D_{31} = 0. \qquad (9.29)$$

Proof

Select a coordinate system where two of the axes (x_2 and x_3) lie in the plane of symmetry $x_1 = 0$. The coefficients A_{ij} are then unchanged under the symmetry operation $x_1 \to -x_1$. Under this same operation, F_1 changes into $F'_1 = -F_1$ and U_1 into $-U_1$, while the other components stay unchanged. Equation 9.22 for the components F_1 and F'_1 is then:

$$F_1 = -\eta\,(A_{11}U_1 + A_{12}U_2 + A_{13}U_3) \qquad \text{and} \qquad F'_1 = -\eta\,(A_{11}(-U_1) + A_{12}U_2 + A_{13}U_3).$$

Since the two expressions above must have equal and opposite values, it follows that: $A_{12} = A_{13} = 0$.

By using the other components of **F**, one shows in the same manner that the coefficients A_{21} and A_{31} vanish. Quite different results are obtained for the coefficient C_{ij}, relating the components G_j of the applied torque to those of the velocity. Indeed, **G** is a pseudo-vector with the same kind of symmetry as a rotational velocity. It therefore does not change sign under a reflection in a plane normal to it, but does become inverted for a symmetry relative to a plane parallel to its own direction. Thus, when x_1 is changed into $-x_1$, G_1 changes into G_1, G_2 into $-G_2$, and G_3 into $-G_3$. By an argument similar to that given above, one obtains Equations 9.27 and 9.28. In the case of the tensor D_{ij}, relating the torque **G** and the rotational velocity vector $\mathbf{\Omega}$, the same components as for A_{ij} vanish, since both of these pseudo-vectors are characterized by the same symmetry.

Bodies with three mutually perpendicular planes of symmetry

When the body has three mutually orthogonal planes of symmetry (e.g., ellipsoid, parallelepiped, or any body having a similar level of symmetry . . .), every coefficient C_{ij}, and consequently B_{ij}, is zero with respect to any coordinate system in accordance with Equations 9.27 and 9.28. There is a total decoupling between translational and rotational motion; only a torque can cause the object to rotate (if it is non-zero when computed with respect to an axis of rotation passing through the point of intersection of the planes of symmetry).

In the same way, according to Equation 9.26, the matrix A_{ij} is diagonal relative to a set of coordinate axes perpendicular to the planes of symmetry of the solid.

(i) ***A cylindrical rod with circular cross-section falling in a viscous fluid***

Such a body has an axis of revolution (assumed to be pointing in the z-direction) in addition to three mutually perpendicular axes of symmetry. Not only does it fall without spinning in a viscous fluid, but its motion is determined by only two coefficients which characterize the viscous friction parallel and normal to the axis (e.g. $A_{zz} = \lambda_{\parallel}$ and $A_{xx} = A_{yy} = \lambda_{\perp}$). If the ratio of these two coefficients is known, it is possible to determine the angle at which the rod falls relative to the vertical as a function of the angle of inclination of the rod.

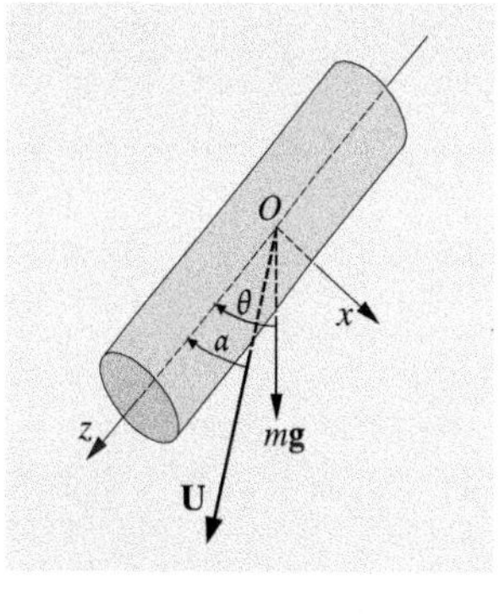

Figure 9.8 *A cylindrical rod falling obliquely in a viscous fluid*

Proof

Consider a long uniform rod, of circular cross-section, with center of symmetry O (Figure 9.8). Let the z-axis be along the axis of the rod, and let the x-axis, perpendicular to it, be located in the vertical plane of symmetry. If the rod is of uniform density, the force of gravity acts at the center of symmetry, and there is no torque which could cause rotation. In this coordinate system, then:

$$F_z = -\eta\,\lambda_{\parallel}\,U_z, \qquad F_x = -\eta\,\lambda_{\perp}\,U_x, \qquad F_y = -\eta\,\lambda_{\perp}\,U_y. \tag{9.30}$$

As indicated below (Section 9.4.3), it can be shown, by a calculation valid in the limit of very slender objects, that: $\lambda_{\perp} = 2\,\lambda_{\parallel}$ (Equations 9.71a-b): the resistance to motion parallel to the long dimension is half that in the normal direction. Let us define α as the angle of the trajectory relative to the axis of the rod, and θ the angle of inclination of this axis to the vertical. When the rod moves with constant velocity, we can write:

$$F_z = -mg\cos\theta = -\eta\,\lambda_{\parallel}\,U_z \tag{9.31a}$$

and:

$$F_x = -mg\sin\theta == -\eta\,\lambda_{\perp}\,U_x = -2\,\eta\,\lambda_{\parallel}\,U_x, \tag{9.31b}$$

where m represents the mass of the rod minus the contribution of the Archimedes buoyancy.

Then:
$$\tan\alpha = U_x/U_z = (\tan\theta)/2.$$

The angle of deviation $\theta-\alpha$ of the trajectory relative to the vertical direction then satisfies:

$$\tan(\theta-\alpha) = \frac{(\tan\theta - \tan\alpha)}{1 + \tan\theta\,\tan\alpha} = \frac{\tan\theta}{\left(2 + \tan^2\theta\right)}. \tag{9.32}$$

For $\alpha = 0$ and $\alpha = \pi/2$, the rod falls vertically; one is back to the case where the applied force (the weight of the rod) is perpendicular to a plane of symmetry of the solid. The maximum value of the trajectory angle $(\theta-\alpha)$ occurs when $\tan\theta = \sqrt{2}$, so that: $\tan(\theta-\alpha) = 1/(2\sqrt{2})$, corresponding to $(\theta-\alpha) \approx 19.5°$.

(ii) ***Case of a cube or a sphere***

In the case of a cube, planes parallel to the cube faces and passing through the center of symmetry, are mutually orthogonal symmetry planes, and the tensor A_{ij} is diagonal in the corresponding coordinate system. Since these planes are all equivalent, the three corresponding eigenvalues are equal, and we can write:

$$[\mathbf{A}] = \lambda\,[\mathbf{I}],$$

where $[\mathbf{I}]$ is the identity matrix. For pure translational motion, the viscous friction force is colinear with the velocity in any possible direction, so that Equation 9.19b becomes:

$$\mathbf{F} = -\eta\,C\,L\,\mathbf{U}, \tag{9.33}$$

where L is a length characteristic of the solid, and C a constant. The force $\mathbf{F}$ acts at the center of symmetry of the object. This result holds equally for a sphere of radius R (= L), with $C = 6\pi$ (as we show in Section 9.4.2). Moreover, since this force acts at the center of symmetry of the solid, it cannot induce rotation. Thus, a body with spherical symmetry, such as a cube, or a regular tetrahedron, always falls vertically in a viscous fluid at low Reynolds numbers, irrespective of the initial orientation of its faces relative to the vertical (the density must, however, be uniform throughout the body). We estimate further down (Section 9.4.3) the viscous drag force in such cases.

In the case of pure rotation, the tensor D_{ij}, given the relationship between the torque and the rotational angular-velocity vector, is also proportional to $[\mathbf{I}]$ (provided that the axis of rotation passes through the center of symmetry). Accordingly:

$$\mathrm{G} = -\eta\, L^3\, D\,\Omega. \tag{9.34}$$

For a sphere of radius $R = L$, it will be shown that $D = 8\pi$ (Section 9.4.3).

Translational–rotational coupling for a body devoid of planes of symmetry

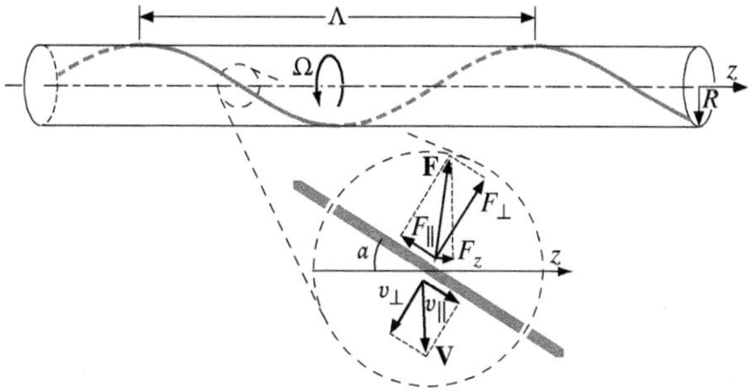

Figure 9.9 *Motion of a helix spinning in a viscous fluid*

In this case, some of the coefficients which we have previously considered as vanishing can actually play a very important role. Consider, for example, the motion of an object with helical symmetry around an axis oriented in the z-direction (Figure 9.9). Since the x-y plane is not a plane of symmetry, the coefficients B_{zz} and C_{zz} do not vanish. A rotation of the helix with angular velocity Ω_z thus induces a propulsive force $B_{zz}\,\Omega_z$ in the z-direction, parallel to the axis of rotation, its direction dependent on whether the spiral is right- or left-handed. Per unit length of the helix, B_{zz} has the value:

$$B_{zz} = 2\pi\,\lambda_{\parallel} R^2/\Lambda. \tag{9.35}$$

Proof

We demonstrate this result by analyzing the local forces acting on an element of length of the helix, assuming it to be similar to a short, straight rod moving with azimuthal velocity $\Omega\, R$ around the axis of rotation (where R is the radius of the helix). We assume that the pitch Λ of the helix is large relative to the radius R; locally, such an element of length makes therefore a small angle $\alpha = 2\pi\, R/\Lambda$ with the z-axis. The corresponding velocity components $v_{\parallel} \approx \Omega\,\alpha R = 2\pi\,\Omega\, R^2/\Lambda$ and $v_{\perp} = \Omega\, R$, respectively parallel and perpendicular to the rod, induce the respective force components $F_{\parallel}$ and $F_{\perp}$. These forces can be expressed per unit length as:

$$F_{\parallel} = -\eta\,\lambda_{\parallel}\, v_{\parallel} = -2\pi\,\eta\,\lambda_{\parallel}\,\Omega\, R^2/\Lambda \qquad \text{and} \qquad F_{\perp} = -\eta\,\lambda_{\perp}\, v_{\perp} = -2\,\lambda_{\parallel}\,\eta\,\Omega\, R.$$

These forces, in turn, can be projected into a component parallel to the z-axis, and a component normal to the radius.

As seen in the inset of Figure 9.9, the two projections of $F_{\parallel}$ and $F_{\perp}$ in the z-direction are opposite to each other. If we compute their projection in the z-direction, keeping in mind that the angle $\alpha = 2\pi\ R/\Lambda$ is small, we find that they have a non-zero resultant with magnitude:

$$F_z = \left|\alpha F_\perp - F_\parallel\right| = 2\pi\,\eta\,\lambda_\parallel\,\Omega\,R^2/\Lambda,$$

per unit length of the helix. The orientation of these two components and, as a result, the force F_z are reversed if either the direction of rotation or the pitch of the helix is changed. The component F_z has the same value for every element of the helix. There is consequently a macroscopic driving force in the z-direction proportional to the velocity and to the viscosity. We thus recover Equation 9.35 for the coefficient of coupling B_{zz} between the angular velocity Ω and the driving force per unit length of the helix.

The orientations of the components of $F_\parallel$ and $F_\perp$ normal to the radius vary continuously all along the helix. They therefore have a zero resultant, but they do contribute to a resistive torque with value:

$$G_z = \left(F_\perp\ \cos\alpha + F_\parallel\ \sin\alpha\right) R \approx -2\,\eta\,\lambda_\parallel\,\Omega\,R^2.$$

Since the distance from the point of application of these forces to the axis of rotation is R, the coefficient of coupling D_{zz} between the angular velocity and the resistive torque is therefore: $D_{zz} = 2\ \lambda_\parallel R^2$.

9.3.3 Propulsion at low Reynolds numbers

Propulsion by the rotation of a helix

The previous model allows us to explain the motion of certain types of bacteria: thus, *Escherichia coli* propels itself by rotating its tail (flagellum) by means of a molecular motor around an articulation located at one end of its body. The method of propulsion is completely different from the action of propellers on a boat. In the latter case, which corresponds to a flow at high Reynolds number, the effect of propulsion is due to the circulation of the velocity of the fluid around the blades of the propeller. This causes the appearance of a Magnus force perpendicular to the velocity of the blade, which is analogous to the lift on the wings of an airplane, as seen in Section 7.5.2. There results an overall propulsion force proportional to the square of the velocity of rotation: the efficiency of such a classical propeller would be much smaller at low Reynolds numbers, since this lift force is missing.

The scallop theorem

A significant point in the propulsion at low Reynolds numbers is a property known as the *scallop theorem*. Let us assume that we have a miniature version of this bivalved mollusk, of a size small enough so that the Reynolds number Re is smaller than 1, which alternately opens and closes by suction or ejection of a liquid. Such flows would be reversible with respect to a change of direction of the motion of the walls; the model would then undergo merely a sequence of forward and backward movements of same amplitude and no overall motion would occur.

In real life, in contrast to the case of our model, this type of propulsion takes place in seawater at a Reynolds number of several thousands, where inertial effects dominate and reversibility no longer applies. More precisely, scallops move by opening their shell slowly, and closing it rapidly. Then, the emission of a rearward jet (and thus of momentum) in the second stage is not balanced by the momentum oriented in the forward direction, thus resulting in a net propulsion.

As seen previously, such forces are proportional to the velocity of the walls. If we denote by $e(t)$ the maximum opening of the shell, the instantaneous propulsive force will vary in the manner: $F(e) = k(e)\ \mathrm{d}e/\mathrm{d}t$. The total impulse resulting from a change of the opening from a value e_1 to a value e_2 would then be $\Delta P = \int_{e_1}^{e_2} k(e)de$. The contributions of the phases of opening and closing are thus opposite, and this is independent of the velocities of motion in each phase (as long as $Re \ll 1$). More generally, the propulsion of microorganisms at low Reynolds number cannot result from a series of motions of the same amplitude in alternating directions.

Some methods to avoid reversibility

Let us go back to the example of bacteria or other microorganisms made up of a body and a flagellum, already mentioned above in the propulsion of *E. coli*. In contrast to that case, let us assume that the flagellum moves inside a plane. If we create a stationary wave on the flagellum with, for example, a point fixed at the place where it ties in with the main body, the average propulsion force is zero: we have effectively a sequence of symmetric deformations in both directions. On the other hand, the net force would be non-zero for a progressive wave (such results were demonstrated experimentally by applying alternating forces via magnetic ball-bearings along model flagella).

Still referring to flagella, irreversibility can result from their mechanical properties and, specifically, from their elasticity. Let us assume that we excite a flexible, elastic flagellum by means of a sinusoidal motion of one of its ends (this model mimics in a simplified manner the propulsion of certain microorganisms). The dynamics of each element is the result, not only of the hydrodynamic force due to its motion relative to the fluid, but also from the elastic forces created by its deformation. This results in the fact that the geometry, and thus the forces, are not the same at a given position, during the back and forth phases of the motion. A net driving force can thus appear.

In such a situation, the object is not passive; the motion of its different points is constrained, e.g. its length must remain constant, although it can exert forces on the fluid. In the experiment of Figure 9.2, we would also violate reversibility by replacing the spot of dye with a flexible thread that, depending on the phase of its motion, could be stretched or folded.

Let us now leave the case of flagella, and come back to that of an articulated object. The scallop theorem tells us that there can be no propulsion if there is only one degree of freedom (the opening and closing of the model shell). On the other hand, propulsion has been demonstrated in the case of objects including two articulations (e.g, a set of three articulated stems): in this case, with an appropriate sequence of individual motions of the articulations, bringing back the object to its initial state without requiring exactly inverse motion, propulsion can be. This is also possible for a chain of three spheres by using an appropriate sequence of changes of their relative position: one can thus model the propulsion of certain types of microorganisms by the deformation of their bodies.

9.4 Constant-velocity motion of a sphere in a viscous fluid

9.4.1 The velocity field around a moving sphere

Assume that we have a fluid at rest at infinity, with a velocity field described in spherical, polar coordinates (r, θ, φ), such that the polar, z-axis ($\theta = 0$), points along the direction of a sphere, of radius R, moving with uniform velocity $\mathbf{U}$ (Figure 9.10a). Because of the rotational symmetry of the system around the polar axis z, the velocity field is axially symmetric: therefore, the velocity component v_φ is zero, and the other two components v_r and v_θ are independent of φ. As a result of calculations carried out in detail below, one obtains for the two components v_r and v_θ:

$$v_r = U\cos\theta\left(\frac{3R}{2r} - \frac{R^3}{2r^3}\right) \quad (9.36a) \qquad \text{and:} \qquad v_\theta = -U\sin\theta\left(\frac{3R}{4r} + \frac{R^3}{4r^3}\right). \quad (9.36b)$$

These velocity components vanish at infinity and coincide with those of the sphere at its surface ($r = R$) as required. The above solution is only valid when the Reynolds number is small compared to unity, and for a sphere moving along at constant velocity U.

The most striking aspect of the above result is the slow decrease, as $1/r$, of the flow velocity with the distance r from the center of the sphere. We contrast this result, valid at low Reynolds numbers, with the much faster decrease, as $1/r^3$, of the velocity near a sphere in potential flow, discussed in Section 6.2.4.

This slow decrease of the perturbation in velocity results from the ineffectiveness of diffusion in transporting, away from the sphere, the momentum imparted to the fluid by viscous friction forces. We can retrieve this variation as $1/r$ by means of a simple physical argument: assume that the velocity decreases as $r^{-\alpha}$ far from the sphere; the momentum flux resulting from diffusion can be expressed in terms of gradients of the velocity components, varying therefore as $r^{-\alpha-1}$. The integral of this flux over a sphere of radius r thus varies as $r^{-\alpha+1}$; but it must be constant, independent of r, and its magnitude must correspond to the total frictional force on the sphere. Accordingly, we find that $\alpha = 1$, so that the corresponding velocity field varies indeed as $1/r$.

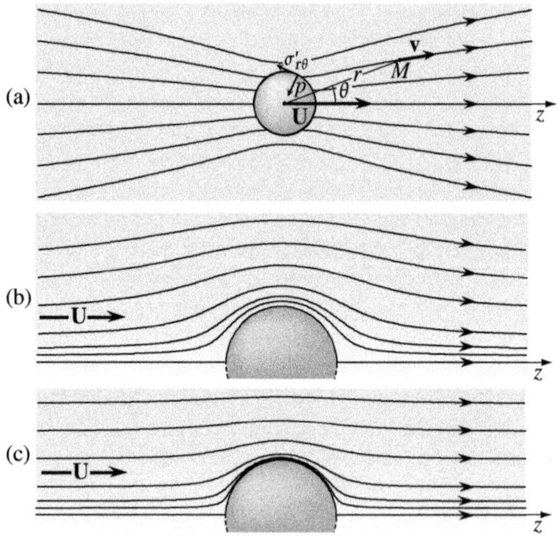

Figure 9.10 *(a) Streamlines of the flow around a sphere moving with constant velocity* **U** *in a fluid at rest. The components of the normal and tangential stresses acting at a point on the surface of the sphere are also shown on the figure. (b) Streamlines of the flow of a viscous fluid with uniform velocity* **U** *at infinity and Re* $\ll$ *1. (c) Potential flow of an ideal fluid around the same sphere (Section 6.2.4)*

For a sphere at rest in a flow of uniform velocity **U** far from it, the velocity **V** of the fluid is obtained by subtracting from the velocity **U** the components given by Equations 9.36 a-b:

$$V_r = U\cos\theta - v_r = U\cos\theta\left(1 - \frac{3R}{2r} + \frac{R^3}{2r^3}\right), \tag{9.37a}$$

and:

$$V_\theta = -U\sin\theta - v_\theta = -U\sin\theta\left(1 - \frac{3R}{4r} - \frac{R^3}{4r^3}\right). \tag{9.37b}$$

In this frame of reference, where the sphere is at rest, the streamlines return much more slowly to the configuration of uniform velocity as we move away from the sphere than for a potential flow (Figure 9.10b & c).

We compute below step-by-step the velocity components given by Equations 9.36a & b. The result is based initially on a guess of a trial function for the distribution of pressure around the sphere; this distribution is then used to calculate the velocity field obeying the required boundary conditions, from which we evaluate the arbitrary constants of the trial function. Since we have shown above that solutions for the velocity field are unique, the resultant solution must be the correct one.

Calculation of the pressure field

The pressure field $p(\mathbf{r})$ is a harmonic function since it satisfies $\nabla^2 p = 0$ (Equation 9.9). We can therefore expand $p(\mathbf{r})$ in spherical polar coordinates (r, θ, φ) as a linear combination of a term in $1/r$, and of its successive derivatives relative to the various coordinates; each of these terms is a solution of Laplace's equation $\nabla^2 f = 0$, and corresponds to the multipole field resulting from a charge, a dipole, a quadrupole, etc. The leading terms of this expansion are:

$$\phi_0 \propto \frac{A}{r}, \qquad \boldsymbol{\phi}_1 \propto \nabla\left[\frac{1}{r}\right] = -\frac{\mathbf{r}}{r^3} \quad \text{and} \quad [\boldsymbol{\phi}_2] \quad \text{such that} \quad \phi_{2ij} \propto \left(\frac{\delta_{ij}}{r^3} - 3\,\frac{x_i x_j}{r^5}\right).$$

Assume that we neglect the hydrostatic pressure term and that the remaining components of p vanish at infinity (this only appears as an additive constant): no constants, or terms containing positive powers of r, appear in $p(\mathbf{r})$. We thus use, as a trial function, the simplest of these terms that obeys the symmetry of the problem. The pressure field must be in the form $\phi\,\eta\,U$, since p is proportional to U and to η (Section 9.2.4). ϕ is the linear combination of the basis functions introduced above. The only term compatible with the scalar form of p is the component of ϕ_1 parallel to U, i.e., $\partial/\partial z(1/r) = -\cos\theta/r^2$ (indeed, the terms of the form $(\sin\theta\cos\varphi)/r^2$ and $(\sin\theta\sin\varphi)/r^2$ do not display the axial symmetry around the polar axis, chosen in the direction of U). On the other hand, the term ϕ_0 would imply a radial pressure gradient and thus a source of flow which does not exist. We can therefore write:

$$p = C\,\eta\,U\frac{\cos\theta}{r^2} = -C\,\eta\,\mathbf{U}\cdot\nabla\left(\frac{1}{r}\right) = -C\,\eta\,\nabla\cdot\left(\frac{\mathbf{U}}{r}\right). \tag{9.38}$$

The forces acting on the sphere are thus proportional to the velocity and to the viscosity. We now investigate whether there exists a velocity field corresponding to this pressure distribution and which obeys the Stokes equation and the boundary conditions at the wall of the sphere. If this was not the case, we would have to add into $p(\mathbf{r})$ higher-order terms in the expansion (this the case for a body having a shape more complex than that of a sphere).

Vorticity field corresponding to the distribution of pressure

Starting with the Stokes equation as written in Equation 9.8, and combining it with Equation 9.38, we obtain:

$$-C\,\eta\nabla\{\nabla\cdot(\mathbf{U}/r)\} = -C\,\eta\,\nabla\times\{\nabla\times(\mathbf{U}/r)\} - C\,\eta\,\nabla^2\,(\mathbf{U}/r) = -\eta\,\nabla\times\boldsymbol{\omega}. \tag{9.39}$$

But, we have:

$$\nabla^2\cdot\left(\frac{\mathbf{U}}{r}\right) = \mathbf{U}\cdot\nabla^2\left(\frac{1}{r}\right) = 0, \quad (9.40) \quad \text{so that:} \quad \boldsymbol{\omega} = C\,\nabla\times\left(\frac{\mathbf{U}}{r}\right) + \nabla g(r). \quad (9.41)$$

where $g(\mathbf{r})$ is a function obeying Laplace's equation, as can be seen by taking the divergence of Equation 9.41. However, only the component ω_φ is non-zero, since $v_\varphi = 0$ and $\mathbf{v}$ is independent of φ. The unknown function $g(\mathbf{r})$ must therefore be of the form $\alpha\,\varphi + \beta$ where α and β are constants. But $\boldsymbol{\omega}$ is independent of φ just as $\mathbf{v}$; the constant α must then be zero. Moreover, as g appears only through its gradient, we can also take $\beta = 0$. Thus, the function g vanishes identically. Using the vector identity $\nabla\times(m\,\mathbf{A}) = m\,(\nabla\times\mathbf{A}) + (\nabla m)\times\mathbf{A}$, one obtains:

$$\boldsymbol{\omega} = C\,\nabla\times\left(\frac{\mathbf{U}}{r}\right) = -C\,\mathbf{U}\times\nabla\left(\frac{1}{r}\right), \quad (9.42) \qquad \text{i.e.:} \quad \omega_\varphi = CU\frac{\sin\theta}{r^2} \quad (9.43)$$

Evaluation of the stream function Ψ from the vorticity field

If we now introduce the Stokes stream function Ψ (Section 3.4.3), such that:

$$v_r = \frac{1}{r^2\sin\theta}\frac{\partial\Psi}{\partial\theta} \quad (9.44) \qquad \text{and:} \quad v_\theta = -\frac{1}{r\sin\theta}\frac{\partial\Psi}{\partial r}. \quad (9.45)$$

We then obtain:
$$\omega_\varphi = \frac{1}{r}\left(\frac{\partial(r\,v_\theta)}{\partial r} - \frac{\partial v_r}{\partial\theta}\right) = -\frac{1}{r\sin\theta}\frac{\partial^2\Psi}{\partial r^2} - \frac{1}{r^3}\frac{\partial}{\partial\theta}\left(\frac{1}{\sin\theta}\frac{\partial\Psi}{\partial\theta}\right). \quad (9.46)$$

Substituting Equation 9.46 into Equation 9.43, we obtain:

$$-\frac{1}{r\sin\theta}\frac{\partial^2\Psi}{\partial r^2} - \frac{1}{r^3}\frac{\partial}{\partial\theta}\left(\frac{1}{\sin\theta}\frac{\partial\Psi}{\partial\theta}\right) = C\,U\frac{\sin\theta}{r^2}. \quad (9.47)$$

One can achieve separation of variables by assuming $\Psi = U\,\sin^2\theta\,f(r)$, a form justified by the fact that the z-axis must be a streamline. Equation 9.47 then becomes, after factoring out the terms in $\sin\theta$:

$$-\frac{1}{r}\frac{\partial^2 f}{\partial r^2} + \frac{2f}{r^3} = \frac{CU}{r^2}. \quad (9.48)$$

Solutions to the homogeneous form of the above equation can be written as L'/r and $(M'r^2)$, where L' and M' are the required arbitrary constants, while a particular solution is $CU\,r/2$. Letting then: $L = L'/U$ and $M = M'/U$, one obtains:

$$\Psi = U\,\sin^2\theta\left(\frac{L}{r} + M\,r^2 + \frac{C\,r}{2}\right). \quad (9.49)$$

Calculation of the velocity field

The components of the velocity field can now be obtained from Equations 9.44 and 9.45:

$$v_r = U\cos\theta\left(\frac{C}{r} + \frac{2L}{r^3} + 2M\right) \quad (9.50a) \quad \text{and:} \quad v_\theta = -U\sin\theta\left(\frac{C}{2r} - \frac{L}{r^3} + 2M\right). \quad (9.50b)$$

The constants are determined by using the boundary conditions:

for $r\to\infty$, $\mathbf{v}\to 0$, so that $M = 0$;
for $r = R$, one must have: $v_r = U$ for $\theta = 0$, and $v_\theta = -U$ for $\theta = \pi/2$.

Thus: $C = 3R/2$ and $L = -R^3/4$.

One then obtains the velocity components given by Equations 9.36 as well as the following equations for the pressure and vorticity fields ($\mathbf{n}$ is the unit vector along the radius vector $\mathbf{OM}$):

$$p = \frac{3}{2}\eta\,U\,R\frac{\cos\theta}{r^2} = \frac{3}{2}\eta\,R\frac{\mathbf{U}\cdot\mathbf{n}}{r^2} \quad (9.51) \quad \text{and} \quad \omega_\varphi = \frac{3}{2}U\,R\frac{\sin\theta}{r^2} = \frac{3}{2}R\frac{(\mathbf{U}\times\mathbf{n})_\varphi}{r^2}. \quad (9.52)$$

9.4.2 Force acting on a moving sphere: the drag coefficient

Case of a fluid of infinite volume

By using the above results, one finds that the total drag force **F** on a sphere of radius R and velocity **U** in a fluid of viscosity η has the value:

$$\mathbf{F} = -6\,\pi\,\eta\,R\mathbf{U}. \tag{9.53}$$

Experimentally, we find that this expression, known as the *Stokes law*, is closely obeyed up to Reynolds numbers of the order of unity (even though it has been derived under the much more restrictive assumption $Re \ll 1$). The drag force **F** has, moreover, the dimensional form given in Section 9.2.4 (Equation 9.19b).

Proof

The normal stress, due to the pressure at the surface of the sphere, can be written:

$$\sigma_r = \left[-p + 2\eta\left(\frac{\partial v_r}{\partial r}\right)\right]_{r=R} = \frac{3}{2}\frac{\eta\,U\cos\theta}{R}. \tag{9.54a}$$

Let us emphasize that the contribution due to the velocity gradient vanishes at the surface of the sphere. The effect of this is that the normal stress has a maximum along the z-axis. Moreover there appears a tangential stress associated with the viscosity which, as stated in Appendix A-2, Chapter 4, has the form:

$$\sigma_{r\theta} = \eta\left[\frac{1}{r}\frac{\partial v_r}{\partial \theta} + \frac{\partial v_\theta}{\partial r} - \frac{v_\theta}{r}\right]_{r=R} = \frac{3}{2}\frac{\eta\,U\sin\theta}{R}. \tag{9.54b}$$

One observes that $\sigma_{r\theta}$ is largest at a right angle to the polar axis. All the other terms of the stress tensor vanish; the resultant force per unit area can therefore be written at every point on the sphere:

$$\frac{d\mathbf{F}}{dS} = [\sigma]\cdot\mathbf{n} = \sigma_{rr}\,\mathbf{e}_r + \sigma_{r\theta}\,\mathbf{e}_\theta, \quad \text{i.e:} \quad \frac{d\mathbf{F}}{dS} = -\frac{3}{2}\frac{\eta\,U\cos\theta}{R}\mathbf{e}_r + \frac{3}{2}\frac{\eta\,U\sin\theta}{R}\mathbf{e}_\theta = -\frac{3}{2}\frac{\eta\,\mathbf{U}}{R}. \tag{9.55}$$

The force $d\mathbf{F}/dS$ is then independent of θ and φ and its orientation is exactly opposite to **U** since the two components of **U** in the radial direction, and normal to it, are respectively (U cos θ) and ($-U$ sin θ). The total drag force is finally simply obtained by multiplying the value of $d\mathbf{F}/dS$ by the area $4\pi\,R^2$ of the sphere.

To take into account the effect of gravity, one adds to the pressure p the contribution $-\rho_f\,g\,z$ of the hydrostatic pressure. By integrating over the surface of the sphere, this term results in the Archimedes buoyancy $(-\rho_f\,V_{\text{sphere}}\,g)$.

Application to the determination of the terminal velocity for a falling sphere in a viscous fluid at low Reynolds numbers

When a sphere of radius R and density ρ_s sediments in a fluid of infinite extent, of density ρ_f and of viscosity η, it reaches a terminal velocity which results from the equilibrium between the Stokes drag force and the weight of the sphere, corrected by Archimedes buoyancy. This velocity can be written:

$$V_{\text{terminal}} = \frac{2}{9}\frac{(\rho_s - \rho_f)\,g\,R^2}{\eta}. \tag{9.56}$$

For a glass bead 1 mm in diameter ($\rho_s = 2.5 \times 10^3$ Kg/m^3) sedimenting in glycerine ($\rho_f \approx 10^3$ Kg/m^3, $\eta \approx 1$ Pa.s), one has $V_{\text{terminal}} \approx 10^{-3}$ m/s; the Reynolds number for this flow is $Re = 10^{-3}$ and the condition $Re \ll 1$ is, therefore, well obeyed.

We note that, in contrast to the mere determination of its functional dependence by dimensional analysis, the exact calculation of the force exerted on a sphere by a low Reynolds number flow (a problem which appears, at first glance, particularly simple because of the symmetries involved), is quite complicated. This underscores the great importance, in fluid mechanics, of the search for approximate solutions illustrating the underlying principles, and leading to expressions, such as Equation 9.19b, in terms of dimensionless parameters. The evaluation of the exact numerical coefficients, determined by the specific shapes of the flow, can often only be carried out by numerical calculations or by a series of experimental tests.

This relationship between the terminal velocity of the bead and the viscosity of a fluid is used in some viscometers; there, the velocity V_{terminal} is determined from the fall time of a calibrated bead over a known distance. Practical viscometers of this type use vertical tubes for which the diameter is not large relative to that of the bead. As will be seen below, this leads to significant correction factors in the resulting frictional force. The viscosity is determined by calibrating the device by means of liquids of known viscosity.

The drag coefficient can now be computed from the Stokes force, using Equation 9.20 in the form:

$$C_d = \frac{F}{(\pi R^2)\frac{1}{2}\rho\, U^2} \qquad \text{so that:} \qquad C_d = \frac{24}{Re} \tag{9.57}$$

where $Re = (2\ U\ R/\nu)$. This drag coefficient displays the variation as $1/Re$ predicted on the basis of dimensional analysis in Section 9.2.4. It has also the same form as obtained earlier in Section 4.5.3 for the case of Poiseuille flow.

The effect of walls

The presence of plane or cylindrical walls, located parallel or perpendicular to the motion of a sedimenting sphere, increases the effect of friction. The significant effect of the Stokes force in the case of these obstacles must be emphasized.

As an example, for a sphere located in a tube with radius 10 times than of the sphere, there is a 20% increase in the force compared to the case of an infinite volume. This increase is essentially the effect of the wall of the tube, as well as, to a lesser degree, of the upward counter-flow of the liquid displaced by the moving sphere, which also contributes to the drag. The magnitude of this latter effect is determined by the ratio of the area of the particle, projected onto a plane perpendicular to its motion, to that of the tube section here, this ratio is proportional to the *square* of the ratio of the radius of the particle to that of the tube while the first effect was proportional to this ratio.

Figures 9.11a–b display streamlines around a sphere moving along the axis of a cylindrical tube in reference frames respectively fixed relative to the sphere (Figure 9.11a) and to the tube (Figure 9.11b). In the second figure, we see clearly the existence of a recirculation due to a rising flow (counterflow) near the wall of the tube, opposite to the downward motion of the sphere (and of the fluid close to it). In both these cases, the streamlines are symmetric relative to the horizontal diameter plane of the sphere. This would no longer be the case at higher Reynolds numbers.

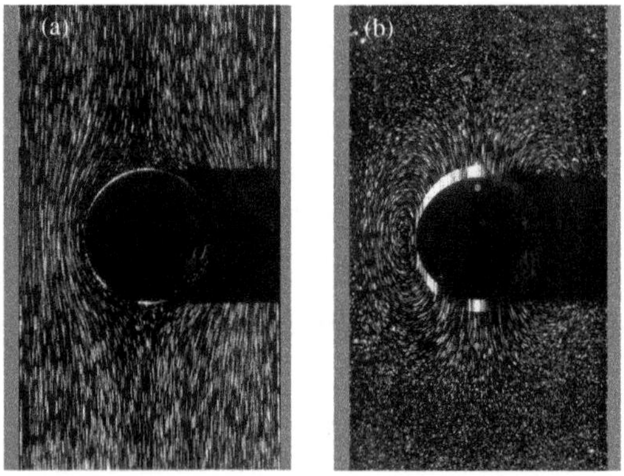

Figure 9.11 *Visualization of the flow resulting from the sedimentation of a sphere along the axis of a vertical, cylindrical tube filled with glycerol, and of diameter* 163 mm, *double that of the sphere. The Reynolds number is* $Re = 0.1$. *The streamlines are visualized by means of small magnesium particles illuminated by a plane light sheet (the illumination comes from the left of the figure, and the dark patch at the right of the sphere corresponds to its shadow). (a) The camera moves in such a way that it is fixed relative to the sphere. (b) The camera is fixed relative to the tube. During the exposure time, the sphere has moved through a small fraction of its diameter (plates courtesy of M. Coutanceau)*

In the case of a sphere moving perpendicular to a flat plate, the drag force when the sphere is very close to the plate has been determined in Section 8.1.6, Equation 8.33. In this case, Equation 9.53 for the drag force is multiplied by a factor R/h, in which h is the minimum distance between the sphere and the wall, such that:

$$\mathbf{F} = -6\pi\,\eta\,R\mathbf{U}\,(R/h).$$

We note specifically that, because of the reversibility of flows at low Reynolds numbers, this equation applies whether the sphere is getting closer to or farther from the walls.

We can tie in this equation, valid for $h \ll R$, to Equation 9.53, which is applicable in the limit $h \gg R$, so as to obtain:

$$F = -6\pi\,\eta\,R\left(1+\frac{R}{h}\right)U.$$

The force resulting from this equation (*a priori* valid in both limits $R/h \gg 1$ and $R/h \ll 1$) remains within 6.5% of the exact value (determined from a calculation involving a series expansion) for intermediate values of the ratio R/h.

9.4.3 Generalization of the solution of the Stokes equation to other experiments

The Stokes problem can be generalized to other geometries such as a sphere in the presence of walls, a rotating sphere, a set of spheres, ellipsoidal objects (of which the cylindrical rod, treated in Section 9.3.2, is a limiting case) or, even more simply, a case where the solid sphere is replaced by a fluid. We discuss now a few of these problems.

Rotating sphere

In general terms, any instantaneous motion of a solid can be decomposed into a translation and a rotation. Aside from the translational motion of the sphere, we will now study the flow resulting from its rotation about its own axis in a fluid at rest far away (Figure 9.12). We find that the azimuthal component of the velocity field has the value:

$$v_\varphi(r,\theta) = \Omega\,R^3\sin\theta/r^2. \tag{9.58}$$

The torque due to forces opposing this rotation is in the z-direction, so that:

$$\mathbf{G} = -8\pi\,\eta\,R^3\,\mathbf{\Omega}, \tag{9.59}$$

a form which we had predicted by dimensional analysis in Section 9.3.2 (Equation 9.34).

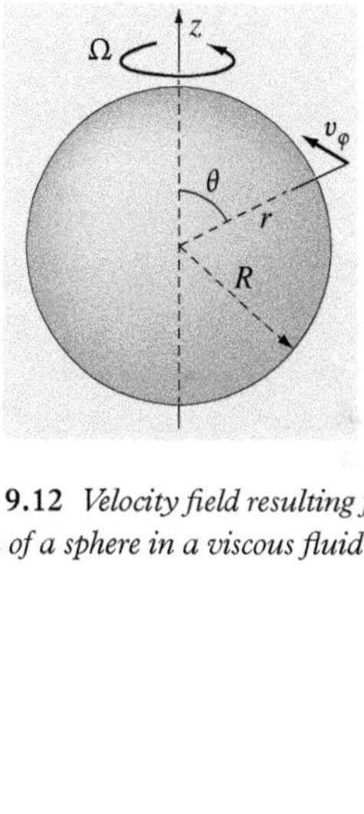

Figure 9.12 *Velocity field resulting from the rotation of a sphere in a viscous fluid*

Proof

Given the symmetry of the problem, the only non-zero component of the velocity is v_φ which depends only on r and the polar angle θ and which is such that $v_\varphi(r=R) = \Omega\,R\sin\theta$. Also, as a consequence of symmetry, the pressure must be independent of φ. The component in the φ direction in the equation of motion in spherical coordinates (Appendix 4A.3, Chapter 4) can then be written:

$$\eta\left(\frac{1}{r}\frac{\partial^2(r\,v_\varphi)}{\partial r^2} + \frac{1}{r^2}\frac{\partial^2 v_\varphi}{\partial\theta^2} + \frac{\cot\theta}{r^2}\frac{\partial v_\varphi}{\partial\theta} - \frac{v_\varphi}{r^2\sin^2\theta}\right) = 0.$$

Using the expansion $v_\varphi(r,\theta) = f(r)\,g(\theta)$, where $g(\theta) = \sin\theta$ in order to satisfy the boundary conditions at the surface of the sphere, this equation can be simplified into:

$$\frac{\partial^2 f}{\partial r^2} + \frac{2}{r}\frac{\partial f}{\partial r} - \frac{2f(r)}{r^2} = 0.$$

By looking for a solution of the form $f \propto r^\alpha$, one finds that $\alpha = 1$ or -2; only the second value allows one to obey the condition of zero velocity at infinity. Equation 9.58 is then obtained by choosing the multiplying factor so as to obey the boundary conditions on the sphere. One computes from $v_\varphi(r, \theta)$ the value of the only non-vanishing component of the stresses exerted on the sphere ($r = R$) by means of the equation:

$$\sigma_{\varphi,r} = \eta\left(\frac{\partial v_\varphi}{\partial r} - \frac{v_\varphi}{r}\right) = -3\eta\ \Omega\ \sin\theta.$$

The total resistive torque is obtained by multiplying $\sigma_{\varphi r}$ by the distance $R \sin \theta$ and integrating over the entire surface of the sphere. One then retrieves Equation 9.59.

The respective influence of two spheres

One examines the case of two identical spheres, sedimenting in a fluid of infinite extent, but sufficiently close to each other so that the velocity field resulting from the motion of one sphere affects the motion of the other. The sedimentation velocity can be evaluated by adding the individual values of the sedimentation velocity of one sphere to that resulting from the motion of the other. By using the property of reversibility of flows at low Reynolds numbers, discussed in Section 9.2.3, one shows that the resulting velocities of the two spheres must be identical. The straight line joining their centers thus does not rotate during the fall, regardless of its original direction, and the distance between their centers remains constant. The combination of the two spheres thus behaves like a solid of revolution with a plane of symmetry perpendicular to its axis. The inverse of the sedimentation velocity V_s (normalized by its value V_{s0} for one sphere in an infinite medium) is shown in Figure 9.13 as a function of $d/2R$, for the two cases where the straight line joining the centers of the spheres is either vertical or horizontal.

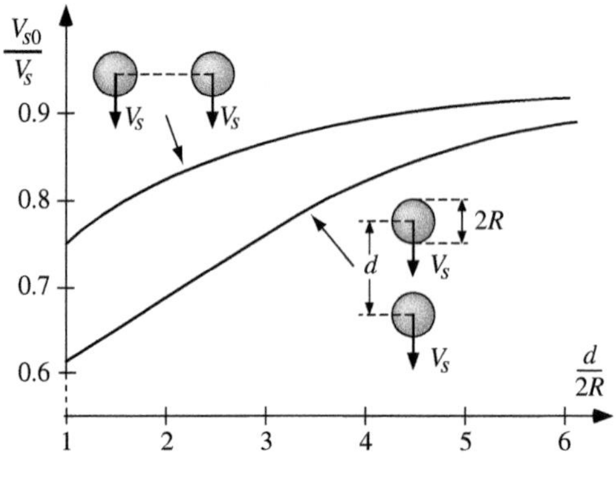

Figure 9.13 *Comparison between the sedimentation velocity V_s of two identical spheres, with either vertical or horizontal separation between them. V_{s0} is the velocity of an isolated sphere*

Let us assume that the distance d between the centers is sufficiently large relative to the radius R so that we can neglect the terms in $1/d^3$ in the velocity field resulting from the effect of each of the spheres on the other (Equations 9.36). When the line joining the centers is vertical, each sphere induces in the vicinity of the other one a velocity component v_i of the order of $(3/2)\, V_s R/d$ parallel to the sedimentation velocity $\mathbf{V}_s$. The relative velocity of the spheres with respect to the fluid is then smaller than V_s by the amount v_i; it must also be equal to the terminal velocity V_{s0} of an isolated sphere in order to insure the equilibrium between the weight of the sphere and the viscous drag. One thus has:

$$V_{s0} = V_s - v_i \approx V_s\left(1 - \frac{3}{2}\frac{R}{d}\right), \tag{9.60}$$

so that:

$$\frac{V_{so}}{V_s} = \left(1 - \frac{3}{2}\frac{R}{d}\right). \tag{9.61}$$

For the case where the line joining the centers is horizontal, one obtains:

$$\frac{V_{so}}{V_s} = \left(1 - \frac{3}{4}\frac{R}{d}\right). \tag{9.62}$$

Both the above equations are only valid when the distance between the spheres is sufficiently large. The curves on Figure 9.13 were obtained instead by an exact calculation.

Drop of fluid in motion within another immiscible fluid

Consider a fluid sphere of viscosity η_i with its center fixed within another, external, non-miscible fluid with viscosity $\eta_e = \lambda\eta_i$ moving at velocity **U** at infinity (Figure 9.14). In order to determine the external velocity field, the general form of Equations 9.50 (taking the value $M = 1/2$ to get the velocity at infinity) is used. Within the drop, a calculation similar to that carried out above provides a velocity field of the form:

$$v_r = U\cos\theta\left(A + Br^2\right) \quad (9.63\text{a}) \qquad \text{and} \qquad v_\theta = -U\sin\theta\left(A + 2Br^2\right). \quad (9.63\text{b})$$

where A and B are constants evaluated by means of the boundary conditions requiring that the velocity and the shear stresses be continuous at the surface of the sphere. One notes that the r-dependence of the two velocity fields is quite different; this is understandable since, on the one hand, the distribution of vorticity inside the drop is not the same as that obtained in Equations 9.36a–b and, on the other hand, the boundary conditions are also different. In particular, no terms of the type $1/r^n$ $(n > 0)$ may appear in the inner velocity field since the velocity at the center of the sphere must be finite.

One obtains, for each of the two velocity fields:

- external:

$$v_r = U\cos\theta\left[1 - \frac{3 + 2\lambda}{2\,(1 + \lambda)}\frac{R}{r} + \frac{1}{2\,(1 + \lambda)}\frac{R^3}{r^3}\right], \tag{9.64a}$$

$$v_\theta = -U\sin\theta\left[1 - \frac{3 + 2\lambda}{4\,(1 + \lambda)}\frac{R}{r} - \frac{1}{4\,(1 + \lambda)}\frac{R^3}{r^3}\right]. \tag{9.64b}$$

- internal:

$$v_r = -U\cos\theta\frac{\lambda}{2\,(1 + \lambda)}\left[1 - \left(\frac{r}{R}\right)^2\right], \tag{9.65a}$$

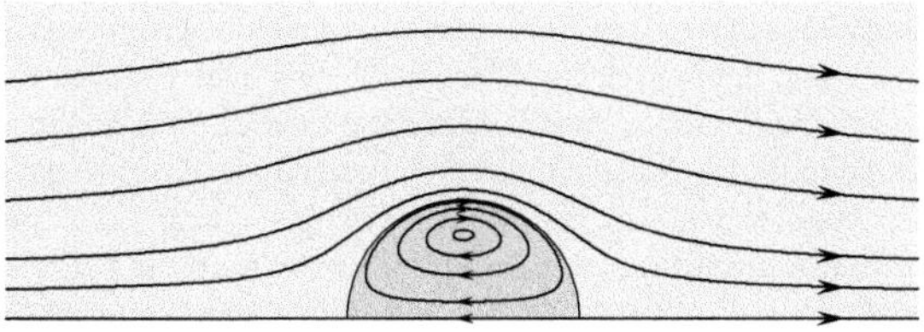

Figure 9.14 *Flow patterns outside and inside a liquid sphere of fixed center placed inside another fluid of uniform velocity **U** at infinity (configuration comparable to that of Figure 9.10b)*

$$v_\theta = U \sin\theta \frac{\lambda}{2(1+\lambda)} \left[1 - 2\left(\frac{r}{R}\right)^2\right]. \tag{9.65b}$$

The Stokes force acting on the sphere is then:

$$\mathbf{F} = 2\pi\, \eta_e\, R\mathbf{U} \frac{3+2\lambda}{1+\lambda}. \tag{9.66}$$

In the case of a sphere moving at a velocity $\mathbf{U}$ in a fluid at rest, a minus sign must be added in front of the right hand side of Equation 9.66. This result reduces to Stokes law (Equation 9.53) in the limit where λ approaches zero (the inner fluid is infinitely viscous), and to the expression for a gas bubble within the fluid, in the opposite limit, as λ tends to infinity:

$$\mathbf{F} = 4\pi\, \eta_e\, R\mathbf{U}. \tag{9.67}$$

The change in the numerical coefficient from 6π to 4π results from the fact that the boundary conditions are different in the two cases. In a real experiment, values intermediate between 6π and 4π are obtained if surfactants are present in the liquid and get attached to the interface, making it more rigid. This is often the case in real life for a bubble rising through water containing even small amounts of impurities.

Frictional force on an object of arbitrary shape, the concept of a Stokeslet

At a large distance $L \gg R$ from a sphere moving in a fluid at rest, the terms of order greater than $1/r$ in Equations 9.36a-b for the velocity field become negligible, so that:

$$v_r = \frac{3}{2} U R \frac{\cos\theta}{r}, \tag{9.68a}$$

and:

$$v_\theta = -\frac{3}{2} U R \frac{\sin\theta}{2r}. \tag{9.68b}$$

By combining these equations with the expression $F = 6\,\pi\,\eta\,R\,U$ of the force exerted by the sphere on the fluid (opposing the Stokes drag force), they become:

$$v_r = \frac{F}{4\pi\,\eta} \frac{\cos\theta}{r}, \tag{9.69a}$$

and:

$$v_\theta = -\frac{F}{4\pi\,\eta} \frac{\sin\theta}{2r}. \tag{9.69b}$$

By using Equation 9.51 we can also express the pressure at large distances as a function of F by:

$$p = \frac{F}{4\pi} \frac{\cos\theta}{r^2}. \tag{9.70}$$

It can be shown that Equations 9.69a–b and 9.70 apply to a body of arbitrary shape, provided that it has three finite dimensions (of order of magnitude denoted by R for the largest one), that the distance r is large compared to these latter dimensions ($r \gg R$), and that the axis $\theta = 0$ of the coordinate system corresponds to the direction of the force $\mathbf{F}$. These equations remain valid even if the body lacks spherical symmetry and/or that the direction of the force $\mathbf{F}$ is not parallel to that of the velocity $\mathbf{U}$ of the body.

It may seem surprising that it is the force exerted on the fluid which determines the velocity at large distances, and not the velocity of the body. As a matter of fact, the latter velocity results in boundary conditions at a short distance and their effect is no longer observed far from the body. In contrast, the force $\mathbf{F}$ exerted by the body on the fluid generates momentum per unit time. In order to ensure that the flow field is stationary, this contribution must be compensated by an outward momentum flux far away from the body: $\mathbf{F}$ must then be equal to the overall value of this flux through a surface surrounding the body, such as a sphere. This value must be independent of the radius of the sphere of integration provided the latter is much greater than R. In the Stokes regime, momentum transport is achieved only by viscous diffusion (we will see further on that this is true only if $r/R \ll 1/Re$): the local flux of momentum is then determined by the gradients of the components of the velocity, which vary as $1/r^2$ if the velocity varies as $1/r$. The integral of the flux of the sphere of radius r should be of the order of magnitude of the product of the surface $4\pi r^2$ by the local flux varying as $1/r^2$ and can thus be a constant (such would not be the case for any other dependence of the velocity as a function of r). The velocity field at large distances is thus determined by the local diffusive flux $\mathbf{F}$ of the momentum and not by the flow near the body.

In fact, this velocity field would be the same if the force $\mathbf{F}$ was applied at a single point, in which case Equations 9.69a–b would be valid at all distances from that point: one refers to such a velocity and pressure field as a *Stokeslet*. Actually, the velocity field resulting from the motion of a large number of particles is well modeled, except close to the particles, by that created by a set of Stokeslets, each coincident with the center of the particles: simulation techniques based on this approach are known as *Stokesian dynamics*.

Frictional force on an object of arbitrary shape

We have already suggested this problem in Section 9.3.2. For each geometry, we will have different coefficients relating the components of the force to those of the velocity.

The minimum energy dissipation theorem (Section 9.2.3) allows us to evaluate the Stokes drag force acting on a cube of edge $2L$ moving with velocity $\mathbf{U}$ (Figure 9.15). It should first be noted that, as shown above, (Section 9.3.2, example (ii)), the force $\mathbf{F}$ acting on the cube is independent of its orientation relative to $\mathbf{U}$; it is, moreover, always opposite to $\mathbf{U}$. Consider the sphere (S) of radius $L\sqrt{3}$ *circumscribing* the cube (C). Let us now replace the true velocity field around (C), $\mathbf{v}_c(\mathbf{r})$, by a superposition of the velocity $\mathbf{v}_s(\mathbf{r})$ around the sphere (assumed solid) and of a uniform field between (S) and (C) corresponding to the velocity $\mathbf{U}$. This second solution, obeying the boundary conditions on (S) and (C), corresponds to an energy dissipation higher than that of the true solution. By applying the Stokes law to the flow around this sphere, we obtain an *upper limit* for the energy dissipated, namely:

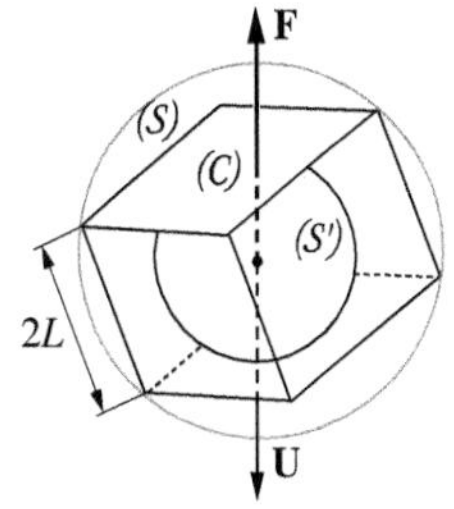

Figure 9.15 *Geometrical illustration for the approximate calculation of the hydrodynamic drag force* $\mathbf{F}$ *on a cube (C) moving at velocity* $\mathbf{U}$

$$-\mathbf{F}.\mathbf{U} < 6\pi\,\eta\,\sqrt{3}\,LU^2;$$

F and U are the magnitudes of the force and the velocity, while $-\mathbf{F}\cdot\mathbf{U}$ is the true rate of energy dissipation around the cube (C). The right-hand term of the inequality represents the power dissipated by a sphere of radius $L\sqrt{3}$, while the rate of dissipation between the sphere and (C) is zero, since the velocity is uniform. Consider now similarly the sphere (S'), of radius L, *inscribed* within (C); the rate of dissipation due to (S') is lower than that obtained for the solution corresponding to uniform flow of velocity $\mathbf{U}$ between (S') and (C) and to the unknown flow pattern around (C). Accordingly, this sets a *lower limit* to the dissipation: $6\pi\,\eta\,L\,U^2$. Combining these two inequalities, one concludes that the force F, acting on the cube, is bounded by:

$$6\pi\,\eta\,L\,|U| < |F| < 6\pi\,\eta\,\sqrt{3}\,L\,|U|.$$

It should be noted that these two limits bracket the exact solution rather closely. In general terms, the force acting on a finite-size object, of maximum linear dimension $L_{\max}$, is very nearly the same as that acting on a sphere of diameter $L_{\max}$.

The correspondence with the spherical case discussed in the text at the right allows us to evaluate the orders of magnitude of the sedimentation velocities of macromolecules of polymers which have, in solution, a configuration of a ball, or of colloidal aggregates; these are non-compact objects for which we can still define a *hydrodynamic radius* R_h compatible with the application of the Stokes equation. This radius R_h remains of the order of magnitude of the radius of the sphere circumscribing the object. One can find the value of this hydrodynamic radius by light scattering techniques.

Forces on long cylindrical objects

For a long rod of radius R and length L, aligned parallel ($F_\parallel$) or perpendicular ($F_\perp$) to the force of gravity, a detailed calculation leads to the equations:

$$F_\parallel \simeq -\frac{2\pi\eta LU}{\mathrm{Log}\left(\frac{L}{R}\right) - \frac{3}{2} + \mathrm{Log}2} \quad (9.71a) \quad \text{and:} \quad F_\perp \simeq -\frac{4\pi\eta LU}{\mathrm{Log}\left(\frac{L}{R}\right) - \frac{1}{2} + \mathrm{Log}2}. \quad (9.71b)$$

The ratio of these values of the forces is quite close to two: this is the result which we have used for computing both the angle from vertical at which a cylindrical rod falls, and the force resulting from the rotation of a helix (Section 9.3.2). The values of the two forces are quite close to the force on a circumscribed sphere of radius $L/2$ as is also the case for other geometrical shapes. This is another example of the long range of hydrodynamic interactions at low Reynolds numbers: as a result, the details of the shape of a moving object influence very weakly the flow velocity field far away from this object or the forces on it.

9.5 Limitations of the Stokes description at low Reynolds numbers

9.5.1 Oseen's equation

The proof of the Stokes equation at the beginning of this chapter is based on the assumption that the inertial terms $(\mathbf{v}\cdot\nabla)\mathbf{v}$ and the non-stationary term $\partial\mathbf{v}/\partial t$ are negligible relative to the viscous terms in the equation of motion. We will show, starting from the case of flow around a sphere, that these assumptions are no longer obeyed at a sufficiently great distance from the sphere. The Stokes equation is then no longer valid and needs to be replaced by Oseen's equation.

Kinetic energy of the fluid far from the sphere

At a large distance from the sphere ($r \gg R$), the values of the velocity components (Equations 9.36a-b) can be approximated by:

$$v_r \approx \frac{3}{2}\frac{R}{r}U\cos\theta \quad (9.72a) \quad \text{and:} \quad v_\theta \approx -\frac{3}{4}\frac{R}{r}U\sin\theta. \quad (9.72b)$$

One infers a lower limit for the kinetic energy of the fluid e_k per unit volume by writing:

$$e_k = (1/2)\,\rho\,v^2\,(r) > (9/32)\,\rho\,U^2\left(R^2/r^2\right).$$

The kinetic energy $(\mathrm{d}E_k/\mathrm{d}r)$ in the volume between the spheres of radii r and $r + \mathrm{d}r$ thus satisfies:

$$\mathrm{d}E_k = e_k\,4\pi\,r^2\,\mathrm{d}r > (9\pi/8)\,\rho\,U^2\,R^2\,\mathrm{d}r. \quad (9.73)$$

Integration over the entire physical space then gives an infinite total energy. Therefore, Stokes equation cannot be valid at a large distance from the sphere.

Convection and acceleration effects far from the sphere – Oseen's equation

Assume a sphere ($\mathcal{S}$) moving at a velocity $\mathbf{U}$ within a viscous fluid which is at rest at infinity; the Reynolds number is taken equal to: $Re = (2\rho\,U\,R/\eta) \ll 1$. Let us now evaluate the order of magnitude of the various terms of the Navier–Stokes equation at a large distance $L \gg R$ from the sphere. The order of magnitude of the velocity at this distance is $v \approx U(R/L)$ (Equations 9.72a–b). In the reference frame of a fixed observer, located at the distance L from $(\mathcal{S})$, the velocity field of the fluid induced by the motion of $(\mathcal{S})$ is not stationary, even if the velocity of $(\mathcal{S})$ is constant: the velocity which is given by Equations 9.36a–b depends indeed on the distance of the observer to the sphere $(\mathcal{S})$ which varies due to the motion of the latter. One has then:

$$\frac{\partial \mathbf{v}}{\partial t} = -(\mathbf{U}\cdot\nabla)\,\mathbf{v}, \qquad \text{so that:} \qquad \rho\left|\frac{\partial \mathbf{v}}{\partial t}\right| \approx \rho\, U\left|\frac{\partial \mathbf{v}}{\partial r}\right| \approx \rho\, U\left(U\frac{R}{L}\right)\frac{1}{L} = \rho\frac{U^2 R}{L^2}.$$

Proof

Let us call O and O' the positions of the center of the sphere respectively at times t and $t + \mathrm{d}t$. If we define the radius vector $\mathbf{r} = \mathbf{OM}$ with $r \gg R$, the radius vector at time $t + \mathrm{d}t$ will be $\mathbf{O'M} = \mathbf{O'O} + \mathbf{r} = -\mathbf{U}\,\mathrm{d}t + \mathbf{r}$. The change in velocity at the fixed point M between t and $t + \mathrm{d}t$ will then be: $\mathbf{v}(\mathbf{r} - \mathbf{U}\,\mathrm{d}t) - \mathbf{v}(\mathbf{r}) = -(\mathbf{U}\cdot\nabla)\,\mathbf{v}\,\mathrm{d}t$. By writing that this change represents $\partial\mathbf{v}/\partial t\,\mathrm{d}t$, one obtains the order of magnitude needed.

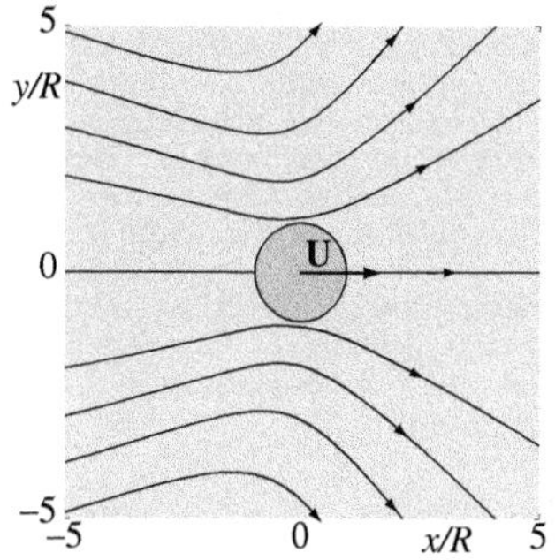

Figure 9.16 *Asymmetric flow around a sphere moving with constant velocity (Re = 0.5). The velocity field was obtained solving Oseen's equation in a reference frame where the fluid is at rest at infinity. This flow configuration should be compared to that in Figure 9.10a which corresponds to a Reynolds number much smaller than one. (Plate courtesy of E. Guazzelli, IUSTI)*

The term $\rho\,\partial\mathbf{v}/\partial t$ thus decreases as $1/L^2$, while the viscous stress term is of the order of:

$$\eta\left|\nabla^2\mathbf{v}\right| \approx \eta\left(U\frac{R}{L}\right)\frac{1}{L^2} \approx \eta\frac{UR}{L^3},$$

and which thus varies as $1/L^3$. The ratio of these two terms is then of the order of $\rho UL/\eta$, and increases with the distance. Therefore, at a distance L of the sphere such that:

$$\frac{L}{R} \approx \frac{1}{Re} \qquad \text{or:} \qquad L \approx \frac{\nu}{U},$$

one has $\rho UL/\eta \approx 1$ and the assumption of quasi-stationarity is no longer valid. On the other hand, still for a fluid at rest at infinity, the convection transport term $(\mathbf{v}\cdot\nabla)\,\mathbf{v}$ satisfies:

$$|\rho\,(\mathbf{v}\cdot\nabla)\,\mathbf{v}| \approx \rho\,\frac{U^2R^2}{L^3} \ll \eta\left|\nabla^2\mathbf{v}\right| \approx \eta\frac{UR}{L^3}.$$

It varies therefore as $1/L^3$ and remains always negligible if $Re \ll 1$.

Let us now analyze the case of a motionless sphere within a fluid which moves at constant velocity $\mathbf{U}$ at infinity. In this case, in contrast, the non-stationary term $\partial\mathbf{v}/\partial t$ vanishes identically for a fixed observer, whatever the location of the latter. At a sufficiently large distance the velocity $\mathbf{v}$ of the fluid tends toward $\mathbf{U}$ and $(\mathbf{v}\cdot\nabla)\mathbf{v}$ is very nearly $(\mathbf{U}\cdot\nabla)\mathbf{v}$. It is then the convective term which is of order $\rho U^2R/L^3$ and is no longer negligible at large distances relative to the viscous dissipation term.

Thus, in an infinite volume of fluid, the Stokes equation is only an approximation valid sufficiently close to the solid; because of the weak decrease in the velocity field, the corresponding errors can become sizable. Far from the sphere, one then needs to replace, as a first approximation, the Stokes equation by *Oseen's equation*:

$$\rho\,(\mathbf{U}\cdot\nabla)\,\mathbf{v} = -\nabla p + \eta\nabla^2\mathbf{v}. \tag{9.74a}$$

If Oseen's equation gives a better description for the flow than Stokes equation far from the sphere, it is not correct near the sphere because its estimate of the non-linear terms or of the acceleration is incorrect. More complicated methods (matched asymptotic expansions) are needed to tie together the two types of solutions in the region of the distances $R \ll L \ll R/Re$. The actual value of $\mathbf{F}$ is intermediate between those predicted by the Stokes equation (Equation 9.53) and by Equation 9.75 and satisfies approximately:

$$F \simeq -6\pi\,\eta R U \sqrt{1 + \frac{3}{16} Re} + O(Re^2).$$

Going back to the first case where the solid moves at constant velocity in a fluid at rest at infinity; one just needs to add to the Stokes equation the time-varying term $\rho\, \partial\mathbf{v}/\partial t = -(\mathbf{U} \cdot \nabla)\, \mathbf{v}$ obtained above, so that:

$$\rho\,(-\mathbf{U} \cdot \nabla)\, \mathbf{v} = -\nabla p + \eta \nabla^2 \mathbf{v}. \qquad (9.74b)$$

Using these equations, one obtains the corrected value for the drag force $\mathbf{F}$:

$$F = -6\pi\, \eta\, R\, U \left(1 + \frac{3}{16} Re\right) + O(Re^2), \qquad (9.75)$$

where $$Re = 2RU/\nu.$$

In contrast with the Stokes flows, the resulting velocity fields are asymmetric relative to the diametral plane perpendicular to the flow. Figure 9.16 above displays a moving sphere within a fluid at rest at infinity. The streamlines are closer together behind the sphere than in front of it (the vorticity is more concentrated downstream). In fact, for the length scales $L \approx R/Re$, the vorticity, locally created by the motion of the sphere, does not diffuse rapidly enough to spread out equally between the upstream and downstream regions, and is dragged preferentially to the rear. We shall see in Chapter 10 (Section 10.7) that, at high Reynolds numbers, or at large distances $L \gg R/Re$, the asymmetry is such that the velocity gradients are restricted to a narrow wake downstream of the body (this is however only the case for well-shaped aerodynamic bodies, behind which the wake remains laminar).

The previous analysis remains valid far away from a solid of arbitrary shape, provided that all three of its dimensions are finite. The dominant term of the velocity field is, indeed, still of the order of $1/r$ at large distances from the solid, and the previous approximations remain applicable.

9.5.2 Forces on an infinite circular cylinder in a uniform flow ($Re \ll 1$)

Assume now an infinitely long cylinder of radius R located in a flow of velocity $\mathbf{U}$ perpendicular to its axis. The crux of this problem is the *Stokes paradox*: there exists no solution of the Stokes equation which obeys, simultaneously, the boundary conditions at the surface of the cylinder and at infinity. A physical explanation of this mathematical result is that two-dimensional flows governed solely by viscosity have a much slower decrease in the velocity and pressure fields than those around a sphere. By using the same kind of argument as in Section 9.4.1, one finds a variation of the velocity as Log r and of the pressure as $1/r$ (instead of $1/r$ and $1/r^2$ for a sphere). The calculation of the kinetic energy carried out in Section 9.5.1 then indicates that the kinetic energy between two cylinders of radii r and $r + \mathrm{d}r$ increases with r: one can therefore not deal with such a flow in the Stokes approximation, which is indeed only valid, as shown above, at short distances.

Assume that the velocity $v(r)$ varies as $Uf(r)$ as a function of r for $r \gg R$ (R being the radius of the cylinder): the viscous stresses (as well as the changes in pressure) will then decrease as $\eta U \mathrm{d}f/\mathrm{d}r$. The flux of momentum must, this time, be calculated over cylinders of radius r and unit length (and no longer over spheres): the product $(2\pi r)\eta\; U \mathrm{d}f/\mathrm{d}r$ is proportional to the force F per unit length and must be independent of r. $r\,\mathrm{d}f/\mathrm{d}r$ must therefore be a constant, leading to $f(r) \propto \mathrm{Log}\, r$, $p(r) \propto 1/r$ and $F \propto \eta U$. With such a variation of $f(r)$, the kinetic energy per unit length between two cylinders of radii r and $r + dr$ must vary as $U^2 r\, \mathrm{Log}^2(r)$ and increases indeed with r.

It is thus Oseen's equation which, while not valid near the cylinder (just as in the case of a sphere), provides a first approximation for the drag force per unit length:

$$\mathbf{F} = \frac{4\pi\eta U}{(1/2) - \gamma - \mathrm{Log}(Re/8)} \cong \frac{4\pi\eta U}{2 - \mathrm{Log}\, Re}, \qquad (9.76)$$

where $Re = 2UR/\nu$ and $\gamma \cong 0.577$ is Euler's constant.

At distances small enough so that $Ur/\nu \ll 1$, the flow still obeys the Stokes equation. Seeking the solution for zero velocity at the cylinder boundary corresponding to the value of the force given by Equation 9.76, we find the stream function (Lamb, 1911):

$$\Psi = \frac{U}{2\,(2 - \text{Log}\, Re)} \left(2\, r\, \text{Log}\, \frac{r}{R} - r + \frac{R^2}{r} \right) \sin\varphi,$$

with $Re = 2\ RU/\nu$ and the radial and tangential components of the velocity:

$$v_r = \frac{1}{r}\frac{\partial \Psi}{\partial \varphi} = \frac{U}{2\,(2 - \text{Log}\, Re)} \left(2\, \text{Log}\, \frac{r}{R} - 1 + \frac{R^2}{r^2} \right) \cos\varphi \qquad (9.77a)$$

and

$$v_\varphi = -\frac{\partial \Psi}{\partial r} = -\frac{U}{2\,(2 - \text{Log}\, Re)} \left(2\, \text{Log}\, \frac{r}{R} + 1 - \frac{R^2}{r^2} \right) \sin\varphi. \qquad (9.77b)$$

The term varying as Log r thus becomes dominant when one moves far enough from the cylinder so that $R^2/r^2 \ll 1$. Just like in the case of the sphere, a more exact calculation requires matched asymptotic expansions using the solutions of the Stokes equation near the cylinder and Oseen's equation further away. We would find of course the same results for a moving cylinder within a fluid at rest (within a change of sign for the drag force and an additive term for the components of the cylinder velocity).

It is interesting to compare Equation 9.76 to Equation 9.71b for the force per unit length on a cylinder of finite length L. The two equations are similar but, again within a multiplicative coefficient, the ratio L/R is replaced by $1/Re$ for the infinite cylinder, so that L is replaced by ν/U. This suggests that ν/U represents a maximum distance beyond which the different parts of the length of the cylinder no longer influence one another.

In Section 9.5.1, ν/U is also the distance for which the non-stationary or convective terms of the Stokes equation become significant.

9.6 Dynamics of suspensions

The last two Sections 9.6 and 9.7 deal with flows at low Reynolds numbers in the case of suspensions and of porous media. They are distinct, first of all, from the previous sections because one deals with a large number of solid particles distributed in disorderly fashion within a flow; in the case of suspensions they are free to move, while they are fixed in porous media. Moreover, we are interested in average laws governing the flows, and not on the detailed velocity field around each particle. These topics are related to the extent that the laws for very concentrated suspensions tie in with those for porous media.

The problem of particles in suspension is of great practical importance. It covers the sedimentation of particles, the flow of suspensions (clay, drilling muds, cements, extruded foods such as varieties of pasta, etc.), as well as the behavior of *fluidized beds*; the latter, of large industrial importance, correspond to solid particles levitated by an upward injection of a fluid from the bottom of a container, resulting in a suspension of separated particles.

The behavior of suspensions vary greatly depending on the size L of the particles involved. In the case of very small Brownian particles (Section 1.3.1), effects due to thermal motion are very significant. For large particles, hydrodynamic effects dominate. The relative importance of these two mechanisms is measured by the *Péclet number*, Pe, defined in Chapter 2 (Equation 2.16) as:

$$Pe = U\,L/D,$$

where U is the flow velocity and D the Brownian diffusion coefficient for the particles. Here, the Péclet number is taken equal to the ratio of the diffusion time for spherical particles over a distance of the order of their radius R, to the convection time for particles due to the flow over the same distance R. In Chapter 1, the Brownian diffusion coefficient D for particles has been shown to satisfy (Equation 1.48):

$$D = \frac{k_B\,T}{6\pi\,\eta\,R},$$

where R is the radius of the particles, and η is the viscosity of the fluid. The characteristic time τ_D for Brownian diffusion of the particles over a distance R is $\tau_D \approx R^2/D$. The convection time of particles at the characteristic velocity $U \approx \dot{\gamma} R$ of the flow is $\tau_C \approx R/U \approx 1/\dot{\gamma}$ ($\dot{\gamma}$ is the shear rate of the flow). Thus, the Péclet number satisfies:

$$Pe = \frac{\tau_D}{\tau_C} = \frac{6\pi\, \eta\, \dot{\gamma}\, R^3}{k_B\, T}. \tag{9.78}$$

A priori, the change in behavior of a particle from Brownian for $Pe \leq 1$ to non-Brownian for $Pe \gg 1$ depends on the value of the shear rate $\dot{\gamma}$. However, because of the large exponent in the factor R^3, it is essentially the size of the particles which determines the crossover between the two types of behavior. In practical terms, we consider that the crossover between Brownian and non-Brownian particles occurs when their size is of the order of 1 μm.

Justification

Consider particles in water at ambient temperature. The limit $Pe \approx 1$ corresponds to a diameter of particles of the order of 1 μm for shear rates of $\dot{\gamma} \approx 1\,\mathrm{s}^{-1}$. The following coincidence allows one to remember the crossover between the two domains: particles 1 μm in diameter suspended in water at ambient temperature and in a flow with shear rates of 1 s^{-1}, have a Brownian diffusion coefficient $D \approx 1\ \mu\mathrm{m}^2/\mathrm{s}$, a corresponding characteristic time $\tau_D \approx R^2/D \approx 1$ s, and a Péclet number of the order of one. Taking the micrometer and the second as fundamental units for these different magnitudes, one summarizes this property by:

$$2R \approx D \approx \dot{\gamma} \approx \tau_D \approx Pe \approx 1. \tag{9.79}$$

An additional factor must be considered in the discussion of the hydrodynamic motion of small objects: it corresponds to the *van der Waals* forces between particles or, if the particles carry electric charges and the solvents have polar molecules, to the electrostatic forces. The corresponding (colloidal) interactions play a dominant role when particles come into very close proximity (at distances typically less than a hundred nanometers).

9.6.1 Rheology of suspensions

Dilute suspensions behave as homogeneous Newtonian fluids with a viscosity η greater than the viscosity η_0 of the fluid alone. The viscosity η obeys an equation derived by Einstein in 1905, in his famous paper on the theory of Brownian motion:

$$\eta = \eta_0\,(1 + 2.5\,C). \tag{9.80}$$

where $C\,(\ll 1)$ is the volume fraction occupied by the particles (the ratio of the total volume occupied by the spheres to the total volume of the suspension). This result is remarkably simple and general and does not depend on whether the particles are Brownian or non-Brownian nor on the type of flow. The only assumption on which it rests is that of a lack of hydrodynamic interactions between particles: this allows one to consider only the perturbation to the flow due to the presence of a single particle, and to sum this perturbation over all particles. Such a summation implies that the global contribution to the viscosity is proportional to the volume fraction C. The complete proof requires a computation of the increase in the dissipated energy of the suspension due to the presence of particles and is not straightforward.

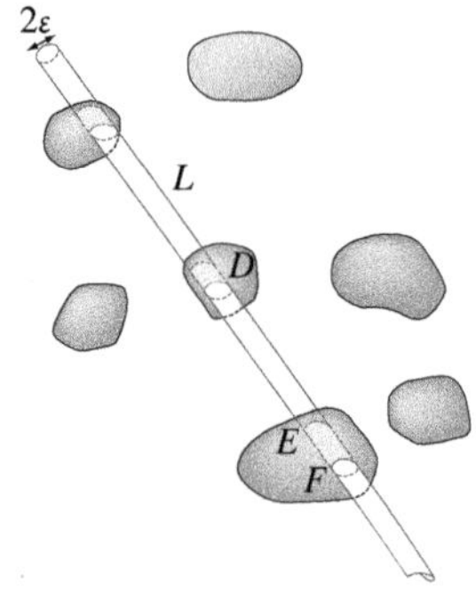

Figure 9.17 *Schematic diagram of a suspension of particles, and the manner of estimating its viscosity*

Qualitative justification for Equation 9.80

Imagine the transfer of momentum along a line L drawn at random in the suspension (Figure 9.17). Such transport is diffusive in the fluid region DE between two particles, but occurs quasi-instantaneously along the segment EF within a solid particle, which effectively accelerates the transfer. Overall, everything goes on as if the length of the line was shortened by the effect of the particles through which it passes. According to a theorem in stereology outlined below, the fraction of length located inside solid particles, along a line passing at random through the suspension, is equal to the volume fraction C of the particles. The ratio of the lengths DE and DF is then, on average:

$$\frac{DE}{DF} = 1 - \frac{EF}{DF} = 1 - C.$$

Writing the expression of the diffusion time τ_D of the momentum along the line DF in two different ways, one obtains:

$$\tau_D = \frac{DF^2}{\eta/\rho} = \frac{DE^2}{\eta_0/\rho} = \frac{[DF\,(1-C)]^2}{\eta_0/\rho}.$$

The first expression involves the overall viscosity η of the suspension of fluid and particles; in the second, one only includes the viscosity of the fluid, while taking into account the fact that diffusion through the fluid occurs only over the distance DE. Expanding to first order in C, we find $\eta_0 = (1 + 2\,C)$, agreeing reasonably with the exact result.

Relationship between the particle concentration and the intercepted fraction of a randomly drawn line: Along a line element L drawn at random through the heterogeneous suspension of particles (Figure 9.17), imagine that one constructs a cylinder of infinitesimal radius ε (where ε is much smaller than any characteristic dimension of the particles). The volume fraction of the above cylinder and the fraction of the length of a line L which is located inside these particles are equal: in the calculation of the volume ratio, the same factor $\pi\varepsilon^2$ appears indeed in the numerator and the denominator. Since the cylindrical tube was randomly drawn through the suspension, the volume fraction of solid matter within the tube equals the average volume fraction in the entire system, provided that the length L is much greater than the size of individual particles and that the suspension is homogeneous.

The variation predicted by Equation 9.80 is correct up to a volume fraction C of a few percent. For these values, the average distance between the particles is five to ten times their radius. Under these conditions, it is reasonable to neglect the effect of the presence of adjacent particles on the motion of a single particle, in spite of the slow decrease of the hydrodynamic perturbations that they generate (these perturbations are proportional to the ratio of the radius of the particle to the distance from one particle to its neighbor). For greater concentrations, the terms neglected in the previous approximation become larger than the first order correction of the viscosity. One must then take into account the hydrodynamic interactions between particles. As a first approximation, only the contribution of the interactions between pairs of particles need be considered; it is then summed over all pairs of particles in the suspension, giving a term in the viscosity of the suspension proportional to the square of the volume fraction C:

$$\eta_0 = \left(1 + 2.5\,C + k\,C^2\right). \tag{9.81}$$

However, despite the above simplifying assumptions, the calculation leading to this formula is rather elaborate (partly because of the slow decrease of the velocity perturbations resulting from the presence of particles). Moreover, the coefficient k of the C^2 term depends both on the nature of the flow to which the suspension is subjected, and on the Brownian motion of the particles (the value of k ranges between 5.2 and 7.6): these variations reflect the influence of the flow on the spatial arrangement of the particles (the formation of aggregates, the effect of the orientation for ellipsoidal particles, etc.). Thus, the viscosity of a suspension subjected to an elongational flow is not identical to that of the same suspension subjected to a shear flow because of the effects of the temporary formation of chains of particles. The viscosity depends therefore on the nature and the velocity of the flow to which the suspension is subjected, and can also be a function of time (see Section 4.4.3).

9.6.2 Sedimentation of particles in suspension

Sedimentation of dilute suspensions

The study of the influence of the flow induced by a moving sphere on the motion of another sphere (Section 9.4.3), has shown that this mutual interaction increases the sedimentation velocity of both spheres. For a suspension of a large number of particles, this effect is however hidden by the counterflow at velocity $\mathbf{v}_c$ around each particle (Figure 9.18a). Such flows are visible in Figure 9.11b: they ensure the conservation of the rate of flow of a fluid, assumed to be incompressible, throughout any cross-section perpendicular to the motion of the particles.

The sedimentation velocity V_S of a dilute suspension of spheres with volume fraction C is smaller than the velocity V_{S0} of an isolated sphere. In an idealized geometry, where sedimentation occurs toward an infinitely distant plane, we would obtain:

$$V_S(C) = V_{S0}\,(1 - 6.5\,C). \tag{9.82}$$

This equation is a first-order expansion in the concentration C and takes into account interactions between a pair of particles. A second-order term can be evaluated, but only through a much more complicated calculation which involves assumptions about the relative distribution of particles which is practically unknown. In addition, the calculations can no longer be restricted to nearest neighbor interaction due long-range effects (also resulting from the slow decrease of the velocity with distance) as $1/r$.

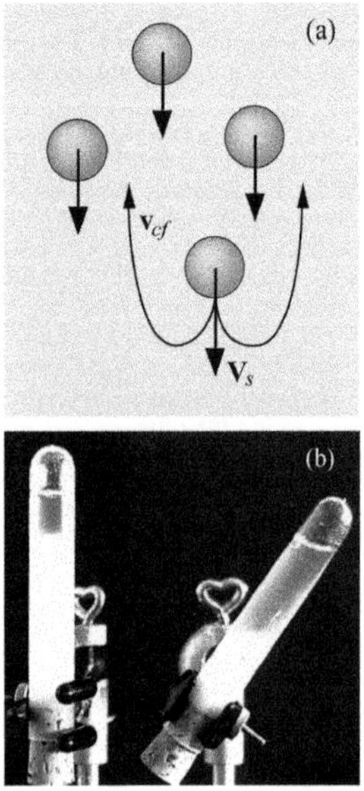

Figure 9.18: *(a) Local counterflow effect during the sedimentation of a suspension in a vertical tube. (b) Boycott effect, when sedimentation occurs in a tube at an angle to the vertical. The two tubes contain a suspension of the same concentration and were flipped at the same time (plates courtesy L. Petit)*

The effect of heterogeneities in the concentration, Boycott effect

When heterogeneities exist in the concentration of particles in a horizontal plane, there results a horizontal gradient of the average density in the suspension. As shown in Section 4.5.5, such a gradient induces a recirculation motion: depending on the local vertical component of the velocity, sedimentation will then be accelerated or slowed down. Such motion can be favored by the geometry of the walls (e.g., for a spherical wall).

A process with similar origin is the *Boycott effect,* discovered by A.E. Boycott while he observed the sedimentation of red blood cells in a tube tilted from the vertical: this occurs much more rapidly in this situation than when the tube is vertical (Figure 9.18b). On the one hand, in the vertical tube, one has only the sedimentation velocity predicted by Equation 9.82 and slowed down by counterflow. On the other hand, in a tilted tube, the component of the gravitational force transverse to the axis of the tube leads to a transverse segregation of particles toward the lower part of each cross-section. One then observes a horizontal gradient of the concentration, and thus of the density, leading to a recirculation flow. The latter is directed downward in the region where the particles are grouped: it drags them along and accelerates markedly the sedimentation.

Sedimentation of concentrated suspensions

Except for a global translational motion, the high concentration limit of such flows is very similar to the flow in a porous medium, where the overall particles are fixed and the fluid flows between them (Section 9.7.3).

Moreover, at high concentration, the velocity V_S varies more rapidly with C than the linear dependence predicted by Equation 9.82. One frequently uses the empirical equation of Richardson-Zaki: $V_S(C) = V_{S0}\,(1-C)^n$ ($n \approx 5.5$) valid at low values of Re and for $C \leq 0.5$.

At high concentrations C, the interactions between particles increase, and their relative motion becomes negligible as C approaches a limiting value C_m; C_m represents the concentration of a compact aggregate of particles falling under its own weight in the fluid.

During the sedimentation of a bed of similarly sized particle, rather abrupt variations are frequently observed in the concentration profiles (as shown in the pictures of Figure 9.19). For sufficiently high concentrations ($C > 20\%$) such experiments display an abrupt trailing front in the concentration. Trailing particles move faster than those deeper within the suspension, since the slowing effect due to nearby particles is less significant; they can therefore catch up to the main pack even if variations in individual velocities due to the dispersion of

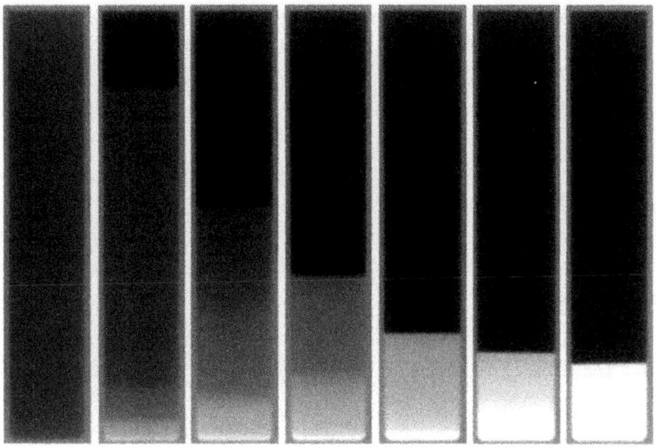

Figure 9.19 *Sedimentation of a suspension of particles observed by means of an X-ray scanner for times increasing from left to right. The gray levels correspond to the concentration of particles which is a maxium in the brightest regions. (Plate courtesy F. Auzerais)*

particle sizes are significant. There is also, in the lower part of the images of Figure 9.19, a leading front moving up against the motion of the particles: this front corresponds to the upper edge of the motionless sediment formed at the bottom of the container. There are also intermediate fronts due to the non-monotonic change of the flow rate $C\,V_S(C)$ as a function of C: this latter feature corresponds to the formation of a shock wave, which has some analogies with traffic jams on busy highways.

The flow rate $C\,V_S(C)$ of the particles starts from zero for $C = 0$ and displays a maximum for an intermediate concentration before dropping back to zero for the concentration of the static sediment. Such a variation leads to the appearance of discontinuities in C (one cannot have a continuous variation because the low concentration layers would catch up with the slower more concentrated ones below them).

9.7 Flow in porous media

9.7.1 A few examples

A porous medium is a piece of solid matter, often characterized in terms of its cavities, or pores, which occur either as an interconnected network, or as isolated hollows. One sees on the plates of Figure 9.20 two examples of such materials observed in scanning and optical microscopes. In the majority of practical instances, the size of the pores is sufficiently small and the flows sufficiently slow that the condition of low Reynolds number applies. The study of these materials leads to three types of problems, depending on the nature of the phases present within the pores.

A first category is that of flows for which the pores are saturated (i.e. completely filled) with a single fluid phase, as in the case of a soil fully saturated with water.

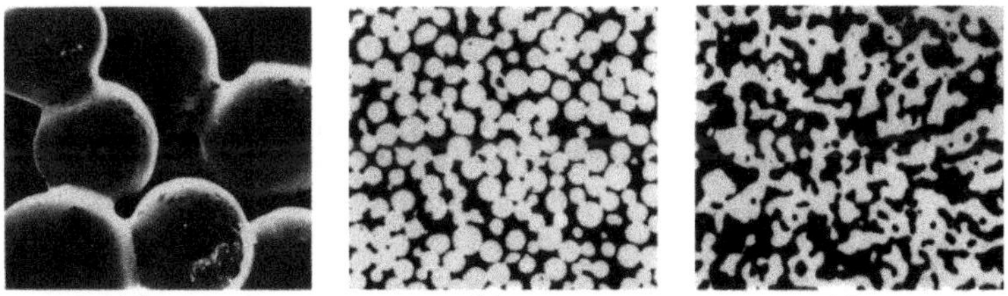

(a) (b) (c)

Figure 9.20 *(a) Scanning electron microscope image of a bronze powder sintered by heating it to a temperature of* 700 °C (500×). *(b) Cross-section of this same sintered material* (100×)*: the pore space appears in black on the image. (c) Cross-section of a sample of less regular cobalt powder: the distribution of pores, again shown in black, is more heterogeneous (Plates courtesy J.-P. Jernot). The porosity of such media, assumed homogeneous on a larger scale, can be estimated from the fraction of the cross-sectional surface corresponding to the pores (Section 9.7.2)*

In a second class of problems, two or more immiscible fluids coexist in the cavities. There are then a large number of interfaces between the different fluids. Each interface corresponds to the presence of a meniscus: capillary effects (see Section 1.4) must therefore be taken into account in order to characterize the respective flows of the different phases. Such flows are encountered in numerous situations: soils partly saturated with water (the second phase is high-humidity air) or mixtures of water and oil in petroleum-bearing rocks; they are discussed briefly in Section 9.7.6.

Finally, the transport of solid particles is very important in filtration problems, either on the surface of the filter or within its bulk. This generally affects the fluid flow, and leads to time-dependent hydrodynamic properties of the medium (by clogging of the filter).

9.7.2 Parameters characterizing a porous medium

Porosity

It is defined by the ratio:

$$\phi = \frac{\text{volume of the pores}}{\text{total volume}} = 1 - C, \tag{9.83}$$

where C is the packing fraction defined in Section 1.1.1. It can be shown that, for an isotropic, homogeneous medium, ϕ is also equal to the fraction ϕ_s of the surface corresponding to the pores in a plane cross-section of the material, as well as to the fraction ϕ_L of the length located inside the pores for a line randomly passing through the medium.

The equality of ϕ_S and ϕ is demonstrated by taking a given cross-section S and moving it in the perpendicular z-direction through a distance Δz: one thus sweeps out a volume $S\Delta z$ within which the pore volume is, by definition, $\phi S\Delta z$. Moreover, one can also calculate this pore volume by integrating in the z-direction the surface of each cross-section located inside the pores, which is equal to $\phi_S S$. By equating the result $\phi_S S\Delta z$ of this integration to the previously obtained value, one finds: $\phi = \phi_S$. The second equation is derived in the same manner as the one relating the concentration of particles in a suspension and the fraction of the length of a randomly selected line within these particles (see Section 9.6.1 and Figure 9.17).

Specific area

The *specific area*, denoted by S_V hereafter, is the surface of the walls of the pores per unit volume of a porous medium. It has the dimension of an inverse length.

For the case of pores of simple geometry (i.e. when their walls are almost smooth), the quantity S_V is of the order of the reciprocal of the local size of the pores. For example, for a cylinder of length L and diameter d, one finds:

$$S_V = \frac{(\pi\, d)\, L}{(\pi\, d^2/4)\, L} = \frac{4}{d}. \tag{9.84}$$

S_V can also be determined by counting, on the image of a cross-section of the medium, the number of points of intersection per unit length of the walls of the pores with a straight line crossing the porous medium. Going back to Figure 9.17, we see that the element of a cylinder that cuts the particles intercepts a total area of their walls equal to the product of the number of intercepts n by each segment, multiplied by the surface of the base of the cylinder S and by a numerical factor F which expresses the effect of the inclination relative to the axis of the cylinder of the elementary surfaces intercepted. The ratio of the surface $(n\,S\,F)$ to the volume $(S\,L)$ is then equal to $F\,n/L$. For an isotropic medium, it can be shown that $F = 2$: the specific area S_V is then equal, in that case, to twice the number of interceptions of the walls of the pores of the medium by a random line of unit length.

Classically, if a molecular species is allowed to be progressively adsorbed in the medium which has been previously well evacuated, the completion of the first adsorbed molecular monolayer can be detected by a step in the observed pressure. If a represents the size of the adsorbed molecules and V_a the total volume of the molecules adsorbed in the volume V of the porous medium, one has:

$$S_V = \frac{V_a}{(V a)}. \tag{9.85}$$

The specific area is an essential paramenter for all processes involving adsorption and catalysis. Such processes depend indeed strongly on the area available within the material for chemical or physico-chemical reactions. For the case where the grains are themselves porous, we must distinguish between the effects of intragranular porosity, which has the greatest contribution to the effects of adsorption, and of intergranular porosity, which determines largely the value of the total porosity.

For many applications, one characterizes frequently the specific area of a medium by dividing S_V by its density, giving thus the surface of the pores for unit mass of the medium. This ratio reaches several million m^2/Kg for activated charcoal or divided catalysts, and of the order of 1000 m^2/Kg for a set of grains of the order of a micrometer.

Parameters such as the porosity or the specific area are traditionally obtained by averaging over a certain representative elementary volume ($\mathcal{REV}$) of the medium defined in Section 1.3.2. The size of this volume must be at least ten pore lengths for very homogeneous materials; it is much greater, and even sometimes impossible to define, for heterogeneous media, or for those whose structure displays a very broad range of characteristic length scales.

Tortuosity

The transport properties of porous media (the flow of fluids or passage of an electric current) involve other geometrical characteristics than the specific area and the porosity. The topology of the medium (the number of pores to which every pore is connected), the complexity of the paths through the pore space, the characteristic sizes of the pores and of the channels which join them, also have a significant effect. Moreover, one must take into account dangling arms (the railroad dead end sidings in the flow), which are particularly significant in highly heterogeneous media with small numbers of pores. To take into account these different factors, one defines a parameter T, called the *tortuosity*, in terms of the effective electrical conductivity σ_p of the porous medium, intrinsically an insulator, when saturated with a conducting fluid of known conductivity σ_f:

$$T = \phi \, \frac{\sigma_f}{\sigma_p}. \tag{9.86}$$

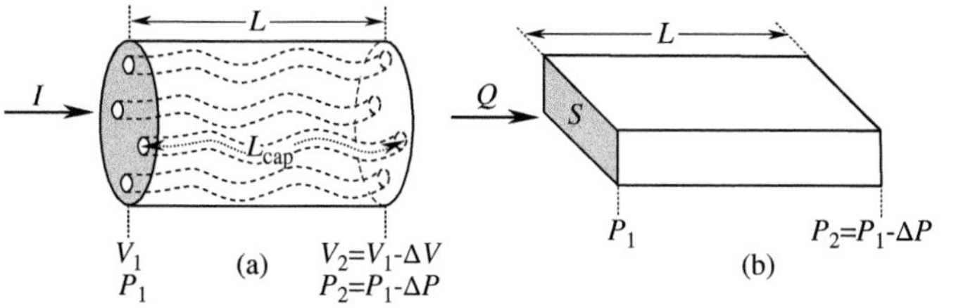

Figure 9.21 *(a) Model of the geometry of the porous medium in the form of a network of wavy capillaries. This model allows defining the tortuosity of the porous medium on the basis of its electrical conductivity obtained by applying a potential difference ΔV and measuring the value I of the electrical current through the medium. (b) Parameters involved in the flow of a fluid through a porous medium*

In order to appreciate the practical significance of the tortuosity, a porous medium is modeled by means of a system of wavy capillaries (Figure 9.21a). One finds in this case:

$$T = \left(\frac{L_{cap}}{L}\right)^2, \tag{9.87}$$

where L_{cap} is the curvilinear length of the capillaries, and L the length of the sample. For wavy capillaries, L_{cap} is greater than L, and the tortuosity is greater than one; in contrast, T equals one if all capillaries are straight. From a general theoretical point of view, T corresponds to the scattering of the electric field by the obstacles in the medium.

Proof

Let us calculate the effective electrical conductivity σ_p of a porous medium, saturated with a conducting fluid. Here, ΔV is the electrical potential difference between two cross-sections of area S of the medium (Figure 9.21a), and I the electric current passing through it; one has:

$$\Delta V = (1/\sigma_p)\ (L/S)\, I. \tag{9.88}$$

Similarly for each individual capillary, one can write:

$$\Delta V = (1/\sigma_f)\ (L_{cap}/S_{cap})\, I_{cap}, \tag{9.89}$$

where L_{cap} represents the average length of the capillaries. The ratio of these two equations satisfies:

$$\frac{\sigma_p}{\sigma_f} = \frac{L}{L_{cap}} \frac{I}{I_{cap}} \frac{S_{cap}}{S} = \left(\frac{L}{L_{cap}}\right)^2 \phi, \tag{9.90}$$

because $I = N\, I_{cap}$ and $\phi = (N\, S_{cap}\, L_{cap})/(S\, L)$ where N is the number of capillaries in the surface S. By using the definition of Equation 9.86, one finds Equation 9.87 for the tortuosity. This model will be used below (Section 9.7.4) to calculate the flow rate Q of the fluid induced through such a medium by a pressure difference ΔP.

9.7.3 Flow in saturated porous media–Darcy's Law

One-dimensional, low velocity flow

One is interested here in saturated porous media for which the pore space is completely filled with a single fluid, assumed to be Newtonian and incompressible. If the flow is slow enough so that the Reynolds number based on the pore size and the local velocity is much smaller than one, the pressure gradients under stationary conditions will be proportional to the flow velocities in the pores (Poiseuille's law applied to each pore). This proportionality, valid in the case of each individual pore, is maintained after averaging the flow rate and the pressure gradients over a volume large relative to the size of the pores. For a homogeneous sample of length L and constant cross-section S (Figure 9.21b), and for a pressure gradient oriented along the length, the flow rate Q satisfies therefore (*Darcy's law*):

$$Q = \frac{K}{\eta}\, S \frac{\Delta P}{L}. \tag{9.91}$$

The constant of proportionality K is the *permeability,* which is one of the characteristics of the porous medium. As shown below (Section 9.7.4), it has the dimensions of a surface area and is of the order of magnitude of the cross-section of an individual pore. A common unit for the permeability is the Darcy ($\cong 1\ \mu m^2$) which is very appropriate for the orders of magnitude encountered in natural porous media (see Table 9.1).

Table 9.1 – A few typical values of the permeability for porous media.

Medium	Permeability (Darcy)
Soil	0.3 – 15
Brick	0.005 — 0.2
Limestone	0.002 — 0.05
Sandstone	0.0005 — 5
Cigarette	1000
Glass fibres	20 — 50
Sand	20 — 200
Silica powder	0.01 — 0.05

Darcy equation generalized to three dimensions

In three dimensions, and taking into account gravity, Equation 9.91 can be generalized for an isotropic medium to:

$$\mathbf{v}_s = \frac{Q\, \mathbf{n_u}}{S} = -\frac{K}{\eta}\,(\nabla p - \rho\, \mathbf{g}), \tag{9.92}$$

where $\mathbf{n_u}$ is a unit vector. Each component v_{si} of $\mathbf{v}_s$ corresponds to the rate of flow Q_i through a unit area with its normal in the direction of the corresponding axis i. The velocity $\mathbf{v}_s$ is known as the *Darcy velocity* (also called *superficial* or *filtration velocity*).

If K and η are constant, one has, following Equation 9.92:

$$\nabla \times \mathbf{v}_s = 0, \quad \text{where:} \quad \mathbf{v}_s = -\nabla \Phi, \tag{9.93}$$

with:

$$\Phi = \frac{K}{\eta}\,(p + \rho g z).$$

The real average velocity $\mathbf{v}_p$ within the pores, often called interstitial velocity, can in fact be much greater than $\mathbf{v}_s$ since only part of the total volume of the medium is available for the transport of fluid. Thus, for a set of parallel capillaries of total porosity ϕ, $\mathbf{v}_p$ will satisfy $\mathbf{v}_p = \mathbf{v}_s/\phi$. In fact, the average flow rate per unit cross-section of the material perpendicular to the capillaries is, on the one hand, equal to $\mathbf{v}_s$ and, on the other hand, equal to the product of $\mathbf{v}_p$ by the fraction of the cross-sectional surface made up by the pores; this is equal to ϕ as seen in Section 9.7.1. For a homogeneous but anisotropic porous medium, such as a layered medium, one generalizes Equation 9.92 by substituting a tensor for the scalar K.

If the fluid is incompressible, the velocity field $\mathbf{v}_s$ satisfies $\nabla \cdot \mathbf{v}_s = 0$, so that:

$$\nabla^2 \Phi = 0. \tag{9.94}$$

The superficial velocity field $\mathbf{v}_s$ can thus be derived from a potential Φ with vanishing Laplacian, just as in the case of the velocity field of an ideal fluid. But, because of the small size of the pores, the flow of a viscous fluid in a porous medium is one of the cases where the effect of the viscosity is greatest and the behavior of the fluid is very different from that of an ideal fluid. The explanation for this paradox comes from the fact that $\mathbf{v}_s$ is not a local velocity, but a *macroscopic* velocity field defined by averaging over a volume large compared to that of the pores. Viscosity effects, which are significant at small length scales, are then averaged out.

Applying Darcy's Law to the flow in a porous embankment

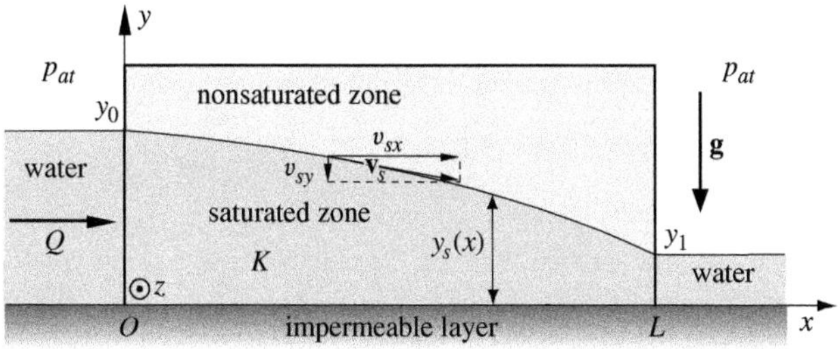

Figure 9.22 *Flow between two water tables at different levels ($y = y_0, y_1$) through a porous embankment bounded by the planes $x = 0$, $x = L$. The geometry is assumed to be two-dimensional*

As an example, let us analyze the flow through a porous embankment of permeability K and length L which separates two water volumes at different levels y_0 and y_1 (Figure 9.22). One assumes a stationary regime in which the water level $y_s(x)$ within the medium is constant in time, with a flow rate Q per unit distance along z constant in each cross-section ($x =$ constant) and independent of x. If the slope of the free surface is sufficiently small inside the porous medium (*Forcheimer's approximation*), we will show below that the dependence of the water level on distance satisfies:

$$\frac{y_0^2 - y_s^2(x)}{y_0^2 - y_1^2} = \frac{x}{L}. \tag{9.95}$$

One obtains a parabolic profile, just as for a water jet freely flowing out of a container (not withstanding the complete difference in the nature of the forces involved). The flow rate Q satisfies:

$$Q = K\, g\, \frac{y_0^2 - y_1^2}{2\, L\, \nu}. \tag{9.96}$$

Proof

The part of the porous medium which is not saturated by water is filled with air; if one neglects the effects of the surface tension, the water pressure just below the surface $y_s(x)$ is therefore equal to the atmospheric pressure p_{atm}. Let us now write the equality of the pressure between two points on the surface located at x and $x + \delta x$. One obtains $p(x, y_s(x)) = p(x + \delta x, y_s(x + \delta x))$ whence, by expanding:

$$\frac{\partial p}{\partial x}\,\delta x + \frac{\partial p}{\partial y}\,\delta y_s = 0. \tag{9.97}$$

If $\partial y_s(x)/\partial x$ is sufficiently small, one can consider that the component v_{sy} of the surface velocity $\mathbf{v}_s$ is much smaller than the component v_{sx} and that the pressure gradient $\partial p/\partial y$ in the y-direction reduces to the hydrostatic pressure gradient $-\rho\, g$ (these assumptions are quite similar to those used for quasi-parallel flows). One then obtains, from Equation 9.97, the horizontal pressure-gradient:

$$\frac{\partial p}{\partial x} = -\frac{\mathrm{d}y_s(x)}{\mathrm{d}x}\,\frac{\partial p}{\partial y} = \rho g\,\frac{\mathrm{d}y_s(x)}{\mathrm{d}x}. \tag{9.98}$$

Applying Darcy's law (Equation 9.92) allows one to calculate the horizontal component $v_{sx}(x)$ of the flow rate velocity, which has the value: $-(K/\eta)\ \partial p/\partial x$ ($\partial p/\partial x$ and, consequently, $v_{sx}(x)$ are both independent of y). The flow rate $Q = v_{sx}(x)\ y_s(x)$ (constant with x) is then:

$$Q = -\rho\, g\,\frac{K}{\eta}\; y_s(x)\;\frac{\mathrm{d}y_s(x)}{\mathrm{d}x} = -\frac{K\,g}{\nu}\; y_s(x)\;\frac{\mathrm{d}y_s(x)}{\mathrm{d}x}. \tag{9.99}$$

Equation 9.96 and the profile $y_s(x)$ are then obtained by integrating Equation 9.99 respectively between 0 and L and between 0 and x. Combining these last two results provides finally Equation 9.95.

Non-linear flow regimes in porous media

Equation 9.92 is valid at low flow velocities such that the Reynolds number *Re* (calculated by taking the size of the pores as the length scale) is much less than one. At higher velocities (and values of *Re* of a few units), the relationship between pressure and flow rate is no longer linear. It can be represented by a power law:

$$|\boldsymbol{\nabla} p| \propto v^n, \tag{9.100}$$

with the exponent n of the order of 2. This dependence indicates the effect of non-linear convective terms in the equation of motion of the fluid. In this case, there is still no turbulence in the pores: even with a laminar flow, the great changes in magnitude and direction of the velocity from one pore to another can render significant the effect of terms such as $(\mathbf{v}\cdot\boldsymbol{\nabla})\mathbf{v}$. This represents a case opposite to that of parallel flows, where such terms can remain negligible even for Reynolds numbers significantly greater than one.

At higher Reynolds numbers, recirculation zones can also appear in the fluid behind the grains that make up the porous medium. Such zones are of the same nature as those described in Sections 2.4.2 and 2.4.3 while discussing the evolution of flow behind a cylinder or a sphere as a function of the Reynolds number.

A two-dimensional model of a porous medium: the Hele-Shaw cell

A *Hele-Shaw cell* (Figure 9.23 below) consists of two parallel plates very close to one another and separated by spacers of size L and thickness $a \ll L$, equal to the separation between the plates, acting as obstacles. This system is used to represent the velocity fields of two-dimensional potential flows of ideal fluids around an obstacle: an example corresponding to flow around a cylinder is shown in Chapter 6 (Figure 6.5).

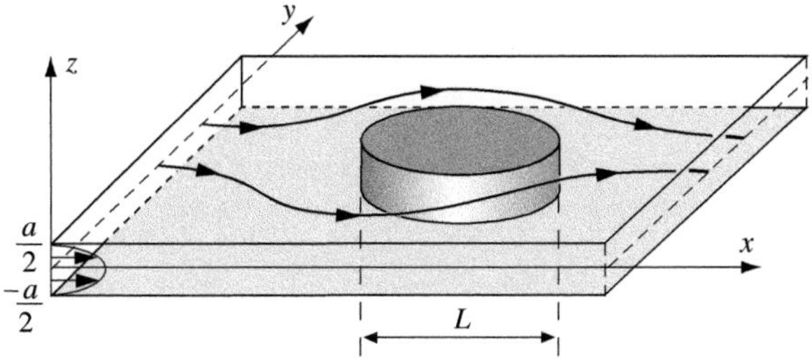

Figure 9.23 *Geometry of a Hele-Shaw cell with an obstacle of length L, large relative to its thickness a.*

Just as for the case of porous media, this result appears at first glance paradoxical. Given the proximity of the plates, viscous friction forces will play an essential role on the characteristics of the flow. In fact, just as in the case of porous media, it is the average velocity field over the gap between the plates which can be derived from a potential, not the *local velocity*.

Between the plates, the flow is nearly parallel to the x-y-plane and one therefore has: $v_z \approx 0$. This can be explained on the basis of the equation of the conservation of mass, which reads: $\partial v_x/\partial x + \partial v_y/\partial y + \partial v_z/\partial z = 0$. Because the length scales in the z-direction (of order a) are very small relative to those in the x- or y- directions (of order L), one can indeed estimate that the order of magnitude of the component v_z is:

$$v_z \approx (a/L)\, v_{x,y} \ll v_{x,y}. \tag{9.101}$$

The great difference between the length scales normal and parallel to the plane of the plates further allows us to write the following inequalities , also valid replacing v_x by v_y:

$$\frac{\partial^2 v_x}{\partial x^2} \approx \frac{\partial^2 v_x}{\partial y^2} \ll \frac{\partial^2 v_{x(y)}}{\partial z^2}. \tag{9.102}$$

Moreover, the inertial terms $\rho\,|(\mathbf{v}\cdot\nabla)\mathbf{v}|$ are negligible relative to those $|\eta\nabla^2\mathbf{v}|$ since the flow occurs at low Reynolds numbers. Keeping in mind Equations 9.101 and 9.102, the Navier–Stokes equation reduces to:

$$\eta\,\frac{\partial^2 \mathbf{v}_\parallel}{\partial z^2} = -\nabla_\parallel p, \tag{9.103}$$

with: $p = p\,(x, y)$. The subscript $\parallel$ indicates that the equation applies to the component of the vectors (velocity or pressure gradient) in the x-y plane. One separates the variables by writing:

$$\mathbf{v}_\parallel\,(x, y, z) = \mathbf{v}_\parallel\,(x, y, 0)\, f(z), \tag{9.104}$$

where, from Equation 9.102, the dependence of $\mathbf{v}_\parallel$ on x and y is very slow relative to that of $f(z)$ on z. From Equations 9.103 and 9.104, one obtains: $f(z) = (1 - 4\,z^2/a^2)$.
Finally, the velocity field between the plates obeys:

$$\mathbf{v}_\parallel\,(x, y, 0) = -\left(a^2/2\eta\right)\nabla_\parallel p \quad (9.105) \qquad \text{and:} \quad v_z\,(x, y, 0) = 0. \tag{9.106}$$

Thus the direction of the velocity $\mathbf{v}$ is everywhere parallel to that of ∇p; moreover, this direction does not change along the direction of the thickness of the fluid layer, even if the *magnitude* of $\mathbf{v}$ is highly dependent on z. Equation 9.105 thus represents, for a Hele-Shaw cell, the equivalent of Darcy's law for porous media. From Equation 9.105, one deduces:

$$\nabla_{\parallel} \times \left[\mathbf{v}_{\parallel}(x, y, 0)\right] = \nabla_{\parallel} \times (\nabla p) = 0. \tag{9.107}$$

The streamlines in a plane of constant z are thus the same for all values of z and are identical to those for two-dimensional potential flow of an ideal fluid around obstacles with geometry identical to that of the spacers. Near the spacers, there is a boundary condition of zero velocity, but its effect is limited to a distance from the boundary of the order of the thickness a.

The Hele-Shaw cell can be used experimentally as a *model porous medium*. One can thus simulate the flow in the porous embankment, discussed above, by means of a Hele-Shaw cell for which the plates are vertical, with the two sides located between two water volumes of different levels. The free surface in the cell will be parabolic, as predicted by Equation 9.95. One can also model the spatial variations in the permeability by varying spatially the distance a between the plates.

9.7.4 Simple models of the permeability of porous media

Permeability of a system of wavy capillaries

Simple porous media, where the pores are well interconnected and of relatively uniform cross-section, are frequently modeled as a system of wavy capillaries with individual diameters d (Figure 9.21a). This approach has already been used in Section 9.7.2 in order to evaluate the electrical conductivity. As shown below, the equivalent permeability of this system of capillaries is:

$$K = \frac{\phi\, d^2}{32} \frac{1}{T}. \tag{9.108}$$

Therefore for a given porosity, the permeability varies as the square of the diameter of the channels: thus, the pressure drop for a given flow rate increases very rapidly as the size of the pores decreases, even if the total pore volume remains constant. This result is quite different from that obtained for the electrical conductivity σ_p of the same porous medium filled with a fluid of given electrical conductivity σ_f. In this latter case, the ratio σ_p/σ_f for circular channels is proportional to the porosity ϕ and independent of d.

Proof

For a single capillary, the dependence of the flow rate δQ on the pressure drop Δp is given by Poiseuille's equation (Equation 4.78):

$$\delta Q = \frac{\pi}{128\,\eta} \frac{\Delta P}{L_{\mathrm{cap}}} d^4. \tag{9.109}$$

Let N be the number of capillaries in the surface S normal to the direction of the flow, and K is the permeability for flows in that direction. The total flow rate per unit area through the system of cross-section S is given by:

$$Q = N\,\delta Q = N \frac{\pi\, d^4}{128\,\eta} \frac{\Delta P}{L_{\mathrm{cap}}}. \tag{9.110}$$

According to Darcy's law (Equation 9.91), Q also satisfies: $Q/S = (K/\eta)(\Delta P/L)$. Taking into account the equation for the porosity $\phi = N(\pi\, d^2/4)\, L_{cap}/(SL)$, one finds Equation 9.108.

For this model, as well as that discussed in the previous section, it is assumed that all capillaries, whether straight or wavy, are on average parallel to one particular direction, an assumption corresponding to a highly anisotropic porous medium. But one can also make a simple estimate of the permeability of an isotropic porous medium by assuming that it consists of three sets of mutually perpendicular capillaries. For a pressure gradient directed along one of these sets, only that particular one contributes to fluid transport, reducing the effective permeability by a factor of three relative to the previous case. One then has:

$$K = \frac{\phi\, d^2}{96}\,\frac{1}{T}. \tag{9.111}$$

The Kozeny-Carman equation

Equation 9.111 involves a pore diameter d which is almost impossible to define unambiguously, even if sections of the material of interest are available. The *Kozeny–Carman equation* allows one to relate the permeability K of a porous medium to experimentally measured, physical parameters, here, the porosity and the specific area of the material. The specific area S_V is used in this approximation to evaluate the diameter of the pores, which is only possible for pores of simple and well connected geometry. For a system of capillaries (Figure 9.21a), the permeability K satisfies, for example:

$$K = \frac{1}{6}\,\frac{\phi^3}{{S_V}^2\, T}. \tag{9.112}$$

The constant 1/6 appears when the medium is modeled as a set of three mutually perpendicular systems of parallel capillaries. An interesting feature of the Kozeny-Carman equation is the fact that for porous grain packings or compressed powders, $6T$ can be replaced by a nearly constant experimental value of the order of five. This result, however, is only valid for grains and packings with simple geometries and when the size of the pores does not vary too much.

Proof

The specific area S_V (the surface of the pore walls per unit volume of the medium) for a set of capillaries is:

$$S_V = \frac{N\,(\pi\, d)\, L_{cap}}{S\, L}, \tag{9.113}$$

where N is the number of capillaries over the cross-section S of the porous medium. The porosity ϕ of the medium can be written as:

$$\phi = \frac{N\,(\pi\, d^2/4)\, L_{cap}}{S\, L}. \tag{9.114}$$

From the two preceding equations we then find, for the specific area as a function of the porosity:

$$S_V = \frac{4}{d}\,\phi, \tag{9.115a}$$

so that:

$$d = \frac{4\,\phi}{S_V}. \tag{9.115b}$$

By substituting this last value for d into Equation 9.111, one then obtains Equation 9.112:

Moreover, one can substitute for the quantity S_V in Equation 9.112 the parameter $S'_V = S_V/(1-\phi)$ representing the area of the pore walls per unit volume of the solid. The second expression is of particular interest when one deals with identical grains of simple shape, because S'_V then represents the surface-to-volume ratio of the individual grains. For a porous composition of spherical grains of diameter D: $S'_V = 6D$ and one obtains what is known as *Ergun's equation*:

For a packing of spherical grains of diameter $D = 100$ μm and porosity $\phi = 0.4$ (corresponding to a packing fraction of 60%), one finds $K \approx 10^{-11}$ m^2.

$$K = \frac{\phi^3 D^2}{180\,(1-\phi)^2}. \tag{9.116}$$

9.7.5 Relationship between the electrical conductivity and the permeability of porous media

One uses here a different approach for evaluating the permeability of a porous medium: it is applicable to a larger set of real media, for which the geometries of the pores and the flow channels are more complex. In this approach, one attempts to establish relationships between the permeability and the electrical conductivity of the medium saturated by a conducting fluid. The latter is indeed also related, albeit in a manner different from the permeability, to the structure and aperture of the channels which cross the medium.

Model of porous medium with two length scales

The porous medium is represented as a system of particles of characteristic size a (which is also the order of magnitude of the length of the channels), for which the surfaces are separated by a characteristic distance d (representing the minimum diameter of the channels) (Figure 9.24). This approach has some common points with the models of capillaries that we have described previously (it is in fact simpler as it does not involve explicitly the tortuosity); it is however characterized by the introduction of the two length scales a and d instead of assuming long tubes.

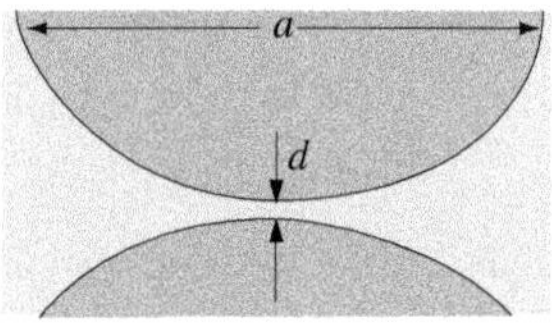

Figure 9.24 *Schematic diagram of a porous medium, showing channels of minimum aperture d between grains of diameter a. While this schematic representation is two-dimensional, one can imagine that this is a cross-section through two grains and that, above and below the plane of the figure, another pair of grains forms the remaining boundaries of the channel*

One evaluates now the relationship between the permeability of the porous medium, its electrical conductivity σ_p and that σ_f of the fluid saturating the medium. Instead of making separate use of the parameters σ_p and σ_f, one introduces the formation factor F (greater than or equal to one) of the medium, which is equal to their ratio so that:

$$F = \sigma_f/\sigma_p \tag{9.117}$$

The dimensions a and d can be related to F and to the permeability K by:

$$d \approx \sqrt{\frac{\sigma_f}{\sigma_p}}\,\sqrt{K} = \sqrt{FK} \quad (9.118a) \qquad \text{and:} \quad a \approx \frac{\sigma_f}{\sigma_p}\sqrt{K} = F\sqrt{K}. \quad (9.118b)$$

Proof

One first evaluates F in the geometry of Figure 9.24. Considering that the cross-sections S and S_{cap} to be used in Equations 9.88 and 9.89 are respectively $S \approx a^2$ and $S_{cap} \approx d^2$ and taking $I = I_{cap}$ and $L = L_{cap} = a$, one obtains:

$$F = \frac{\sigma_f}{\sigma_p} \approx \frac{a^2}{d^2} \tag{9.119a}$$

Moreover, according to Equation 9.109 and taking again $L_{cap} \approx a$, the flow rate through single channels is of the order of $\delta Q \approx (1/\eta)(\Delta P/a)d^4$. Using Darcy's law (Equation 9.91) with, again, $S=a^2$ and with $Q = \delta Q$, one obtains:

$$K \approx \frac{d^4}{a^2}. \tag{9.119b}$$

By combining the two previous equations, one retrieves Equations 9.118a–b.

In the present discussion, Darcy's law was applied to a cube of size a, which is then implicitly considered to be an elementary representative volume on which one can define macroscopic variables such as K and F. In spite of this approximation, which is difficult to justify for very heterogeneous porous media, we will see further down that the equations obtained can be used on natural rocks. This shows that the purpose of the scaling laws of Equations 9.118 and 9.119 is to predict the dependence of the permeability and electrical conductivity on the characteristic lengths of the structure of the medium, without attempting to determine the numerical values of these coefficients.

We explore two different ways of using Equations 9.118a–b. In the first, one determines a characteristic opening d_c for the channels by means of an auxiliary physical measurement; in the second, one uses porous media made up of grains of known and minimally variable size a but with different packing fractions, and thus porosity, as can be obtained by changing the degree of sintering.

Using the conductivity-permeability-size relationship of channels

In this approach, pioneered by Katz and Thompson, a characteristic value of d is estimated by a technique similar to *mercury porosimetry*, to be described in more detail in Section 9.7.6. The characteristic diameter d_c, thus determined, is well adapted for use in Equation 9.118a: the electrical (or hydraulic) conductivity of a given path through the sample is determined by the narrowest openings (those for which the drop of the electrical potential, or that of the pressure, are the highest). The overall conductivity of several channels in parallel is, in contrast, determined by those through which flow is easiest, i.e. those through which the electrical current or fluid flow rate is largest. It is thus the *narrowest* part of these *easiest* paths which controls the transport properties: one will see that it is, specifically, this narrowest dimension which is determined by porosimetry measurements.

One thus expects, by taking the square of Equation 9.118a, to obtain $d_c^2 \approx FK$. Remarkably, this equation is obeyed by samples with permeability varying over a huge range (from 1 to 10^8) and with very different porosities and mineral compositions. These measurements have resulted in the following quantitative relationship, obeyed within a factor of two:

$$K = \frac{\sigma_p}{\sigma_f} \frac{d_c^2}{226}. \tag{9.120}$$

The type of analysis carried out here and based on the concept of a critical size d_c for channels has been applied to other transport problems, such as electrical transport in very heterogeneous solids, or hydraulic transport in fissured media. The underlying concept is that for a system for which one parameter varies greatly from one point to another in the medium, (the electrical resistance of a network of conducting elements with greatly varying individual resistances), the overall conductivity is minimally affected by those elements which are excellent conductors: they only act as local short circuits. The same is true for the elements of the network which are poor conductors: they contribute very little to the overall conductivity because they are generally short-circuited by elements which are better conductors. These are the elements with intermediate resistance but which are just sufficiently numerous to create a continuous path through the sample which determine the conductivity of the overall system.

Using the conductivity-permeability-size relationship of grains

Equation 9.118b has a more restricted domain of application; it can be used, however, to make predictions for sedimentary rocks compacted over long geological periods or granular media welded by sintering with no significant change in the grain size. Systematic experiments have been carried out on model porous media obtained by high-temperature sintering, during different time lapses, of glass beads of the same diameter (a few hundred microns); as the level of sintering increases, the size of the channels d decreases while their length a stays constant. Samples prepared from beads of different diameters have also been compared. The measurements lead to the following equation for the permeability K as a function of the square of the grain size and the form factor:

$$K = 14.1\, a^2 \left(\frac{\sigma_p}{\sigma_f}\right)^2 = 14.1\, \frac{a^2}{F^2}. \tag{9.121}$$

This relation has the form of Equation 9.118b squared. These results are, however, only valid for moderate levels of sintering, corresponding to porosities $\phi \geq 10$ %. For lower porosities, and until the sintered material is no longer permeable, the effect of dangling arms (unconnected pores) alters the validity of the model.

9.7.6 Flow of immiscible fluids in a porous medium

This type of flow occurs when two or more immiscible fluids (or a liquid and a gas) are simultaneously present in a porous environment, a situation occurring in oil-bearing rocks, as well as in the hydrology of non-saturated soils. This is the case for the enhanced recovery of petroleum (oil and water mixture) and in the experiment of mercury porosimetry described above. One must then take into account the viscous pressure drop due to flow in each of the phases, as well as the differences of capillary pressure determining the curvature of the interfaces of the two fluids. The relative importance of these two phenomena can be measured by the dimensionless *capillary number* $Ca = \eta v/\gamma$. This number characterizes the ratio of the viscosity effects and of the capillary pressure and has been already defined in Section 8.2.2.

We will first discuss the more classical approach, that of the model of relative permeabilities: it can be used when the viscous pressure gradients play an important part, and the distribution of the two phases within the porous material is relatively homogeneous. This approach uses Darcy's law by continuing to use macroscopic variables (defined by averaging over an elementary representative volume) which characterize the distribution and movement of the fluids. We will then discuss a few problems associated with the action of capillary forces and several cases where they play a dominant role. In this case, the observed phenomena are quite different depending on whether the fluid initially present is displaced by a fluid which is more (*imbibition* process) or less (*drainage* process) wetting for the porous medium.

Relative permeability of a porous medium

Let us analyze the penetration of water in a cylindrical sample initially saturated with oil (Figure 9.25). The first important parameter is the *saturation* S_w of the porous material with water, defined as the fraction of the volume of the pores filled with water. Similarly, one defines a saturation S_o in oil (if the pores are completely saturated and no other fluid is present, one has $S_o + S_w = 1$). The parameters S_w and S_o, as well as the respective local flow velocities $\mathbf{v}_{sw}$ and $\mathbf{v}_{so}$ for water and oil, are averages over an elementary representative volume, and vary as one moves farther from the initial injection phase. In the center there will be a region where the two phases, water and oil, are continuous. In contrast, at each of the two ends, only one of the respective phases, water or oil, is continuous.

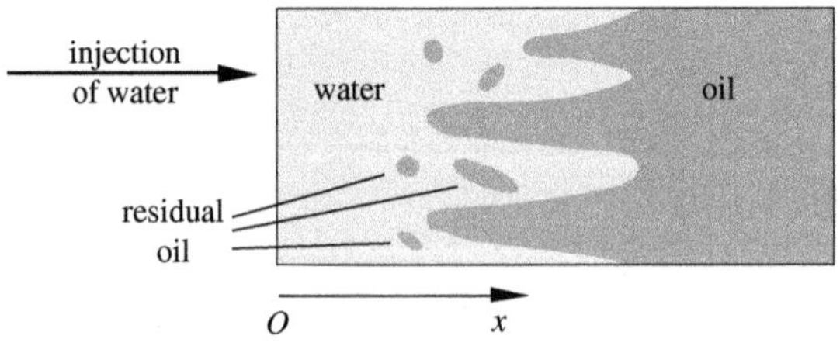

Figure 9.25 *Schematic representation of the displacement of oil by water in a porous medium*

At a given distance x from the injection face, one defines coefficients k_{rw} and k_{ro}, referred to as the average relative permeabilities:

$$\frac{Q_w}{A}\,\mathbf{n}_w = \mathbf{v}_{sw} = -K\,\frac{k_{rw}}{\eta_w}\,(\nabla p_w - \rho_w\,\mathbf{g}) \tag{9.122a}$$

and:

$$\frac{Q_o}{A}\,\mathbf{n}_o = \mathbf{v}_{so} = -K\,\frac{k_{ro}}{\eta_o}\,(\nabla p_o - \rho_o\,\mathbf{g}), \tag{9.122b}$$

where K is the average Darcy permeability such as would be measured in the presence of a single phase. Q_w and Q_o are the flow rates of water and oil, A the cross-section of the sample, $\mathbf{n}_w$ and $\mathbf{n}_o$ are unit vectors, η_w, ρ_w, η_o and ρ_o are the respective viscosities and densities for water and oil. This approach attempts to generalize Darcy's law in the simplest manner possible: one takes into account the presence of two fluids by merely introducing two multiplying numerical factors k_{rw} and k_{ro}. The pressures p_w and p_o within the water and oil can be different because of the capillary pressure jump across the oil–water interface. Equations 9.122a–b include the changes in pressure due to the hydrostatic pressure gradients, which are different in the oil and water phases. These equations make the implicit assumption that there is a homogeneous distribution of water and oil on the scale of the representative volume over which S_w, Q_w and Q_o are defined. It is difficult for this condition to be obeyed in heterogenous porous media, or in the presence of hydrodynamic instabilities due, for example, to viscosity differences between the two fluids.

It is normally assumed, particularly in the oil industry, that the parameters k_{rw} and k_{ro} are functions of S_w only and do not depend on the prior history of the flow. It is then possible to determine experimentally k_{rw} and k_{ro} as a function of S_w by measuring, as a function of time, the amounts of water and oil flowing out of a sample during the injection process, as well as the time-dependence of the pressure difference between the ends. This much simplified assumption is generally not obeyed, particularly if one compares the first injection in a sample initially saturated with a single fluid, with invasion resulting from several injections of one or the other fluid. Therefore, routine measurements of relative permeability are always carried out with a precise specific experimental procedure so as to provide comparable results from one material to another.

Figure 9.26 indicates the typical functional dependence of k_{rw} and k_{ro} on the saturation S_w. One observes that k_{rw} vanishes at a finite, non-zero value of the saturation S_{wi}: this indicates that the water films or droplets, still present at yet lower values of the saturation, no longer form a continuous path through the sample, and can therefore no longer be set in

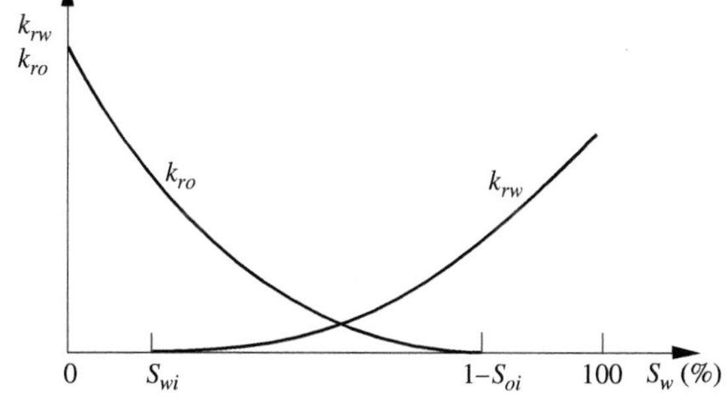

Figure 9.26 *Typical variation of the relative permeabilities k_{rw} for water and k_{ro} for oil as a function of the relative saturation S_w of water in the pore space. The residual saturations in water and oil are S_{wi} and S_{oi}*

motion by small pressure gradient (irreducible water-saturation). Quite similarly, there must be a minimum oil saturation $1-S_{oi}$ for a continuous path of oil to exist, and thus get a flow of oil under a pressure gradient. The concept of relative permeability, in spite of its empirical nature, allows one to account for the profiles of invasion fronts when one fluid displaces another within a porous medium. It further occurs in numerous other situations, where the porous medium is partially saturated with water, like in soil science and in agronomics.

Effects of wetting in two-phase flows in porous media

One considers here the case where the capillary number Ca is small enough so that the capillary effects become significant, or even dominate. One assumes that the porous medium is initially completely saturated with a fluid (or completely empty); another fluid is then injected into it. As mentioned above, there will be two types of such processes: drainage and imbibition.

Porosimetry represents a special case of drainage in which the non-wetting fluid is mercury, and the wetting fluid is represented by vacuum. Because of this, the capillary pressure difference is equal to the pressure of the mercury.

An example of drainage: porosimetry Fluid already present is replaced by a second fluid which is less wetting for the porous medium. The term *mercury porosimetry* designates a technique for characterizing the size of pores in a material where an excess pressure is used to force non-wetting mercury into a porous sample, initially empty of fluid. In order to overcome the capillary forces and penetrate into a cylindrical pore of diameter d, one needs to apply an excess pressure, Δp, varying as a function of d as:

$$\Delta p = \frac{4\,\gamma\cos\theta}{d}, \tag{9.123}$$

where θ is the wetting angle, and γ the surface tension of the mercury; if the meniscus is a spherical cap making an angle θ with the walls, its radius of curvature is indeed $R = d/(2\cos\theta)$ and Δp is given by Young-Laplace's equation (Equation1.55).

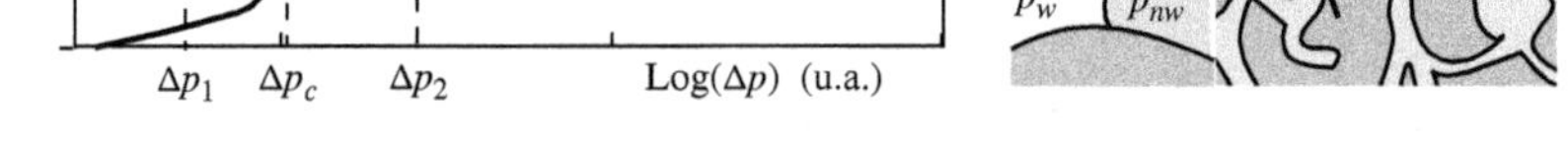

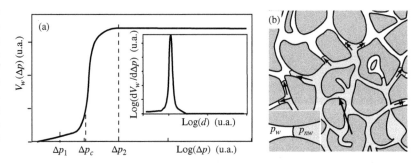

Figure 9.27 *(a) Measuring the porosity: variation of the pore volume V_w invaded by mercury as a function of the injection pressure Δp. Inset: estimate of the histogram of the pore sizes d, obtained by taking the derivative of $V_w(\Delta p)$. (b) Schematic representation, on the scale of a few pores, of the process of porosimetry by the penetration of non-wetting mercury (light gray) in a sample initially empty (white areas). Insert: enlarged image of the meniscus inside a connecting channel.*

In practical terms, we measure the change in the injected volume of mercury in the sample as a function of the applied excess pressure (Figure 9.27a): the injection velocity is kept very low so that the effect of the viscous forces remains negligible, and only the capillary forces affect the measurement.

Figure 9.27b represents schematically, at the scale of a few pores, the distribution of the non-wetting phase (*nw*, in this case: mercury) when it displaces the wetting phase (*w*, in this case, the vacuum) and the corresponding interface menisci. For a given excess pressure Δp, the only invaded pores are those connected to the inlet face of the porous medium by a path along which the diameters of the channels and pores are all greater than the diameter $d(\Delta p)$ obtained from Equation 9.123: the interface menisci can go through the narrow parts a channel only if their diameter is greater than $d(\Delta p)$. Similarly, a large pore completely surrounded by inlets is only invaded if the excess pressure is sufficient to enter these inlets. Finally, at any time, the progression of the front occurs through the nearest, largest-diameter pore.

At small values Δp_1 of the excess pressure Δp, the fluid *nw* only invades the largest pores near the inlet surface, and the menisci are stopped in the first narrow channels they encounter: the invaded volume V_w is then small ($\Delta p \approx \Delta p_1$ on Figure 9.27a). As Δp increases further, the penetration is deeper and deeper. For a critical value Δp_c, there appears a first path going all the way through the porous medium. Physically, $d_c(\Delta p_c)$ is the opening of the narrowest passage along the easiest path going through the porous medium: it is, effectively, this path which determines the pressure Δp_c to be applied in order to travel along the entire path. Many pores of size $d > d_c$ are not invaded if they are not connected to the injection face, and the fraction of the pore volume invaded is small.

One detects the appearance of this path by measuring the electrical conductivity between the inlet and outlet of the sample; it corresponds to the instant when this conductance no longer vanishes. It is therefore this value d_c which is used in Equation 9.120 to evaluate the permeability.

If Δp is increased further, the invasion around this critical path increases, and other paths appear: the value of V_w increases then quite rapidly before reaching saturation when almost every pore is invaded, as is the case for $\Delta p \approx \Delta p_2$ in Figure 9.27a.

In elementary interpretations of porosimetry measurements, the invaded volume for a given excess pressure Δp is assumed to represent the total volume of pores with a size larger than $d(\Delta p)$, obtained from Equation 9.123. One then takes the derivative of the curve with respect to Δp and converts the injection pressures into diameters, again by means of Equation 9.123: this provides the estimation of the distribution of the pore volumes as a function of their diameter (insert in Figure 9.27a). The previous discussion of the physical mechanism of invasion shows, however, that these assumptions are very approximate, even if through are quite frequently used: first of all, the volume V_w is largely determined by the volume of the pores, while $d(\Delta p)$ indicates the diameter of the inlet channels. On the other hand, we have seen that, at a given injection pressure, an important fraction of the large channels of diameter greater than $d(\Delta p)$ remain empty.

The previous description of the very slow invasion process of the porous medium by a non-wetting fluid is a particular example of the general phenomenon of *percolation*. This concept was first introduced for the study of a filter initially made up of interconnected, open channels which are gradually and randomly blocked. In this case, the percolation threshold corresponds to the moment when the permeability vanishes. The situation of mercury porosimetry corresponds, in contrast, to the case of a porous medium which is gradually unblocked by increasing the pressure Δp, until a pressure threshold Δp_c is reached, at which point the mercury passes through the entire porous medium, and the electrical conductivity of the entire sample is no longer zero.

The process of percolation belongs to the family of critical phenomena, characterized by the fact that certain properties (e.g., here the electrical conductivity or permeability) obey universal laws as a function of the distance to the threshold (here, $\Delta p - \Delta p_c$).

Mercury porosimetry is not the only drainage process which can take place in porous media. For example, millions of years ago, a drainage process occurred when oilfields were formed by the filling of early rock formations by oil, replacing the ocean water previously there. Moreover, in certain cases, the existing reservoir rocks have become preferentially wettable by oil: the recovery of petroleum, by injection of water, then becomes a process of drainage. In this case, since the displaced fluid (water or oil) is incompressible, residual fluid can remain at the end of the process.

Imbibition Imbibition is the operation inverse to drainage: a fluid wetting effectively the surface of pores, displaces spontaneously a less wetting fluid that initially fills the medium.

This is the case, for example, of the penetration of water in a porous medium easily wet by water, and initially filled with air (the non-wetting fluid). If, in contrast, one starts with a porous medium saturated with water and lets it empty under the influence of gravity, the penetration of the air which begins to replace the water is a drainage process.

When petroleum production occurs from a reservoir rock, the petroleum is generally displaced by water (injected or coming from a greater depth). If the rocks are preferentially wetted by water rather than by the oil, as is frequently the case, one has also imbibition.

In imbition processes, the smallest pores are the first ones invaded by the wetting fluid (the opposite of drainage): the capillary forces favour indeed the invasion and are strongest in the narrowest channels.

Continuous films of wetting fluid are frequently found along the walls of a porous medium: they play a key role in the phenomenon of imbibition, by allowing the penetration of wetting fluid well ahead of the average displacement front. The non-wetting fluid, on the contrary, remains in the form of isolated residual drops after the passage of the displacement front.

The phenomena of drainage and imbibition have an important place in the science of soils. However, for particles of small size (clays) a significant fraction of the water is chemically bonded to the particles of the soil.

EXERCISES

All these exercises correspond to Newtonian Stokes flows at $Re \ll 1$ where the inertia of the fluid is negligible.

1) Torque on a cylinder and a cross in a shear flow

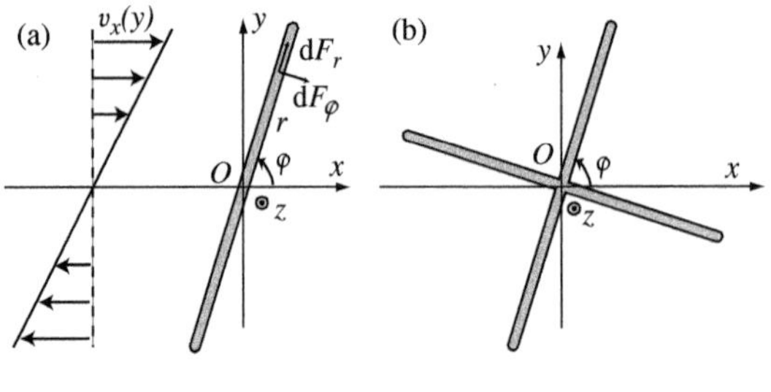

Consider a solid cylinder (length $2L$, diameter $\ll L$) free to rotate around its center O in the plane $z = 0$ (Figure (a)) and placed in a parallel shear flow $v_x(y)$ at $Re \ll 1$ (the shear rate is $S = \partial v_x / \partial y$) The inertia of the cylinder and gravity are neglected. From Equation 9.30, estimate the local components dF_r and dF_φ of the hydrodynamic forces per unit length as the angle $\varphi(t)$ of the cylinder relative to the x-direction varies. Show that the total torque with respect to O is: $G_O = -(2/3)\ \lambda_\parallel\ \eta\ L^2 (S \sin^2\varphi + \dot{\varphi})$. What is the value of G_O? Compute the variation $\varphi(t)$ and the limit of φ at long times.

Two cylinders are rigidly fixed at right angles at the point O in the plane $z = 0$ (see Fig (b)). What is the expression for the torque? How is the motion of the two cylinders related to the vorticity?

2) **Sedimentation of a sphere**
A solid sphere of density ρ_s and radius R is dropped into an infinite bath of viscous fluid of density ρ_f and viscosity η with a zero initial velocity (we take into account the inertia of the sphere while neglecting that of the fluid). Determine the time dependence of the vertical velocity v_z of the sphere and of its height z; what is the corresponding time constant τ? What is the variation of $z(t)$ with time in the two limits $t \ll \tau$ and $t \gg \tau$? If $d = 1\text{mm}$, $\rho_s = 2000\ \text{kg/m}^3$ and $\rho_f = 1000\ \text{kg/m}^3$, compute τ for $\eta = 1$ Pa.s and the corresponding distance to reach the limiting velocity V_{terminal} within 1%. How does this result change if $\eta = 10^{-3}$ Pa.s? Is the condition $Re \ll 1$ still satisfied?

3) **Stokes flow from a three-dimensional point-source**
Show that the radial flow velocity $v_r(r, \theta)$ from a point-like source O in the plane solid wall $\theta = \pi/2$ satisfies $v_r = f(\theta)/r^2$ (use Figure 4A.3 and formula from Appendix 4A.3). Assuming that the pressure p goes to zero at infinity as $1/r^\alpha$: what must be the value of α? Assuming that $f(\theta) = A\cos^2\theta$, compute the flow velocity v_r as a function of the flow rate q through a half sphere of radius r; by writing the components of the equation of motion, show that: $p = -\eta\, q\,(1 - 3\cos^2\theta)/(\pi r^3)$. How does the sign of the radial pressure gradient vary with θ ?

4) **Torque on two concentric rotating spheres**
One wants to extend the computation of the torque on a rotating sphere (Equation 9.59) to two spheres with the same center O and radii $R_1 < R_2$ rotating around the same z-axis at angular velocities Ω_1 and Ω_2. The interval between the two spheres is filled by a fluid of viscosity η. Show that the velocity satisfies: $v_\varphi(r,\theta) = (Ar^\alpha + B\,r^\beta)g(\theta)$. Compute the velocity fields when one sphere rotates while the other is fixed. What happens to these fields if $R_2 \to \infty$ and $\Omega_2 = 0$, or $R_1 \to 0$ and $\Omega_1 = 0$? Compute the torque on the remaining sphere in both configurations. What are the velocity and the torque if $R_2 - R_1 = \delta \ll R_1$ and $\Omega_2 = 0$? How can the problem of two spheres rotating at different velocities be solved using the above results?

5) **Permeability of a layered medium**
Consider a porous medium made up of alternate layers of thicknesses h_1 and h_2 and permeabilities k_1 and k_2. What is the equivalent permeability k of the medium at a scale large compared to the thickness of the layers for flows parallel ($k_\parallel$) (case (a)) and perpendicular ($k_\perp$) (case (b)) to the layers? How do these results change if $k_1 \gg k_2$? We consider now the particular case where each layer is a random packing of beads of diameters d_1 (resp. d_2) for a layer of type 1 (resp. 2): what are the ratios ϕ_1/ϕ_2 and k_1/k_2 between the porosities and the permeabilities of the layers and the ratio $k_\parallel/k_\perp$? If the pores are filled with a same electrically resistive fluid (while the beads are insulating), is the effective electrical conductivity different for currents parallel and perpendicular to the layers?

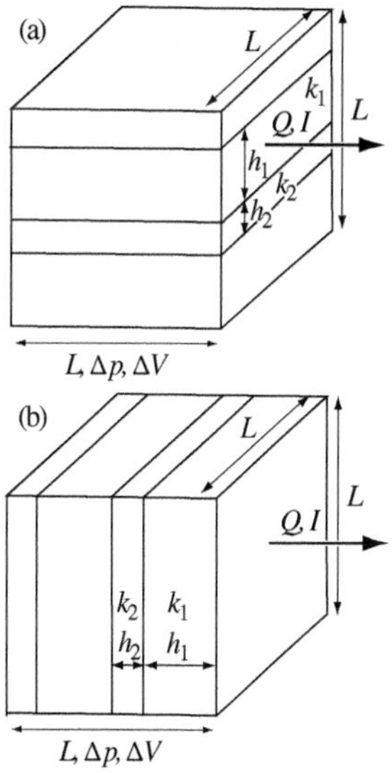

6) **Porous embankment with two fluids of different densities**

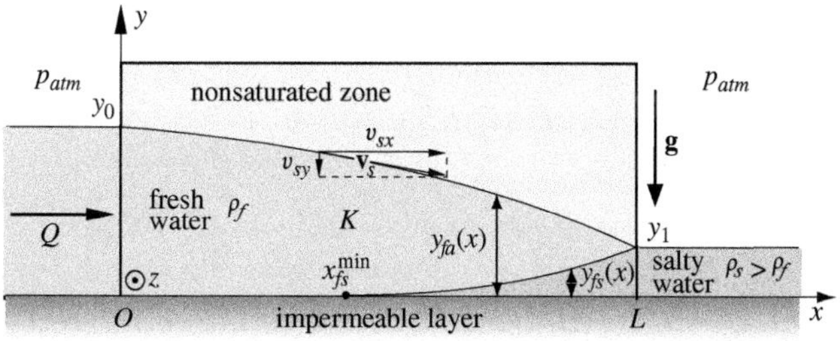

We generalize the model of a porous embankment studied in Section 9.7.3 (Figure 9.22) by assuming: (see figure above) have fresh (f) and salty (s) water respectively upstream and downstream of the embankment with densities ρ_f and $\rho_s > \rho_f$ (as is the case near the sea). The salty water penetrates into the embankment and remains static thereafter. Its interface $y_{fs}(x)$ with the fresh water starts at $y = y_1$ for $x = L$ (see figure); $y_{fa}(x)$ is the interface between the upper fluid and air with $y_{fa}(L) = y_1$. We consider a stationary regime in which the profiles y_{fa} and y_{fs} are independent of time. Show that $(\rho_f - \rho_s)/\rho_s = (y_{fa} - y_1)/(y_{fs} - y_1)$ (neglecting the interdiffusion of the two fluids, which are assumed to remain separated, and ignoring the vicinity of the point $x = L$ where the approximations made in Section 9.7.3 are no longer valid). Compute the flow rate Q of fresh water (per unit length along z) as a function of the slope $\partial y_{fs}/\partial x$ of the fresh-salty water interface. Calculate the distance $L - x_{fs}^{min}$ of salt water penetration into the embankment as a function of Q.

Coupled Transport. Laminar Boundary Layers

10

The combined effects of diffusion and convection and, ultimately of chemical reactions, involve length scales which characterize the relative effectiveness of these mechanisms. The most classical instance is that of the boundary layer around a solid body which appears in laminar flows at high Reynolds numbers: far from the object, and so long as the incident flow is not turbulent, the viscous terms in the equation of motion are negligible, and the flow has the same profile as for an ideal fluid. The connection with the condition of zero velocity along solid stationary walls occurs over the region, known as the boundary layer, which is thinner the higher the Reynolds number. This chapter thus complements the study of the potential flow of ideal fluids in Chapter 6.

After the introduction, the structure of the boundary layers is discussed in Section 10.2 by comparing the effects of diffusion (coming from the viscosity) and convection: this introduces the idea of different characteristic length scales for motion perpendicular and parallel to the surface of an obstacle. The self-similar nature of boundary layer flows is then demonstrated in Section 10.3 and 10.4. The effect of a pressure gradient in the direction of the flow is considered in Section 10.5: it can induce a boundary layer separation which plays an essential role in aerodynamics as shown in Section 10.6. Section 10.7 deals with laminar wakes behind an obstacle: they can indeed be considered as boundary layers evolving in the absence of a solid wall. The occurrence of temperature or concentration gradients near the surface of a body leads to that of thermal or mass-transfer boundary layers (coupled to the fluid velocity ones). Section 10.8 deals with these problems which have a number of applications in thermal or chemical engineering: illustrations of such properties are provided in two cases of boundary layers involving the concentration of ions in the neighborhood of electrochemical electrodes. Finally, in Section 10.9, combustion problems are introduced in two limiting cases, that of a diffusive flame and of a premixed flame. In these two instances, in addition to the effects of convection and diffusion, one must consider the chemical reaction in the combustion, and the corresponding thermal effects.

10.1 Introduction

In laminar flow at high Reynolds number ($Re = UL/\nu \gg 1$) around a solid body, the viscosity terms in the equation of motion (Equation 4.30) are only to be taken into account within a thin region around the obstacle known as a *boundary layer*.

Vorticity created near the walls is convected in the wake (Figure 10.1a below) downstream of the solid: it will be seen in Section 10.7 that, in this region, the velocity gradients

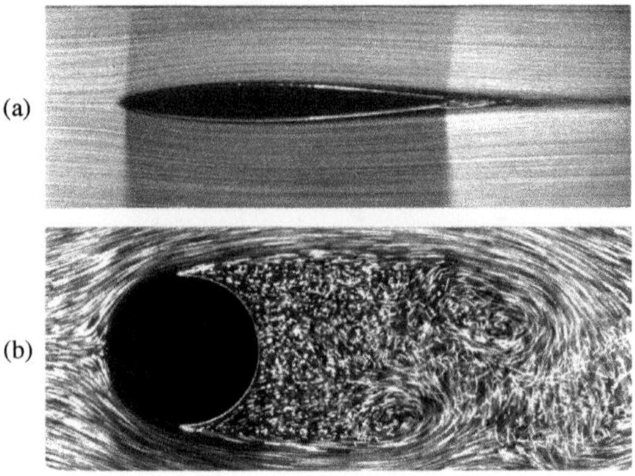

Figure 10.1 *(a) Cross-section of a boundary layer and a laminar wake along and behind a wing in uniform flow at zero angle of incidence; (b) boundary layer, separated from the surface of a non-streamlined object, with a large associated wake. In this case, the size of the wake is of the same order of magnitude as the lateral dimension of the obstacle (courtesy of H. Werlé, ONERA)*

are also localized within a small part of the total volume of the flow. This justifies, *a posteriori*, the study of the flow of ideal fluids, since the corresponding velocity profiles apply almost everywhere: the effects of viscosity are significant only within the boundary layer near a solid wall or, downstream of an obstacle, in the wake. The concept of a boundary layer thus provides the link between two important domains of fluid mechanics: the study of the velocity fields of ideal fluids in potential flow, carried out in Chapter 6, and the determination, of practical interest, of the flow of viscous fluids at finite Reynolds numbers.

This concept, due to Ludwig Prandtl (1905), must be appropriately adapted in a number of real experimental situations:

The present chapter deals only with laminar boundary layers, within which the velocity field varies only slowly with time. The turbulent boundary layers are discussed in Section 12.5.5.

- flows which are turbulent upstream of a body and/or in the boundary layer: convective momentum-transport then plays an important role and dominates the diffusive transfer which normally occurs in a laminar boundary layer. The velocity profiles are significantly altered in these *turbulent boundary layers*;
- for solid bodies not well streamlined: the boundary layer only exists along a portion of the surface of the object, and a turbulent *wake* of width comparable to that of the obstacle appears downstream. This corresponds to the phenomenon of *boundary layer separation* illustrated in Figure 10.1b. In that case, the downstream flow bears little resemblance to that of an ideal fluid; both the energy dissipation and the drag force on the object are greatly increased.

10.2 Structure of the boundary layer near a flat plate in uniform flow

One considers a uniform laminar flow, of velocity **U**, parallel to a semi-infinite flat plate with its edge parallel to the z-axis and perpendicular to the plane of the figure (Figure 10.2a below). If the velocity is sufficiently high, the influence of the plate will not be noticeable upstream of its leading edge: velocity gradients will indeed not have had time to diffuse an appreciable distance from the edge of the plate before being carried along downstream by the flow. As a result, near the edge, the velocity gradients, and the vorticity, are confined very close to the surface of the plate. Just as in the problem of a flat plate set in motion

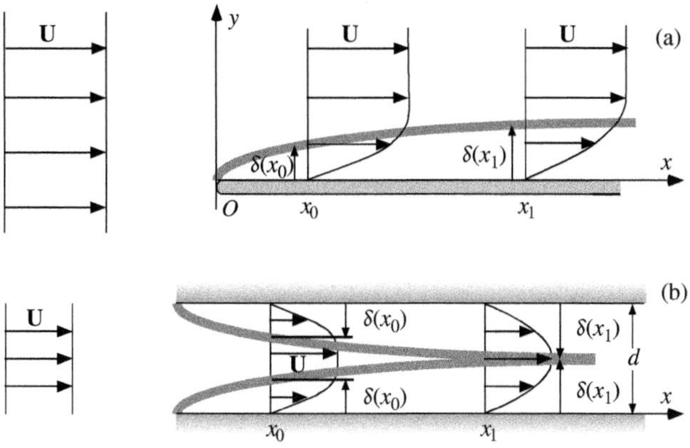

Figure 10.2 *(a) Growth of the boundary layer along a flat semi-infinite plate, with its edge O normal to the plane of the figure, located in otherwise uniform flow with velocity* **U**. *(b) Entry length effect for flow between two semi-infinite, parallel plates. We find again a Poiseuille-type flow after a distance x_1. In both cases, the scale in the y-direction is exaggerated relative to that in the x-direction*

parallel to itself, discussed in Section 2.1.2, the velocity gradients attenuate with time; indeed, the spatial distribution of vorticity, initially localized along the wall, spreads out by viscous diffusion over distances of the order of $\delta \approx \sqrt{\nu t}$, where ν is the kinematic viscosity of the fluid.

However, in contrast to the problem discussed in Chapter 2, the fluid is here simultaneously carried along parallel to the plate with a velocity of the order of magnitude of the external flow velocity, U. In the reference frame of the moving fluid, the order of magnitude of the time t for moving through a distance x_0 downstream from the edge of the plate is: $t \approx x_0/U$. Substituting this value in the equation $\delta \approx \sqrt{\nu t}$, one finds that, at a distance x_0 from the edge, the velocity gradients are localized within a distance from the surface of the plate Figure 10.2a:

$$\delta(x_0) \approx \sqrt{\frac{\nu\, x_0}{U}}; \qquad (10.1a)$$

$\delta(x_0)$ thus represents the thickness of the *boundary layer* over which the transition occurs, between the ideal-fluid flow far from the plate and the flow very close to it. This flow is governed by the viscosity which requires the condition of zero-velocity at the wall. One therefore has:

$$\frac{\delta(x_0)}{x_0} \approx \sqrt{\frac{\nu}{U\, x_0}} \approx \frac{1}{\sqrt{Re_{x_0}}} \ll 1, \qquad (10.1b)$$

in which Re_{x0} is a local Reynolds number defined by taking the distance x_0 to the plate edge as the local length scale. Thus, as Re_{x0} tends toward infinity, the maximum thickness of the boundary layer becomes very small in comparison with the characteristic global length of the plate. This result explains why, at an increasingly large Reynolds number, the external, non-turbulent flow-region of a viscous fluid behaves more and more like an ideal fluid: the thickness of the layer where viscous effects are observed tends indeed toward zero.

This is not to say that the effects of such a layer, even infinitely thin, can be neglected. On the contrary, from the results derived in Section 5.3.1, one finds that the energy dissipation due to a finite change in the velocity over an infinitely thin layer is itself infinite. This is known as a *singular perturbation* which describes the approach to the limit $\delta = 0$.

We will see in Section 10.4.4 that the edge of the boundary layer is not a streamline, and that the rate of flow within it increases as $\sqrt{x_0}$. Moreover, if Re_{x_0} exceeds a certain value, the boundary layer itself becomes unstable and turbulent: the estimates made above are then no longer valid, since the momentum transport by turbulent convection causes δ to increase much more rapidly with x_0 than in the laminar case.

A phenomenon closely related to the boundary layer is the *entry length effect* which delays the establishment of a stationary flow profile downstream of the sharp entrance of a channel.

Figure 10.2b (above) illustrates schematically this process, for the simple case of flow between two semi-infinite plates perpendicular to the plane of the figure and separated by a distance d. At a short distance x_0, downstream from the edges O and O' marking the entrance ends of the two plates, the velocity between them is almost uniform with magnitude equal to the upstream velocity U; this profile is referred to as a *plug flow*. The transition to the zero-velocity condition at the surface of the plates occurs once again over a thin boundary layer of local thickness $\delta(x_0)$. Further downstream, the thickness $\delta(x)$ increases, and the two boundary layers ultimately merge at a distance x_1 from the opening such that:

$$\frac{x_1}{d} \approx \frac{U\,d}{\nu} \approx Re_d,$$

where Re_d is the Reynolds number based on the velocity U and the separation d between the plates. The length x_1 is then that required for the establishment, between the two plates, of the parabolic, stationary velocity profile discussed in Section 4.5.3: x_1 becomes greater as Re_d increases (1000 times the diameter for $Re = 1000$). For a cylindrical tube, the same kind of behavior is qualitatively observed.

The estimate of x_1 on the right is obtained by means of Equation 10.1a, writing that $\delta(x_1) \approx d/2$.

10.3 Equations of motion within the boundary layer – Prandtl theory

10.3.1 Equations of motion near a flat plate

Let us examine a two-dimensional, stationary flow in the x-y plane near a flat plate $y = 0$, for a potential external flow $\mathbf{U}(x)$ assumed to be in x-direction, parallel to the plate. The results derived below remain valid even for curved walls, provided that the radius of curvature is large relative to the thickness δ of the boundary layer.

The characteristic distance in the direction x is locally, at a given point M, of the order of its distance x_0 from the edge of the plate (Figure 10.3). Normal to the flow direction, the characteristic length is the local thickness of the boundary layer $\delta(x_0)$, which is very small relative to x_0 (Equations 10.1 a-b). The entire discussion which follows is based on the existence of these two very different length-scales, one in the direction parallel to the wall, the other perpendicular to it. Under these assumptions, the equation of motion in the x-direction becomes:

$$v_x \frac{\partial v_x}{\partial x} + v_y \frac{\partial v_x}{\partial y} = U(x)\,\frac{\partial U(x)}{\partial x} + \nu \frac{\partial^2 v_x}{\partial y^2}. \qquad (10.2)$$

while the pressure satisfies:

$$\partial p/\partial y = 0 \qquad (10.3) \qquad \text{and:} \qquad p(x) + \rho U^2(x)/2 = \textit{constant}. \qquad (10.4)$$

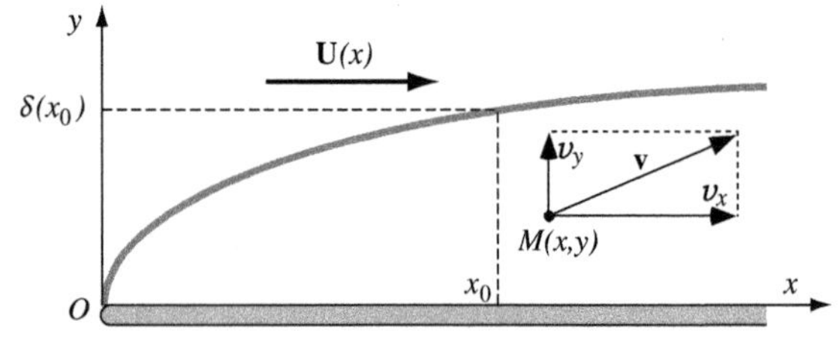

Figure 10.3 *Boundary layer along a semi-infinite plane plate, located in an external velocity field U(x) parallel to the plate; δ(x0) is the order of magnitude of the local thickness of the boundary layer at a point x0 downstream from the edge of the plate*

Equation 10.3 is similar to the absence of the pressure gradient normal to the streamlines for parallel flow, or rather, in the present case, quasi-parallel. Just as in Chapter 9, the pressure is corrected relative to the hydrostatic pressure $p_0 = \rho\, \mathbf{g}{\cdot}\mathbf{r}$ and the notation p represents implicitly the difference $p - p_0$.

Proof

Because the flow of interest is two-dimensional and incompressible, the equation of the conservation of mass can be written:

$$\frac{\partial v_x}{\partial x} + \frac{\partial v_y}{\partial y} = 0, \tag{10.5}$$

where v_x and v_y are the components of the velocity $\mathbf{v}$ of the fluid in the respective x and y directions. Equation 10.5 indicates that the component of velocity perpendicular to the wall is smaller by an order of magnitude than the component parallel to the wall, so that:

$$v_y \approx v_x \frac{\delta(x_0)}{x_0} \approx \frac{v_x}{\sqrt{Re_{x_0}}} \ll v_x, \tag{10.6}$$

since $\delta(x_0) \approx x_0/\sqrt{Re_{x_0}} \ll x_0$. There results:

$$\frac{\partial^2 v_x}{\partial y^2} \approx \frac{v_x}{\delta^2(x_0)} \gg \frac{v_x}{x_0{}^2} \approx \frac{\partial^2 v_x}{\partial x^2} \quad (10.7a) \qquad \text{and:} \qquad \frac{\partial^2 v_y}{\partial y^2} \approx \frac{v_y}{\delta^2(x_0)} \gg \frac{v_y}{x_0^2} \approx \frac{\partial^2 v_y}{\partial x^2}. \quad (10.7b)$$

These inequalities allow one to write the Navier-Stokes equations in the simpler form:

$$v_x \frac{\partial v_x}{\partial x} + v_y \frac{\partial v_x}{\partial y} = -\frac{1}{\rho}\frac{\partial p}{\partial x} + \nu \frac{\partial^2 v_x}{\partial y^2} \quad (10.8a) \qquad \text{and:} \qquad v_x \frac{\partial v_y}{\partial x} + v_y \frac{\partial v_y}{\partial y} = -\frac{1}{\rho}\frac{\partial p}{\partial y} + \nu \frac{\partial^2 v_y}{\partial y^2}. \quad (10.8b)$$

The term $v_y\, \partial v_x/\partial y$ has been retained in Equation 10.8a: it is indeed of the same order of magnitude as $v_x\, \partial v_x/\partial x$ because:

$$v_y \frac{\partial v_x}{\partial y} \approx v_x \frac{\delta(x_0)}{x_0} \frac{v_x}{\delta(x_0)} = \frac{v_x^2}{x_0} \approx v_x \frac{\partial v_x}{\partial x}.$$

The small value of v_x is compensated by the large value of the gradient $\partial v_x/\partial y$ in the direction normal to the wall, along which the length scales are much smaller. In the same manner, the terms $v_y\, \partial v_y/\partial y$ and $v_x\, \partial v_y/\partial x$ are of the same order of magnitude in Equation 10.8b. In order to go from Equation 10.8a to Equation 10.8b, $\partial p/\partial x$ is replaced by $\partial p/\partial y$, and v_y substituted for v_x in the three other terms; since v_y is very small compared to v_x, these three terms are, each, smaller by an order of magnitude than the corresponding terms in Equation 10.8a. Changes in pressure in the y-direction will therefore have a negligible effect on the velocity profile, when compared to the variations in the x-direction which appear in Equation 10.8a. One can therefore write Equation 10.8b as $\partial p/\partial y = 0$ (Equation 10.3), so that:

$$p = p(x). \tag{10.9}$$

At the edge of the boundary layer, at a distance $\delta(x_0)$ from the wall, the term $v_x(\partial v_x/\partial x)$ corresponding to the convective transport of momentum and the term $\nu(\partial^2 v_x/\partial y^2)$ for momentum transport by viscous diffusion are of the same order of magnitude. Indeed, $v_x(\partial v_x/\partial x)$ is of the order $(U^2(x_0)/x_0)$, while $\nu(\partial^2 v_x/\partial y^2)$ satisfies, for $y \approx \delta(x_0)$:

$$\nu \frac{\partial^2 v_x}{\partial y^2} \approx \nu \frac{U(x_0)}{\delta^2(x_0)} \approx \nu \frac{U(x_0) Re_{x_0}}{x_0^2} \approx \frac{U^2(x_0)}{x_0}.$$

Outside the boundary layer, viscous effects are negligible, and Bernoulli's equation (Equation 10.4) can be applied just as in the case of an ideal fluid; taking the derivative of this equation relative to x leads to:

$$\frac{\partial p}{\partial x} + \rho\, U(x) \frac{\partial U(x)}{\partial x} = 0. \tag{10.10}$$

Equation 10.2 is finally obtained by combining Equations 10.8 and 10.10.

10.3.2 Transport of vorticity in the boundary layer

The general equation for transport of vorticity in a Newtonian fluid (Equation 7.41) can be written:

$$\frac{\partial \boldsymbol{\omega}}{\partial t} + (\mathbf{v}\cdot\nabla)\boldsymbol{\omega} = (\boldsymbol{\omega}\cdot\nabla)\mathbf{v} + \nu\, \nabla^2 \boldsymbol{\omega}. \tag{10.11}$$

It can be simplified in the present case of a two-dimensional flow, where only the component ω_z is non-zero and the flow is stationary. As a result, the term $(\boldsymbol{\omega}\cdot\nabla)\,\mathbf{v}$ vanishes, so that one obtains:

$$v_x \frac{\partial \omega_z}{\partial x} + v_y \frac{\partial \omega_z}{\partial y} = \nu \frac{\partial^2 \omega_z}{\partial y^2}. \tag{10.12}$$

This equation expresses the balance between the transport of vorticity by convection and by diffusion. Changes in ω_z associated with the stretching of vorticity tubes do not indeed occur as a consequence of the two-dimensional character of the flow; these properties have been discussed in Chapter 7 as part of the study of vorticity.

10.3.3 Self-similarity of the velocity profiles in the boundary layer for the case of uniform, constant, external velocity

As already seen above, the length scales x_0 and $\delta(x_0)$ parallel and normal to the wall are very different. The same is true of the corresponding velocity scales $v_x(\approx U)$ and $v_y\left(\approx U/\sqrt{Re}\right)$. Moreover, there is no unique characteristic length scale applicable at every point in the flow. *Two* very different *local* length scales are therefore used:

- a distance x_0, measured from the edge of the plate parallel to the direction of flow;
- the corresponding local thickness $\delta(x_0) \approx \sqrt{\frac{\nu x_0}{U}} \approx \frac{x_0}{\sqrt{Re_{x_0}}}$ of the boundary layer, in the y-direction normal to the wall.

The Navier-Stokes equation is then rewritten using a system of dimensionless variables (denoted using *'primes'*) defined using the two above length scales and the corresponding velocity scales:

- U parallel to the plate;
- $\frac{U}{\sqrt{Re_{x_0}}}$ normal to it (Equation 10.6).

One has then:

$$x' = \frac{x}{x_0}, \quad (10.13a) \qquad y' = \frac{y}{\delta(x_0)} = \frac{y\sqrt{Re_{x0}}}{x_0}, \quad (10.13b)$$

$$v'_x = \frac{v_x}{U}, \quad (10.13c) \qquad \text{and} \qquad v'_y = \frac{v_y\sqrt{Re_{x0}}}{U}. \quad (10.13d)$$

One considers first the case where the velocity $U(x)$ outside the boundary layer is independent of x. One has then: $\partial U(x)/\partial x = 0$ and, consequently, $\partial p/\partial x = 0$ in agreement with Equation 10.10. Equations 10.2 and 10.5 then become:

$$\frac{\partial v'_x}{\partial x'} + \frac{\partial v'_y}{\partial y'} = 0 \quad (10.14) \qquad \text{and:} \qquad v'_x \frac{\partial v'_x}{\partial x'} + v'_y \frac{\partial v'_x}{\partial y'} = \frac{\partial^2 v'_x}{\partial y'^2}. \quad (10.15)$$

Mathematically, the right-hand side of Equation 10.15, representing viscous momentum-transport, plays a role as important as that of the convective terms, a fact indicative of the true physical situation. This is not the case in the limit of an infinite Reynolds number in the dimensionless Navier-Stokes equation, obtained by using a single length scale as in Section 4.2.4. Indeed, in that instance, the equation reduces simply to Euler's equation (Equation 4.31) for ideal fluids. By using the two different length-scales, normal and parallel to the wall, one takes into account properly the velocity gradient (and thus of the viscous forces) which is quite large near the walls within the boundary layer.

Let us now show that, still for a constant external velocity U, there are solutions for Equations 10.14 and 10.15 which are functions, not of x' and y' individually, but of the single dimensionless variable:

$$\theta = \frac{y'}{\sqrt{x'}} = \frac{y}{\sqrt{\nu x/U}}. \quad (10.16)$$

The corresponding variations of the velocity components are of the form:

$$\frac{v_x}{U} = f\left(\frac{y}{\sqrt{\nu x/U}}\right) = f(\theta) \quad (10.17) \qquad \text{and:} \qquad \frac{v_y}{U} = \sqrt{\frac{\nu}{Ux}}\, h\left(\frac{y}{\sqrt{\nu x/U}}\right) = \sqrt{\frac{\nu}{Ux}}\, h(\theta). \quad (10.18)$$

Such a profile is known as *self-similar*: the dependence of the component v_x of the velocity with the distance y from the surface of the plate is always the same, within a scale factor $\sqrt{\nu x/U}$, for any distance x from the edge of the plate.

Proof

Equations 10.14 and 10.15 have solutions of the form:

$$v'_x = f(x', y') \quad (10.19) \qquad \text{and:} \qquad v'_y = g(x', y'). \quad (10.20)$$

One then has:

$$v_x = U f\left(\frac{x}{x_0}, \frac{y}{\delta(x_0)}\right) \quad (10.21)$$

and:

$$v_y = \sqrt{\frac{\nu U}{x_0}}\, g\left(\frac{x}{x_0}, \frac{y}{\delta(x_0)}\right) \quad (10.22) \qquad \text{or:} \qquad v_y = \sqrt{\frac{\nu U}{x}}\, h\left(\frac{x}{x_0}, \frac{y}{\delta(x_0)}\right), \quad (10.23)$$

with $h = g\sqrt{x/x_0}$. The components v_x and v_y depend *a priori* on the two variables x/x_0 and $y/\delta(x_0)$. However, the functions f and h cannot involve independently these two variables, but only in a combination which does not use x_0. The choice of the length-scale x_0 parallel to the plate is indeed arbitrary while the solution of the problem is independent of x_0 and cannot therefore depend on this choice. One selects therefore the variable $\theta = y'/\sqrt{x'}$ which is the simplest combination that does not involve x_0.

The concept of self-similarity, and the method just used, apply very generally to a wide variety of situations. An example of a self-similar velocity profile has indeed already been

The self-similarity property of a velocity profile in the boundary layer is observed whenever the outer velocity profile $U(x)$ has no characteristic length-scale; an example of such a flow has just been presented in the case of a flat plate. Yet, another example occurs in the family of flows of the form $U(x) = Cx^m$ corresponding to a flow near the angle of intersection of two planes, mentioned in Section 6.6.2(iv): it is discussed in Section 10.5.2 in terms of the behavior of the boundary layer in the presence of a pressure gradient in the external flow.

found in Section 2.1.2, for the flow induced by the motion of a plate in a direction parallel to its own plane. In that case, the velocity profile depends only on the dimensionless parameter $y/\sqrt{\nu t}$, involving the distance y from the plate and the elapsed time t. The variable t is analogous, for a non-stationary problem, to the distance x from the edge of a plate in the present case of a stationary boundary layer. A similar example is discussed in Section 10.7, regarding the shape of the transverse velocity profile in a laminar wake behind a solid obstacle. One also finds self-similar profiles for jets and for mixing layers between two adjacent fluids at different velocities, both in the laminar and turbulent regimes.

10.4 Velocity profiles within boundary layers

10.4.1 Blasius equation for uniform external flow along a flat plate

One looks for the differential equation obeyed by the velocity field within the boundary layer, along a flat plate, again located in a uniform flow of velocity $\mathbf{U}$, parallel to its plane: for that purpose, the velocity component $v_x(x, y)$ (Equation 10.17) is expressed as a function of the reduced variable θ, introduced in Equation 10.16, and of the magnitude U of the velocity:

$$v_x(x, y) = U\, f(\theta) \qquad \text{and:} \qquad \theta = \frac{y}{\sqrt{\nu x/U}}.$$

Substituting these expressions into Equations 10.2 and 10.5, which hold within the boundary layer, one obtains the equation known as the *Blasius equation*:

$$f''(\theta) = -\frac{1}{2} f'(\theta) \int_0^\theta f(\xi)\, \mathrm{d}\xi. \tag{10.24}$$

Proof

one uses the incompressibility condition, Equation 10.5, in order to express the variable y as a function of x and θ. One obtains:

$$\frac{\partial v_x}{\partial x} = -\frac{\partial v_y}{\partial y} = U f'(\theta) \frac{\partial \theta}{\partial x} = -U f'(\theta) \frac{\theta}{2x}. \tag{10.25}$$

Moreover:

$$\frac{\partial v_y}{\partial \theta} = \frac{\partial v_y}{\partial y} \frac{\partial y}{\partial \theta} = \frac{\partial v_y}{\partial y} \sqrt{\frac{\nu x}{U}}. \tag{10.26}$$

By combining these two equations, one obtains:

$$\frac{\partial v_y}{\partial \theta} = \sqrt{\frac{\nu U}{x}}\, \theta\, \frac{f'(\theta)}{2}. \tag{10.27}$$

The results, by integrating by parts with respect to θ at constant x (ξ being the variable of integration between 0 and θ):

$$v_y = \sqrt{\frac{\nu U}{x}} \left(\frac{1}{2}\, \theta\, f(\theta) - \frac{1}{2} \int f(\xi) \mathrm{d}\xi \right). \tag{10.28}$$

The constant of integration which determines the limits of the integral appearing in the expression for v_y is evaluated below by using the boundary conditions. The terms in Equation 10.2 are obtained by replacing v_y by the preceding expression:

$$U\frac{\partial U}{\partial x} = 0, \quad (10.29) \qquad v_x\frac{\partial v_x}{\partial x} = \frac{U^2}{x}\left(-\theta\, f(\theta)\,\frac{f'(\theta)}{2}\right), \quad (10.30)$$

$$v_y\frac{\partial v_x}{\partial y} = \frac{U^2}{x}\, f'(\theta)\left(\frac{1}{2}\theta f(\theta) - \frac{1}{2}\int f(\xi)\,\mathrm{d}\xi\right) \quad (10.31) \qquad \text{and:} \qquad \nu\frac{\partial^2 v_x}{\partial y^2} = \frac{U^2}{x} f''(\theta). \quad (10.32)$$

One obtains then:

$$f''(\theta) = -\frac{1}{2}\, f'(\theta)\int f(\xi)\,\mathrm{d}\xi \quad (10.33)$$

by combining the different terms in Equation 10.2 that we have just solved, and cancelling out the factor U^2/x. This eliminates the variable x from the equation and confirms the self-similarity property (θ is the only remaining variable). The limits of integration in Equations 10.28 and 10.33 are determined by using the boundary conditions at the surface of the plate and at infinity. One has indeed:

$$\frac{v_x(y=0)}{U} = f(\theta=0) = 0, \quad (10.34a) \qquad \frac{v_y(y=0)}{\sqrt{\nu\, U/x}} = \frac{v_y(\theta=0)}{\sqrt{\nu\, U/x}} = 0 \quad (10.34b)$$

and

$$\lim_{y\to\infty}\frac{v_x}{U} = \lim_{\theta\to\infty} f(\theta) = 1. \quad (10.34c)$$

The condition of Equation 10.34b determines the limits of integration in Equation 10.28 for v_y, which becomes:

$$v_y = \sqrt{\frac{\nu\, U}{x}}\left(\frac{1}{2}\,\theta\, f(\theta) - \frac{1}{2}\int_0^\theta f(\xi)\,\mathrm{d}\xi\right). \quad (10.35)$$

In the same way, Equations 10.34a-b determine the limits of the integral in Equation 10.33: it takes therefore the form of Equation 10.24.

10.4.2 Velocity profile: the solution of Blasius's equation

Figure 10.4 displays, using the dimensionless quantities $f = v_x/U$ and $\theta = y/\sqrt{\nu\, x/U}$, the variation of the longitudinal velocity component as a function of the distance to the plate as obtained by a numerical integration of Equation 10.24; at small distances, the variation is linear with a slope $\mathrm{d}f/\mathrm{d}\theta(0) = 0.332 \cong 1/3$. Between $\theta = 3$ and $\theta = 5$, there is a transition toward the limiting value $\theta = 1$ at large distances ($v_x/U = 0.99$ for $\theta = 5$). This result confirms the importance of the concept of a boundary layer: as one moves away from the plate, one recovers quickly the external uniform flow. As will be shown below, these characteristics of the velocity profile in the figure can be inferred simply from the form of the Blasius

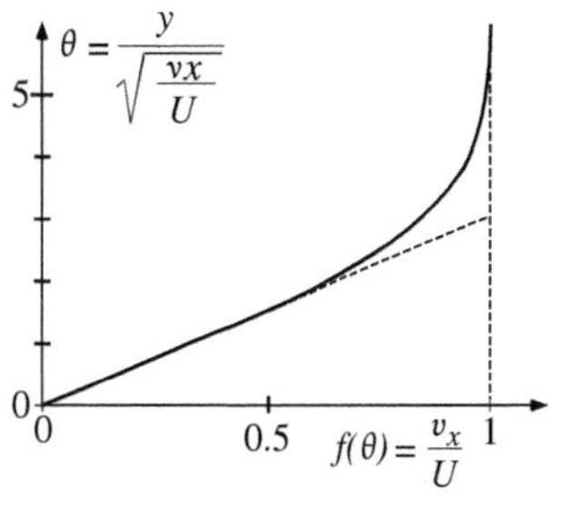

Figure 10.4 *Variation of the normalized velocity component parallel to the plate as a function of the normalized distance* $\theta = \sqrt{\nu\, x/U}$

equation (Equation 10.24). For flows in turbulent boundary layers, the velocity approaches more slowly its asymptotic value (see Section 12.5.5).

Proof

The derivative of Equation 10.24 can be written:

$$f'''(\theta) = -\frac{1}{2}\left(f'(\theta)\, f(\theta) + f''(\theta) \int_0^\theta f(\xi)\, d\xi\right). \tag{10.36}$$

For $\theta = 0$, one has $f(0) = 0$ (Equation 10.34a) and, according to Equation 10.24, $f''(0) = 0$ so that, from Equation 10.36: $f'(0) = 0$. Thus, at small values of θ, the variation of the velocity component v_x parallel to the plate as a function of the distance to the plate is very linear; it can be expressed in the form of a partial expansion:

$$\frac{v_x(\theta)}{U} = f(\theta) \approx \theta\, f'(0) + b\, \theta^4 + O(\theta^5). \tag{10.37}$$

Substituting this expansion into Equation 10.36 and writing the equality of terms of order θ, one finds:

$$b = -\frac{1}{48}\, f'^2(0). \tag{10.38}$$

Thus:

- the concavity of the profile $f(\theta)$ is oriented upstream of the flow because $f''(\theta) < 0$ (Figure 10.4);
- even for a value of θ as large as 2, the correction $b\,\theta^3$ relative to a linear variation is only of the order of $f'(0)/6 \cong 0.055$.

In the opposite limit, for large values of θ, $f(\theta)$ approaches 1 (Equation 10.34c) and $\int_0^\theta f(\xi)\, d\xi$ becomes of the order of θ. The Blasius equation (Equation 10.24) thus approaches the simplified form:

$$f''(\theta) \approx -\frac{1}{2}\, \theta\, f'(\theta). \tag{10.39}$$

By integrating this expression one finds that $f'(\theta)$ is of the order of $K e^{-\theta^2/4}$. Thus, the profile $f(\theta)$ approaches exponentially its asymptotic value 1: as soon as θ takes on a value of 4 or 5, $f'(\theta)$ is indeed only of the order of a few thousandths. By combining the two preceding limits ($\theta \to 0$ and $\theta \to \infty$), one predicts that $f(\theta)$ shifts abruptly from a linear dependence to its asymptotic value. This result is well obeyed by the variation displayed in Figure 10.4.

Let us determine now the order of magnitude of the value of $f'(0)$ of the slope at the origin on the basis of the approximate expansion in Equation 10.37. This expansion predicts indeed a very linear initial variation followed by a sudden rounding (corresponding to the bend in the profile of Figure 10.4) beyond which $f(\theta)$ would reach a maximum for a value θ_m obtained by setting to zero the derivative of Equation 10.37 with respect to θ:

$$\theta_m = \left(12/f'(0)\right)^{1/3}. \tag{10.40}$$

The maximum corresponding value of $f(\theta)$ should be close to 1: this asymptotic value is indeed reached very quickly, i.e. exponentially, by the velocity profile beyond the bend. One has, as a first approximation:

$$f(\theta_m) = \theta_m f'(0) - \frac{1}{48} f'^2(0)\, \theta_m^4 \approx 1.$$

Substituting for θ_m in this equation the value given by Equation 10.40, one obtains:

$$f'(0) \approx 0.29 \qquad \text{and:} \qquad \theta_m \approx 3.44.$$

This estimation of $f'(0)$ is of the same order of magnitude as the exact value $f'(0) = 0.332$ mentioned above.

10.4.3 Frictional force on a flat plate in uniform flow

For flow in the x-direction, the frictional force on a surface with its normal in the y-direction, having length L in the x-direction and unit width, has the value:

$$F_{\text{total}} = \frac{4}{3}\, \rho U^2 L \sqrt{\frac{1}{Re_L}}. \tag{10.41}$$

The drag coefficient C_d is defined by normalizing the force per unit area by the product of the dynamic pressure ($1/2\ \rho\ U^2$) and of the surface area in contact with the fluid ($2L$), so that:

$$C_d = \frac{F_{\text{total}}}{\rho U^2 L} = \frac{1.33}{\sqrt{Re_L}}. \tag{10.42}$$

This variation of the drag coefficient C_d as $1/\sqrt{Re_L}$ is slower than the variation as $1/Re$ obtained for flow at low Reynolds numbers, where the transport is controlled by viscous diffusion (Equation 9.20). The transport of momentum by viscous diffusion is in fact confined within the small thickness of the boundary layer, and enhanced by the presence of velocity gradients significantly higher than those existing when the Reynolds number is small compared to unity. In contrast, the variation is faster than for turbulent flows (Section 12.5.4): in the latter case, convective transport dominates, the force is proportional to the square of the velocity, and C_d is approximately independent of the Reynolds number (Equation 12.74).

Proof

The frictional force per unit area at a distance x from the edge is the component σ_{xy} of the stress:

$$\sigma'_{xy} = \eta \left[\frac{\partial v_x}{\partial y}\right]_{y=0} = \eta\ U f'(0)\, \frac{\partial \theta}{\partial y} = \eta\ U f'(0) \sqrt{\frac{U}{\nu x}} \tag{10.43}$$

or, introducing a term of the form ρU^2, having the dimensions of a pressure:

$$\sigma'_{xy} = \rho U^2 f'(0) \sqrt{\frac{\nu}{U x}}. \tag{10.44}$$

In order to obtain the total force exerted by the fluid on the plate, σ'_{xy} is integrated with respect to x over both sides of the flat plate of length L. One finds:

$$F_{\text{total}} = 2\, \rho U^2 f'(0) \sqrt{\frac{\nu}{U}} \int_0^L \frac{dx}{\sqrt{x}} = 4 \rho U^2 f'(0) \sqrt{\frac{\nu L}{U}}. \tag{10.45}$$

Equation 10.41 is then obtained by defining a Reynolds number $Re_L = UL/\nu$ using, as a characteristic length, the size L of the plate in the direction of the flow and taking $f'(0) \approx 1/3$.

10.4.4 Thicknesses of the boundary layers

We have shown that, in the presence of a flow with velocity U, the thickness of the boundary layer varies as $\sqrt{\nu x/U}$. In order to evaluate more precisely the constant of proportionality, we take the thickness of the boundary layer δ as equal to the value y for which $f(\theta) = v_x(y)/U$ has a given value. For instance, $\delta_{0.99}$ (corresponding to $v_x/U = 0.99$) is given by:

$$\delta_{0.99} = 5\sqrt{\frac{\nu x}{U}}. \tag{10.46}$$

The choice of the value of f is arbitrary so other more universally applicable definitions based on a physical effect have been introduced. Their value, however, only differs from that of $\delta_{0.99}$ by a numerical constant.

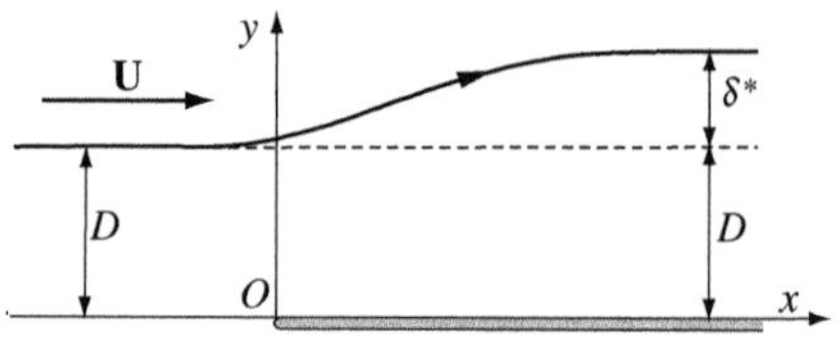

Figure 10.5 *Evaluation of the thickness of the boundary layer along a flat plate, in terms of the displacement δ^* of the streamlines external to the boundary layer (the vertical scale is greatly expanded)*

(i) **The displacement thickness δ^***
A more universally accepted definition of the boundary layer corresponds to the displacement δ^* of the streamlines of the external potential flow. This displacement varies with the distance from the leading edge of the plate in the direction of the flow. In order to evaluate the quantity δ^*, we write the integral for the flow rate within a stream tube of unit width along the direction perpendicular to the plane of Figure 10.5, and of thickness D upstream of the plate (D must be considerably larger than δ^*):

$$\int_0^D U\,\mathrm{d}y = \int_0^{D+\delta^*} v_x\,\mathrm{d}y = \int_0^{D+\delta^*} (v_x - U)\,\mathrm{d}y + \int_0^{D+\delta^*} U\,\mathrm{d}y. \tag{10.47}$$

If D is sufficiently large, we can consider that $(v_x - U)$ is 0 beyond the distance $(D+\delta^*)$, leading to the equation:

$$\int_0^{D+\delta^*} (v_x - U)\,\mathrm{d}y = \int_0^{\infty} (v_x - U)\,\mathrm{d}y. \tag{10.48}$$

If the flow is uniform, Equation 10.47 then becomes:

$$U D - U(D+\delta^*) = \int_0^{\infty} (v_x - U)\,\mathrm{d}y, \tag{10.49}$$

so that:

$$\delta^*(x) = \int_0^{\infty} \left(1 - \frac{v_x(x,y)}{U}\right)\mathrm{d}y. \tag{10.50}$$

And one obtains, by numerical integration:

$$\delta^* = 1.73\sqrt{\frac{\nu x}{U}}. \tag{10.51}$$

(ii) **Momentum thickness δ^{**}**

Quite similarly to the displacement of the streamline, the thickness of a boundary layer can be defined in terms of the change in the momentum inside a stream tube or on the basis of changes in the kinetic energy. Thus:

$$\delta^{**} = \frac{\text{upstream flux of momentum} - \text{flux at distance } x}{\rho U^2}, \tag{10.52a}$$

so that:

$$\delta^{**} = \int_0^\infty \frac{v_x\,(U - v_x)}{U^2}\,\mathrm{d}y. \tag{10.52b}$$

For uniform external flow one obtains, numerically:

$$\delta^{**}(x) = 0.66\,\sqrt{\frac{\nu x}{U}}. \tag{10.53}$$

The product $\rho U^2 \delta^{**}$ is equal to the frictional force on one side of the wall between the edge and the distance x (i.e., to half the value for $L = x$ given by Equation 10.41, which corresponds to using both sides).

10.4.5 Hydrodynamic stability of a laminar boundary layer – transition to turbulence

So far, Reynolds numbers have been defined in this chapter by using, as length-scales, distances in the direction of the flow. One has thus used: $Re_{x_0} = Ux_0/\nu$ and $\mathrm{Re}_L = UL/\nu$, where x_0 and L are respectively the local distance to the edge of the wall, and the total length in the direction of the flow: the conditions $Re_{x_0} \gg 1$ and $Re_L \gg 1$ must be obeyed so that the thickness δ of the boundary layer can be small relative to x_0 and L. In contrast, the stability of a laminar boundary layer is determined by the value of a local Reynolds number using the thickness of the boundary layer as the length scale. Similarly, in order to determine whether the flow is laminar or turbulent in a tube, one takes, as the characteristic length in the Reynolds number, the diameter of the tube and not its length. For example, one defines:

$$Re_{\delta_{0.99}} = \frac{U\ \delta_{0.99}}{\nu} \propto \sqrt{\frac{Ux}{\nu}}. \tag{10.54}$$

This number increases as the square root of the distance x from the edge of the plate. Accordingly, even at high flow velocities, hydrodynamic instabilities appear only beyond an appreciable distance x from the leading edge. These first take the form of regular two-dimensional oscillations of the velocity in the x-y plane and, later, of three-dimensional fluctuations. Beyond a threshold amplitude of the fluctuations, turbulent regions appear. At still higher velocities, the whole boundary layer displays rapid fluctuations of the instantaneous local velocity.

We will discuss in Section 12.5.5 the characteristics of these *turbulent boundary layers*: above a small distance from the plate, momentum transport no longer results from the viscosity, but from the turbulent fluctuations of the velocity.

10.5 Laminar boundary layer in the presence of an external pressure gradient: boundary layer separation

10.5.1 Simplified physical treatment of the problem

Let us assume that the velocity of the external potential flow $U(x)$ decreases with the distance x downstream of the leading edge of the plate as would be the case, for example, in a divergent flow. Outside the boundary layer, the pressure $p(x)$ increases with distance since the pressure gradient $\partial p/\partial x$ in this direction obeys Bernoulli's equation (Equation 10.4):

$$\frac{\partial p}{\partial x} = -\rho U \frac{\partial U}{\partial x} > 0. \tag{10.55}$$

Moreover, because, according to Equation 10.3, variations in pressure in the transverse direction are negligible, the longitudinal pressure-gradient is constant within the boundary layer. Thus, in the low-velocity regions near the wall, the dynamics of fluid elements results from two opposing effects: on the one hand, the positive pressure gradient $\partial p/\partial x$ slows down their motion; on the other hand, momentum transfer by viscous diffusion from higher-velocity regions tends to accelerate them. If the velocity gradient $\partial U/\partial x$ is sufficiently large in magnitude, there will be a reversal of the direction of flow near the wall. This phenomenon characterizes *boundary layer separation* (see Figures 10.1b, 10.6 and 10.7). In the opposite case of a positive downstream velocity gradient, the corresponding pressure-gradient $\partial p/\partial x$ is negative, the fluid near the wall is accelerated, and the boundary layer thins out, leading to stability.

10.5.2 Self-similar velocity profiles – flows of the form $U(x) = C\,x^m$

In Section 6.6.2 (iv), the complex potential $f(z) = C\,z^{m+1}$ has been seen to describe the flow along two solid half-planes intersecting at an angle $\pi/(m+1)$. Along the plane $\varphi = 0$, the velocity reduces to its radial component:

$$v_r = C\,(m+1)\,r^m,$$

in which the radius vector r is equal, in the plane, to the distance x from the vertex of the angle. Thus, assuming a velocity dependence $U(x) = C\,x^m$ just outside the boundary layer, one can analyze the behavior of the boundary layer for flow around planes intersecting at a sharp angle. One derives first the differential equation obeyed by the velocity field within the boundary layer; in this case, it is the analog of the Blasius equation written above for uniform external flows.

The Falkner–Skan equation

One follows a procedure similar to that used in the case of uniform flow (Section 10.4.1). The combination of the dimensionless variables x/x_0 and $y\,/\sqrt{\nu U(x_0)/x_0}$, which does not depend on the local distance x_0 from the vertex, is:

$$\theta = y\sqrt{\frac{U(x)}{\nu x}} = y\sqrt{\frac{C\,x^{m-1}}{\nu}}. \tag{10.56}$$

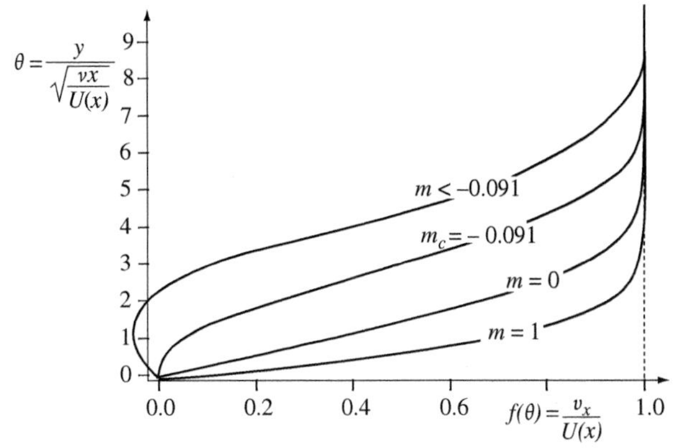

Figure 10.6 *Shape of the velocity profiles within a boundary layer for external flows of the type $U(x) = C\,x^m$; $v_x(x, y)$ is the component of velocity parallel to the wall. The solution $m = 0$ is that displayed in Figure 10.4*

Expressing the different terms of Equation 10.2, in terms of the variable θ, one finds the differential equation satisfied by the ratio $v_x(x,y)/U(x)$ of the velocity component parallel to the plane, to the external flow velocity $U(x)$:

$$m\,(1 - f^2(\theta)) + f''(\theta) = -\frac{m+1}{2}\, f'(\theta) \int_0^\theta f(\xi)\,\mathrm{d}\xi. \tag{10.57}$$

This equation, sometimes known as the *Falkner-Skan equation*, reduces, as expected, to the Blasius equation (Equation 10.24) for the specific case $m = 0$.

Velocity profiles within the boundary layer

By solving numerically the Falkner–Skan equation, one obtains the velocity profiles of the type shown in Figure 10.6: the larger and more positive the exponent m, the thinner the corresponding boundary layer. As m decreases, the boundary layer thickness increases and, for a small negative critical value $m_c = -0.0905$, the velocity gradient at the wall vanishes. For even more negative values of m, the direction of the flow near the wall is reversed, a phenomenon known as boundary layer separation introduced above in Section 10.5.1.
The critical angle α_c, corresponding to the appearance of a flow reversal, is equal to:

$$\alpha_c = \frac{\pi}{m_c + 1} = 198°.$$

There is, in this case, a deviation by an angle equal to 198°–180° = 18° of the velocity of the fluid close to the surface before and after the vertex. Figure 10.7 illustrates the type of flow observed when the angle exceeds this value.

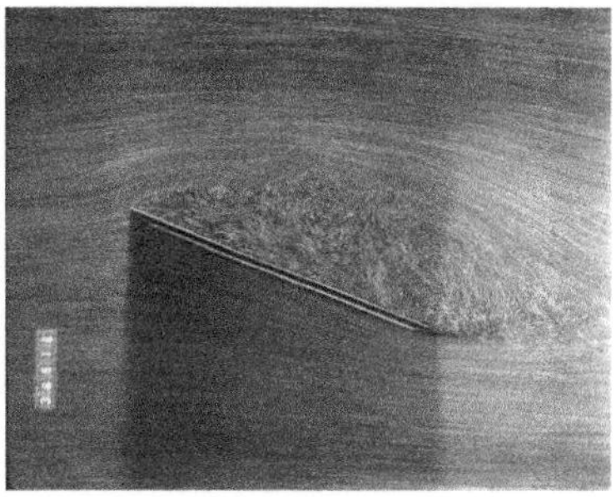

Figure 10.7 *Generalized boundary layer separation for flow at the leading edge of an inclined plate. The figure displays an angle of incidence greater than 20°, a value close to the limit of 18° beyond which boundary layer separation occurs (plate courtesy of ONERA)*

The larger the absolute value of $|m|$ (with $m < 0$), the slower $f(\theta)$ reaches its asymptotic value of unity. In the limiting case where $m = -1$, the velocity has no asymptotic limit. Physically, this corresponds to a divergent flow from the vertex of an angle between two planes ($U \propto 1/x$), and a boundary layer no longer exists. For $m > 0$, the boundary layer is thinner than for $U = constant$ and, when $m = 1$ (flow toward a stagnation point), its thickness remains constant with the distance away from the point of incidence (Figure 10.8a below).

One does not observe in Figure 10.8a any local flow reversal in the divergence region upstream of the stagnation point: this is due to the lack of a solid wall creating a layer of slow fluid flow easily stopped by the pressure gradient opposite to it. Instead, if a flat plate is put normal to the wall (Figure 10.8b), recirculation zones appear within the angles which they form.

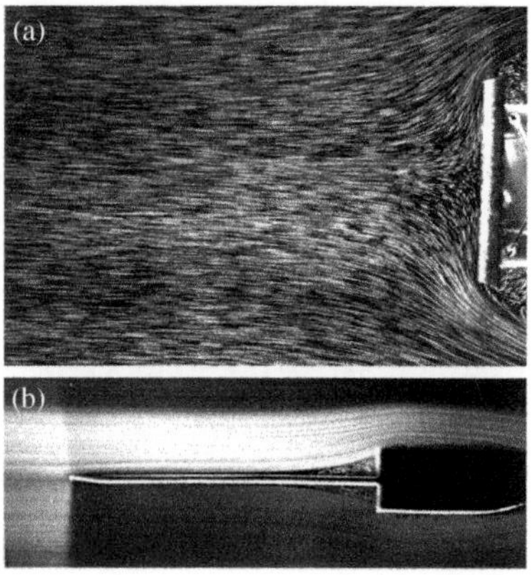

Figure 10.8 *Flow toward a plane with a stagnation point; (a) the absence of a solid wall prevents the formation of a boundary layer in the region where the flow diverges; (b) if a wall is inserted, there appears two vortex regions resulting from the boundary layer separation on each side of this wall (plates courtesy, H. Werlé, ONERA)*

Approximate calculation of the condition for boundary layer separation

Let us go back to the Falkner-Skan equation (Equation 10.57), in the form:

$$m\ (1-f^2(\theta))+f''(\theta)=-\frac{m+1}{2}f'(\theta)\int_0^\theta f(\xi)\,d\xi. \qquad (10.58)$$

For $\theta=0$: $f(0)=0,$ (10.59) so that: $m+f''(0)=0.$

Thus, at small values of θ:

$$f(\theta)=\theta\, f'(0)-\frac{m}{2}\,\theta^2+O(\theta^3). \qquad (10.60)$$

For large values of θ, instead:

$$f(\theta)\to 1 \qquad (10.61) \qquad \text{and, consequently:} \qquad \int_0^\theta f(\xi)\,d\xi\to\theta.$$

Therefore, Equation 10.58 approaches the limiting form:

$$f''(\theta)=-\frac{m+1}{2}\,\theta\, f'(\theta), \qquad (10.62) \qquad \text{so that:} \qquad f'(\theta)\approx C\ e^{-\frac{(m+1)\,\theta^2}{4}}; \qquad (10.63)$$

$f'(\theta)$ then approaches 0 more slowly for $m<0$ ($\partial U/\partial x<0$) than for $m>0$ ($\partial U/\partial x>0$). This is in agreement with the larger thickness of the boundary layer for $m<0$ than for $m>0$ (Figure 10.6). The connection is made again between the dependence predicted by Equation 10.63, for $\theta\to\infty$, and that given by Equation 10.60, for small θ values.

- When $m>0$, $f''(0)$ is negative; the curvature of the velocity profile is therefore directed toward the left on Figure 10.6 and there is no point of inflection. In order to reach the asymptotic value faster than in the case $U=$ *constant*, $f'(0)$ must be larger than in this latter case for a same magnitude of the velocity outside the boundary layer.
- If m is negative, $f'(0)$ is, in contrast, smaller, and the curvature of the velocity profile, near the point O, is directed toward the right. For $|m|$ sufficiently large, we have therefore a reversal of the direction of flow near the wall ($f'(0)<0$).

Let us estimate now the critical value m_c at which $f'(0)=0$. We have, in this case, for small θ:

$$f(\theta)=-m_c\frac{\theta^2}{2}+O(\theta^3).$$

Moreover, $f(\theta)$ must then reach the limiting value of 1 when θ is of the order of a few units. By assuming in the previous expression that $f=1$ for $\theta=5$, we obtain the order of magnitude $m_c=-1/12\cong-0.083$ (the exact value is -0.0905).

10.5.3 Boundary layers of constant thickness

We examine several situations for which the boundary layer along a plate maintains a constant thickness. A first way to obtain this result is to compensate the thickening of a boundary layer, resulting from the diffusion of vorticity, by a velocity component directed toward the wall.

Flow toward a stagnation point

We see, in the example of Figure 10.8a above, that flow toward a stagnation point has such a component. In the idealized case of a stagnation point in two dimensions, the stream function outside the boundary layer is $\Psi = kxy$, with corresponding velocity components $v_x = kx$ and $v_y = -ky$ (for a stagnation point located at $x = 0$ and $y = 0$). The thickness $\delta(x)$ then obeys the equation:

$$\delta(x) \approx \sqrt{\frac{\nu x}{v_x(x)}} \approx \sqrt{\frac{\nu}{k}}, \tag{10.64}$$

so that δ is independent of distance: there is then a compensation between the opposing effects of convective transport toward the wall, and diffusion away from it.

Flow in the presence of suction

One encounters a related case by considering the effect, on the Blasius problem, of suction perpendicular to the plate, with a normal velocity component at the wall: $v_y = -V$ independent of x (Figure 10.9). Such a technique has been suggested for limiting the boundary layer separation on cars and airplane wings (Section 10.6). Assume now that there exists a solution of the equations of motion corresponding to a boundary layer of thickness and velocity profile independent of x. In this case, v_x and v_y depend only on y and the velocity component v_y, normal to the wall, satisfies $\partial v_y/\partial y = 0$ as a result of the condition of incompressibility. We thus have $v_y = constant = -V$ throughout the volume of the fluid. For the transport of vorticity in the stationary regime, Equation 10.12 can be written:

$$-V\,\frac{\partial \omega_z}{\partial y} = \nu\,\frac{\partial^2 \omega_z}{\partial y^2}, \tag{10.65}$$

since the term $\partial \omega_z/\partial x$ is zero. Let us integrate this equation once, obtaining:

$$-V\,(\omega_z - \omega_0) = \nu\,\frac{d\omega_z}{dy}, \tag{10.66}$$

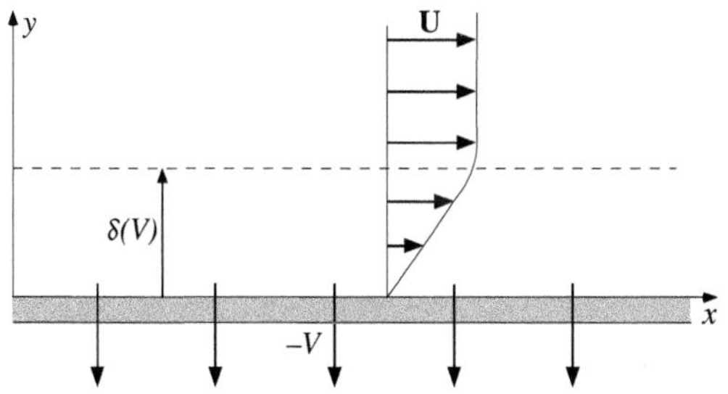

Figure 10.9 *Boundary layer in the presence of suction at a uniform velocity −V along the wall*

where ω_0 represents the vorticity at a large distance from the plate (where $d\omega_z/dy \approx 0$), and equals zero if the flow outside the boundary layer is uniform. Integrating again and choosing the constants of integration such that $v_x = U$, far from the plate, and $v_x = 0$ for $y = 0$, we obtain:

$$\omega_z = \frac{\partial v_x}{\partial y} = \frac{U\,V}{\nu}\,e^{-\frac{V\,y}{\nu}} \qquad \text{and:} \qquad v_x = U\left(1 - e^{-\frac{V\,y}{\nu}}\right).$$

We find that, as initially assumed, the boundary layer has a constant thickness $\delta(V) = \nu/V$; this reflects the fact that, at the edge of the boundary layer, the convective and diffusive transport terms in Equation 10.65, respectively of order $V\omega_z/\delta$ and $\nu\,\omega_z/\delta^2$, balance each other.

Periodic flows with zero-average translational velocity

A first example, already mentioned in Section 4.5.4, is that of flow near a plate oscillating parallel to its own plane with angular frequency ω (not to be mistaken for the vorticity). The effect of the motion of the plane decays exponentially over a distance $\delta_\omega = \sqrt{\nu/\omega}$; here, δ_ω is the order of magnitude of the distance over which changes in velocity diffuse during the period $2\pi/\omega$ of the oscillation. The similar situation of a boundary layer occurring on a disk in uniform rotation used in polarographic measurements will be seen in Section 10.8.2.

10.5.4 Flows lacking self-similarity – boundary layer separation

Figure 10.10 *Velocity profile in a high-angle divergent channel. One observes the occurrence of a stagnation zone followed by a recirculation region (plate courtesy of H. Werlé, ONERA)*

For the self-similar flows of the form $U = C\,x^m$ discussed above, the direction of flow reverses over the entire boundary wall as m crosses the value m_c. In a number of real experimental cases of flows with velocity decreasing in the downstream direction the boundary layer separates only at a certain point, known as the *separation point*, beyond which a recirculation zone appears. Such is the case for flows around bodies lacking aerodynamic shape (Figure 10.1b), or having a divergent-flow profile (Figure 10.10). Thus, for velocity dependences of the kind $U(x) = U_0 - \alpha\,x$, there appears a natural characteristic length U_0/α for the flow. The distance of the separation point relative to the beginning of the zone of negative velocity gradient is determined by this length and has virtually no dependence on the Reynolds number *Re*.

Proof

Very generally, in accordance with Equations 10.14 and 10.15, the velocity field $v_x(x, y)$ inside the boundary layer obeys the equation:

$$\frac{v_x(x, y)}{U(x)} = f\left(\frac{x}{L}, \frac{y}{\delta_L}\right), \tag{10.67}$$

where x and y are local coordinates respectively parallel and normal to the boundary wall, L is the length characteristic of the flow in the direction parallel to the wall, and δ_L is the corresponding thickness of the boundary layer. When the thickness δ_L is small compared to L, the external velocity profile $U(x)$ is independent of the Reynolds number (the effect of the boundary layer region on the external flow is negligible). The condition $[\partial v_x/\partial y]_{y=0} = 0$, which gives the position of the separation point, becomes:

$$f'(x/L, 0) = 0. \tag{10.68}$$

Thus, the value of the ratio x/L, which determines the position of the separation point, is independent of the Reynolds number if it is large enough so that the boundary layer is well established upstream.

10.5.5 Practical consequences of boundary layer separation

Flow regions where recirculation occurs are generally very unstable. The minimum value of the Reynolds number at which instabilities can be easily amplified decreases then to a few tens. A turbulent region of considerable width, in which there is significant energy dissipation appears therefore behind the separation point. The drag force (the component of the viscous force parallel to the direction of flow) then increases substantially; such an effect is observed for bodies which lack an *aerodynamic* profile, as illustrated in Figure 10.1b. In contrast, the drag force is quite low for a streamlined body from which the boundary layer does not separate, and for which the resultant wake is very narrow (Figure 10.1a).

Let us emphasize the difference between the appearance of turbulence associated with boundary layer separation, which causes generally significant energy dissipation, and the transition towards turbulence of the boundary layer itself, which also increases the dissipation, but can delay the onset of the separation. In some cases, one even observes that the turbulent boundary layers, which had separated while they were still laminar, actually reattach. These effects are studied in Chapter 12, devoted to turbulent flows (Section 12.5.5. and 12.5.6).

10.6 Aerodynamics and boundary layers

The aerodynamics of land-based and airborne vehicles is an important field for the application of the concepts of laminar and turbulent boundary layers: even if other factors also play a significant role, the control of the boundary layers represents an important economic issue for the improvement of performance and energy consumption in such vehicles. This significant subject lies beyond the scope of a general discussion; we will confine ourselves to a schematic presentation beginning with an airplane wing which, as a first approximation, can be considered a long object with two-dimensional geometry. In contrast, all three dimensions are of the same order of magnitude for cars, and three-dimensional effects then play an essential role.

10.6.1 Control of boundary layers on an airplane wing

Separation phenomenon

The lift on airplane wings has been discussed in Section 6.6.3 and 7.5.2: more precisely, it represents the Magnus force due to the circulation of velocity around the wing. In order to ensure equilibrium, the lift force F_L must balance the weight of the airplane: from Equation 7.75a, the product $U^2 C_z$, where C_z is the lift coefficient, must then remain a constant. From Equation 6.106, C_z increases linearly with the angle of incidence α for small values of α so that $U^2\alpha$ must also be a constant. In order to reduce the distance required for an airplane to land or to take off, one must then also decrease its velocity during the corresponding phases of flight while increasing the angle of incidence α of the wing.

At moderate values of α, the streamlines follow the profile of the wing (Figure 10.11a) so that the boundary layer remains attached. If α is increased beyond a critical angle α_c, one observes a separation of the boundary layer on the upper surface of the wing, and the appearance of a long, turbulent wake (Figure 10.11b). The pressure on the upper surface of the wing increases, the lift decreases rapidly and the drag force increases, as seen in Figure 10.12 below: when the lift does not balance the weight any more, the airplane *stalls* and the angle of incidence must be reduced to regain control.

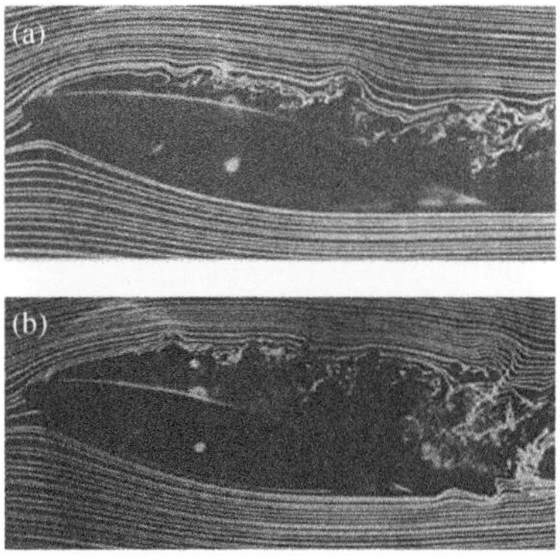

Figure 10.11 *Visualization of the flow around the profile of an airplane wing at two different angles of incidence with and without boundary layer separation: (a) below the angle of separation; (b) above that angle (plates from "Illustrated experiments in fluid mechanics", NCFMF, MIT Press)*

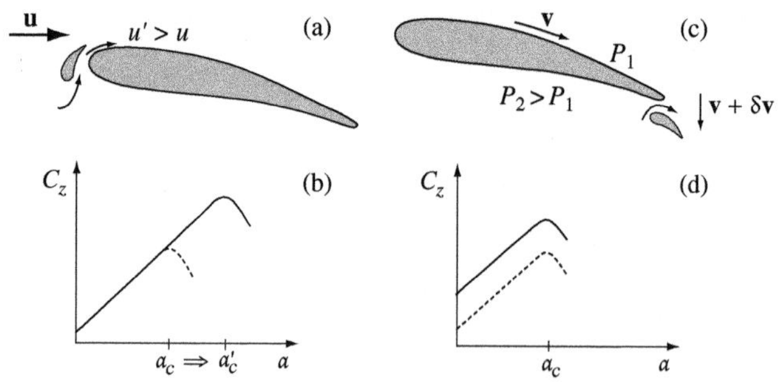

Figure 10.12 *(a) Leading-edge wing flap on an airplane. (b) Functional dependence of the lift coefficient C_z on the angle of incidence, with (solid line) and without (dashes) a leading-edge wing flap. (c) Trailing-edge wing flap, which results in a significant increase in the circulation around the wing for a given velocity and angle of incidence. (d) Comparison of the dependence of the lift coefficient C_z on the angle of incidence α with (solid line) and without (dashes) a trailing-edge wing flap*

Control of the boundary layer separation on airplane wings

The value of α at which this stalling phenomenon appears must then be increased in order to achieve a large enough lift when the airplane flies at a low velocity during take-off and landing. Two complementary strategies currently used in practice are:

- increase the critical angle of incidence α_c by means of *leading-edge wing flaps*;
- increase the lift coefficient C_z for a given angle α by means of *trailing-edge wing flaps*.

A leading-edge wing flap such as the one illustrated in Figure 10.12a, results in an increase of the critical angle α_c, which appears in the dependence of the lift coefficient C_z on α (Figure 10.12b). Due to the effect of the flap, air from below the wing is injected tangentially along the upper surface; it thus *"regenerates"* the boundary layer across the top of the wing by increasing the velocity of the fluid near the wall. At large angles of incidence, the effect of the inverse pressure-gradient is thus reduced, and the critical angle of incidence α_c correspondingly increases.

A trailing-edge wing flap, when extended (Figure 10.12c), causes an increase, at a given velocity, in the circulation around the cross-section of the wing; as a result, the curve $C_z(\alpha)$ is translated upwards, as shown in Figure 10.12d. On very large airliners, the flaps themselves have further flaps attached to them (up to three successive stages on the Airbus A340 or the Boeing 747). One thus often see several sets of flaps extending one behind the other, with the rearmost one almost vertical. These systems increase quite considerably the drag on the wing: they are therefore used exclusively on takeoff and landing, at airplane speeds too low for the lift to be adequate under the normal configuration of the wing. The principle behind the action of trailing-edge wing flaps is a combination of two different effects:

- the existence of a gap between the flap and the main portion of the wing allows for the *"regeneration"* of the boundary layer, by inducing a flow from the lower face of the wing to the upper one (similar to that mentioned in the previous section for leading-edge flaps). This avoids boundary layer separation at the flaps, in spite of their high angle of incidence relative to the main flow;
- the flaps create a significant downward deviation, $\delta\mathbf{v}$, of the flow velocity at the rear of the wing and in the wake, resulting in a major increase of the circulation, and of the corresponding lift.

Other methods for controlling the separation of boundary layers

Many other methods have been used or studied to delay boundary layer separation; some of these are variations on the preceding ones. One can, for example, instead of using leading-edge wing flaps, change the shape of the wing itself in the area of the leading edge to achieve a comparable result. Another example is the use of small vanes perpendicular to the surface of the wing and oblique with respect to the mean flow: these generate vortices parallel to the flow, which bring toward the wall a small amount of high velocity air. These *vortex generators* have been used practically to delay the boundary layer separation but they increase the drag.

10.6.2 Aerodynamics of road vehicles and trains

Specific features of the aerodynamics of cars

In practical applications, one tries to minimize the drag force. The latter is associated mainly to the pressure drop between the upstream and downstream regions, in contrast to the airplane, for which it is the frictional force which dominates. Also, in contrast to airplanes, the lift force does not have to compensate the weight of the vehicle: it must instead be oriented downward in order to ensure good contact of the tires with the road, and to improve handling but cannot be so high as to cause friction and wear on the tires.

The drag force F_D is characterized by a coefficient C_x varying from greater than 0.5 for older cars to less than 0.3 for newer ones. The decrease of this C_x is a major consideration for car manufacturers because of its impact on fuel efficiency. They must, however, also take into account the comfort of the vehicle, which often implies opposite constraints.

The coefficient C_x is defined as $C_x = |F_D| / (\rho U^2 A / 2)$ where A is the projected area of the vehicle's shape onto a plane normal to the velocity **U**. This definition is derived from that of the drag coefficient C_D (Equation 9.20) by substituting the area A for the square of the characteristic length.

Figure 10.13 shows the relative contribution to the drag of the various parts of a vehicle: if the rear of the vehicle plays the essential role, the contribution of the additional elements (such as the wheels, the side-view mirrors and the door handles) is not negligible. This drag is the result of the combined frictional (viscous or, more often, turbulent) and pressure forces: it is the latter which, generally, have the largest contribution.

In contrast to the wing of an airplane, a car has a width and height comparable to its length. The flow of air around the car's body thus has a pronounced three-dimensional nature, which strongly influences the aerodynamic forces on the vehicle.

We will center our discussion on the rear of the vehicle which, according to Figure 10.13, contributes approximately three-quarters of the drag force. This is indeed where the boundary layer separations are localized: these lead to the formation of recirculation and vortex zones which are both sources of energy dissipation.

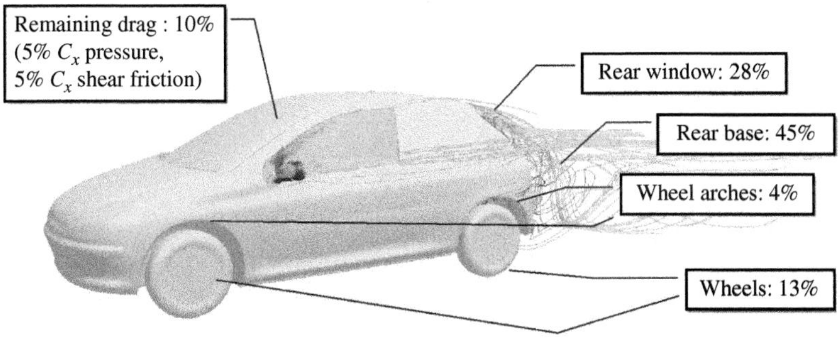

Figure 10.13 *Respective contributions to the drag of the different elements of a car (courtesy of J-L. Aider, PSA Peugeot-Citroën)*

Velocity field behind a motor vehicle – Ahmed body

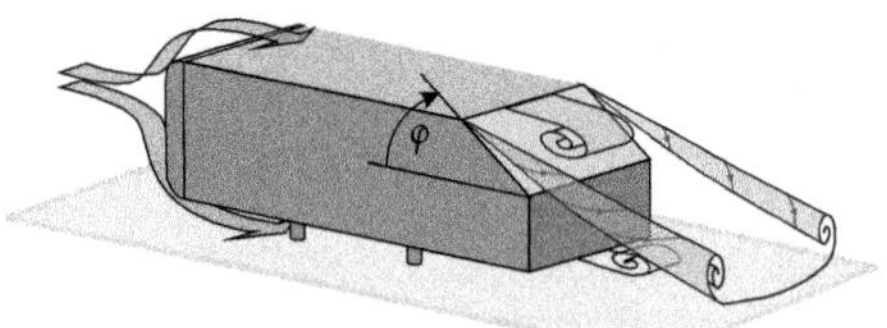

Figure 10.14 *Schematic diagram for the flow around a three-dimensional, non-aerodynamic body (Ahmed body), which models in a simplified manner the flow around a motor vehicle. The structures which contribute to the downstream dissipation are the circulation of the velocity, transverse to the main flow at the level of the rear window, and of the rear base, as well as the longitudinal vortices emitted from the two rear-corners of the roof (as indicated by A. Brunn et al.)*

As a model for the structure of this flow, aerodynamic engineers have introduced a simple standard model: the *Ahmed body* (Figure 10.14). This shape allows them to reproduce experimentally or numerically, in a reasonably realistic manner, the structure of the flow behind a motor vehicle, and to vary effectively the parameters such as the angle φ of the rear window. The velocity field behind this body combines two types of structures:

- **A pair of counter-rotating axial vortices** parallel to the average velocity and emitted at the level of the lateral edges of the model profile, reminiscent of the vortices at the wingtips of airplanes (Figure 7.27c) or of those created behind a sphere (Figure 2.11). Figure 10.15 displays a numerical simulation of the flow behind an

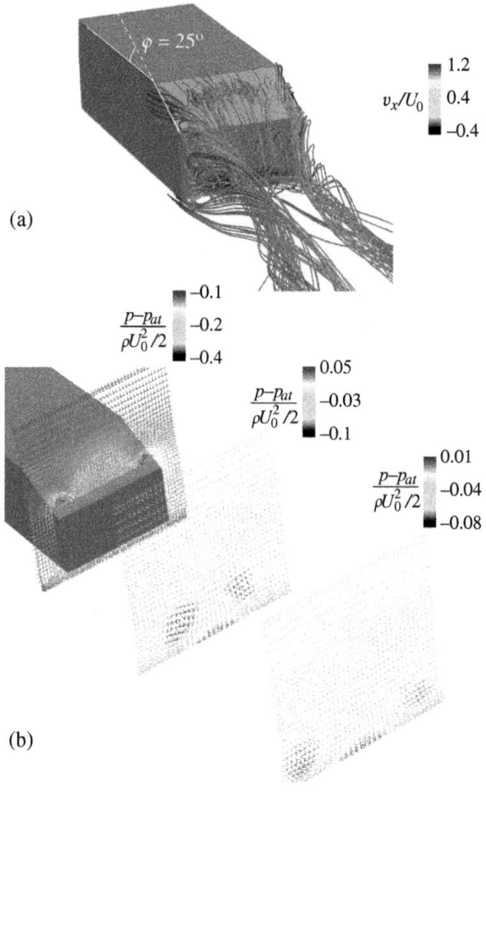

Figure 10.15C *Large-scale simulation (see Section 12.3.4) of the flow behind an Ahmed body with angle $\varphi = 25°$ in a flow of average velocity $U_0 = 40$ m/s upstream of the body (the flow is from left to right). Two longitudinal vortices are visible on the simulation. (a) Streamlines: the color (see in color plate section) corresponds to the normalized component v_x/U_0 of the velocity in the direction of the average flow. (b) Pressure change $(p - p_{at})/(\rho U_0^2/2)$ relative to the normalized atmospheric pressure (colors) and velocity field (vectors). The vortices are displayed as regions of low pressure (in blue on Figure (b)). They represent an important contribution to the drag force (courtesy M. Minguez, R. Pasquetti, and E. Serre)*

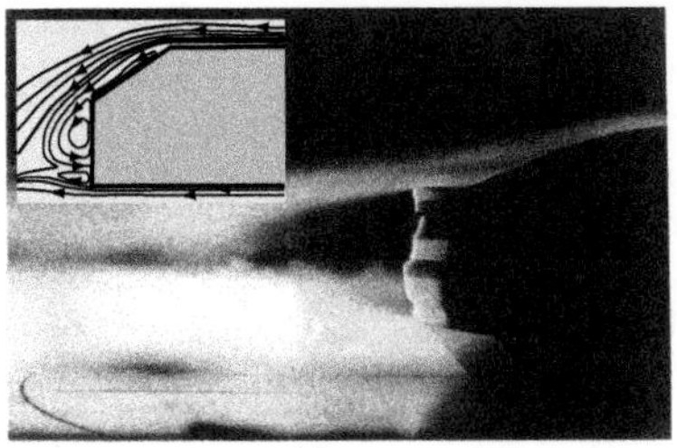
(a)

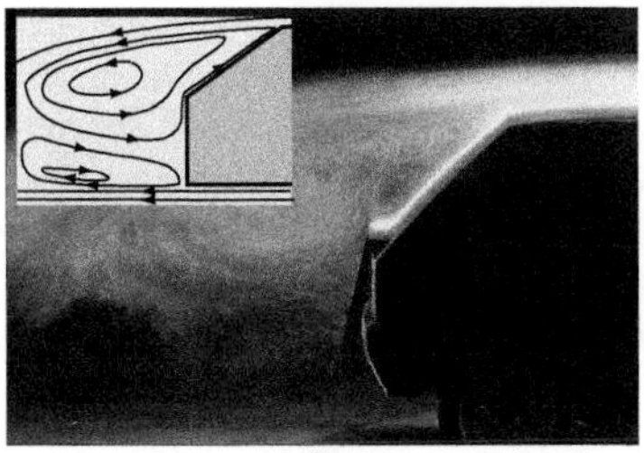
(b)

Figure 10.16 *Visualizations in a wind tunnel of the wakes of vehicles (a) with a fast-back profile (Renault Laguna, plate courtesy Renault), and (b) with a more steeply inclined rear window, hatchback design (Citroën AX, plate courtesy PSA Peugeot-Citroën). The vehicles are stationary and the fluid flows from right to left. One sees a large difference between the widths in cases (a) and (b) of the recirculation zones made visible in the vertical symmetry plane by a light sheet. Insets: recirculation flows in the symmetry plane of two Ahmed bodies with comparable tilt angles: (a) $\varphi = 25°$, (b) $\varphi = 35°$. Just like in the visualizations, the recirculation zones are much larger in case (b), where the boundary layer separates just behind the edge of the roof*

Ahmed body (with $\varphi = 25°$). The large magnitude of these vortices indicates the strongly *three-dimensional* character of the flow.

- **Recirculation flows of axis normal to the symmetry plane of the body** (located in the center part of the rear of the vehicle). These flows are similar to those behind a *two-dimensional* obstacle (inclined ramp or downward step) and are strongly dependent on the tilt angle.

Figure 10.16 shows the influence of this angle by means of wind-tunnel visualizations and numerical simulations of the flow in the vertical symmetry plane, both behind real vehicles and an Ahmed body.

In the vehicle of Figure 10.16a, there is also a slight rise of the profile of the bodywork at the junction between the rear window and the rear base: there is, as a result, a high pressure region at this point creating a force with components oriented downward (increasing the adherence), and forward (decreasing the drag). The height of this rise must not be too large in order to avoid creating a new downstream turbulent region which would increase the drag.

- For small tilt angles of the rear window (case *(a)*), the recirculation zone is either absent or very small: the boundary layer is either attached or weakly separated. In contrast, there is a region of strong recirculation below the vertical rear base and a second small recirculation of opposite direction underneath this first region.
- For high angles of inclination, the boundary layer separates from the top of the rear window and a large recirculation zone appears, covering also part of rear base with a second, weaker, recirculation zone lower down.

The transition between these two regimes occurs around $\varphi = 25°$. This value is of the same order of magnitude as the limit angle ($\approx 18°$) which corresponds to the maximum of the drag and, also, as the angle beyond which the boundary layer separates above a corner between two planes (Section 10.5.2).

The two types of recirculation (longitudinal and transverse) both contribute to the drag force but in a different manner depending on the value of the angle φ. At small values of φ, like in Figure 10.16a, for instance, the contribution of the transverse recirculation is smaller but the effect of axial vortices is strongly magnified. Overall, the resulting dependence of the drag force on the angle is not monotonic (one frequently observes high values around $\varphi = 30°$); in some cases, moreover, values of $\varphi < 10°$ or $\varphi > 35°$ (fast-back) yield acceptable and comparable values of the drag force.

While trying to reduce the drag, one must avoid that, at high speeds, the negative lift which is essential for safe driving, be too small or, even, vanish. On the other hand, sport vehicles of prestigious brands were known for having a poor road handling capability at low speeds, but excellent at higher ones, because of the increase of the negative lift.

10.6.3 Aerodynamics of other land-based vehicles

Drag force on a high-speed train

For a car, the frictional forces at the level of the roof or on the sides are small compared to the pressure forces. This is no longer the case for a truck or a high-speed train (known in France as "TGV"). In contrast to the case of a car, the pressure forces at the front and at

The flow pattern at the rear of a TGV is strongly turbulent because the boundary layer from the roof continues toward the rear and there is no recirculation at the scale of the overall height. Just as for a car, two longitudinal vortices are created.

Semi-trailer aerodynamics:
Without precautions, a turbulent region appears in the gap behind the truck and the trailer of a semi-trailer, creating an additional drag. Placing a baffle between the roof of the truck and the top of the trailer eliminates this recirculation zone. The more continuous aerodynamic profile obtained reduces the perturbations to the flow and the drag.

the rear represent only 10% of the aerodynamic drag, while 30% of this drag is the friction which results from the turbulent boundary layers on the sides and on the roof. There is also a significant contribution from the bogies (40–50%) as well as the pantographs (10–20%) which generate significant turbulence along the roof. Recirculation flows also appear in the space between the cars of the train or between two sections of the train and are sources of dissipation which play an important role.

Finally, for the aerodynamics of TGV's a further constraint comes from the fact that the train must be able to move in either direction: the front and the rear of a high-speed train must then have the same geometry.

10.6.4 Active and reactive control of the drag force and of the lift

Trailing edge wing flap represent an example of *open-loop active control* of the lift: the pilot opens more or less the flaps, depending on the speed of the airplane, and its distance to the airport.

The *closed-loop reactive control* is the object of current research without having yet major industrial applications: the drag can be reduced, for instance, by placing, at the back edge of the roof of an Ahmed body, motorized vortex generators controlled by a sensor.

As seen above, streamlining the upper part of a truck, improving the shape of a rear-view mirror, or of the door handle, can significantly reduce drag: this is known as a *passive control* because such changes are made permanently without any additional energy.

In the case of *active control*, there is a direct action on the flow depending on its instantaneous characteristics and/or on its environment. Because it is the driver who regulates this action as a function of the indications on his dashboard and the characteristics of the environment, one speaks of an *open-loop control.* In a further stage, the resulting action will be controlled by a computer connected to sensors: one will then have a *closed-loop control* also known as a *reactive control.*

This method is rarely used in practice. It had, however, found an application in sailing: the *turbosail* of the wind-driven catamaran "*Moulin à Vent*" developed by Malavard and Cousteau is a very thick wing placed vertically. It is oriented relative to the direction of the wind just like an ordinary sail, and the separation of the boundary layer is prevented by suction of air through a slit located in the region where the separation would take place. In this way, a large circulation of the velocity (and thus horizontal lift forces) can be generated around the "sail."

A possible active control of the separation of boundary layers consists in the suction of low velocity air near a wall or on the contrary, in its acceleration by tangential jets. This allows one to avoid the situation where the adverse pressure-gradient, due to the decrease in the speed of the external flow, might reverse the direction of the flow of the slower fluid: one can thus slow down the separation of the boundary layer even at high angles of incidence. The suction and the injection of air require, however, additional energy: the suction would then only be applied if there is a risk of separation.

Another possible application of active control is the reduction of the drag associated with the emission of Bénard–von Karman vortices behind a cylinder perpendicular to a flow (Section 11.1.2). This emission can be controlled by imposing rotations in alternating directions of the cylinder around its axis. The change in the drag depends on the ratio of the frequency of rotation to that of the vortex emission: at high values of this ratio, the drag as well as the length of the vortex street decreases downstream of the obstacle.

10.7 Wake and laminar jet

10.7.1 Equation of motion of the wake

Qualitative characteristics

One studies the flow behind a motionless solid of comparable characteristic scale a in all directions and placed in a fluid of uniform upstream velocity $\mathbf{U}$ (Figure 10.17 below). The flow is assumed to be laminar and stationary throughout the region: this implies having a Reynolds number $Re = Ua/\nu$ small enough so as to avoid the occurrence of instabilities in the wake (e.g., less than a few tens for a sphere); the case of wakes at high Reynolds numbers will be discussed in Section 12.4. However, as seen in Section 9.5.1, even if Re is small

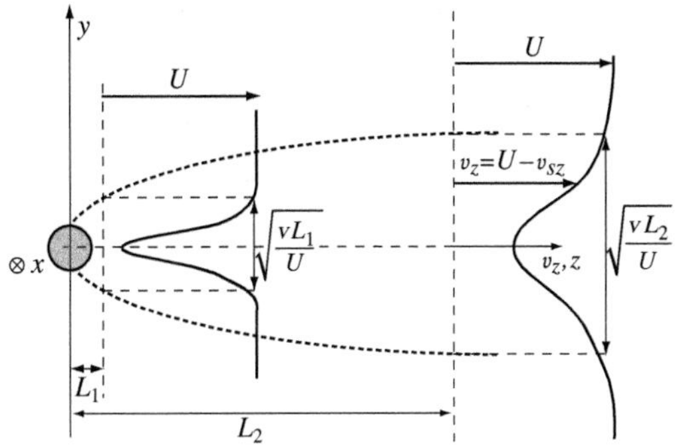

Figure 10.17 *Wake downstream of a solid body located in a flow with velocity* **U** *at infinity. The curves display the transverse profiles of the velocity v_z at two different distances L_1 and L_2 from the body*

compared to unity, the convective terms become dominant at a sufficiently large distance L from the body ($L \gg a/Re$).

One analyzes the perturbation of the average flow **U,** at a distance in the downstream direction sufficiently large compared to the characteristic size of the body so that:

- the exact shape of the body is not relevant;
- Oseen's equation (Equation 9.74a) applies;
- the convection term of this equation dominates.

One uses an approach similar to that of Section 10.2 (Figure 10.2). The transit over the body of a volume of fluid reduces there locally, because of the condition of zero velocity at the walls, the velocity compared to the value U (Figure 10.17). This reduction, as well as the associated vorticity, is initially localized over a distance of the order of the size of the body. The time taken by this volume of fluid to arrive at a distance L downstream of the obstacle is $\Delta t \approx L/U$. During this time interval, the vorticity (and, therefore, the velocity gradients) have diffused over a distance transverse to the flow of the order:

$$e \approx \sqrt{\nu\, \Delta t} \approx \sqrt{\frac{\nu\, L}{U}}. \tag{10.69}$$

They are thus concentrated within a *wake* of thickness e, which increases as $\sqrt{L}$ with distance (Figure 10.17) and, as such, more slowly than the distance L from the body. As already discussed at the beginning of this chapter, this kind of profile is typical of flows where the transport of momentum and vorticity by convection occurs in a particular direction (the transport in the normal direction being uniquely diffusive).

Relative to the body, the thickness e of the wake at distance L is seen to subtend an angle:

$$\alpha \approx \frac{e}{L} \approx \sqrt{\frac{\nu}{U\, L}} = \sqrt{\frac{\nu}{U\, a}}\sqrt{\frac{a}{L}} = \frac{1}{\sqrt{Re_a}}\sqrt{\frac{a}{L}}. \tag{10.70}$$

The Reynolds number Re_a uses the size a of the body as the characteristic length-scale. Thus, even if the Reynolds number is not very large, the angle α becomes small ($\alpha \ll 1$) when the distance L obeys:

$$\frac{L}{a} \gg \frac{1}{Re_a}. \tag{10.71}$$

The velocity gradients and the vorticity are therefore confined in a small part of the fluid space. Compared to the case of a boundary layer near a plate, the velocity is not equal to zero along the axis but is reduced by an amount $\mathbf{v}_s = \mathbf{U} - \mathbf{v}$ which decreases as one gets further away from the body.

Equation of motion in the wake

Taking into account the assumptions made above, the component in the z-direction of the equation of motion can be put in the form:

$$U \frac{\partial v_{sz}}{\partial z} = \nu \left(\frac{\partial^2 v_{sz}}{\partial x^2} + \frac{\partial^2 v_{sz}}{\partial y^2} \right). \tag{10.72}$$

This equation has the form of a thermal diffusion equation (Equation 1.17) in two dimensions (where the time variable is replaced by z/U).

Proof

Equation 10.71 is also the condition required for Oseen's equation (Equation 9.74) discussed in Section 9.5.1 to be valid. This equation can be rewritten for $\mathbf{v}_s = \mathbf{U} - \mathbf{v}$ in the form:

$$-\rho\,(\mathbf{U} \cdot \nabla)\,\mathbf{v}_s = -\nabla p - \eta\,\nabla^2 \mathbf{v}_s; \tag{10.73}$$

$\mathbf{v}_s$ is used in these proofs rather than $\mathbf{v}$, because $\mathbf{v}_s$ becomes zero at large distances, while $\mathbf{v}$ approaches the constant value $\mathbf{U}$. Taking into account the great difference between the length scales e along x and y and L in the z-direction one obtains, in the same manner as for the equations for the boundary layers:

$$v_{sx} \approx v_{sy} \approx v_{sz}\, e/L \ll v_{sz}. \tag{10.74}$$

By taking the different components of Equation 10.73 along the x-, y- and z-directions, the terms which control $\partial p/\partial x$ and $\partial p/\partial y$ are found to be smaller by an order of magnitude in e/L than those which determine $\partial p/\partial z$. The changes in pressure in the directions perpendicular to the velocity of the fluid can thus be neglected, both inside and outside the wake. Moreover, if gravity is also neglected, the pressure outside the wake (where the velocity of the fluid is quite small) is uniform as soon as one is sufficiently far away from the moving body. One can thus also neglect changes in pressure in the direction of the external flow. Combining the two above results, one concludes that the pressure inside the wake can be assumed to be uniform as was already the case for laminar boundary layers.

Moreover:
$$\frac{\partial^2 v_{sz}}{\partial z^2} \approx \frac{e^2}{L^2} \frac{\partial^2 v_{sz}}{\partial y^2} \ll \frac{\partial^2 v_{sz}}{\partial x^2}, \frac{\partial^2 v_{sz}}{\partial y^2}. \tag{10.75}$$

Using these results, the component in the z-direction of Equation 10.73 yields Equation 10.72.

Wake behind a body with all dimensions finite

Using Equation 10.72, it will be shown that:

$$v_{sz} = U - v_z = \frac{Q\ U}{4\pi\ \nu\ z}\ e^{-\frac{U\,r^2}{4\nu\,z}}, \tag{10.76}$$

where:

$$Q = \iint v_{sz}\,\mathrm{d}x\,\mathrm{d}y \qquad (10.77) \qquad \text{and:} \qquad r^2 = \left(x^2 + y^2\right);$$

Q is the deficit in the rate of flow of fluid through the wake; this deficit will be shown below to be independent of the distance z. The deficit in the velocity along the axis thus decreases as $1/z$ with distance. Moreover, Q will be related to the drag force on the body in Section 10.7.2

Proof

The following Equation 10.78 is the axially symmetric solution of Equation 10.72 which satisfies the condition $v_{sz} = 0$ at infinity:

$$v_{sz} = \frac{C\,U}{\pi\,\nu\,z}\,\mathrm{e}^{\frac{-U(x^2+y^2)}{4\nu\,z}} = \frac{C\,U}{\pi\,\nu\,z}\,\mathrm{e}^{\frac{-U\,r^2}{4\,\nu\,z}}, \qquad (10.78)$$

in which C is a constant of integration. The rate of flow Q across the wake is then given by:

$$Q(z) = \int 2\pi\,r\,v_{sz}(z,\,r)\,\mathrm{d}r \approx 4C \int_0^{\infty} \mathrm{e}^{-\theta^2}\,\mathrm{d}(\theta^2) \qquad \text{with:} \qquad \theta^2 = \frac{U\,r^2}{4\,\nu\,z},$$

so that:

$$Q\,(z) = 4C.$$

Q is therefore independent of z; Equation 10.76 is obtained by replacing in Equation 10.78 C by its value.

Let us consider a reference frame which moves at velocity **U** and where the fluid is at rest at infinity and the object has a velocity –**U**: in such a case, the velocity in the wake is $-v_{sz}$ and the rate of flow Q is merely the opposite of the rate of flow of the fluid localized in the wake. This flow is compensated by a flow in the direction of the object distributed over the entire space.

One observes in Equation 10.76 that, along paraboloids obeying the equation $Ur^2/4\nu\,z = constant$, the velocity deficit v_{sz} is a constant fraction of its maximum value $Q\,U/4\pi\,\nu z$ on the z axis.

These paraboloids are, at the same time, stream tubes in the reference frame which is at rest relative to the fluid at infinity. Using Equation 10.76, the deficit in the rate of flow $Q(r_0(z))$ through a cross-section of radius $r_0(z)$ of the surfaces obeys:

$$\int_0^{r_0(z)} 2\pi\,v_{sz} r\,\mathrm{d}r = \int_0^{r_0(z)} \frac{Q\,U}{2\nu\,z}\,\mathrm{e}^{-\frac{U\,r^2}{4\nu\,z}}\,r\,\mathrm{d}r$$

$$= \int_0^{r_0(z)} Q\,\mathrm{e}^{-\frac{U\,r^2}{4\nu\,z}}\,\mathrm{d}\left(\frac{U\,r^2}{4\,\nu\,z}\right)$$

$$= Q\left(1 - \mathrm{e}^{-\frac{U\,r_0^2(z)}{4\nu\,z}}\right).$$

This flux is thus independent of z on a paraboloid ($r_0^2(z)/z = constant$).

Wake behind an infinitely long cylinder

For a two-dimensional flow behind a cylinder which is infinitely long in the x-direction, one finds a variation as $\sqrt{z}$ of the width of the wake as a function of the distance z from the cylinder parallel to the flow. By ignoring derivatives with respect to z in Oseen's equation and by integrating, one finds a velocity deficit:

$$v_{sz} = U - v_z = Q\,\sqrt{\frac{U}{4\pi\,\nu\,z}}\;\mathrm{e}^{-\frac{U\,y^2}{4\,\nu\,z}}. \qquad (10.79)$$

Q represents here again the volume flux within the wake (per unit length in the z direction); it is independent of z, just as in the preceding case. In contrast, the decrease of the velocity maximum as a function of distance is definitely slower, *i.e.* as $1/\sqrt{L}$ instead of $1/L$.

10.7.2 Drag force on a body – relationship with the velocity in the wake

In the preceding discussion (Section 10.7.1), the flow rate Q of fluid across the wake was unknown. For a body located in a flow of uniform velocity **U** (Fig.10.18 below), Q will be shown now to be related to the drag force **F** by:

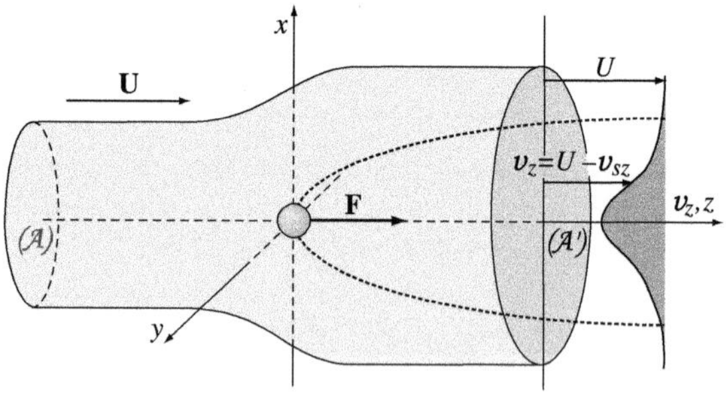

Figure 10.18 *Estimate of the drag force on an axially symmetric object from the flow rate within the wake. Far from the body, the fluid is in motion at a velocity* **U**

$$\mathbf{F} = \iint \rho\, \mathbf{U}\, (U - v_z)\, \mathrm{d}x\, \mathrm{d}y = \rho \mathbf{U} Q. \tag{10.80}$$

This equation is, however, only valid in this form for bodies having either a spherical symmetry or that of a regular polyhedron for which the force **F** is always parallel to the velocity **U**.

Thus, for the specific case of a sphere and a Reynolds number $Re = UR/\nu \ll 1$:

$$\mathbf{F} = 6\pi\, \eta\, R\mathbf{U}, \qquad \text{so that:} \qquad Q = 6\pi\, \nu\, R. \tag{10.81}$$

Unexpectedly, the deficit of the rate of flow Q in the wake does not depend on the velocity U: an increase in U leads to an increase in v_{sx}, but the latter is exactly compensated by the corresponding decrease of the width e of the wake.

Thus, in Equation 10.76, the component v_{sz} of the velocity deficit in the wake obeys (again for a sphere with $Re \ll 1$):

$$v_{sz} = U - v_z = \frac{3\, R\, U}{2\, z}\ \mathrm{e}^{-\frac{U(x^2+y^2)}{4\,\nu\, z}}, \tag{10.82}$$

As indicated above, using Oseen's equation does not contradict the assumption $Re \ll 1$, used to derive Equation 10.81, but Equation 10.82 is only valid if $z/R \gg 1/Re$.

where z is the downstream distance to the center of the sphere.

The problem of a sphere with velocity **U** moving within a fluid at rest at infinity is equivalent, within a relative translation, to the preceding case of a motionless sphere in a fluid with velocity –**U**. In that case, Equation 10.82 represents directly the component v_z in the z-direction of the velocity of the fluid resulting from the motion of the sphere. On the other hand, the drag force is in the direction opposite to the velocity **U**.

Proof of Equation 10.80

Consider the case of a sphere in uniform flow where the velocity deficit v_{sz} satisfies Equation 10.76. According to this equation, v_{sz} is largest up to distances $r = \sqrt{x^2 + y^2}$ from the axis of the order of $\sqrt{\nu z/U}$ and decreases exponentially as a function of r thereafter. Outside the wake (for instance upstream of the body), the perturbation $\mathbf{v}_s$ to the average velocity U decreases as $1/L^2$ with the distance L from the body (instead of $1/L$ in the wake). Far from the body, one has an approximately radial flow with a velocity of order Q/L^2 directed outward, which compensates the volume rate Q of the flow in the wake (one has a similar velocity field in Figure 9.16). The conservation of the flow rate through sections $\mathcal{A}$ and $\mathcal{A}'$

of the same stream tube located respectively far upstream and downstream of the body can be written (Figure 10.18):

$$\iint_{(\mathcal{A})} v_z(x, y)\, \mathrm{d}x\, \mathrm{d}y = A\, U = \iint_{(\mathcal{A}')} v_z(x, y)\, \mathrm{d}x\, \mathrm{d}y. \tag{10.83}$$

In fact, if the cross-section $\mathcal{A}$ of area A is sufficiently far upstream of the body, the contribution of the radial flow (which decreases as $1/L^2$) to the integral over section $\mathcal{A}$ will be negligible relative to that of the wake (velocity deficit in the wake decreasing as $1/L$).
In the same way, due to the stationarity of the flow, the equation of the conservation of momentum in the volume of the stream tube bounded by $\mathcal{A}$ and $\mathcal{A}'$ (Equation 5.13) reduces to:

$$\iint_{(S)} \rho\, v_z v_j n_j\, \mathrm{d}S = \iint_{(S)} \sigma_{zj} n_j\, \mathrm{d}S + (-F_z), \tag{10.84}$$

in which $\mathcal{S}$ is the whole surface bounding the volume $\mathcal{V}$. The only volume force that appears is a force $(-F_z)$, opposite to the drag force, which must be applied to the body in order to maintain it in equilibrium.
Again, on the right side of Equation 10.84 the term σ_{ij} includes the components associated with both viscous and pressure forces. Outside the wake, one can neglect the effect of viscous forces. Regarding the contribution of the integral over the section $\mathcal{A}'$, only σ'_{xy} is non-zero and it does not contribute to the integral, so that:

$$\iint_{(S)} \sigma'_{zj}\, n_j\, \mathrm{d}S = 0\,.$$

The integral $\iint (-p)\, n_z\, \mathrm{d}S$ (contribution from the pressure) is also zero. Outside the wake, one can indeed apply Bernoulli's equation:

$$p + \rho U^2/2 = \textit{constant};$$

U and, consequently, p are therefore constant. Inside the wake, the streamlines are almost parallel; the pressure thus does not vary along the direction perpendicular to the streamlines. It has therefore the same constant value inside the wake as outside. The integral $\iint_{(S)} (-p)\, n_z\, \mathrm{d}S$ reduces then to $-p \iint_{(S)} n_z\, \mathrm{d}S$, which is zero over the entire closed surface.
Regarding now the integral on the left-hand side of Equation 10.84, the contribution of the walls of the stream tube vanishes. This integral then is reduced to the following difference between the fluxes of momentum through the cross-sections $\mathcal{A}$ and $\mathcal{A}'$:

$$\iint_{(\mathcal{A}')} \rho v_z^2\, \mathrm{d}S - \iint_{(\mathcal{A})} \rho v_z^2\, \mathrm{d}S = \iint_{(\mathcal{A}')} \rho v_z^2\, \mathrm{d}S - \rho\, A\, U^2,$$

where A is the area of the cross-section $\mathcal{A}$. Multiplying Equation 10.83 by ρU provides a value of $\rho\, A\, U^2$ to be substituted in the previous equation in order to obtain, after a change of sign, the following form for Equation 10.84:

$$\rho \iint_{(\mathcal{A}')} v_z\, (U - v_z)\, \mathrm{d}x\, \mathrm{d}y = F_z. \tag{10.85}$$

This equation simplifies to Equation 10.80 when one is sufficiently far from the body, so that $U - v_z \ll U$ and one can take in the first factor $v_z \approx U$.

One observes that Equation 10.85 implies that the integral $\rho\iint_{(\mathcal{A}')} v_z\,(U - v_z)\,dx\,dy$ is constant and that, therefore, if $v_z \cong U$, $Q = \rho\iint_{(\mathcal{A}')} (U - v_z)\,dx\,dy$ is also independent of the chosen cross section $\mathcal{A}'$. The constant value of the flow rate Q and Equation 10.85 itself are thus the result of the combination of the conservation of momentum and mass.

The relationship between the *drag force* and the *flow rate* of fluid within the wake has a very general physical meaning not limited to laminar flows. Behind an aerodynamically shaped body, the wake has a very small width, and the rate of flow inside the wake is correspondingly small (the velocity deficit $(U - v_z)$ is of the order of magnitude of U). One has then, according to Equation 10.80, a resulting low drag force. In contrast, behind an object lacking aerodynamic shape, the boundary layer separates, becomes unstable and disperses rapidly. This results in additional dissipation, and is generally associated to the appearance of turbulence downstream of the obstacle (as is the case in Figure 10.1b). The drag force is then greatly increased.

10.7.3 Two-dimensional laminar jet

The problem of the laminar jet corresponds to the injection of a viscous fluid into the same fluid at rest ($U = 0$); here, x will be the direction of the injection and y that of the width of the jet which has a structure invariant in the direction z. Just as in the case of the boundary layer and the wake, the velocity gradients are localized within the width e of the jet, which increases gradually as $e \approx \sqrt{\nu t}$ by transverse viscous diffusion as a function of the transit time t of the fluid particles (and therefore also of the distance x). Proceeding as in the case of the viscous wake, one shows then that the integral $\rho\iint v_x^2\,dy$ (corresponding to the left-hand term of Equation 10.85 with $U = 0$) is independent of x, provided that the integral along y is evaluated over the entire cross-section of the jet (and over a unit distance in the z-direction). One must then have $v_x^2(x)\,e(x) \approx M/\rho \approx$ *constant* (here, M is the total flux of momentum over a unit distance along z, and ρ is the density). The velocity v_x will then decrease as a function of the distance x. More precisely, the variation of the width can be rewritten in the differential form:

$$\frac{\partial(e^2)}{\partial t} \approx \nu \approx \frac{\partial(e^2)}{\partial x}\,v_x \approx \frac{\partial(e^2)}{\partial x}\sqrt{\frac{M}{e\,\rho}}. \qquad (10.86)$$

Since ν is constant, one obtains then, by integrating with respect to x:

$$e(x) \approx x^{2/3}\nu^{2/3}\left(\frac{\rho}{M}\right)^{1/3} \qquad (10.87) \qquad \text{and, also:} \quad v_x(x) \approx x^{-1/3}\nu^{-1/3}\left(\frac{M}{\rho}\right)^{2/3}. \qquad (10.88)$$

As might have been expected, the exponents differ from those obtained in the case of the wake: however, just like for the wake, the ratio $e(x)/x$ decreases with distance as $x^{-1/3}$. The assumption of a strongly localized vorticity at great distance thus still applies.

A detailed calculation indicates that the velocity profiles $v_x(x,y)$ are self-similar with:

$$v_x(x,\,\xi) \approx x^{-1/3}\,\nu^{-1/3}\left(\frac{3M}{32\,\rho}\right)^{2/3}\frac{1}{\cosh^2\xi}, \qquad (10.89a) \qquad \text{where:} \quad \xi \approx \frac{y}{x^{2/3}}\left(\frac{M}{48\,\rho\nu^2}\right)^{1/3}. \qquad (10.89b)$$

By computing from these two equations: $v_x(x) = v_x(x,\,\xi = 0)$ and $e(x) = y(x)$ for $\xi = 1$, one retrieves the dependence on x, ν and M/ρ predicted by Equations 10.87 and 10.88. For a three-dimensional jet, one must integrate along y and z in order to compute the flux of momentum. Its constant value leads to $v_x \propto e(x)^{-1}$. By proceeding as for Equation 10.86, one obtains $\partial e/\partial x = constant$. In this case, the *boundary layer approach* used in the present chapter is no longer valid, because the ratio e/x no longer approaches zero at large distances.

10.8 Thermal and mass boundary layers

The concept of a boundary layer is useful not only for characterizing the flow regimes and forces acting on an obstacle for a large range of Reynolds numbers, but also for computing heat and mass transfer between a solid body and the surrounding fluid. Such transfer is affected by the flow around the object, as we instinctively know whenever we blow along the surface of a hot body in order to cool it. Indeed, some of the most important problems associated with the re-entry of a space vehicle into the earth's atmosphere are related to the heat exchange in the thin boundary layer around the nose of the rocket. In this situation, heat is generated by the aerodynamic flow itself in the neighborhood of the obstacle. The concept of a *thermal boundary layer* is closely related to that of the hydrodynamic boundary layer discussed previously. Such a layer results from a combination of *longitudinal heat convection* by the velocity field, and *transverse conduction* by diffusion. Nonetheless, its structure is strongly affected by the fact that it coexists with a velocity boundary layer: more precisely, it depends on the relative effectiveness of diffusive heat and momentum transfers, measured by means of the Prandtl number $Pr = \nu/\kappa$, defined in Chapter 2, Section 2.3.2. Electrochemical reactions at plane electrodes provide several examples of *mass boundary layers:* these will be governed by the *Schmidt number*, $Sc = \nu/D$.

The concepts of thermal and mass boundary layers have numerous practical applications in the domain of heat or mass transfer: for instance, when the thermal or mass boundary layers are thin relative to the velocity boundary layer, the measurement of the characteristics of heat or mass transfer allows for a determination of the velocity gradient near the walls.

10.8.1 Thermal boundary layers

Consider again the geometry of Figure 10.2a with a temperature difference between a value T_0 at the flat plate and the uniform temperature T_1 of the fluid far from the wall. Once the temperature and velocity profiles have become stationary, the temperature differs appreciably from T_1 only over a thin region near the wall. In order to study the structure of this *thermal boundary layer*, one must add to the Navier-Stokes equation, which results in the Blasius velocity profile, another equation accounting for heat transport. Such an equation is derived, just like the equation of motion for a fluid, by writing that the variation $\mathrm{d}T/\mathrm{d}t$ of the temperature of an element of fluid, in a reference frame moving with the fluid element, satisfies the heat-conduction equation (Equation 1.17). Using Equation 3.2, it then takes on the form:

$$\frac{\mathrm{d}T}{\mathrm{d}t} = \frac{\partial T}{\partial t} + (\mathbf{v}\cdot\nabla)T = \kappa\,\nabla^2 T. \tag{10.90a}$$

Just as for a velocity boundary layer, the spatial derivatives in the direction of flow are smaller by an order of magnitude than those in the direction normal to the wall. One can therefore write:

$$\frac{\mathrm{d}T}{\mathrm{d}t} = \frac{\partial T}{\partial t} + v_x\frac{\partial T}{\partial x} + v_y\frac{\partial T}{\partial y} = \kappa\,\frac{\partial^2 T}{\partial y^2}, \tag{10.90b}$$

neglecting the diffusive term $\kappa(\partial^2 T/\partial x^2)$ relative to $\kappa(\partial^2 T/\partial y^2)$. The second and third convective terms of the equation play, for heat conduction, the same role as those of the

Navier-Stokes equation for the momentum. The temperature must satisfy the boundary conditions:

$$T(x, y = 0) = T_0 \quad \text{for } x > 0, \qquad T(x, y) = T_1 \quad \text{for } y \text{ large.}$$

The form of Equations 10.90a-b implies the following assumptions:

- there are no sources of heat such as those which would be observed in the presence of high-speed flows and strong velocity gradients near solid walls (this is the case for the re-entry of the rocket mentioned above). The case of combustion will be discussed in Section 10.9;
- the effect of Archimedes' buoyancy on the warmer, less dense fluid is neglected;
- finally, the viscosity is assumed to be constant with the temperature.

When the above assumptions are satisfied, the thermal problem can be solved, in principle, once the velocity field is given; there is indeed no back coupling from the variation of the thermal field on the hydrodynamic behavior because of the lack of coupling between the temperature and velocity fields of the fluid. Let us now discuss the structure of the boundary layers observed at different values of the Prandtl number $Pr = \nu/\kappa$.

Prandtl number much greater than one

This is the case for insulating and/or high viscosity liquids. Transverse heat conduction is then quite inefficient compared to the diffusion of momentum: however, it is this transverse conduction which determines the growth of thermal boundary layers as a function of distance downstream. The local thickness of the layer corresponds indeed to the distance from the wall at which diffusive and convective transport phenomena are equally effective (this result was already pointed out, for the velocity boundary layer, in Section 10.3.1).

The thickness $\delta_\theta(x)$ of the thermal boundary layer $\mathcal{T}$ must, therefore, be small compared to that, $\delta(x)$, of the velocity boundary layer $\mathcal{V}$ (Figure 10.19): one can then assume that, within the thickness of the thermal boundary layer $(0 < y < \delta_\theta)$, the velocity varies linearly with y, so that:

$$v_x \simeq U \frac{y}{\delta(x)}, \tag{10.91}$$

where $\delta(x) = \sqrt{\nu x/U}$ is the thickness of the velocity boundary layer.

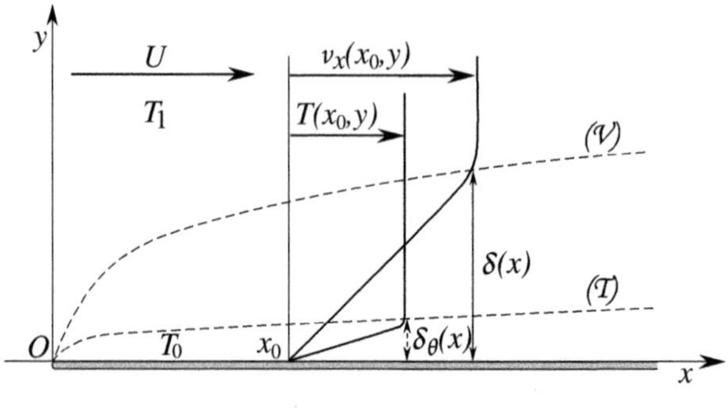

Figure 10.19 *Thermal ($\mathcal{T}$) and velocity ($\mathcal{V}$) boundary layers for a Prandtl number much greater than one*

These assumptions lead to the result that the ratio δ_θ/δ is independent of x with:

$$\left(\frac{\delta_\theta}{\delta}\right)^3 \approx \frac{\kappa}{\nu}, \quad (10.92a) \qquad \text{i.e.:} \quad \frac{\delta_\theta}{\delta} \approx \left(\frac{\kappa}{\nu}\right)^{1/3} = \left(\frac{1}{Pr}\right)^{1/3} \ll 1. \quad (10.92b)$$

Thus, δ_θ also increases with distance as $\sqrt{x}$ and, using Equations 10.1a and 10.92b: $\delta_\theta \approx \left(\kappa\, \nu^{1/2}\right)^{1/3} \sqrt{x/U}$.

Justification

The thickness $\delta_\theta(x)$ is of the order of $\sqrt{\kappa\, t_\theta(x)}$ where $t_\theta(x)$ is the time taken by the fluid at the edge of the thermal boundary layer $(y = \delta_\theta)$ to cover the distance x. Using Equation 10.91, the corresponding velocity is $v_x(y = \delta_\theta) \simeq U(\delta_\theta/\delta)$: the assumption that δ_θ/δ does not vary with x implies that $v_x(y = \delta_\theta)$ is also independent of x, so that:

$$t_\theta \approx \frac{x\ \delta(x)}{U\ \delta_\theta(x)} \quad (10.93a) \qquad \text{and:} \qquad \delta_\theta(x) \approx \sqrt{\kappa\, t_\theta(x)} \approx \sqrt{\kappa\, \frac{x\ \delta(x)}{U\ \delta_\theta(x)}}. \quad (10.93b)$$

By dividing term by term, Equation 10.93b by Equation 10.1a, one obtains Equation 10.92b which justifies, a posteriori, the hypothesis of a ratio δ_θ/δ constant with x.

Like in Equation 10.2, the two convective terms $v_x\, \partial T/\partial x$ and $v_y\, \partial T/\partial y$, and the diffusive term $\kappa\, \partial^2 T/\partial y^2$ in the thermal transport equation (Equation 10.90b) are of the same order of magnitude at the edge of the thermal boundary layer $(y = \delta_\theta)$. Just as the velocity boundary layer, the thermal one corresponds to the boundary between the regions where the diffusive transport of heat is dominant $(y < \delta_\theta)$ and those where convective transport dominates $(y > \delta_\theta)$.

Proof

Equation 10.5 for the conservation of mass can be rewritten using Equation 10.91, leading to:

$$\frac{\partial v_y}{\partial y} = -\frac{\partial v_x}{\partial x} \approx U \frac{y}{\delta^2(x)} \frac{d\delta}{dx}, \qquad \text{so that, by integration:} \qquad v_y \approx U \frac{y^2}{\delta^2} \frac{d\delta}{dx}. \quad (10.94)$$

The order of magnitude of the different terms of Equation 10.90b for the thermal transport along the edge of the thermal boundary layer $(y = \delta_\theta(x))$ can be evaluated using the same approach as in Section 10.3.1; for the temperature, one uses the reduced variable $\theta(y) = (T - T_0)\,/\,(T_1 - T_0)$ (note that this definition of θ differs from that used at the beginning of this chapter). One obtains first, using Equation 10.94:

$$v_y \frac{\partial \theta}{\partial y} \approx U \frac{\delta_\theta^2}{\delta^2} \frac{d\delta}{dx} \frac{\theta(\delta_\theta)}{\delta_\theta(x)} \approx U \frac{\delta_\theta}{\delta^2} \frac{d\delta}{dx}, \quad (10.95a)$$

because $\partial\theta/\partial y \approx \theta(\delta_\theta)\,/\,\delta_\theta(x)$ and, at the edge of the thermal boundary layer $(y = \delta_\theta(x))$, one has $T(\delta_\theta) = T_1$ so that $\theta(\delta_\theta) \approx 1$. In the same way, the second convective term $v_x(\partial\theta/\partial x)$ and the diffusive term $\kappa(\partial^2\theta/\partial y^2)$ satisfy respectively, still for $y = \delta_\theta(x)$:

$$v_x \frac{\partial \theta}{\partial x} \approx U \frac{\delta_\theta}{\delta} \frac{\theta}{\delta_\theta} \frac{d\delta_\theta}{dx} \approx \frac{U}{\delta} \frac{d\delta_\theta}{dx} \quad (10.95b) \quad \text{and:} \quad \kappa \frac{\partial^2 \theta}{\partial y^2} \approx \kappa \frac{\theta(\delta_\theta)}{\delta_\theta^2} \approx \frac{\kappa}{\delta_\theta^2}. \quad (10.95c)$$

One used above the estimate $\partial\theta/\partial x \approx \partial\theta/\partial y\,(\mathrm{d}\delta_\theta/\mathrm{d}x) \approx \theta/\delta_\theta\,(\mathrm{d}\delta_\theta/\mathrm{d}x)$: it corresponds to the assumption that the ratio between the distances in the y- and x-directions along which there is a significant change of θ is of the order of $\mathrm{d}\delta_\theta/\mathrm{d}x$. The two convective terms in Equations 10.95a-b are of the same order of magnitude due to the fact that, δ_θ/δ being independent of x, one has $\mathrm{d}\delta_\theta/\mathrm{d}x \approx (\delta_\theta/\delta)\,\mathrm{d}\delta/\mathrm{d}x$. By substituting for δ_θ in Equations 10.95a and 10.95c its expression as a function of δ from Equation 10.92b, one obtains respectively (again for $y = \delta_\theta(x)$):

$$v_y\,\frac{\partial\theta}{\partial y} \approx U\left(\frac{\kappa}{\nu}\right)^{1/3}\frac{1}{\delta}\frac{\mathrm{d}\delta}{\mathrm{d}x} \approx U\left(\frac{\kappa}{\nu}\right)^{1/3}\frac{1}{x}, \quad \text{and:} \quad \kappa\frac{\partial^2\theta}{\partial y^2} \approx \frac{\kappa}{\delta^2}\left(\frac{\nu}{\kappa}\right)^{2/3} \approx \frac{U}{x}\frac{\kappa}{\nu}\left(\frac{\nu}{\kappa}\right)^{2/3},$$

which are indeed of the same order of magnitude.

Using the preceding results, one estimates the thermal flux between the solid and the fluid:

$$Q \propto k\,\Delta T\,Re^{1/2}\,Pr^{1/3}. \tag{10.96}$$

Here, $\Delta T = T_0 - T_1$ and the Reynolds number Re is taken equal to UL/ν where L is the total length of the plate in the direction of the flow.

Proof

As mentioned above, the diffusive thermal heat-exchange is dominant within the thermal boundary layer. The thermal flux $\mathrm{d}Q$ through a surface element of length $\mathrm{d}x$ in the direction of the velocity U and of unit length in the z-direction thus obeys:

$$\mathrm{d}Q = k\,\frac{T_0 - T_1}{\delta_\theta}\,\mathrm{d}x = k\,(T_0 - T_1)\,\frac{Pr^{1/3}}{\delta}\,\mathrm{d}x. \tag{10.97}$$

The thermal conductivity k, which accounts for the diffusion of heat in the fluid, was defined in Section 1.2.1. By integrating Equation 10.97 over the total length L of the plate, and writing $\Delta T = T_0 - T_1$ we find the global flux, which leads to Equation 10.96.

Equation 10.96 for the variation of the thermal flux as a function of Pr and Re is the key result of this calculation: it is frequently rewritten by replacing Q by the dimensionless parameter $Nu = Q/(k\Delta T)$. This *Nusselt number Nu* is the ratio between the actual flux Q in the presence of flow, and the flux which would be obtained under the same conditions of geometry and temperature without any convection. One then obtains:

$$Nu = C\,Re^{1/2}Pr^{1/3}. \tag{10.98}$$

This is a relation which an engineer would find upon looking up in tables for the solution for a heat-transfer problem. For this specific case of flow parallel to a warm plate, the coefficient of proportionality C is approximately 0.34. The exponents 1/2 and 1/3 of Re and Pr are characteristic of thermal transfers for Prandtl numbers very large compared to one.

Prandtl number much smaller than unity

This is a situation found only for liquid metals (e.g., $Pr = 0.01$ for mercury). In this limit, the viscous boundary layer is, in contrast to the preceding case, much thinner than the thermal one; thus, the velocity is uniform and of value U throughout the thickness δ_θ. In this case, one can use the arguments from Section 10.2, replacing the kinematic viscosity ν by the thermal diffusivity κ; one might also carry out the same calculation as for $Pr \gg 1$, but taking $v_x = U$ in the thermal boundary layer. One then has:

$$\delta_\theta(x) \approx \sqrt{\frac{\kappa x}{U}} \qquad \text{and:} \qquad \frac{\delta(x)}{\delta_\theta(x)} \approx \sqrt{\frac{\nu}{\kappa}}. \tag{10.99}$$

Carrying out the same calculation as in the above section leads to a dependence of the Nusselt number of the type $Nu \approx Re^{1/2} Pr^{1/2}$. Thus the variation with respect to the velocity as $Re^{1/2}$ is preserved, but the interchange in the relative thickness of the velocity and temperature boundary layers causes the exponent for the Prandtl number to change from 1/3 to 1/2.

Prandtl number of order unity

Physically, such a situation corresponds to the case of gases for which the thermal conductivity and momentum diffusion coefficients are of the same order of magnitude. In this case, the growth of the thermal boundary layer occurs at the same rate as that of the viscous one, and the two thicknesses are of the same order of magnitude (this is in fact equivalent to taking the limit at $Pr = 1$ of the two equations obtained above).

An application of heat exchange laws between a solid and a flowing fluid: the hot-wire anemometer

This device, described in Section 3.5.3, makes use of a wire heated by an electrical current, and cooled by a transverse flow, to measure the velocity of the flow. The temperature of the wires is determined by the equilibrium between the Joule heating of the wire and the cooling by the forced convection and by thermal diffusion. Only the component U_n of the velocity of the fluid normal to the wire contributes significantly to the cooling. Experimental measurements show that the Nusselt number, which corresponds to thermal heat-exchange between the wire and the flow, obeys the equation:

$$Nu = 0.57\, Pr^{1/3}\, Re^{1/2} + 0.42\, Pr^{1/5}. \tag{10.100}$$

When both Pr and Re are large, the right hand side of the equation can be reduced to the first term which is the same as for a laminar boundary layer near a wall. This result may seem, at first, surprising because these devices are often used in turbulent flows: the device is however sufficiently small relative to the size of the vortices present in the flow that, at the scale of its diameter, the flow appears as laminar. The fact that the device is not a plate parallel to the flow but a circular cylinder with stagnation points accounts for the second term which is independent of the Reynolds number.

The Nusselt number Nu is proportional to the ratio between the thermal power $P = R I^2$ generated in the wire by electrical heat, and the temperature difference $T_w - T_0$ between the wire and the distant fluid. The temperature T_w of the wire is measured by its electrical resistance, which obeys $R = R_0(1 + a(T_w - T_0))$ where R_0 is the resistance of the wire at temperature T_0. The electrical resistance R of the hot wire obeys the following experimental law as a function of the velocity U_n normal to the wire:

$$\frac{R I^2}{R - R_0} = A + B\, U_n^{0.5}. \tag{10.101}$$

This is in agreement with the variation of Nu given by Equation (10.100), since $R - R_0 \propto T - T_0$ and $Re \propto U_n$.

10.8.2 Concentration boundary layers, polarography

Concentration boundary layer resulting from an electrode imbedded in a wall

Concentration boundary layers appear when a chemical reaction occurs at a solid wall, with absorption or emission of one of the components of a flowing mixture. This is particularly the case for electro-chemical reactions induced by a metallic electrode placed within the flow. This phenomenon is well illustrated by the polarographic technique which allows for

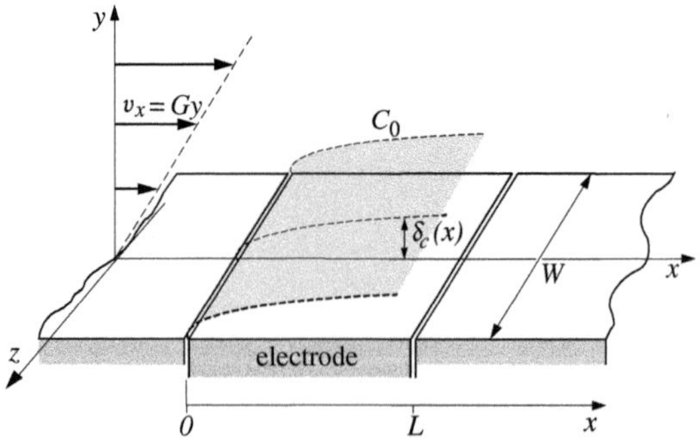

Figure 10.20 *Formation of the concentration boundary layer (indicated by the dashed curves) near an electrode at which an oxidation–reduction reaction is induced*

An example is the ferrocyanide–ferricyanide reaction in the presence of a low-resistivity solution (e.g., KCl): the latter eliminates almost completely parasitic potential differences due to electrical currents flowing in the bulk of the solution.

the measurement of the velocity gradient in a liquid near a solid wall: it uses a controlled, electro–chemical oxidation–reduction reaction within a flowing reagent.

The reaction is induced by injecting a current into a small size electrode (of length L parallel to the flow, and width W), embedded in the wall near which one wants to measure the velocity gradient (Figure 10.20). Assume, for example, that the solution contains initially only ferrous ions Fe^{2+} (in general in the form of complex $Fe(CN)_6^{4-}$) with concentration C_0, which are oxidized into Fe^{3+} ions by the injection of an electric current. The oxidation reaction, decreasing the concentration C_w of Fe^{2+} ions at the wall, creates a concentration gradient relative to the value C_0; this, in turn, induces a diffusive flux of ions toward the wall, $\mathcal{J} = -D_m(\partial C/\partial y)$, compensated by the influx of ions from the upstream convective current (D_m is the molecular diffusion coefficient of the ions). The concentration gradient increases with the electric current density (controlled by the polarization potential) until the concentration C_w at the wall drops to zero. At that point, all the Fe^{2+} ions reaching the electrode by diffusion and convection are oxidized, any attempt to increase further the current by increasing the electrical potential difference triggers other oxidation–reduction reactions. This boundary condition is equivalent to the condition of constant temperature or zero velocity at the walls for thermal and velocity boundary layers: just as in these cases, the thickness δ_c of the concentration boundary layer is the distance at which the diffusive and convective fluxes become comparable.

In most cases, the molecular diffusion coefficient D_m is much smaller than the kinematic viscosity ν. The problem is therefore very similar to that of a thermal boundary layer, for a Prandtl number very large compared to unity. The *Schmidt number* $Sc = \nu/D_m$, introduced in Section 2.3.2 (Table 2.1), plays the role of the Prandtl number Pr; values of Sc of the order of 1,000 are common for many liquids. The dynamics of exchanges is, therefore, analogous to that discussed in Section 10.8.1 for the thermal boundary layer (with $Pr \gg 1$): the concentration boundary layer is, in the present case also, much thinner than the velocity one, and its thickness δ_c obeys the equivalent of Equation 10.92b:

$$\frac{\delta_c(x)}{\delta(x)} \approx \left(\frac{D_m}{\nu}\right)^{1/3} \approx \left(\frac{1}{Sc}\right)^{1/3} \ll 1. \qquad (10.102)$$

The difference between the present experiment and the thermal problem is that the electrode covers only a small portion of the length of the solid wall in the direction of the flow (in the example discussed in Section 10.8.1, the plate was, in contrast, assumed to be heated throughout its length). The momentum boundary layer, initiated far upstream, is already highly developed at the electrode: one therefore assumes that its thickness, and consequently the velocity gradient $G = \partial v_x/\partial y$ at the wall, are constant over the length of the electrode.

As shown below, the thickness δ_c of the concentration boundary layer varies as:

$$\delta_c(x) \approx \left(\frac{D_m x}{G}\right)^{1/3}, \qquad (10.103)$$

where x is the distance from the upstream edge of the electrode. The concentration boundary layer thus grows as $x^{1/3}$, instead of $x^{1/2}$ in the example of Section 10.8.1: this difference arises because the velocity gradient G normal to the wall is, as mentioned above, constant over the active region of the electrodes, instead of decreasing with the distance x. Such concentration boundary layers can occur even in established flows where a velocity boundary layer is absent provided $\delta_c(x)$ satisfies:

$$\frac{\delta_c(x)}{x} \approx \left(\frac{D_m}{G\,x^2}\right)^{1/3} \ll 1.$$

Proof

The thickness $\delta_c(x)$ of the boundary layer is calculated by locating the origin, $x=0$, at the upstream edge of the electrode, and letting C_0 be the concentration of ions in the incident solution outside the boundary layer. Just like for the equation of thermal transport (Equation 10.90a), the transport equation for the concentration C (here, of ions) is shown to be, in the most general case:

$$\frac{\mathrm{d}C}{\mathrm{d}t} = \frac{\partial C}{\partial t} + (\mathbf{v}\cdot\nabla)C = D_m\,\nabla^2 C. \qquad (10.104a)$$

In a stationary regime, and using the usual approximations for boundary layers, the equation becomes:

$$v_x\frac{\partial C}{\partial x} + v_y\frac{\partial C}{\partial y} = D_m\,\frac{\partial^2 C}{\partial y^2}. \qquad (10.104b)$$

Like in Section 10.8.1, the thickness δ_c of the concentration boundary layer is taken equal to the distance from the wall at which the convective-transport and diffusive-transport terms become comparable. Since the velocity gradient G is assumed to be constant over the entire surface of the electrode, one has a parallel flow with $v_y=0$ and $v_x=Gy$: the second convective term $v_y(\partial C/\partial y)$ is therefore itself zero. On the other hand, at the edge of the concentration boundary layer $(y=\delta_c(x))$, $v_x=G\,\delta_c(x)$. Thus, the first convective term of Equation 10.104b has an order of magnitude:

$$v_x\frac{\partial C}{\partial x} \approx v_x\,\frac{\partial C}{\partial y}\,\frac{\partial\delta_c(x)}{\partial x} \approx G\,\delta_c(x)\frac{C_0}{\delta_c(x)}\,\frac{\partial\delta_c(x)}{\partial x} \approx G\,C_0\frac{\partial\delta_c(x)}{\partial x}. \qquad (10.105)$$

Like in Equation 10.95b, one takes $\partial C/\partial x \approx \partial C/\partial y\ (\mathrm{d}\delta_c(x)/\mathrm{d}x) \approx (C_0/\delta_c(x))(\mathrm{d}\delta_c(x)/\mathrm{d}x)$. The diffusive term $D_m(\partial^2 C/\partial y^2)$ is of the order of $D_m\,C_0/\delta_c{}^2$ since the characteristic scale on which the concentration varies in the y-direction is δ_c. By writing that the estimates of the diffusive and convective terms obtained above are equal, one obtains, by removing the factor C_0:

$$\delta_c^{\,2}\,\frac{\partial\delta_c(x)}{\partial x} \approx \frac{D_m}{G},$$

from which one infers Equation 10.103 by integration with respect to x.

We now show that a detailed analysis of the mass transfer within the concentration boundary layer leads to the measurement of the transverse velocity gradient G near the solid wall.

Measurement of a velocity near a wall by a polarographic method

In a practical experiment, the potential difference ΔV between the probe electrode and a reference electrode (maintained, by construction, at a fixed potential with respect to the solution) is kept constant. After stabilization, the current I at the electrodes is measured for several values of ΔV (Figure 10.21). This current is a result of a transfer of electrical charges between the electrode and the ions with which it comes into contact.

In the *polarogram* $I=f(\Delta V)$ of Figure 10.21, a minimal voltage is required to induce the oxidation $Fe^{2+} \rightarrow Fe^{3+} + e^-$ (or, more exactly, the reaction $Fe(CN)_6^{4-} \rightarrow Fe(CN)_6^{3-} + e^-$). Beyond this value, the current increases with ΔV at the same time as the concentration C_w along the wall decreases. The limiting current is reached when the concentration C_w of ions at the wall is zero and all the Fe^{2+} ions brought by diffusion to the boundary layer have been oxidized. At still higher values of ΔV, the current again increases, either because other oxidation–reduction reactions are induced, or because electrolysis of the solvent occurs. The limiting value I_L of the current is related to the velocity gradient $G = \partial v_x/\partial y$ near the wall by:

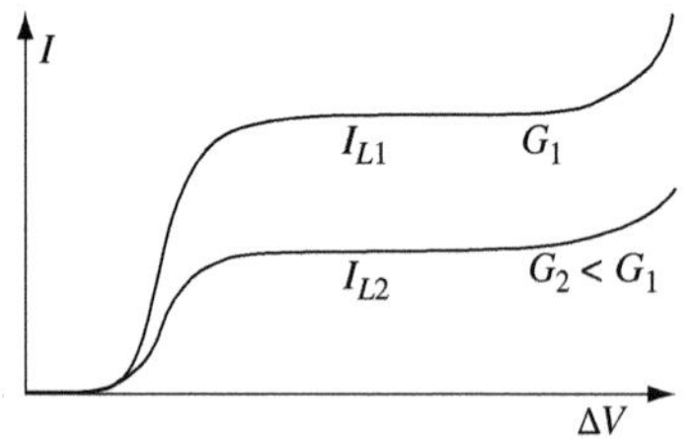

Figure 10.21 *Variation of the electrical current I in the measuring electrode as a function of the potential difference ΔV relative to a reference electrode in an oxidation–reduction reaction. The two curves correspond to different values of the velocity gradient G at the wall. The potential differences depend on the ionic reaction used, and are of the order of a few hundred millivolts. The current intensities depend on the area of the electrodes, and can vary from a fraction of a microampere to several milliamperes*

$$I_L \propto G^{1/3}; \tag{10.106}$$

therefore, the measurement of I_L allows one to determine G. This variation of the signal as $G^{1/3}$ complicates the use of this measurement, but the readings are consistent and the perturbation of the flow is quite small. The measurement of the velocity gradient at the wall is particularly reliable.

Proof

The motion of the ions is a result of the combination of three mechanisms:

(i) migration of active ions due to the electric field (determined by an electrical potential ϕ);
(ii) molecular diffusion due to the concentration of active ions created by the reaction at the level of the electrode;
(iii) convection due to the flow near the electrode.

The flux $\mathbf{J}$, in moles per unit area and unit time, of the active ion (Fe^{2+}) with molecular concentration C can then be expressed as:

$$\mathbf{J} = \frac{\beta F}{RT} D_m C \, \nabla\phi - D_m \nabla C + C\mathbf{v}, \tag{10.107}$$

where: D_m = molecular diffusivity of the reagent;
β = number of active electrons per molecule of reagent, with $\beta = 1$ for the case of Fe^{2+} ions;
F = Faraday's constant ($F = \mathcal{N}_A\, e = 96{,}500$ coulombs).

At the level of the wall ($y = 0$), C and $\mathbf{v}$ vanish, and only the flux associated with molecular diffusion is non-zero. When the concentration C_w of ions at the wall is zero, this diffusive flux is equal to: $\mathcal{J} = -D_m\, \partial C/\partial y \approx D_m C_0/\delta_c$; there results an electrical current density:

$$j(x) \approx \beta F\, D_m \frac{C_0}{\delta_c}. \tag{10.108}$$

The total current per unit width of the electrode (normal to the flow) thus equals:

$$I = -\beta F D_m C_0 \int_0^L \frac{\mathrm{d}x}{\delta_c(x)}, \tag{10.109}$$

so that, by substituting for δ_c in Equation 10.103:

$$I \propto \beta F C_0 D_m^{2/3} G^{1/3} L^{2/3}, \tag{10.110}$$

where L is the length of the electrode in the direction of the flow. One then finds the variation $I \propto G^{1/3}$ of Equation 10.106.

If the electro-chemical reaction involved is sufficiently fast, the electro-chemical probes can carry out non-stationary measurements (fluctuations of the parietal velocity gradient in the presence of instabilities or turbulence). However, while this measurement of the parietal velocity gradient by means of the electrical current I is straightforward in a stationary regime, in a non-stationary one, the transfer functions relating the different variables of interest are more complex.

Mass and velocity boundary layers on a rotating disk electrode

Velocity-field near a rotating disk Measurements of the kind just described above are frequently performed by means of *rotating-disk electrodes*: these use a circular electrode located, close to the axis of rotation, at the surface of a disk rotating with constant angular velocity Ω_0 in an electrolyte solution (Figure 10.22). The disk acts as a centrifugal pump: the fluid undergoes axial suction toward the disk, and is then ejected radially. A useful feature of this system is the presence of a velocity boundary layer of constant thickness over the surface with $\delta = \sqrt{\nu/\Omega_0}$. Moreover, the axial velocity v_z is also independent of r and φ, and is of the order of $-{\Omega_0}^{3/2}\nu^{-1/2}z^2$ (this is known as a *uniformly accessible electrode*). Besides, the radial and tangential components are proportional to r.

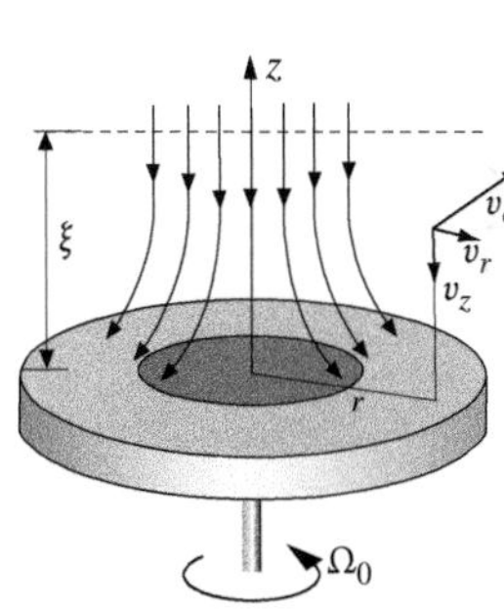

Figure 10.22 *Schematic diagram of a rotating disk electrode*

Justification

One uses a system of cylindrical coordinates, assuming a stationary flow and axial symmetry around the axis of rotation. The effect of gravity is neglected, so that the pressure is constant outside the velocity boundary layer: this pressure is therefore assumed to be constant everywhere (by continuity), just as in previous cases. Throughout this example, we consider a fixed laboratory frame, in which the electrode rotates, and we do not therefore need to introduce Coriolis forces into the equations of motion.

At the wall of the electrode ($z = 0$), the velocity of the fluid is strictly orthoradial with $v_\varphi(r, 0) = \Omega_0 r$, $v_r(r, 0) = 0$ and $v_z(r, 0) = 0$; v_r and v_φ vanish for distances z greater than a few thicknesses of the boundary layer δ. Within the boundary layer, the radial component of the flow is the result of an equilibrium between the centrifugal force and the radial viscous stress: let us now evaluate this equilibrium for a cylinder of fluid of height of the order of δ in the z-direction and with a base of area: $r\,\mathrm{d}r\,\mathrm{d}\varphi$. Using the simplifying assumption that the velocities v_r and v_φ vanish outside of the boundary layer ($z \geq \delta$), the only viscous stress to be taken into account is that along the disk $\tau_0 \approx \eta\,[\partial v_r/\partial z]_{z=0}$. Simplifying by $r\,\mathrm{d}r\,\mathrm{d}\varphi$, the equilibrium of radial forces on the cylinder can then be written in the form:

$$\left(\rho\,\Omega_0^2 r\right)\delta \approx \eta\left[\frac{\partial v_r}{\partial z}\right]_{z=0}, \tag{10.111a}$$

so that, near the wall:

$$v_r \approx \frac{\Omega_0^2}{\nu} r\,z\,\delta. \tag{10.111b}$$

In order to estimate the velocity v_z, one uses the condition of incompressibility:

$$\frac{1}{r}\frac{\partial}{\partial r}(r\,v_r) + \frac{\partial v_z}{\partial z} = 0, \quad \text{so that, near the wall:} \quad v_z \approx -\frac{\Omega_0^2}{\nu} z^2 \delta. \tag{10.112}$$

In order to estimate the thickness δ, one considers again that it corresponds to the distance at which there is an equilibrium between the transverse convection of momentum (at velocity v_z) and the viscous diffusion of the projection $\rho\,\mathbf{v}_\parallel$ of the momentum on the disk (with components ρv_r and ρv_φ). This process is comparable to the suction boundary layer (Section 10.5.3). The projection $\mathbf{v}_\parallel$ of the velocity must therefore obey:

$$\nu\,\frac{\partial \mathbf{v}_\parallel}{\partial z} \approx \nu\,\frac{\mathbf{v}_\parallel}{\delta} \approx -v_z(\delta)\,\mathbf{v}_\parallel, \qquad \text{i.e.:} \qquad \delta \approx \frac{\nu}{|v_z(\delta)|}.$$

Combining this result with Equation 10.112 for $z=\delta$, one retrieves the relation:

$$\delta = \sqrt{\nu/\Omega_0}. \tag{10.113}$$

The thickness δ of the velocity boundary layer is thus constant over the surface of the disk. Substituting this value for δ in Equations 10.111b and 10.112, one obtains:

$$v_r \approx \Omega_0^{3/2}\nu^{-1/2}r\,z. \tag{10.114}$$

For $z \approx \delta$, v_r is of the order of the tangential velocity $v_\varphi \approx \Omega_0\,r$, and:

$$v_z = -\alpha_0\,z^2 = -\Omega_0^{3/2}\nu^{-1/2}z^2, \quad (10.115a) \qquad \text{with:} \qquad \alpha_0 = \Omega_0^{3/2}\nu^{-1/2}. \quad (10.115b)$$

The value of v_z at the surface of the disk is then constant. The complete analytical calculation gives a similar equation valid to second order in z:

$$v_z = -c\,\Omega_0^{3/2}\nu^{-1/2}\,z^2, \quad (10.116a) \qquad \text{with:} \qquad c = 0.51. \quad (10.116b)$$

Equations 10.114 and 10.115 are only valid inside the boundary layer ($z < \delta$). Outside this, v_z approaches a constant value as the distance z increases and the velocities v_r and v_φ approach zero.

Mass transfer toward a "rotating-disk electrode"

One assumes that the central region of the rotating disk is taken up by an electrode of radius R small relative to the radius of the disk (in order to be able to ignore edge effects) at the same level as the disk itself. One also assumes that the angular velocity Ω_0 is constant and that a stationary regime has been reached both for the distribution of velocity and for that of the concentration. Finally, one assumes that, just as in the case of the velocity, the concentration C is only a function of z. The equation for the transport of concentration then becomes:

$$v_z(z)\,\frac{\mathrm{d}C}{\mathrm{d}z} = D_m\frac{\mathrm{d}^2C}{\mathrm{d}z^2}, \tag{10.117}$$

with boundary conditions $C = C_0$ for $z \to \infty$ and $C = 0$ at $z = 0$. Just as for the other cases discussed in this section, the thickness δ_c of the concentration boundary layer is taken as the distance from the wall where the convective and diffusive terms are of the same order of magnitude with $C \approx C_0$. The order of magnitude of the thickness of the mass boundary layer is then:

$$\delta_c \approx \left(\frac{D_m}{\alpha_0}\right)^{1/3} \approx D_m^{1/3}\,\nu^{1/6}\,\Omega_0^{-1/2}, \quad (10.118a) \qquad \text{so that:} \qquad \frac{\delta_c}{\delta} \approx \left(\frac{D_m}{\nu}\right)^{1/3} \propto Sc^{-1/3}. \quad (10.118b)$$

This variation as $Sc^{-1/3}$ of the ratio between the thicknesses of the two boundary layers is then the same as for polarography. The total electrical current I corresponding to this ionic transfer is:

$$I \approx \beta F \left(\pi R^2\right) D_m^{2/3} \nu^{-1/6} \Omega_0^{1/2} C_0. \qquad (10.119)$$

The exact result, known as the *Levich equation*, from the name of a contemporary Russian physical-chemist, is identical, within an added numerical factor 0.62. Given the viscosity of the fluid, Equation 10.119 then allows one to measure the diffusion coefficient, D_m, of the ions.

Justification

As the coordinates r and φ appear neither in the equation nor in the boundary conditions, one assumes therefore that, just as the velocity, the concentration C is independent of r. One uses then Equation 10.115a to estimate the two sides of Equation 10.117 for $z = \delta_c$ with:

$$v_z \frac{\partial C}{\partial z} \approx \alpha_0 \, \delta_c^{\,2} \frac{C_0}{\delta_c} \approx \alpha_0 \, \delta_c \, C_0, \quad (10.120\text{a}) \qquad \text{and:} \qquad D_m \frac{\mathrm{d}^2 C}{\mathrm{d}z^2} \approx D_m \frac{C_0}{\delta_c^{\,2}}. \quad (10.120\text{b})$$

Writing that these two terms are of the same order of magnitude, and replacing α_0 by its value, one recovers Equations 10.118a-b.

At the electrode ($z = 0$), just like in the preceding example of polarography, the electrical current density j corresponds completely to diffusion, so that:

$$j = \beta \, F \, D_m \left.\frac{\partial C}{\partial z}\right]_{z=0} \approx \beta \, F \, D_m \frac{C_0}{\delta_c}. \qquad (10.121)$$

It is then sufficient to substitute for δ_c in this latter equation its value given by Equation 10.118a and to multiply the result by the area πR^2 of the electrode to obtain Equation 10.119.

The preceding result corresponds to the limiting case where the electrochemical reaction at the wall is instantaneous. In a more general case, we would need to take into account the time constant of the chemical reaction, as well as the time needed for the different reagents to reach the wall. This is specifically the case for slow reactions such as corrosion.

10.8.3 Taylor dispersion

In the two examples of the thermal boundary layer and of polarography, one dealt with a longitudinal convective transport of a scalar quantity (the thermal energy or the mass) and with a transverse diffusive transport. The changes in temperature and concentration then remained localized in a boundary layer which is thin relative to the global width of the flow.

In Taylor dispersion, one still has a longitudinal convective and diffusive transport, but it can extend over the entire area of the flow, and one no longer has a boundary layer. The generic process corresponding to Taylor dispersion is the spreading of a small spot of dye injected into a small diameter tube inside which a fluid flows. A two dimensional illustration of this process is shown in Figure 10.23: a narrow band of dye is injected initially at the entrance of a channel made up of two parallel planes separated by a distance a, and between which there is a parabolic Poiseuille velocity profile. The dots represent molecules carried along by the local flow and undergoing, at the same time, a Brownian motion representing

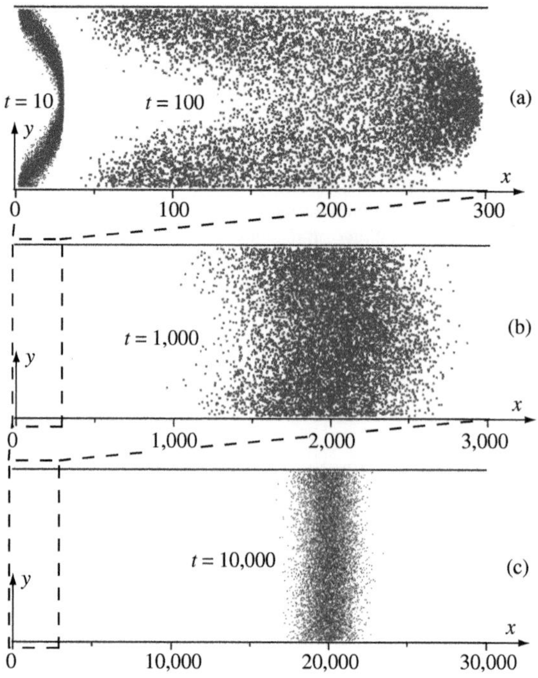

Figure 10.23 *Schematic diagram of the spreading of a band of tracer particles in a Poiseuille flow between parallel plates, at four successive instants of time ($t = 10$, 100, 1000 and 10000 time steps). The scale in the x-direction is contracted by a factor of 10 as one proceeds from one graph to the one lower down (numerical Monte-Carlo simulation by the authors)*

molecular diffusion. The Péclet number $Pe = a\,U/D_m$ defined in Equation 2.16, Section 2.3.2, and which characterizes the relative importance of the convective and diffusive transport processes, plays here a key role (U = average velocity, a = channel opening).

Without flow, we would only have molecular diffusion, both parallel and normal to the walls of the channel: the spreading at the end of a time interval t, in the x-direction is then:

$$\Delta x \approx \sqrt{D_m t}. \tag{10.122}$$

In the presence of a flow, at short time intervals, the band of dye is deformed following the local motion of the fluid, and takes on a parabolic form indicative of the velocity profile (curve $t = 10$ in Figure 10.23a); the velocity of the dye particles varies indeed according to their distance from the walls. In that case, the distance over which the spot of dye spreads out in the x-direction of the flow is:

$$\Delta x \approx U\,t, \tag{10.123}$$

For flow between two planes, one should take into account the factor 3/2, corresponding to the ratio between the maximum velocity and the average velocity. Also, at very short times t, molecular diffusion in the direction of the flow becomes again dominant: this is the case, in accordance with Equations 10.122 and 10.123, if: $\sqrt{D_m t} > U\,t$. or, equivalently: $t < D_m/U^2$.

where U is the average velocity of the flow.

At longer times ($t = 100$ and $t = 1{,}000$ in Figure 10.23a & b), molecular diffusion spreads out the dye between the planes. Then, a distribution which is almost uniform in the direction y is reached for $t \gg \tau_D$, where τ_D is the characteristic diffusion time over a distance a with:

$$\tau_D = \frac{a^2}{D_m}; \tag{10.124}$$

this corresponds to Figure 10.23c ($t = 10{,}000$).

Moreover, one observes experimentally (or numerically) that the variation with x of the average over y of the concentration is quite close to a Gaussian function of width Δx increasing as the square-root of the elapsed time, with:

$$\Delta x \approx \sqrt{D_{Taylor}\, t}\,. \tag{10.125}$$

One has therefore a *macroscopically diffusive spreading* (superimposed on the average displacement at velocity U) characterized by D_{Taylor} (coefficient of the Taylor dispersion). One estimates D_{Taylor} by assuming that at the time $t = \tau_D$ corresponding to the transition between the convective and diffusive spreading, Equations 10.123 and 10.125 are simultaneously obeyed, so that:

$$\Delta x \approx \sqrt{D_{Taylor}\,\frac{a^2}{D_m}} \approx U\,\frac{a^2}{D_m}, \qquad \text{and :} \qquad D_{Taylor} \approx \frac{a^2\,U^2}{D_m}\,. \tag{10.126}$$

A somewhat surprising result is the decrease of the dispersion as D_m increases, (with Ua remaining constant), in contrast to the case of pure molecular diffusion. Indeed, when the transverse molecular diffusion increases, the concentration of dye is homogeneized more rapidly in the y direction and spreading due to velocity variations in the y direction is reduced.

For two parallel planes, Taylor's calculation, complemented by Aris, yields:

$$D_{Taylor} = D_m + \frac{a^2\,U^2}{210\;D_m}, \quad (10.127a) \qquad \text{or:} \qquad \frac{D_{Taylor}}{D_m} = 1 + \frac{Pe^2}{210}\,. \quad (10.127b)$$

For a cylindrical capillary of diameter d, we have:

$$D_{Taylor} = D_m + \frac{d^2\,U^2}{192\;D_m}, \quad (10.128a) \qquad \text{or:} \qquad \frac{D_{Taylor}}{D_m} = 1 + \frac{Pe^2}{192}\,. \quad (10.128b)$$

here $Pe = d\,U/D_m$. At small values of the Peclet number ($Pe \ll 1$), molecular diffusion dominates and one recovers the molecular diffusion coefficient D_m.

Taylor dispersion plays an important role in the motion and mixing of chemical reagents in the network of small channels found in micro-fluidic experiments. It also represents a practical method to determine the molecular diffusion coefficient D_m.

Proof

In the geometry of Figure 10.23, the convection-diffusion equation (Equation 10.104a) becomes:

$$\frac{\partial C}{\partial t} + U_0\left(1 - \frac{4y^2}{a^2}\right)\,\frac{\partial C}{\partial x} = D_m\,\frac{\partial^2 C}{\partial y^2}, \tag{10.129}$$

where $v_x(y) = U_0\,(1 - 4y^2/a^2)$ represents the Poiseuille velocity profile between the two planes $y = \pm a/2$ ($U_0 = 3U/2$). The second derivative $\partial^2 C/\partial x^2$ can be neglected provided the variations of the concentration in the x-direction are sufficiently slow (length of flow channel along x large compared to the distance a). The derivative $\partial C/\partial t$ is then estimated in the reference frame moving at the average velocity of the fluid (here, $2U_0/3$), by means of the changes of coordinates:

$$x_1 = x - 2U_0\,t/3, \qquad \text{and} \qquad t_1 = t$$

and of the equation:

$$\left(\frac{\partial C}{\partial t}\right)_x = \left(\frac{\partial C}{\partial t}\right)_{x_1} - \frac{2\,U_0}{3}\left(\frac{\partial C}{\partial x_1}\right)_t. \tag{10.130}$$

In this reference frame, in this reference frame, the concentration profile is independent of time with $(\partial C/\partial t)_{x_1}$=0: an observer moving at the velocity of the flow does not indeed explore the concentration profile like a static observer would do. Moreover, the variations of the convective flux of the dye parallel to the average flow must be balanced by the transverse diffusive transport so that Equation 10.129 becomes:

$$U_0\left(\frac{1}{3}-\frac{4y^2}{a^2}\right)\frac{\partial C}{\partial x_1}=D_m\frac{\partial^2 C}{\partial y^2}. \tag{10.131}$$

Finally, $(\partial C/\partial x_1)_t$ must be independent of y because of the transverse homogeneization of the concentration and of the slow variation of $C(x_1, t_1)$ with x_1. By integrating Equation 10.131 between 0 and y, taking into account the fact that C must be an even function of y, one finds:

$$C(x_1, y) = C(x_1, 0) + \frac{a^2 U_0}{3D_m}\frac{\partial C}{\partial x_1}\left(\frac{y^2}{2}-\frac{y^4}{a^2}\right). \tag{10.132}$$

The average tracer flux $\mathcal{J}$ per unit area convected in the direction of the flow through a cross-section $x_1 = constant$ is given by:

$$a\,\mathcal{J} = 2\int_0^{a/2} C(x_1, y)\, U_0\left(\frac{1}{3}-\frac{4y^2}{a^2}\right)\mathrm{d}y; \tag{10.133}$$

the choice of the constant $C(x_1, 0)$ is arbitrary because, in Equation 10.133, it is multiplied by an integral equal to zero. Replacing $C(x_1, y)$ by Equation 10.132, and the maximum velocity U_0 by its expression $3U/2$ as a function of the average velocity, one finds:

$$\mathcal{J} = -\frac{a^2\, U^2}{210\, D_m}\frac{\partial C}{\partial x_1}. \tag{10.134}$$

This corresponds, as expected, to the expression: $\mathcal{J} = -D_{Taylor}\,(\partial C/\partial x_1)_t$ of a diffusive flux with a coefficient D_{Taylor} equal to the U^2 term in Equation 10.127a. A more elaborate approach is required to recover the additive term D_m in Equations 127a-b.

10.9 Flames

We will be using a very simplified form of the chemical reactions by assuming that a single global reaction accounts for the overall production of thermal energy and the disappearance of the reagent: actually, the "simple" case of the H_2-air mixture involves more than a dozen reversible reactions between eight chemical species; for the frequently used mixture of air and kerosene, 1800 reactions between 220 species must be taken into account. The problem of flames is thus already formidable even in its simplest forms!

The study of flames belongs at the same time to fluid mechanics, molecular-transport phenomena, and chemical kinetics. It does not deal directly with boundary layers of the type discussed in most of this chapter: howevever, like them, flames involve simultaneously convection, thermal diffusion and mass processes, each having its own characteristic time constant. Moreover, flames involve a combustion chemical reaction, frequently localized within a thin region, and characterized by its own specific time constant. Finally, the *Navier-Stokes equation* must take into account the changes in density due to thermal dilation and to the production of the various chemical species.

We will not attempt here to give a complete overview of the rather complicated problem of flames and combustion and will refer the reader to specialized resources cited in the references: we wish only to give, by means of a few examples, an idea of the relationship between these phenomena and the transport processes studied in the remainder of this book.

10.9.1 Flames, mixing and chemical reactions

Diffusion and premixed flames

A flame brings into play a strongly exothermal chemical reaction between a *combustible gas* (e.g., natural gas) and an *oxidizing gas* (generally the oxygen in the air) within the flow of fluids. A flame can only appear if there exists a region where these reagents are present in a suitable mixture: several types of flames may be observed depending on the geometry of the mixing region and the mechanism of the mixture.

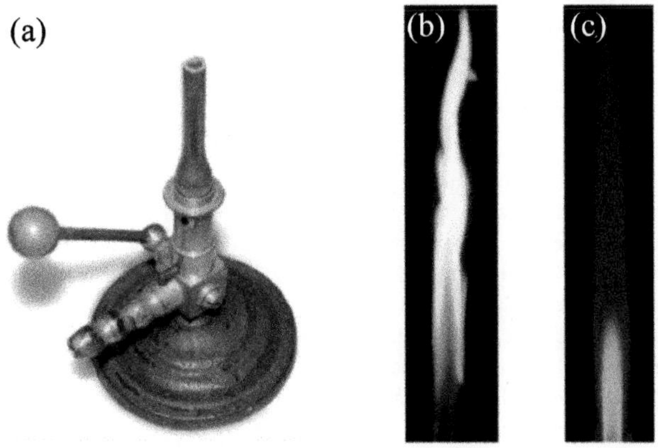

Figure 10.24C *(a) Bunsen burner (collection: Powerhouse Museum, Sidney); (b) diffusion-type flame (closed ferrule); (c) premixed flame (open ferrule); ((b-c): excerpts from a document by A. J. Fijałkowski)*

The behavior of a Bunsen burner illustrates the very different types of flames observed, depending on the conditions of the mixing of the two components. This device is made up of a vertical tube in which the combustible natural gas is injected; a lateral opening on the side, adjustable by rotating the ferrule, allows for the insertion of a selectable amount of air (Figure 10.24a).

If the ferrule is closed, the pure combustible gas exits from the burner, and the ambient air is used for oxidation: if the flow is laminar, the mixture takes place by molecular diffusion, and the flame appears in this mixing region. One then has a *non-premixed diffusion flame* at the interface between the combustible and pure oxidizing gases. In this case, combustion is frequently incomplete, and there appear soot particles, which glow when they are heated to high temperatures. This explains the brilliant region above the flame in Figure 10.24b. This case is discussed in Section 10.9.2

If the ferrule is wide open, air coming in through the lateral opening mixes with the gas inside the tube: it is this mixture which reacts (generally as it exits at the upper opening) when it attains a region where the temperature is high enough (Figure 10.24c). It will be shown in Section 10.9.3 that, in such a *premixed flame*, the reaction takes place within a thin layer separating the fresh gases usptream and the burnt gases downstream. As seen in the figure, not only the geometry but also the color of the flame differs from the preceding case of the non-premixed flame. The reaction is more complete because the mixture is closer to a stoichiometric composition. One does not then see any soot particles. The blue color visible on Figure 10.24c (see in color plate section) corresponds to the region of highest temperature in the flame.

Characteristic reaction time—Damköhler number

Obtaining a mixture of combustible and oxidizing gases of suitable composition is not the only condition for creating a flame. The speed of the reaction depends not only on the fraction (per unit mass) of the fresh reactive species, but also on the reaction kinetics: the latter is characterized in simple cases by a time constant τ_R which varies very rapidly with temperature according to an Arrhenius-type law:

$$\tau_R(T) = \tau_0 \, e^{E/RT} \tag{10.135}$$

(E is the energy of activation of the combustion reaction, and τ_0 an elastic collision time for the gas). This very rapid change of the speed of the reaction as a function of temperature is an essential characteristic of flames, as compared to many reactions in chemical engineering. In the neighborhood of the maximum temperature T_B in the flame ($T_B \approx 2{,}000$ K), one has:

$$\tau_R(T_B) = \tau_0 \, e^{E/RT_B} = \tau_0 \, e^{\beta}, \quad (10.136) \qquad \text{where:} \quad \beta = E/RT_B. \quad (10.137)$$

Typically β is of order of magnitude 10 and τ_R of the order of a few 10^{-4} s.

Evaluation of the flame temperature T_B. One assumes a simple reaction scheme in the form $F \to B + Q$ accounting for the transition from the initial fresh products F (the combustible and oxidizing gases) to the final burnt combustion products B (specifically carbon dioxide and water vapor) with a resultant exothermal energy Q. The maximum temperature T_B reached in the flame is thus estimated by assuming (adiabatic hypothesis) that the energy created by the reaction of a given volume is used to heat the corresponding volume of reagent gas, i.e.: $Q = C_P\,(T_B - T_F)$, where T_F is the temperature of the unburnt gases. Generally, flames do not appear spontaneously: a local rise in temperature must be induced by means of an external method (match, spark, etc.) in order to initiate the reaction, which, then, maintains itself thanks to the exothermal energy that it generates.

In order to estimate the influence of this characteristic time, one frequently uses a *Damköhler number Da* equal to the ratio τ_A/τ_R where τ_A is the characteristic time for bringing reagents together. Depending on the type of the flame (*diffusion* or *premixed*), the time-constant τ_A is a characteristic time for diffusion or convection. Large values of the Damköhler number indicate rapid combustion, localized within a very thin layer; this layer separates the oxidizing and combustible gases for a diffusion flame, and the mixture of fresh and burnt gases for a premixed flame. In contrast, a small or zero Damköhler number corresponds to a "frozen" chemical reaction with a passive interpenetration of the species induced by diffusion or convection.

In the following, one always assumes that the Damköhler number is large (short reaction intervals). This assumption is valid for cars and rockets engines and gas turbines (premixed flames), and for the flames in wildfires or torches (diffusion flames).

Laminar and turbulent flames

The idealized Bunsen burner corresponds to a laminar flow regime: this is seldom the general case, and the nature of the flow greatly influences the characteristics of the flames. In a turbulent regime, laminar diffusion is replaced by transport through turbulent velocity fluctuations (Section 12.2.3), which are much more effective. These play an essential role, for example, in the combustion occurring in the cylinders of car engines.

In a turbulent regime, there is a very strong coupling between the changes in volume due to temperature fluctuations, the composition of the mixture, and the fluctuations in the velocity of the gases; this problem only be treated very approximately. On the other hand, the velocity fluctuations lead to deformations of the surface of the flame (*wrinkled* flame).

In the following sections, the discussion is restricted to laminar flows and to simple geometries for which it is possible to define characteristic time constants in simple terms.

10.9.2 Laminar diffusion flames

In addition to the Bunsen burner with a closed ferrule, the flame of a candle is another example of a diffusion flame. In this case, the reactive gases created by pyrolysis at the level of the wick meet the oxygen of the air during a rising motion resulting from the thermal convection of the hot gases. Mixing is achieved here through molecular diffusion. Diffusion flames are also present in numerous applications using jets or combustible droplets. For a diffusion flame, the Damköhler number is taken equal to $Da = \tau_d/\tau_R$ where $\tau_d = l^2/D_m$ is the characteristic time-constant for the diffusion of fresh gases over the width l of the flame.

A simple case: the Burke-Schumann (1928) model

This model of a diffusion flame is based on several, drastically simplifying, assumptions, but gives a good first-approximation of the physics of the behavior of such flames.

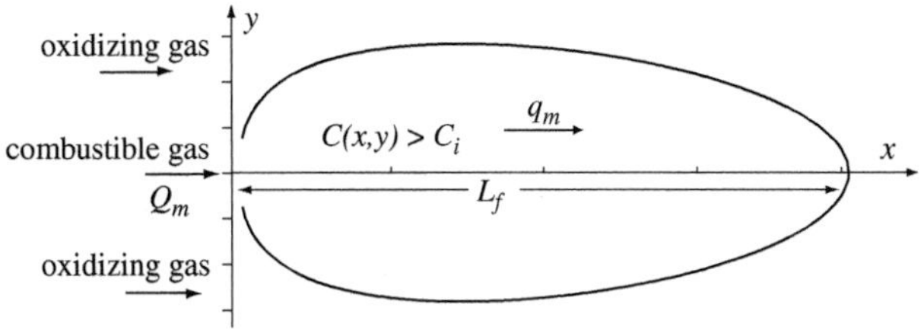

Figure 10.25 *Schematic diagram of a diffusion-type flame according to the Burke-Schumann model. The solid-line contour has been calculated on the basis of Equation 10.139 on the next page*

The flow is assumed to be parallel ($v_y = v_z = 0$), invariant in the z-direction and stationary (Figure 10.25): the combustible gas is injected into the flowing, oxidizing one by means of a slit normal to the plane of the figure. The mass flux $q_m = \rho\, v_x$ per unit length in the y- and z-directions is assumed to be independent of y and z, and to have the same value in the two fluids. The viscosity of the fluids is not taken into account and the molecular-diffusion coefficient, D_m, and the coefficient of thermal diffusion, κ, are assumed to be equal to each other. There is then a coupling between a molecular (and thermal) transverse diffusion and a longitudinal convection: mathematically, this problem is analogous to the transport of momentum within a laminar wake (Section 10.7.1 and 10.7.2).

There are other possible geometries for diffusion-type flames, e.g., those appearing at the interface between a jet of oxidizing gas and a combustible one incident from opposite directions.

In the case where there is no chemical reaction (e.g., if $\tau_R = \infty$) , there is a passive transverse diffusion of the combustible gas into the oxidizing one: then, at a large enough distance x from the injection-point ($x = 0$), the concentration field satisfies (see proof below):

$$C(x,y) = Q_m \sqrt{\frac{1}{4\pi\,\rho\, q_m\, D_m\, x}}\; e^{\frac{-q_m y^2}{4\rho D_m x}}, \qquad (10.138)$$

where Q_m is the total mass flow rate of the combustible gas injected per unit length through a slit in the z-direction. As x increases, the concentration profile $C(x, y)$ in the y-direction also widens, and the maximal value $C(x, 0)$, along the axis decreases as $1/\sqrt{x}$.

Proof

As the flow is parallel ($v_y = 0$), Equation 3.25 for the conservation of mass in a stationary regime requires that $\rho\, v_x = q_m$ be independent of x: note that the velocity v_x can vary as the temperature (and thus the density) changes. Taking into account the similarity between this problem and the transport of momentum in a laminar wake, one uses Equation 10.72 while ignoring derivatives with respect to z, replacing the viscosity ν by D_m, and multiplying the equation by the density ρ. One obtains then for the spatial variations of the concentration $C(x, y)$:

$$\rho\, v_x\, \frac{\partial C}{\partial x} = q_m\, \frac{\partial C}{\partial x} = \rho\, D_m\, \frac{\partial^2 C}{\partial y^2}.$$

An equation of the same type is obtained for the variation of temperature by using $\rho\,\kappa = k/C_p$ instead of ρD_m. Assume that $\rho\, D_m = constant$ and that the distance x is large enough so that the width of the injection-slit is negligible compared to that of the wake: the solution (Equation 10.79) of the equation of the two-dimensional laminar wake can be used to obtain:

$$C(x, y) = \left(\int_{-\infty}^{\infty} C(x,y)\, \mathrm{d}y\right) \sqrt{\frac{q_m}{4\pi\rho\, D_m\, x}}\; e^{\frac{-q_m y^2}{4\rho\, D_m\, x}}.$$

Equation 10.138 results from this relation since, like q_m, the mass flow rate $Q_m = \int_{-\infty}^{\infty} \rho\, C\, v_x\, \mathrm{d}y = q_m \int_{-\infty}^{\infty} C\, \mathrm{d}y$ must be independent of x; the integral over y of the concentration can therefore be replaced by the ratio Q_m/q_m.

The assumption of a very large Damköhler number used here avoids to have to introduce a chemical reaction term.

Assume now that the reaction time is short relative to the diffusion time constant ($Da \gg 1$): in this case, the reaction takes place in a very thin layer considered as the contour of the flame. One assumes that the reaction occurs at a distance $y_i(x)$ for which $C(x, y)$ has a value C_i (e.g., this might be the value corresponding to a stoichiometric composition of the mixture). The concentration profile (Equation 10.138) is also assumed to be valid between $y = 0$ and y_i. The profile $y_i(x)$ is obtained by taking $y = y_i$ and $C = C_i$ in Equation 10.138 and by taking then the logarithm of the square of the equation:

$$y_i^2 = \frac{2\, \rho\, D_m\, x}{q_m} \left[\mathrm{Log} \frac{Q_m^2}{4\pi\, \rho\, C_i^2\, q_m\, D_m} - \mathrm{Log}\, x \right]. \tag{10.139}$$

The length L_f of the flame is the value of x for which the concentration on the axis ($y_i = 0$) is equal to C_i (Equation 10.139 no longer gives a real value beyond this point).

One therefore has:

$$L_f = \frac{Q_m^2}{4\pi\, \rho\, q_m\, D_m\, C_i^2}. \tag{10.140}$$

The ratio Q_m/q_m is constant, independent of x, and represents the width h_0 of the slit through which the pure combustible gas is injected with $C_i = 1$; L_f is then proportional to the rate of flow of the combustible gas as well as to the reciprocal of the molecular diffusion coefficient D_m (or rather, here, of the product $\rho\, D_m$). Equation 10.140 can be rewritten in an alternative form: if v_0 and ρ_0 are the velocity and the density inside the injection slit, $q_m = \rho\, v_0$ and $Q_m = \rho\, v_0 h_0$, so that:

$$L_f \approx \frac{v_0\, h_0^2}{4\pi\, D_m^0\, C_i^2} \tag{10.141a}$$

or:

$$\frac{L_f}{h_0} \approx \frac{Pe_0}{4\pi\, C_i^2}, \tag{10.141b}$$

where $Pe_0 = v_0\, h_0 / D_m^0$ is the Péclet number at the point of the injection.

More general cases

The Burke-Schumann model does not include the motion of the hot fluid rising by convection, which can play a very important role. It can be taken into account by introducing the *Froude number* $Fr_0 = \rho\, v_0^2/(\Delta\rho\, g\, d)$ of the flow ($\Delta\rho$ being of the order of magnitude of the changes in the density). One can then generalize Equation 10.141b in the form of a scaling law:

$$\frac{L_f}{h_0} \propto f\left(\frac{\rho\, v_0^2}{\Delta\rho\, g\, d}\right) \frac{v_0\, h_0}{D_m^0} = f(Fr_0)\, \frac{v_0\, h_0}{D_m^0}, \tag{10.142}$$

where h_0 can be replaced by the diameter d_0 in the case of a circular tube. Such scaling laws are approximately valid for slow flows such that the time-constant τ_R can be considered as negligible compared to the other characteristic times. The case of fast flows is much more complicated.

10.9.3 Premixed flames

Deflagration flames and detonation

Let us now discuss an elementary model of a premixed flame, assuming, for simplicity, that it results from the injection of a mixture of fresh gases in an open tube. In a first step, one assumes that the tube is initially filled with a static mixture and is then lit at one end: as the combustion progresses, one observes a rearward propagation of the flame which separates the fresh and cold gases from the burnt hot ones. The flame progresses in this manner until all the fresh gases in the tube have been used up. The appearance of a thin *flame front* separating the fresh gases from the burnt ones results from the large value of *Da* (a condition already used to define a *flame surface* in the preceding case).

There are two regimes for the propagation. In the case of a deflagration flame, the velocity u_d is of the order of a meter per second, the pressure in the flame can be considered as constant: this is the regime which is studied in the following. In contrast, there is frequently a transition towards a *detonation* regime (also called *explosive regime*), with a velocity of the order of the speed of sound: such a process is not considered in this book which deals mostly with incompressible flows.

Practically, gases are injected, at a finite flow rate, into a tubular burner: when the velocity v_f of the gases, at the level of the flame-front, is equal to the absolute value of u_d, it compensates exactly the velocity of burning, and one has a plane stationary flame at the outlet of the tube (Figure 10.26a).

For $|v_f| > u_d$, the flame front is conical (Figures 10.24c, 10.26b and, below, 10.28); in this case, the velocity u_d has to be equal to the absolute value of the velocity component $\mathbf{v}_f \sin\alpha$ normal to the front (α is the half-angle of the cone) so that the flame front remains stationary. The corresponding value of α is then given by the equation:

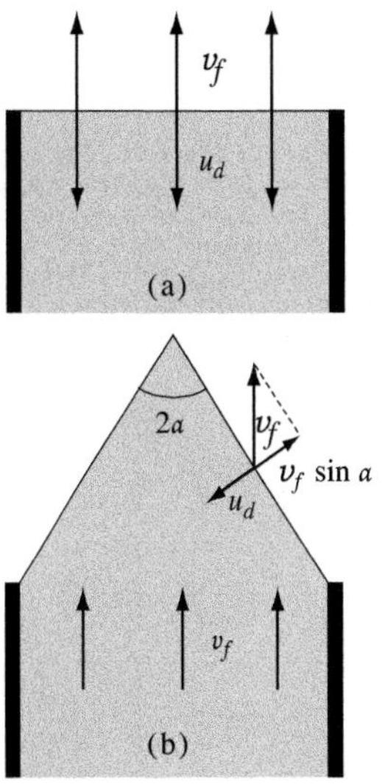

Figure 10.26 *Premixed flames. (a) The flame is plane when the velocities u_d for the propagation of the deflagration flame and v_f of the fresh gases are equal and opposite. (b) For $|v_f| > u_d$, the flame is conical with a half-angle α function of the ratio $|v_f|/u_d$*

$$u_d = |v_f| \sin\alpha. \tag{10.143}$$

In the case of a Bunsen burner where $|v_f| < u_d$, the flame is "swallowed": it moves into the burner which can cause great damage to the tube. To avoid such an accident, devices arresting the flame are placed in some burners at the exit of the mixture of fresh gases.

The greater the velocity of the gases, the smaller the angle of the flame tip.

Propagation of a one-dimensional flame

One considers now a plane flame, using in reference frame static relative to the flame front in which the flow, concentration and temperature profiles can be considered as stationary. Figure 10.27 indicates the variation of these different quantities as a function of the normalized distance x/δ_κ to the flame-front.

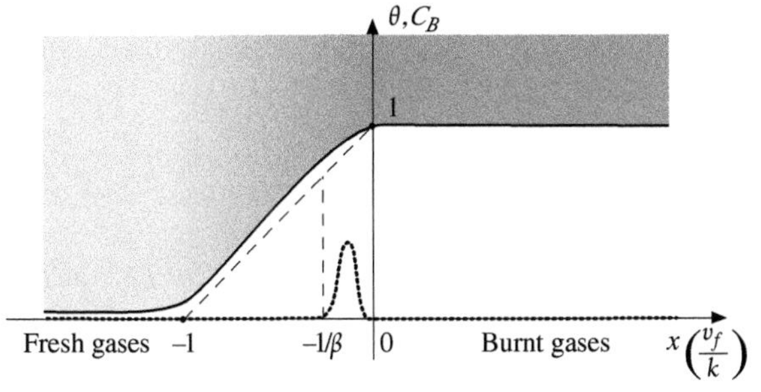

Figure 10.27 *Variations with distance to the flame front, parallel to the mean flow, of the normalized temperature and relative concentration (solid curve) and of the rate of reaction (dotted curve) in a reference frame fixed relative to a one-dimensional flame front. The width of the reaction region is smaller than the distance over which the temperature and concentration vary. The rate of reaction is in arbitrary units. The horizontal scale is normalized by the width of the temperature front $x/\delta_\kappa \approx x/\delta_{Dm} \approx x\, v_f/\kappa$ (Equation 10.146)*

We use a reduced temperature θ defined by:

$$\theta = (T - T_F)/(T_B - T_F), \quad (10.144)$$

which varies continuously from zero to one, just like C_B, in the transition zone between fresh and burnt gases. One also assumes, for simplicity, in Figure 10.27, that the variations of θ and C_B as a function of the distance x are the same: this will be justified further down. The concentration C_F of fresh gases is then equal to $1 - C_B$ and therefore decreases from one to zero as x increases.

Both the reduced temperature θ and the relative concentration C_B of the burnt gases increase continuously between the two (upstream and downstream) limiting values, 0 and 1.

Surprisingly, the combustion reaction takes place only within a very narrow region where one has simultaneously a small (but sufficient) amount of reagents, and a high temperature; it is also in this region that one has the strongest emission of light, as seen in Figure 10.28. As shown in Figure 10.27, there are more reagents available upstream, but the temperature is not sufficient for the reaction to take place. The rise in temperature due to the combustion propagates through a nearest neighbor mechanism (e.g. thermal diffusion), toward the region of fresh gases located just upstream: the combustion reaction can then start there, in its turn, when the threshold temperature is reached. The flame-front then propagates between nearest neighbors in the direction of the fresh gases. The diffusion of burnt gases coming from the reaction zone can also play a role in the propagation of the rise in temperature.

Characteristic distances within a one-dimensional flame

One uses again the reference frame of the flame-front in which flow is stationary, and assumes that the variables of the problem (temperature T, density ρ, average velocity v) depend only on the distance x. The subscripts F and B will refer to the values of these variables for fresh and burnt gases, respectively upstream or downstream of the flame-front. Finally, we will assume that the pressure is constant.

The equation of conservation of mass for the flow can be written in the form:

$$\rho_F v_F = \rho_B v_B \quad (10.145a) \qquad \text{with:} \quad \frac{\rho_B}{\rho_F} = \frac{n_F}{n_B}\frac{T_F}{T_B}, \quad (10.145b)$$

in the limit of an ideal gas, and denoting by n_F/n_B the ratio of the initial number of molecules to the final one, in the chemical reaction. Therefore, the density of the gas decreases and the velocity of the flow increases as one passes from the region of fresh gases to that of the burnt ones.

This change in the velocity is the origin of the variation of the orientation of the streamlines at the flame front observed in the conical flame of Figure 10.28 below. In this case, the velocity component normal to the front varies as predicted by Equations 10.145a-b, while the tangential component remains constant.

A very important fact is the very similar variations of the temperature and relative concentration of burnt gases with the distance x: this similarity results from the very similar values of the thermal and molecular diffusion coefficients κ and D_m for gases (Section 1.3.2 and 2.3.2). The width of the region over which this variation takes place is of the order of:

$$\delta_\kappa \approx \delta_D \approx \frac{\kappa}{v_f} \approx \frac{D_m}{v_f}, \quad (10.146)$$

where v_f is the typical velocity of the gas mixture inside the front.

Figure 10.28C *Visualization of a premixed flame illuminated by a plane light sheet. The white region indicates the boundary between the fresh gases and the burnt ones. The fresh gases have been seeded with small refractory grains of caesium oxide so as to visualize the streamlines: these are deflected as they cross the flame front (plate courtesy J. Quinard and G. Searby, IRPHE)*

Justification

In the reference frame of the flame front, where the flow and the thermal transfer are stationary, the normalized temperature θ depends only on x and the partial derivatives relative to time vanish. One can therefore write the equation for thermal balance (Equation 10.90a) in the form:

$$\rho\, C_p\, v_x \frac{\mathrm{d}\theta}{\mathrm{d}x} = k \frac{\mathrm{d}^2\theta}{\mathrm{d}x^2} + Q_R, \quad (10.147a) \qquad \text{so that:} \quad v_x \frac{\mathrm{d}\theta}{\mathrm{d}x} = \kappa \frac{\mathrm{d}^2\theta}{\mathrm{d}x^2} + \frac{Q_R}{\rho\, C_p}, \quad (10.147b)$$

where C_p is the heat capacity per unit mass, Q_R the thermal energy created by the reaction per unit volume and unit time and v_x the local velocity, which depends on x. Regarding the relative concentration C_B, it satisfies:

$$v_x \frac{\mathrm{d}C_B}{\mathrm{d}x} = D_m \frac{\mathrm{d}^2 C_B}{\mathrm{d}x^2} + Q_B, \quad (10.148)$$

where Q_B corresponds to the production of gases burnt by the reaction. When the pressure is not too high, we can assume that the values of D_m and κ for gases are close to each other (the Lewis number $Le = D_m/\kappa$ is of the order of one). Moreover, C_B and θ both vary between 0 and 1 from the downstream and upstream ends of the flame and, again, the terms Q_R and Q_B have the same dependence on x since the generation of thermal energy accompanies the production of burnt gases. The changes in C_B and θ are thus described by effectively identical equations with similar boundary conditions, and accordingly can only have similar values.

Outside the (narrow) region of the reaction, the terms Q_R and Q_B can be considered to vanish. The convection and diffusion terms in Equation 10.147b (and, therefore, also in Equation 10.148) must also be equal. We thus obtain $v_f\, \theta/\delta_\kappa \approx \kappa\, \theta/{\delta_\kappa}^2$ where the velocity v_f is chosen to be of the same order of magnitude as that corresponding to the inflection point of the curve around the value of $\theta \approx 0.5$, so that we can infer the order of magnitude (Equation 10.146).

Another essential characteristic length-scale for this problem is the width δ_R of the region where the reaction takes place. This obeys:

$$\delta_R \approx \frac{\delta_\kappa}{\beta} \approx \frac{\delta_D}{\beta}, \qquad (10.149)$$

where β, defined in Equation 10.137, determines the kinetics of the reaction. The estimate $\beta \approx 10$ given above confirms the low value of δ_R relative to δ_D and δ_κ .

Justification

One assumes that the term of Equation 10.148 describing the production of burnt gases is proportional to the reciprocal of the characteristic time constant $\tau_R(T)$ of the reaction and to the relative concentration $(1 - C_B)$ of fresh gases available for combustion; then, in accordance with Equation 10.136:

$$Q_B \propto (1 - C_B)\, e^{-\beta^{T_B/T}} \approx (1 - C_B)\, e^{-\beta/\theta}. \qquad (10.150)$$

Assuming that $T_0 \ll T_B$ leads to: $\theta \approx T/T_B$ from Equation 10.144. Equation 10.150 indicates therefore the need, already discussed above, to have, at the same time, a sufficient injection of fresh gases and a temperature high enough to induce the reaction. As already seen above, $(1 - C_B)$ varies approximately as $(1 - \theta)$ as a function of x, and thus much slower than the exponential term $e^{-\beta/\theta}$. The width δ_R of the reaction region can then be assumed to correspond to a variation of the exponential by a factor of the order of 1/e: the corresponding change of θ is approximately $1/\beta$ (Figure 10.28). Since the variation of θ is approximately one over the distance δ_κ, one obtains the estimate (Equation 10.149) for the width δ_R of the reaction zone. In this last region, one shows that the diffusion-term dominates relative to the convection one, so that Equations 10.147b and 10.148 reduce to an equilibium *reaction–diffusion*, while there is an equilibrium between *convection* and *diffusion* in the part of the region of thickness δ_κ where the reactions are negligible.

Propagation velocity of a premixed flame

As shown above, the propagation of a premixed flame occurs between nearest neighbors: the combustion reaction taking place at a given location brings the adjacent layer, with contains fresh gases, to the ignition temperature T_B. During the characteristic duration τ_R

of the reaction (Equation 10.136), a layer of thickness $e = \sqrt{\kappa \tau_R}$ is heated. The propagation velocity of the burning, u_d, thus has an order of magnitude:

$$u_d \approx e/\tau_R \approx \sqrt{\kappa/\tau_R}. \quad (10.151)$$

For a typical value of κ, 5×10^{-5} m^2/s at $T_B \approx 2{,}200$ K with $\tau_R \approx 3 \times 10^{-4}$ s, which corresponds to a mixture of methane and air, one obtains $u_d \approx 0.4$ m/s et $e \approx 0.12$ mm. For an air-hydrogen mixture, the time-constant τ_R would be shorter ($\tau_R \approx 8 \times 10^{-6}$ s for $T_B \approx 2{,}400$ K) which leads to a higher velocity $u_d \approx 3$ m/s, and to $e \approx 0.02$ mm.

10.9.4 Instability of a plane, premixed flame

This instability results from the refraction of the streamlines, already observed in Figures 10.26b and 10.28, in the case where the gas flow is at an angle relative to the front of a premixed flame. One considers a flame front initially plane and stationary, to which is applied an infinitesimal sinusoidal perturbation (Figure 10.29). The streamlines crossing the flame-front are closer together (and thus the velocity is higher) in the part of the front which is deformed in the direction of the burnt gases, and the velocity is lower in the region where the deformation is directed toward the fresh gases.

Like in Figure 10.27, the normal component of the local velocity varies as one crosses the front (Equation 10.145a), while the tangential component is continuous; one must, however, recover a constant velocity far enough upstream and downstream of the front.

Thus, in the regions of increasing velocity (Figure 10.29), the velocity v_f of the gases has a lower absolute value than the intrinsic velocity u_d in the direction opposite to the front: this front is no longer locally stationary, and the deformation is amplified. In the regions where the velocity is decreasing, one has, in contrast, $|v_f| < |u_d|$ but the deformation (which is in the opposite direction) is also amplified.

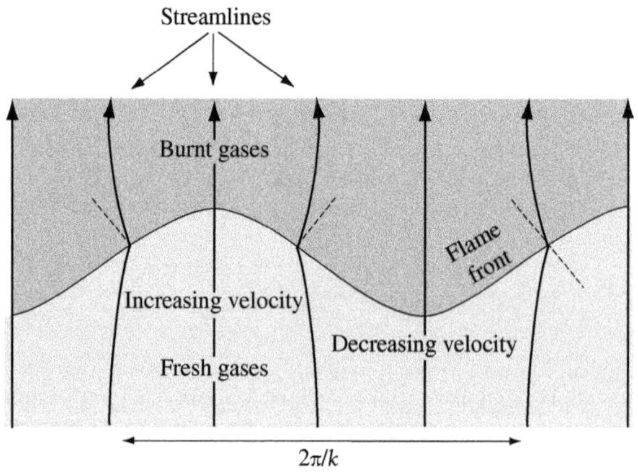

Figure 10.29 *Schematic diagram for the Landau-Darrieus instability: a modulation of the transverse profile of a plane flame is amplified by the refraction of the streamlines*

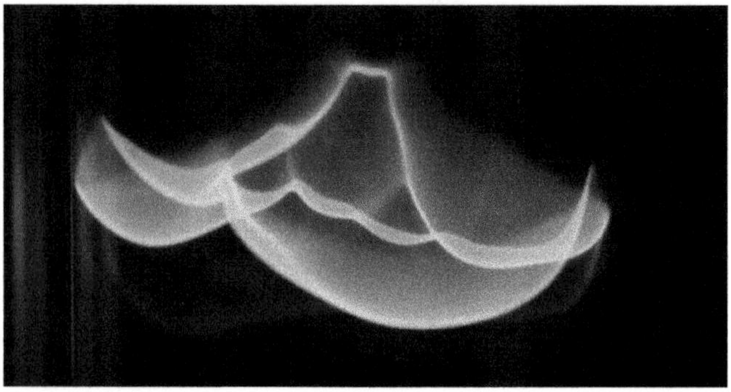

Figure 10.30C *Landau-Darrieus instability of a premixed methane-air flame-front, propagating freely from top to bottom within a tube of diameter* 14 cm, *when no acoustic resonance occur (plate courtesy of J. Quinard, G. Searby, B. Denel, and J. Graña-Otero, IRPHE)*

One also observes these instabilities for flames propagating in a static mixture as shown in Figure 10.30 above.

The development of these instabilities can be slowed by gravitational effects (for a vertical flame and a stabilising density-difference between the two sides of the flame front): this effect will be particularly sensitive to long wavelengths, while the diffusion phenomena prevent the development of short-wavelength deformations. We thus see a preferential development of deformations with characteristic intermediate values.

....

EXERCISES

1) **Stationary dissolution boundary layer near a free surface of a falling film**
Consider a Newtonian liquid film of constant thickness h and of height L flowing along a vertical wall $y = 0$; the flow is stationary, invariant along z and unidirectional ($v_z = 0$, $v_y = 0$). The velocity profile is parabolic with $v_x = V_0 y/h\ (2-y/h)$ for $y \leq h$ (Section 8.2.1) and the Reynolds number is defined as $Re = V_0 h/\nu$. The film surface is in contact with a gas which diffuses into the liquid with a diffusion coefficient D_g. The concentration of gas at the free surface ($y = h$) is assumed in the following to be constant with $C(x, h) = C_0$. If the thickness δ_c of the diffusion layer is small compared to h, show that the concentration $C(x,y)$ of dissolved gas satisfies, in the stationary regime: $V_0\, \partial C/\partial x - D_g\, \partial^2 C/\partial y^2 = 0$. Using a self-similarity assumption (discuss it) with a reduced variable $u = (h-y)/\sqrt{x}$, show that the equation becomes $-(V_0\ u/2)\partial C/\partial u - D_g\ \partial^2 C/\partial u^2 = 0$, and compute the profile $C(x, y)$. What is the scaling law satisfied by the thickness δ_c of the boundary layer? What is the condition on the value of D_g for the transport equation established above to be valid? How does the transverse concentration gradient $\partial C/\partial y$ at the surface ($y = h$) vary with x?

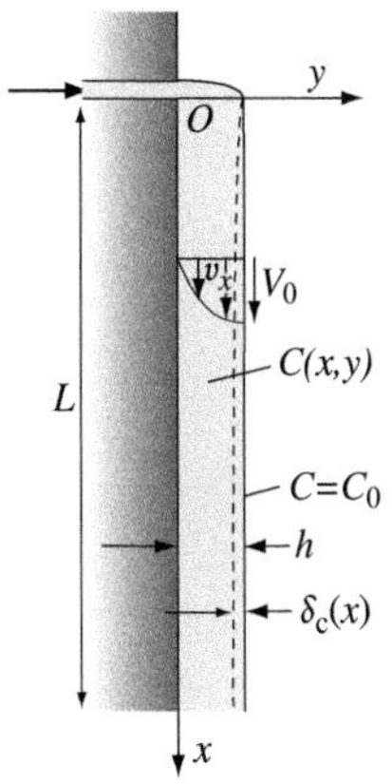

2) **Boundary layer for a cylindrical Couette flow with radial aspiration**
We consider two coaxial cylinders of axis z, of radii R_1 and $R_2 > R_1$, with porous walls bounding a Newtonian fluid. The outside cylinder rotates at an angular velocity Ω and the inner one is fixed. The tangential slip velocity of the fluid at the walls and the component v_z are assumed to be zero. The fluid is sucked uniformly though the walls toward the z-axis with a total absolute flow rate Q per unit length along z. We assume that stationary flow is reached around the cylinder, with a velocity depending only on r (gravity is neglected).

Show that the components v_r and v_φ of the fluid velocity satisfy $v_r(r) = -\alpha/r$ and $-\alpha\, \partial(r v_\varphi)/\partial r = \nu\left(r^2 \partial^2 v_\varphi/\partial r^2 + r\, \partial v_\varphi/\partial r - v_\varphi\right)$. What is the value of α ? Show that there are power-law stationary solutions $v_\varphi(r) \propto r^\gamma$; what are the possible values of γ ? Explain why α/ν represents a characteristic Reynolds number of the flow. Show that $v_\varphi = (R_2^2\, \Omega)/(r)\ ((r/R_2)^{2-\alpha/\nu} - (R_1/R_2)^{2-\alpha/\nu})/(1-(R_1/R_2)^{2-\alpha/\nu})$ and that, for $\alpha/\nu \gg 1$, and except close to $r = R_1$, v_φ corresponds to a potential flow. Compute the order of magnitude of the distance δ at which viscous-momentum transport is of the same order as the radial, convective, one. Compare these results to the characteristics of the boundary layer on a flat plate with aspiration.

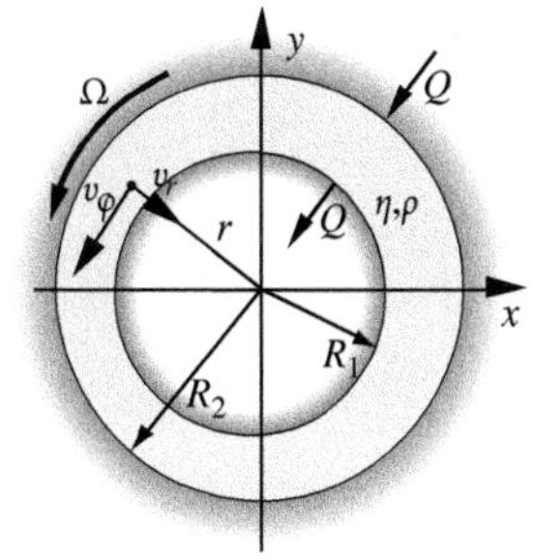

3) **Transient flow around a rotating cylinder**
A solid circular cylinder of radius R and infinite length is initially at rest in an infinite Newtonian fluid; it is set in rotation at $t = 0$ at a constant angular velocity Ω around its axis Oz the rotation generates a velocity field $v_\varphi(r, t)$ ($v_r = v_z = 0$) with a Reynolds

number $Re = \Omega R^2/\nu \gg 1$. Discuss qualitatively, as in Section 10.2, the development of the velocity field after the rotation has started, at times t such that $\sqrt{\nu t} \ll R$. Show that $v_\varphi(r, t)$ satisfies a diffusion equation. Assuming stationary flow and a power-law variation of v_φ with r, show that the velocity profile should be $v_{\varphi S}(r) = \Omega R(R/r)$. What would be the corresponding stationary distribution $\omega_{rS}(r)$ of the z-component of the vorticity and the total energy? $(\omega_z(r, t) = (1/r)\,\partial(r\,v_\varphi)\partial r)$. Can this flow be reached in a finite time? Studying now the variation with time, we look for a self-similar solution: $v_\varphi(r,t) = v_S(r)\,f(u)$ in which $u = r^2/t$. Show that $f = e^{-r^2/4\nu t}$ if $R \ll \sqrt{\nu t}$. Compute the corresponding local vorticity ω_z and the total vorticity in the fluid.

Hydrodynamic instabilities

11

Instabilities in fluid mechanics are part of a broad set of phenomena occurring in many areas of science (changes of phase in the physics of condensed matter, buckling in the mechanics of solids, etc.). Instabilities often correspond to bifurcations observed when a control parameter (e.g., temperature difference in the case of thermal convection) exceeds a certain threshold. They generally imply a reduction in the symmetry of the system characterized by an order parameter. For the simplest instabilities, the amplitude increases in a continuous and reversible manner when the control parameter goes beyond the threshold. Other instabilities display hysteresis or lead to a complete loss of symmetry. We introduce first the formalism used by Landau in describing instabilities (Section 11.1), starting with a mechanical example; we then apply this formalism to the case of flow around a cylinder, which has already been described qualitatively in Section 2.4.2. We examine then the classical Rayleigh–Bénard convective instability (Section 11.2) induced by a vertical temperature gradient: it occurs beyond a threshold temperature difference determined by the balance between the forces due to the density gradient which lead to the flow, and the diffusive effects which oppose it. We discuss then, in Section 11.3, the similarities of this instability to, and differences from, other closed-flow instabilities governed by centrifugal forces (the Taylor–Couette instability) or by gradients in the surface tension (the Bénard–Marangoni instability). This last case allows one to introduce the concept of a subcritical instability. In Section 11.4, the Kelvin–Helmholtz instability for parallel flows at different velocities represents an example of the case of open flows; the effect of the form of the velocity profiles of these flows is then described before discussing the stability of Couette and Poiseuille flows.

11.1 A global approach to instability: the Landau model

When an instability causes a transition from a certain type of flow regime to another, a simple but effective approach to determine the overall local velocity field consists of characterizing this transition by means of macroscopic variables; this can be, for example, a Fourier component of the flow velocity if the instability displays a periodicity in the space or time domains. Such models were initially developed by the Russian physicist, L. Landau, to describe phase transitions at a threshold (critical point) of systems in thermodynamic equilibrium.

Often, in physics, the investigation of the neighborhood of a transition point is a rich source of information on the different regimes on both sides of the threshold. In order to characterize a physical system near a transition, one uses a series expansion of the *order parameter* (a variable characterizing the state of the transition) as a function of a *control*

The Landau model was initially developed to explain second-order phase transitions. It has been subsequently applied to hydrodynamic instabilities, assuming that the local interactions may be averaged out by means of a *mean field approximation*. This will be the case for several of the hydrodynamic instabilities described here. In contrast, for a transition such as the critical point of fluids, this assumption, which is equivalent to that of the simple *van der Waals* model, is not valid for real fluids, because of strong critical fluctuations.

Physical Hydrodynamics. Second Edition. Etienne Guyon *et al.*

parameter representing the distance from the threshold. As an example, for the critical point of a liquid–gas transition (*van der Waals* model), the control parameter is the relative deviation, $\varepsilon = (T - T_c)/T_c$, of the temperature from the critical value T_c; the order parameter is the difference in density between the liquid and the gas and is non-zero only for temperatures less than T_c, at which liquid and gas can coexist. This approach will be extended below to fluid flows which are non-equilibrium systems, starting with the example of flow downstream of a cylinder. A simple analog mechanical model of such instabilities displaying the essential characteristics of "Landau-style" approaches is first presented.

11.1.1 A simple experimental model of a mechanical instability

Consider a solid ball of mass m free to roll without friction along the inside of a circular ring of radius R, spinning at angular velocity Ω around a vertical axis (Figure 11.1a). The location of the equilibrium position of the ball-bearing is studied as a function of the angular velocity Ω (this position is characterized by the angle θ relative to the vertical). In the reference frame of the ring, the total energy of the ball-bearing is its potential energy: the sum of a gravitational term and a term corresponding to the centrifugal force, so that:

$$E_p = mg\,R\,(1-\cos\theta) - \frac{m\,\Omega^2 R^2}{2}\sin^2\theta. \tag{11.1}$$

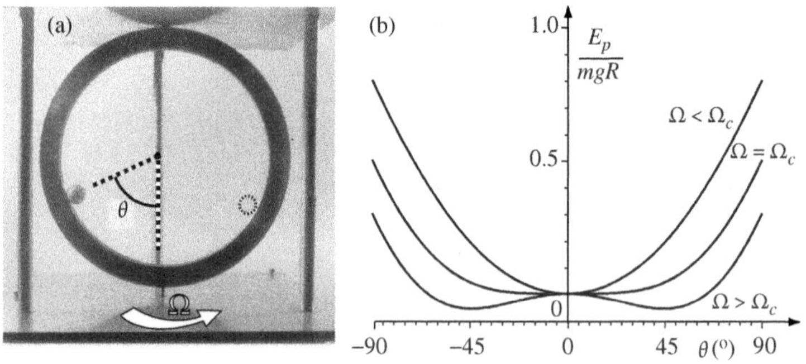

Figure 11.1 *(a) Experimental model of the mechanical instability for a ball-bearing in a ring rotating around a vertical axis (plate courtesy C. Rousselin). (b) Normalized potential energy E_p/mgR of the ball-bearing, resulting from gravity and the centrifugal force as a function of angular position θ for three normalized rotational velocities: $\Omega/\Omega_c = 0.6$ (upper curve), $\Omega/\Omega_c = 1$ (central curve) and $\Omega/\Omega_c = 1.2$ (lower curve); Ω_c is the critical angular velocity $(g/R)^{1/2}$. Depending on the rotational velocity, we have either a single stable equilibrium position, or two stable ones and one unstable*

Justification

The potential energy E_p is equal to the work done by the components of the forces tangential to the ring, which act on the ball bearing, as it moves from a zero angle to an angle θ. The normal components are compensated by the force of reaction from the ring. For a given angle θ, the tangential components are respectively $-g\sin\theta$ and $m\,\Omega^2 R\sin\theta\cos\theta$, so that:

$$F_t = -mg\,\sin\theta + m\,\Omega^2 R\,\sin\theta\cos\theta. \tag{11.2}$$

Using the convention of zero total energy when $\theta = 0$, Equation 11.1 is obtained by calculating the integral $E_p = -\int_0^\theta F_t\,(\theta)\,R\,\mathrm{d}\theta$.

Figure 11.1b displays the variation of the energy E_p normalized by mgR as a function of θ for three different angular velocities. The equilibrium position of the ball-bearing corresponds to the extrema of E_p for which we have also $F_t = 0$ since $F_t = -\,dE_p/d\theta$. We find a unique equilibrium position when $\Omega \leq \Omega_c$ and two stable and one unstable one for $\Omega > \Omega_c$. The threshold value is:

$$\Omega_c = \sqrt{g/R}. \tag{11.3}$$

Justification

The value $\theta = 0$ is always a solution for $F_t = 0$. The other values of the angle for which F_t vanishes must obey the equation $\cos\theta = g/(\Omega^2 R)$: it provides a physical value of θ only if $g/(\Omega^2 R) \leq 1$, which takes one back to the previous threshold.

Depending on the value of Ω, one obtains the following results:

- For $\Omega \leq \Omega_c$, the variation of the energy $E_p(\theta)$ as a function of θ has a unique minimum corresponding to the stable equilibrium at $\theta = 0$.
- For $\Omega > \Omega_c$, the position $\theta = 0$ corresponds to a maximum, thus to an unstable equilibrium. There are two symmetrical, stable equilibrium positions $\pm\,\theta_e$, such that $\cos\theta_e = g/(\Omega^2 R)$. These two positions then coincide at $\theta = 0$ for $\Omega = \Omega_c$.

In the classical vocabulary of Landau-type models, the variation as a function of Ω of the angles θ corresponding to a stable equilibrium, displayed schematically in Figure 11.2, is the *bifurcation diagram* for the system. In this mechanical transition, the equilibrium-angle θ is the *order parameter*, and Ω the *control parameter*. Beyond the threshold Ω_c, the system selects one of the solutions $+\theta$ or $-\theta$ (either of the branches of the stability diagram). This choice is accompanied by a reduction in the symmetry of the entire system (ring+solid ball), known as *symmetry breaking*: one loses indeed the symmetry relative to the vertical axis as soon as the ball moves away from the location $\theta = 0$.

In the limit $\Omega \to \Omega_c$ and for small values of θ, the variation of the potential energy may be written as a function of the variable $\Omega - \Omega_c$ in the form of the following so-called *Landau expansion*:

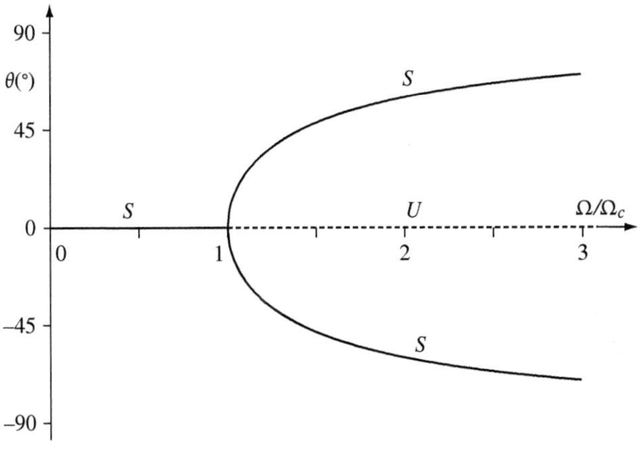

Figure 11.2 *Variation of the angular position of the solid ball in the experimental model of mechanical instability as a function of the angular velocity Ω. Ωc is the threshold of the instability, the branches S correspond to stable equilibria, and the branch U to an unstable equilibrium*

There is another analogy between the mechanical model and phase-transitions: near the threshold Ω_c, after undergoing a small displacement from its equilibrium position, the ball-bearing returns very slowly to it. This is the result of the very flat shape of the curve corresponding to the variation of the potential energy with θ. In the same manner, near the critical point of the liquid, there is a divergence of the observed during of the fluctuation (e.g., of the density): this is referred to as *critical slowing down*.

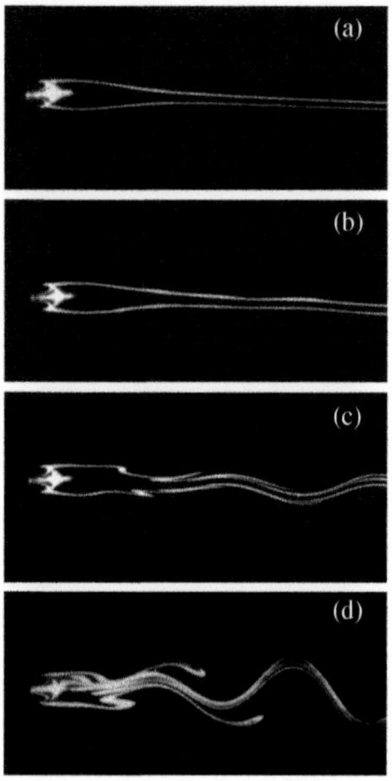

Figure 11.3 *Flow past a cylinder for various small deviations (Re − Re$_c$) of the Reynolds number) from its threshold value for vortex generation: (a) (Re − Re$_c$) = −0.4, no vortices present; (b) (Re − Re$_c$) = 0.3, light waviness; (c) (Re − Re$_c$) = 1, oscillations are amplified; (d) (Re − Re$_c$) = 5.2, generation of alternating vortices downstream of the cylinder. (Plates courtesy of C. Mathis and M. Provansal)*

In a more complete analysis, the study of the variation of the amplitude (Re) leading to equations (11.6,7) should not be restricted to one point downstream of the obstacle: the time variation of the complete spatial profile of the amplitude (global mode) should be analyzed. But the shape of this profile varies with (Re −Re_c), an effect neglected in the simple Landau model: one uses then instead the related *Ginzburg–Landau* model, which takes into account the spatial variations of the order parameter.

$$E_p = m R^2 \left[-\theta^2 \, \Omega_c \, (\Omega - \Omega_c) + \theta^4 \left(\frac{\Omega_c^2}{8} \right) \right]. \qquad (11.4)$$

Beyond the threshold, the solutions take on the simple form:

$\theta(\varepsilon) = \pm 2\, \varepsilon^{\beta}$ (11.5a) with: $\varepsilon = (\Omega - \Omega_c)/\Omega_c$ (11.5b) and an exponent: $\beta = 1/2$.

The exponent β, known as the critical exponent, characterizes the increase in the angle θ (*order parameter*) as a function of the angular velocity Ω (*control parameter*), beyond the threshold.

11.1.2 Flow around a cylinder in the neighborhood of the vortex-generation threshold

The generation of alternating vortices downstream of a cylinder (Section 2.4.2) is a first application of the Landau model in fluid mechanics. One selects as the *control parameter* the ratio:

$$\varepsilon = \frac{Re - Re_c}{Re_c}. \qquad (11.5)$$

This parameter characterizes the relative deviation of the Reynolds number Re from the critical value Re_c corresponding to the transition between the regimes of stable flow and periodic emission of vortices. The amplitude A_y of the oscillation of the transverse velocity due to vortex emission will be used as the *order parameter* of the transition. It can be measured by the laser anemometry technique (Section 3.5.3) in the high velocity regions, and is non-zero when the Reynolds number is larger than the critical value Re_c.

One first analyzes the evolution of the flow downstream of the cylinder: we consider Reynolds numbers Re slightly larger than the threshold Re_c above which one observes a periodic emission of vortices, and thus a periodic change in the velocity. The series of photographs in Figure 11.3 displays visualizations of the flow by a dye at Reynolds numbers close to the threshold value Re_c. So long as the amplitude A_y of the velocity oscillations is small enough so that the linear analysis (carried out earlier) remains valid, one has:

$A_y = 0$ for $\varepsilon < 0$ (11.6) and: $(A_y)^2 \propto \varepsilon = \dfrac{Re - Re_c}{Re_c}$ for $\varepsilon > 0$. (11.7)

This variation, plotted in Figure 11.4a below, is similar to that obtained for the mechanical system described in Section 11.1.1. The symmetry is broken when the flow choses the side of the cylinder where the first vortex is emitted.

11.1.3 Time-dependent evolution of the instabilities in the Landau model

In this section, the model is applied step by step to the specific case of vortex shedding generated by the flow around a cylinder. One assumes that, superimposed on the laminar flow (stable for $Re < Re_c$), there is a set of instability modes characterized by an index k. Their amplitude $A_k(t)$ is, for a given mode, assumed to be of the form $e^{\sigma_k t}$ where $\sigma_k = \sigma_{kr} + i\, \sigma_{ki}$ is the complex coefficient for the exponential growth of the mode. The imaginary part σ_{ki} characterizes the frequency of the oscillations corresponding to the mode; the real part indicates the rate of growth ($\sigma_{kr} > 0$), or of decrease ($\sigma_{kr} < 0$), of this mode:

- for $Re < Re_c$, all modes (and, therefore, every perturbation) decay exponentially (σ_{kr} is negative for all values of k);
- for $Re = Re_c$, σ_{kr} is negative for all modes, save for a *marginally stable* mode m such that $\sigma_{mr} = 0$;
- for $Re > Re_c$, σ_{kr} is negative for most modes, but a range of modes exists for which $\sigma_{kr} > 0$. The dominant mode for which σ_{mr} is most positive is characterized by the suffix m: in the following, one refers to σ_{mr} as σ_m. This parameter characterizes here the development of the vortex emission.

One considers now the evolution of the time-average $|A|^2 = <|A(t)|^2>$. This average must be calculated over a time long relative to the period $T \approx 2\pi/\sigma_i$ of the oscillation, but sufficiently short compared to the time-constant $\tau \approx 1/\sigma_m$ of its growth or decay. These two conditions define the domain of the validity of the Landau equation and can only be simultaneously obeyed if $\sigma_i \gg \sigma_m$ (one denotes by σ_i a typical imaginary component of the complex frequency σ); this condition is met near the threshold where σ_m becomes zero. In this limit, the velocity increases exponentially as $A_y(t) \propto e^{\sigma_m t}$. The square of the magnitude of the amplitude is then a solution of the equation:

$$\Phi(A) = 2\,\sigma_m |A|^2. \qquad (11.8)$$

The variation of the amplitude $A(t)$ of the perturbation predicted by Equation 11.8 is, however, not valid for long times because this amplitude must be bounded. To ensure this saturation, Landau altered Equation 11.8 by including an additional term in the power-law expansion of the right-hand term, so that:

$$\frac{d}{dt}|A(t)|^2 = -\Phi(A), \quad (11.9a) \quad \text{where:} \; -\Phi(A) = 2\,\sigma_m|A|^2 - 2\,b\,|A|^4. \quad (11.9b)$$

The third-order term would contain, indeed, a periodic factor of vanishing average value. The next non-vanishing term is of fourth order, and its amplitude must be negative ($b > 0$) so as to avoid the divergence of the solution at very long times.

One thus obtains : $$\frac{d}{dt}|A|^2 = 2\,\sigma_m|A|^2 - 2\,b\,|A|^4. \qquad (11.10)$$

Equation 11.10 can be transformed to describe the evolution of the amplitude $|A(t)|$:

$$\frac{d\,|A|}{dt} = \sigma_m\,|A| - b\,|A|^3. \qquad (11.11)$$

This equation is suitable for describing an instability leading to a stationary flow, such as the Rayleigh–Bénard instability discussed in Section 11.2.
For an instability generating a periodic flow, such as vortex emission, a more general form involving the complex amplitude $|A(t)|\,e^{i\varphi}$ and a complex value $\sigma = \sigma_m + i\,\sigma_i$ is needed:

$$\frac{dA}{dt} = \sigma\,A - b\,|A|^2 A = (\sigma_m + i\,\sigma_i)\,A - b\,|A|^2 A. \qquad (11.12)$$

Writing Equation 11.11 in this way, for the magnitude, one obtains a new equation for the evolution of the phase:

$$\frac{d\varphi}{dt} = \sigma_i. \qquad (11.13)$$

Here, $\sigma_i = \omega$ represents the angular frequency of vortex generation.

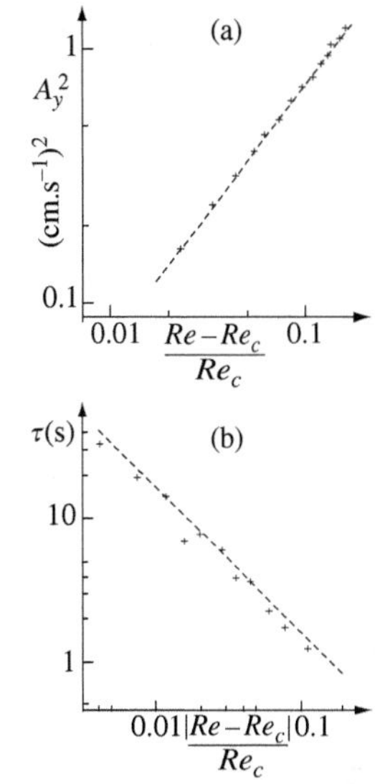

Figure 11.4 *(a) Variation of the square of the steady state amplitude A_y of velocity oscillations induced by the Bénard–von Kármán vortices behind a 6 mm. diameter cylinder as a function of the normalized difference $\varepsilon = (Re - Re_c)/Re_c$ from the threshold Re_c. The dashed line corresponds to a variation of A_y^2 as $(Re - Re_c)$ (b) Variation of the characteristic time $\tau = 1/\sigma_r$ for the relaxation of the perturbation below the threshold Re_c as a function of the absolute normalized difference $|\varepsilon|$. The dashed line corresponds to a variation as $(Re_c - Re)^{-1}$ (results courtesy C. Mathis and M. Provansal)*

The form of the potential in (11.9b) is the same as that found for the expansion (Equation 11.4) of the total energy in the mechanical example of the spherical ball. The order parameter θ, in that example, corresponded to the variable A in the present case. However, in the example of the spherical ball, the time dependence was omitted since only the equilibrium position was of interest.

In the present case, Equation 11.12 accounts for the different characteristics of vortex generation downstream of a cylinder:

- The solution stationary in amplitude has a magnitude $|A_{eq}| = \sqrt{\sigma_m/b}$. Expanding σ_m in a power-series of the Reynolds number, one obtains: $\sigma_m = k' (Re - Re_c) + O(Re - Re_c)^2$, in which k' is positive and equal to the inverse of the characteristic time-constant of the problem. The stationary amplitude of the oscillation satisfies then Equation 11.11 so that:

$$|A_{eq}| = \sqrt{\frac{\sigma_m}{b}} \propto \sqrt{Re - Re_c}. \tag{11.14}$$

These results allow one to draw, for vortex generation, a diagram identical to that in Figure 10.2 provided that θ is replaced by $|A_{eq}|$, and Ω by Re. The second branch of the diagram indicates the possibility of generating the initial vortex on either of the two sides of the cylinder (initial breaking of the symmetry of the flow).

The variation of the magnitude $|A_{eq}|$ of the order parameter as a function of the control parameter ε of the instability must be of the same type as that illustrated by Figure 11.2. Figure 11.4a, on the previous page, displays the results of measurements of the velocity by laser anemometry downstream of a cylinder: the squared amplitude $|A_{eq}|^2$ is indeed proportional to $Re - Re_c$ above the threshold Re_c (the variation has a unit slope in the logarithmic coordinates used). On the other hand, $|A_{eq}|^2$ vanishes for $Re < Re_c$.

- The time-constant $1/\sigma_m$ which characterizes the dynamics of the instability obeys:

$$\frac{|A|}{\mathrm{d}\,|A|\,/\mathrm{d}t} = \frac{1}{\sigma_m} \propto \frac{Re_c}{Re - Re_c}. \tag{11.15}$$

The Landau equation allows one to describe precursor effects observed just below the threshold; a small perturbation of the flow is sufficient in this region to trigger an oscillation which then decays exponentially as a function of time.

Above the threshold, $1/\sigma_m$ accounts for the rate of growth of the instability; below it, the negative value of $1/\sigma_m$ refers to the damping of the perturbation, which might consist in a transient vortex emission. Figure 11.4b indicates that the characteristic time $\tau = 1/\sigma_m$ is indeed proportional to $|Re_c - Re|^{-1}$ below the threshold.

The above predictions regarding the variation of $1/\sigma_m$ are in agreement with the experimental measurements of the flow beyond an obstacle. This is shown by Figure 11.5 which displays, for various negative values of the parameter ε, the time variation of the oscillations of the transverse component of the velocity. These oscillations are generated by a brief perturbation upstream of the obstacle: they correspond to an emission of vortices in the wake downstream of the cylinder which persists for a limited time, increasing

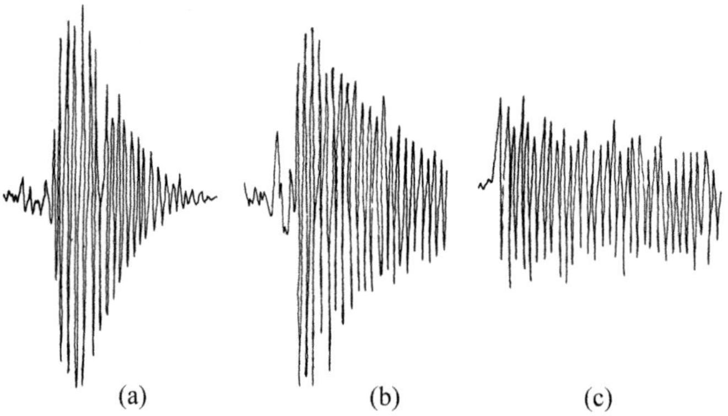

Figure 11.5 *Dependence of the amplitude of oscillation of the velocity in response to an initial impulse at different distances below the threshold Re_c. One observes that the frequency is quite insensitive to the value of ε, but that the relaxation time τ becomes quite long near the threshold $Re_c = 47$. The three curves correspond to values of $\varepsilon = (Re_c - Re)/Re_c$ respectively equal to (a) 0.2, (b) 0.057 and (c) 0.027. (Docs. courtesy C. Mathis and M. Provansal)*

as ε gets closer to 0. Moreover, the frequency of these oscillation is close to that (σ_i) of the natural oscillations above the threshold: σ_i does not display indeed a singular behavior for $Re = Re_c$.

Experimentally, the characteristic time-constant of the exponential relaxation of the oscillation varies as $|Re_c - Re|^{-1}$ as predicted by Equation 11.15 and shown in Figure 11.4b.

The divergence of this relaxation time-constant τ, when the Reynolds number equals Re_c, is the *critical slowing*: it is observed too, as noted previously, in second-order phase transitions. More precisely, the experimental variation of τ can be expressed as a function of the viscous diffusion time d^2/ν (d is the diameter of the obstacle, and the kinematic viscosity ν of the liquid) by:

$$\tau = \frac{1}{\sigma_m} \simeq 5 \frac{d^2}{\nu\,|Re - Re_c|}. \tag{11.16}$$

The critical Reynolds number determined from the variation of τ with Re is the same as that deduced from the variation of the amplitude $|A_{eq}|$ with Re.

11.2 The Rayleigh-Bénard instability

In this section, we discuss a hydrodynamic instability known as the *Rayleigh-Bénard (R-B) instability*, caused by buoyancy forces resulting from spatial variations of the temperature (and therefore of the density) of the fluid. This instability occurs in a fluid located between two horizontal plates and subjected to a vertical temperature gradient. An essential characteristic of such *thermal convection phenomena* is the strong coupling between the temperature and velocity fields in the fluid.

One first seeks to determine the conditions of appearance of this instability from the equations of motion and transport and from the local velocity field. It is then shown that, just like the generation of vortices behind a cylinder, this instability can also be described by Landau's global formalism.

The name of this instability associates those of the French physicist Henri Bénard who first described the formation, above a temperature threshold, of periodical convective structures of a liquid layer heated from below, and of Lord Rayleigh who interpreted these effects in terms of thermal expansion. However, the instability observed by Bénard on a layer of liquid with a free upper surface was due mainly to gradients of surface tension (it resulted actually from the Bénard–Marangoni effect discussed in Section 11.3.1).

We discussed in Section 4.5.5 a thermal convection flow resulting due to a horizontal temperature gradient which also created Archimedes' buoyancy forces. In that case one did not deal with an instability since no temperature threshold was needed to create the flow (the fluid velocity increased linearly with the temperature gradient).

11.2.1 Convective thermal transport equations

The coupling between the temperature and the velocity of the fluid appears in the two governing equations of the problem: the equation of heat transport and the Navier-Stokes equation. In the first one, the convective transport terms correspond to the temperature variations induced by the flow. In the second equation, the term containing the variations of the density with temperature corresponds to the driving force generating the flow of the fluid.

In the presence of a flow at velocity $\mathbf{v}$ and without volume thermal sources, the thermal-transport equation already discussed in Chapter 10 (Equation 10.90a) is written:

$$\frac{\mathrm{d}T}{\mathrm{d}t} = \frac{\partial T}{\partial t} + (\mathbf{v} \cdot \nabla)\, T = \kappa\, \nabla^2 T. \tag{11.17}$$

Actually, this equation merely represents the application, to a moving fluid particle, of the equation for thermal transport derived in Chapter 1 (Equation 1.17). The derivative $\mathrm{d}T/\mathrm{d}t$ represents the change in temperature of a particle followed along its trajectory: this is the Lagrangian derivative appearing throughout this volume in different terms associated with a particle of fluid (velocity, temperature, concentration of tracer particles, etc.). The derivative $\partial T/\partial t$ represents the temperature variation with time at a given fixed point. The next term $(\mathbf{v}\cdot\nabla)\,T$ in the equation for $\mathrm{d}T/\mathrm{d}t$ describes the thermal transport due to the fluid flow.

The velocity field **v** appearing in the diffusion–convection equation (Equation 11.17) often represents a forced flow: one has then a *forced convection*. The present discussion deals with the opposite case of *free convection*, where the flow results from spatial variations in the temperature: this corresponds, for example to the circulation of hot air above a heater.

The second equation involved is the Navier-Stokes equation (Equation 4.30); after dividing by ρ, it can be written:

$$\frac{\partial \mathbf{v}}{\partial t} + (\mathbf{v} \cdot \nabla)\, \mathbf{v} = -\frac{1}{\rho} \nabla p + \nu \nabla^2 \mathbf{v} + \mathbf{g}. \qquad (11.18)$$

The origin of the coupling between the velocity of the fluid and the spatial temperature variations is the variation of the density ρ of the fluid with the temperature T: these variations influence indeed the term $-(1/\rho)\ \nabla p$ of the Navier-Stokes equation.

11.2.2 Stability of a layer of fluid in the presence of a vertical gradient of temperature

Let us consider a layer of fluid located between two fixed horizontal planes each at a different temperature. The density of the fluid decreases as its temperature increases: therefore, if the temperature of the lower plate is below that of the upper plate, the resultant density gradient is stable (the denser fluid remains at the bottom). This configuration occurs in the atmosphere of certain cities when the layers of hot air are above the layers of cooler air: this so-called *temperature inversion* does not allow for the proper removal of air pollution (Figure 11.6) and a layer of smog results in the upper atmosphere. This smog reinforces the process by absorbing solar radiation and preventing the heating of the lower atmospheric layers.

Here and throughout this section, the density of the fluids is assumed to decrease in a monotonic way with temperature. This is not always the case and, in particular, liquid water has a density maximum at 4° C. This density anomaly explains the fact that, below the surface of frozen lakes and rivers, one find a layer of liquid water at 0° C in equilibrium with solid ice, and then, further down, a denser layer of liquid water at 4° C. The stratification of this layer is stable: it is the presence of this stable liquid layer that allows for a survival of marine life.

The opposite configuration of a fluid heated from below is unstable because the fluid particles of lower density are below those with higher density. However, in contrast to the case of a horizontal gradient, a motion of the fluid only appears when the temperature difference is higher than a certain value, known as the *instability threshold*. This phenomenon is known as the *Rayleigh–Bénard instability*.

Figure 11.6C *Temperature-inversion layer and accumulation of "smog" above downtown Los Angeles (plate courtesy M. Luethi)*

11.2.3 Description of the Rayleigh–Bénard instability

Let us first describe the experimental results observed in the configuration of Figure 11.7: T_1 is the temperature of the upper plate $(y=a)$ and $T_2 > T_1$ is that of the lower plate $(y=0)$.

- So long as the temperature difference $\Delta T = T_2 - T_1$ is lower than a critical value ΔT_c, the heat exchanges are strictly diffusive (as in the case where the fluid is heated from above).
- For $\Delta T > \Delta T_c$ the fluid is set into motion: parallel counter-rotating rolls appear simultaneously throughout the cell (Figure 11.8). Their diameter is comparable to the distance between the plates. The velocity of the fluid in the rolls varies continuously and reversibly as the temperature difference $(\Delta T - \Delta T_c)$ varies above the threshold.
- For sufficiently high temperature differences above ΔT_c, further thresholds corresponding to the onset of non-stationary phenomena can be identified: they are detected from the variations with time of local temperatures measured by thermocouples. We observe the existence of temperature fluctuations coupled with fluctuations of the velocity field. The different sequences of instabilities which ultimately lead to turbulence have been extensively studied in relation to their mathematical properties.

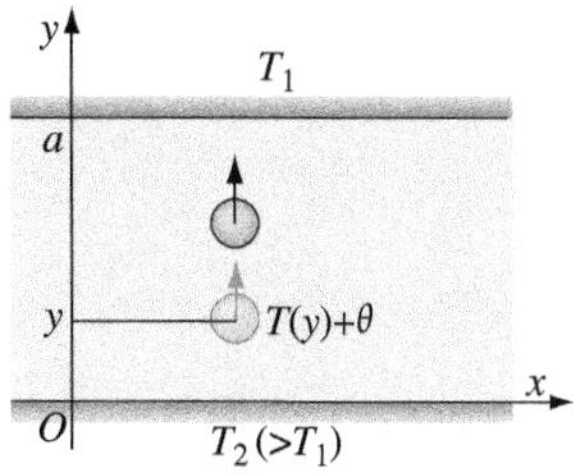

Figure 11.7 *Schematic diagram of the onset of the Rayleigh–Bénard instability. A local fluctuation θ of the temperature of a fluid particle (shaded) initiates an upward convective motion. Once the particle has started moving (black), its difference of temperature with the surrounding fluid increases with the displacement and is stronger if the thermal exchange with this fluid is small: this reinforces the effect of the initial fluctuation*

Several ways of estimating the threshold ΔT_c will now be discussed: this will make us more familiar with the various approaches of this type of problem.

11.2.4 Mechanism for the Rayleigh–Bénard instability and corresponding orders of magnitude

Qualitative mechanism of the instability

The mechanism can be understood on the basis of the diagram in Figure 11.7.

- Let us assume that an initial temperature fluctuation θ relative to the local equilibrium temperature $T(y)$ occurs within a small fluid particle.
- As a result of the Archimedes' buoyancy, the particle starts to move vertically: if θ is positive, the motion is upward.
- If the particle does not cool down too fast during this latter motion, its density contrast with respect to the surrounding fluid (which is increasingly cold) rises and, consequently, Archimedes' buoyancy increases as the particles rise. This effect reinforces the initial temperature perturbation, and amplifies the convection.

Figure 11.8 *System of convective rolls in a circular layer of oil between two plates maintained at different temperatures (view from the top). The white lines, where light is focused, correspond to the regions separating adjacent rolls when the fluid is rising. The defects and branches of the network of rolls are the result of the conditions to be satisfied at the circular boundary; the rolls connect indeed preferentially at right angles to the side walls (plate courtesy of V. Croquette, P. Legal, A. Pocheau, CEA Saclay)*

This positive feedback loop allows the convective instability to maintain itself: for this, the effect of stabilizing mechanisms such that the thermal conductivity (which causes the temperature perturbation to spread), and the kinematic viscosity (which attenuates the velocity perturbations) must not be too large. The conditions for the onset of the instability are thus governed by the relative values of the time constants for thermal heat exchange and for momentum exchange which occur in this problem. A very important parameter is the Prandtl number $Pr = \nu/\kappa$, defined in Section 2.3.2, which characterizes the relative diffusivity of momentum and temperature. The qualitative analysis presented in the next section corresponds to a Prandtl number much greater than one like, for example, for very viscous oils. In this case, the viscous diffusion time constant for momentum over a given distance is much shorter than that for thermal diffusion (these time constants vary respectively as $1/\nu$ and $1/\kappa$). The fluid velocity is then a *fast* variable adjusting rapidly to the variations of the vertical forces due to changes in the density; in contrast, the temperature is a *slow* variable. The thermal equilibrium of particles displaced from their static equilibrium point is then assumed to control the vertical force, and, as a result, the amplification of the instability. In the opposite case $Pr \ll 1$ (for example for mercury), this conclusion is no longer valid. However, both theory and experiments show that the value of the instability threshold remains the same.

Physical criterion for the instability in the case of a Prandtl number much greater than unity

The stability of the layer of liquid is determined by the value of the *Rayleigh number*:

$$Ra = \frac{\alpha\,\Delta T\,g\,a^3}{\nu\,\kappa}, \tag{11.19a}$$

where ΔT is the temperature difference between the plates and α characterizes the change in density of the fluid as a function of temperature, as:

$$\rho = \rho_0\,(1 - \alpha\,(T - T_0)), \tag{11.19b}$$

T_0 is a reference temperature and ρ_0 the corresponding density. In order for the convective instability to arise, Ra must be greater than its critical value Ra_c: its experimental value is 1,708 when the fluid is located between two rigid horizontal plates. The critical temperature difference ΔT_c varies very rapidly as a function of the distance a between the plates: for a thickness twice as small, ΔT_c is increased by a factor of eight.

Order of magnitude

For an experiment carried out using silicone oil with $\nu \approx 10^{-4}$ m^2/s, $\kappa = 10^{-7}$ m^2/s, $\alpha = 10^{-3}$ K^{-1} and $a = 10^{-2}$ m, one finds a value $\Delta T_c \approx 1.7$ K.

Proof

Let us analyze the motion of a spherical liquid particle with radius r_0. Assuming that, as a consequence of a fluctuation, the particle acquires a velocity v (positive if it is directed upward). Is this velocity fluctuation subsequently attenuated or does it continue to be amplified? Because of its vertical motion, the particle reaches regions where the temperature is different from its initial value: if $Pr \gg 1$ the particle does not reach immediately a thermal equilibrium while, in contrast, its velocity takes immediately an equilibrium value. Let us now evaluate the temperature difference δT acquired relative to the surrounding fluid. The characteristic time-constant τ_Q for thermal relaxation obeys:

$$\tau_Q = A\,r_0^2/\kappa, \tag{11.20}$$

where A is a geometrical constant. During this time interval, the particle has moved through a distance $\delta y = v\,\tau_Q$ and the differences δT and $\delta\rho$ between its temperature and density and those of the surrounding fluid become respectively:

$$\delta T = \frac{\partial T}{\partial y}\,\delta y = A\,v\frac{r_0^2}{\kappa}\frac{\Delta T}{a} \quad (11.21) \quad \text{and:} \quad \delta\rho = -\rho_0\,\alpha\,\delta T = -A\,\rho_0\,\alpha\,v\frac{r_0^2}{\kappa}\frac{\Delta T}{a}, \quad (11.22)$$

where α is defined in Equation 11.19b. Accordingly, the driving force on the liquid sphere (the Archimedes' buoyancy) is:

$$F_m = -\frac{4}{3}\pi\, r_0^3\, \delta\rho\, g = \frac{4}{3}\pi\, A\, \rho_0\, \alpha\, g\, v \frac{r_0^5}{\kappa}\frac{\Delta T}{a}. \qquad (11.23)$$

The velocity v of the particle then increases if the driving force is greater than the Stokes viscous force $F_{visc} = -\,6\pi\,\eta\, r_0\, v$ calculated in Chapter 9 (Equation 9.53). The instability criterion thus becomes:

$$F_m > F_{visc}, \qquad (11.24a) \qquad \text{i.e.:} \quad \frac{4}{3}\pi\, A\, \rho_0\, \alpha\, g\, v \frac{r_0^5}{\kappa}\frac{\Delta T}{a} > 6\pi\,\eta\, r_0\, v \qquad (11.24b)$$

The value of the velocity v of the particle does not appear in the instability criterion because of the linearity of the equations derived. In order to determine the stationary amplitude of the fully developed convection, additional non-linear terms must be included.

The greater the spatial extent of the perturbation, the greater its instability. Taking as a maximum size $r_0 = a/2$ and substituting for η/ρ_0 the kinematic viscosity ν, the condition for instability becomes:

$$Ra = \frac{\alpha\, \Delta T\, g\, a^3}{\nu\,\kappa} > \frac{72}{A} = Ra_c.$$

Replacing the liquid sphere by a liquid cylinder, so as to respect the symmetry of the rolls generated, simply amounts to remove a factor a in Equations 11.22 and 11.23: the force on a cylinder depends indeed only logarithmically on its diameter, as seen in Section 9.5.2. The final result (Equations 11.24) is the same within a geometrical factor.

One recovers therefore the form of the experimental condition for instability; the numerical value of Ra_c is different, particularly because the stiffness of the walls is not taken into consideration in the simplified model.

11.2.5 Two-dimensional solution of the Rayleigh–Bénard problem

Approximate calculation of the stability threshold

Experimentally, above the instability threshold, one observes parallel rolls with nearly circular cross-section: the fluid in each roll rotates in the opposite direction relative to its adjacent neighbor. Together with this observation, one assumes, on the other hand, that the liquid layer extends to infinity in the two horizontal directions, x and z, that the velocity field is independent of z and that $v_z = 0$. As an approximation, the vertical component v_y of the convective velocity is taken equal to the periodic function:

$$v_y\,(x, t) = v_{y0}\,(t)\, \cos kx. \qquad (11.25)$$

One takes $k = \pi/a$ as the value of the wave vector: this value takes into account the fact that the rolls are circular and of a diameter of the order of a. This formulation corresponds to an *analysis in terms of modes*, in which one looks for Fourier components of the solution of wave vectors k. The amplitude $v_{y0}\,(t)$ plays the role of an order parameter of the Rayleigh–Bénard instability, just like the angle θ in the mechanical example of Section 11.1.1. Equation 11.25 is only an approximation since it does not obey the boundary conditions along the horizontal walls with ordinates $y = \pm\, a/2$; it does obey however the incompressibility condition and provides a good approximation for the critical Rayleigh number. The equations (11.18) and (11.17) for vertical transport of the velocity and of the thermal energy can be written as:

The component of Equation 11.18 corresponding to the horizontal velocity component v_x has not been written; it is however needed for satisfying the incompressibility condition if one takes into account the variation of v_y with y, in order to reproduce more realistically than by Equation 11.25 the velocity field of the rolls.

$$\frac{\partial v_y}{\partial t} + v_y \frac{\partial v_y}{\partial y} = \nu \left(\frac{\partial^2 v_y}{\partial x^2} + \frac{\partial^2 v_y}{\partial y^2} \right) - \frac{1}{\rho}\frac{\partial p}{\partial y} - g, \qquad (11.26a)$$

and:
$$\frac{\partial T}{\partial t} + v_y \frac{\partial T}{\partial y} = \kappa \left(\frac{\partial^2 T}{\partial x^2} + \frac{\partial^2 T}{\partial y^2} \right). \qquad (11.26b)$$

One assumes now that the system is close to the instability threshold and that one can use the *Boussinesq approximation:* this implies that the variations of the parameters of the fluid as a function of the temperature can be neglected, except for the density. Moreover, the deviations θ of the temperature from the profile below the threshold and the velocity components v_x and v_y are assumed to be small, which allows one to neglect second-order terms. In this case, Equations 11.26a–b can be rewritten in the form:

In Equations 11.27a–b, only first-order terms in v and θ have been retained and second-order ones were neglected: as a result, there is still a convective term in the equation for thermal transport, but none in the Navier–Stokes equation (it would indeed be of second order in v).

$$\frac{\partial v_y}{\partial t} = \nu \left(\frac{\partial^2 v_y}{\partial x^2} + \frac{\partial^2 v_y}{\partial y^2} \right) + \alpha g \theta \quad (11.27a) \text{ and: } \frac{\partial \theta}{\partial t} = \kappa \left(\frac{\partial^2 \theta}{\partial x^2} + \frac{\partial^2 \theta}{\partial y^2} \right) + v_y \frac{\Delta T}{a}. \quad (11.27b)$$

The two symmetrical Equations 11.27a–b both include a diffusive term inhibiting the development of the instabilities and a time-dependent term. It is the coupling terms, $v_y\,(\Delta T/a)$ and $\alpha g \theta$, between the variables v_y and θ which account for the onset of the convective motion.

Proof

To lowest order in the perturbation expansion, the solution of the two coupled equations (Equations 11.26a–b) yields $v_y = 0$, and a linear change in the temperature as a function of the y-coordinate. These solutions correspond to the state of the system in the absence of convection, below the critical Rayleigh number. In this regime, where thermal transport is purely diffusive, the velocity, pressure and temperature obey:

$$v_y = 0 \quad (11.28a), \qquad p_0 = constant - \rho_0\, g\, y \quad (11.28b)$$

and:
$$T_0 = T_2 + \frac{(T_1 - T_2)\, y}{a} = T_2 - \frac{\Delta T}{a} y. \qquad (11.28c)$$

The solution of the perturbation expansion to the next order corresponds to the appearance of the convection: if one remains infinitesimally close to the threshold, the velocity v_y is infinitesimally small to the first-order with respect to the distance from the threshold. The velocity and pressure fields can therefore be written in the form:

$$T(x,y,t) = T_0(y) + \theta(x,t), \quad (11.29a) \qquad p(x,y,t) = p_0(y) + \delta p(x,t) \quad (11.29b)$$

and:
$$\rho(x,y,t) = \rho_0(y) + \delta\rho(x,t). \qquad (11.29c)$$

Equation 11.26a can be rewritten, replacing p and ρ by Equations 11.29b–c and keeping only the first-order terms in v_y, $\delta\rho$ and δp. One obtains then, by using the expressions of Equations 11.28a-b-c which correspond to the diffusive regime:

$$\frac{\partial v_y}{\partial t} = \nu \left(\frac{\partial^2 v_y}{\partial x^2} + \frac{\partial^2 v_y}{\partial y^2} \right) + \frac{\delta\rho}{\rho_0^2} \frac{\partial p_0}{\partial y} - \frac{1}{\rho_0} \frac{\partial(\delta p)}{\partial y}. \qquad (11.30)$$

Combining Equations 11.19b, 11.28b and 11.29b–c leads to:

$$\frac{\delta\rho}{\rho_0^2}\frac{\partial p_0}{\partial y} = \alpha\, g\, \theta. \qquad (11.31)$$

Using this value in Equation 11.30, and further assuming that δp is only a function of x and t, the equation simplifies to Equation 11.27a. Similarly, by using Equations 11.28c and 11.29a, in Equation 11.26b, one obtains 11.27b.

One now looks for the stability condition, using only solutions of Equations 11.27a–b for which the variation of the amplitude $v_y(x, t)$ has the form of Equation 11.25 with a time dependence:

$$v_{y0} = v_0\, e^{\sigma t}. \qquad (11.32)$$

Here, the real parameter σ has the dimensions of a reciprocal time. The exponential dependence $e^{\sigma t}$ indicates that an infinitesimal velocity perturbation increases exponentially with time ($\sigma > 0$) if ΔT is greater than ΔT_c, or undergoes exponential damping ($\sigma < 0$) in the opposite situation. The value $\sigma = 0$ corresponds to the threshold of the instability ($\Delta T = \Delta T_c$). This type of variation was already encountered during the discussion of the Landau model for flow beyond a cylinder (where σ is a complex variable) in Section 11.1.3. We then find that this threshold value ΔT_c obeys:

$$\frac{\Delta T_c}{a} = |\nabla T|_c = \frac{\nu\,\kappa\, k^4}{\alpha\, g}, \qquad (11.33a) \quad \text{so that:} \quad Ra_c = \frac{\alpha\,\Delta T_c\; g\; a^3}{\nu\,\kappa} = \pi^4, \qquad (11.33b)$$

where the Rayleigh number Ra is again defined by Equation 11.19a, with $k = \pi/a$.

In the neighborhood of Ra_c, the growth rate σ, defined by Equation 11.32, satisfies:

$$\sigma = \left(\frac{Ra - Ra_c}{Ra_c}\right)\left(\frac{\nu\,\kappa}{\nu + \kappa}\right)k^2. \qquad (11.34)$$

For $Ra \to Ra_c$, the growth of perturbations (velocity, temperature) becomes infinitesimally slow. This phenomenon known as *critical slowing* was already discussed in connection with Figure 11.4b. The continuous transition from real negative values of σ to positive ones, as the Rayleigh number goes through its critical value, is known as the *stability exchange principle*. It characterizes the continuous transition from a *thermodynamic state* of thermal diffusive exchange below the threshold, to a state of *dynamic convective order* above it.

The value $\pi^4 \approx 97$ in Equation 11.33b is much smaller than the experimental value $Ra_c = 1{,}708$ for a fluid layer bounded by two rigid plates. This difference is hardly surprising if one keeps in mind the gross approximations which have been used; the goal, here, was only to obtain the correct dimensional form for Ra.

Proof

The variations of the velocity field given by Equations 11.25 and 11.32 are associated to variations of the temperature given by the similar expression:

$$T(x, y, t) = T_0(y) + \theta\,(x, t) = T_0(y) + \theta_0 \cos(kx) e^{\sigma t}. \qquad (11.35)$$

The same factor $\cos(kx)$ is found in the two equations for v_y and T because the extrema of the vertical velocity and the temperature are in phase with the rising, or descending, fluid columns. By substituting these expressions of v_y and T in the equations of motion, Equations 11.27a–b, and by factoring out $\cos(kx)\, e^{\sigma t}$, one obtains:

In the more frequent case of a liquid which is a poor thermal conductor, the Prandtl number $Pr = \nu/\kappa$ is much greater than one. From Equation 11.32, the prefactor for the growth rate (or the rate of decrease) is then of the order of κk^2: it is thus governed by the thermal conductivity. When $Ra = 0$, a perturbation of wave vector $k \approx 1/a$ close to that of the rolls has a thermal relaxation time constant $\tau_Q = a^2/\kappa$. This justifies the assumption made above according to which, when Pr is much greater than unity, the thermal conductivity is responsible for the damping of the rolls. The velocity variable which has a faster intrinsic response is slaved to the evolution of the temperature field. On the other hand, when Pr is much smaller than one, the viscous diffusion of the velocity perturbations governs the attenuation of the convective motion.

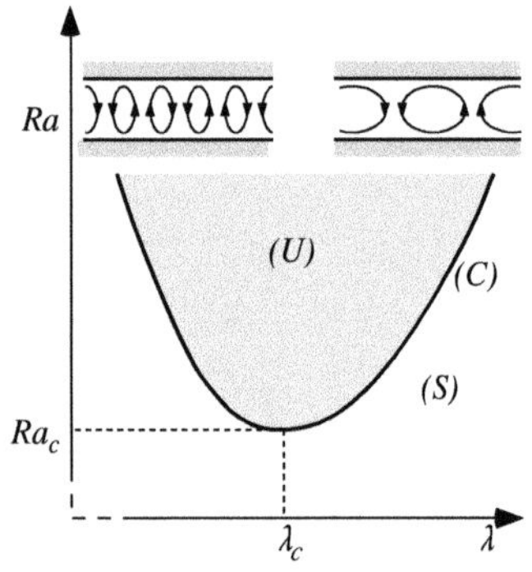

Figure 11.9 *Stability diagram for the Rayleigh–Bénard instability, obtained by selecting the wavelength $\lambda = 2\pi/k$ of the system of rolls, and the Rayleigh number Ra, as control parameters. The so-called marginal stability curve (C), corresponding to a growth rate $\sigma = 0$, separates the unstable region (U) from the stable one (S) (i.e. stable with respect to infinitesimal perturbations). Configurations of rolls corresponding to wavelengths λ both lower and higher than the critical value λ_c are illustrated schematically*

Let us now analyze qualitatively the increase in the critical Rayleigh number seen in Figure 11.9 as λ goes away from the value λ_c. For rolls with a narrow width along x, viscosity effects due to counter-flows within the same roll delay the onset of convection; moreover, thermal convection from the edge of one roll to the next tends more efficiently to uniformize the temperature. On the other hand, for very wide rolls, convective transport is quite ineffective and there is a significant energy dissipation by viscous friction at the level of the horizontal plates.

$$\sigma v_0 = -\nu k^2 v_0 + \alpha g \theta_0 \quad (11.36) \quad \text{and:} \quad \sigma \theta_0 = -\kappa k^2 \theta_0 + v_0 (\Delta T/a). \quad (11.37)$$

The term $\alpha g \theta_0$ corresponds to the modulation of the Archimedes' buoyancy of the liquid, due to the periodic variation of the temperature in a horizontal plane. The compatibility condition of these two equations, homogeneous in v_0 and θ_0 is:

$$\left(\nu k^2 + \sigma\right)\left(\kappa k^2 + \sigma\right) - \alpha g (\Delta T/a) = 0. \quad (11.38)$$

Equation11.33a is then obtained by setting $\sigma = 0$ in this expression and Equation 11.33b by assuming that $k = \pi a$, as mentioned at the beginning of the present section.

Equation 11.38 first allows one to calculate the growth rate σ of the instability as a function of the difference $\Delta T - \Delta T_c$, assumed to be small enough so that the terms in σ^2 can be neglected. Equation 11.34 is then obtained by using Equation 11.33a for ΔT_c because $(Ra - Ra_c)/Ra_c = (\Delta T - \Delta T_c)/\Delta T_c$.

The stability domain as a function of the wavelength

The complete solution of the linear, two-dimensional problem, which takes into account (as opposed to the above calculation) the variations of the velocity in the y-direction and the velocity component v_x yields the *stability diagram* in Figure 11.9. The solid curve (C) separates a region (S), where the system is stable relative to linear perturbations of the stationary problem, from an unstable region (U); in the latter the system is linearly unstable, i.e., unstable with respect to perturbations of infinitesimal amplitude. The minimum of the curve occurs at a value $\lambda_c \approx 2a$ of the wavelength in the case of boundary conditions of zero velocity on the rigid horizontal plates; this value agrees with the assumption made in the simpler calculations carried out above.

Amplitude variation of the instability as a function of the distance to the threshold

The linear analysis discussed above provides information on the value of the threshold and the nature of the solutions in the neighborhood of Ra_c. However, it does not predict the long term evolution of the flow for $Ra > Ra_c$ and, more specifically, the amplitude of the spatial variations of the velocity: the exponential variation in Equation 11.32 would indeed lead to a divergence over a long period of time. In this problem, the non-linear changes of the base flow determine the final limit to the growth of the amplitude. One predicts then theoretically that the limiting amplitude of the velocity components v_i ($i = x, y$) obeys:

$$v_i^2 = \frac{1}{\gamma_i} \frac{\Delta T - \Delta T_c}{\Delta T}, \quad (11.39)$$

in which γ_i is proportional to the constant b in Equation 11.9b. This result agrees with experimental measurements of the velocity by laser Doppler anemometry carried out on Rayleigh–Bénard experiments. These measurements show that the velocity v_x at long times increases continuously with $\Delta T - \Delta T_c$ beyond the threshold as:

$$v_x \propto [\Delta T - \Delta T_c]^\beta, \quad (11.40)$$

with an exponent $\beta = 1/2$ (Figure 11.11a two pages below).

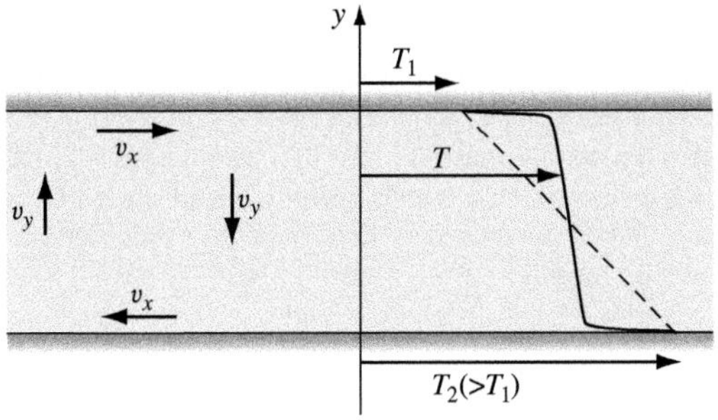

Figure 11.10 *Temperature profile between two horizontal plates, heated in the presence of a convective flow of vertical and horizontal velocity components v_x and v_y when the velocity is high enough for thermal boundary layers to appear in the vicinity of the walls*

Justification

The physical origin of these non-linear effects is described schematically in Figure 11.10. The convective thermal heat exchange reduces the temperature difference in the central region of the fluid layer: the largest temperature changes occur then in the thermal boundary layers near the walls, where the velocity v_y is small, and where thermal transport is due mainly to the diffusive heat conduction.

The lower thermal gradient $(\nabla T)_0$ in the central region leads to a reduction of the local Rayleigh number based on this gradient and, therefore of the velocity. More quantitatively, one assumes that:

$$(\nabla T)_0 \approx \left(1 - \gamma_i\, v_i^2\right) \nabla T, \tag{11.41}$$

in which ∇T is the mean gradient $\Delta T/a$ applied to the system; the presence of the quadratic term v_i^2 takes into account the fact that $+v_i$ and $-v_i$ reduce the gradient by the same amount. By taking as the order of magnitude of the temperature gradient in the center region $(\nabla T)_0$ the threshold value $\Delta T_c/a$, one obtains then Equation 11.39.

A dimensionless macroscopic parameter, frequently used to characterize the development of the instability above the threshold, is the Nusselt number Nu, already encountered in Section 10.8.1 and defined as:

$$Nu = \frac{\textit{measured heat flux}}{\textit{heat flux by thermal conductivity in the absence of convection}}. \tag{11.42}$$

The term in the denominator can be determined by heating the upper plate of the cell: in that case, the density stratification is stable, and thermal heat exchange occurs uniquely through the thermal conductivity (neglecting radiative transfers). In agreement with the definition in Equation 11.42, one has:

$$Nu = 1 \qquad \text{for } \Delta T \leq \Delta T_c. \tag{11.43a}$$

Moreover, the experimental measurements of the heat flux, from one plate to the other, in the presence of convection indicate a variation of Nu of the type:

$$Nu - 1 \propto (\Delta T - \Delta T_c) \qquad \text{for } \Delta T \geq \Delta T_c. \tag{11.43b}$$

Physical explanation: The vertical heat flux is the same at every level between the plates: in the central region, it is dominated by convection and corresponds to the average value of the product θv_y in a z-x-plane. As v_y and θ each vary as $(\Delta T - \Delta T_c)^{1/2}$ (Equation 11.40), one has a resultant variation of Nu as $(\Delta T - \Delta T_c)$. One notes that, in the regions where $v_y > 0$, convection carries warm fluid ($\theta > 0$) upward; on the other hand, when $v_y < 0$, the flow carries cold fluid ($\theta < 0$) downward: in both cases the product θv_y is positive, and the two contributions add up together.

11.2.6 The Landau model applied to Rayleigh–Bénard convection

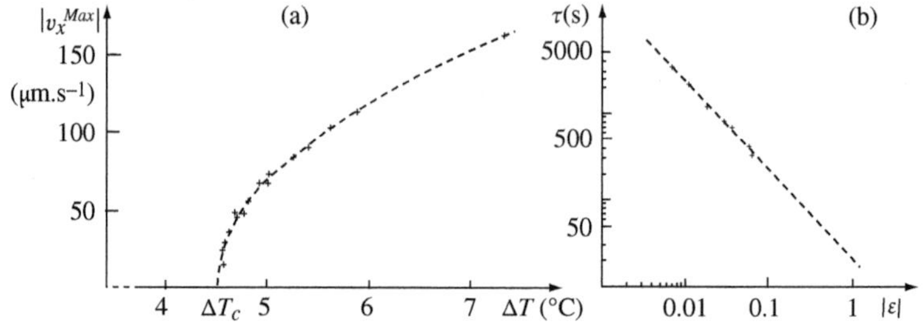

Figure 11.11 *(a) Variation of the maximum horizontal velocity component in a system of rolls as a function of the temperature difference* ΔT. *The dashed line corresponds to a variation as* $\sqrt{\Delta T - \Delta T_c}$. *(b) Variation of the relaxation time-constant* τ *of the system of rolls with the ratio* $\varepsilon = (Ra\text{-}Ra_c)/Ra_c$ *characterizing the deviation from the threshold. The dashed line corresponds (in logarithmic coordinates) to a variation* $\tau \propto \varepsilon^{-1}$ *(from docs. courtesy of J.E. Wesfreid, M. Dubois, P. Bergé)*

In the Rayleigh–Bénard problem, one has a broken symmetry like for the solid ball in the rotating hoop. This corresponds to the choice of the phase φ in the function $\cos(kx + \varphi)$ which describes the structure of the system of rolls. For a given position of the rolls, there are two possible directions of rotation which satisfy the inversion of the direction of rotation from one roll to the next (φ changed into $\varphi+\pi$). These two solutions are equivalent to the two symmetrical equilibrium positions of the ball above the instability threshold.

Detailed models of the R-B instability allow for the calculation of the complete velocity and temperature fields for a given temperature difference: one can then, in principle, determine the coefficients of the Landau model from these results.

The variation as $(\Delta T - \Delta T_c)^{1/2}$ of the velocity components in Equations 11.39 and 11.40 suggests that the global behavior of the Rayleigh–Bénard instability may be described by means of the Landau model introduced in Section 11.1. Since the convective flow created by this instability is stationary, Equation 11.11 can be used to describe it. One can choose, for example, the Rayleigh number Ra (or the temperature difference ΔT) as a control parameter and, as the order parameter $|A|$, the maximum $|v_i^{Max}|$ of the absolute value of one of the velocity components in the cell.

One expects then a variation of $|v_x|$ and $|v_y|$ as a function of Ra (or ΔT) of a type similar to those displayed in Figures 11.2 and 11.4a. Figure 11.11 displays indeed an experimental variation as $\sqrt{\Delta T - \Delta T_c}$ of the maximum $|v_x^{Max}|$ of the absolute value of the velocity component v_x for $\Delta T > \Delta T_c$. For $\Delta T < \Delta T_c$, the velocity $|v_x^{Max}|$ is zero. The second branch of the curves in Figure 11.2 would correspond to a system of rolls shifted by one unit: this would be equivalent to inverting the direction of rotation of each roll, because they alternate in direction.

Another proof of the correctness of the Landau description of this instability is given by the relaxation of $|v_x^{Max}|$ after a sudden change in ΔT ($> \Delta T_c$) toward a new value satisfying Equation 11.14. Experimental measurements of the characteristic time-constant τ of this relaxation show that it varies as the inverse of the distance from the threshold (Figure 11.11b) in agreement with Equation 11.15.

11.2.7 Evolution toward turbulence above the convection threshold

Sufficiently above the thermal convection threshold, one reaches a turbulent regime in which the velocity field has a random time and space component. This does not imply, however, that the macroscopic roll structures disappear. In the case of R-B convection, as well as for other instabilities described further down, there are several possible "scenarios" for the transition from an initial stationary instability to a state of weak turbulence. The scenarios actually observed depend on several experimental parameters (geometry and size of the box, Prandtl number, ...).

In the case of an instability in "small boxes", i.e., containing a small number of convection rolls, one might observe at first *a two frequency scenario.* In order to detect these, one can, for example, carry out a spectral analysis of the temporal dependence of the temperature measured by local sensors placed in the cell. In this case, if the temperature difference is increased further above the linear threshold ΔT_c, there will appear a first unstable frequency f_1 above a new threshold $\Delta T_{c1} > \Delta T_c$: it will correspond, for example, to a periodic modulation of the rolls. Then, a second frequency f_2 incommensurate with the first one appears for $\Delta T_{c2} > \Delta T_{c1}$. Finally, for a value $\Delta T_{c3} > \Delta T_{c2}$, the flow becomes unpredictable while no third frequency appears. One has a transition toward *weak turbulence,* in contrast with the classical description in which turbulence is expected to appear as a result of the combination of an infinite set of unstable frequencies.

A second scenario, known as that of *frequency division,* obtained under different experimental conditions but still in small containers, consists in the appearance of a first unstable frequency f_1, followed by a second $f_2 = f_1/2$, a third $f_3 = f_2/2$, and so forth. The thresholds corresponding to each of these *period doublings* become closer and closer, and converge toward a value Re_t where one observes a transition to full turbulence.

The situation is quite different in "large containers": turbulent puffs appear when one is sufficiently above the linear instability threshold.

This *intermittent* turbulent scenario has similarities with the subcritical transition mechanisms toward turbulence, which will be considered in Section 11.4.3. It is also encountered in the case of the instability of the *Taylor–Couette* rolls (Section 11.3.2).

11.3 Other closed box instabilities

Closed box instabilities develop in a space of finite dimensions, with no mean flow, as was the case, for example, for the Rayleigh–Bénard instability, and in contrast to the Bénard–von Karman instability described in Section 2.4.2.

11.3.1 Thermocapillary Bénard-Marangoni instability

The so-called *Bénard-Marangoni instability* appears when one heats the lower face of a thin layer of fluid with a free upper surface: it results from the forces induced on the free surface by the gradients of the surface tension discussed in Section 8.2.4. This effect is superimposed on the Rayleigh-Bénard mechanism, which is due to a density gradient.

In a volume of fluid bounded by two solid horizontal plates, the Rayleigh–Bénard instability mechanism is the only one at work.

Beyond a critical value of the temperature difference, hexagonal flow-cells appear between the bottom of the liquid layer and the free surface (Figure 3.17). The fluid rises in the center of the hexagons and flows down along the edges (Figure 11.12 below). The size of the hexagons is of the same order of magnitude as the thickness of the fluid layer.

Qualitatively, the mechanism of the instability is the following: one assumes that, initially, the temperature rises slightly by an amount θ at a point on the free surface, thus reducing the local surface tension. As a result, the fluid at the surface is expelled radially toward the colder region outward where the surface tension is higher. In order to conserve the flow rate, warmer fluid rises from the lower region of the cell: this additional thermal energy reinforces the initial perturbation. The driving force of the instability is therefore the differences between the surface tensions in the regions of different temperatures at the surface of

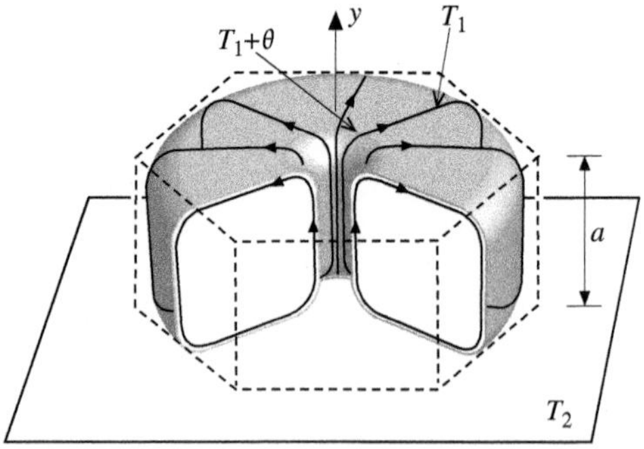

Figure 11.12 *Schematic diagram of the flow occurring in cells with a Bénard–Marangoni instability (see also the visualization (top view) shown in Figure 3.17)*

the liquid. Just as in the case of the Rayleigh–Bénard instability, the stabilizing processes are the thermal conductivity and the viscosity which, respectively, smooth out the temperature variations and slow down the motion of fluid particles; they also determine the sizes of the cells (these are of the order of their thickness).

Threshold of the Bénard–Marangoni instability

The dimensionless parameter, which controls this instability, is the *Marangoni number, Ma*, which satisfies:

$$Ma = \frac{b\ \gamma\ \Delta T\ a}{\eta\ \kappa}, \tag{11.44}$$

where $b = -(1/\gamma)(\partial\gamma/\partial T)$ represents the relative rate of variation of the surface tension γ with the temperature T, as defined in Equation 1.54, and a is the thickness of the fluid layer (flows resulting from these surface tension gradients were already discussed in Section 8.2.4).

Proof

One assumes here that the characteristic length scale for the variations of the velocity and temperature in the horizontal direction is the same as that in the vertical direction (i.e., the thickness a of the fluid layer). This implies that the cross-section of a convection cell by a vertical plane is roughly a square as shown by Figure 11.12.

The principle of this calculation is very similar to that used for the Rayleigh–Bénard instability. The only difference comes from the fact that, for the Marangoni effect, the driving force (Equation 8.66) is localized on the free surface of the fluid. One perturbs an element of the fluid by applying to it a velocity v_c and analyzing whether or not this velocity decreases as a function of time. In the two previous examples the stability criterion was obtained by comparing the forces on elements of fluid with a size r_0 of the order of $a/2$: the present discussion uses elements of the same size and with one face coinciding with the free surface of the fluid. Evaluating then, like in Section 11.2.4, the temperature difference induced between the element and the surrounding fluid provides Equation 11.21 in the equivalent form:

$$\delta T \simeq A\, \boldsymbol{v}_c \frac{a^2}{\kappa} \frac{\Delta T}{a}. \tag{11.45}$$

The driving force F_m on a volume element is then related to the absolute value of the derivative $\mathrm{d}\gamma/\mathrm{d}T$ of the surface tension with respect to the temperature by

$$F_m \simeq \frac{a}{2}\left|\frac{\mathrm{d}\gamma}{\mathrm{d}T}\right|\delta T \simeq \boldsymbol{v}_c\frac{a^3}{\kappa}\left|\frac{\mathrm{d}\gamma}{\mathrm{d}T}\right|\frac{\Delta T}{a}. \tag{11.46}$$

The face of the fluid element which coincides with the interface has been assumed to be square with two sides perpendicular to the convection flow v_c; one then obtains Equation 11.46 by assuming that the temperature difference between these two sides is of order δT, and that F_m is the difference between the surface tension forces on them. The order of the magnitude of the damping force F_{visc} is, as in the Rayleigh–Bénard case:

$$F_{visc} \approx \eta\, v_c\, a.$$

The condition for the growth of the instability can then be written:

$$F_m > F_{visc}, \quad \text{so that :} \quad \boldsymbol{v}_c\,\frac{a^3}{\kappa}\left|\frac{\mathrm{d}\gamma}{\mathrm{d}T}\right|\frac{\Delta T}{a} > \eta\,\boldsymbol{v}_c\,a. \tag{11.47}$$

The ratio F_m/F_{visc} is then equal to the Marangoni number given by Equation 11.44: it is therefore the dimensionless combination which determines the possible occurrence of the instability.

The ratio of the Rayleigh number to the Marangoni number obeys:

$$\frac{Ra}{Ma} = \frac{\alpha\,\rho\,g\,a^2}{|\mathrm{d}\gamma/\mathrm{d}T|} = \frac{|\delta\rho|\,g\,a^2}{|\delta\gamma|}, \tag{11.48}$$

where $|\delta\rho|$ and $|\delta\gamma|$ are the absolute values of the respective changes of the density and surface tension resulting from a same temperature difference δT. This ratio characterizes the relative influence of Archimedes' buoyancy and of the surface tension forces, due to temperature variations in a horizontal layer with a free upper surface. The ratio Ra/Ma is then equivalent here to the Bond number introduced in Chapter 1 (Equation 1.64). For a fixed temperature difference, the ratio Ra/Ma varies as a^2: surface tension effects are therefore dominant for small thicknesses of the fluid layer, and those due to gravity for larger thicknesses.

In a weightless environment, the effect of Archimedes' buoyancy disappears and only the Marangoni effect remains. It may thus be interesting to carry out experiments with large single crystals grown in a zero gravity environment, where convection motions due to gravitational forces, which usually disturb the regular growth of crystals, are absent. Very large crystals, or those normally very difficult to grow (e.g., protein crystals), have been grown in a spaceship environment: however, the Marangoni effect may then play a dominant role.

There exists another difference between the Rayleigh-Bénard and Bénard-Marangoni instabilities than that between the definitions of Ra and Ma: in the case of the Rayleigh-Bénard instability, there is a symmetry between the fluid rising to the top and descending to the bottom (Section 11.2.6). In the problems of the Bénard-Marangoni instability, there is no such symmetry between top and bottom: the instability creates then hexagonal structures (Figure 11.12) where the liquid rises in the center and spreads out toward the corresponding edges where it flows downward; in contrast to the Rayleigh-Bénard instability, this configuration cannot be transformed by a simple translation into one in which the liquid rises along the edges and descends in the center.

Behavior above the threshold

The asymmetry mentioned just above is visible in the stability diagram of Figure 11.13a below: this diagram displays the variation, as a function of Ma, of the vertical component v_y of

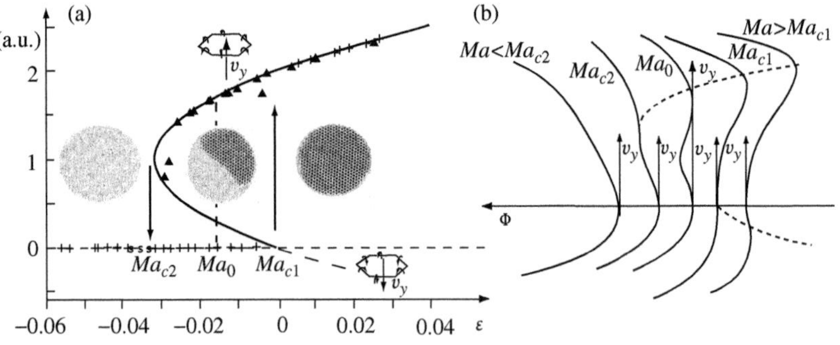

Figure 11.13 *(a) Variation as a function of the normalized Marangoni number $\varepsilon = (Ma - Ma_{c1})/Ma_{c1}$ of the convective velocity for the Bénard–Marangoni instability of a thin liquid layer heated from below (according to H. Swinney et al.). In contrast with the model mechanical instability (Figure 11.2a) or the Rayleigh–Bénard one, the stability diagram is not symmetrical relative to the axis $v_y = 0$. The images in the "insets" display typical observations in the stable and unstable regions (left and right, respectively) and for $Ma = Ma_0$ (in the center). The + symbols correspond to data obtained by increasing the value of Ma from small numbers, and the ▲ to values obtained by decreasing Ma starting from larger numbers $Ma > Ma_{c1}$. (b) Free energy curves $\Phi(v_y)$ associated with values of Ma corresponding to the different regimes. The curves have been shifted horizontally with respect to one another, and the amplitude of the secondary minimum has been exaggerated to improve the readability (all v_y axes correspond to a zero value of Φ). The two dashed curves indicate the tendency for non-zero values of v_y to evolve, corresponding to minima in Φ*

the velocity at the center of the hexagons. When the temperature difference ΔT, and thus *Ma*, is increased, starting from $\Delta T = 0$ ($Ma = 0$), thermal transfer results first from *thermal conduction* without flow. When the critical threshold Ma_{c1} is reached, there is a sudden transition to a regime of *thermal convection* with a finite value of the velocity v_y increasing continuously as a function of *Ma*. If, instead, *Ma* is decreased progressively from a value $Ma > Ma_{c1}$, one can then go down to a value $Ma_{c2} < Ma_{c1}$ before the disappearance of the convection cells. At values intermediate between Ma_{c2} and Ma_{c1}, one can induce a transition from the thermal conductivity regime to the convection one by applying a perturbation to the static fluid layer: the amplitude required to trigger the transition becomes smaller as one gets closer to the critical value Ma_{c1}.

Extension of the Landau model to the Bénard–Marangoni instability

In order to account for the asymmetry of the instability diagram, we need to modify the Landau model by adding into Equation 11.9b for the free energy, terms corresponding to odd powers of v_y. The velocity v_y plays here the role of A and, as seen just above, the values of v_y and $-v_y$ are not equivalent. One uses thus the following expression:

$$\Phi\left(v_y\right) = 2k' \left(Ma_c - Ma\right) v_y^2 + c\, v_y^3 + 2\, b\, v_y^4. \tag{11.49}$$

Figure 11.13b displays the variations $\Phi(v_y)$ resulting from this equation: for $Ma < Ma_{c2}$, the curves have a single minimum corresponding to $v_y = 0$: no convection cells appear and one is in the thermal conduction domain. For $Ma_{c2} < Ma < Ma_{c1}$, Φ has two minima: one is at $v_y = 0$ and the other at a finite value of v_y, even at the threshold Ma_{c2} for which the minimum is merely a point of inflection. In this domain, depending on the previous history of the system and the perturbations it has undergone, one observes either a stable state, or a convective state with hexagonal cells. Moreover, transitions from one state to the other (generally to the convective regime) can be induced by perturbations of sufficiently large amplitude to overcome the intermediate maximum of the energy curve. There is also, on the other hand, a value Ma_0 for which (as seen in the center inset in Figure 11.13a), one can have coexistence of the two states, because the two minima of Φ have the same value. Finally for $Ma > Ma_{c1}$, one has an energy maximum (unstable equilibrium) at $v_y = 0$. One almost always observes in this case flows with liquid rising in the center of the cells (right-hand inset). Flows in the opposite direction correspond to a shallower energy minimum (lower, dashed branch on Figure 11.13a) and are usually difficult to observe.

One notes analogies between this hysteretic transition and first-order thermodynamic transitions, such as those between the liquid and gas states. By heating a well degassed volume of water in a container with smooth walls, water can indeed be kept in its liquid state well above the normal boiling point: boiling can be triggered by introducing a few bubbles, e.g., with the help of a porous stem. This phenomenon of *delayed boiling* is equivalent to the hysteresis in Figure 11.13a. In the same way, the liquid and gas phases may coexist at a temperature which is a unique function of the external pressure (determined in this case by a "Maxwell construction"): this is equivalent to the existence of stable and unstable states for $Ma = Ma_0$.

An instability displaying the characteristics of the Bénard–Marangoni instability (hysteresis, the ability to trigger the instability by perturbations, the possible coexistence of the two states,...) is referred to as *subcritical*. Instabilities such as the Rayleigh–Bénard instability, or that leading to the emission of vortex streets (Section 2.4.2), are instead called *supercritical*.

11.3.2 Taylor–Couette instability

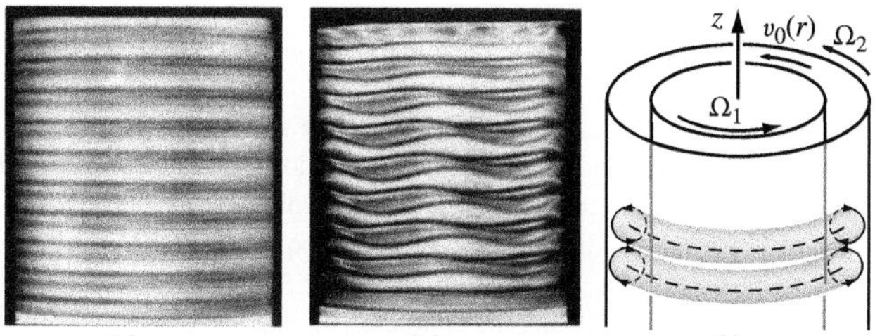

Figure 11.14 *Pictures of a system of Taylor–Couette rolls visualized by means of reflecting particles suspended in the fluid. (a) Just above the threshold Ta_c; (b) further above the threshold we observe wave-like rolls (plates courtesy of D. Andereck and H. Swinney); (c) schematic view of the rolls appearing in the Taylor–Couette instability for flow between two cylinders with different rotational velocities Ω_1 and Ω_2*

Couette flow between two concentric cylinders has been discussed in Section 4.5.6. The stationary velocity field obtained at low rotational velocities is a purely tangential flow, of velocity profile given by Equation 4.110. If the inner cylinder is fixed and the outer cylinder rotated, the flow is stable up to a velocity at which it becomes turbulent without any appearance of characteristic structures in the flow: this is the problem that Maurice Couette had studied in his doctoral thesis, in 1901. On the other hand, if the inner cylinder is rotated while maintaining the outer cylinder fixed, a fluid circulation with a toroidal rolls geometry appears above a critical value Ω_c of the rotational angular velocity Ω; the plane of the tori is horizontal and their cross-section has a diameter equal to the distance a between the cylinders (Figure 11.14). This type of structure was first described by G. I. Taylor in 1923: the rolls are easily visualized if the liquid contains small flat particles which reflect light (see Section 3.5.1). The rolls can be studied quantitatively by means of velocity gradient probes located on the walls as discussed in Section 10.8.2. They can also be studied by laser Doppler anemometry or by the PIV techniques described in Section 3.5.4). The resultant velocity field is the sum of the unperturbed tangential flow and of a toroidal advective motion.

This Taylor–Couette instability has a close correspondence with the Rayleigh–Bénard one. The radial gradient of the centrifugal force, due to the greater angular momentum of the fluid closer to the axis, replaces, in this problem, the effect of the Archimedes' buoyancy

	Rayleigh–Bénard Instability	Taylor–Couette Instability	Bénard–Marangoni Instability
Viscous-damping force	$F_{visc} = \eta\, v_c\, a$	$F_{visc} = \eta\, v_c\, a$	$F_{visc} = \eta\, v_c\, a$
Driving-force	$F_{Archimedes'\ buoyancy}$ $\rho_0\, \alpha\, g\, \dfrac{a^5}{\kappa}\dfrac{\Delta T}{a}\, v_c$	$F_{centrifugal\ force}$ $\rho_0\, a^2\, \dfrac{\Omega^2 R}{a}\dfrac{a^5}{v}\, v_c$	$F_{suface\ tension}$ $\dfrac{a^3}{\kappa}\dfrac{d\gamma}{dT}\dfrac{\Delta T}{a}\, v_c$
Relaxation time-constant of the perturbation with the surrounding fluid	a^2/κ	a^2/v	a^2/κ
Characteristic parameter of the instability	$Ra = \dfrac{\alpha\,\Delta T\, g\, a^3}{v\,\kappa}$	$Ta = \dfrac{\Omega^2\, R\, a^3}{v^2}$	$Ma = -\dfrac{(d\gamma/dT)\,\Delta T\, a}{\eta\,\kappa}$
Critical value for the onset of the instability	$Ra_c = 1{,}708$ $k_c = 3.11/a$	$Ta_c = 1{,}712$ $k_c = 3.11/a$	$Ma_c = 80$

Table 11.1 *Table comparing the characteristic parameters of the Rayleigh–Bénard, Taylor–Couette and Bénard–Marangoni instabilities.*

The kinematic viscosity v occurs squared in the Taylor number. It first appears once in the evaluation of the time constant τ_v for a velocity perturbation to reach equilibrium with the surrounding fluid by viscous friction (in the case of the Rayleigh–Bénard instability, the corresponding time constant is that for thermal equilibrium). v comes into play a second time in the calculation of the viscous–friction force which opposes the driving force. The v^2 denominator in the Taylor number thus replaces the product $v\,\kappa$ in the Rayleigh number.

due to the unstable stratification of the density. Also, the stabilizing effect of the viscosity (allowing for a fluid element displaced from its initial position to reach a velocity equilibrium with the surrounding fluid), replaces that of the thermal conductivity.

One finds for this instability a dimensionless number equivalent to the Rayleigh number for the Rayleigh–Bénard one by writing an equilibrium of forces similar to that used in this former case. One finds then (see below) the characteristic number:

$$Ta = \frac{\Omega^2 a^3 R}{v^2}, \tag{11.50}$$

known as the *Taylor number* (a is the distance between the cylinders and $R \gg a$ their average radius; Ω is the rotational angular velocity of the inner cylinder). Table 11.1 indicates the parallels between the parameters and the forces characteristic of these two instabilities, and of the Bénard–Marangoni instability discussed just above (geometrical constants are not given in this table).

Proof

Like for the Rayleigh–Bénard instability, one assumes that a *radial, convective velocity perturbation v_c, normal* to the tangential mean flow is applied to a sphere of fluid with a small radius r_0 and density ρ_s. The momentum of the sphere is then $p = m\, v_c$, ($m = (4\pi/3)r_0^3$ is the mass of this fluid sphere) and is lost by viscous damping after a characteristic time τ_v. Because the viscous damping force acting on the sphere is $F_{visc} = 6\pi\, \eta\, r_0\, v_c$, one find then:

$$\frac{4}{3}\pi\, \rho\, r_0^3 \frac{dv_c}{dt} = -6\pi\, \eta\, r_0\, v_c, \qquad \text{so that :} \qquad \frac{1}{\tau_v} = \frac{1}{v_c}\frac{dv_c}{dt} = A\frac{v}{r_0^2},$$

where A is a geometrical numerical coefficient. This expression for the viscous-damping time τ_ν is, moreover, equivalent to that of the thermal diffusion time in Equation 11.20. After a time τ_ν, the particle has moved by a distance: $\delta r \approx v_c \tau_\nu$. The order of magnitude of the driving force F_m correspond then to the variation of the centrifugal force $m\,\Omega^2(r) r$ over the distance δr. The local rotational angular velocity $\Omega(r)$ varies from a value Ω on the inner cylinder to zero on the outer cylinder ($r = R + a$). One takes then as the order of magnitude for F_m:

$$F_m = \frac{4}{3}\pi \rho r_0^3 \frac{\partial}{\partial r}(\Omega^2(r))\,\delta r \approx B\, r_0^3 \rho \frac{\Omega^2 R}{a}\, v_c\, \tau_\nu = \frac{B}{A}\frac{\rho}{\nu} v_c\, \Omega^2 \frac{r_0^5\, R}{a},$$

where B is a geometrical constant. The instability condition equivalent to Equations 11.24a–b is thus:

$$F_m > F_{visc}, \qquad \text{i.e.:} \qquad \frac{B}{A}\frac{\rho}{\nu} v_c\, \Omega^2 \frac{r_0^5\, R}{a} > 6\pi\, \rho\, \nu\, r_0 v_c.$$

Just as in the case of the Rayleigh–Bénard instability, the largest perturbations obey most easily the latter condition with an upper limit $r_0 \approx a/2$. Neglecting geometrical coefficients, one finds indeed that the ratio F_m/F_{visc} is equal to the Taylor number given in Equation 11.50.

The critical wave vector k_c corresponds to the instability modes involving the smallest Taylor number: just as in the case of the Rayleigh–Bénard instability, it is of the order of π/a which corresponds, for both instabilities, to a diameter of the rolls comparable to the distance between the solid walls. The stability curve in the plane (Ta, λ) is also very similar to that obtained in the (Ra, λ) plane for the Rayleigh–Bénard instability (Figure 11.9).

11.3.3 Other centrifugal instabilities

The Taylor–Couette instability is part of broader class of instabilities resulting from centrifugal force effects for fluids in rotation, or for curved flows.

For example, the *Görtler instability* appears in a boundary layer developing along a curved concave wall (flow parallel to the plane of curvature): it develops in a Blasius-like velocity profile (Section 10.4.2) instead of a Couette profile for the Taylor–Couette instability. Beyond a threshold velocity, convection rolls appear with their axis parallel to the mean flow.

The *Dean instability* is observed in developed flows of the Poiseuille-type in curved channels such as the one with rectangular cross-section displayed in Figure 11.15a below. For any flow velocity, two recirculation cells first appear at the two ends of the cross-section: these correspond to a *Dean secondary flow* described in Section 7.7.1 and already shown in Figure 7.42 for a cross-section of different aspect ratio. The actual *Dean instability* corresponds to the appearance of initially smaller cells localized near the concave wall of the channel between two secondary flow cells (Figure 11.15a below): in contrast to the secondary flow, these new cells only appear above a threshold flow velocity.

The Dean instability shows up beyond a critical value Dn_c of the *Dean number, Dn,* which represents the ratio between the centrifugal and viscous forces. The Dean number, Dn, is defined as:

$$Dn = (U_m D_h/\nu)\sqrt{D_h/R_c},$$

where U_m is the average velocity in a section, R_c is the radius of curvature of the channel, and D_h is a length characterizing the hydraulic aperture. Figure 11.15b displays the development of the instability as Dn increases.

In a channel with a square cross-section, the Dean instability results in the appearance, near the concave walls of the channel, and near the plane of symmetry of the flow, of two additional small cells rotating in opposite directions, and superimposed on the secondary flow shown schematically in Figure 7.42.

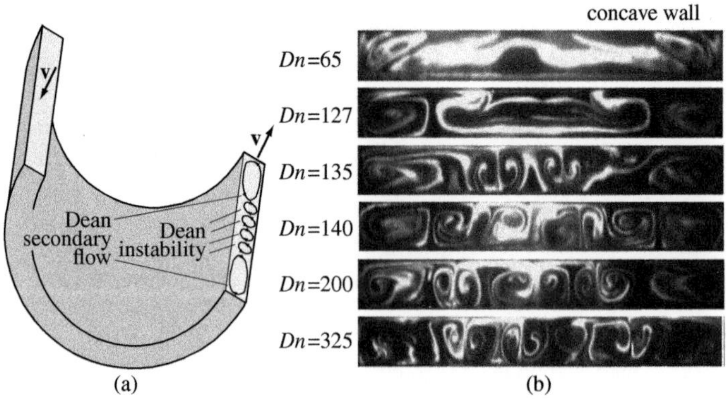

Figure 11.15 *(a) Schematic diagram of the secondary flow and of the Dean instability in an elongated rectangular cross-section of a curved tube. (b) Visualization by the LIF technique (Section 3.5.2) in a flow section for six values of the Dean number Dn of a fluorescent dye flow injected upstream (the angle between the entry and measurement cross-sections is 180°) (plates courtesy of H. Fellouah, C. Castelain, A. Ould El Moctar, H. Peerhossaini)*

The closed box type instabilities (Rayleigh–Bénard, Bénard–Marangoni, Taylor–Couette) discussed above are sometimes referred to as *absolute instabilities*: beyond the instability threshold, they develop indeed without propagating. In open flows such as jets, other kinds of instabilities known as *convective* (not to be confused with the *thermal convection instabilities* discussed previously) may, in addition, be encountered: perturbations in the flow develop (or attenuate, if they happen to be below threshold) within a finite-size region, as they are carried along (convected), by the average motion of the fluid.

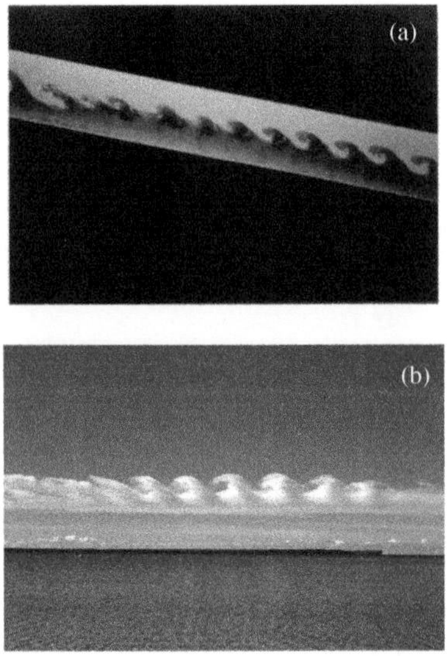

Figure 11.16C *(a) Experimental visualization of an instability at the interface between two immiscible liquids in a parallelepipedic cell initially horizontal and tilted downward to the right. (plate courtesy O. Pouliquen). (b) Kelvin–Helmholtz instability in a layer of clouds above the bay of Jervis, New South Wales, Australia (plate courtesy G. Goloy)*

11.4 Instabilities in open flows

In this section, one first discusses in detail the instability of a shear flow (*Kelvin–Helmholtz instability)* for which the threshold value is independent of viscosity effects. This discussion allows one to review a whole series of concepts related to ideal fluids, to vorticity, and to waves. Next, after having discussed the influence of the shape of velocity profiles, one considers the case of Poiseuille and Couette flows which demonstrates the richness and complexity of the mechanisms of the transition toward turbulence.

11.4.1 Kelvin–Helmholtz instability

Wind blowing parallel to a free surface of water may induce the formation of waves: for a force 7 wind on the Beaufort scale (i.e., approximately 60 km/h) these will amplify into breaking waves. This phenomenon is a manifestation of the Kelvin–Helmholtz instability. This instability can also be observed in the experiment of Figure 11.16a: a parallelepipedic cell, containing two superimposed non-miscible liquids, is tilted from an initially horizontal position. By resetting the cell to a horizontal position, one creates a shear flow of the fluids at the level of the interface: one observes that the interface becomes wavy as it returns to a horizontal position. Such effects occasionally appear on a much larger scale, in an atmosphere in which clouds are present (Figure 11.16b).

In the following discussions, the effect of viscosity is neglected: this assumption is valid if the thickness of the viscous boundary layer is small relative compared to the perturbation to be amplified. This is the case for a flow at a velocity high enough so that the corresponding Reynolds number is much greater than one.

Physical explanation for the Kelvin–Helmholtz instability

The shear flow results in a velocity gradient which is assumed to be localized in the neighborhood of the interface between the two fluids (left-side of Figure 11.17 below). A deformation of the interface reduces the distance between the streamlines along the convex face and increases therefore the velocity v in the term $\rho v^2/2$ of Bernoulli's equation (Equation 11.58). On the other side of the interface (the concave face), both the speed and $\rho v^2/2$ decrease because the streamlines spread apart. This contrast between the values of the $\rho v^2/2$ terms on either side of the interface results in an amplification of its deformation.

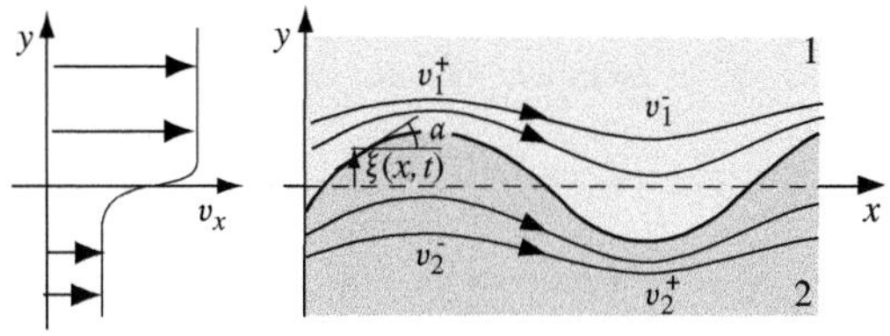

Figure 11.17 *Deformation of the interface between two superimposed fluids, moving parallel to each other at different velocities in the x-direction*

Justification

If the interface remained motionless, according to Equation 11.58, a pressure difference oriented so as to amplify the deformation should appear. Here, the interface is free and, absent surface tension, the pressure is the same on both sides: the difference between the values of the $\rho v^2/2$ terms will then be compensated in Bernoulli's equation (Equation 11.58) by a term $\partial\Phi/\partial t$ corresponding to an acceleration of the fluid. Therefore, only deformation modes of the interface for which the displacement is in the same direction as the acceleration will be amplified.

Such structures of parallel vortices generated at the interface of two liquid jets of differing velocities can be observed at high Reynolds numbers where turbulence can also occur. These *mixing layers* represent an example of *turbulent coherent structures* which will be discussed in Section 12.7.2.

Instability model in a linear regime

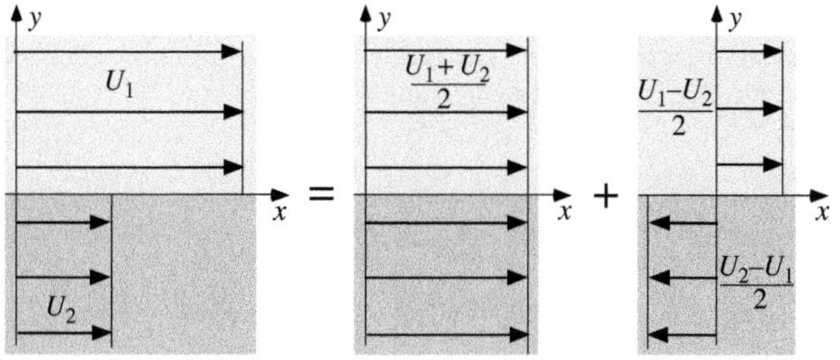

Figure 11.18 *Flow of two superimposed fluids in contact, moving parallel to each other at different velocities and representation of the velocity field of this flow as the superposition of an overall translation and a relative flow of zero average velocity*

These processes are modelled as the flow of two superimposed fluids (1) and (2), moving parallel to each other at different velocities U_1 and U_2 in the x-direction (Figure 11.18). This flow is the superposition of a global translation at velocity $(U_1 + U_2)/2$, and of a flow antisymmetric relative to a horizontal plane, $y = 0$, and at a velocity $\pm\, U/2$ with $U = U_1 - U_2$. As seen in Section 7.4.2, this is equivalent to assuming the presence of an infinitesimal layer of vorticity at $y = 0$.

One also assumes (a general result for two-dimensional flows) that the first instability which develops corresponds to a two-dimensional perturbation characterized by the height $\xi(x, t)$ of the interface above a base plane $y = 0$ (Figure 11.17). In the reference frame moving at the global translational velocity $(U_1 + U_2)/2$, the velocity fields $\mathbf{v}_1$ et $\mathbf{v}_2$ in each liquid can be written in the form:

$$\mathbf{v}_1 = \nabla\left[\frac{Ux}{2} + \Phi_1(x, y, t)\right], \quad (11.51a) \qquad \mathbf{v}_2 = \nabla\left[-\frac{Ux}{2} + \Phi_2(x, y, t)\right]. \quad (11.51b)$$

The first term in each equation between the brackets is the velocity potential of the unperturbed flow; the second term is the potential for the velocity perturbation resulting from the deformation $\xi(x, t)$. The calculation is carried out in a linear approximation valid if the amplitude $\xi(x, t)$ of the perturbations is small relative to their wavelength: this implies that the angle α with the x–axis of the tangent at the interface (Figure 11.17) is small, so that:

$$\alpha \approx \tan\alpha = \frac{\partial \xi}{\partial x} \ll 1. \tag{11.52}$$

The corresponding velocity perturbations are also assumed to be small relative to the velocities $\pm U/2$ of the basic flow. One now takes, for the coupled variables ξ, Φ_1, and Φ_2 characterizing this perturbation, solutions of the form:

$$\frac{\xi}{A} = \frac{\Phi_1}{B_1 e^{-ky}} = \frac{\Phi_2}{B_2 e^{ky}} = e^{ikx+\sigma t}. \tag{11.53}$$

One will then look for relations between these several variables resulting from the boundary conditions and the equations of motion.

Justification

The terms $B_1 e^{-ky}$ and $B_2 e^{ky}$ are required by the structure of Laplace's equation, in the same way as they were used in the study of surface waves carried out in Section 6.4.1. One also needs to associate to the sinusoidal variation in the x-direction an exponential variation along the perpendicular y-axis with a characteristic decay length equal to the wavelength. The "–" and "+" signs ensure that the physical solution corresponds to an attenuation of the wave in the two upper and lower half-spaces respectively. The term $e^{\sigma t}$ indicates that one looks for modes varying exponentially with time like in the Rayleigh–Bénard problem. One cannot exclude the existence of an imaginary component in σ: for surface waves of stationary amplitude, this is indeed the only non-zero component.

A first relation between Φ_1, Φ_2, and the amplitude $\xi(x, t)$ is provided by the boundary conditions at $y = 0$ for the velocity component of the fluids *normal to the interface*:

$$v_{1y} = \frac{\partial \Phi_1}{\partial y} = \frac{\partial \xi}{\partial t} + \frac{U}{2}\frac{\partial \xi}{\partial x}, \quad (11.54a) \quad \text{and:} \quad v_{2y} = \frac{\partial \Phi_2}{\partial y} = \frac{\partial \xi}{\partial t} - \frac{U}{2}\frac{\partial \xi}{\partial x}. \quad (11.54b)$$

Applying Equations 11.53 in the above equations and simplifying by the factor $e^{ikx+\sigma t}$, one then obtains:

$$kB_1 + \left(\sigma + ik\frac{U}{2}\right)A = 0 \quad (11.55a) \quad \text{and:} \quad kB_2 - \left(\sigma - ik\frac{U}{2}\right)A = 0. \quad (11.55b)$$

Proof

The normal velocity components at the interface $[v_{1\perp}]$, $[v_{2\perp}]$ must be continuous and equal to the component $(\partial\xi/\partial t)\cos\alpha$ of its velocity (the *tangential* components may, however, be different because the fluids are assumed to be ideal). Calculating these normal components for the two fluids, taking the projection of the velocity of each fluid on the normal to the boundary and taking also $\cos\alpha = 1$, the previous equation becomes:

$$v_{iy} - v_{ix}\,\alpha = \frac{\partial \xi}{\partial t} \quad (i = 1, 2), \quad (11.56a) \quad \text{so that:} \quad v_{iy} \cong \frac{\partial \xi}{\partial t} + v_{ix}\frac{\partial \xi}{\partial x} \quad (i = 1, 2). \quad (11.56b)$$

By taking $v_{1x} = U/2$ in Equations 11.56a–b, we obtain Equations 11.54a–b by using Equations 11.51a–b for the velocity fields $\mathbf{v}_1$ and $\mathbf{v}_2$ and retaining only first order terms.

The second term in Equation 11.56b is the Lagrangian derivative of the position $\xi(x, t)$ of the interface. It is obtained by following a fluid particle located close to the interface.

The approximation $\cos\alpha \cong 1$ introduces a second-order error in α while the other terms in Equations 11.54a–b are of first-order in α.

Case where the interfacial surface tension and the differences in density are neglected

In a first step, the effects of capillarity and of the difference in density between the two liquids are neglected. In this case, the pressures on each side of the interfaces are equal, so that:

$$p_1(x, y, t) = p_2(x, y, t), \qquad \text{for} \quad y = \xi \tag{11.57}$$

Moreover, Bernoulli's equation (Equation 5.36) allows one to write, for each of the two fluids (assuming they have equal densities $\rho_1 = \rho_2 = \rho$):

$$p_i + \rho\frac{\partial \Phi_i}{\partial t} + \rho g y + \frac{1}{2}\rho v_i^2 = \textit{constant} \quad (i = 1, 2). \tag{11.58}$$

$$\text{Therefore}: \left(\frac{\partial \Phi_1}{\partial t}\right)_{y=0} + \frac{U}{2}\left(\frac{\partial \Phi_1}{\partial x}\right)_{y=0} = \left(\frac{\partial \Phi_2}{\partial t}\right)_{y=0} - \frac{U}{2}\left(\frac{\partial \Phi_2}{\partial x}\right)_{y=0}. \tag{11.59}$$

Justification

By subtracting Equations 11.58 and taking $y = \xi$, the term $\rho g y$ cancels out as well as the term p_i (from Equation 11.57). In the equation obtained in this way, the terms including $v_i{}^2$ can be expressed by means of Equations 11.51a–b: by writing that those of terms of the resulting expression which are of first-order relative to the perturbations Φ_i (and thus vary as $e^{ikx+\sigma t}$) are equal, one obtains Equation 11.59. The condition $y = \xi$ has been replaced by that of $y = 0$, a substitution justified by the fact that the effect of variations over the infinitesimal distance ξ would be second-order. Note that subtracting Equation 11.58 eliminates possible components of p and ξ proportional to $e^{ikx+\sigma t}$. Equation 11.59 expresses the dynamic equilibrium for this problem (it plays the role of an equation of motion), while Equations 11.54a–b express kinematic conditions.

It follows that the growth rate σ is related to the wave vector k by:

$$\sigma = \pm\, k\frac{U}{2}. \tag{11.60}$$

This relationship between the growth rate and the wave vector k, which corresponds to a dispersion relation for a wave, show that there is always an unstable mode. Equation 11.60 implies also that, if the surface tension and gravitational forces are absent, the modes with a short wavelength always display the greatest amplification.

Proof

Substituting Equation 11.53 into Equation 11.59, one obtains, after having cancelled out $e^{ikx+\sigma t}$, and taking again $y = 0$:

$$\left(\sigma + ik\frac{U}{2}\right)B_1 - \left(\sigma - ik\frac{U}{2}\right)B_2 = 0. \tag{11.61}$$

Equations 11.55a–b and 11.61 represent a system of homogeneous, linear equations with unknown coefficients A, B_1, B_2. These equations will only be compatible if the determinant of the matrix of these coefficients vanishes, so that:

$$k\left(\sigma + ik\frac{U}{2}\right)\left(\sigma + ik\frac{U}{2}\right) + k\left(\sigma - ik\frac{U}{2}\right)\left(\sigma - ik\frac{U}{2}\right) = 0, \quad (11.62)$$

whence:

$$\sigma^2 = \left(k\frac{U}{2}\right)^2. \quad (11.63)$$

Equation 11.60 results immediately. One observes that the value of the growth rate is obtained by a dimensionally correct combination of the wave vector k and of the velocity U.

Effects of the interfacial tension and of differences in the density

Capillary forces and gravity limit respectively the growth of small-wavelength perturbations and of those of long wavelengths (if the heavier fluid is below the lighter one). The previous calculations will now be altered in order to take into account these effects.

- The pressure boundary condition (Equation 11.57) is replaced by:

$$(p_1)_{y=\xi} = (p_2)_{y=\xi} + \gamma\frac{\partial^2\xi}{\partial x^2} = (p_2)_{y=\xi} - \gamma k^2 \xi. \quad (11.64)$$

- In Bernoulli's equation, one introduces the hydrostatic term $\rho_i g \xi$ $(i = 1, 2)$ with $\rho_1 \neq \rho_2$.

The condition for the compatibility of the three equations is written like in the previous case, and leads this time to the instability condition:

$$\frac{U^2 \rho_1 \rho_2}{(\rho_1 + \rho_2)^2} > c_0^2 = \frac{g}{k}\frac{\rho_2 - \rho_1}{\rho_2 + \rho_1} + \frac{\gamma k}{\rho_2 + \rho_1}, \quad (11.65)$$

where c_0 is the velocity of the surface waves in the absence of flow ($U = 0$). The stabilizing effects of the density difference $(\rho_1 - \rho_2)$ (first term) and of the surface tension γ are clearly apparent. The minimum value of c_0, $c_{0_{min}}$, and the critical wavelength kc associated with it, satisfy the equations:

$$c_{0\,min}^2 = \sqrt{\frac{4\gamma g(\rho_2 - \rho_1)}{(\rho_1 + \rho_2)^2}} \quad (11.66a) \qquad \text{and:} \qquad k_c = \sqrt{\frac{g(\rho_2 - \rho_1)}{\gamma}}. \quad (11.66b)$$

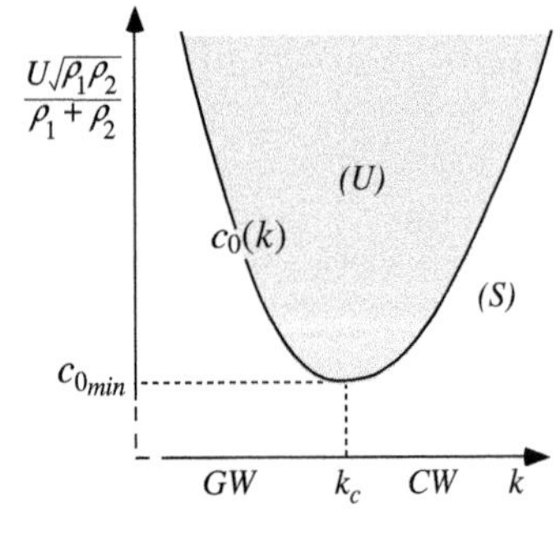

Figure 11.19 *Stability diagram for the Kelvin–Helmholtz instability as a function of the wave vector k. Along the solid-line (representing the marginal-stability curve), the velocity $c_0(k)$ of the surface waves is given by the right-hand side of Equation 11.65. (U) represents the domain of instability; GW and CW are those regions in which the stabilizing effects of gravity and capillary forces respectively dominate*

Note that k_c is the reciprocal of the capillary length introduced in Section 1.4.4.

The set of threshold values of $U\sqrt{\rho_1\rho_2}/(\rho_1 + \rho_2)$ corresponding to different wave vectors is represented by the solid curve on Figure 11.19. This curve represents the limit of linear stability of the problem analogous to the curve displayed in Figure 11.9 for the Rayleigh–Bénard instability. The minimum of the curve (and thus of the velocity U required to observe the instability) corresponds to a *critical wave vector* k_c for which gravitational and capillary effects are of comparable magnitude. The inner region (U) is the domain where perturbations with a given wave vector are amplified. In the region (S) outside this curve, gravity and capillary waves excited by external perturbations are damped by viscous effects and their amplitude decreases exponentially with time, as soon as the excitation stops.

Experimental measurements

This curve can be determined experimentally by using a water-filled channel above which air is blown parallel to the free surface, at a selectable velocity. By means of a movable plate, located at one end of the channel, a local surface oscillation of adjustable frequency is "forced", simultaneously determining the wave vector. For each driving frequency, the minimum wind velocity at which an instability appears is measured: this minimum is the velocity for which the amplitude of the waves increases as a function of the distance from the point of excitation. Such measurements lead to the determination of the equivalent of the stability curve shown in Figure 11.19 (with the wave vector k replaced by the driving frequency).

Kelvin–Helmholtz instability in a non-linear regime

The above linear analysis does not apply to large amplitude regimes for which a description in terms of the exponential growth of a single wave vector mode is insufficient. In order to understand the subsequent evolution of the instability, one uses a discrete representation of the velocity discontinuity as a system of parallel vortex lines (a method previously used in Section 7.4.2). Such techniques are frequently used for numerical calculations: the computation of the non-linear evolution of the Kelvin–Helmholtz flow was first carried out as early as in the1930s. Let us consider, for example, the behavior of four, parallel, vortex lines A, B, C and D: these lines are part of a sheet of linear vortices (of core perpendicular to the plane of the figure) with a contour displaying a bump at B. In order to study the evolution of the shape of this bump, one looks at the effect, on the vortex at B, of two vortices placed symmetrically at A and C (Figure 11.20a).

As a consequence of the deformation of the shape of the boundary, the field induced by these vortices has a horizontal component which causes the vortex at B to move closer to that at C; similarly, the vortex at D also comes closer to C. Thus, the vorticity is increased in the neighborhood of C while, around A, it is weakened. As a result of the additional flow due to this vorticity, the interface becomes steeper, and takes on the form shown in Figure 11.20b (it is referred to as a *type-N wave*, a reference to the asymmetric form of that letter). Finally, the surface can break, just like ocean waves.

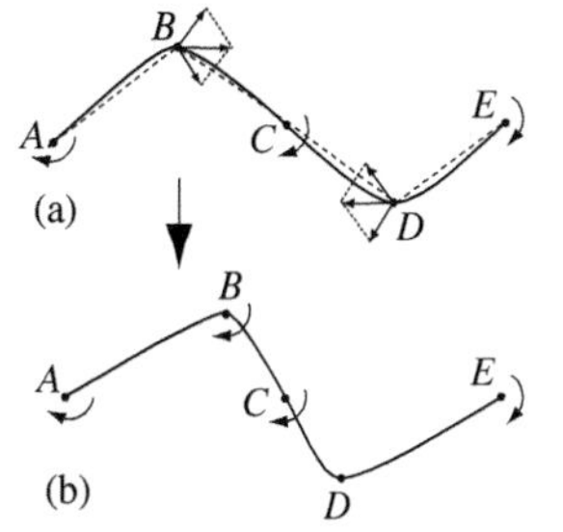

Figure 11.20 *Evolution of a set of four identical vortex filaments, normal to the plane of the figure, and which are part of a vortex sheet; (a) initial shape of the vortex sheet after a bump has appeared on it; (b) evolution of the shape of the sheet resulting from the interaction between the velocity fields of the vortices*

11.4.2 Role of the shape of the velocity profile for open flows

For the Kelvin–Helmholtz instability discussed just above, the basic shear flow displays an inflection point: such flows were shown to be always unstable, absent the effects of gravity, capillarity and viscosity. We will now show that, on the other hand, flows for which the velocity profile does not display an inflection point are stable, provided the effect of viscosity is neglected: this is, for example, the case for Poiseuille-type flows, or those inside a boundary layer.

However, for viscous fluids, even flow profiles without an inflection point can be unstable. This may seem paradoxical since viscosity had always appeared previously as a stabilizing parameter: in contrast, in this case, momentum transfers due to the viscosity are the driving force for the instability for high values of the Reynolds number.

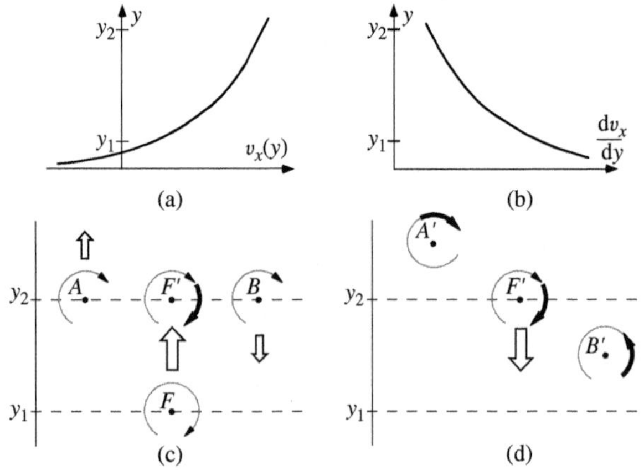

Figure 11.21 *Evolution of the perturbation in the parallel flow of an ideal fluid for the case where the variation of the fluid velocity $v_x(y)$ with the distance y, normal to the flow has no points of inflection: (a) Velocity profile for $v_x(y)$. (b) Variation of the derivative dv_x/dy of the velocity profile as a function of y. (c) Perturbation of the flow due to the displacement of an element of fluid from F to F′. (d) Velocity field resulting from the elements of fluid A and B, displaced to their new positions A′ and B′. On the schematic diagrams (c) and (d), the fine-print arrows represent the local vorticity, those in bolder print the additional vorticity representing the local difference between the displaced element of fluid and the local vorticity field*

Justification

Consider now a two-dimensional flow with a velocity parallel the x-direction: the exact velocity profile $v_x(y)$ is not known but one assumes that its curvature d^2v_x/dy^2 (negative in the case of Figure 11.21a) does not change sign for any value of y. The vorticity $\omega_z = -dv_x/dy$ varies then monotonically with the coordinate y (Figure 11.21b).

One displaces now an element of fluid by a small distance $y_2 - y_1$ from point F to point F' (Figure 11.21c). If viscosity effects are neglected, the vorticity is only carried along, and cannot diffuse, as previously seen in Section 7.3.1. One assumes therefore that the fluid particle retains its initial vorticity when it reaches the point F': it appears, relative to the equilibrium profile, as a little vortex with an effective vorticity (bold-face arrow on the figure) proportional to the difference between the values of ω_z at points F and F' (fine-print arrows). This vortex, in turn, acts on the elements of fluid A and B, located near F': A and B are then displaced in opposite directions toward the points A' and B', also retaining their own initial vorticity (Figure 11.21d). As a result, each of them is out of equilibrium with the surrounding fluid: this process is thus equivalent to the appearance of two opposite vortices at A' and B' (their effective vorticity is once again displayed by the bold-face arrow on the figure). The direction of the velocity that these vortices induce at F' is such that it pushes back toward F the fluid element considered initially.

A velocity profile without a point of inflection is therefore stable relative to an infinitesimal perturbation in the limit that the fluid can be considered as ideal, so that only convection effects can affect the vorticity. This conclusion clearly no longer applies when one includes effects of the viscous diffusion of the vorticity.

11.4.3 Sub-critical instabilities for Poiseuille and Couette flows

The Kelvin–Helmholtz instability displays supercritical characteristics, at least for sufficiently deep fluid layers, and develops thereafter without displaying hysteresis effects. It will be seen that this is not the case for open flows such as those of the Poiseuille and Couette type. Just as in the Bénard–Marangoni instability (Section 11.3.1), the instabilities of these flows can be subcritical and perturbations of high enough amplitude can cause a sudden transition toward an unstable regime, starting from rest, below a linear threshold. For such flows, instabilities can show up by the appearance of structures which propagate with the flow (one then refers to a *convective regime* of the instabilities) or which appear in a given region and persist there (an *absolute regime*).

Poiseuille flow in a circular tube

In his study of the transition toward turbulence in a cylindrical tube discussed in Section 2.4.1, O. Reynolds was the first to show evidence of such effects; this problem is however still poorly understood. The *subcritical* nature of these instabilities is made more clear by the fact that the critical Reynolds number corresponding to their occurrence has been shown to be infinite, if the fluctuations in the flow are infinitesimal. A finite amplitude perturbation is then needed to trigger instabilities leading to turbulence. Measurements by means of an experimental set-up similar to that of Reynolds have shown that, in the range $1{,}500 < Re < 1{,}750$, one can induce localized perturbations in the flow with a length of about 20 diameters: these *turbulent puffs* coexist with the Poiseuille flow and decay faster with distance as Re is farther below 1,750. Between $Re = 1{,}750$ and $Re = 2{,}200$ (typically), there are *turbulent plugs* (already observed by Reynolds) of constant length which propagate at a velocity of the same order as the average flow. When the Reynolds number increases further, these *turbulent plugs* end up occupying the entire flow. The critical Reynolds numbers for the spontaneous transition to turbulence can vary from a few thousands to about 10^5 depending on external conditions in the experiment (vibrations, conditions at the entry point, roughness of the walls, etc.). Current studies using PIV techniques and the introduction of controlled perturbations, allow for a detailed analysis of these unstable structures.

Poiseuille flow between parallel plates

On the other hand, for Poiseuille flow between two parallel plates, there exists a well-defined threshold for the appearance of infinitesimal instabilities ($Re_c = 5{,}900$): this is known as the *linear threshold*, since linear approximations can be used to predict the evolution of such perturbations. However, it is equally possible for *finite amplitude* perturbations to be amplified for a Reynolds number (of order 1,000) much lower than the above value. In fact, experimentally, the transition to well-developed turbulence throughout the volume occurs in a manner similar to the case of a circular tube.

Plane Couette flow

Plane Couette flow also displays subcritical characteristics. There is a non-linear threshold $Re_{NL} \approx 1{,}500$, representing the minimum velocity for large amplitude perturbations to be amplified ($Re = \Delta U\, h/\nu$, where h is the distance between the planes, and ΔU their relative velocity). As one approaches this threshold from higher values, one observes a continuous increase in the amplitude required to create a permanent turbulent structure. A minimum spatial extent of these perturbations is also required. At higher Reynolds numbers, the entire flow volume can be made turbulent. Experiments displaying these properties have been carried out between two parallel flat plates moving in opposite directions at the same absolute velocity: the velocity in the median plane is zero. This make the observation of the perturbations, which are then at rest, easier.

These results display analogies with first-order phase transitions (such as the transition from a liquid to gaseous state near a critical point). The transition from one state to another depends in this case, in addition to the amplitude of the perturbations, on a critical radius for nucleation of the unstable (turbulent) phase which grows from the stable state. They can be put in relation with the analysis of the B-M instability presented in Figure 11.13.

12 Turbulence

This book concludes with the most challenging aspect of the mechanics of fluids: turbulence which we have already encountered in various manifestations throughout this book. In Chapter 2, we have seen that the flow behind a cylindrical obstacle become turbulent at Reynolds numbers of a few hundreds. In Chapter 7, we have studied the physics of vortices, which are important elements of turbulence. Boundary layer separation (Chapter 10) produces turbulent wakes downstream of bodies placed within a flow. Finally, the study of hydrodynamic instabilities in Chapter 11 forms a beginning to the study of turbulence.

After a brief historical overview (Section 12.1), we derive (Section 12.2) the basic equations governing turbulence, by splitting up the local instantaneous velocity field into a mean flow and fluctuations; we apply the same procedure to other variables (vorticity, pressure, energy). The fluctuations are correlated and thus have a non vanishing contribution: the resulting turbulent transport mechanism plays a key role, both for energy and momentum transfer and for the transfers of thermal energy and solutes. We present then (Section 12.3) several empirical relationships between the transport of momentum by turbulent fluctuations and the gradient of the mean flow. We also discuss (Section 12.4) their application to free flows such as jets, wakes, or mixing layers.

In Section 12.5, we tackle turbulent flows bounded by solid walls. Momentum transport involves viscosity only in the immediate vicinity of the wall (viscous sublayer); then it takes place by means of increasingly larger vortices as one moves away from the wall (inertial layer).

Kolmogorov's theory (Section 12.6) applies, on the contrary, to homogeneous turbulence and uses the variations of the fluctuating quantities (velocity, vorticity ...) as a function of the size of the vortices to which they correspond: it describes, within three-dimensional flows, the transfer of the energy from the mean flow, and of its large-scale components, toward those corresponding to smaller spatial scales where the effect of viscosity becomes significant. One thus derives scaling laws characteristic of the transport, and encounters again vorticity amplification by the stretching of vortices already discussed in Chapter 7.

Finally, in Section 12.7, we will discuss two-dimensional flows, for which the mechanism of the stretching of vortices is non-existent, as well as the case of large, coherent, turbulent structures.

12.1 A long history

Turbulence makes up an essential chapter in the study of fluid flows. From the times of antiquity, it has been the object of imaged descriptions. The classical quotation "We never bathe twice in the same river", attributed to Heraclitus, refers to the everchanging

Physical Hydrodynamics. Second Edition. Etienne Guyon *et al.*

character – non-stationary and unpredictable – of turbulent flows. In *"De natura rerum"*, Lucretius gives a description of the surface of the sea, alternately calm, and very agitated. He introduces, following Epicurus, the idea of *"clinamen"* (in English, *deviation*), which refers to changes of the trajectories of *atoms* which, amplified, lead to the build-up of bodies in an unexpected manner, in the same way as the appearance of the turbulent motion of a fluid.

In the previous chapter, we have already discussed this *sensitivity to initial conditions*, as we studied hydrodynamic instabilities, which which lead to an unpredictable behavior.

Moreover, turbulent flows lead to the appearance of easily recognized geometrical forms, such as the vortices and filamentary structures drawn by *Leonardo da Vinci*, and which became associated with the modern idea of *coherent structures*. The Bénard–von Karman vortex-streets, discussed in Section 2.4.2, as well as the mixing-layer structures at the interface of two fluids moving at different velocities (Section 11.4.1), are examples of such structures. *Leonardo da Vinci* also gives a first description of transport within a turbulent flow, for which we borrow the transcription in the reference book by Uriel Frisch on this subject:

> *"Where the turbulence of water is generated;*
> *where the turbulence of the water maintains for long;*
> *where the turbulence comes to rest."*

We will see in Section 12.6 that this can be related to the *energy cascade*, known as *Kolgomorov's cascade*, which extends from the scale of large vortices to that of smaller and smaller ones, and to which we have referred while studying the Rankine vortex in Section 7.3.2. Finally, we have seen in Section 2.4.1 how, as early as 1880, Reynolds studied the transition between laminar and turbulent flows.

We might continue with this historical presentation... to state, in conclusion, that the problem of predicting turbulent flows, which has been the object of many studies as early as the end of the nineteenth century, remains to this day incompletely understood.

12.2 The fundamental equations

12.2.1 Statistical description of turbulent flows

Turbulent flows display random fluctuations of the velocity. For this reason, their study resembles the study of gases of particles in statistical mechanics. Rather than attempt to describe the velocity at every point **x** and at any instant of time t, one needs to focus on the probability of obtaining a given velocity at an ensemble of carefully chosen points. In practical terms, a complete determination of such probability distributions is impossible, and one is limited to the determination of moments of this distribution, such as the average or the mean-square deviation, as well as the correlation functions between the velocity fluctuations at neighboring points and times.

The variables describing a turbulent flow (velocity, pressure, temperature, composition...) can then be split up into average values, and fluctuations related to the turbulence, of zero-average value. Let us apply this decomposition, known as *Reynolds decomposition*, to the component v_i of the velocity in the form:

$$v_i = \bar{v}_i + v'_i \qquad \text{with:} \qquad \overline{v'_i} = 0. \tag{12.1}$$

We are now interested in deriving the laws which affect the variations of the average value $\bar{v}_i$ and those of the correlations between the components v'_i of the fluctuations. Strictly speaking, in order to determine these average values, one must create a large number N of configurations of the same flow with identical initial conditions and geometries, and carry out an *ensemble average* over the N values obtained for all these configurations:

$$\bar{v}_i(\mathbf{x}, t) = \lim_{N\to\infty} \frac{1}{N} \sum_1^N v_i^\alpha(\mathbf{x}, t), \tag{12.2}$$

where the index α varies from 1 to N and characterizes the specific configuration. This procedure is the only possible one when the flow is time-dependent, and one is interested in the temporal evolution of $\bar{v}_i(\mathbf{x}, t)$ (e.g., like for turbulence resulting from the motion of a grid through a fixed volume of fluid). Practically, this approach can however only be applied to flows easily reproducible, and for which the evolution time (e.g., the relaxation of the turbulent velocity toward zero), is not too long.

Such ensemble averages are not required if one deals with a statistically stationary flow. In such flows, the values of the velocity and the pressure fluctuate as a function of time, but their statistical properties (their probability distribution, average value, etc.) remain constant. One then assumes that turbulence satisfies the hypothesis of *ergodicity*: if one waits long enough, the flow passes through all possible states, and the time elapsed in each of these states is proportional to its probability. One infers from this the following definition for the component $\bar{v}_i$ of the average velocity of the flow at a given point **x:**

$$\bar{v}_i(\mathbf{x}) = \lim_{T\to\infty} \frac{1}{T} \int_{t_0}^{t_0+T} v_i(\mathbf{x}, t)\, \mathrm{d}t. \tag{12.3}$$

In order to obtain a significant value of the average, the averaging time T must be large relative to the characteristic duration τ of the slowest fluctuations in the velocity associated to turbulent motions.

If the mean flow is not rigorously stationary but evolves slowly, e.g. as a result of a variation in the external parameters, one can still define an average $\bar{v}_i(\mathbf{x}, t_0)$ for a given time t_0 by adapting the definition of Equation 12.3: for this, the averaging time T must be small relative to the characteristic time T_1 for the large scale evolution of the global flow. The combination of these two conditions for T implies that the duration τ of the slowest turbulent fluctuations must be much shorter than the time T_1: the mean flow must therefore be almost stationary, and one can then assume that the ergodicity assumption is valid.

In addition to the time averages of the components of the velocity and of the other fluctuating variables (pressure, temperature, etc.), one will also use the higher order moments of these variables (or the averages of products of these variables), which indicate the relationship between their variations at different locations and times. The study of such *space and time correlations* from instantaneous measurements of the velocity at two different locations is one of the major aspects of the experimental studies of turbulence, in relation with the development of statistical models.

The experimental devices used have been, among others, hot-wire and laser-doppler anemometers (Section 3.5.3). Other, more recent, techniques, such as laser-induced fluorescence (Section 3.5.2) or the velocimetry by particle images (Section 3.5.4), provide important new additional information on the geometry of turbulence.

12.2.2 Derivatives of average values

When an ensemble average is used, one finds, by taking a space or time derivative of Equation 12.2 and interchanging the order of the operations of summation and differentiation:

$$\frac{\partial \bar{v}_i}{\partial x_j} = \overline{\frac{\partial v_i}{\partial x_j}} \tag{12.4}$$

and:

$$\frac{\partial \bar{v}_i}{\partial t} = \overline{\frac{\partial v_i}{\partial t}}. \tag{12.5}$$

When a time average is used instead, one obtains the same results for the space derivatives by applying the same procedure to Equation 12.3. For the time-derivative, Equation 12.5

is valid only if the condition (discussed above) of a characteristic time τ of the fluctuations much smaller than that for the evolution of the flow is satisfied.

Proof

The time-derivative of the average velocity obeys, by splitting v_i into $\bar{v}_i + v'_i$:

$$\overline{\frac{\partial v_i}{\partial t}} = \lim_{\delta t \to 0} \overline{\left(\frac{v_i(t_0 + \delta t) - v_i(t_0)}{\delta t} \right)} = \lim_{\delta t \to 0} \left(\frac{\bar{v}_i(t_0 + \delta t) - \bar{v}_i(t_0)}{\delta t} + \frac{\overline{v'_i(t_0 + \delta t)} - \overline{v'_i(t_0)}}{\delta t} \right). \tag{12.6}$$

The two averages of the fluctuations of the term on the right-hand side will cancel out if the averaging time T is much greater than τ. On the other hand, one can assume that the term on the left-hand side equals $\overline{\partial v_i / \partial t}$ if T is sufficiently small relative to the global evolution time T_1. Taking $\delta t = T$, one then obtains a significant value for this derivative.

12.2.3 Governing equations of turbulent flows

Reynolds Equation

The Navier–Stokes equation continues to apply to the instantaneous velocity v_i so long as the length scale of the smallest turbulent motion of particles in the fluid is large relative to the mean-free-path of individual molecules (the validity of this assumption is discussed further down). One also assumes, as in the remainder of this book, that the fluid is incompressible.

The equation of motion, known as *Reynolds equation*, obeyed by the average value $\bar{v}_i$ of the velocity, is as follows:

$$\frac{\partial \bar{v}_i}{\partial t} + \bar{v}_j \frac{\partial \bar{v}_i}{\partial x_j} + \overline{v'_j \frac{\partial v'_i}{\partial x_j}} = -\frac{1}{\rho} \frac{\partial \bar{p}}{\partial x_i} + \nu \frac{\partial^2 \bar{v}_i}{\partial x_j^2} + f_i. \tag{12.7}$$

In Equation 12.7, one sums over the indices j which appear twice in a term.

This so-called Reynolds equation differs only in the term $\overline{v'_j \, \partial v'_i / \partial x_j}$ from that obtained by replacing simply, in the Navier–Stokes equation, p and v_i by their averages $\bar{p}$ and $\bar{v}_i$.

Proof

Using the splitting of Equation 12.1 in the Navier–Stokes equation normalized by the density ρ, one obtains:

$$\frac{\partial}{\partial t}\left(\bar{v}_i + v'_i\right) + \left(\bar{v}_j + v'_j\right) \frac{\partial}{\partial x_j}\left(\bar{v}_i + v'_i\right) = -\frac{1}{\rho} \frac{\partial}{\partial x_i}\left(\bar{p} + p'\right) + \nu \frac{\partial^2}{\partial x_j^2}\left(\bar{v}_i + v'_i\right) + f_i. \tag{12.8}$$

As above, the Einstein convention of implicit summation over the index j is used; Equation 12.8 is then averaged over time. All terms in which the fluctuations p' or v'_i appear only once vanish when they are averaged. Only the cross-product $\overline{v'_j \, \partial v'_i / \partial x_j}$ and the terms which only include the average values of v_i and p are non-zero, resulting in Equation 12.7.

This chapter deals only with incompressible turbulence and the incompressibility condition, $\nabla \cdot \mathbf{v} = 0$, applies both to the average and fluctuating fields. These two conditions can be written as:

$$\frac{\partial}{\partial x_j}\left(\bar{v}_j\right) = 0 \tag{12.9}$$

and:

$$\frac{\partial}{\partial x_j}\left(v'_j\right) = 0, \tag{12.10}$$

where Equation 12.10 is valid at every instant of time.

Just like the usual Navier–Stokes equation, Equation 12.7 can be transformed into an equation for momentum conservation:

$$\rho \frac{\partial \bar{v}_i}{\partial t} = \frac{\partial}{\partial x_j} \left(\overline{\sigma_{ij}} - \rho\, \bar{v}_i\, \bar{v}_j - \rho\, \overline{v'_i v'_j} \right) + \rho f_i, \tag{12.11a}$$

where:

$$\bar{\sigma}_{ij} = -\bar{p}\, \delta_{ij} + \rho\, \nu \left(\frac{\partial \bar{v}_i}{\partial x_j} + \frac{\partial \bar{v}_j}{\partial x_i} \right). \tag{12.11b}$$

Proof

Equations (12.11) are obtained by multiplying Equation 12.7 by ρ, rewriting the cross-product by using Equation 12.10:

$$\overline{v'_j \frac{\partial}{\partial x_j} v'_i} = \frac{\partial}{\partial x_j} \overline{v'_i v'_j} - \overline{v'_i \frac{\partial}{\partial x_j} v'_j} = \frac{\partial}{\partial x_j} \overline{v'_i v'_j},$$

and carrying out this same operation on the term $\bar{v}_j\, \partial \bar{v}_i / \partial x_j$.

The term:

$$\tau_{ij} = -\rho\, \overline{v'_i\, v'_j}. \tag{12.11c}$$

in Equation 12.11a is the *Reynolds tensor.* The average $\overline{\Pi_{ij}}$ of the momentum flux $\Pi_{ij} = \rho\, v_i\, v_j + p\, \delta_{ij} - \sigma'_{ij}$ introduced in Section 5.2.2 (Equation 5.11) obeys:

$$\overline{\Pi_{ij}} = -\overline{\sigma_{ij}} + \rho\, \bar{v}_i\, \bar{v}_j + \rho\, \overline{v'_i\, v'_j} = -\overline{\sigma'}_{ij} + \bar{p}\, \delta_{ij} + \rho\, \bar{v}_i \bar{v}_j + \rho\, \overline{v'_i v'_j}. \tag{12.12}$$

Again, Equations 12.11a and 12.12 obeyed by the averages $\bar{v}_i$ and $\bar{p}$, are virtually identical to the equations obtained by replacing, in the corresponding Equations 5.6 and 5.11, v_i and p by their average values. The additional term, fundamental for turbulent transport, is the Reynolds tensor.

The meaning of the Reynolds tensor

The term $\tau_{ij} = -\rho\, \overline{v'_i v'_j}$, characterizing the correlation between the instantaneous components of the fluctuating velocity, accounts for the momentum transport by means of the turbulent fluctuations. It is also known as the *turbulent stress tensor* and is a symmetric tensor. It has therefore three diagonal components τ_{ii}, and three off-diagonal components $\tau_{i \neq j}$ which play a particularly important role: for a turbulent flow in a cylindrical tube, or between two planes, they ensure the transport towards the walls of the momentum parallel to the mean flow (Section 12.5.2). In this way, they are similar to the off-diagonal components of the viscosity stress tensor discussed in Chapter 4.

Turbulent transport of thermal energy or matter

Up to this point, we have attempted to derive equations obeyed by the average velocity of the turbulent flow which controls, consequently, the momentum transport. One can handle in the same way the problem of thermal transport in turbulent flow. Thermal transport by turbulent fluctuations of the velocity has some common features with the kinetic-theory model of gases discussed in Chapter 1 (Section 1.3.2). In this last case, the transfer was associated with the adiabatic displacement of the molecules resulting from their thermal random motion within a temperature gradient. Here, this thermal agitation is replaced by the turbulent velocity fluctuations.

In the general case, the equation of thermal transport in a flowing fluid in the absence of a heat source is:

$$\frac{\partial T}{\partial t} + v_j \frac{\partial T}{\partial x_j} = \kappa \frac{\partial^2 T}{\partial x_j^2}, \qquad (12.13)$$

where v_j is the local velocity of the fluid and $\kappa = k/(\rho C_p)$ is its *thermal diffusivity* (C_p is the thermal heat capacity per unit mass, ρ the density, and k the thermal conductivity). Just as in the case of the velocity, we frequently have, in a turbulent flow, fluctuations of the temperature T, which we can split up in the form $T = \overline{T} + T'$. The space and time derivatives of $\overline{T}$ are considered, just as in the case of the velocity, as equal to the average values of the derivatives. By splitting both T and the velocity $v_i = \bar{v}_i + v'_i$, and averaging the thermal transport equation, one obtains an equation equivalent to the Reynolds equation:

$$\frac{\partial \bar{T}}{\partial t} = -\frac{\partial}{\partial x_j}\left(\bar{v}_j\, \bar{T} + \overline{v'_j\, T'}\right) + \kappa \frac{\partial^2 \bar{T}}{\partial x_j^2}. \qquad (12.14)$$

The average $\overline{v'_j\, T'}$ plays a role equivalent to the Reynolds tensor and accounts for the global thermal transfer due to the transport of temperature fluctuations by those of the velocity. From Equation 12.14, one infers, as above, the following expression of the thermal flux $\mathbf{Q}_T$ within the turbulent flow:

$$Q_{Tj} = -k\frac{\partial \bar{T}}{\partial x_j} + \rho\, C_p\, \bar{v}_j \bar{T} + \rho\, C_p\, \overline{v'_j\, T'}. \qquad (12.15)$$

One considers now, instead of thermal transport, that of a soluble component (ion, dye, radioactive tracer, chemical pollutant, etc.) with a concentration C expressed as the mass of solute per unit volume of solvent. In the same way, one obtains the following equation for the mass flux $\mathbf{Q}_m$ of the solute per unit area and time, in the absence of a source term (e.g., a chemical reaction):

$$Q_{mj} = -D_m \frac{\partial \bar{C}}{\partial x_j} + \bar{v}_j\, \bar{C} + \overline{v'_j\, C'}. \qquad (12.16)$$

The coefficient D_m is the molecular diffusivity associated with the quantity involved: for turbulent flows, the corresponding term contributes generally weakly to Q_{mj}. This is the idea which is used naturally while stirring a cup of coffee after adding cream!

The ratio Pe_t between the turbulent and molecular diffusion terms can be considered, whether one refers to a thermal transfer or a mass transfer, as a *turbulent Péclet number*; using the same definition as in Section 2.3.2. Its order of magnitude is $Pe_t \approx 10^4$, for the previous example.

Example: Let us compare, for an ionic solution, the orders of magnitude of the flux components associated with the different mechanisms contributing to $\mathbf{Q}_m$. The typical value of D_m for a simple ion is $D_m = 10^{-9}$ m^2/s. Let us assume that, for an average value $\overline{C}$ of the concentration, there is an average concentration-gradient corresponding to a variation of the order of $10^{-1}\, C$ across a container of size 0.1 m. Let us now generate in this container, for example with a large spoon, a turbulent mixing flow which results in turbulent velocity fluctuations of the order of 10^{-2} m/s, and concentration fluctuations of the order of $10^{-1}\, \overline{C}$. The orders of magnitude of the mass transport terms by turbulent fluctuations and by molecular diffusion (if the fluid had remained at rest) are respectively $\overline{v'\, C'} \approx 10^{-3}\, \overline{C}$ m/s and $D_m \partial \overline{C}/\partial x_j \approx 10^{-7}\, \overline{C}$ m/s. Turbulent diffusion is clearly much more efficient than molecular diffusion.

12.2.4 Energy balance in a turbulent flow

The kinetic energy of a turbulent flow is the sum of the energy of the mean flow and of the turbulent fluctuations. Effectively, when we average $\left(\bar{v}_i + v'_i\right)^2$, the cross terms $\bar{v}_i v'_i$ in the expansion of the square average out to zero, such that $\overline{v_i^2} = \bar{v}_i^2 + \overline{v_i'^2}$. We will thus write out separately the balance equations for these two components in order to understand the change in the total kinetic energy.

Changes in the kinetic energy of the mean flow in the absence of gravity

To calculate this balance, one multiplies the Navier–Stokes equation by $\bar{v}_i$, sums over the different components and then takes averages (the volume force f_i is assumed to be zero). By splitting up each velocity component as in Equation 12.1, dividing by ρ in order to obtain the energy per unit mass, and eliminating those terms where the fluctuations only occur in a single factor so that their average is zero, one obtains:

$$\frac{\partial}{\partial t}\left(\frac{\bar{v}_i^2}{2}\right) + \bar{v}_i\,\bar{v}_j\frac{\partial\,\bar{v}_i}{\partial x_j} = -\bar{v}_i\frac{\partial\bar{p}}{\partial x_i} - \bar{v}_i\,\overline{v_j'\frac{\partial v_i'}{\partial x_j}} + \nu\,\bar{v}_i\frac{\partial^2\bar{v}_i}{\partial x_j^2}. \qquad (12.17)$$

Using the incompressibility condition (Equations 12.9 and 12.10), Equation 12.17 can be rewritten in the form:

$$\frac{\partial}{\partial t}\left(\frac{\bar{v}_i^2}{2}\right) + \bar{v}_j\frac{\partial}{\partial x_j}\left(\frac{\bar{v}_i^2}{2}\right) = \frac{\partial}{\partial x_j}\left(-\bar{v}_i\overline{v_i'v_j'} - \frac{\bar{p}\,\bar{v}_j}{\rho} + \nu\,\bar{v}_i\frac{\partial\bar{v}_i}{\partial x_j}\right) - \nu\left(\frac{\partial\bar{v}_i}{\partial x_j}\right)^2 + \overline{v_i'v_j'}\,\frac{\partial\,\bar{v}_i}{\partial x_j}. \qquad (12.18)$$

On the left-hand side:

- the term $\partial/\partial t\left(\bar{v}_i^2/2\right)$ corresponds to the time variation of the average energy. It is non-zero if the mean flow is not stationary.
- the term $\bar{v}_j\,\partial/\partial x_j\left(\bar{v}_i^2\right)$ corresponds to the convection by the mean flow of the spatial gradient of the average kinetic energy. By grouping this term with the first one, one obtains a kind of Lagrangian derivative but constructed by means of the average velocity and not of the total velocity.

The right-hand side of the equation includes the following contributions of the various mechanisms:

- the first three terms (grouped within the same parentheses) represent the divergence of the flux terms from different origins (turbulent fluctuations, pressure, viscosity). Their integral over a control volume corresponds to the work done by the corresponding stresses.
- the term $-\nu\left(\partial\,\bar{v}_i/\partial x_j\right)^2$ expresses, by reference to Section 5.3.1, the loss of the energy in the mean flow by viscous dissipation (this term is obtained by replacing v_i by $\bar{v}_i$ in Equation 5.26).
- the last term $\overline{v_i'v_j'}\,\partial\,\bar{v}_i/\partial x_j$ of the equation represents the transfer of energy between the mean flow and the turbulent fluctuations: it involves at the same time the gradient of the mean flow and the Reynolds tensor $\tau_{ij} = -\rho\,\overline{v_i'\,v_j'}$ for the transport of momentum by turbulent fluctuations.

Changes in the kinetic energy of the turbulent fluctuations in the absence of gravity

One proceeds exactly as for obtaining Equations 12.17 and 12.18, but the Navier–Stokes equation is multiplied by v_i' instead of $\bar{v}_i$ before summing over the indices i and j and averaging. By proceeding just as before, to regroup the different terms of the equations thus obtained, one obtains:

$$\frac{\partial}{\partial t}\left(\frac{\overline{v_i'^2}}{2}\right)+\bar{v}_j\frac{\partial}{\partial x_j}\left(\frac{\overline{v_i'^2}}{2}\right)=\frac{\partial}{\partial x_j}\left(-\frac{\overline{v_i'^2 v_j'}}{2}-\frac{\overline{p' v_j'}}{\rho}+\nu\,\overline{v_i'\frac{\partial v_i'}{\partial x_j}}\right)-\nu\overline{\left(\frac{\partial v_i'}{\partial x_j}\right)^2}-\overline{v_i' v_j'}\frac{\partial \bar{v}_i}{\partial x_j}. \quad (12.19)$$

This equation is similar in form to Equation 12.18, replacing the average values by their fluctuating components. On the left hand side:

- the term $\partial\left(\overline{v_i'^2}/2\right)/\partial t$ characterizes the lack of stationarity of the energy of the turbulent fluctuations;
- the term $\bar{v}_j\,\partial\left(\overline{v_i'^2}/2\right)/\partial x_j$ indicates, as in Equation 12.18, a convection by the mean flow but, here, of the gradient in the energy of the turbulent fluctuations. We could also group it with the first term in order to obtain a particular term of the Lagrangian derivative (based on the convection by the mean flow).

On the right-hand side:

- the first three terms, all in the form of a divergence, also represent, after integrating over an element of volume, the work done by the turbulent pressure and viscous stresses. These stresses are fluctuating but yield non-zero contributions, because they are multiplied by fluctuating velocity components;
- the term $-\nu\overline{\left(\partial v_i'/\partial x_j\right)^2}$ expresses, again by reference to Section 5.3.1, the loss of energy of the turbulent fluctuations by viscous dissipation.
- the term $-\overline{v_i' v_j'}\,\partial\,\bar{v}_i/\partial x_j$ is the same as that in Equation 12.18, but it has the opposite sign: this confirms its role in the exchange of energy between the mean flow and the fluctuations.

We will see further on that the velocity gradients associated with vortices (particularly those of small size) are much greater than those associated with the mean flow, except very close to the walls (Section 12.6.1). The term $-\nu\overline{\left(\partial v_i'/\partial x_j\right)^2}$ for the viscous dissipation of turbulent fluctuations in Equation 12.19 is then much greater than the corresponding term $-\nu\left(\partial \bar{v}_i/\partial x_j\right)^2$ for the mean flow. Regarding the relative order of magnitude of the last two terms in Equation 12.18, one shifts from one to the other by substituting in $-\nu\left(\partial \bar{v}_i/\partial x_j\right)^2$ the factor $-\nu\left(\partial \bar{v}_i/\partial x_j\right)$ (viscous stressses associated with the mean flow) by $\overline{v_i' v_j'}$ (turbulent stresses). Here, also, the turbulent stresses are generally dominant except, as will be seen, very close to the walls.

12.2.5 Transport of the vorticity in a turbulent flow

Balance of the vorticity of the turbulent fluctations

Just as for the velocity and pressure fields, one splits up the vorticity $\boldsymbol{\omega} = \nabla \times \mathbf{v}$ into $\boldsymbol{\omega} = \bar{\boldsymbol{\omega}} + \boldsymbol{\omega}'$. The fluctuation terms ω'_i correspond to combinations of the gradients of the components of the fluctuations of the velocity: in absolute value, they are generally much greater than the terms $\bar{\omega}_i$ (see Section 12.6.1).

One is thus interested only in the equation of conservation of the components ω'_i, or rather, since their average value is zero, in the quantity $\omega_i'^2/2$ (summed over i), known as the *enstrophy*. The equation of conservation of the enstrophy $\omega'^2/2 = \sum_i \omega'^2_i/2$ of the turbulent fluctuations is then derived by starting from Equation 7.42 for the transport of vorticity and proceeding as in the case of the energy and the momentum. Keeping only the dominant terms, one obtains:

$$\frac{\partial}{\partial t}\left(\frac{\omega'^2}{2}\right)+\bar{v}_j\frac{\partial}{\partial x_j}\left(\frac{\omega'^2}{2}\right)=\overline{\omega'_i\,\omega'_j\frac{\partial v_i'}{\partial x_j}}-\nu\overline{\left(\frac{\partial \omega_i'}{\partial x_j}\right)^2}. \quad (12.20a)$$

Proof

One multiplies Equation 7.42 by ω'_i and averages, using the same splitting as previously for v_i and neglecting the average vorticity $\bar{\omega}_i$ relative to ω'_i. The different terms become:

$$\overline{\omega'_j v_j \frac{\partial \omega_i}{\partial x_j}} = \bar{v}_j \frac{\partial}{\partial x_j}\left(\frac{\omega'^2_j}{2}\right) + \overline{v'_j \frac{\partial}{\partial x_j}\left(\frac{\omega'^2_i}{2}\right)} + \overline{\omega'_j v'_j}\frac{\partial \bar{\omega}_i}{\partial x_j},$$

where:

$$\overline{v'_j \frac{\partial}{\partial x_j}\left(\frac{\omega'^2_i}{2}\right)} = \frac{\partial}{\partial x_j}\left(\overline{v'_j \frac{\omega'^2_i}{2}}\right),$$

$$-\overline{\omega'_i \omega_j \frac{\partial v_i}{\partial x_j}} = -\overline{\omega'_i \omega'_j}\frac{\partial \bar{v}_i}{\partial x_j} - \overline{\omega'_i \omega'_j \frac{\partial v'_i}{\partial x_j}},$$

$$\nu\, \overline{\omega'_i \frac{\partial^2 \omega_i}{\partial x_j^2}} = \nu\, \overline{\omega'_i \frac{\partial^2 \omega'_i}{\partial x_j^2}} = \nu \frac{\partial}{\partial x_j}\overline{\left(\omega'_i \frac{\partial \omega'_i}{\partial x_j}\right)} - \nu\, \overline{\left(\frac{\partial \omega'_i}{\partial x_j}\right)^2} = \frac{\nu}{2}\frac{\partial^2 \overline{\omega'^2_i}}{\partial x_j^2} - \nu\, \overline{\left(\frac{\partial \omega'_i}{\partial x_j}\right)^2}.$$

Terms such as $\overline{\omega'_i \omega'_j}\, \partial \bar{v}_i / \partial x_j$ which include the derivative of terms already averaged are then considered negligible relative to terms of the type $\overline{\omega'_i \omega'_j\, \partial v'_i / \partial x_j}$. Taking these remarks into account, one finds the two dominant terms of the right-hand term of Equation 12.20a. For a quasi-stationary and homogeneous flow, the left-hand side of this equation vanishes, and it becomes:

$$\overline{\omega'_i \omega'_j \left(\partial v'_i / \partial x_j\right)} = \nu\, \overline{\left(\partial \omega'_i / \partial x_j\right)^2}. \tag{12.20b}$$

Equation 12.20a-b plays for the vorticity (and, through it, for the angular momentum) the role of the energy balance equation (Equation 12.19) for the kinetic energy. The term $-\nu\, \overline{\left(\partial \omega'_i / \partial x_j\right)^2}$ represents the vorticity dissipation by viscous diffusion. The second term $\overline{\omega'_i \omega'_j \left(\partial v'_i / \partial x_j\right)}$ results from the term $\omega_j\, \partial v_i / \partial x_j$ of Equation 7.42: this term corresponds to variations of the turbulent vorticity due to the stretching of the vorticity tubes by velocity fluctuations. In order to compensate for viscous diffusion, this term must be positive: there is thus, on average, more stretching than compression of the vorticity tubes. Thus, the larger ones are continuously being stretched into tubes of smaller cross-section: this process continues up to the point where the losses due to viscosity come into equilibrium with the increase in vorticity.

A similar equilibrium between stretching of the vorticity and viscous diffusion was discussed in Section 7.3.2, in the case of vortices created by emptying a container.

This particular model applies only to three-dimensional flows: for a purely two-dimensional one, the vorticity is always perpendicular to the plane of the flow, and the stretching term vanishes identically. In Section 12.7.3, this is shown to influence considerably the properties of the turbulence of two-dimensional flows.

The effectiveness of this stretching process is well demonstrated experimentally for turbulent flows between parallel disks turning in opposite directions around the same axis. Vortex filaments with a very localized core appear sporadically. However, these vortices become unstable and disappear soon after they have appeared. Such filaments are also observed in numerical simulations.

Divergence of the vorticity at a finite time in the absence of viscosity

In the absence of viscosity, the amplification of the vorticity by stretching would lead to a divergence of this vorticity within a finite time-interval. Let us go back to Equation 12.20a while neglecting the effect of viscosity. Dimensionally, the right hand term is of order ω^3. In a symbolic manner, the vorticity thus obeys the equation:

$$\frac{1}{2}\frac{d\omega^2}{dt} = A\,\omega^3,$$

where A is a constant of the order of unity. This equation has a solution of the form:

$$\frac{1}{\omega_0} - \frac{1}{\omega} = A\,t,$$

where ω_0 is the vorticity at an initial time $t = 0$, so that:

$$\omega = \frac{\omega_0}{1 - A\,\omega\,t} = \frac{\omega_0\,t_0}{t_0 - t}, \quad (12.21a) \qquad \text{with:} \qquad t_0 = \frac{1}{A\,\omega_0}. \quad (12.21b)$$

In this model, the vorticity ω diverges after a time t_0, known as *the catastrophe time*, which is of the same order of magnitude as the time it takes to flip, by rotation, the original vortex packet. Whether the effect of viscosity can inhibit such a divergence from occurring within a finite time-interval remains an open question.

12.3 Empirical expressions for the Reynolds tensor and applications to free flows

12.3.1 Closure of the Reynolds equation

The set of Equations 12.7 and 12.9 includes more unknowns than equations: in addition to the four unknowns $\bar{v}_i$ and $\bar{p}$ one has indeed the six components of τ_{ij} (the correlation functions). One finds the equations obeyed by τ_{ij} by multiplying Equations 12.7 by v'_i, and then taking their average. However, the resulting equations then contain higher order moments, which bring into play three fluctuating components. One might iterate once more this process, which would cause the appearance, within the same time-scale, of fourth order moments! One can then never obtain in this way a *closed* system containing as many equations as unknowns. This problem, due to the presence of a nonlinear term $(\mathbf{v}\cdot\nabla)\mathbf{v}$ in the equations of motion, is essential in the study of turbulence.

One has therefore to find approximate solutions by looking for possible expressions for τ_{ij} as a function of the average components of the flow velocity: this allows then to *close* the equations and to solve the averaged equations of motion. The major shortcoming of these approaches is that they are not based on rigorous theoretical grounds: they do not involve indeed physical variables with well defined values and clear significance, but they introduce adjustable parameters which need to be determined for each flow. Nonetheless, these equations, initially intended to be theoretical models of turbulence, remain very useful for the practical characterization of turbulent flows.

In such an approach, one seeks to relate statistical quantities (essentially first and second-order moments) obtained by averaging operations, rather than derive equations relating instantaneous values, valid at every instant of time and every point in space. One looks more specifically for relationships between second-order moments (such as the Reynolds tensor), and first-order moments, such as the gradient of the average velocity.

12.3.2 Eddy viscosity

This concept was in 1875 by Boussinesq in his "*Essai sur la théorie des eaux courantes*" (Essay on the theory of running waters).

In this approach, one accounts for the increase in friction and momentum transport in a turbulent flow by introducing a viscosity ν_t which acts in addition to the classical viscosity of fluids. In this approach, one writes, in analogy with the viscosity tensor, the tensor $\tau_{ij} = -\rho\, \overline{v'_i v'_j}$ for the turbulent stresses in the form:

$$\tau_{ij} = \rho\, \nu_t \left(\frac{\partial \bar{v}_i}{\partial x_j} + \frac{\partial \bar{v}_j}{\partial x_i} \right). \tag{12.22}$$

This concept has however no solid physical basis: this *eddy* (or *turbulent*) *viscosity* ν_t is independent of the properties of the fluid, but depends instead upon the nature of the flow as well as on the point at which it is measured. In fact, defining an eddy viscosity is equivalent to replace the problem of the variation of τ_{ij} as a function of the characteristics of the flow by that of the variation of ν_t.

Several models have been developed to estimate the eddy viscosity. A first one, discussed below, uses the concept of a mixing length. Another approach is the k-ε model, often used in numerical simulations and where ν_t is estimated from the energy of the turbulent fluctuations and of the energy being dissipated.

12.3.3 Mixing length

This approach was suggested by Prandtl in 1903: it is inspired by the kinetic theory of gases, and provides an estimate of the eddy viscosity ν_t. For that, one introduces a scaling length which plays the same role as the mean free path of molecules, and assumes that the fluctuating components of the velocity play the same role as the velocity of thermal motion (see Section 1.3).

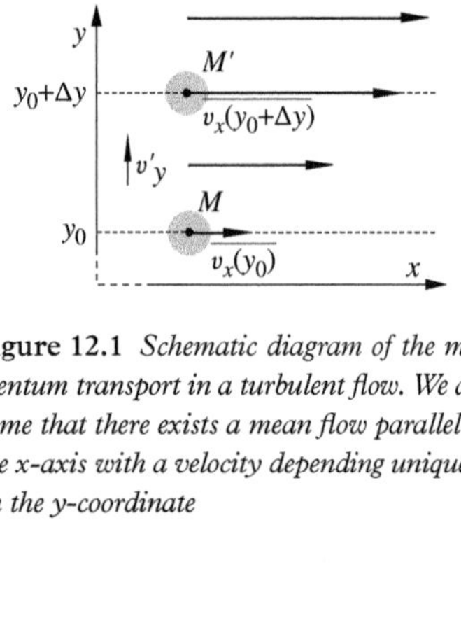

Figure 12.1 *Schematic diagram of the momentum transport in a turbulent flow. We assume that there exists a mean flow parallel to the x-axis with a velocity depending uniquely on the y-coordinate*

The gradient of the average velocity $\partial \bar{v}_x/\partial y$ is assumed to be positive (Figure 12.1); one also assumes that the fluid particles initially in the plane $y = y_0$ and moving toward the plane $y = y_0 + \Delta y$ under the action of a velocity fluctuation $v'_y > 0$ retain their initial velocity component in the x-direction which is of the order of $\bar{v}_x(y_0)$. The arrival of such a particle in the plane $y = y_0 + \Delta y$, where the average velocity $\bar{v}_x(y_0 + \Delta y)$ is higher appears then as a negative velocity fluctuation of the order of magnitude:

$$v'_x \approx \bar{v}_x(y_0) - \bar{v}_x(y_0 + \Delta y) = -\Delta y \frac{\partial \bar{v}_x}{\partial y}.$$

This displacement gives then an instantaneous contribution to the momentum transport $\rho\, v'_x v'_y \approx -\rho \Delta y\, v'_y (\partial \bar{v}_x/\partial y) < 0$; the same argument can be applied to a particle coming from above (e.g., going from the plane $y_0+\Delta y$ to the plane y_0) by replacing Δy by $-\Delta y$. In this case, one has $v'_y < 0$ and the instantaneous contribution to $\rho\, v'_x v'_y$ is again negative. There is therefore a correlation between the signs of v'_x and v'_y due the fact that the fluctuations v'_x are determined from v'_y and the spatial variations of the average velocity: one has therefore $\rho\, \overline{v'_x v'_y} < 0$ in a one-dimensional flow with a velocity gradient $\partial \bar{v}_x / \partial y > 0$.

Averaging over the various trajectories and possible velocities of an element of fluid, one obtains for the corresponding component of the Reynolds tensor:

$$\tau_{xy} = -\rho\, \overline{v'_x v'_y} = \rho \frac{\partial \bar{v}_x}{\partial y} \overline{\Delta y\, v'_y}, \quad (12.23a) \qquad \text{so that:} \quad \nu_t = \overline{\Delta y\, v'_y}. \quad (12.23b)$$

In the case of the mixing length model, ν_t is written in the form:

$$\nu_t = u^* \ell, \tag{12.24a}$$

where ℓ is the *mixing length*: in terms of Equation 12.23b and the preceding discussion, ℓ can be interpreted as the average distance over which the velocity of an element of fluid remains correlated to its initial value and ensures an effective transport of the momentum. The *characteristic velocity* u^* is of the same order of magnitude as the amplitude of the transverse velocity fluctuations $\sqrt{\overline{v'^2_y}}$.

Values of the mixing length

The values of ℓ and u^* used in practice vary from one kind of flow to another. Thus, we will see in Section 12.4 that, for free flows such as jets or turbulent wakes, the mixing length can be considered constant in the direction transverse to the mean flow, and of the order of the average width of this flow. For such flows, we generally relate the velocity u^* to the transverse profile $\bar{v}_x(y)$ of the average velocity by:

$$u^* = \ell \, | \partial \bar{v}_x / \partial y | . \tag{12.24b}$$

For flow near a wall (such as in a channel or for a turbulent boundary layer near a plane wall), the local value of ℓ is generally taken as proportional to the distance from the wall, particularly when this is the only available characteristic length. The velocity u^* is, in turn, directly related to the stress at the wall. In Section 12.5, it will be seen that one can then predict the logarithmic variation of the average velocity as a function of the distance to the wall which is observed experimentally. However, this agreement is essentially a reflection of dimensional constraints: at a given distance y from the wall, y is indeed the only characteristic length involved.

For quasi-parallel flows, a slightly more general definition, due to von Karman, relates ℓ to the curvature of the velocity profiles with: $\ell = -k | \partial \bar{v}_x / \partial y | / | \partial^2 \bar{v}_x / \partial y^2 |$.

Such simple assumptions cannot however be made for every flow, especially when there are scaling lengths and characteristic times. This represents one of the major problems of these models.

Validity of the analog between the mixing length model and the kinetic theory of gases

As just stated, the mixing length approach has similarities with the mean free path introduced in Sections 1.3.2 and 2.2.1 to calculate the viscosity (or the diffusivity) on the basis of the kinetic theory of gases: the velocity u^* plays, for momentum transport, the same role as the thermal motion velocity in kinetic theory. The essential difference is that, for the microscopic calculation of viscosity, there is a *scale separation* between small scales of random motions due to thermal agitation and the averaging distance used to calculate the viscosity. Such a separation does not exist in turbulent motion, for which the largest turbulent scales are of the same order of magnitude as the average motion. This lack of separability of scales in turbulent motion presents a major obstacle to attempts to relate the kinetic theory of gases to the study of turbulence.

12.3.4 Other practical approaches to turbulence

Many other models have been suggested to deal with turbulence, particularly in the perspective of numerical simulations.

The increase in computing power now allows one to simulate numerically turbulent flows by integration of the Navier–Stokes equation. Such *direct numerical simulations (DNS)* require, however, quite long calculation times on very powerful computers: they are thus limited, probably for another few years, to simple geometries and to moderate Reynolds numbers *Re*.

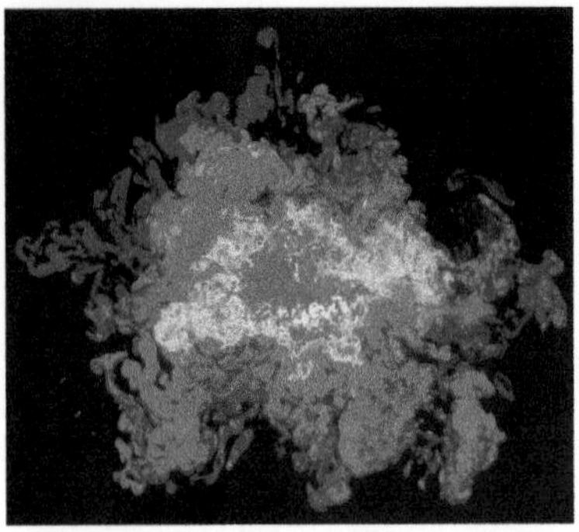

Figure 12.2C *Cross-sectional view of a turbulent jet lighted by a plane of light perpendicular to its axis and injected into a volume of the same fluid. The jet contains a fluorescent dye while the external fluid is pure. The colours indicate the concentration of the dye, increasing from blue to red (courtesy H.E. Catrakis et P.E. Dimotakis)*

For most practical applications, other approaches must be used. An important obstacle to this type of simulation comes from the great disparity between the size of large scale motions and that of the smallest vortices; this scale is smaller as *Re* is greater (Section 12.6.1). In order to simulate these at the same time as the global flow, the number of points that must be included in the simulation mesh (and the corresponding length of the calculation) increases very rapidly with *Re*.

However, the mean flow only exchanges energy and momentum with the largest vortices, and its geometry only affects these vortices. This has led to the development of *large eddy simulation (LES)* methods for which only motions of large and medium scale are directly simulated. Momentum and energy exchange at the level of small scale turbulence are, for their part, described by models similar to that of eddy viscosity.

12.4 Free turbulent flows: jets and wakes

The eddy viscosity models frequently provide good results for *free* turbulent flows such as jets, mixing layers or wakes behind a body. We will present a parallel discussion of the cases of the wake and jet: the latter is demonstrated visually in Figure 12.2 by means of the technique of *laser induced fluorescence (LIF)* described in Section 3.5.2.

12.4.1 Basic properties of two-dimensional and turbulent jets and wakes

Equations of motion

Let us first discuss the cases of a two-dimensional turbulent wake behind a cylindrical obstacle in a uniform, non-turbulent flow at velocity $\mathbf{U}$ (Figure 12.3a) and of a jet emitted into a stationary fluid from a slit, very long in the z-direction (Figure 12.3b). For both cases, the downstream distance x is assumed to be large enough so that the flow is insensitive to the exact geometry of the obstacle or slit. The perturbation to the external flow of velocity $\mathbf{U}$ (with $\mathbf{U} = 0$ for the jet) is also assumed to be limited to a region of finite characteristic transverse dimension $e(x)$: this half-width of the wake (or the jet) is assumed to be small relative to x. Finally, these flows are not influenced by the walls, so that the viscous terms $\eta \partial^2 \bar{v}_i / \partial x_j^2$ are neglected in the equation of motion throughout the volume of the fluid.

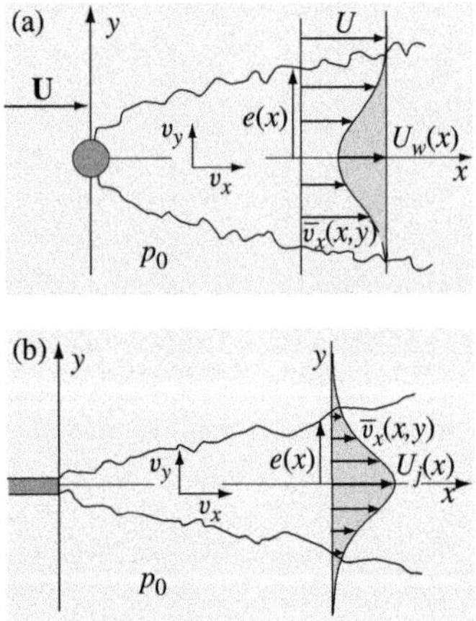

Figure 12.3 *Schematic diagram of the velocity profiles (a) in a wake behind an infinitely long cylinder, and (b) in a jet emitted from a slit, virtually infinitely long in the z-direction*

In the wake, the *velocity deficit* $U - \bar{v}_x$ is assumed to be small relative to U. On the other hand, the conservation equation (Equation 12.9) and the condition $e(x) \ll x$ imply that the transverse component $\bar{v}_x(x, y)$ of the average velocity is $\ll U - \bar{v}_x$ (or $\ll \bar{v}_x$ for the jet). For both cases the velocity and pressure p_0 are constant outside the perturbation zone ($y \gg e(x)$).

One assumes that the flow is statistically stationary in the sense discussed in Section 12.2.1 and invariant in the z-direction. Assuming also that the volume forces vanish, and neglecting the viscous terms, the equations for momentum transport (Equation 12.11) can be written in the form:

$$\bar{v}_x \frac{\partial \bar{v}_x}{\partial x} + \bar{v}_y \frac{\partial \bar{v}_x}{\partial y} = -\frac{1}{\rho}\frac{\partial \bar{p}}{\partial x} - \frac{\partial}{\partial y}\left(\overline{v'_x v'_y}\right) - \frac{\partial}{\partial x}\left(\overline{v'^2_x}\right), \tag{12.25a}$$

and:

$$\bar{v}_x \frac{\partial \bar{v}_y}{\partial x} + \bar{v}_y \frac{\partial \bar{v}_y}{\partial y} = -\frac{1}{\rho}\frac{\partial \bar{p}}{\partial y} - \frac{\partial}{\partial x}\left(\overline{v'_x v'_y}\right) - \frac{\partial}{\partial y}\left(\overline{v'^2_y}\right). \tag{12.25b}$$

Equation 12.25a can be simplified, for the wake, to:

$$U\frac{\partial \bar{v}_x}{\partial x} = -\frac{\partial}{\partial y}\left(\overline{v'_x v'_y}\right), \tag{12.26a}$$

and, for the jet, to:

$$\frac{\partial}{\partial x}\left(\bar{v}_x^2\right) + \frac{\partial}{\partial y}\left(\bar{v}_x\bar{v}_y\right) = -\frac{\partial}{\partial y}\left(\overline{v'_x v'_y}\right). \tag{12.26b}$$

Equation (12.26a) is very similar to the 2D version of Equation 10.72 valid for a laminar jet: the viscous stress $\eta \partial v_x/\partial y$ is simply replaced by the corresponding component $-\rho\,\overline{v'_x v'_y}$ of the Reynolds tensor.

These equations express the equilibrium between the variations of the longitudinal momentum flux of the mean flow and the transverse momentum flux due both to fluctuations and to the average transverse flow.

Justification:
Several inequalities allow one to neglect some of the terms of Equations 12.25a–b.

- Equation 12.9, for the conservation of the flow, can be rewritten in the form $\partial(U-\bar{v}_x)/\partial x = \partial\bar{v}_y/\partial y$ for the wake: one obtains then the order of magnitude of the average transverse velocity $\bar{v}_y \approx (U-\bar{v}_x)e(x)/x \ll Ue(x)/x$. For the jet, Equation 12.9 leads to: $\bar{v}_y \approx \bar{v}_x e(x)/x \ll \bar{v}_x$.
- The small value of the velocity deficit ($\bar{v}_x - U \ll U$) implies that, for the wake: $\bar{v}_x\partial\bar{v}_x/\partial x \cong U\partial\bar{v}_x/\partial x$: just as in the case of a laminar wake, this term will dominate (because of the factor U) the other term $\bar{v}_y\partial\bar{v}_x/\partial y$ which is then neglected in Equation 12.26a. In contrast, for the jet, the two terms on the right-hand side of Equation 12.25b must be kept: one shows that their sum is equal to the left-hand side of Equation 12.26b by expanding the latter and using the conservation equation (Equation 12.9).
- Just as in the case of laminar quasi-parallel flows, the terms on the left-hand side of Equation 12.25b are negligible.
- In contrast with the average components of the velocity, the different components of the Reynolds tensor are of the same order of magnitude; however, their derivatives with respect to x are negligible relative to the derivatives $\partial\left(\overline{v'_x v'_y}\right)/\partial y$ and $\partial\left(\overline{v'^2_y}\right)/\partial y$ with respect to y because $e(x) \ll x$. Taking into account the previous assumptions, Equation 12.25b becomes then:

$$-(1/\rho)\,\partial\bar{p}/\partial y - \partial\left(\overline{v'^2_y}\right)/\partial y = 0, \tag{12.27a}$$

so that, by integrating between the interior and exterior of the wake, or of the jet:

$$\bar{p}(x,y) = p_0 - \rho\,\overline{v'^2_y}(x,y). \tag{12.27b}$$

One infers from Equation 12.27b the value $(1/\rho)\partial p/\partial x = -\,\partial\overline{v'^2_y}/\partial x$ of the pressure gradient in the x-direction: this term is negligible because it is smaller by an order of magnitude than the term $\partial\left(\overline{v'_x v'_y}\right)/\partial y$ in Equation 12.25a. One recovers then Equations 12.26a-b.

One encounters in other turbulent flows such variations of the pressure in the direction transverse to the mean flow: even if they do not influence it, they contribute to differentiating the flows from parallel or quasi-parallel laminar flows, for which these transverse gradients are considered to vanish.

Conservation of momentum and of flow rate

Let us now integrate Equations 12.26a–b relative to y over an infinite transverse distance (in practical terms, sufficiently large to contain the whole perturbed zone).

- ***For the turbulent wake***, one obtains first, having divided Equation 12.26a by U:

$$\int_{-\infty}^{+\infty} -\frac{\partial \bar{v}_x(x,y)}{\partial x}\,\mathrm{d}y = \frac{1}{U}\int_{-\infty}^{+\infty} \frac{\partial}{\partial y}\left(\overline{v'_x v'_y}\right)\mathrm{d}y = 0$$

($\overline{v'_x v'_y}$ vanishes indeed outside the wake). This equation can be rewritten in the form:

$$\frac{\partial Q(x)}{\partial x} = 0 \qquad (12.28a) \qquad \text{with:} \qquad Q(x) = \int_{-\infty}^{\infty} (U - \bar{v}_x(y))\,\mathrm{d}y. \qquad (12.28b)$$

The conservation of $Q(x)$ is only approximate, and results from the conservation of momentum. Proceeding as in the derivation of Equation 10.85 for a laminar wake, one finds that the quantity actually constant with x is $\rho \int_{-\infty}^{\infty} \bar{v}_x (U - \bar{v}_x)\,\mathrm{d}y$. This condition simplifies to $\rho\, U\, Q(x) = constant$ only if $U - \bar{v}_x \ll \bar{U}_x$ as has been, indeed, assumed.

The rate of flow $Q(x) = \int_{-\infty}^{\infty} (U - \bar{v}_x(y))\mathrm{d}y$ associated with the velocity deficit $U - \bar{v}_x$ in the wake is then constant along it. $Q(x)$ is also equal, in absolute value, to the flow rate in the reference frame moving at velocity $\mathbf{U}$, and it is defined per unit length in the z-direction.

Writing the conservation of momentum between the upstream and downstream regions of the body, one shows, like in Chapter 10 for a laminar wake (Equation 10.80), that there is the following relationship between the drag force F_D per unit length of the cylinder and the constant value Q of the flow rate $Q(x)$.

$$F_D = \rho U Q. \qquad (12.29)$$

- ***For the turbulent jet***, the integral of Equation 12.26b can be written:

$$\rho \int_{-\infty}^{+\infty} \frac{\partial}{\partial x}\left(\bar{v}_x^2\right)\mathrm{d}y = -\rho \int_{-\infty}^{+\infty} \frac{\partial}{\partial y}\left(\bar{v}_x \bar{v}_y\right)\mathrm{d}y - \rho \int_{-\infty}^{+\infty} \frac{\partial}{\partial y}\left(\overline{v'_x v'_y}\right)\mathrm{d}y = 0.$$

The two integrals of the right-hand side are zero, because the average velocities and the fluctuations vanish outside the jet. One thus obtains:

$$\frac{\partial \phi_j(x)}{\partial x} = 0, \qquad (12.30a) \qquad \text{where:} \qquad \phi_j(x) = \int_{-\infty}^{+\infty} \rho\, \bar{v}_x^2(x,y)\,\mathrm{d}y. \qquad (12.30b)$$

The flux of momentum over the width of the jet has then a value ϕ_j independent of the distance x: this requires an increase in the flow rate $Q(x) = \int_{-\infty}^{+\infty} \bar{v}_x(x,y)\,\mathrm{d}y$ with x since the average velocity $\bar{v}_x$ is decreasing. The jet "sucks in" more and more external fluid as it moves along.

If the self-similarity assumption is valid, Equations 12.31a–b are rigorously exact since the flow rate Q_w in the wake and the flux ϕ_j of the momentum in the jet obey:

$$Q_w = \int_{-\infty}^{+\infty} (U - \bar{v}_x(x,y))\,dy = U_w(x)\,e(x) \int_{-\infty}^{+\infty} f(\xi)\,d\xi = constant \quad \text{and:}$$

$$\phi_j = \int_{-\infty}^{+\infty} \rho \bar{v}_x^2(x,y)\,dy = \rho U_j^2(x)\,e(x) \int_{-\infty}^{+\infty} f^2(\xi)\,d\xi = constant.$$

Estimate of the variation of the two-dimensional widths of the wake and turbulent jet as a function of distance

Let us take as respective characteristic velocities $U_w(x)$ and $U_j(x)$ in the wake and in the jet, the components of the velocity $U - \bar{v}_x(x,0)$ and $\bar{v}_x(x,0)$ along the axis $y = 0$. Estimating the integrals in Equations 12.28b and 12.30b using these characteristic values then leads to:

$$U_w(x)\,e(x) \approx constant \qquad (12.31a) \qquad \text{and:} \qquad U_j^2(x)\,e(x) \approx constant. \qquad (12.31b)$$

Let us now estimate the terms of the equations of motion Equations 12.26a–b by assuming that the component $\overline{v'_x v'_y}$ varies as $U_w{}^2$ or $U_j{}^2$. One further assumes that the characteristic distances for the variation of the averages of the velocity components and of the corresponding Reynolds tensor along the x- and y-directions are respectively x and $e(x)$. By equating the estimates found in this manner, one obtains respectively:

$$\frac{U U_w(x)}{x} \approx \frac{U_w^2(x)}{e(x)} \qquad (12.32a) \qquad \text{and:} \qquad \frac{U_j^2(x)}{x} \approx \frac{U_j^2(x)}{e(x)}. \qquad (12.32b)$$

Combining Equations 12.31a–b and 12.32a–b, one obtains:

for the ***wake:*** $e(x) \propto \sqrt{x}$ (12.33a) and: $U_w(x) \propto \frac{1}{\sqrt{x}}$, (12.33b)

and for the ***jet:*** $e(x) \propto x$ (12.34a) and: $U_j(x) \propto \frac{1}{\sqrt{x}}$. (12.34b)

Moreover, one infers from Equations 12.33a–b and 12.34a–b that the Reynolds number $Re_w = U_w(x)\ e(x)/\nu$ for the wake is independent of the distance x, while that of the jet $(Re_j = U_j(x)\ e(x)/\nu)$ increases with that distance. One also notes that the variations of $e(x)$ and $U_w(x)$ are the same for laminar and turbulent wakes, while they are different in the case of a jet (for a laminar jet: $e(x) \propto x^{2/3}$ and $U_j(x) \propto x^{-1/3}$).

12.4.2 Self-similar velocity fields in two-dimensional jets and wakes

Equations of motion for self-similar jets and wakes

In order to determine quantitatively the velocity profiles $\bar{v}_x(x,y)$, the equations of motion are rewritten with an assumption of self-similarity of the same nature as that used for the boundary layers and laminar wakes in Chapter 10 (Sections 10.3.3 and 10.7).

For this purpose, the velocity deficit $U - \bar{v}_x$ for the wake and the velocity $\bar{v}_x$ for the jet, normalized by U_w or U_j, are assumed to depend solely on the normalized transverse distance $\xi = y/e(x)$ and not separately on y and x, respectively with:

$$U - \bar{v}_x(x,y) = U_w(x)\, f(\xi) \quad (12.35a) \qquad \text{and:} \qquad \bar{v}_x(x,y) = U_j(x)\, f(\xi). \quad (12.35b)$$

U_w and U_j are defined like in the previous case for $\xi = 0$, which implies: $f(0) = 1$.

One also assumes that similar conditions are obeyed by the transverse component $\overline{v'_x v'_y}$ of the Reynolds tensor (normalized by the same characteristic velocities):

$$\overline{v'_x v'_y}(x,y) = U_w^2(x)\, g(\xi) \quad (12.36a) \qquad \text{and:} \qquad \overline{v'_x v'_y}(x,y) = U_j^2(x)\, g(\xi). \quad (12.36b)$$

The component of the transverse velocity $\bar{v}_y(x,\xi)$ in the jet may be computed from the incompressibility condition (Equation 12.9) and the definition (Equation 12.35b), so that:

$$\bar{v}_y = \frac{1}{2}\frac{de(x)}{dx} U_j(x)\left[2\,\xi\, f(\xi) - \int_0^{\xi} f(u)\, du\right]. \quad (12.39)$$

This component then has also a self-similar form.

- For the ***turbulent wake***, the equation of motion, Equation 12.26a, is shown to become:

$$U\,\frac{de(x)}{dx}\,[f + \xi f'] = U_w(x)\, g' = \frac{K}{e(x)} g', \quad (12.37)$$

where f' and g' are derivatives with respect to the variable ξ and $K = Q_w / \left(\int_{-\infty}^{+\infty} f(\xi)\, d\xi\right) = cst$. Equation 12.37 can be obeyed, when x and ξ vary independently, only if $e(x)\ de(x)/dx = constant$. One then recovers Equations 12.33a–b.

Experimentally, self-similarity is obeyed at a few hundred diameters downstream of the cylinder; the width of the wake and the velocity deficit also vary respectively as $\sqrt{x}$ and $1/\sqrt{x}$.

The constant of proportionality between $e(x)$ and $\sqrt{x}$ can have an arbitrary value. If it is changed, the value of ξ corresponding to a given value of y is also changed so that Equation 12.37 continues to be obeyed.

- For the ***turbulent jet***, the equation of motion (Equation 12.26b) becomes:

$$\frac{\alpha}{2}\left[f' \int_0^{\xi} f(u)\, du + f^2\right] = \frac{\alpha}{2}\frac{\partial^2}{\partial \xi^2}\left[\int_0^{\xi} f(u)\, du\right]^2 = g', \quad (12.38a)$$

so that:

$$\alpha = \frac{de(x)}{dx} = constant. \tag{12.38b}$$

This last equation is equivalent to Equation 12.34a, and confirms that $U_j(x) \propto 1/\sqrt{x}$ according to Equation 12.31b. Experimentally, the self-similarity condition is obeyed as soon as the distance x exceeds about ten times the initial diameter of the jet.

Application of eddy viscosity and mixing length models

The explicit determination of the transverse flow profile requires the choice of a closure relation between $\overline{v'_x v'_y}(x,y)$ and the mean velocity $\bar{v}_x(x,y)$. One uses for this purpose the concepts of eddy viscosity and of mixing length defined by Equations 12.22 and 12.24a-b.

Identifying by means of Equations 12.35 and 12.36 the terms $\overline{v'_x v'_y}(x,y)$ and $-\nu_t\, \partial\bar{v}_x/\partial y$, leads respectively to: $\nu_t \propto U_w(x)e(x)$ and $\nu_t \propto U_j(x)e(x)$ for the wake and the jet. One then has a constant eddy viscosity ν_t independent of x in the case of the wake, while it increases as $\sqrt{x}$ for the jet. The mixing length $\ell(x)$ is, in both these cases, proportional to the width $e(x)$. This suggests that the size of the vortices which ensure the momentum transport is proportional to the local width of the wake or jet.

Justification

Let us equate, for the case of the wake, the two terms $-\nu_t\, \partial\bar{v}_x/\partial y = \nu_t\, (U_w(x)/e(x))\, f\{'(\xi)$ and $\overline{v'_x v'_y}(x,y) = U_w^2(x)\, g(\xi)$. One obtains $\nu_t = U_w(x)\, e(x)\, g(\xi)/f'(\xi)$ which has therefore the expected dependence on x. The mixing length ℓ obtained by combining Equations 12.24a-b is related to ν_t by $\nu_t = \ell^2\, |\partial\bar{v}_x/\partial y|$ so that: $\ell^2 = e^2(x)\ g(\xi)/f'^2(\xi)$. One has therefore: $\ell(x) \propto e(x)$. Using this procedure, the same result is obtained for the jet.

Profile of the average velocity in a two-dimensional wake

The eddy viscosity ν_t is now assumed to be constant as a function of ξ in addition to being independent of x, as seen above. Equation 12.27 is then identical to that used for a laminar wake, merely replacing ν by ν_t. The profile of the velocity deficit is then given by the same Equation 10.79 obtained in that previous case:

$$U - \bar{v}_x = Q\sqrt{\frac{U}{4\pi\nu_t x}}\, \mathrm{e}^{-\frac{U\,y^2}{4\nu_t\,x}}. \tag{12.40}$$

Experimentally, sufficiently far downstream from a cylinder placed within a flow, the agreement with the predictions of this equation is acceptable, specifically in the central region of the flow (Figure 12.4a on the next page).

The difference from the laminar wake is the large value of the eddy viscosity ν_t. This viscosity is generally much greater than ν when the Reynolds number is high. One finds experimentally that $\nu_t \cong 0.017\, U d$, where d is the diameter of the cylinder.

Equation 12.40 and the expression for ν_t in the text at right imply that: $e(x) \propto \sqrt{\nu_t x/U} \propto \sqrt{x\,d}$. For this reason, $\xi = y/\sqrt{x\,d}$ is chosen as the vertical coordinate in Figure 12.4a.

Profile of the average velocity for a two-dimensional jet

By assuming, just as for the wake, that the eddy viscosity is independent of y over the width of the jet, the component $\bar{v}_x(y)$ of the average velocity obeys:

$$\bar{v}_x(y) = \frac{U_j(x)}{\cosh^2\left(\dfrac{y\, Re_t}{x\,\sqrt{2}}\right)}; \tag{12.41}$$

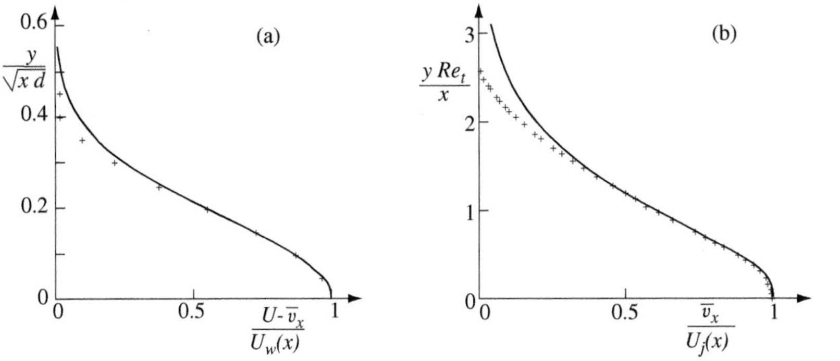

Figure 12.4 *(a) Comparison of the experimental measurements (+) and the theoretical predictions for Equation 12.40 (solid line) for the variation of the normalized velocity deficit $(U - \bar{v}_x)/U_w(x)$ with the normalized transverse distance $\xi = y/\sqrt{x\,d}$ in a turbulent wake behind a cylinder (from Townsend, 1956). (b) Comparison of the experimental measurements (+) and the theoretical predictions of Equation 12.41 (solid line) for the normalized velocity profile $\bar{v}_x/U_j(x)$ in a plane turbulent jet as a function of the normalised transverse distance Re_t/x (experiments of Heskestad (1965); from Pope, 2000)*

$Re_t = U_j(x)\,e(x)/\nu_t(x)$ is the turbulent Reynolds number defined on the basis of the eddy viscosity ν_t, the velocity $U_j(x)$ at $y = 0$ and the width $e(x)$. Since, as seen above, $\nu_t \propto U_j(x)\,e(x)$, Re_t is therefore constant and independent of x.

Experimentally, Equation 12.41 is satisfied in the central region of the jet but with significant deviations when $\bar{v}_x/U_j$ becomes smaller than 0.25 (Figure 12.4b). Also, still experimentally, Re_t, as well as the ratio $\alpha = e(x)/x$, are both independent of the rate of flow of the jet.

Proof

By introducing the eddy viscosity coefficient and using Equations 12.35b and 12.36b, the function $g(\xi)$ is found to obey: $g(\xi) = \overline{v'_x v'_y}/U_j^2 = -(\nu_t\,\partial\bar{v}_x/\partial y)/U_j^2 = -(1/Re_t)f'(\xi)$. Replacing g by this expression in Equation 12.38a, one obtains the differential equation obeyed by $f(\xi)$ which has the solution:

$$f(\xi) = 1\Big/\cosh^2\left(\xi\sqrt{\alpha Re_t/2}\right).$$

Just as for the wake, no precise quantitative definition had been given up to now for $e(x)$ or α: the value $\alpha = e(x)/x = 1/Re_t$ selected here gives a reasonable order of magnitude for the width of the jet, since $f(1)$ (corresponding to $y = e(x)$) is then of the order of 0.6. Equation 12.41 results immediately.

12.4.3 Three-dimensional axially symmetric turbulent jets and wakes

For the three-dimensional case of an axially symmetric wake or jet, the conservation of the flow rate is obeyed in the first instance, and that of the momentum, in the second. The calculation of Q and ϕ_j (Equations 12.28b and 12.30b) involves however an integration in both the y- and z-directions, rather than only along y. Equations 12.31a-b are then replaced by:

$$U_w(x)e^2(x) \approx constant \quad (12.42a) \qquad \text{and:} \qquad U_j^2(x)e^2(x) \approx constant. \quad (12.42b)$$

Combining these equations with Equations 12.36a-b, which remain valid, one obtains for the *axially symmetric wake:*

$$e(x) \propto x^{1/3} \quad (12.43a) \qquad \text{and:} \qquad U_w(x) \propto x^{-2/3}, \quad (12.43b)$$

while, for the *jet:*

$$e(x) \propto x \quad (12.43c) \qquad \text{and:} \qquad U_j(x) \propto \frac{1}{x}. \quad (12.43d)$$

In conclusion, the *turbulent* (or *eddy*) *viscosity* and *mixing length* models provide an acceptable approximate expression for the velocity profiles as a function of distance in quasi-parallel turbulent flows far from solid walls, such as jets and wakes (this is also the case in turbulent mixing layers). More precisely, at each distance x, the mixing lengths are proportional to the transverse dimension of the flows.

We will now discuss flows close to walls for which the effect of viscosity needs to be taken into account.

12.5 Flows near a solid wall

12.5.1 Qualitative properties of turbulent flows in the presence of a wall

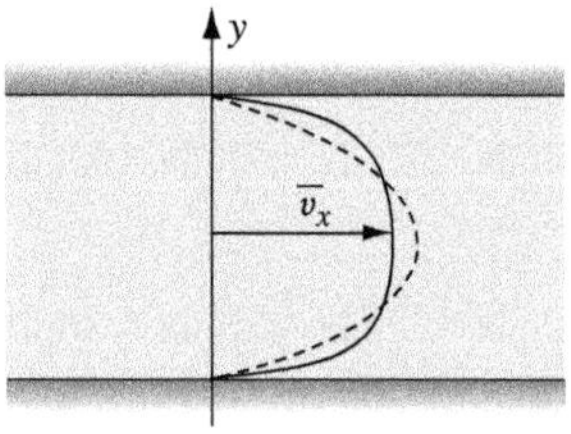

Figure 12.5 *Average velocity profile for flow in laminar (dashed line) and turbulent (solid line) regimes within a conduit. The horizontal velocity scale has been in each case divided by the average velocity in order to make the comparison easier*

This problem, of great practical importance, is found in numerous examples of flows within a conduit and near obstacles. This occurs, coupled with mass transfer, in chemical engineering flows and, coupled with heat transfer, in thermal flows. In aerodynamics, it occurs in the evaluation of pressure drops and frictional forces associated with turbulent flows.

Let us start off by considering a fully developed flow in a cylindrical conduit or between two parallel plates. The properties of such flows differ quite considerably from those of a laminar flow. Thus, the profile for the average velocity $\bar{v}_x$ inside the conduit is, in particular, much flatter in the central region than the parabolic profile which characterizes viscous laminar flows (Figure 12.5). The velocity gradient at the wall is, in contrast, much higher. On the other hand, the pressure drop between the ends of the conduit is no longer proportional to the flow rate Q, as in the case of a laminar flow, but varies approximately as Q^2.

Just as for the free turbulent flows in Section 12.4, momentum transfer is essentially convective and associated with velocity fluctuations by means of the Reynolds tensor of the turbulent stresses. This is the case throughout the flow except for a fine *viscous sublayer* near the wall: the latter results from the vanishing velocity and fluctuations at the wall which do not occur in free flows. The thickness of this sublayer (i.e. the distance at which there is a transition from viscous diffusive to convective transport) represents a second characteristic length scale in addition to the global transverse size of the flow.

Finally, the velocity profile $\bar{v}_x$ no longer depends on the geometry of the flow at distances small relative to the global scale where one observes, in addition to the viscous sublayer, *an inertial sublayer* where the convective transport dominates. These two sublayers appear both in a conduit and in a turbulent boundary layer.

12.5.2 Stationary turbulent flows parallel to a plane wall

Equations of motion

Let us consider a turbulent flow where the average velocity is parallel to the x-direction along a plane $y = 0$; let us further assume that the flow is well established, and that the

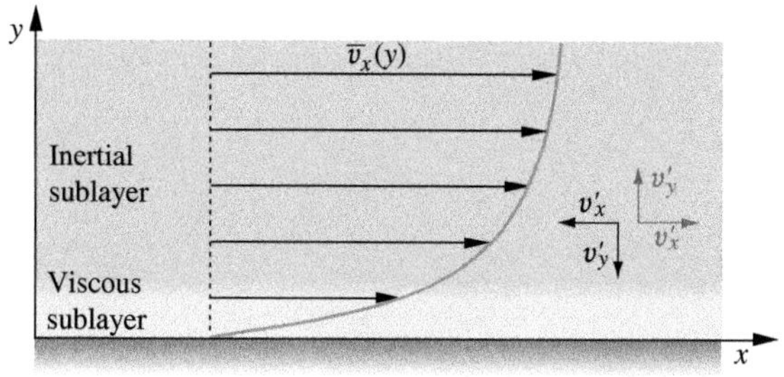

Figure 12.6 *Schematic diagram of the turbulent velocity field for a flow parallel to a plane wall*

statistical properties of the turbulent fluctuations are independent of the distance x along the flow Figure 12.6. The component $\bar{v}_y$ of the average velocity then vanishes; the longitudinal component $\bar{v}_x$ and the moments of the components of the velocity fluctuations (such as $\overline{v'_y{}^2}$ and $\overline{v'_x v'_y}$) are then only functions of y. The components of Equation 12.11 for the transport of momentum in a stationary regime can then be written:

$$\frac{\partial}{\partial y}\left(\nu \frac{\partial \bar{v}_x}{\partial y} - \overline{v'_x v'_y}\right) - \frac{1}{\rho}\frac{\partial \bar{p}}{\partial x} = 0 \quad (12.44a) \qquad \text{and:} \qquad \frac{\partial}{\partial y}\left(\overline{v'^2_y}\right) + \frac{1}{\rho}\frac{\partial \bar{p}}{\partial y} = 0. \quad (12.44b)$$

When the turbulent fluctuations vanish (laminar flows), one recovers the Navier–Stokes equations for parallel flows.

According to Equation 12.44b: $\bar{p} = p_0 - \rho\,\overline{v'^2_y}$, where the pressure p_0 at the wall is only a function of x ($\overline{v'^2_y}$, just as every component of the velocity and its fluctuations, vanishes at the wall). Since $\overline{v'^2_y}$ is independent of x, the derivatives $\partial\bar{p}/\partial x = \partial\bar{p}_0/\partial x$ must then equal. These gradients are independent of x since the terms in Equation 12.44a depend only on y.

By integrating Equation 12.44a relative to y, one obtains then:

$$\rho\,\nu \frac{\partial \bar{v}_x}{\partial y} - \rho\,\overline{v'_x v'_y} = \frac{\partial \bar{p}}{\partial x}\,y + \tau_0. \qquad (12.45)$$

- At very small distances from the wall, where the velocity fluctuations and the term containing the pressure gradient are negligible, Equation 12.45 reduces to:

$$\tau_0 = \rho\,\nu\,\frac{\partial \bar{v}_x}{\partial y}, \qquad (12.46)$$

  τ_0 represents the stress on the wall, which reduces to the viscous stress because of the condition of zero velocity. The stress τ_0 depends not only on the average velocity of the flow, but also on the Reynolds number and the roughness of the wall. Taking into account the condition of zero-average velocity at $y = 0$, one then has, by integrating:

$$\bar{v}_x = \frac{\tau_0\,y}{\rho\,\nu}. \qquad (12.47)$$

- As one gets further away from the wall, the convective transport of momentum by the turbulent fluctuations substitutes for the transport by viscous diffusion (the distance for this transition is evaluated further down), and Equation 12.45 becomes:

$$-\rho\,\overline{v'_x v'_y} = \frac{\partial \bar{p}}{\partial x}\,y + \tau_0. \qquad (12.48a)$$

Let us discuss the condition for the term $(\partial \bar{p}/\partial x)$ to be negligible in the case of a flow between two parallel plates separated by a distance $2h$: as the mid-plane $y = h$ is the plane of symmetry of the flow, $\rho\, \overline{v'_x v'_y}$ vanishes there because the momentum has no reason to be transported toward one plate rather than the other. One then has $\tau_0 = -h\,(\partial \bar{p}/\partial x)$ which expresses the equilibrium between the forces due to the pressure gradient and the stress on the two walls. The term involving $(\partial \bar{p}/\partial x)$ in Equation 12.48a is then negligible if y is small compared to h. This assumption is assumed to remain valid for other geometries.

In the inertial sublayer which we introduced above, where simultaneously Equation 12.48a is satisfied and y is small relative to the transverse global distance h of the flow, the term $\partial \bar{p}/\partial x$ is negligible and:

$$\tau_0 = -\rho\, \overline{v'_x v'_y}. \tag{12.48b}$$

Because of this equality, one can define a characteristic velocity u^*, known as the *frictional velocity*, by:

$$\tau_0 = \rho\, u^{*\,2}, \tag{12.49}$$

which represents a characteristic velocity of the turbulent fluctuations.

- Finally, a third region, known as the *external flow region*, corresponds to larger values of y, going up to the transverse global size: there, one must use the complete form of Equation 12.48a.

Evaluation of the velocity profile by means of the mixing-length model

We will now determine the velocity profile in the inertial and viscous sublayers, and the distance at which there is a transition between the viscous and convective transport mechanisms. In a first approach, we will use the concept of mixing length, which has already been applied to the case of jets and turbulent wakes in Section 12.4.

- In the viscous sublayer, the velocity profile is linear and given by Equation 12.47.
- In the inertial sublayer, one obtains by combining Equations 12.49 and 12.24b:

$$\tau_0 = \rho\, u^{*2} = \rho\, u^* \ell\, \frac{\partial \bar{v}_x}{\partial y}. \tag{12.50a}$$

In the present case of a uniform flow parallel to a plane wall, one considers that the mixing length ℓ is proportional to the distance y from the wall so that $\ell = \kappa y$. This assumes that, in the range of length-scales for which the equation $\tau_0 = -\rho\, \overline{v'_x v'_y}$ is valid, the structure of the flow and the momentum transport at a given distance from the wall are affected neither by vortices of size much greater than this distance, nor by the layer very close to the wall where the viscosity dominates. The distance y from the wall then represents the only significant length-scale. This also implies that the momentum transport takes place over a distance of the order of y. Equation 12.50a then becomes:

$$\frac{\partial \bar{v}_x}{\partial y} = \frac{u^*}{\kappa y}, \tag{12.50b}$$

so that:

$$\bar{v}_x = \frac{u^*}{\kappa}\, \mathrm{Log}\, y + \mathit{constant}. \tag{12.51}$$

One finds experimentally that the value of $1/\kappa$ is approximately 2.5 for a large range of turbulent flows in the presence of a wall: this parameter κ is known as the *von Karman constant* (and should not be confused with the thermal diffusivity κ).

The order of magnitude of the viscous sublayer

According to Equation 12.50b, the velocity gradient should diverge as y approaches 0. This increases the viscosity term $\rho\nu(\partial \bar{v}_x/\partial y) \cong \rho\, \nu\, u^*/(\kappa y)$ in Equation 12.45. This term becomes of the order of magnitude of the total stress $\tau_0 = \rho\, u^{*2}$ at a distance $y \approx \nu/(\kappa u^*)$. At smaller distances, the viscous stress dominates, and remains equal to $\rho\, u^{*2}$. The length ν/u^* then represents the order of magnitude of the thickness of the viscous sublayer; beyond this distance, one gets into the inertial sublayer.

Common characteristics of turbulent flows near a wall

- As soon as one gets to a small distance from the wall, momentum transport is accounted for by the turbulent velocity fluctuations (inertial sublayer). This mechanism is much more effective than the viscosity, and the flux of momentum is larger than in a laminar flow.
- In the immediate neighborhood of the wall, where $v'_x = v'_y = 0$, the transport of momentum results solely from viscous diffusion (viscous sublayer). The momentum flux is then: $\rho\, \nu\, \partial\bar{v}_x/\partial y\,(0)$ and it must take the same large value as in the inertial sublayer: the velocity gradient $\partial\bar{v}_x/\partial y\,(0)$ is then also higher than if the entire flow had been laminar.
- While studying the viscous and inertial sublayers, we have not used the global geometry of the flow. Experiments show indeed that, in numerous cases, it has no effect on the flow in these two regions, but only on the external flow on a larger scale.
- The pressure is not constant along a perpendicular to the wall, in contrast to the laminar case (but the pressure gradient in the direction of the flow remains constant).

12.5.3 Turbulent flow between two parallel plates

Let us now analyze more quantitatively the case of flows within channels with parallel walls. We will use a more dimensional approach taking advantage of the specific features of the flow without referring to the mixing length model.

Equations of motion in scaled coordinates

The flow between two parallel plates is symmetrical relative to the plane $y = h$ parallel to the plates and half-way between them (Figure 12.7); one discusses therefore below only the part of the velocity profile corresponding to $0 \le y \le h$, the remaining region ($h \le y \le 2h$) being obtained by symmetry. Because of the same symmetry argument, the derivative $\partial\bar{v}_x/\partial y$ and the convective flux $\rho\, \overline{v'_x v'_y}$ must both vanish at $y = h$. Equation 12.45 then becomes:

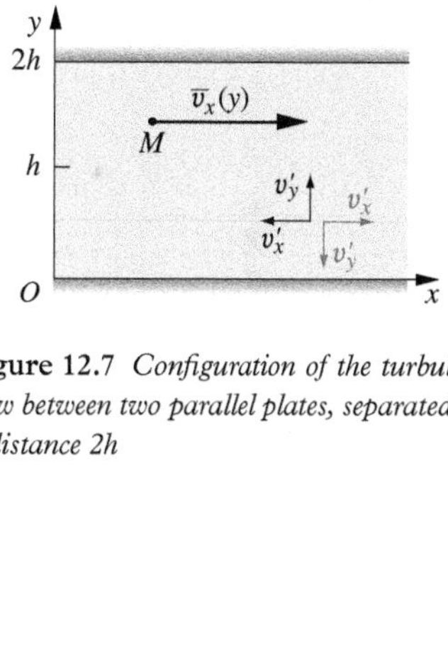

Figure 12.7 *Configuration of the turbulent flow between two parallel plates, separated by a distance 2h*

$$\tau_0 + \frac{\partial\bar{p}}{\partial x}\,h = 0 \qquad (12.52a) \qquad \text{or:} \qquad \rho\, u^{*2} = -\frac{\partial\bar{p}}{\partial x}\,h. \qquad (12.52b)$$

By dividing the two terms of Equation 12.45 by ρ, and using Equation 12.52b, one obtains:

$$\nu\frac{\partial\bar{v}_x}{\partial y} - \overline{v'_x v'_y} = u^{*2}\,\frac{h-y}{h}. \qquad (12.53)$$

There are two characteristic length scales in this problem: the distance $2h$ between the plates and the thickness ν/u^* of the viscous sublayer. Substituting for the stress τ_0 its value as a function of the frictional velocity u^*, Equation 12.47 becomes:

$$\frac{\bar{v}_x}{u^*} = \frac{u^* y}{\nu}. \qquad (12.54)$$

In order to calculate the velocity profile in the inertial sublayer, one introduces two different dimensionless variables related to the two characteristic scales h and ν/u^*:

$$Y = \frac{y}{h} \qquad (12.55a) \qquad \text{and:} \qquad y^+ = y\,\frac{u^*}{\nu}, \qquad (12.55b)$$

for which the ratio is given by a Reynolds number:

$$\frac{y^+}{Y} = \frac{u^* h}{\nu} = Re^*. \tag{12.55c}$$

Either variable may be used, depending on whether one is interested in the structure in the center of the flow or near the walls: in the latter case, the velocity profile is not influenced by the global structure of the flow, but the viscosity can play a role, and the variable y^+ is then more suitable. Far from the walls, in contrast, the viscosity no longer affects the velocity profile and the variable Y is better suited. In laminar flow, one did not need to distinguish between the two dimensionless variables, because the viscosity plays a dominant role throughout the flow.

For the average velocity $\bar{v}_x$, one uses in the same way two different dimensionless variables. For studying the flow near the walls, one uses the ratio $\bar{v}_x/u^*$ (this normalization appears already in Equation 12.54). If, on the other hand, one is interested in the velocity field in the center of the flow, the normalized velocity deficit $(\bar{v}_x - U_0)/u^*$ will be used instead (the relevant natural velocity is then the velocity U_0 in the symmetry plane $y = h$). By analogy with the solution of the dimensionless Navier–Stokes equations, one assumes that the two scaled average velocity variables can be written as functions of Y and y^+ in the form:

$$\bar{v}_x(Y) - U_0 = u^* f_1(Y) \qquad \text{for:} \qquad \left(y \gg \frac{\nu}{u^*},\ Re^* \gg 1\right). \tag{12.56}$$

In the same way, near the wall:

$$\bar{v}_x(y^+) = u^* f(y^+) \qquad \text{for:} \qquad (y \ll h,\ Re^* \gg 1). \tag{12.57}$$

Moreover, one has $f(0) = f_1(1) = 0$. The two functions f and f_1 are in principle universal, so long as there is a sufficiently long time left for the turbulent flow to develop and lose the "memory" of its initial configuration. More precisely, one must be sufficiently far away from the upstream end of the conduit or of the wall so that the entry features of the flow have disappeared.

Inertial sublayer – logarithmic variation of the velocity

In this approach, the inertial sublayer is the domain of values of y such that Equations 12.56 and 12.57 may be valid simultaneously; this implies the double inequality:

$$\frac{\nu}{u^*} \ll y \ll h \qquad \text{which requires that:} \qquad Re^* = \frac{u^* h}{\nu} \ggg 1. \tag{12.58}$$

Typically, such a domain with $30\ \nu/u^* < y < 0.1\ h$ exists only if $Re^* > 10^3$. In this region, the convective momentum transport plays an essential role. Equations 12.56 and 12.57 must, of course, yield in this domain the same value of the average velocity for a given value of y (this represents an example of a *matched asymptotic expansion* approach, very often used in fluid mechanics). By taking derivatives of the two equations with respect to y one must, first of all, obtain the same function: multiplying the result by y, one obtains then:

$$Y\,\frac{\partial f_1(Y)}{\partial Y} = y^+\,\frac{\partial f(y^+)}{\partial y^+}. \tag{12.59}$$

The functions f and f_1 must be universal and have a unique respective dependence on y^+ and on Y. However, it is possible to vary independently these variables by changing, for example, Re^* to keep one of them constant (and, consequently, to keep constant the corresponding term in Equation 12.59). The two sides of the equation must be equal to the same constant written as $1/\kappa$, so that:

$$f_1(Y) = \frac{1}{\kappa}\,\mathrm{Log}\,Y + C \quad \text{or:} \quad \bar{v}_x(y) - U_0 = u^*\left(\frac{1}{\kappa}\mathrm{Log}\,\frac{y}{h} + C\right) \tag{12.60a}$$

and:

$$f(y^+) = \frac{1}{\kappa}\,\mathrm{Log}\,y^+ + C' \quad \text{or:} \quad \bar{v}_x(y) = u^*\left(\frac{1}{\kappa}\mathrm{Log}\,\frac{y\,u^*}{\nu} + C'\right). \tag{12.60b}$$

The parameter κ is the von Karman constant already defined in Equation 12.50b. Since the two Equations 12.60a and 12.60b must be identical in the intermediate region joining them, one has:

$$\frac{U_0}{u^*} = \frac{1}{\kappa}\,\mathrm{Log}\,Re^* + C' - C. \tag{12.61}$$

The above equation allows one to determine the maximum velocity, once the pressure gradient and the width of the channel are known.

Velocity profiles for turbulent flows in a conduit

Figure 12.8 displays the variation of the normalized average velocity as a function of the scaled distance y^+ from the wall, for flow in a circular tube of diameter D (symbols "o"). The corresponding measurements in a turbulent boundary layer are also indicated (symbols "Δ") and will be discussed in Section 12.5.5.

The use of a logarithmic horizontal scale for the variable y^+ demonstrates a linear variation (accordingly, exponential as a function of Log y^+), for values of $y^+ \geq 25$ and up to a distance y of the order of $0.1D$; $v_x(y)/u^*$ then obeys Equation 12.60b with $C' \approx 5.0$ and $1/\kappa = 2.5$ (dashed line), so that:

$$\frac{\bar{v}_x}{u^*} = 2.5\,\mathrm{Log}\,\frac{y\,u^*}{\nu} + 5. \tag{12.62}$$

We also see in the figure that, at short distances $y^+ \leq 5$ within the viscous sublayer, there is good agreement between the experimental points and the theoretical value $\bar{v}_x/u^* = y^+$

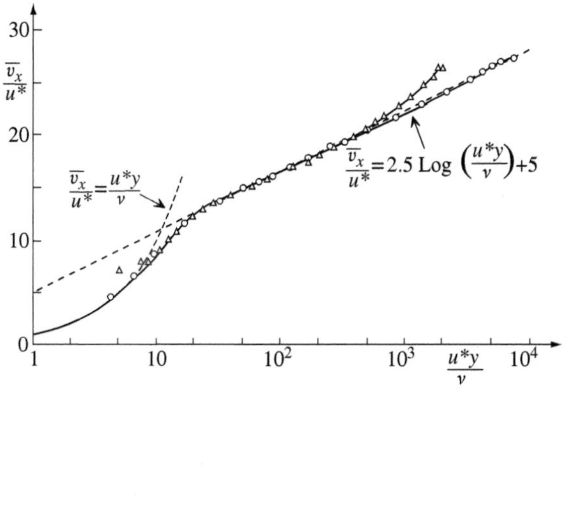

Figure 12.8 *Profile for the variation of the average scaled velocity $\bar{v}_x/u^*$ as a function of the scaled distance to the wall $y^+ = u^*y/\nu$ for turbulent flows in a cylindrical tube* 200 mm *in diameter (symbols "o") and in a boundary layer resulting from uniform flow near a plane wall (symbols "Δ") (according to Patel, 1965)*

We find graphically that the intersection of the two limiting equations, Equations 12.54 and 12.60b, corresponds to $y^+ = u^* y/\nu = 11$ (dashed lines in Figure 12.8 above). We must then have:

$$\bar{v}_x / u^* = 11 = 2.5 \,\mathrm{Log}\, 11 + C'.$$

We verify then that: $C' = 11 - 2.5 \,\mathrm{Log}\, 11 \approx 5$.

As Re^* increases, the range of distances to the wall corresponding to the inertial sublayer, where the logarithmic dependence applies, becomes broader. More precisely, its upper limit is always a fraction of h with the same order of magnitude while its lower limit decreases (it is of the order of $10\nu/u^*$).

(Equation 12.54) and, beyond that, a transition region toward the logarithmic dependence (the linear dependence appears as a curved line in Figure 12.8 because of our using a horizontal, logarithmic scale).

For $y^+ = 5$, the two terms for momentum transport obey:

$$-\frac{\overline{v'_x v'_y}}{u^{*2}} \approx 0.1 \,\frac{\partial\,(\bar{v}_x / u^*)}{\partial y^+}. \tag{12.63}$$

Thus in the viscous sublayer, the velocity fluctuations are not negligible, but the momentum transport due to the viscosity is dominant relative to the convective transport. For $y^+ = 11$, corresponding to the intersection of the two curves representing Equations 12.54 and 12.60b (see above at left), the two terms are equivalent.

Relationship of the average velocity to the maximum velocity in turbulent flows

In order to describe the central region of the flow, one must add an additional term $W(y/h)$ to the right-hand side of Equation 12.60a:

$$\bar{v}_x\,(y) - U_0 = u^* \left(\frac{1}{\kappa} \mathrm{Log}\,\frac{y}{h} + C + W\left(\frac{y}{h}\right)\right). \tag{12.64}$$

For flow between parallel plates, one finds experimentally that one can choose $W(Y) \propto \sin \alpha Y$ with $C + W(1) = 0$ and $W(0) = 0$. As soon as $y/h < 0.1$, the logarithmic dependence is again valid. In practical terms, one frequently takes as a first approximation $C = 0$ and $W = 0$ in Equation 12.64.

According to these latter assumptions, and using Equations 12.61 and 12.60a, as well as the values $C' = 5$ (given above) and $C = 0$, one obtains:

$$\frac{\bar{v}_x\,(y) - U_0}{u^*} = 2.5 \,\mathrm{Log}\,\frac{y}{h} \quad (12.65) \qquad \text{and:} \qquad \frac{U_0}{u^*} = 2.5 \,\mathrm{Log}\, Re^* + 5. \quad (12.66)$$

By combining these two equations, one obtains:

$$\frac{\bar{v}_x\,(y)}{u^*} = 2.5 \left[\mathrm{Log}\,\frac{y}{h} + \mathrm{Log}\, Re^*\right] + 5. \tag{12.67}$$

By integrating this equation over the half-interval between the two planes, one obtains the average velocity U_m with:

$$\frac{U_m}{u^*} = \frac{1}{h} \int_0^h \bar{v}_x \,\mathrm{d}y = 2.5 \,\mathrm{Log}\, Re^* + 2.5. \tag{12.68a}$$

For a circular tube of diameter D, one obtains a similar equation by taking $Re^* = u^* D/\nu$:

$$\frac{U_m}{u^*} = 2.5 \,\mathrm{Log}\, Re^* - 0.5. \tag{12.68b}$$

Taking $Re^* = 10^3$ in Equations 12.66 and 12.68a, one finds for the flow between plates $U_m/U_0 = 0.88$. This value is notably larger than the ratio 2/3 for a laminar flow between plates (Equation 4.69): this is a confirmation of the fact that a turbulent profile is much flatter than the laminar, parabolic profile. The ratio U_m/U_0 increases with Re^* and, according to Equations 12.66 and 12.68, it tends toward 1 when 2.5 Log $Re^* \gg 1$: this indicates that the velocity profile approaches plug flow (except very close to the walls).

12.5.4 Pressure losses and coefficient of friction for flows between parallel planes and in tubes

Flows between plates and in tubes with smooth walls

Equation 12.68a allows one to determine the relationship between the average velocity and the pressure gradient $\Delta p/L$ parallel to the flow between two plates. In fact, the force $(2\Delta p\, h/L)$ on an element of fluid of unit length in the x- z-directions, which extends across the interval $2h$ between the two plates, must be in equilibrium with the total wall stress $\tau_0 = 2\rho\, u^{*2}$ on these plates.

Therefore: $$u^* = \sqrt{\frac{\Delta p\, h}{\rho L}} \quad (12.69a) \qquad \text{and:} \quad Re^* = \sqrt{\frac{\Delta p\, h^3}{\rho\, \nu^2 L}}. \quad (12.69b)$$

The rate of flow Q between the plates per unit depth in the z-direction is $Q = 2U_m h$. Equation 12.68a can therefore be transformed into the relation between the flow rate and the pressure gradient:

$$Q = 2U_m h = 2\sqrt{\frac{\Delta p\, h^3}{\rho L}}\left(2.5\,\text{Log}\sqrt{\frac{\Delta p\, h^3}{\rho\, \nu^2 L}} + 2.5\right). \quad (12.70)$$

This equation indicates that the rate of flow Q between the plates is approximately proportional to the square root of the pressure-gradient $\Delta p/L$ because of the slow variation of the logarithmic term. The flow velocity therefore increases slower as a function of the pressure drop than for a laminar flow for which $Q \propto \Delta p/L$.

For the case of a tube of diameter D with smooth walls, we would similarly obtain $\tau_0 = (D/4)(\Delta p/L) = \rho\, u^{*2}$, so that:

$$u^* = \sqrt{\frac{\Delta p D}{4\,\rho\, L}}, \quad (12.71a) \qquad Re^* = \sqrt{\frac{\Delta p\, D^3}{4\rho\, \nu^2 L}}, \quad (12.71b)$$

and: $$Q = \frac{\pi\, D^2}{4}\sqrt{\frac{\Delta p D}{4\rho L}}\left(2.5\,\text{Log}\sqrt{\frac{\Delta p\, D^3}{4\rho\, \nu^2 L}} - 0.5\right). \quad (12.72)$$

Let us now introduce, still for the case of a tube, a friction factor:

$$C_d = \frac{\Delta p}{\frac{L}{D}\,\frac{\rho\, U_m^2}{2}} = \frac{8\, u^{*2}}{U_m^2}. \quad (12.73)$$

Using Equation 12.68b, this equation becomes:

$$\sqrt{\frac{8}{C_d}} = 2.5\ \text{Log}\left(\frac{u^* D}{\nu}\right) - 0.5 = 2.5\,\text{Log}\left(Re_D\sqrt{C_d}\right) - 3.1, \quad (12.74)$$

By using a base 10 logarithm, as is customary for practical applications, one derives from Equation 12.74, the equation:

$$C_d = \left(2.03\log_{10}\left(\sqrt{C_d}\,Re_D\right) - 1.09\right)^{-2}. \quad (12.75)$$

This equation is very similar to the equation used in engineering for commercial tubes with circular cross-section and smooth walls:

$$C_d = \left(2\log_{10}\left(\sqrt{C_d}\,Re_D/2.51\right)\right)^{-2} = \left(2\log_{10}\left(\sqrt{C_d}\,Re_D\right) - 0.8\right)^{-2}. \quad (12.76)$$

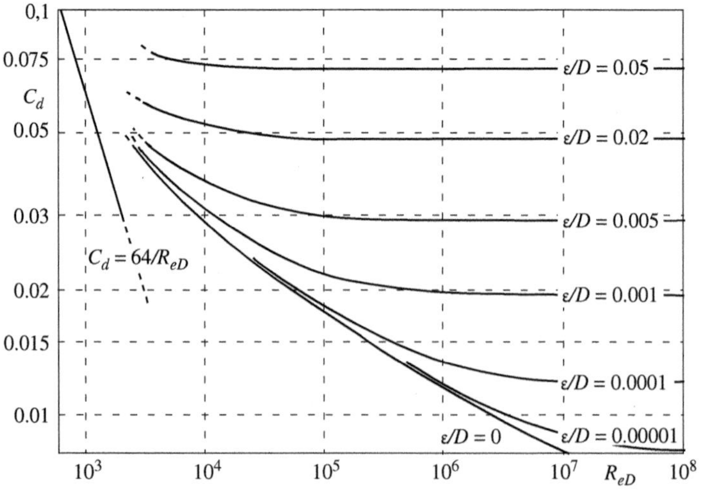

Figure 12.9 *Variation of friction factor C_d (Equation 12.73) as a function of the Reynolds number $Re_D = U_m D/\nu$ for different tubes with circular cross-sections with smooth and rough walls. The roughness is characterized in the case of each curve by the ratio between the characteristic height ε of the surface asperities and the diameter D of the tube. The linear dependence on the left indicates a laminar flow. The lowest curve of the right-hand group corresponds to a tube with smooth walls. The dashed lines indicate the laminar-turbulent transition zone (according to L.F. Moody, 1944)*

where $Re_D = U_m D/\nu$. This variation corresponds to the lowest of the curves on Figure 12.9 ($\varepsilon/D = 0$) where one notes that C_d varies only by a factor 3 when the Reynolds number Re_D varies by a factor 100.

Flows in conduits with rough walls

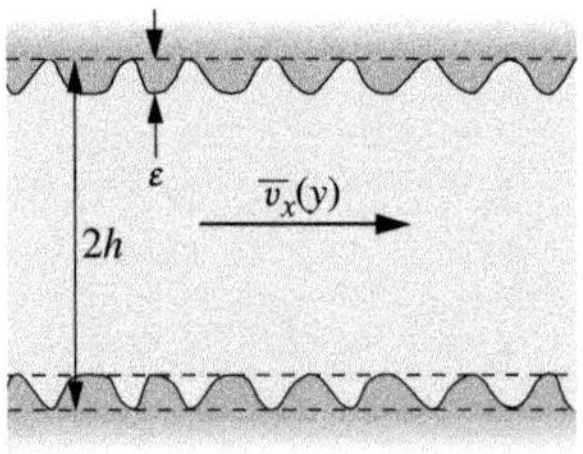

Figure 12.10 *Schematic diagram for a flow between two planes with rough walls, separated by a distance 2h with a thickness ε of asperities along the walls*

In this case of great practical importance, the average height ε of the asperities (Figure 12.10) introduces a new parameter into the problem. If $\varepsilon \ll \nu/u^*$, the presence of the asperities does not change the structure of the viscous sublayer, and the previous results remain valid: the assumption of a *smooth wall* may be retained.

If $\varepsilon \gg \nu/u^*$, it is the scale length ε and no longer the thickness ν/u^* of the viscous sublayer which governs the structure of the velocity field at a small distance from the walls; one then has *rough walls*. The dimensionless variable y^+ must then be replaced by $y_\varepsilon = y/\varepsilon$; but one can only expect simple predictions for distances $y \gg \varepsilon$, such that the details of the local geometry of the asperities no longer affect the flow. If $\nu/u^* \ll \varepsilon \ll y \ll h$, one substitutes then for Equation 12.60b, the equation:

$$\bar{v}_x = \frac{u^*}{\kappa} \operatorname{Log} \frac{y}{\varepsilon}. \tag{12.77}$$

One assumes for simplicity that $\bar{v}_x = 0$ for $y = \varepsilon$, so that there is no additive constant. One obtains then, by assuming that the equation remains valid up to $y = h$ and integrating $\bar{v}_x$ between ε and h:

$$U_m \approx \frac{u^*}{\kappa} \left(\operatorname{Log} \frac{h}{\varepsilon} - 1 \right). \tag{12.78}$$

Proceeding as in the case of Equations 12.72 to 12.76, one finds that the flow rate between the plates is exactly proportional to $\sqrt{\Delta p/L}$, and that the friction factor C_d is constant (without a logarithmic dependence on Re_D).

Figure 12.9 displays the variation of C_d with the Reynolds number Re_D for different values of the ratio ε/D (again for tubes of circular cross-section). The transition to the turbulent regime results in an increase of C_d. Then, as Re_D increases further, one has first a regime where the thickness ν/u^* of the viscous sublayer is large relative to ε: C_d follows in this regime the universal dependence (Equation 12.74) obtained for smooth tubes. At larger values of Re_D such that $\nu/u^* \ll \varepsilon$, C_d has a constant value determined by the roughness of the walls. The weak variation of C_d with Re_D then justifies the use of this coefficient to characterize the wall friction for industrial turbulent flows of fluids with low viscosity.

Experimentally, one obtains, for tubes with rough walls, the following value (independent of the velocity) for the limit of C_d at high Reynolds numbers:

$$C_d = \left(2\log_{10}\frac{\varepsilon/D}{3.7}\right)^{-2} = \left(2\log_{10}\frac{D}{\varepsilon} + 1.14\right)^{-2}. \qquad (12.79)$$

For a laminar flow corresponding to Poiseuille's Equation 4.78, one obtains instead with the same definition of C_d:

$$C_d = \frac{64}{Re_D}. \qquad (12.80)$$

This variation appears at the left in Figure 12.9. The coefficient C_d varies as $1/Re_D$ like for the other viscous flows for which the velocity of the fluid is proportional to the applied pressure gradients. This variation is much faster than that observed for turbulent flows and even for laminar boundary layers (as $1/Re^{1/2}$ in the case of the latter).

12.5.5 Turbulent boundary layers

In Section 12.5.3, we discussed flows between two plates and in a tube, and assumed that they had reached a stationary velocity profile valid over the whole cross-section of the flow and invariant along it. We will now analyze the case of a turbulent boundary layer which develops along a smooth semi-infinite plane located in a uniform flow **U** parallel to the x-axis (Figure 12.11). One assumes that the characteristic distances for the evolution of the velocity of the flow in the x-direction are large relative to those in the y-direction. Just as for the laminar boundary layer discussed in Chapter 10, the flow outside the turbulent boundary layer is considered as potential and with uniform velocity **U**: in this latter case, the boundary between the turbulent region and the external flow at constant velocity fluctuates as a function of time, even though, instantaneously, this boundary is quite clear in experimental visualizations (Figure 12.11). In order to determine the thickness of the boundary layer $\delta(x)$, instantaneous velocity profiles at different times are averaged.

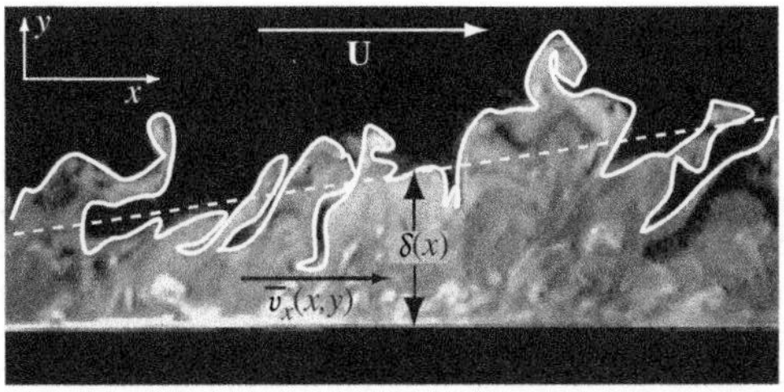

Figure 12.11 *Flow in a turbulent boundary layer. The gray-levels in the image correspond to the intensity of the fluorescence induced by a laser plane perpendicular to the wall (the fluorescent dye is emitted at the wall)*

Variation of the thickness δ as a function of the distance along the wall

The elements of fluid are carried along parallel to the wall at a velocity of the order of U. One assumes that the transverse fluctuations v'_y of the velocity tend to drive away from the plate the boundary of the turbulence: this transverse increase is thus convective, not diffusive, as was the case for the laminar boundary layer. But, in the inertial sublayer: $\rho\, u^{*2} = -\rho\, \overline{v'_x v'_y}$ and experiments indicate that one also often has $\left|\overline{v'_x v'_y}\right| \approx \overline{v'^2_y}$. One can therefore consider that the transverse velocity fluctuations v'_y are of the order of u^*. By using the same kind of qualitative reasoning as in Section 10.2, one obtains:

$$\frac{d\delta(x)}{dx} \approx \frac{u^*}{U}, \qquad (12.81a) \qquad \text{so that:} \qquad \delta(x) \approx \frac{u^*\,x}{U}, \qquad (12.81b)$$

if we take as the origin $x = 0$ the edge of the plate. We will see in fact, further down, that u^* varies only little with the distance x in the direction of the mean flow. The average thickness thus increases proportionately to the distance x and not as $\sqrt{x}$, as it would in a laminar boundary layer.

Laminar and viscous sublayers for a smooth wall

The symbols (Δ) in Figure 12.8 correspond to the experimental variation of the velocity $\bar{v}_x/u^*$ as a function of Log ($y\,u^*/\nu$) in a turbulent boundary layer resulting from uniform flow near a flat plate (u^* is, as in the previous case, related to the stress on the wall by Equation 12.49). The experimental variations are practically identical to those observed for flow in a tube of circular cross-section. We have, just as in this previous case, both a viscous and an inertial sublayer corresponding to variations, respectively linear and logarithmic, of the average velocity as a function of the distance to the wall. The two normalized velocity profiles are almost identical when the distance to the wall is less than 0.1 D for the tube, or small in comparison with the thickness of the boundary layer. Thus:

- in the *viscous sublayer* ($y \ll \nu/u^*$), one has:

$$\bar{v}_x\,(y) \approx \frac{u^{*2}\,y}{\nu} \qquad (12.82) \qquad \text{and:} \qquad \tau\,(x) = \rho\,u^{*2}. \qquad (12.83)$$

- in the *inertial sublayer* ($\nu/u^* \ll y \ll \delta(x)$), the velocity profile y obeys an equation identical to Equation 12.60b:

$$\bar{v}_x\,(y) = u^*\left(\frac{1}{\kappa}\ \text{Log}\ \frac{y\,u^*}{\nu} + C'\right), \qquad (12.84)$$

with the same coefficients $1/\kappa = 2.5$ and $C' \cong 5$. One has also an equation equivalent to Equation 12.60a for the variable $Y = y/\delta(x)$.

Velocity profile at the edge of a turbulent boundary layer in the case of a smooth wall

At high Reynolds numbers, we move away from the logarithmic dependence of the profile as soon as $y > 0.15\ \delta(x)$, approximately. In this transitional region, the flow is alternately potential and turbulent because of the great amplitude of the motion of the edge of the boundary layer seen in Figure 12.11. Equation 12.84 is then no longer valid. The correction term W, which we need to introduce, depends only, as in the case of the two plates, on the variable $y/\delta(x)$. For $y \gg \nu/u^*$, Equation 12.84 can thus be generalized into the form:

$$\frac{\bar{v}_x\,(y)}{u^*} = \frac{1}{\kappa}\ \text{Log}\ \frac{y\,u^*}{\nu} + C' + W\left(\frac{y}{\delta(x)}\right) \qquad (12.85)$$

The function $W(y/\delta(x))$ vanishes in the inertial sublayer ($y \ll \delta$), and its variation near the edge of the layer is similar to that observed at the center of the flows between plates. Writing, by means of Equation 12.85, the condition $\bar{v}_x\,(\delta) = U$ for $y = \delta(x)$, one obtains:

$$\frac{U}{u^*} = \frac{1}{\kappa}\ \text{Log}\ Re^*_\delta + C'', \qquad (12.86)$$

where $C'' = C' + W(1)$ and $Re^*_\delta = u^*\delta(x)/\nu$.

Frictional coefficient along a smooth wall

Defining the friction factor C_d as $C_d = \rho\,u^{*2}/\rho U^2$, Equation 12.86 becomes:

$$C_d = \frac{1}{\left((1/\kappa)\ \text{Log}\ Re^*_\delta + C''\right)^2} \qquad (12.87)$$

C_d varies therefore slowly (logarithmically) with Re^*_δ (and therefore with $\delta(x)$), for a smooth plane wall: this result is in contrast with the variation as $1/Re^{1/2}$ for laminar flows (Equation 10.42).

These results justify *a posteriori* the assumption of a constant value of u^*, made above in order to evaluate the variation of $\delta(x)$ with x. Quantitatively, by using Equation 12.86 and Equation 12.81b, one obtains:

$$\delta(x) \approx \frac{x}{(1/\kappa)\,\mathrm{Log}\,Re^*_\delta + C''}. \tag{12.88}$$

The variation of $\delta(x)$ with x is thus approximately linear in x; more precisely, δ varies as $x^{0.9}$ because of the increase as Log x of the Log Re^*_δ term in the denominator of Equation 12.88. Moreover, the ratio $\delta(x)/x$ depends only very slightly on the external velocity U. Note that δ represents only an average thickness, since the boundary with the external flow is strongly fluctuating and tortuous.

Turbulent boundary layer near a rough surface

Just as in the case of a fully developed flow, the effect of surface roughness is dominant when its depth ε becomes greater than the thickness of the viscous sublayer. In this case, the velocity profile as well as the frictional force and the coefficient C_d depend on the two length scales δ and ε. Just as in Equation 12.79, which corresponds to a fully developed flow in a circular tube, the form of the equations for C_d and for $\delta(x)$ is inferred from Equations 12.87 and 12.88 by substituting for Log $Re_\delta{}^*$ the value Log (δ/ε).

Comparison of the laminar and turbulent boundary layers

In concluding this presentation of the characteristics of turbulent boundary layers, let us summarize the main differences from the laminar boundary layers:

- In a laminar layer, the velocity varies linearly with the distance from the wall in the vicinity of the wall, and then it joins up exponentially with its asymptotic value (Figure 12.12). For the turbulent layer, in normalized coordinates, the initial slope $\partial\bar{v}_x/\partial y$ of the profile in the laminar sublayer is higher than in the laminar case; then, we join the velocity of the external flow more slowly than in the laminar case because of the logarithmic variation in the inertial sublayer.
- The thickness of the boundary layer increases approximately as the distance x from the edge of the turbulent layer (instead of $\sqrt{x}$ in the case of a laminar layer). The convective momentum flux resulting from the turbulent fluctuations in the inertial sublayer is greater than the diffusive flux taking place in a laminar layer. As a result, the friction coefficient C_d is higher. Just as for a fully developed flow, this also explains the large value of the slope $\partial\bar{v}_x/\partial y$ at the wall: in the viscous sublayer, the viscous momentum transport must indeed compensate for the smaller magnitude of the convective transport so that both fluxes match.
- The coefficient C_d at a plane wall varies slower with the Reynolds number for the turbulent layer (as Log $(1/Re)$) than for a laminar layer (as $1/\sqrt{Re}$). C_d is then most often higher in the turbulent case; moreover, C_d also increases with the roughness of the wall at high Reynolds numbers.

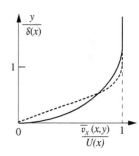

Figure 12.12 *Comparison between the normalized velocity profiles for flows in a laminar (dashes) and turbulent (solid line) regimes in boundary layers in the neighborhood of a smooth flat plate*

12.5.6 Separation of turbulent boundary layers

Another characteristic of turbulent boundary layers, essential in practical applications, is the fact that the gradient $\partial U/\partial x < 0$, required in order to generate the separation, is much

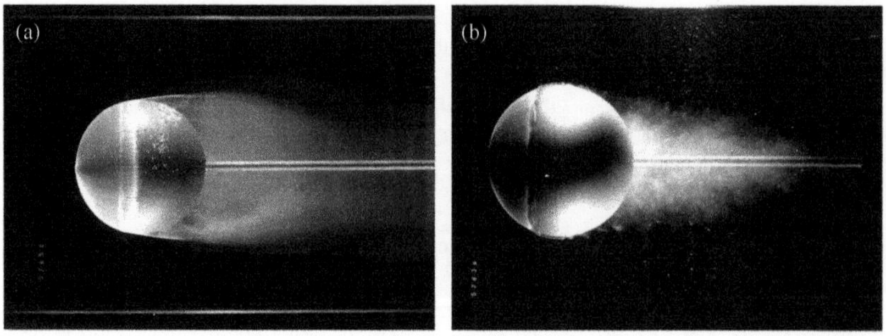

Figure 12.13 *Difference between the locations of the separation points for the case (a) of a laminar boundary layer and (b) of a turbulent one. In the second case, a wire around the sphere upstream of the separation point causes the transition toward a turbulent boundary layer and thus slows down the separation (plates courtesy H. Werlé, ONERA)*

higher than for a laminar layer (Section 10.5.4). This is an indication of the much greater effectiveness of the convective momentum transfer by vortices in the turbulent layer as compared to the purely diffusive transfer in the laminar layer. The transfer of momentum toward the low-velocity regions near the wall is thus more significant, and slows down the inversion of the direction of the flow.

One can thus stabilize the boundary layers by inducing a transition toward turbulence upstream of the normal separation point (e.g., by a small roughness along the surface). The separation of the boundary layer then occurs much farther along than if it had remained laminar. The width of the downstream turbulent wake is then significantly reduced (Figure 12.13b), in comparison with a laminar boundary layer (Figure 12.13a).

The effect of the reduced size of the wake is generally much greater than that of the increase of the frictional force, in the region where the boundary layer has not separated because it became turbulent: the amplitude of this latter effect can be visualized in Figure 12.9 through the difference between the values of C_d for laminar and turbulent flow, in the transition region for a tube of circular cross-section ($Re_D \approx 3000$).

This reduction of the size of the wake causes a significant decrease of the energy dissipation in the wake. The global drag can then be significantly decreased, an important consideration in a number of practical applications.

One of the most spectacular manifestations of this effect is the phenomenon of the *drag crisis*: it is observed at Reynolds numbers of several hundred thousands, even on smooth spheres and circular cylinders. It is the result of a spontaneous transition of the laminar boundary layer toward turbulence: this phenomenon is accompanied by a dramatic reduction of the drag (Figure 12.14) related to the reduction of the size of the wake. For dimpled

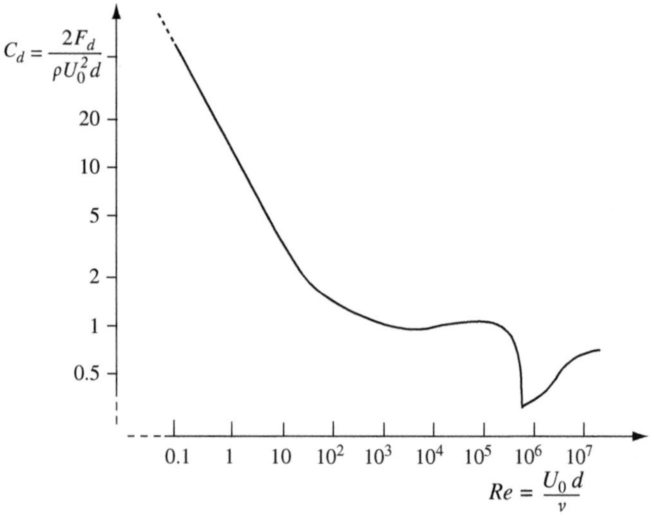

Figure 12.14 *Illustration of the drag-coefficient crisis for a cylinder of circular cross-section: at a Reynolds number of the order of 3×10^5, the drag exerted on a circular cylinder suddenly drops by a factor of the order of 2. This is the result of the sudden decrease in the width of the wake*

surfaces, such as a golf ball, or rough surfaces, such as a tennis ball, this transition occurs earlier. This allows for a reduction of the drag, provided that the velocity required for the transition is exceeded, a result instinctively recognized by strong players.

A related effect is the *reattachment of boundary layers* which sometimes occurs when a boundary layer becomes turbulent just downstream of a separation point Figure 12.15. In this case, the transverse momentum transfer, by turbulent fluctuations, can be sufficient to reaccelerate the fluid near the wall, and suppress the recirculation flow.

The reattachment of the boundary layer along a wall may be observed (Figure 12.15) by looking at the tarpaulin on the side of a passing truck.

Figure 12.15 *Separation and subsequent reattachment of a boundary layer due to its transition to turbulence, the result of a snub-nosed profile (plate courtesy ONERA)*

12.6 Homogeneous turbulence – Kolmogorov's theory

The multiplicity of the characteristic sizes of vortex-like motions is one of the essential features of turbulence. We have already written, in Section 12.2, the balance equations for the exchange of energy and vorticity between the mean flow and the ensemble of these turbulent fluctuations of the velocity. The goal of the theory developed by Kolmogorov, in 1941, is to make these balances more precise, and to study the distribution and the exchange of energy and vorticity among vortices of different sizes.

12.6.1 Energy cascade in a homogeneous turbulent flow

Qualitative description

In the Kolmogorov model of an *energy cascade*, one considers that the energy injected into the mean flow at its largest scale is transmitted to the largest vortices of characteristic size ℓ (Figure 12.16); these transmit then the energy to immediately smaller vortices located in the same region of space and, sequentially, eventually to the smallest vortices, at whose level the energy is dissipated by the viscosity.

The energy transfer between vortices of different sizes in Kolmogorov's theory displays some analogies with the transfer of momentum in a turbulent flow near a wall. The average motion of a homogeneous turbulent flow corresponds to the profile of the global velocity, or to the velocity at a large distance from the walls. The smallest vortices for which viscous dissipation is important correspond to the viscous sublayer very close to the walls. Between these two scales, there exists an intermediate domain, in which the transfer of momentum or energy is dominated by inertial effects, and with characteristics independent both of the viscosity and the global size of the flow.

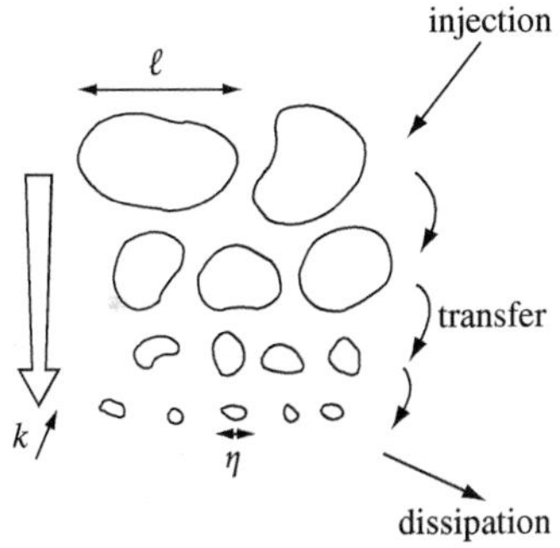

Figure 12.16 *Principle of Kolmogorov's energy-transfer cascade from the largest vortices toward the smallest ones. Note that the vertical scale is defined here in terms of the virtual space of the sizes of the vortices, not in real space: the vortices are actually randomly distributed with no segregation with respect to their size*

In order to justify this description, one considers Equation 12.19 for the transport of the energy of the fluctuations, assuming at this stage that the mean flow is stationary. One assumes further that the flow evolves slowly, as a function of the distance, so that the divergence of the terms involving the flux of energy is small. The balance of the energy of the turbulent fluctuations thus becomes, neglecting these terms and those on the left-hand side:

$$-\nu\overline{\left(\frac{\partial v'_i}{\partial x_j}\right)^2} = \overline{v'_i v'_j}\,\frac{\partial \bar{v}_i}{\partial x_j}. \tag{12.89}$$

This equation indicates that, for these fluctuations, there is equilibrium between the energy dissipated by the viscosity and that received from the mean flow. Let us assume more precisely a fact verified *a posteriori* in Equations 12.92a–b. The velocities associated with the vortices increase with their size; in contrast, the velocity gradients are greater for the small vortices. In this case, for the energy balance (Equation 12.89), the largest vortices contribute, in dominant fashion, to the term $-\overline{v'_i v'_j}\left(\partial \bar{v}_i / \partial x_j\right)$ and the smallest ones to $\nu\overline{\left(\partial v'_i / \partial x_j\right)^2}$.

The image of the Rankine vortex discussed in Section 7.1.2 allows one to understand the origin of this energy exchange: it results from the stretching of the vortex tubes, which appears in the equation for the balance of vorticity (Equation 12.20a) through the term $\overline{\omega'_i \omega'_j\left(\partial v'_i / \partial x_j\right)}$. In a stationary homogeneous flow for which $d\overline{\omega'^2}/dt = 0$, this term must be, on average, positive for Equation 12.20b to be obeyed. As seen in Section 7.3.1, if one stretches a vortex tube of length L and radius R, the circulation of the velocity around the vortex tube (equal to the integral $\pi\omega' R^2$ of the vorticity through a cross-section) and its volume $V = \pi R^2 L$ remain constant as the vortex is deformed. The kinetic energy of rotation of the tube, of the order of $\rho V R^2 \omega'^2$, is, in contrast, proportional to $1/R^2$ and also to L: the energy of the vortex increases therefore as it is stretched.

The key complementary assumption of the above-mentioned Kolmogorov model is that the vortices of a given size receive energy only from those of immediately greater size, and transmit it only to those of immediately smaller size. Let us denote by $\xi (\leq \ell)$ one of the characteristic sizes of the vortices present and by v'_ξ and $(\partial v'/\partial x)_\xi$ the order of magnitude of the velocity fluctuations and of the velocity gradient associated with vortices of size ξ. The present discussion will take place in the space of the size ξ of the vortices and assume that the component v'_ξ increases as ξ increases, while the derivatives $(\partial v'/\partial x)_\xi$ decrease: since, in the following, one assumes also that $(\partial v'/\partial x)_\xi \approx v'_\xi/\xi$, these assumptions imply that, if v'_ξ follows a power law in ξ, the exponent must be less that one. The stretching of the vortex tubes corresponding to a component of size ξ of the turbulent velocity field results only from velocity gradients $(\partial v'/\partial x)_{\xi'}$ created by the components of size $\xi' > \xi$ (the components of size smaller than ξ give a vanishing effect when averaged over the whole vortex). Among these components, those which give the largest values of $(\partial v'/\partial x)_{\xi'}$, and which have the strongest effect correspond to the sizes immediately greater than ξ.

Let us describe, in more precise terms, what we mean by "size immediately larger or smaller". Arguments such as the previous one classify implicitly vortices along a logarithmic scale which is the only one adequate to describe the variations by several orders of magnitude of the size of the vortices in a highly developed turbulence. One divides then the scale of the useful values of Log ξ into equal intervals which correspond, for example, to a change by a factor of two, as one shifts from one length-scale to the neighboring one. The vortices are then distributed among the various intervals: one will say, for example, that vortices with a size of the order of ξ receive energy from vortices of size 2ξ and then pass it on to those of typical size $\xi/2$ (the number two is arbitrary).

Scaling laws and characteristic sizes resulting from the energy-cascade model

Let us call ε the amount of energy per unit mass exchanged during a given time interval by the mean flow with the largest vortices (ε is measured in m^2/s^3). In the cascade process which has just been described, ε is also equal to the energy exchanged by vortices of arbitrary size ξ with those of a neighboring size, so long as the size ξ is large enough so that one can neglect dissipation due to viscosity.

One considers first the level of the large scales of turbulence: one denotes as ℓ the size of the largest vortices, v'_ℓ the associated characteristic velocity and t_ℓ the time taken by a particle of matter to go around the entire vortex with $t_\ell \approx \ell/v'_\ell$; t_ℓ is assumed to represent the order of magnitude of the time taken by the vortex to transfer its initial energy. One then has:

$$\frac{dv_\ell'^2}{dt} \approx -A\frac{v_\ell'^2}{\left(\ell/v'_\ell\right)}, \qquad (12.90a) \qquad \text{i.e.:} \qquad \varepsilon \approx \frac{A\, v_\ell'^3}{\ell}. \qquad (12.90b)$$

A is a dimensionless constant of the order of unity. Equation 12.90b can be verified experimentally by measuring the loss of energy from the turbulence created by the passage of a flow at velocity U through a grid, as a function of the distance to the latter. The experiments confirm approximately the initial assumptions, even when the size of the largest vortices changes. These experimental verifications suggest that Equations 12.90a–b remains valid for smaller vortices, of size $\xi < \ell$, but large enough so that viscous effects remain negligible. Then:

$$\frac{dv_\xi'^2}{dt} \approx A\,\frac{v_\xi'^2\, v_\xi'}{\xi}, \qquad (12.91a) \qquad \text{so that:} \qquad \varepsilon \approx \frac{A\, v_\xi'^3}{\xi}. \qquad (12.91b)$$

One refers to the range of sizes of vortices for which this equation is obeyed as the *inertial subrange*: it is justified by the fact that the time $t_\xi \approx \xi/v'_\xi$ is the only characteristic time associated with a vortex of given size, intermediate between those of the largest vortices and of the smallest (affected by viscosity). This implies that the dynamics of the intermediate-sized vortices is independent of both the global structure of the flow (the largest vortices) and of the viscosity which only affects the smallest ones: this assumption results from the cascade model in which vortices in a given range of sizes interact only with the immediately larger and smaller ones.

Since the rate of transfer of energy ε is the same for all vortices and is therefore independent of ξ, Equation 12.91b implies:

$$v'_\xi \propto \xi^{1/3} \qquad (12.92a) \qquad \text{and:} \qquad (\partial v'/\partial x)_\xi \propto \xi^{-2/3}. \qquad (12.92b)$$

This confirms the assumption that v'_ξ decreases and the gradient $(\partial v'/\partial x)_\xi$ increases as the spatial scale $\boldsymbol{\xi}$ of the vortices decreases.

As the total amount of the energy transferred must be equal to the dissipated energy, ε also represents the rate of dissipation of energy per unit mass. Let us call η the characteristic size of the smallest vortices (this established notation should not be confused with the dynamic viscosity η). More specifically, η represents the size of the vortices for which the energy dissipated by viscosity $\nu\,\overline{\left(\partial v'_i/\partial x_j\right)^2}$ becomes of the order of ε. Substituting for this last expression the approximate value $\nu\,(v'_\eta/\eta)^2$, the length η and the corresponding velocity v'_η must obey the double condition, taking $A = 1$ in Equations 12.91a-b:

$$\varepsilon = \nu \left(\frac{v'_\eta}{\eta}\right)^2 = \frac{v_\eta'^3}{\eta}. \qquad (12.93)$$

Solving this equation, one obtains:

$$\eta = \nu^{3/4}\,\varepsilon^{-1/4} \qquad (12.94) \qquad \text{and:} \qquad v'_\eta = (\nu\,\varepsilon)^{1/4}, \qquad (12.95)$$

where η and v'_η are respectively referred to as the *Kolmogorov length* and *Kolmogorov velocity.* Since the smallest vortices lose their energy very rapidly by viscous dissipation, one can consider that η represents the typical lower limit of the size of vortices. Another way to obtain this same result consists in calculating the Reynolds number $Re_\eta = v'_\eta \eta/\nu$ by taking the Kolmogorov velocity and length as characteristic scales. One finds $Re_\eta = 1$ which confirms that η is the length-scale for which convective and viscous effects become of the same order of magnitude.

The *inertial subrange* of vortex sizes ξ for which scaling laws such as Equation 12.92a-b are valid extends (as an order of magnitude) from η to ℓ. According to the very principle of the energy cascade, this concept is indeed no longer applicable to scales of size greater than ℓ, as seen above, or less than η, for which viscous dissipation becomes important. Experimental results corresponding to this *inertial subrange* will be discussed in Section 12.6.3.

In this same domain of vortex sizes, Equation 12.91b remains valid and the kinetic energy E_ξ per unit mass of fluid, corresponding to vortices of size ξ has the value:

$$E_\xi \propto v_\xi'^2 \propto (\varepsilon\,\xi)^{2/3}. \qquad (12.96)$$

The energy is thus principally associated with the largest vortices.

On the other hand, the *enstrophy* $\omega_\xi'^2$ (the square of the vorticity defined in Section 12.2.5) associated with vortices of size ξ has a dependence $\omega_\xi'^2 \propto (v_\xi'/\xi)^2 \propto \varepsilon^{2/3}\xi^{-4/3}$. The enstrophy is then concentrated in the smallest vortices because it increases as their size decreases.

In addition to the scaling laws which have just been discussed, Kolmogorov has derived a precise general equation, based on the conservation of energy and valid in homogeneous and isotropic turbulence: it involves a third-order moment of the component $v_\parallel$ of the velocity, parallel to the increment $r_\parallel$:

$$\left\langle \left(v_\parallel\left(\mathbf{x}+\mathbf{r}_\parallel\right)-v_\parallel\left(\mathbf{x}\right)\right)^3\right\rangle = -\frac{4}{5}\,\varepsilon\, r_\parallel. \quad (12.98)$$

This equation is known as Kolmogorov's four-fifths law, and is valid for distances r small compared to the size of the largest vortices. It may be used to determine the parameter ε (which plays a key role, as seen above, in Kolmogorov's cascade model).

Kolmogorov cascade and correlations between velocity fluctuations

An equation related to the previous one characterizes the correlation between velocity fluctuations measured at two points separated by an interval $\mathbf{r}$ in a turbulent homogeneous and isotropic flow. One finds that:

$$\left\langle \left(v_\parallel\left(\mathbf{x}+\mathbf{r}\right)-v_\parallel\left(\mathbf{x}\right)\right)^2\right\rangle_{\mathbf{x}} = C\,\varepsilon^{2/3}\,r^{2/3}, \quad (12.97)$$

in which $r = |\mathbf{r}|$ and $v_\parallel$ is the velocity component in the direction of $\mathbf{r}$. The correlation function of the left-hand side is independent of the volume over which the average is taken because of the homogeneity of the flow; it does not depend either on the orientation of $\mathbf{r}$ because of the isotropy. It is understood that this equation is only valid in the same range of distance scales r for which Equation 12.96 is itself valid.

12.6.2 Spectral expression of Kolmogorov's laws

So far, turbulent flow has been considered as a superposition of vortices of different sizes. We describe now another classical method using the Fourier transform of the velocity field to perform a spectral decomposition into sinusoidal components of different wave numbers.

Spectral decomposition of the spatial variation of the velocity

The spectral decomposition of the velocity field is performed by taking the three-dimensional *Fourier transform* of the components v_j'. This amounts to represent the velocity fluctuations as a superposition of sinusoidal waves. One thus defines:

$$v_{\mathbf{k},j}' = \frac{1}{(2\pi)^3}\iiint v_j'(\mathbf{r})\,\mathrm{e}^{-\mathrm{i}\mathbf{k}\cdot\mathbf{r}}\,\mathrm{d}^3\mathbf{r}, \quad (12.99a)$$

and, taking the inverse Fourier transform:

$$v_j'(\mathbf{r}) = \iiint v_{\mathbf{k},j}'\,\mathrm{e}^{\mathrm{i}\,\mathbf{k}\cdot\mathbf{r}}\,\mathrm{d}^3\mathbf{k}. \quad (12.99b)$$

The kinetic energy of the flow can then be computed as a function of the components $v_{\mathbf{k},j}'$ by means Equation 12.99b. The energy per unit mass obeys (summing over the index j and the volume $\mathcal{V}$ corresponding to a unit mass of the fluid):

$$\iiint_{\mathcal{V}} \frac{\overline{v_j'^2}}{2}\,\mathrm{d}^3\mathbf{r} = \iiint \frac{\overline{\left|v_{\mathbf{k},j}'\right|^2}}{2}\,\mathrm{d}^3\mathbf{k}. \quad (12.100)$$

$\left(\overline{\left|v_{\mathbf{k},j}'\right|^2}/2\right)\mathrm{d}^3\mathbf{k}$ thus represents the energy per unit mass associated to the components of the *wave vector* near $\mathbf{k}$, within a volume $\mathrm{d}^3\mathbf{k}$ of the spectral domain. In a homogeneous and isotropic turbulent flow, this energy depends only on the modulus $|\mathbf{k}|$ and not on the direction of $\mathbf{k}$. One can then define a spectral energy density $E(k)$: $E(k)\,\mathrm{d}k$ is the energy associated with the components for which the *modulus of the wave vector* ranges between k and $k + \mathrm{d}k$. As the corresponding volume in the spectral region is $4\pi\,k^2\,\mathrm{d}k$, one obtains then (again summing over j):

$$E(k) = 2\pi\, \rho\, k^2\, \overline{\left|v'_{\mathbf{k},j}\right|^2}. \qquad (12.101)$$

An important problem with the Fourier transform is that the sine wave functions have an infinite extent through the physical space, while turbulent structures have a finite size. One must then consider vortices of different sizes as wave packets, corresponding to an interval Δk of values of k centered around zero; more specifically, the size ξ of the vortex is of the order of $2\pi/\Delta k$. Just as in the case of Heisenberg's uncertainty principle, we cannot perfectly localize the turbulent fluctuations both in real space and in the space of the wave vectors: the larger Δk, the smaller the spatial extent will be.

Other more elaborate techniques such as the *wavelet transform* have also been used. The wavelets, like sinusoidal waves, constitute an orthonormal basis. In contrast to them, they are only non-vanishing in a finite size region which corresponds to the characteristic size of the turbulent fluctuations that they represent.

The viscous energy dissipation can be computed from Equation 5.26 by means of a similar mathematical method. For homogeneous and isotropic turbulence, this dissipation obeys, again for a volume $\mathcal{V}$ of unit mass of fluid:

$$\iiint_{\mathcal{V}} 2\,\nu\, e_{ij}^2\, \mathrm{d}V = 2\nu \int k^2\, E(k)\, \mathrm{d}k. \qquad (12.102)$$

As in the previous case, only the modulus k of the wave vector is involved. The factor k^2 appears while computing the Fourier transforms of the spatial derivatives of the velocity occurring in $e_{ij}{}^2$, leading to the appearance of the different components of $\mathbf{k}$ as multiplying factors.

Spectral tranformation based on variations in time

The PIV techniques described in Section 3.5.4 determine the instantaneous velocity field in a measurement plane, and thus allow one to carry out the Fourier transformations of the velocity components which have just been discussed (at least for two of the components of the velocity). However, the experimental velocity mapping barely amounts to a hundred points in each direction: the range of length-scales covered by these mappings is thus narrow since it extends only from the interval between points up to the global field of view. At high Reynolds numbers, one will then be unable to cover the full domain of length-scales required in order to describe the flow; this region should, indeed, range from the global size of the flow down to scales smaller than the Kolmogorov length η. One cannot then verify precisely scaling laws as in Equations 12.92a-b.

Other techniques allow one, however, to determine $E(k)$ in such cases. One frequently uses a fixed probe to measure one or several components of the velocity and records its output as a function of time. Information on the spatial variation of the velocity at a given time may be deduced from this measurement if individual vortices cross the probe over a time lapse shorter than the characteristic time for their evolution: such measurements require therefore a high mean velocity of the flow (compared to that of the velocity fluctuations). This latter assumption is that of Taylor's *frozen turbulence*. Practically, spectral components $v'_{\omega,j}$ which are functions of the angular frequency ω are computed by applying a Fourier transform to the variations with time. Using the assumption of frozen turbulence, one considers that $v'_{\omega,j}$ coincides with the spectral spatial component $v'_{k_1,j}$, in which k_1 is the component of the wave vector parallel to the velocity $\bar{\mathbf{v}}$ of the mean flow and related to its modulus by $k_1 = \omega/\bar{v}$. Using probes which have a high frequency response, such as hot-wire anemometers (Section 3.5.3), very fast time variations can be measured, allowing for the analysis of vortices over a very broad domain of characteristic length scales.

One cannot measure in this way the components $v'_{k_i,j}$ for every component of the wave vector $\mathbf{k}$, but only for the component k_1 parallel to the mean flow $\mathbf{U}$. Moreover, one measures generally only one or two components of the velocity: one does not, therefore, determine the sum $\overline{|v'_{\mathbf{k},j}|^2}/2$ for every component of $\mathbf{k}$ and all the indices j, but only parameters like:

$$E_1(k_1) = \frac{\overline{\left|v'_{k_1,1}\right|^2}}{2} \qquad (12.103)$$

However, if the turbulent fluctuations are isotropic, $E_1(k_1)$ and $E(k)$ are shown to be related by:

$$E(k) = 2k_1^3\, \frac{\mathrm{d}}{\mathrm{d}k_1}\left(\frac{1}{k_1}\frac{\mathrm{d}E_1}{\mathrm{d}k_1}\right). \qquad (12.104)$$

$E(k)$ can then be evaluated as a function of $E_1(k_1)$. If, as seen further down, $E(k)$ varies with k following a power law, $E_1(k_1)$ also obeys a power law with the same exponent and in the same range of wave vectors as $E(k)$.

In another approach one uses a detector attached outside a body (e.g. plane or boat) moving rapidly within a flow and measures the time variation of its relative velocity with respect to the fluid. These variations are translated to spatial variations by performing a spectral analysis and replacing the frequency ω by the product $U k_1$ (U is the velocity of the probe and k_1 the component of the wave vector parallel to the motion). If the distribution of the turbulent fluctuations is isotropic, $E(\mathbf{k})$ can be deduced from these spectral components (Equation 12.104). The assumption of *frozen turbulence* is valid, in this case, if the probe crosses a turbulent structure in a time lapse shorter that needed for the structure to evolve.

Kolmogorov's similarity laws in wave vector space

We will now describe similarity laws which express, in the wave vector space $\mathbf{k}$, the process of *energy cascade* discussed in real space in Section 12.6.1. Just like above, vortices are considered as a wave packet for which the range of the values of the wave vector k is centerd about zero over a width $\Delta k \approx 2\pi/\xi$. The wave vectors occurring in the description of vortices of size ξ are thus of the order of $1/\xi$, and are greater, the smaller the vortices. We will now derive, by means of dimensional arguments, two similarity laws suggested by Kolmogorov: the first is valid for sufficiently small vortices (thus for large wave vectors) so that they are not influenced by the global structure of the mean flow; the range of wave vectors over which this law is valid includes the smallest vortices for which viscosity cannot be neglected. The second of Kolmogorov's laws is more specific, and only applies in the intermediate *inertial subdomain* where neither the viscosity nor the mean flow play a role.

Kolmogorov's first similarity law

One first examines the case of vortices of size $\xi \approx 1/k \ll \ell$. In this case, the energy density $E(k)$ per unit mass, of dimension L^3/T^2, must depend only on the variables k, ν and ε, of respective dimensions: $1/L$, L^2/T and L^2/T^3, and not on ℓ. The *Kolmogorov length* $\eta = \nu^{3/4}\varepsilon^{-1/4}$ is then the only characteristic scale length in this problem (one recalls that it is the size of vortices for which the energy dissipation by viscosity is of the order of magnitude of the rate ε of transfer of energy from the mean flow).

Kolmogorov's first similarity law assumes then that, again if $k\ell \gg 1$, the wave vector $\mathbf{k}$ is only involved through a function f of the dimensionless combination $k\eta$. There must be, in front of f, a prefactor $\varepsilon^{\alpha}\nu^{\beta}Z$ with the same dimension as $E(k)$ so that: $\alpha = 1/4$ and $\beta = 5/4$. One then obtains:

$$E(k) = \nu^{5/4}\,\varepsilon^{1/4}\,f(k\eta). \tag{12.105}$$

Because of the loss of energy by vortices of size less than η due to viscous dissipation, one predicts (and verifies experimentally) that the function $E(k)$ decreases faster for $k \gg 1/\eta$ than for smaller values of the wave vector.

Kolmogorov's second similarity law: inertial subrange

Let us now analyze the case of vortices of size $\xi \gg \eta$: in contrast to the previous case, viscosity does not play a role in the dissipation of their energy. The energy density $E(k)$ then must be only a function of k, ℓ and ε. Proceeding as in the previous case, one finds:

$$E\,(k) = \varepsilon^{2/3}\,\ell^{5/3}\,g\,(k\ell) \qquad \text{for:} \qquad k \ll 1/\eta; \tag{12.106}$$

k must be slightly greater than $1/\ell$ so that the scales involved are not too much affected by the large-scale structure of the flow.

If ℓ is much larger than η, there exists a range of values of k for which the conditions $k\ell \gg 1$ of Kolmogorov's first law, and $k\eta \ll 1$ of Equation 12.106, are simultaneously obeyed. This range is the *inertial subrange*.

In this domain, $E(k)$ must be independent of ℓ in Equation 12.106, implying that $g(k\ell) \propto (k\ell)^{-5/3}$. On the other hand, $E(k)$ must also be independent of ν in Equation 12.105; substituting for η its value in Equation 12.94, this implies that $f(k\eta) \propto (k\,\eta)^{-5/3}$. These two conditions lead to the same equation:

$$E\,(k) = K_0\,\varepsilon^{2/3}k^{-5/3}. \tag{12.107}$$

One finds experimentally that K_0, known as *Kolmogorov's constant*, is of the order of 1.45, and that it is independent of ℓ and of ν. Equation 12.107 confirms that, as seen above, the energy is dominantly concentrated in the large vortices, and decreases with their size.

Equation 12.107 represents the famous *Kolmogorov* $k^{-5/3}$ *law* which has been well verified experimentally, as shown by Figure 12.17. As said above, this equation is only valid in the range of wave vectors corresponding to the inertial subdomain, i.e. to vortices large enough so that the effects of viscosity are negligible, but small relative to the characteristic size of the mean flow where energy injection occurs. This condition translates into the relationship $\eta \ll 1/k \ll \ell$: this implies that $\ell \ggg \eta$ (in practical terms $\ell > 10^3\eta$ or even $10^4\eta$). Making use of Equation 12.94, this inequality becomes:

Equation 12.107 is equivalent to the equation derived above for the energy E_ξ of vortices of size ξ. Considering that the size ξ of the vortices corresponding to turbulent fluctuations of wave vectors of modulus k is of the order of $1/k$, Equation 12.96 becomes $E_{\xi=1/k} \propto \varepsilon^{2/3}k^{-2/3}$. Discussing again the size of vortices in terms of a logarithmic scale (e.g., energy exchange between vortices and others of double and half their size), one can consider that $E_{\xi=1/k}$ represents, in the space of the wave vectors, the integral of $E(k)$ over an interval $\Delta k \approx k$. One then obtains, from Equation 12.107, the estimate $E_{\xi=1/k} \propto (\varepsilon^{2/3}k^{-5/3})k$, which corresponds well with that deduced from Equation 12.96.

$$\left(\frac{v'_\ell\,\ell}{\nu}\right)^{3/4} = Re_\ell^{\,3/4} \ggg 1, \qquad (12.108)$$

where Re_ℓ is the Reynolds number associated with the velocity and size scales of large vortices. These scales are generally significantly smaller than the global velocity and size scales of the flow (e.g., for turbulence resulting from a grid). The number Re_ℓ is thus generally smaller than the global Reynolds number of the flow: this makes it even more difficult to satisfy experimentally the condition of Equation 12.108.

12.6.3 Experimental verification of Kolmogorov's theory

Verification of the $k^{-5/3}$ law

The verification can be carried out in natural flows. In the atmosphere, one easily reaches a Reynolds number $Re \approx 10^7$, with $\ell = 100$ m, $v'_\ell = 1$ m/s, $\nu \approx 10^{-5}$ m^2/s. Measurements on meteorological towers, or using a probe dragged by a plane flying through the atmosphere, allowed one to verify the variation of $E(k)$ as $k^{-5/3}$ over about three orders of magnitude (vortices with a size varying between a few centimeters and a hundred meters) and also verified the isotropy of the established turbulence. This dependence is also verified on oceanic flows (for instance tidal flows in straits), where the global Reynolds numbers can be as large as 3×10^8. Finally, experiments carried out in the large wind-tunnel at Modane (France) have shown, in a well-controlled manner, variations of this type over a broad range of wave vectors (there, again, over a little less than three orders of magnitude) (Figure 12.17). One notes the very clear appearance of a cut-off effect at large wave vectors, and also for large vortices at small wave vectors.

Verification of Kolmogorov's first law

Such a verification can be carried out in a laboratory setting by means of hot-wire anemometers, which provide very localized measurements. Experimentally, one can demonstrate that, if $k\ell \gg 1$, the ratio $E(k)/(\varepsilon^{1/4}\,\nu^{5/4})$ is a universal function of $k\eta$. In the laboratory, it is quite rare to be able to observe the *inertial subrange*; as mentioned above, this is mostly possible in the atmosphere, in the ocean, or in large wind-tunnels (Figure 12.17). The parameter ε can be determined indirectly, since it is equal to the total energy dissipated by the smallest vortices. One verifies experimentally that, for very various Reynolds numbers and methods of generating flows, the measurements correctly follow a universal law for $k\ell \gg 1$.

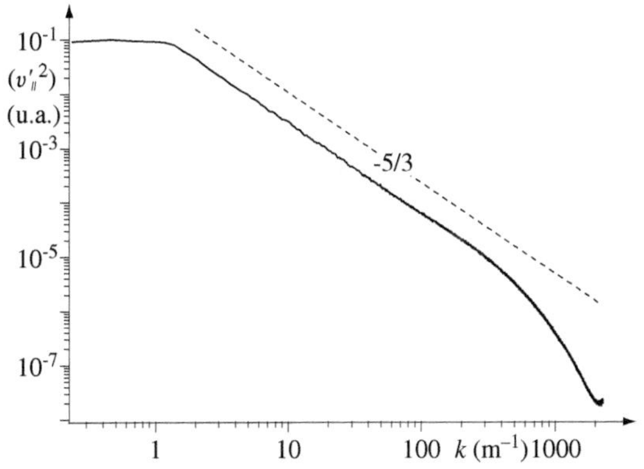

Figure 12.17 *Spatial spectrum, in arbitrary units, of the turbulent fluctuations of the longitudinal component $v'_\parallel$ of the velocity squared,measured in the ONERA S1 wind-tunnel at Modane-Avrieux as a function of the wave vector k. The dashed line corresponds to a variation as $k^{-5/3}$. Experimental parameters: average velocity $\bar{v}_\parallel$ =* 20.6 m/s, $\sqrt{\overline{v'^2_\parallel}}$ = 1.66 m/s, *diameter of the experimental cross-section* $D = 24$ m, *Kolmogorov length* $\eta = 0.31$ mm *($k_\eta = 1/\eta = 3.2 \times 10^3$* m^{-1}*). The spectrum was calculated as a function of the wave vector k from the time fluctuations of the velocity, taking into account the variations of the local velocity of the flow $\bar{v}_\parallel + v'_\parallel$ (from Y. Gagne and Y. Malécot)*

12.7 Other aspects of turbulence

Up to this point, we have considered that turbulence was characterized by the existence of unpredictable flows, with strong non-linear effects and a significant convective transport (this transport is the reason for the great effectiveness of turbulent mixing). We are, however, far from having explored all the possible manifestations of turbulence, which frequently end up beyond this simple description. We will discuss here briefly a few of these situations by making use of what we have been able to learn throughout this textbook, and by referring the reader to more specialized works for a detailed analysis.

12.7.1 Intermittent turbulence

Kolmogorov's theory gives a good overview of isotropic turbulence, and remains a key reference for a first approach to homogeneous turbulence. Nonetheless, large vortices are never in statistical equilibrium in real-life experiments. Since these determine the amplitude of the energy exchange, this implies that there are other temporal variations in the value of ε: experimental measurements show, in fact, that there exist considerable variations of the energy-dissipation ε as a function of time, and from one point to another. Physically, these fluctuations can be correlated to the passage of turbulent structures that one easily observes on the surface of a river. The *turbulent puffs* are closely related to the concept of *intermittency*. The importance of fluctuations in ε leads to take them into account in the consideration of higher-order moments of the velocity where Kolmogorov's theory is deficient; one must then take into account not only the average value, but also the higher-order moments of the probability distribution of ε.

12.7.2 Coherent turbulent structures

One has seen in several instances that turbulence is not incompatible with the existence of large-scale, extremely stable structures, known as *coherent structures*. The red spot in Jupiter's atmosphere is an outstanding example at an extremely high Reynolds number. Assuming a velocity of the order of 100 m/s, a kinematic viscosity of hydrogen of the order of 10^{-5} m^2/s, and a height of the atmosphere of 100 km, one obtains a Reynolds number: $Re \approx 10^{12}$.

In Chapter 2, we showed that vortices with alternating directions of the circulation, the Bénard–von Karman vortex street, appear behind a cylindrical obstacle within a flow (Figure 2.9). Such vortex streets can still be observed at very high Reynolds numbers (Figure 2.10).

One encounters similar structures at the boundary between two plane parallel flows with different velocities: the corresponding Kelvin–Helmholtz instability for parallel flows at different velocities has been discussed in Section 11.4.1. This instability creates a periodic row of vortices with their axes perpendicular to the flow (Figure 11.16) and with the same direction of circulation, in contrast to the Bénard–von Karman vortex street. If, instead of having a flow between walls, the parallel flows are unconfined, the size of the vortices increases approximately linearly as a function of the distance from the point where the jets meet. Such structures create an admixture between the two fluids so that one also speaks of a *mixing layer* instability.

The American physicist A. Roshko showed experimentally that these large two-dimensional vortex structures persist almost identically at every Reynolds number, regardless of how high it is. These large-scale structures are visible in the three images of Figure 12.18 at different Reynolds numbers. On the other hand, finer structures which correspond to smaller-size vortices are more visible for the highest value of Re (Figure 12.18a).

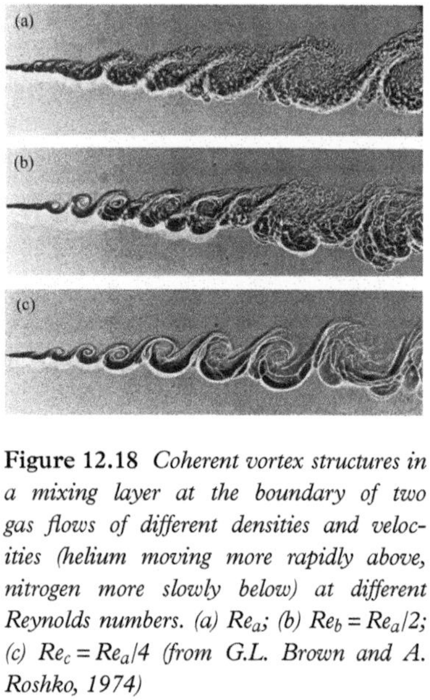

Figure 12.18 *Coherent vortex structures in a mixing layer at the boundary of two gas flows of different densities and velocities (helium moving more rapidly above, nitrogen more slowly below) at different Reynolds numbers. (a) Re_a; (b) $Re_b = Re_a/2$; (c) $Re_c = Re_a/4$ (from G.L. Brown and A. Roshko, 1974)*

In spite of the presence of smaller vortices, it is the large-scale structures which govern the mixing: this mixing has very different characteristics in the regions where the structures roll up (giving the appearance of a cake roll) than in the stretching regions, which also appear simultaneously. This is especially important for two fluids undergoing a chemical reaction, such as in a gas burner (Section 10.9.1 and Figure 10.24b).

Even when the result is unchanged regarding large-scale coherent structures, increasing the Reynolds number leads to the appearance of small three-dimensional structures within the larger ones: the two-dimensional rolls are distorted, then stretched in the direction of the flow before producing smaller vortices. This last result is actually conform to Kolmogorov's theory, which predicts a decrease in the scale of the dissipative structures as the Reynolds number increases. We can thus observe an intermediate inertial regime, at scales smaller than those of the large structues.

12.7.3 Dynamics of vortices in two-dimensional turbulence

Another spectacular characteristic of unconfined mixing layers, such as those displayed in Figure 12.18, is the evolution of vortices as they move. Their velocity is intermediate between that of the two flows, i.e., $1/2(U_1 + U_2)$ as seen in Sections 7.4.2 and 11.4.1. The vortices tend to group themselves in pairs with the same direction of the circulation as they move away from the point where they were created; they end up playing hopscotch, until they combine to produce a third one, larger than the first two.

Because this result involves the large-scale structures, which are essentially two-dimensional, it can be considered as a characteristic of two-dimensional turbulence. In fact, the process of accretion of vortices, which has just been described, is also observed in numerical simulations of two-dimensional turbulence such as those in Figure 12.19. The turbulent structures (visualized by the distribution of vorticity) grow through the coalescence of vortices with the same direction of circulation, as was the case in the mixing layer described above. At the same time, these structures attenuate due to viscous effects, since one does not have a new source of energy after the creation of the vortices. Similar

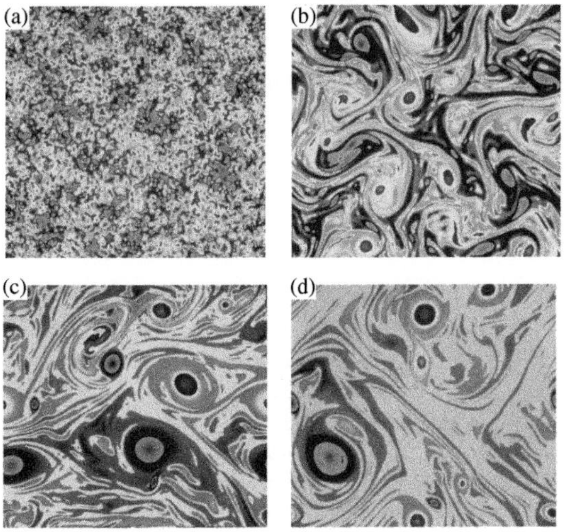

Figure 12.19C *Numerical simulation of the evolution, in time, of a distribution of vorticity, initially random within a two-dimensional turbulent flow, and evolving without further addition of external energy. The color code (see color plate section) is used to represent the vorticity field: red for the strongly positive values, blue for the strongly negative values, gray for low values and yellow for the value zero. The images (a), (b), (c) and (d) correspond respectively to normalized times* $t = 0$, $t = 1$, $t = 3$, $t = 5$ *(plates courtesy M. Farge and J-F. Colonna)*

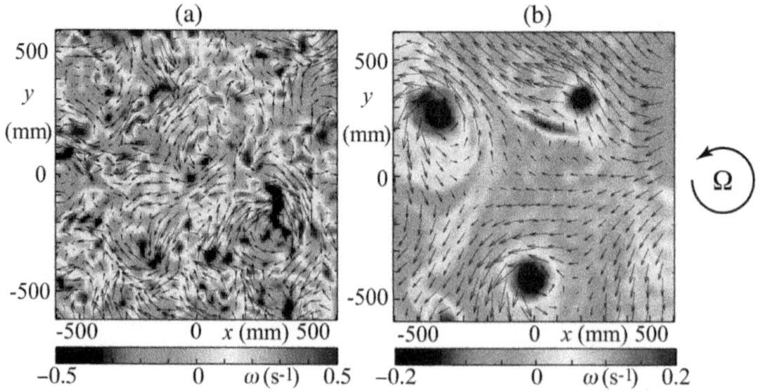

Figure 12.20C *Vorticity distribution in a 13 m diameter tank rotating around a vertical axis normal to the plane of the figure (rotation period = 30 s): (a) after 2 periods (b) after 10 periods. The turbulent flow has been created by the motion, in the plane of the figure, of a grid perpendicular to it. The component of the vorticity along the axis is shown by the colours and the velocity field, by the vectors. The cyclonic vorticity zones (in red) corresponds to a local rotation of same orientation as the applied rotation; the anticyclonic zones of opposite sign are in blue (plates courtesy F. Moisy, C. Morize, M. Rabaud and J. Sommeria)*

results are obtained in model two-dimensional flows, such as can be produced by moving a rod in a soap film.

Globally, two-dimensional turbulent flows differ from three-dimensional ones. In three-dimensional flows, the essential mechanism for the evolution of turbulence is the stretching of the vorticity by the velocity-gradients parallel to the axis (Section 7.3.2). This mechanism is obviously missing in two-dimensional flows since the velocity gradients are then perpendicular to the axis of the vorticity. One then refers to an *inverse cascade* to describe the transfer of turbulence toward large-scale structures; this is in contrast to the Kolmogorov's *direct* cascade, discussed above for three-dimensional flows.

A particularly important example of almost two-dimensional turbulence is that of the turbulence at very large scales, such as in oceans, or in the atmosphere. In fact, one saw in Section 7.6.2 that the Coriolis force uncouples the motion perpendicular to the axis of rotation from that parallel to it. Moreover, the amplitude of the vertical motion perpendicular to the surface is bounded by thermal stratification, and the thickness of the atmosphere. These effects affect very strongly the large-scale turbulent mixing within the oceans, and in the atmosphere.

Experimentally, one observes the coalescence of vortices in the turbulence created in rotational flows, as we can see in Figure 12.20. In this case (like in earth's atmosphere), the cyclonic vorticity (in the same direction as that of the applied rotation) is privileged. Such laboratory experiments are helpful in order to understand these results important in meteorological problems.

EXERCISES

1) **Energy dissipation in a turbulent flow**

We consider a turbulent flow of water assumed to be homogeneous, stationary and isotropic, in which the characteristic size of the largest vortices is $\ell = 0.3$ m and their velocity $v_\ell = 3$ m/s. What is the Reynolds number associated to the largest vortices? What is the energy ε per unit time and mass of the fluid which is transmitted by means of the Kolmogorov energy cascade (described in 12.6.1)? This energy is ultimately turned into heat by viscous dissipation in the smallest vortices. Neglecting external losses, compute the corresponding rise of the temperature after one minute (the specific heat of water is $C_V = 4200$ J/(Kg.K)). What is the order of magnitude of the size of the smallest vortices in such a flow?

2) **Transport of a solute in turbulent flow**
A solute of molecular diffusion coefficient D_m is dropped into a turbulent flow: C and $\mathbf{v}$ are respectively the local values of the mass concentration of the dye and of the velocity, C' and $\mathbf{v}'$, $\overline{C}$ and $\overline{\mathbf{v}}$ their fluctuations and mean values. Using Equation 12.16, show that the equation equivalent to Equation 12.14 for the transport of the solute is:

$$\frac{\partial \overline{C}}{\partial t} = -\nabla \cdot (\overline{\mathbf{v}}\overline{C} + \overline{\mathbf{v}'C'}) + D_m \nabla^2 \overline{\mathbf{C}} \qquad (2)$$

The measured rms fluctuations $\sqrt{\overline{v_z'^2}}$ of a given velocity component v_z and $\sqrt{\overline{C'^2}}$ of the concentration C are respectively $\sqrt{\overline{v_z'^2}} = 1$ m/s and $\sqrt{\overline{C'^2}} = 10^{-3}$ kg/m^3. One measures, in addition, a correlation coefficient equal to 0.5 between the fluctuations C' and v_z'. Compute the mass flux of the solute in the direction z resulting from the turbulent fluctuations. Assuming a molecular diffusion coefficient $D_m = 10^{-9}$ m^2/s and a gradient of the mean concentration of 10^{-2} kg/m^4, compare the corresponding mass flux to that resulting from turbulent fluctuations.

The correlation coefficient of two random variables $a'(t)$ et $b'(t)$ of zero mean value is the ratio: $\dfrac{\overline{a'(t)\,b'(t)}}{\sqrt{\overline{a'^2(t)}}\sqrt{\overline{b'^2(t)}}}$

3) **Decay of a turbulent flow**
We consider a highly turbulent flow of water with a zero mean velocity produced in a container filled with water by moving up and down several times at a high velocity a horizontal grid of area close to that of the container cross-section. The grid is then suddently extracted from the bath at the time $t = 0$ after which the turbulence created by the stirring is left to decay. We call ℓ_0 and v_{ℓ_0} the size and velocity of the largest vortices at the initial time $t = 0$.
In a first step, we assume that the size of the largest vortices involved in the Kolmogorov cascade process remains constant ($\ell = \ell_0$) during the decay until the Reynolds number is too small and flow becomes laminar. Assuming that most of the total kinetic energy is carried by the largest vortices, relate its variation to the parameter ε of the Kolmogorov theory, and obtain the equation satisfied by the velocity $v_\ell(t)$ of the largest vortices. Integrating this equation, show that the velocity v_ℓ would decrease as $1/t$ at long times. Estimate the time t_{10} at which the Reynolds number $Re_\ell = v_\ell\, \ell/\nu$ becomes of the order of 10. Beyond this time, one assumes that decay is controlled by viscous dissipation assumed to be represented by a Stokes force $f \approx c_\nu\, \rho \nu\, \ell\, v_\ell$ on a volume of size ℓ (c_ν is a constant coefficient): show that the velocity decays exponentially with time in this regime and determine the prefactor by assuming that the transition between the two regimes takes place at $t = t_{10}$. Experimental measurements show that, in the first regime, the decay of the velocity of the turbulent fluctuations is slower that predicted above (variation of v_ℓ as $t^{-0.6}$ rather than as $1/t$): assuming that this is due to a variation of ℓ with time, what must be this variation in order to obtain the experimental exponent?

4) **Turbulent plane Couette flow**
We study the parallel, turbulent, Couette flow between two parallel plates $y = 0$ and $y = 2h$ of respective velocities 0 and $2U$ in the direction x and with no applied pressure gradient. Assuming that the mean velocity and the Reynolds tensor satisfy: $\bar{v}_x(y) = U f(\xi)$ and $\overline{v_x' v_y'} = U^2 g(\xi)$ with $\xi = y/h$, show, using the equations of motion (Equation 12.44a), that: $f'' = Re\, g'$ with $Re = U h/\nu$.

The eddy viscosity approach (Equation 12.22), leads to the closure relation: $\overline{v'_x v'_y} = -\nu_t\, \partial \bar{v}_x / \partial y$. One assumes that $\nu_t = \kappa_c U y$ for $y < h$ in which κ_c is a constant ≈ 1: discuss this last assumption by referring to Equations 12.24a and 12.50 used for a turbulent Poiseuille flow. Show that the function $f(\xi)$ satisfies: $f''(\xi + 1/(\kappa_c Re)) + f' = 0$ (for $\xi < 1$) and compute $f(\xi)$. Deduce from this result the value of the friction coefficient defined by $C_d = \bar{\sigma}_{xy} / (\rho U^2/2)$. Show that these two latter expressions can still be used for $Re \ll 1$.

What do the velocity profile and the friction coefficient become at large Reynolds numbers? Write the limit of this expression at short distances from the wall: what is the dominant momentum transport mechanism in this region? What is the limit at large distances and the corresponding dominant transport mechanism?

Solutions to the Exercises

Chapter 4

1) Laminar Poiseuille flow between two coaxial tubes.

As for flow in a rigid circular tube (Poiseuille flow, see Section 4.5.3), we infer from the three components of the equation of motion that the pressure gradient $\partial p/\partial z$ along the z-axis is constant; if $\partial p/\partial z < 0$, we have $\partial p/\partial z = -\Delta p/L$. The r and φ components of the equation of motion express hydrostatic pressure variations, and the z component is $\Delta p/L = -(\eta/r)\, \partial/\partial r\,(r\, \partial v_z/\partial r)$ so that $v_z = -(r^2/4\eta)\Delta p/L + C \operatorname{Log} r + D$. The Log r term is retained since the fluid volume does not include the axis $r = 0$. Applying the boundary conditions $v_z = 0$ for $r = R_1$ and $r = R_2$ gives $C = -\dfrac{\Delta p}{4\eta\, L}\dfrac{R_1^2 - R_2^2}{\operatorname{Log}(R_2/R_1)}$ and $D = -\dfrac{\Delta p}{4\eta\, L}\dfrac{R_2^2 \operatorname{Log}(R_1) - R_1^2 \operatorname{Log}(R_2)}{\operatorname{Log}(R_2/R_1)}$ providing the velocity profile and flow rate. If $R_1 \to 0$, the logarithmic term vanishes, leading to the formula for a single tube. If $R_1 \to R_2$: $Q \simeq \pi\, R_1\, \Delta p\, e^3/(6\eta\, L)$ with $e = R_2 - R_1$. This latter value is equal to the flow rate per unit width between two parallel planes separated by a distance e and multiplied by the perimeter $2\pi\, R$ which represents the effective transverse size.

2) Flow of a Bingham Fluid on a tilted plane.

The x- and y-components of Equation 4.25 are: $\rho g \sin\alpha - \partial p/\partial x + \partial\sigma'_{xy}/\partial y = 0$ and $-\rho g \cos\alpha - \partial p/\partial y = 0$. The boundary conditions are $v_x(0) = 0$ for the velocity, $p(h) = p_0$ for the pressure, and $\sigma'_{xy}(h) = 0$ for σ'_{xy}. The variation of p is purely hydrostatic with $p = p_0 + \rho g(h - y)\cos\alpha$ and $\partial p/\partial x = 0$. Integrating the x component gives: $\sigma'_{xy} = \rho g \sin\alpha (h - y)$. If $\rho g h \sin\alpha \le \sigma_c$, $\sigma'_{xy} \le \sigma_c$(and $\partial v_x/\partial y = 0$) in the whole fluid layer, we have $v_x \equiv 0$. If $\rho g\, h \sin\alpha > \sigma_c$, $\sigma'_{xy} > \sigma_c$(and $\partial v_x/\partial y \ne 0$) for $y < y_c$ and $\sigma'_{xy} \le \sigma_c$(and $\partial v_x/\partial y = 0$) for $y \ge y_c$, we then have $y_c = h - \sigma_c/(\rho g \sin\alpha)$. Integrating in both regions and writing the continuity of the velocity at $y = y_c$, leads to: $v_x = (\rho g \sin\alpha/D)(y_c y - y^2/2)$ for $y \le y_c$ and $v_x = \rho g y_c^2 \sin\alpha/2D$ for $y \ge y_c$; D is the constant appearing in the equation (4.132a) relating the strain and the stress for the Bingham fluid (plug flow in the upper part of the layer). Increasing α increases the thickness of the plug flow domain.

3) Dynamics of a meniscus rising in a cylindrical tube.

For a wetting angle $\theta \ge \pi/2$ (non-wetting fluid), the meniscus does not enter the tube if the latter just touches the surface. If $\theta < \pi/2$, the meniscus rises at a height $h(t)$ with $h(0) = 0$ which saturates at long times to the value $h_\infty = 4\gamma\cos\theta/(\rho g d)$ given by Equation 1.70b. At shorter times, writing the flow rate in the tube as a balance between the capillary, gravity and viscous forces (Equations 1.70a and 4.78) leads to the equation: $h\, dh/dt = (\rho g d^2/(32\eta))(h_\infty - h)$ so that: $-h - h_\infty \operatorname{Log}(1 - h/h_\infty) = \rho g d^2 t/(32\eta)$. At short times, h increases from 0 according to the law: $h \simeq \sqrt{(\gamma d \cos\theta/4\eta)\, t}$; at long times, h reaches its limiting value exponentially with $h_\infty - h = h_\infty \exp\left(-\rho^2 g^2 d^3\, t/(128\, \gamma\, \eta\, \cos\theta)\right)$.

4) Poiseuille flow in an elastic tube.

Equation 4.77 can be rewritten: $Q = -A^2 \partial p/\partial x/(8\pi\eta) =$ constant. The dependence of A on pressure p provides a second relation between $\partial p/\partial x$ and $A(x)$, leading to $\partial A^3/\partial x = -24\pi\,\eta\,QKA_0$. Integrating this equation with the boundary condition $A(L) = A_0$ leads to the variation given in the text. The pressure at the inlet then satisfies $p(0) - p_0 = \dfrac{A(0) - A_0}{A_0 K} = \left(\left(1 + \dfrac{24\pi\,\eta\,QKL}{A_0^2}\right)^{1/3} - 1\right)\Big/ K.$

Chapter 5

1) Momentum conservation using a moving control volume.

If the control volume is moving, the only term to be changed in Equation 5.10 is that corresponding to the flux of momentum through the surface S: while the momentum of the fluid is still $\rho\mathbf{v}$, it is the velocity relative to the reference frame of the control volume **v-w** which determines its flux through the surface S. Equation 5.10 then becomes: $\frac{\mathrm{d}}{\mathrm{d}t}\left(\iiint_{\mathcal{V}} \rho\,\mathbf{v}\,\mathrm{d}V\right) = -\iint_S \left(\rho\mathbf{v}((\mathbf{v}-\mathbf{W}).\mathbf{n}) = p\mathbf{n} - [\sigma'] \cdot \mathbf{n}\right)\mathrm{d}S + \iiint_{\mathcal{V}} \rho\mathbf{f}\,\mathrm{d}V$. Similarly, the mass conservation equation (Equation 3.23) becomes: $\mathrm{d}m/\mathrm{d}t = \mathrm{d}/\mathrm{d}t\left(\iiint_{\mathcal{V}} \rho \mathrm{d}V\right)/\mathrm{d}t = -\iint_S \rho(\mathbf{v}-\mathbf{W})\cdot\mathbf{n}\,\mathrm{d}S$.

2) Principle of a rotating lawn sprinkler.

In order to apply the equation of conservation of momentum, we use a control volume (dotted line) moving with the nozzle at its tangential velocity $W = \Omega\,R$. The relative velocity U of the fluid with respect to the nozzle at its outlet is q/S. The velocity of the fluid in the fixed reference frame is therefore $q/S - \Omega\,R$. If the velocity of rotation is constant, the first side of the momentum conservation equation 5.10 is zero; moreover, the radial inlet part does not contribute to the tangential component of the equation which becomes: $F_t = \rho\,q(q/S - \Omega\,R)$: the forces F_t are localized at the distance $r = R$ of the axis of rotation and their momentum $2F_tR$ must balance the torque Γ so that $\Gamma(\Omega) = 2\rho\,qR\,(q/S - \Omega\,R)$. In the limiting case $\Omega = 0$, we then have: $\Gamma(\Omega) = 2\rho\,q^2R/S$. For $\Gamma = 0$, $\Omega = q/(RS)$. As an engine, the sprinkler supplies a power $\Gamma\,\Omega$ which is largest for $\Omega = q/(2\,RS)$ with a value $\rho q^3/2S^2$ (or, equivalently: $\rho U^3 S/2$).

3) Impact of a circular jet on a plane and a cone.

For $\alpha = 0$, we retain the circular symmetry of the flow. Both in the circular jet and in the layer on the plate beyond the vicinity of the stagnation point, flow remains quasi-parallel with a free surface so that the pressure remains equal to the atmospheric pressure: as in Section 5.4.1, both U, to satisfy Bernoulli's equation, and the total flow rate Q, to insure mass conservation, must be constant with r. Since $Q = 2\pi\,r\,e(r)\,U$, this implies that $e(r) \propto 1/r$. The component of the momentum flux normal to the plane is equal to the force $F_\perp$ on the plane, so that $F_\perp = \pi\,\rho\,U^2R^2\cos\alpha$.

In the case of a conical obstacle, the pressure in the film remains equal to the atmospheric pressure and both the velocity U and the flow rate Q must be independant of the distance h. Here, $Q = 2\,e(h)\,h\tan\beta$ so that $e(h) \propto 1/h$. By symmetry, the force **F** on the cone must be parallel to the x-direction. In order to compute F_x, we must take into account the fluxes of momentum both at the inlet and the outlet (they both have a non-zero component along x). F_x corresponds to their difference with: $F_x = \pi\rho R^2U^2(1 - \cos\beta)$. For $\beta = \pi/2$, we find, as could be expected, the same result as for a plane.

4) Principle of a turbine flow-meter.

In the local (rotating) reference frame, the relative flow velocity $\mathbf{V}_{rel}$ must be parallel to the surface of the blade (zero normal relative velocity). The x component is equal to U (mass-conservation condition) so that the tangential component must be $U\tan\alpha$ (< 0 in the figure). The local tangential velocity in the laboratory frame at the distance r is then $v_\varphi + U \tan\alpha = \Omega\, r + U \tan\alpha$. As in Exercises 1 and 2, if there is no friction, no tangential force (and, therefore, no momentum flux component) is needed to keep the blade rotating and the sum must be zero: this condition may be satisfied simultaneously over the whole blade provided $\tan\alpha \propto r$. One then retains the same velocity U through the device. The angular velocity is therefore $\Omega = -U\,(\tan\alpha)/r$ and is proportional to U, allowing the use of the system as a flow velocity measurement device. If there is a torque due to solid or fluid friction, we have $\Omega\, r + U \tan\alpha \neq 0$ and a torque compensating for the frictional one appears.

Chapter 6

1) The method of images: line source in front of a solid plane surface.

If we use two sources symmetrical with respect to the solid surface $y = 0$, the y-components of the velocities from the two sources cancel out on the surface, while the x-components parallel to it add up: the potential flow resulting from the combined effects of the source and its "image" satisfies the required boundary conditions and provides the actual flow field in the region. From Equation 6.92, the complex potential of a source is $q/(2\pi z)$: after replacing z by $z + ia$ and $z - ia$ and summing the two terms, we obtain: $f(z) = (q/2\pi)\,(\mathrm{Log}\,(z + ia) + \mathrm{Log}\,(z - ia)) = (q/2\pi)\left(\mathrm{Log}\left(z^2 + a^2\right)\right)$ and the complex velocity is therefore $w(z) = q\,z/\left(\pi\left(z^2 + a^2\right)\right)$. Furthermore, on the plane $y = 0$, we find: $v_x(x) = q\,x/\left(\pi\left(x^2 + a^2\right)\right)$. The corresponding local Bernoulli pressure $\delta p = -\rho v_x^2/2$ therefore has the value given in the text. By taking $u = x/a$, we obtain: $F_p = \int_{-\infty}^{+\infty} \delta p(x, 0)\,\mathrm{d}x = \rho\, q^2/(4\pi\, a)$. The plane is attracted by the source because of the reduced local pressure on the plane as a result of Bernoulli's equation. This attractive force remains the same if the source is replaced by a sink since the pressure variation is quadratic in q.

2) Force on a half-cylinder placed in a uniform flow.

The flow velocity around the half-cylinder is identical for $y > 0$ to that around the complete cylinder due to the symmetry of the latter flow with respect to the diametral plane $y = 0$. We can then apply Equations 6.17 which give $v_\varphi = -2U \sin\varphi$ and $v_r = 0$ at the surface of the half-cylinder. If we start by neglecting buoyancy forces and apply the Bernoulli equation for potential flows in the form $p\,(R, \varphi) + \rho_f v_\varphi^2\,(R, \varphi)\,/2 = p_0 + \rho_f\, U^2/2$, we find the relation given in the text (in particular, $p\,(0) = p\,(\pi) = p_0 + \rho_f\, U^2/2$ and $p\,(\pi/2) = p_0 - 3\rho_f\, U^2/2$). By symmetry, only the y-component of the pressure force $\mathbf{F}$ on the upper surface of the half-cylinder, per unit length, is non-zero with: $F_{y(\text{upper surface})} = -\int_0^\pi p(\varphi)\, R \sin\varphi\, \mathrm{d}\varphi = -2p_0 R + (5/3)\,\rho_f U^2 R$. The force on the base of the half-cylinder is: $F_{y(\text{base})} = p\,(0)\, 2R = 2p_0 R + \rho_f U^2 R$ so that the global hydrodynamic force is: $F_{y(\text{total})} = (8/3)\,\rho_f U^2 R$. If buoyancy effects are then introduced, the cylinder will remain fix on the plane until the force $F_{y(\text{total})}$ exceeds the weight of the cylinder corrected by Archimedes buoyancy: $-\pi\, R^2\, g\left(\rho_{hc} - \rho_f\right)/2$; the corresponding threshold velocity is: $U_M = \sqrt{(3\pi\, R g(\rho_{hc} - \rho_f))/(16\rho_f)}$.

3) Gravity waves at the interface between fluids of different densities.

The vertical velocities of each fluid must be zero at the corresponding wall and must be equal at the interface with: $\partial\Phi_1/\partial y\,(0) = \partial\Phi_2/\partial y\,(0)$, $\partial\Phi_1/\partial y\,(h_1) = 0$, $\partial\Phi_2/\partial y\,(h_2) = 0$. Writing the Bernoulli Equation 6.65 in both fluids at the interface, taking derivatives with respect to time after multiplying by ρ_j, and making use of the equality of the pressures at the interface, leads to: $\rho_1 \partial^2\Phi_1/\partial t^2 + \rho_1\, g\partial\Phi_1/\partial y = \rho_2 \partial^2\Phi_2/\partial t^2 + \rho_1\, g\partial\Phi_2/\partial y$ (all derivatives are taken at $y=0$). With the form assumed for $f_j(x,t)$, Laplace's equation $\nabla^2\Phi_j = 0$ becomes: $\partial^2 g_j/\partial y^2 = k^2 g_j$ with solutions which are linear combinations of $e^{\pm ky}$. The boundary conditions at the walls lead to $g_1(y) = \cosh k(y-h_1)$ and $g_2(y) = \cosh k(y+h_2)$ since the derivative of g_j must be zero on the corresponding wall. Inserting these forms of g_1 and g_2 in Bernoulli's equation and in the expression of the continuity of the velocity at the interface leads respectively to: $\rho_1 A_1 \left(\omega^2 \cosh(kh_1) + g\,k \sinh kh_1\right) = \rho_2 A_2 \left(\omega^2 \cosh kh_2 - g\,k \sinh kh_2\right)$ and: $A_1 \sinh kh_1 = -A_2$. Combining these two equations leads to the dispersion relation $\omega(k)$ given in the text. For $k h_j \ll 1$ (shallow-water case), the approximation $\coth k h_j \simeq 1/(k h_j)$ leads to: $c^2 = \omega^2/k^2 = g h_1 h_2 (\rho_2 - \rho_1) / (\rho_1 h_2 + \rho_2 h_1)$. For $h_1 = h_2$, this becomes: $c^2 = g h_j (\rho_2 - \rho_1)/(\rho_1 + \rho_2)$. We then have an additional factor: $(\rho_2 - \rho_1)/(\rho_1 + \rho_2)$ compared to the case of an interface between a liquid and air: the upper factor corresponds to the restoring force, which is smaller for liquids of similar density, and the lower one to the inertia of the fluids, which is instead larger. The velocity of the wave may then be much lower than for a single liquid if $\rho_2 - \rho_1 \ll \rho_1, \rho_2$. For $k h_j \gg 1$ (deep-water case), the approximation $\coth kh_j \simeq 1$ leads to: $c^2 = (g/k)(\rho_2 - \rho_1)/(\rho_1 + \rho_2)$. The velocity is multiplied by the same factor as in the case of a shallow layer.

4) Reflection of a gravity wave due to a variation of depth.

Since the equations of motion and the boundary conditions are the same as those used in Section 6.4.1 for establishing Equation 6.73, the latter will remain valid (taking $\gamma = 0$) with different propagation velocities in the two regions. In the shallow-water limit, we have $c_1 = \sqrt{g h_1}$ and $c_2 = \sqrt{g h_2}$ (if $k_j h_j \ll 1$) leading to wave vectors $k_i = \omega/c_1$, $k_t = \omega/c_2$, $k_r = -\omega/c_1$. We write the continuity of the vertical velocity across the step with: $v_{yi} + v_{yr} = v_{yt}$ or: $\partial\Phi_i/\partial y + \partial\Phi_r/\partial y = \partial\Phi_t/\partial y$. Ideally, we should take x=0 and y=y_0 but, as a first order approximation, the values are taken at $y=0$. Using the expressions for Φ_j given in the text and taking $\sinh k_j(y+h_j) \cong k_j(y+h_j)$ $(k_j h_j \ll 1)$, we obtain, after eliminating some common factors: $A_i k_i^2 h_1 + A_r k_r^2 h_1 = A_t k_t^2 h_2$, so that $A_i + A_r = A_t$ (all products of the form $k^2 h$ have the same value, as $k_j^2 = (\omega/c_j)^2 \approx 1/h_j$). The second boundary condition is provided by the continuity of the flux in the x-direction through the plane $x = 0$. If $kh \ll 1$, the velocity is uniform in the vertical direction so that the conservation equation becomes: $h_1 (\partial\Phi_i/\partial x + \partial\Phi_r/\partial x) = h_2 \partial\Phi_t/\partial x$, leading to $A_i k_i h_1 - A_r k_r h_1 = A_t k_t h_2$ so that: $(A_i - A_r)\sqrt{h_1} = A_t \sqrt{h_2}$. Combining the two conditions leads to: $A_t/A_i = 2\sqrt{h_1}/(\sqrt{h_1} + \sqrt{h_2})$ and $A_r/A_i = (\sqrt{h_1} - \sqrt{h_2})/(\sqrt{h_1} + \sqrt{h_2})$. Increasing the difference between the depths increases the magnitude of the reflected wave compared to the transmitted one.

Chapter 7

1) Vortex and plane boundary.

The flow may be reproduced by adding a second vortex filament of opposite circulation $-\Gamma$ and located at point $(x, -y)$. This image vortex insures that the resulting velocity at the plane $y = 0$ is only tangential. The velocity induced on the vortex by its image is along the x direction and equal to $\Gamma/4\pi y$. It will therefore move, like a vortex pair, at a constant velocity parallel to the plane $y = 0$.

2) Vortex in a corner.

The boundary conditions are satisfied at the planes by replacing them by two vortices of circulation $-\Gamma$ at points $(x, -y)$ and $(-x, y)$ and one with $+\Gamma$ at point $(-x, -y)$. Applying Equation 7.60 and removing the additive constants leads to the given potential. Taking derivatives with respect to z leads to $v_x = (\Gamma/2\pi)\, x^2/(y(x^2+y^2))$ and

$v_y = -(\Gamma/2\pi)\, y^2/(x\,(x^2+y^2))$ from which we obtain immediately the result $dx/x^3 = -dy/y^3$. Integrating this relation leads to: $1/x^2 + 1/y^2 = 1/x_0^2 + 1/y_0^2$. Depending on the sign of Γ, the vortex moves towards $x \to \infty$ or $y \to \infty$ with a limit for the other coordinate equal to: $x_0 y_0/\sqrt{x_0^2 + y_0^2}$.

3) Geostrophic flow in a rotating channel.

Equation 7.131 becomes: $\nu\, \partial^2 v_x/\partial y^2 = (1/\rho)\partial p/\partial x$, $-2\ \Omega\ v_x = (1/\rho)\ \partial p/\partial y$ and $(1/\rho)\ \partial p/\partial z = 0$. $\partial p/\partial x$ is therefore constant and v_x is zero at the two walls, so that we have the symmetrical Poiseuille velocity profile given by Equation 4.66. The velocity field is: $v_x(y) = -(6Q/a^3)(y^2 - a^2/4)$ and the pressure field : $p = p_0 - 12\eta Qx/a^3 + 12(\rho\Omega/a^3)(y^3 - ya^2/4)$ (Q = volume flow rate in the flow section per unit length along z). There results: $p(x, a/2) - p(x, -a/2) = -2\ \rho\ \Omega$. We then obtain: $\partial p/\partial x \ll \partial p/\partial y$ if $12\ \eta\ v_x/a^2 \ll 2\ \rho\ \Omega\ v_x$, i.e. when the Ekman number: $Ek = \nu/\left(a^2\ \Omega\right)$ is $\ll 1$. In this case, we can consider p as constant along x: we then recover a geostrophic flow with a velocity parallel to the isopressure lines.

The appearance of a pressure gradient transverse to the flow is analogous to the Hall effect in a sample of (for instance) a semiconductor subjected to a magnetic field **B:** when an electrical current **I** perpendicular to **B** is injected in the sample, a gradient of the electrical voltage perpendicular (and proportional) both to **B** and **I** appears.

4) Tilt of the free surface of the oceans between two coasts.

We have: $2\ \Omega\ V_x = (-1/\rho)\ \partial p/\partial y$, $0 = (1/\rho)\ \partial p/\partial x$ and $g = (-1/\rho)\ \partial p/\partial z$. Since $p(h(y)) = p_0$, we have the result: $p(y) = p_0 + \rho\, g\ (h - y)$ so that: $2\,\Omega\ V_x = -g\ \partial h/\partial y = -g\ \tan\beta$. For the current numerical values of the different parameters, $h(-L/2) - h(L/2) \cong 3$ m.

In the case of two fluids, the previous relations are valid separately for each of them with: $2\ \Omega\ V_{x1} = -g\ \tan\beta_1$ and $2\ \Omega\ V_{x2} = -g\ \tan\beta_2$. The pressures in fluids 1 and 2 must be equal at their interface $z = h_i(y)$. Integrating with respect to x and y from point $z = h_0$, $y = 0$ where $p = p_0$ leads to: $p_1(y,z) = p_0 + \rho_1 g\ (h_0 - z) - 2\ \Omega\ \rho_1 V_{x1} y$ and to a similar expression for p_2 (replacing the index 1 by 2) Since $p_1 = p_2$ for $z = h_i(y)$, we obtain finally: $\tan\beta_i = (h_i - h_0)/y = -(2\ \Omega/g)\ (\rho_1\ V_{x1} - \rho_2\ V_{x2})/(\rho_1 - \rho_2)$. As a result, the angle β_i is much larger than either β_1 or β_2 if $\rho_2 - \rho_1 \ll \rho_1$ (for sea currents of different salinities, we have: $(\rho_1 - \rho_2)/\rho_1 \approx 10^{-3}$).

5) Modelization of thermal winds.

The equations of motion are: $\partial p/\partial x = -2\rho\,\Omega\, v_y$, $\partial p/\partial y = 2\rho\,\Omega\, v_x$, $\partial p/\partial z = -\rho\, g$. For an incompressible fluid, taking derivatives of the two first equations with respect to z and combining them with the third one leads to $\partial \mathbf{v}/\partial z = -\left(g/\left(2\rho|\mathbf{\Omega}|^2\right)\right) \mathbf{\Omega}\times\nabla\rho$ The dependence on ∇T results immediately (second-order terms associated with the variation of ρ are neglected).

Combining the equation of state of the gas with the expression for $\partial p/\partial z$ provides the relation between dz and d (Log p); the relationship between $\partial\mathbf{v}/\partial z$ and $\nabla\rho$ obtained above leads to that between $\partial\mathbf{v}/\partial(\text{Log } p)$ and ∇T by replacing, in addition, $\nabla\rho$ by $-(\rho/T)(\nabla T)_p$.

The variation $\Delta\mathbf{v}$ of the velocity is given by: $|\Delta\mathbf{v}| = (R\,|\nabla T|\,/2M\,|\mathbf{\Omega}|)\ \text{Log}(p_2/p_1)$ and is of the order of 2 m/s for the current values of the parameters.

Chapter 8

1) Flow of a fluid layer around a horizontal cylinder.

The pressure is constant at the interface ($p = p_{atm}$) so that the pressure variations $p - p_{atm}$ within the film are of the order of $\rho g\, h$ (considering the variations of p perpendicular to the surface, i.e. the velocity). The pressure gradient parallel to the film is then of the order of $\rho g\, h/R$: it is, therefore, one order of magnitude smaller in h/r than the volume force component $\rho\, g\, h \sin\varphi$ and can be neglected. The flow rate q is then computed as in Section 8.2.1, replacing the gravity g by its component $g \sin\varphi$ so that $q = g\ h^3 \sin\varphi/(3\ \nu)$. The expression $\partial h/\partial t$ is then obtained from the mass conservation equation: $\partial h/\partial t = -1/R\ (\partial q/\partial\varphi)$.

Integrating this expression for a Newtonian fluid, we find that the thicknesses at $\varphi = 0$ and $\varphi = \pi$ vary respectively as $h(0,\ t) = h_0/\sqrt{1 +\ t/\tau_f}$ and $h(\pi,\ t) = h_0/\sqrt{1 -\ t/\tau_f}$.

For a fluid of varying viscosity, we obtain for $\varphi = 0$ and $\varphi = \pi$: $h(0, t) = h_0\Big/\sqrt{1 +\ \tau_d(1 - e^{-t/\tau_d})/\tau_f}$ and $h(\pi, t) = h_0\Big/\sqrt{1 -\ \tau_d(1 - e^{-t/\tau_d})/\tau_f}$. For $t \ll \tau_d$, a first order development with respect to t/τ_d gives the same result as if the fluid had not dried. At long times such as $t \gg \tau_d$, we obtain: $h(0, t \gg \tau_d) = h_0/\sqrt{1 +\ \tau_d/\tau_f}$ and $h(\pi, t \gg \tau_d) = h_0/\sqrt{1 -\ \tau_d/\tau_f}$, so that, for $\tau_d < \tau_f$, we reach a constant thickness, larger for $\varphi = \pi$ than for $\varphi = 0$.

2) Spin coating.

The pressure in the fluid film can be considered constant and the flow driven by the centrifugal forces. The radial velocity profile is $v_r(r, z) = (\Omega^2 r/2\nu)\,(2hz\ -\ z^2)$: the flow rate q is the result of integration with respect to z and multiplication by $2\pi r$. Writing the equation of conservation of mass within the cylinder, we obtain: $\partial h/\partial t = -(1/2\pi r)\partial q/\partial r = -\Omega^2/\nu\ \left[2h^2/3 + r h^2 \partial h/\partial r\right]$. The variation of h with t is then independent of r if $\partial h/\partial r = 0$ at any given time. By integration, $h = h_0/\sqrt{1 + t/\tau}$ with $\tau = 3\nu/(4h_0^2\Omega^2)$, resulting numerically in: $h(10\ \text{s}) = 4.6\ \mu\text{m}$, $h(100\ \text{s}) = 1.45\ \mu\text{m}$. In practical cases, the thickness h of the layer is not initially uniform; however, these variations are damped out due to the dependence of the flow rate on h (more fluid flows out of thicker regions than they receive from thinner ones).

3) Rough sphere dropping away from a plane.

For $h_0 \ll \varepsilon$, the dominant contribution comes from the three individual rugosities closest to the contact: the diameter a must be replaced by ε in Equation 8.33 and a factor 3 must be added. For $\varepsilon \ll h_0 \ll a$, Equation 8.33 remains valid provided h_0 is replaced by $h_0 + \varepsilon$. At very large distances, $h_0 \gg a$, the expression $F = -6\pi\eta a\,(1 + a/h_0)\, dh_0/dt$ coincides with Stokes equation for a sphere while, at small distances, it coincides with Equation 8.33. By combining these different components, we obtain the expression for F given in the text. The equation of motion shown in the exercise is then obtained by writing that F balances the apparent weight $(4/3)\,\pi\, a^3\,(\rho_s - \rho_f)g$. This gives the implicit expression of $h_0(t)$ after integration with respect to time. Subtracting the values of this expression for $h_0 = a$ and $2a$ and dividing by its value for $h_0 = 2a$ leads to: $\varepsilon/a = 2\ \exp\,(2 - t_{2a}\,(1 + \text{Log}\,2)/(t_{2a} - t_a))$.

4) Liquid film draining under gravity along a vertical plate.

The pressure is constant within the film and equal to the atmospheric pressure p_{atm}: the flow rate q per unit length through a section x = constant of the film is then given by Equation 8.46b. The equation of conservation of mass at a given height x is $\partial h/\partial t = -\partial q/\partial x$: combining these two equations leads to the partial differential equation given in the text. Replacing h by $f(x/t^\alpha)$ gives: $g\, f^2/\nu\ =\ \alpha\, x/t$. Therefore, one must have $\alpha = 1$ and $h = \sqrt{(\nu/g)\, x/t}$. For the values of x and ν of interest and $t = 1$, 10, 100 and 10^4 s, we obtain respectively $h = 4.5$, 1.4, 0.45 and 0.045 mm.

5) Rayleigh–Plateau instability on a cylindrical fiber.

The difference between the local pressure and the external value p_{atm} is given by Equation 8.86 because, as in Section 8.3.2, only capillary forces need to be taken into account (in principle R_0 should be replaced by $R_0\ +\ e_0$ but the value R_0 is retained since $e_0\ \ll\ R_0$). When inertia is negligible, from Equation 8.25, the expression of the flow rate $Q(z)$ is: $Q(z) \simeq -(e_0^3/3\eta)2\pi R_0\,(\partial p/\partial z)$; substituting in this expression the value of $p(z)$ from Equation 8.86 leads to the equation given in the text. dh/dt and $\tau(k)$ are then obtained by means of the mass conservation equation $\partial Q/\partial z = -2\pi\ R_0\ \partial h/\partial t$. Just as for the jet, the time constant is shortest for $k = 1/R\sqrt{2}$ with $\tau_{min} = 12\,\eta\,R^4/(e_0^3\gamma)$; here, $\tau_{min} = 190$ s and the corresponding wavelength is: $\lambda = 2\pi\ R\sqrt{2} \simeq 4.5\,\text{mm}$. The initial assumption that the effect of gravity is negligible will be valid if the characteristic time τ_{min} is shorter than the characteristic time τ_d for drainage by gravity determined in Exercise 1.

6) Liquid film rising due to the Marangoni effect (Section 8.2.4).

The pressure is again constant within the film due to the small curvature of the surface. The equation of motion then reduces to $\eta\ \partial^2 v_z/\partial x^2 = g$. Integrating with the boundary conditions $v_z(0) = 0$ and $\eta\ (\partial v_z/\partial x)_{x=h} = \sigma^{(\gamma)} = -\ b\,\gamma\,(\partial T/\partial z)$ (see Equation 8.67) leads to $v_z(x) = (g/2\eta h)\,x(x-2h) + \sigma^{(\gamma)}\,x/\eta$. We then obtain $q(h)$ by integration with respect to x. The conservation of mass implies $(\partial q/\partial z)_t = -(\partial h/\partial t)_z$ so that $(dq/dh)(\partial h/\partial z)_t = -(\partial h/\partial t)_z$, because q is only a function of h. The velocity $V(h) = (\partial z/\partial t)_h$ of the point of height h satisfies the relation: $V(h) = -\ (\partial h/\partial t)_z/(\partial h/\partial z)_t$: this equation is obtained by writing that the Lagrangian derivative: $dh/dt = (\partial h/\partial t)_z + V(h)(\partial h/\partial z)_t$ is equal to zero. We have then: $V(h) = dq/dh = \sigma^{(\gamma)}\ h/\eta - \rho g\ h^2/\eta$. The velocity $V(h)$ is therefore constant with time so that the coordinate z of this point is proportional to the time t with $z = V(h)\ t$. The profiles $h_i(z)$ at different times can therefore be deduced from one another by an affine transformation: for example, the values of z_1 need to be multiplied by the ratio t_2/t_1 of the times in order to obtain z_2. The maximum value of $V(h)$ at which $d^2q/dh^2 = 0$ is obtained for $h_{max} = \sigma^{(\gamma)}/(2\rho\,g)$ so that $V(h_{max}) = (\sigma^{(\gamma)})^2 t/(4\eta\ \rho g)$. The presence of such a maximum requires that the thickness $h(z)$ does not decrease continuously to zero: assuming that $h > h_{max}$ for $z = 0$, there would indeed be two values of h corresponding to a same normalized height $x/t\ (= V)$. Mathematically, this implies that h goes abruptly to zero after decreasing continuously from $h(0)$ to a value $h_s > h_{max}$ determined by the shape of the curve $V(h)$. This is formally similar to phenomena such as shock waves in compressible fluids (Section 5.4.4). Experimentally, we observe that variations of h in the direction y also take place due to an instability of the front.

Chapter 9

1) Torque on a cylinder and a cross in a shear flow.

The forces per unit length are determined by the local relative velocity of the cylinder and the fluid, using notations identical to those in Equation 9.30: $dF_\varphi/dr = -2\eta\ \lambda_\parallel\ (r/2L)\left(\dot\varphi + S\sin^2\varphi\right)$ and $dF_r/dr = \eta\ \lambda_\parallel\ (r/2L)\ (S\ \sin\varphi\ \cos\varphi)$. This provides the total torque G_0 after multiplying dF_φ/dr by r and integrating. We have $G_0 = 0$ due to the zero inertia so that: $\varphi(t) = \tan^{-1}\ [\tan\varphi_0/(1 + S(\tan\varphi_0)t)]$ and the rod gets aligned with the flow at long times. For two cylinders at right angles, we add the torques leading to: $G_0 = -(2/3)\ \eta\ \lambda_\parallel\ L^3\ (2\,\dot\varphi + S) = 0$. The angular velocity in the stationary regime is then constant and equal to $-\ S/2 = \omega_z/2$. This is the principle of a simple device measuring the vorticity in two-dimensional flows. More generally, the rotation of a solid body floating at the surface of a flowing liquid layer may be used to estimate the component of the local vorticity perpendicular to the surface (see Section 7.3.1).

2) Sedimentation of a sphere.

Neglecting the inertia of the fluid since $Re \ll 1$, we obtain: $\left[2R^2\rho_s/(9\eta)\right] dv_z/dt = -\left[v_z - 2R^2 g\left(\rho_s - \rho_f\right)/(9\eta)\right]$; leading to: $z(t) = V_{terminal}\left[t - \tau\left(1 - \mathbf{e}^{-t/\tau}\right)\right]$ with $\tau = 2/9\,(R^2/\eta)\,\rho_s$ (see Equation 9.56 for $V_{terminal}$). This terminal velocity is reached within 1% for $\mathbf{e}^{-t/\tau} = 0.01$ or $t = t_c = 2.3\ \log_{10}(100)\ \tau$. At short times,

$z(t) = g\ (\rho_s - \rho_f)\ t^2/(2\ \rho_s)$ (free fall equation); at long times: $z(t) \simeq V_{\text{terminal}}\,(t-\tau)$. For the present numerical values, $\tau = 1.1 \times 10^{-4}$ s and $t_c = 5.1 \times 10^{-4}$ s in the first case and respectively 0.11 s and 0.51 s in the second. The corresponding velocities V_{terminal} are 0.55 mm/s and 550 mm/s and the Reynolds numbers are 5.5×10^{-4} and 550. In the first case, $z(t_c) = 0.2$ μm while the condition $Re \ll 1$ is not satisfied in the second case.

3) Stokes flow from a three-dimensional point-source.

The incompressibility equation reduces to $(1/r^2)\partial(r^2 v_r)/\partial r = 0$ so that $r^2 v_r$ is only a function of θ because of the circular symmetry around the z-axis. From the Stokes equation, $\partial p/\partial r$ must therefore vary as $1/r^4$ and p as $1/r^3$. Multiplying the velocity $A\cos^2\theta/r^2$ by $2\pi r^2 \sin\theta\ d\theta$, integrating over θ between 0 and $\pi/2$ and equating the result to q leads to: $A = 3q/2\pi$. Integrating the θ component of the equation of motion with respect to θ leads to: $p(r,\theta) = (2\eta/r^3)\ (C + 3q\cos^2\theta/2\pi)$. Computing then the r-component of the equation of motion (for instance for $\theta = \pi/2$) gives $C = -q/2\pi$. Both p and the gradient $\partial p/\partial r$ change sign at the angle $\theta = \arccos(1/3^{1/2}) < \pi/2$ although the radial velocity is always of the same sign.

4) Torque on two concentric rotating spheres.

Using the same method as in Section 9.4.3, we find that $f(r) = Ar + B/r^2$; the values of A and B are obtained from the boundary conditions on the two spheres with: $v_\varphi(r,\theta) = \Omega_1 \sin\theta\left[R_1^3 R_2^3/(R_1^3 - R_2^3)\right]\left[1/r^2 - r/R_2^3\right]$ for $\Omega_2 = 0$ (swap the indices 1 and 2 in the other case). For $R_2 \to \infty$ and $\Omega_2 = 0$, we retrieve Equation 9.58 for the flow around a rotating sphere. For $R_1 \to 0$ and $\Omega_1 = 0$, we have solid body rotation of the fluid inside the sphere of radius R_2. The torque is computed as in Section 9.4.3 with: $G = -8\pi\,\eta\,\Omega_1\left[R_1^3 R_2^3/(R_1^3 - R_2^3)\right]$ for $\Omega_2 = 0$ on cylinder 1 and $-G$ on the other (for $\Omega_1 = 0$, $\Omega_1 \to \Omega_2$ and the signs are inverted). For $\delta \ll R_1$, $v_\varphi(r,\theta) = \Omega_1 \sin\theta\, R(r - R_2)/\varepsilon$ and we have $G = -(8\pi/3)\,\eta\,\Omega\,R^4/\delta$. If both spheres rotate, we just add the expressions for the given velocity for $\Omega_1 = 0$ and $\Omega_2 = 0$ (see the superposition of solutions of the Stokes equation in Section 9.2.3).

5) Permeability of a layered medium.

For flow parallel to the layers, the pressure drop Δp is the same along all layers but the flow rate depends on the permeability. Estimating from Darcy's law the flow rates Q_1 and Q_2 in a layer of each type and writing their sum using the global effective permeability $k_{//}$ of the layered medium, we find: $k_{//} = (k_1 h_1 + k_2 h_2)/(h_1 + h_2)$. For flow perpendicular to the layers, the flow rate Q through all layers is the same but the pressure drops Δp_1 and Δp_2 are different. Writing their sum as a function of Q using the effective permeability $k_\perp$ leads to: $k_\perp = (h_1 + h_2)/(h_1/k_1 + h_2/k_2)$. If $k_1 \gg k_2$, $k_{//} = k_1 h_1/(h_1 + h_2)$ and $k_\perp = k_2(h_1 + h_2)/h_2$.

The porosity and tortuosity of random unconsolidated packings of spherical beads of uniform diameter d and prepared using a same procedure is practically independent of their diameter d (see Section 1.1.1). We have therefore $\phi_1/\phi_2 = 1$ and $k \propto d^2$ so that: $k_1/k_2 = (d_1/d_2)^2$ and, using the previous relations, one finds: $k_\parallel/k_\perp = \left(h_1 d_1^2 + h_2 d_2^2\right)\left(h_1 d_2^2 + h_2 d_1^2\right)/((h_1 + h_2)^2 d_1^2 d_2^2)$. In contrast, the electrical conductivity of the packings does not depend on d but only on the porosity and tortuosity and is the same for the two layers. The global conductivities in the parallel and perpendicular configurations are then both equal to this latter common value.

6) Porous embankment with two fluids of different densities.

Just as for a single fluid, we can neglect the effect of the flow on the transverse pressure gradient which is then purely hydrostatic. Computing the pressure profile $p_{fs}(x)$ at the interface $y_{fs}(x)$ separately in the salty stagnant water and in the fresh water and equating the two resulting expressions provides the relation between y_{fs} and y_{fa}. The flow rate Q can be computed in the same way as for a single fluid with $Q = -(kg\rho_f/\mu)(y_{fa} - y_{fs})\partial y_{fa}/\partial x$. Using the relation between y_{fs} and y_{fa} to compute Q as a function of $\partial y_{fs}/\partial x$ leads to: $(y_{fs} - y_1)^2 = 2\nu Q\rho_f(x - L)/\left(g\,k\left(\rho_f - \rho_s\right)\right)$. The profile $y_{fs}(x)$ is therefore parabolic. The penetration distance $L - x_{fs}{}^{\min}$ is obtained by taking $y_{fs} = 0$ in the previous expression so that: $L - x_{fs}^{\min} = k\,g\,y_1^2\left(\rho_s - \rho_f\right)/(2\,\nu\,Q\,\rho_f)$. The slope $\partial y_{fs}/\partial x$ of the interface increases (and $L - x_{\mathrm{fs}}^{\min}$ decreases) with the viscous pressure gradient $\partial p/\partial x$ and, therefore, with Q. For a given pressure gradient $\partial p/\partial x$, the slope must be larger at small values of the density contrast $\rho_s - \rho_f$: this allows one to keep constant the hydrostatic pressure terms which are the product of the two and balance the viscous pressure.

Chapter 10

1) Stationary dissolution boundary layer near a free surface of a falling film.

Since the flow velocity is parallel to the x-axis, the local flux $\mathbf{J}(x,y)$ of dissolved gas has a convective x-component: $\mathcal{J}_x^{conv} = C(x,y)\,v_x(y)$ and a diffusive y-component: $\mathcal{J}_y^{diff} = -D_g\,\partial C(x,y)/\partial y$. The diffusive x-component can be neglected because: $\partial C/\partial x \ll \partial C/\partial y$. If D_g is low, the gas penetrates only over a distance from the surface $\delta_c \propto (D_g)^{1/2}$ which is assumed to be $\ll h$ (this will be discussed below): the velocity $v_x(y)$ in this layer can be taken equal to V_0 so that $\mathcal{J}_x^{conv} \simeq C(x,y)\,V_0$. Writing the stationarity condition $\nabla\cdot\mathbf{J} = 0$ in this penetration layer leads then to the equation given in the text. Replacing $\partial C/\partial x$ by $-u(\partial C/\partial u)/(2x^{1/2})$ and $\partial^2 C/\partial y^2$ by $(1/x)(\partial^2 C/\partial u^2)$ provides the reduced variable form of the equation. Two successive integrations with the boundary conditions $C = 0$ for $y \to -\infty$ and $C = C_0$ for $y = h$ lead to: $C = C_0\left[1 - \mathrm{erf}\left((h - y)\sqrt{V_0/(4\,D_g\,x)}\right)\right]$. The characteristic distance δ_c over which the concentration varies from C_0 to 0 is therefore $\sqrt{4D_g x/V_0}$ and represents the thickness of the gas diffusion boundary layer in this problem. The above hypothesis of a small penetration distance can be expressed by $\delta_c(L) \ll h$ which requires a Péclet number $Pe = V_0 h/D_g \ll L/h$. The transverse gradient $\partial C/\partial y$ at the free surface $(y = h)$ satisfies: $(\partial C/\partial y)_{y=h} = -(\partial C/\partial u)_{u=0}/\sqrt{x} = -C_0\sqrt{V_0/(\pi\,D_g x)}$; it therefore decreases as $1/\sqrt{x}$.

2) Boundary layer for a cylindrical Couette flow with radial aspiration.

The absolute flow rate Q per unit length along z and through a cylinder of radius r is $Q = -2\pi\,r\,v_r$ and is constant so that: $\alpha = Q/2\pi$. Replacing v_r by $-\alpha/r$ in the φ component of the Navier–Stokes equation (see Appendix 4A.2) provides the expression given in the text since v_φ depends only on r. Replacing v_φ by $C\,r^\gamma$ and simplifying by $C\,\nu\,r^\gamma$ leads to: $\gamma = -1$ and $\gamma = 1 - \alpha/\nu$. The Darcy velocity through the inner porous wall is $v_s = Q/(2\pi R_1) = -\alpha/R_1$ so that $\alpha/\nu = |v_s|R_1/\nu$ represents the Reynolds number associated to v_s and R_1. Assuming that $v_\varphi(r)$ is of the form: $C_1/r + C_2\,r^{1-\alpha/\nu}$ and using the boundary conditions on the two cylinders provides then the given expression. For $\alpha/\nu \gg 1$, the exponent $2 - \alpha/\nu$ is negative and the terms containing R_1/R_2 are the only important ones in the fraction, except when $r \simeq R_1$. One has then $v_\varphi(r) \propto 1/r$ (a vortex potential flow), except for $r \to R_1$. The thickness δ of the layer where the vorticity is localized is estimated by writing that, at a distance δ from the inner cylinder $(r \simeq R_1)$, the viscous transport term $\eta\,\partial v_\varphi/\partial r \approx \eta\,v_\varphi/\delta$ is of the same order as the convective one: $\rho\,v_\varphi\,v_r \approx \rho\,v_\varphi\,\alpha/R_1$ so that:

$\delta \approx \nu R_1/\alpha = \nu/v_s$ where v_s is the suction velocity. This value is the same as for a flat plate with aspiration but the variation with distance is a power law rather than an exponential one.

3) Transient flow around a rotating cylinder.

At short times, fluid motion is localized in a thin layer of thickness $\delta \approx \sqrt{\nu t}$ (Section 2.1.2). The variation of $v_\varphi(r,t)$ is then governed by the φ component of the equation of motion in cylindrical coordinates (see Appendix 4A.2) which reduces to a pure diffusion equation. The r component of the equation $\partial p/\partial r = \rho\, v_\varphi^2/r$ shows that the pressure gradient balances the centrifugal force. In the stationary regime, taking $v_\varphi = C\, r^\gamma$ leads to: $\gamma^2 = 1$. Eliminating $\gamma = 1$ (fluid at rest at infinity) leads to $v_{\varphi S} = C/r$ with $C = \Omega\, R^2 (v_{\varphi S} = \Omega R$ for $r = R)$ and $\omega_{zS} = 0$. The corresponding kinetic energy per unit volume is $\rho\, \Omega^2\, R^4/r^2$ and its integral over the whole fluid volume diverges at large distances: this stationary flow cannot then be reached in a finite time. Inserting the expression $v_\varphi = v_{\varphi S}\, f(u)$ into the equation of motion and simplifying by $v_{\varphi S}$ and r^2/t leads, after integration, to: $f = C\,\mathrm{e}^{-u^2/4\nu t} + D$ with $D = 0$ (finite total kinetic energy) and $C = 1$ if $R \ll \sqrt{\nu t}$. The vorticity satisfies: $\omega_z = (1/r)\ \partial(r v_\varphi)/\partial r = -(\Omega R^2/(2\nu t))\mathrm{e}^{-r^2/4\nu t}$ and it decays over a distance $\propto \sqrt{\nu t}$: if $R \ll \sqrt{\nu t}$, the total vorticity $\int_R^\infty 2\pi\, r\, \omega_z\, dr$ is constant and equal to $2\pi R^2 \Omega$ (circulation of the velocity around the cylinder).

Chapter 12

1) Energy dissipation in a turbulent flow.

The Reynolds number $Re = v_\ell\, \ell/\nu$ is $9\ 10^5$ and the corresponding energy ε dissipated per unit time and unit mass satisfies: $\varepsilon = v_\ell^3/\ell = 90\, W/kg$. The temperature rise ΔT per minute (duration $\tau_M = 60$ s) then satisfies: $C_v\ \Delta T = \tau_M\ \varepsilon$ so that: $\Delta T \cong 1.3$ K. Finally, the size η of the smallest vortices given by Equation 12.94 is: $\eta = \nu^{3/4}\varepsilon^{-1/4} \simeq 10\ \mu$m.

2) Transport of a solute in turbulent flow.

The mass conservation equation of the solute can be written: $\partial \overline{C}/\partial t + \nabla \cdot \mathbf{Q}_m = 0$. Using Equation 12.16 as the equation for $\mathbf{Q}_m$, we obtain the equation given in the text.

For a correlation coefficient of 0.5, the turbulent transport term associated to the fluctuations of the velocity and concentration is: $\overline{\mathbf{v}'\ C'} = 0.5\sqrt{\overline{v_z'^2}}\ \sqrt{\overline{C'^2}} = 5 \times 10^{-4}$ kg/(m^2.s). On the other hand, the mass flux due to molecular diffusion is $-D_m\ \nabla \overline{C} \approx 10^{-11}$ kg/(m^2.s) which is considerably smaller than the turbulent flux.

3) Decay of a turbulent flow.

The rate of variation with time of the kinetic energy must be equal to the energy transmitted per unit time in the energy cascade by vortices of a characteristic size to the next smaller ones. As a result, we have, per unit mass: $(1/2)\ dv_\ell^2/dt \simeq -v_\ell^3/\ell$ or $dv_\ell/v_\ell^2 \simeq -dt/\ell$. After integrating, we obtain: $v_\ell(t) = v_{\ell 0}/(1 + v_{\ell 0}\, t/\ell)$ so that, at long times: $v_\ell(t) \simeq \ell/t$ and, $Re_\ell \simeq (\ell^2/\nu)/t = \tau_\nu/t$. The characteristic diffusion time over the distance ℓ is τ_ν and $v_{\ell o}$ is the value of v_ℓ at $t = 0$. These approximations are valid if $v_{\ell_0}\, t/\ell \gg 1$ or: $v_{\ell_0}/v_\ell \gg 1$.

For a very large initial Reynolds number $Re_{\ell_0} \gg 10$, the latter condition is satisfied as Re decreases towards 10: the time t_{10} for reaching $Re = 10$ is then of the order of $\tau_\nu/10$ while Re_ℓ and v_ℓ decrease as 1/t. In the final viscous dissipation regime, the variation of the velocity of fluid particles of size $\simeq \ell$ satisfies: $-c_\nu\, \rho\nu\, \ell v = \rho\, \ell^3 \mathrm{d}v/\mathrm{d}t$ or $-c_\nu\ (\nu/\ell^2)\ v = \mathrm{d}v/\mathrm{d}t$ so that $v_\ell = K\,\mathrm{e}^{-c_\nu t/\tau_\nu}$ (c_ν is a constant). Assuming that the transition from the turbulent regime occurs at $t = t_{10}$, we then have: $v_\ell = (\nu/10\ell)\ \mathrm{e}^{-c_\nu(t - t_{10})/\tau_\nu}$.

If $v_\ell \propto t^{-0.6}$ in the first regime, assuming that the equation $dv_\ell/v_\ell^2 \simeq -dt/\ell$ remains valid, leads to $\ell \propto t^{0.4}$: the size of the largest eddies increases slowly with time until it is limited by the container size.

4) Turbulent plane Couette flow.

Due to the absence of applied pressure gradient ($\partial\bar{p}/\partial x = 0$), Equation 12.44a for parallel flows becomes: $\nu\ \partial\bar{v}_x/\partial y - \overline{v'_x v'_y} = \bar{\sigma}_{xy}/\rho = cst$, in which $\bar{\sigma}_{xy}$ is the total stress at the wall. Using dimensionless parameters and taking derivatives with respect to x in order to eliminate the constant leads to $f'' = Re\, g'$.

Rewriting the definition of ν_t in dimensionless coordinates: $g = -(\nu_t/Uh)\, f'$ and using the value: $\nu_t = \kappa_c\, U\, y$ leads to $g = -\kappa_c\, \xi\, f'$; the expression of ν_t is similar to that derived for turbulent Poiseuille flows (Equations 12.24a and 12.50) in which u^* is replaced by U and the constant κ by κ_c which may also be expected to be ≈ 1. Taking the derivative of this equation with respect to ξ and combining the result with the relation between f' and g' leads to the differential equation for f given in the text.

Dividing by $(\xi + (1/(\kappa_c\, Re))\, f'$ and integrating once leads to: $(\xi + 1/(\kappa_c\, Re))\, f' = C_1$. Dividing by $(\xi + 1/(\kappa_c\, Re))$ and integrating with $f = 0$ and 1 respectively for $\xi = 0$ and 1 gives: $f(\xi) = \mathrm{Log}(1 + \kappa_c\, Re\, \xi)/\mathrm{Log}(1 + \kappa_c Re)$. Using $\bar{\sigma}_{xy}/\rho = \nu\ \partial\bar{v}_x/\partial y\,(0)$ leads to: $C_d = 2\,\kappa_c/\mathrm{Log}(1 + \kappa_c\, Re)$.

For $Re \ll 1$, we have $f(\xi) = \xi$ and $C_d = 2/Re$: both results correspond to the variations for laminar Couette flows.

For $Re \gg 1$, C_d varies slowly (logarithmically) with Re like for a turbulent Poiseuille flow between smooth plates.

For $Re \gg 1$ and for $\xi\, Re \ll 1$ (i.e. $y \ll \nu/U$), we have $f(\xi) \simeq \kappa_c Re\, \xi/\mathrm{Log}(1 + \kappa_c Re)$: this corresponds to a linear variation $f(\xi) \approx yU/\nu$, similar (replacing again u^* by U) to that found in the viscous sublayer of Poiseuille flows close to the wall (Equation 12.54).

For $Re\, \xi \gg 1$, the variation of the velocity with y/h is logarithmic as in the inertial sublayer of Poiseuille turbulent flows. A remarkable feature is that the expression for the velocity profile established above is valid at all distances from the wall and at all Reynolds numbers.

Bibliography

General references

Batchelor, G.K. (2000) *An Introduction to Fluid Mechanics*. Cambridge University Press.
Faber, T.E. (1995) *Fluid Dynamics for Physicists*. Cambridge University Press.
Feynman, R.P., Leighton, R. and Sands, M. (1964) *The Feynman Lectures on Physics*. Volume 2, Chapters 40 "The flow of dry water" and 41 "The flow of wet water." Addison-Wesley.
Kundu, P.K., Cohen, L.M. and Dowling, D.R. (2011) *Fluid Mechanics*, 5th edition, Academic Press.
Lamb, H. (1993) *Hydrodynamics*. Cambridge University Press.
Landau, L.D. and Lifschitz, E.M. (1987) *Fluid Mechanics*. 2nd edition. Chapters 1 to 4. Pergamon Press.

Books and multimedia documents giving an approach of fluid mechanics based on images

Guyon, E., Hulin, J-P. and Petit, L. (2011) *Ce que disent les fluides, la science des écoulements en images*, 2nd edition. Belin.
Guyon, E. and Guyon, MY. (2014). "Taking fluid mechanics to the general public" *Annual Review of Fluid Mechanics*. **46**, 1–22.
Homsy, G. (2008) *Multimedia Fluid Mechanics (MMFM)* second edition. Cambridge University Press.
Paterson, A.R. (1983) *A First Course in Fluid Dynamics*. Cambridge University Press.
Samimy, K.S., Breuer, K.S., Leal, L.G. and Steen, P.H. (2004) *A Gallery of Fluid Motion*. Cambridge University Press.
Shapiro, A.H. (1961–1972) *National Committee for Fluid Mechanics Films* (NCFMF) and its companion book: *Illustrated Experiments in Fluid Mechanics* (IEFM, 1974). MIT Press. For internet access to all these films use Shapiro and NCFMF as search keywords.
Tritton, D.J. (1988) *Physical Fluid Dynamics*. 2nd edn. Oxford Science Publications.
Van Dyke, M. (1982) *An Album of Fluid Motion* (AFM). Parabolic Press.

History of hydrodynamics

Darrigol, O. (2008) *Worlds of Flows: A History of Hydrodynamics from the Bernoullis to Prandtl*. Oxford University Press.

In addition to the above general references, we provide below, chapter by chapter, more specialized ones which are arranged as follows:

- References relevant to the chapter: review articles in scientific journals or fluid mechanics textbooks discussing in greater depth some content of the chapter.
- Specific references dealing with specific ideas discussed in a particular section of the chapter.
- Films from the NCFMF series: see above.

Chapter 1

Adamson, W.A. and Gast, A.P. (1997) *Physical Chemistry of Surfaces*. John Wiley & Sons.
Bird, B.R., Stewart, W.E. and Lightfoot, E.N. (2006) *Transport Phenomena*. John Wiley & Sons.
Carslaw, H.S. and Jaeger, J.C. (1959) *Conduction of Heat in Solids*. Oxford University Press.
Egelstaff, P.A. (1994) *An Introduction to the Liquid State*. Oxford University Press.
Guyon, E. et al. (2010) *Matière et matériaux. De quoi est fait le monde*. Belin.
McQuarrie, D. (1976) *Statistical Mechanics*. Harper and Row.
Reif, F. (1972) *Statistical Physics*. McGraw-Hill. See bibliography of Chapter 8.

Specific references

Barker, J. and Henderson, D. (1981) "Numerical simulations of two-dimensional phase transitions." *Scientific American.* **180**, Nov. 130–138.

Charmet, J., Cloitre, M., Fermigier, M., Guyon, E.; Jenffer, P. ; Limat, L. ; Petit, L. (1986). "Application of forced Rayleigh scattering to hydrodynamic measurements." *IEEE Journal of Quantum Electronics.* **QE-22**, 1461–1468.

Betrencourt, C., Guyon, E. and Giraud, L.G. (1980) "Teaching physics out of a bag of marbles." *European Journal of Physics.* **1**, 206–211.

Pieranski, P. (1984) "An experimental model of a classical many body problem." *American Journal of Physics.* **52**, 68–73.

Chapter 2

Specific references

Koplik, J., Banavar, J. and Willemsen, J.F. (1988) "Molecular dynamics of Poiseuille flow and moving contact lines." *Physical Review Letters.* **60**, 1282–1285.

Provansal, M., Mathis, C. and Boyer, L. (1987) "Bénard-von Kármán instability: transient and forced regimes." *Journal of Fluid Mechanics.* **182**, 1–22.

Miedzik, J., Gumowski, K., Goujon-Durand, S., Jenffer, P. and Wesfreid, J.E. (2008) "Wake behind a sphere in early transitional regimes." *Physical Review E.* 77, 055308.

Reynolds, O. (1883) "An experimental investigation on the circumstances which determine whether the motion of water shall be direct or sinuous and the law of resistance in a parallel channel." *Philosophical Transactions of the Royal Society.* **174**, 935–982.

Rott, N. (1990) "Note on the history of the Reynolds number." *Annual Review of Fluid Mechanics.* **22**, 1–20.

Chapter 3

Adrian, R.J. (1991) "Particle imaging techniques for experimental fluid mechanics." *Annual Review of Fluid Mechanics.* **23**, 261–304.

Merzkirch, W. (1979) *Flow Visualization.* Academic Press.

Specific references

Bartol, I.K., Kruger, K.S., Stewart, W.S. and Thomson, J.T. (2009) "Hydrodynamics of pulsed jetting in juvenile and adult brief squid Lolliguncula brevis: evidence of multiple jet 'modes' and their implications for propulsive efficiency." *Journal of Experimental Biology.* **212**, 1889–1903.

Taylor, G.I. (1934) "The formation of emulsions in definable fields of flow." *Proceedings of the Royal Society A.* **146**, 501–523.

NCFMF film

Kline, S.J. (1966) *Flow Visualization* (IEFM, p. 34).

Lumley, J.L. (1963) *Deformation of Continuous Media* (IEFM, p. 11).

Chapter 4

Byron Bird, R. and Hassager, O. (1987) *Dynamics of Polymeric Liquids.* John Wiley & Sons.

Coussot, P. (2005) *Rheometry of Pastes, Suspensions and Granular Materials.* John Wiley & Sons Ltd.

Larson, R.G. (1999) *The Structure and Rheology of Complex Fluids.* Oxford University Press.

Oswald, P. (2009) *Rheophysics: The Deformation and Flow of Matter.* Cambridge University Press.

Specific references

Allain, C., Cloitre, M. and Perrot, P. (1997) "Experimental investigation and scaling laws of die swelling in semi-dilute polymer solutions." *Journal of Non-Newtonian Fluid Mechanics.* 73, 51–66.

Berret, J-F., Porte, G. and Decruppe J-P. (1997) "Inhomogeneous shear flows of wormlike micelles." *Physical Review E.* **55**, 1668–1676.

Charlaix, E., Kushnick, A.P. and Stokes, J.P. (1989) "Experimental study of the dynamic permeability of porous media." *Physical Review Letters.* **61**, 1595–1598.

Cottin-Bizonne, C., Cross, B., Steinberger, A. and Charlaix, E. (2005) "Boundary slip on smooth hydrophobic surfaces." *Physical Review Letters.* **94**, 056102.

Reiner, M. (1964) "The Deborah number." *Physics Today.* 17 n°1, 62.

NCFMF films

Markovitz, H. (1964) *Rheological Behaviour of Fluids* (IEFM, p. 18).
Trefethen, L. (1967) *Surface Tension in Fluid Mechanics* (IEFM, p. 26).

Chapter 5

Middleman, S. (1995) *Modeling Axisymmetric Flows: Dynamics of Films, Jets and Drops*. Academic Press.
Milne-Thomson, L.M. (1996) *Theoretical Hydrodynamics*. Dover Publications.

NCFMF film

Shapiro, A.H. (1962) *Pressure Fields and Fluid Acceleration* (IEFM, p. 39).

Chapter 6

Lighthill, M.J. (2001) *Waves in Fluids*, 2nd edition. Cambridge University Press.
Stoker, J.J. (1957) *Water Waves, The Mathematical Theory with Applications*. John Wiley & Sons.

Specific references

Davies, R.M. and Taylor, G.I. (1950) "The mechanics of large bubbles rising through extended liquids and through liquids in tubes." *Proceedings of the Royal Society of London A*. **200**, 375–390.

NCFMF film

Bryson, A.E. (1964) *Waves in Fluids* (IEFM, p. 105).

Chapter 7

Childress, S. (1981) *Mechanics of Swimming and Flying*. Cambridge University Press.
Cushman-Roisin, B. (1994) *Introduction to Geophysical Fluid Dynamics*. Prentice Hall.
Greenspan, H.P. (1990) *The Theory of Rotating Fluids*. Breukelen Press.
Holton, J.R. (2004) *An Introduction to Dynamic Meteorology*, 4th edition, Academic Press.
Saffman, P.J. (1995) *Vortex Dynamics*. Cambridge University Press.

Specific references

Cortet, P-P., Lamriben, C. and Moisy, F. (2010) "Viscous spreading of an inertial wave beam in a rotating fluid." *Physics of Fluids*. **22** 086603.
Donnelly, R.J., Glaberson, W.I. and Parks, R. (1967) *Experimental Superfluidity*. University of Chicago Press.
Donnelly, R.J. (1993) "Quantized Vortices and Turbulence in Helium II." *Annual Review of Fluid Mechanics*. **25**, 327–371.
Godoy-Diana, R., Aider, J-L. and Wesfreid, J.E. (2008) "Transitions in the wake of a flapping foil." *Physical Review Letters E*. 77, 016308.
Guyon, E., Kojima, H., Veitz, W. and Rudnick, I. (1972) "Persistent current states in rotating superfuid He." *Journal of Low Temperature Physics*. **9**, 187–193.
Williams, G. and Packard, R.E. (1980) "A technique for photographing vortex positions in rotating superfluid He." *Journal of Low Temperature Physics*. **39**, 553–577.

NCFMF films

Shapiro, A.H. (1961) *Vorticity* (IEFM, p. 63).
Fultz, D. (1969) *Rotating Flows* (IEFM, p. 143).

Chapter 8

De Gennes, P.G., Brochard, F. and Quéré, Y. (2004) *Capillarity and Wetting Phenomena: Drops, Bubbles, Pearls, Waves*. Springer.

Specific references

Cazabat, A.M. and Cohen Stuart, M. (1986) "Dynamics of wetting: effects of surface roughness." *Journal of Physical Chemistry*. **90**, 5845–5849.

Fermigier, M. and Jenffer, P. (1991) "An experimental investigation of the dynamic contact angle." *Journal of Colloid and Interface Science*. **146**, 226–241.

Mora, S., Abkarian, M., Tabuteau, H. and Pomeau, Y. (2011) "Surface instability of soft solids under strain." *Soft Matter*. 7, 10612–10619.

Ribe, N.M. (2004) "Coiling of viscous jets." *Proceedings of the Royal Society of London A*. **460**, 3223–3239.

Trouton, F.T. (1906) "On the coefficient of viscous traction and its relation to that of viscosity." *Proceedings of the Royal Society of London A*. 77, 426–440.

Chapter 9

Bear, J. (1989) *Dynamics of Fluids in Porous Media*. Dover Publications.

Dullien, F.A. (1992) *Porous Media; Fluid Transport and Pore Structure*. Academic Press.

Guazzelli, E. and Morris, J.F. (2012) *A Physical Introduction to Suspension Dynamics*. Cambridge University Press.

Happel, J. and Brenner, H. (1983) *Low Reynolds Number Hydrodynamics (with special applications to particulate media)*. Kluwer Academic Publishers.

Hinch, E.J. (1988) "Hydrodynamics at Low Reynolds Number: a brief and elementary introduction" in *Disorder and Mixing*. Guyon, E., Nadal, J.P. and Pomeau, Y. eds., N.A.T.O. A.S.I. E series, Kluwer Academic Publishers. **152**, 43–55.

Moffatt, K. (1977) "Six lectures on fluid dynamics" in: *Fluid Dynamics, les Houches 1973*. Balian, R. and Peube, J.L. eds., Gordon and Breach, 149–234.

Scheidegger, A.E. (1979) *The Physics of Flow Through Porous Media*. University of Toronto Press.

Specific references

Katz, A.J. and Thompson, A.H. (1986) "Quantitative prediction of permeability in porous rocks." *Physical Review B*. **34**, 8179–8181.

Lliboutry, L.A. (1987) *Very Slow Flows of Solids*. Ch. 7. M. Nijjhoff Publishers.

Purcell, E.M. (1977) "Life at low Reynolds number." *American Journal of Physics*. **45**, 3–11.

Tabeling, P. (2005) *Introduction to Microfluidics*. Oxford University Press.

Wong, P., Koplik, J. and Tomanic, J.P. (1984) "Conductivity and permeability of rocks." *Physical Review B*. **30**, 6606–6614.

NCFMF film

Taylor, G.I. (1964) *Low Reynolds Number Flows* (IEFM, p. 47).

Chapter 10

Levich, V.G. (1962) *Physicochemical Hydrodynamics*. Prentice Hall.

Prandtl, L. and Tietjens, O.G. (1957) *Fundamentals of Hydro- and Aerodynamics*. Dover Publications.

Probstein, R.F. (1994) *Physicochemical Hydrodynamics, An Introduction*. 2nd edition, Wiley-Blackwell.

Schlichtling, H. and Gersten, K. (2000) *Boundary Layer Theory*. 8th edition, Springer.

Specific references

Comte-Bellot, G. (1976) "Hot-wire anemometry". *Annual Review of Fluid Mechanics*. **8**, 209–231.

Freymuth, P., Bank, W. and Palmer, M. (1984) "First experimental evidence of vortex splitting." *Physics of Fluids*. **27**, 1045–1046.

Minguez, M., Pasquetti, R. and Serre, E. (2008) "High-order large-eddy simulation of flow over the 'Ahmed body' car model." *Physics of Fluids*. **20**, 095101.

Quinard, J., Searby, G., Denet, B. and Graña-Otero, J. (2011) "Self turbulent flame speed." *Flow, Turbulence and Combustion*. **89**, 231–247.

Werle, H. (1974) *The Uses of a Hydrodynamic Wind Tunneling Space Research*. ONERA publication # 156, p. 43; ibid. publication #1303, p. 343.

NCFMF film

Abernathy, F.H. (1968) *Fundamentals of Boundary Layers* (IEFM, p. 75).

Chapter 11

Berge, P., Pomeau, Y. and Vidal, C. (1987) *Order within Chaos*. Wiley-VCH.
Charru, F. (2011) *Hydrodynamic Instabilities*. Cambridge University Press.
Lin, C.C. (1955) *The Theory of Hydrodynamic Stability*. Cambridge University Press.
Manneville, P. (2010) *Instabilities, Chaos and Turbulence: An Introduction to Nonlinear Dynamics and Complex Systems*, 2nd edition, Imperial College Press.
Mutabazi, I.,Wesfreid, J.E. and Guyon, E. (2006) *Dynamics of Spatio-temporal Structures*. Springer.

Specific references

Andereck, C.D., Liu, S.S. and Swinney, H.L. (1986) "Flow regimes in a circular Couette system with independently rotating cylinders," *Journal of Fluid Mechanics*. **164**, 155–183.
Fellouah, H., Castelain, C., Ould El Moktar, A. and Peerhossani, H. (2006) "A criterion for detection of the onset of Dean instability," *European Journal of Mechanics. B/Fluids*. **46**, 505–531.
Guyon, E. and Pieranski, P. (1974) "Convective instabilities in nematic liquid crystals." *Physica*. **73**, 184–194.
Heslot, F., Castaing, B. and Libchaber, A. (1987) "Transitions to turbulence in helium gas." *Physical Review. A* **36**, 5870–5873.
Libchaber, A., Fauve, S. and Laroche, C. (1983) "Two-parameter study of the routes to chaos." *Physica D*. **7**, 73–84.
Schatz, M.F., Van Hook, S.J., McCormick, W.D., Swift, J.B. and Swinney, H.L. (2008) "Onset of surface-tension-driven Bénard instability," *Physical Review Letters*. **75**, 1938–1941.
Thorpe, S.A. (1969) "Experiments on the instability of stratified shear flows: immiscible fluids." *Journal of Fluid Mechanics*. **39**, 25–48.
Thorpe, S.A. (1971) "Experiments on the instability of stratified shear flows: miscible fluids." *Journal of Fluid Mechanics*. **46**, 299–319.

NCFMF film

Mollo-Christensen, E.L. (1972) *Flow Instabilities* (IEFM, p. 113).

Chapter 12

Davidson, P.A. (2004) *Turbulence: An Introduction for Scientists and Engineers*. Oxford University Press.
Frisch, U. (1996) *Turbulence: The legacy of A.N. Kolmogorov*. Cambridge University Press.
Lesieur, M. (1997) *Turbulence in Fluids*. Kluwer Academic Publishers.
Pope, S.B. (2000) *Turbulent Flows*. Cambridge University Press.
Tennekes, M. and Lumley, J.M. (1972) *A First Course in Turbulence*. Cambridge University Press.

Specific references

Catrakis, H.J. and Dimotakis, P.E. (1996) "Mixing in turbulent jets: scalar measures and isosurface geometry." *Journal of Fluid Mechanics*. **317**, 369–406.
Kahlerras, H., Malecot, Y. and Gagne, Y. (1998) "Intermittency and Reynolds number," *Physics of Fluids*. **10**, 910–921.
Moisy, F., Morize, C., Rabaud, M. and Sommeria, J. (2011) "Decay laws, anisotropy and cyclone–anticyclone asymmetry in decaying rotating turbulence." *Journal of Fluid Mechanics*. **666**, 5–35.

NCFMF film

Stewart, L.W. (1966) *Turbulence* (IEFM, p. 82).

Index